MATHEMATICAL CENTERS IN THE THALASSIC AGE

1. Syracuse	Abdera	10	(Democritus)
2. Crotona	Alexandria	25	(Euclid, Heron, Ptolemy, Pappus, Menelaus and others)
3. Elea	Athens	8	(Plato, Theaetetus)
4. Rome	Byzantium	11	(Proclus)
5. Tarentum	Chalcedon	12	(Xenocrates)
6. Cyrene	Chalcis	23	(Iamblichus)
7. Elis	Chios	16	(Hippocrates)
8. Athens	Cnidus	20	(Eudoxus)
9. Stagira	Crotona	2	(Pythagoras)
10. Abdera	Cyrene	6	(Theodorus, Eratosthenes)
11. Byzantium	Cyzicus	14	(Callippus)
12. Chalcedon	Elea	3	(Parmenides, Zeno)
13. Nicaea	Elis	7	(Hippias)
14. Cyzicus	Gerasa	24	(Nicomachus)
15. Pergamum	Miletus	19	(Thales)
16. Chios	Nicaea	13	(Hipparchus)
17. Samos	Perga	22	(Apollonius)
18. Smyrna	Pergamum	15	(Apollonius)
19. Miletus	Rhodes	21	(Eudemus, Geminus)
20. Cnidus	Rome	4	(Boethius)
21. Rhodes	Samos	17	(Pythagoras, Conon, Aristarchus)
22. Perga	Smyrna	18	(Theon)
23. Chalcis	Stagira	9	(Aristotle)
24. Gerasa	Syene	26	(Eratosthenes)
25. Alexandria	Syracuse	1	(Archimedes)
26. Syene	Tarentum	5	(Pythagoras, Archytas, Philolaus [?])

A History of Mathematics

Carl B. Boyer

Professor of Mathematics
Brooklyn College

A History
of Mathematics

JOHN WILEY & SONS, INC.   New York London Sydney

To the Memory of My Parents

    Howard Franklin Boyer

    and

    Rebecca Catherine (Eisenhart) Boyer

Preface

Numerous histories of mathematics have appeared during this century, many of them in the English language. Some are very recent, such as J. F. Scott's *A History of Mathematics*[1]; a new entry in the field therefore should have characteristics not already present in the available books. Actually, few of the histories at hand are textbooks, at least not in the American sense of the word, and Scott's *History* is not one of them. It appeared, therefore, that there was room for a new book—one that would meet more satisfactorily my own preferences and possibly those of others.

The two-volume *History of Mathematics* by David Eugene Smith[2] was indeed written "for the purpose of supplying teachers and students with a usable textbook on the history of elementary mathematics," but it covers too wide an area on too low a mathematical level for most modern college courses, and it is lacking in problems of varied types. Florian Cajori's *History of Mathematics*[3] still is a very helpful reference work; but it is not adapted to classroom use, nor is E. T. Bell's admirable *The Development of Mathematics*.[4] The most successful and appropriate textbook today appears to be Howard Eves, *An Introduction to the History of Mathematics*,[5] which I have used with considerable satisfaction in at least a dozen classes since it first appeared in 1953. I have occasionally departed from the arrangement of topics in the book in striving toward a heightened sense of historical-mindedness and have supplemented the material by further reference to the contributions of the eighteenth and nineteenth centuries especially by the use of D. J. Struik, *A Concise History of Mathematics*.[6]

The reader of this book, whether a layman, a student, or a teacher of a course in the history of mathematics, will find that the level of mathematical background that is presupposed is approximately that of a college junior or senior, but the material can be perused profitably also by readers with either stronger or weaker mathematical preparation. Each chapter ends with a set of exercises that are graded roughly into three categories. Essay questions that are intended to indicate the reader's ability to organize and put into his own words the material discussed in the chapter are listed first. Then follow relatively easy exercises that require the proofs of some of the theorems mentioned in the chapter or their application to varied situations. Finally,

---

[1] London: Taylor and Francis, 1958.
[2] Boston: Ginn and Company, 1923–1925.
[3] New York: Macmillan, 1931, 2nd edition.
[4] New York: McGraw-Hill, 1945, 2nd edition.
[5] New York: Holt, Rinehart and Winston, 1964, revised edition.
[6] New York: Dover Publications, 1967, 3rd edition.

there are a few starred exercises, which are either more difficult or require specialized methods that may not be familiar to all students or all readers. The exercises do not in any way form part of the general exposition and can be disregarded by the reader without loss of continuity.

Here and there in the text are references to footnotes, generally bibliographical, and following each chapter there is a list of suggested readings. Included are some references to the vast periodical literature in the field, for it is not too early for students at this level to be introduced to the wealth of material available in good libraries. Smaller college libraries may not be able to provide all of these sources, but it is well for a student to be aware of the larger realms of scholarship beyond the confines of his own campus. There are references also to works in foreign languages, despite the fact that some students, hopefully not many, may be unable to read any of these. Besides providing important additional sources for those who have a reading knowledge of a foreign language, the inclusion of references in other languages may help to break down the linguistic provincialism which, ostrich-like, takes refuge in the mistaken impression that everything worthwhile appeared in, or has been translated into, the English language.

The present work differs from the most successful presently available textbook in a stricter adherence to the chronological arrangement and a stronger emphasis on historical elements. There is always the temptation in a class in history of mathematics to assume that the fundamental purpose of the course is to teach mathematics. A departure from mathematical standards is then a mortal sin, whereas an error in history is venial. I have striven to avoid such an attitude, and the purpose of the book is to present the history of mathematics with fidelity, not only to mathematical structure and exactitude, but also to historical perspective and detail. It would be folly, in a book of this scope, to expect that every date, as well as every decimal point, is correct. It is hoped, however, that such inadvertencies as may survive beyond the stage of page proof will not do violence to the sense of history, broadly understood, or to a sound view of mathematical concepts. It cannot be too strongly emphasized that this single volume in no way purports to present the history of mathematics in its entirety. Such an enterprise would call for the concerted effort of a team, similar to that which produced the fourth volume of Cantor's *Vorlesungen über Geschichte der Mathematik* in 1908 and brought the story down to 1799. In a work of modest scope the author must exercise judgment in the selection of the materials to be included, reluctantly restraining the temptation to cite the work of every productive mathematician; it will be an exceptional reader who will not note here what he regards as unconscionable omissions. In particular, the last chapter attempts merely to point out a few of the salient characteristics of the twentieth century. In the field of the history of mathematics perhaps nothing is more to be desired than that there should appear a latter-day Felix Klein who would complete for our century the type of project Klein essayed for for the nineteenth century, but did not live to finish.

A published work is to some extent like an iceberg, for what is visible constitutes only a small fraction of the whole. No book appears until the author has lavished time on it unstintingly and unless he has received encouragement and support from others too numerous to be named individually. Indebtedness in my case begins with the many eager students to whom I have taught the history of mathematics, primarily at Brooklyn College, but also at Yeshiva University, the University of Michigan, the University of California (Berkeley), and the University of Kansas. At the University of Michigan, chiefly through the encouragement of Professor Phillip S. Jones, and at Brooklyn College through the assistance of Dean Walter H. Mais and Professors Samuel Borofsky and James Singer, I have on occasion enjoyed a reduction in teaching load in order to work on the manuscript of this book. Friends and colleagues in the field of the history of mathematics, including Professor Dirk J. Struik of the Massachusetts Institute of Technology, Professor Kenneth O. May at the University of Toronto, Professor Howard Eves of the University of Maine, and Professor Morris Kline at New York University, have made many helpful suggestions in the preparation of the book, and these have been greatly appreciated. Materials in the books and articles of others have been expropriated freely, with little acknowledgment beyond a cold bibliographical reference, and I take this opportunity to express to these authors my warmest gratitude. Libraries and publishers have been very helpful in providing information and illustrations needed in the text; in particular it has been a pleasure to have worked with the staff of John Wiley and Sons. The typing of the final copy, as well as of much of the difficult preliminary manuscript, was done cheerfully and with painstaking care by Mrs. Hazel Stanley of Lawrence, Kansas. Finally, I must express deep gratitude to a very understanding wife, Dr. Marjorie N. Boyer, for her patience in tolerating disruptions occasioned by the development of yet another book within the family.

Brooklyn, New York   CARL B. BOYER
January 1968

Contents

A History of Mathematics

CHAPTER I

Primitive Origins

Did you bring me a man who cannot number his fingers?

From the Book of the Dead

Mathematicians of the twentieth century carry on a highly sophisticated intellectual activity which is not easily defined; but much of the subject that today is known as mathematics is an outgrowth of thought that originally centered in the concepts of number, magnitude, and form. Old-fashioned definitions of mathematics as a "science of number and magnitude" are no longer valid, but they do suggest the origins of the branches of mathematics. Primitive notions related to the concepts of number, magnitude, and form can be traced back to the earliest days of the human race, and adumbrations of mathematical notions can be found in forms of life that may have ante-dated mankind by many millions of years. Darwin in *Descent of Man* (1871) noted that certain of the higher animals possess such abilities as memory and imagination, and today it is even clearer that the abilities to distinguish number, size, order, and form—rudiments of a mathematical sense—are not exclusively the property of mankind. Experiments with crows, for example, have shown that at least certain birds can distinguish between sets containing up to four elements.[1] An awareness of differences in patterns found in their environment is clearly present in many lower forms of life, and this is akin to the mathematician's concern for form and relationship.

At one time mathematics was thought to be directly concerned with the world of our sense experience, and it was only in the nineteenth century that pure mathematics freed itself from limitations suggested by observations of nature. It is clear that originally mathematics arose as a part of the everyday life of man, and if there is validity in the biological principle of the "survival of the fittest," the persistence of the human race probably is not unrelated to the development in man of mathematical concepts. At first the primitive notions of number, magnitude, and form may have been related to contrasts

[1] See Levi Conant, *The Number Concept. Its Origin and Development* (1923). Cf. H. Kalmus, "Animals as Mathematicians," *Nature*, **202** (1964), 1156–1160.

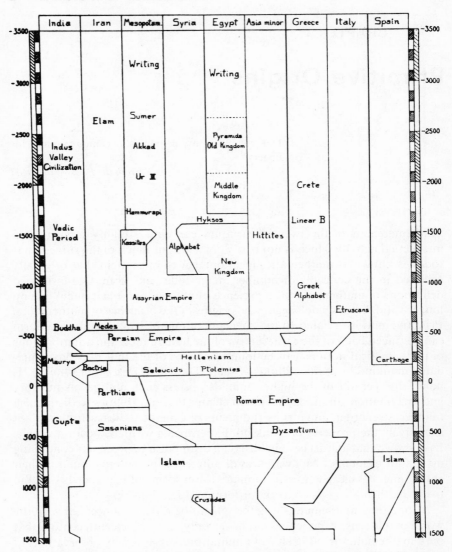

Chronological scheme representing the extent of some ancient and medieval civilizations. (Reproduced, with permission, from O. Neugebauer, *The Exact Sciences in Antiquity*.)

rather than likenesses—the difference between one wolf and many, the inequality in size of a minnow and a whale, the unlikeness of the roundness of the moon and the straightness of a pine tree. Gradually there must have arisen, out of the welter of chaotic experiences, the realization that there are samenesses; and from this awareness of similarities in number and form both science and mathematics were born. The differences themselves seem to point to likenesses, for the contrast between one wolf and many, between one sheep and a herd, between one tree and a forest, suggests that one wolf, one sheep, and one tree have something in common—their uniqueness. In the same way it would be noticed that certain other groups, such as pairs, can be put into one-to-one correspondence. The hands can be matched against the feet, the eyes, the ears, or the nostrils. This recognition of an abstract property that certain groups hold in common, and which we call number, represents a long step toward modern mathematics. It is unlikely to have been the discovery of any one individual or of any single tribe; it was more probably a gradual awareness which may have developed as early in man's cultural development as his use of fire, possibly some 300,000 years ago. That the development of the number concept was a long and gradual process is suggested by the fact that some languages, including Greek, have preserved in their grammar a tripartite distinction between one and two and more than two, whereas most languages today make only the dual distinction in "number" between singular and plural. Evidently our very early ancestors at first counted only to two, any set beyond this level being stigmatized as "many." Even today many primitive peoples still count objects by arranging them into bundles of two each.

The awareness of number ultimately became sufficiently extended and vivid so that a need was felt to express the property in some way, presumably at first in sign language only. The fingers on a hand can be readily used to indicate a set of two or three or four or five objects, the number one generally not being recognized at first as a true "number." By the use of the fingers on both hands, collections containing up to ten elements could be represented; by combining fingers and toes, one could mount as high as twenty. When the human digits were inadequate, heaps of stones could be used to represent a correspondence with the elements of another set. Where primitive man used such a scheme of representation, he often piled the stones in groups of five, for he had become familiar with quintuples through observation of the human hand and foot. As Aristotle had noted long ago, the widespread use today of the decimal system is but the result of the anatomical accident that most of us are born with ten fingers and ten toes. From the mathematical point of view it is somewhat inconvenient that Cro-Magnon man and his descendants did not have either four or six fingers on a hand.

Although historically finger counting, or the practice of counting by fives and tens, seems to have come later than countercasting by twos and threes, the quinary and decimal systems almost invariably displaced the binary and ternary schemes. A study of several hundred tribes among the American

2

Indians, for example, showed that almost one third used a decimal base and about another third had adopted a quinary or a quinary-decimal system; fewer than a third had a binary scheme, and those using a ternary system constituted less than 1 per cent of the group. The vigesimal system, with twenty as a base, occurred in about 10 per cent of the tribes.[2]

Groups of stones are too ephemeral for preservation of information; hence prehistoric man sometimes made a number record by cutting notches in a stick or a piece of bone. Few of these records remain today, but in Czechoslovakia a bone from a young wolf was found which is deeply incised with fifty-five notches. These are arranged in two series, with twenty-five in the first and thirty in the second; within each series the notches are arranged in groups of five. Such archaeological discoveries provide evidence that the idea of number is far older than such technological advances as the use of metals or of wheeled vehicles. It antedates civilization and writing, in the usual sense of the word, for artifacts with numerical significance, such as the bone described above, have survived from a period of some 30,000 years ago. Additional evidence concerning man's early ideas on number can be found in our language today. It appears that our words "eleven" and "twelve" originally meant "one over" and "two over," indicating the early dominance of the decimal concept. However, it has been suggested that perhaps the Indo-Germanic word for eight was derived from a dual form for four, and that the Latin *novem* for nine may be related to *novus* (new) in the sense that it was the beginning of a new sequence. Possibly such words can be interpreted as suggesting the persistence for some time of a quaternary or an octonary scale, just as the French *quatre-vingt* of today appears to be a remnant of a vigesimal system.

3   Man differs from other animals most strikingly in his language, the development of which was essential to the rise of abstract mathematical thinking; yet words expressing numerical ideas were slow in arising. Number *signs* probably preceded number *words*, for it is easier to cut notches in a stick than it is to establish a well-modulated phrase to identify a number. Had the problem of language not been so difficult, rivals to the decimal system might have made greater headway. The base five, for example, was one of the earliest to leave behind some tangible written evidence; but by the time that language became formalized, ten had gained the upper hand. The modern languages of today are built almost without exception around the base ten, so that the number thirteen, for example, is not described as three and five and five, but as three and ten. The tardiness in the development of

---

[2] W. C. Eels, "Number Systems of North American Indians," *American Mathematical Monthly*, **20** (1913), 293. Cf. also D. J. Struik, "Stone Age Mathematics," *Scientific American*, **179** (December 1948), 44–49.

language to cover abstractions such as number is seen also in the fact that primitive numerical verbal expressions invariably refer to specific concrete collections—such as "two fishes" or "two clubs"—and later some such phrase would be adopted conventionally to indicate all sets of two objects. The tendency for language to develop from the concrete to the abstract is seen in many of our present-day measures of length. The height of a horse is measured in "hands," and the words "foot" and "ell" (or elbow) have similarly been derived from parts of the body.

The thousands of years required for man to separate out the abstract concepts from repeated concrete situations testify to the difficulties that must have been experienced in laying even a very primitive basis for mathematics. Moreover, there are a great many unanswered questions relating to the origins of mathematics. It usually is assumed that the subject arose in answer to man's practical needs, but anthropological studies suggest the possibility of an alternative origin. It has been suggested[3] that the art of counting arose in connection with primitive religious ritual and that the ordinal aspect preceded the quantitative concept. In ceremonial rites depicting creation myths it was necessary to call the participants onto the scene in a specific order, and perhaps counting was invented to take care of this problem. If theories of the ritual origin of counting are correct, the concept of the ordinal number may have preceded that of the cardinal number. Moreover, such an origin would tend to point to the possibility that counting stemmed from a unique origin, spreading subsequently to other portions of the earth. This view, although far from established, would be in harmony with the ritual division of the integers into odd and even, the former being regarded as male, the latter as female. Such distinctions were known to civilizations in all corners of the earth, and myths regarding the male and female numbers have been remarkably persistent.

The concept of whole number is one of the oldest in mathematics, and its origin is shrouded in the mists of prehistoric antiquity. The notion of a rational fraction, however, developed relatively late and was not in general closely related to man's systems for the integers. Among primitive tribes there seems to have been virtually no need for fractions. For quantitative needs the practical man can choose units that are sufficiently small to obviate the necessity of using fractions. Hence there was no orderly advance from binary to quinary to decimal fractions, and decimals were essentially the product of the modern age in mathematics, rather than of the ancient period.

Statements about the origins of mathematics, whether of arithmetic or **4** geometry, are of necessity hazardous, for the beginnings of the subject are

[3] See A. Seidenberg, "The Ritual Origin of Counting," *Archive for History of Exact Sciences*, **2** (1962), 1–40.

older than the art of writing. It is only during the last half-dozen millennia, in a career that may have spanned thousands of millennia, that man has been able to put his records and thoughts in written form. For data about the prehistoric age we must depend on interpretations based on the few surviving artifacts, on evidence provided by current anthropology, and on a conjectural backward extrapolation from surviving documents. Herodotus and Aristotle were unwilling to hazard placing origins earlier than the Egyptian civilization, but it is clear that the geometry they had in mind had roots of greater antiquity. Herodotus held that geometry had originated in Egypt, for he believed that the subject had arisen there from the practical need for re-surveying after the annual flooding of the river valley. Aristotle argued that it was the existence of a priestly leisure class in Egypt that had prompted the pursuit of geometry. We can look upon the views of Herodotus and Aristotle as representing two opposing theories of the beginnings of mathematics, one holding to an origin in practical necessity, the other to an origin in priestly leisure and ritual. The fact that the Egyptian geometers sometimes were referred to as "rope-stretchers" (or surveyors) can be used in support of either theory, for the ropes undoubtedly were used both in laying out temples and in realigning the obliterated boundaries. We cannot confidently contradict either Herodotus or Aristotle on the motive leading to mathematics, but it is clear that both men underestimated the age of the subject. Neolithic man may have had little leisure and little need for surveying, yet his drawings and designs suggest a concern for spatial relationships that paved the way for geometry. Pottery, weaving, and basketry show instances of congruence and symmetry, which are in essence parts of elementary geometry. Moreover, simple sequences in design, such as that in Fig. 1.1, suggest a sort of applied

FIG. 1.1

group theory, as well as propositions in geometry and arithmetic. The design makes it immediately obvious that the areas of triangles are to each other as squares on a side, or, through counting, that the sums of consecutive odd numbers, beginning from unity, are perfect squares. For the prehistoric period there are no documents, hence it is impossible to trace the evolution of mathematics from a specific design to a familiar theorem. But ideas are like hardy spores, and sometimes the presumed origin of a concept may be only the reappearance of a much more ancient idea that had lain dormant.

The concern of prehistoric man for spatial designs and relationships may have stemmed from his aesthetic feeling and the enjoyment of beauty of form, motives that often actuate the mathematician of today. We would like to think that at least some of the early geometers pursued their work for the sheer joy of doing mathematics, rather than as a practical aid in mensuration; but there are other alternatives. One of these is that geometry, like counting, had an origin in primitive ritualistic practice. The earliest geometrical results found in India constituted what were called the *Sulvasutras*, or "rules of the cord." These were simple relationships that apparently were applied in the construction of altars and temples. It is commonly thought that the geometrical motivation of the "rope-stretchers" in Egypt was more practical than that of their counterparts in India; but it has been suggested[4] that both Indian and Egyptian geometry may derive from a common source—a proto-geometry that is related to primitive rites in somewhat the same way in which science developed from mythology and philosophy from theology. We must bear in mind that the theory of the origin of geometry in a secularization of ritualistic practice is by no means established. The development of geometry may just as well have been stimulated by the practical needs of construction and surveying or by an aesthetic feeling for design and order. We can make conjectures about what led men of the Stone Age to count, to measure, and to draw. That the beginnings of mathematics are older than the oldest civilizations is clear. To go further and categorically identify a specific origin in space or time, however, is to mistake conjecture for history. It is best to suspend judgment on this matter and to move on to the safer ground of the history of mathematics as found in the written documents that have come down to us.

BIBLIOGRAPHY

Conant, Levi, *The Number Concept. Its Origin and Development* (New York: Macmillan, 1923).

Eels, W. C., "Number Systems of North American Indians," *American Mathematical Monthly*, **20** (1913), 293.

Kalmus, H., "Animals as Mathematicians," *Nature*, **202** (1964), 1156–1160.

Menninger, Karl, *Zahlwort und Ziffer: Eine Kulturgeschichte der Zahlen*, 2nd ed. (Göttingen: Vandenhoeck & Ruprecht, 1957–1958, 2 vols.).

[4] A. Seidenberg, "The Ritual Origin of Geometry," *Archive for History of Exact Sciences*, **1** (1962), 488–527.

Seidenberg, A., "The Ritual Origin of Counting," *Archive for History of Exact Sciences*, **1** (1962), 488–527.

Seidenberg, A., "The Ritual Origin of Counting," *Archive for History of Exact Sciences*, **2** (1962), 1–40.

Smeltzer, Donald, *Man and Number* (New York: Emerson Books, 1958).

Smith, D. E., *History of Mathematics* (Boston: Ginn, 1923–1925, 2 vols.; paperback ed., New York: Dover, 1958).

Smith, D. E., and Jekuthiel Ginsburg, *Numbers and Numerals* (Washington, D.C.: National Council of Teachers of Mathematics, 1958).

Struik, D. J., "Stone Age Mathematics," *Scientific American*, **179** (December 1948), 44–49.

EXERCISES

1. Describe the type of evidence on which an account of prehistoric mathematics is based, citing some specific instances.
2. What evidence, if any, is there that mathematics began with the advent of man? Do you think that mathematics antedates man?
3. List evidences from language for the use at some time of bases other than ten.
4. What are the advantages and disadvantages of the bases two, three, four, five, ten, twenty, and sixty? Do you think that these influenced early man in his choice of a base?
5. If you had to choose a number base, which would it be? Why?
6. Which do you think came first, number names or number symbols? Why?
7. Why are there few traces of scales from six to nine?
8. What do you think were the first plane and solid geometric figures to be consciously and systematically studied? Why?
9. Which do you think was more influential in the rise of early geometry, an interest in astronomy or a need for surveying? Explain.
10. Which of the following time divisions was prehistoric man likely to notice: the year, the month, the week, the day, the hour? Explain.

CHAPTER II

Egypt

Sesostris... made a division of the soil of Egypt among the inhabitants... If the river carried away any portion of a man's lot, ... the king sent persons to examine, and determine by measurement the exact extent of the loss... From this practice, I think, geometry first came to be known in Egypt, whence it passed into Greece.

Herodotus

It is customary to divide the past of mankind into eras and periods, with **1** particular reference to cultural levels and characteristics. Such divisions are helpful, although we should always bear in mind that they are only a framework arbitrarily superimposed for our convenience and that the separations in time they suggest are not unbridged gulfs. The Stone Age, a long period preceding the use of metals, did not come to an abrupt end. In fact, the type of culture that it represented terminated much later in Europe than in certain parts of Asia and Africa. The rise of civilizations characterized by the use of metals took place at first in river valleys, such as those in Egypt, Mesopotamia, India, and China; hence we shall refer to the earlier portion of the historical period as the "potamic stage." Chronological records of the civilizations in the valleys of the Indus and Yangtze rivers are quite unreliable, but fairly dependable information is available about the peoples living along the Nile and in the "fertile crescent" of the Tigris and Euphrates rivers. Before the end of the fourth millennium B.C. a primitive form of writing was in use in both the Mesopotamian and Nile valleys. There the early pictographic records, through a steady conventionalizing process, evolved into a linear order of simpler symbols. In Mesopotamia, where clay was abundant, wedge-shaped marks were impressed with a stylus upon soft tablets which then were baked hard in ovens or by the heat of the sun. This type of writing is known as cuneiform (from the Latin word *cuneus* or wedge) because of the shape of the individual impressions. The meaning to be transmitted in cuneiform was determined by the patterns or arrangements of the wedge-shaped impressions. Cuneiform documents had a high degree of permanence; hence many thousands of such tablets have survived from antiquity, many of them

9

Reproduction (*top*) of a portion of the Moscow Papyrus showing the problem on the volume of a frustum of a square pyramid, together with hieroglyphic transcription (*below*).

dating back some 4000 years. Of course, only a small fraction of these touch on themes related to mathematics. Moreover, until about a century ago the message of the cuneiform tablets remained muted because the script had not been deciphered. In the 1870s significant progress in the reading of cuneiform writing was made when it was discovered that the Behistun Cliff carried a trilingual account of the victory of Darius over Cambyses, the inscriptions being in Persian, Medean, and Assyrian. Knowledge of Persian consequently supplied a key to the reading of Assyrian, a language closely related to the older Babylonian. Even after this important discovery, decipherment and analysis of tablets with mathematical content proceeded slowly, and it was not until the second quarter of the twentieth century that awareness of Mesopotamian mathematical contributions became appreciable, largely through the pioneer work of Fr. Thureau-Dangin in France and Otto Neugebauer in Germany and America.[1]

[1] See, for example, O. Neugebauer, *Vorgriechische Mathematik* (Berlin: Springer, 1934). For a more general account in English see his *The Exact Sciences in Antiquity* (1957).

Egyptian written records meanwhile had fared better than Babylonian in **2** one respect. The trilingual Rosetta Stone, playing a role similar to that of the Behistun Cliff, had been discovered in 1799 by the Napoleonic expedition. This large tablet, found at Rosetta, an ancient harbor near Alexandria, contained a message in three tongues: Greek, Demotic, and Hieroglyphic. Knowing Greek, Champollion in France and Thomas Young in England made rapid progress in deciphering the Egyptian hieroglyphics (that is, "sacred carvings"). Inscriptions on tombs and monuments in Egypt now could be read, although such ceremonial documents are not the best source of information concerning mathematical ideas. Egyptian hieroglyphic numeration was easily disclosed. The system, at least as old as the pyramids, dating some 5000 years ago, was based, as we might expect, on the ten-scale. By the use of a simple iterative scheme and of distinctive symbols for each of the first half-dozen powers of ten, numbers over a million were carved on stone, wood, and other materials. A single vertical stroke represented a unit, an inverted wicket or heel bone was used for 10, a snare somewhat resembling a capital letter C stood for 100, a lotus flower for 1000, a bent finger for 10,000, a burbot fish resembling a polywog for 100,000, and a kneeling figure (perhaps God of the Unending) for 1,000,000. Through repetition of these symbols the number 12,345, for example, would appear as

Sometimes the smaller digits were placed on the left, and sometimes the digits were arranged vertically. The symbols themselves occasionally were reversed in orientation, so that the snare might be convex toward either the right or the left.

Egyptian inscriptions indicate familiarity with large numbers at an early date. A museum at Oxford has a royal mace more than 5000 years old on which a record of 120,000 prisoners and 1,422,000 captive goats appears.[2] These figures may have been exaggerated, but from other considerations it is nevertheless clear that the Egyptians were commendably accurate in counting and measuring. The pyramids exhibit such a high degree of precision in construction and orientation that ill-founded legends have grown up around them. The suggestion, for example, that the ratio of the perimeter of the base of the Great Pyramid (of Khufu or Cheops) to the height was consciously set at 2π is clearly inconsistent with what we know of the geometry of the Egyptians.[3] Nevertheless, the pyramids and passages within them were so precisely oriented that attempts are made to determine their age from the known rate of change of the position of the polestar.

[2] J. E. Quibell, *Hierakonpolis* (London: B. Quaritch, 1900). See especially Plate 26B.

[3] Noel F. Wheeler, "Pyramids and Their Purpose," *Antiquity*, **9** (1935), 5–21, 161–189, 292–304.

The Egyptians early had become interested in astronomy and had observed that the annual flooding of the Nile took place shortly after Sirius, the dog-star, rose in the east just before the sun. By noticing that these heliacal risings of Sirius, the harbinger of the flood, were separated by 365 days, the Egyptians established a good solar calendar made up of twelve months of thirty days each and five extra feast days. But this civil year was too short by a quarter of a day, hence the seasons advanced about one day every four years until, after a cycle of about 1460 years, the seasons again were in tune with the calendar. Inasmuch as it is known through the Roman scholar Censorinus, author of *De die natale* (A.D. 238), that the calendar was in line with the seasons in A.D. 139, it has been suggested through extrapolation backward that the calendar was instituted in the year 4241, just three cycles earlier. More precise calculations (based on the fact that the year is not quite $365\frac{1}{4}$ days long) have modified the date to 4228, but other scholars feel that the backward extrapolation beyond two cycles is unwarranted and suggest instead an origin around 2773 B.C.

3   There is a limit to the extent of mathematical information that can be inferred from tombstones and calendars, and our picture of Egyptian contributions would be sketchy in the extreme if we had to depend on ceremonial and astronomical material only. Mathematics is far more than counting and measuring, the aspects generally featured in hieroglyphic inscriptions. Fortunately we have other sources of information. There are a number of Egyptian papyri that somehow have survived the ravages of time over some three and a half millennia. The most extensive one of a mathematical nature is a papyrus roll about 1 foot high and some 18 feet long which now is in the British Museum (except for a few fragments in the Brooklyn Museum). It had been bought in 1858 in a Nile resort town by a Scottish antiquary, Henry Rhind; hence it often is known as the Rhind Papyrus or, less frequently, as the Ahmes Papyrus in honor of the scribe by whose hand it had been copied in about 1650 B.C.[4] The scribe tells us that the material is derived from a prototype from the Middle Kingdom of about 2000 to 1800 B.C., and it is possible that some of this knowledge may have been handed down from Imhotep, the almost legendary architect and physician to the Pharaoh Zoser, who supervised the building of his pyramid about 5000 years ago. In any case, Egyptian mathematics seems to have stagnated for some 2000 years after a rather auspicious beginning.

---

[4] There are two good English editions, one edited by T. E. Peet and published in London in 1923, the other by A. B. Chace et al. and published in two volumes at Oberlin, Ohio, in 1927–1929. Volume I of the latter contains an extensive general account of Egyptian mathematics by R. C. Archibald, a translation with commentary of the Ahmes Papyrus, and a very extensive bibliography of articles on Egyptian mathematics.

The numerals and other material in the Rhind Papyrus are not written in the hieroglyphic forms described above, but in a more cursive script, better adapted to the use of pen and ink on prepared papyrus leaves and known as hieratic ("sacred," to distinguish it from the still later demotic or popular script). Numeration remains decimal, but the tedious repetitive principle of hieroglyphic numeration has been replaced by the introduction of ciphers or special signs to represent digits and multiples of powers of ten. Four, for example, usually is no longer represented by four vertical strokes, but by a horizontal bar; and seven is not written as seven strokes, but as a single cipher ⟍ resembling a sickle. In hieroglyphic the number twenty-eight had appeared as ∩∩||||, but in hieratic it is simply ₌⅄. Note that the cipher ₌ for the smaller digit eight (or two fours) appears on the left rather than on the right. The principle of cipherization, introduced by the Egyptians some 4000 years ago and used in the Rhind Papyrus, represented an important contribution to numeration, and it is one of the factors that makes our own system in use today the effective instrument that it is.

Men of the Stone Age had no use for fractions, but with the advent of more **4** advanced cultures during the Bronze Age the need for the fraction concept and for fractional notations seems to have arisen. Egyptian hieroglyphic inscriptions have a special notation for unit fractions—that is, fractions with unit numerators. The reciprocal of any integer was indicated simply by placing over the notation for the integer an elongated oval sign. The fraction

$\frac{1}{8}$ thus appeared as ⁅||||, and $\frac{1}{20}$ was written as ∩∩. In the hieratic notation,

appearing in papyri, the elongated oval is replaced by a dot, which is placed over the cipher for the corresponding integer (or over the right-hand cipher in the case of the reciprocal of a multidigit number). In the Ahmes Papyrus, for example, the fraction $\frac{1}{8}$ appears as ⩷, and $\frac{1}{20}$ is written as ⅄̇. Such unit fractions were freely handled in Ahmes' day, but the general fraction seems to have been an enigma to the Egyptians. They felt comfortable with the fraction $\frac{2}{3}$, for which they had a special hieratic sign ⁊; occasionally they used special signs for fractions of the form $n/(n + 1)$, the complements of the unit fractions. To the fraction $\frac{2}{3}$ the Egyptians assigned a special role in arithmetic processes, so that in finding one third of a number they first found two thirds of it and subsequently took half of the result! They knew and used the fact that two thirds of the unit fraction $1/p$ is the sum of the two unit fractions $1/2p$ and $1/6p$; they were also aware that double the unit fraction $1/2p$ is the unit fraction $1/p$. However, it looks as though, apart from the fraction $\frac{2}{3}$, the Egyptians regarded the general proper rational fraction of the form m/n not as an elementary "thing," but as part of an uncompleted process. Where today we think of $\frac{3}{5}$ as a single irreducible fraction, Egyptian scribes thought of it as reducible to the sum of the three unit fractions $\frac{1}{3}$ and $\frac{1}{5}$ and $\frac{1}{15}$. To facilitate the reduction of "mixed" proper fractions to the sum of unit fractions, the Rhind Papyrus opens with a table expressing $2/n$ as a sum of unit fractions for all odd values of n from 5 to 101. The equivalent of $\frac{2}{5}$ is given as $\frac{1}{3}$ and $\frac{1}{15}$; $\frac{2}{11}$ is written as $\frac{1}{6}$ and $\frac{1}{66}$; and $\frac{2}{15}$ is expressed as $\frac{1}{10}$ and $\frac{1}{30}$. The last item in the table decomposes[5] $\frac{2}{101}$ into $\frac{1}{101}$ and $\frac{1}{202}$ and $\frac{1}{303}$ and $\frac{1}{606}$. It is not clear why one form of decomposition was preferred to another of the indefinitely many that are possible. At one time it was suggested that some of the items in the $2/n$ table were found by using the equivalent of the formula

$$\frac{2}{n} = \frac{1}{\dfrac{n+1}{2}} + \frac{1}{\dfrac{n(n+1)}{2}}$$

or from

$$\frac{2}{p \cdot q} = \frac{1}{p \cdot \dfrac{p+q}{2}} + \frac{1}{q \cdot \dfrac{p+q}{2}}.$$

Yet neither of these procedures yields the combination for $\frac{2}{15}$ that appears in the table. Recently it has been suggested[6] that the choice in most cases was dictated by the Egyptian preference for fractions derived from the "natural" fractions $\frac{1}{2}$ and $\frac{1}{3}$ and $\frac{2}{3}$ by successive halving. Thus if one wishes to express $\frac{2}{15}$ as a sum of unit fractions, he might well begin by taking half of $\frac{1}{15}$ and then seeing if to the result, $\frac{1}{30}$, he can add a unit fraction to form $\frac{2}{15}$; or he could use the known relationship

$$\frac{2}{3} \cdot \frac{1}{p} = \frac{1}{2p} + \frac{1}{6p}$$

to reach the same result $\frac{2}{15} = \frac{1}{10} + \frac{1}{30}$. One problem in the Rhind Papyrus specifically mentions the second method for finding two thirds of $\frac{1}{5}$ and asserts that one proceeds likewise for other fractions. Passages such as this indicate that the Egyptians had some appreciation of general rules and methods above and beyond the specific case at hand, and this represents an important step in the development of mathematics. For the decomposition of $\frac{2}{5}$ the halving

[5] A list of fractional decompositions of $2/n$ from $n = 5$ to $n = 101$ is given in B. L. van der Waerden, *Science Awakening* (1961) and in Kurt Vogel, *Vorgriechische Mathematik*, Vol. 1, *Vorgeschichte und Ägypten* (ca. 1958). A clear-cut explanation of Egyptian fractions appears also in O. Neugebauer, *The Exact Sciences in Antiquity*. All three works give excellent accounts of Egyptian mathematics.

[6] See Neugebauer, *Exact Sciences in Antiquity*, pp. 74 ff.

procedure is not appropriate; but by beginning with a third of $\frac{1}{5}$ one finds the decomposition given by Ahmes, $\frac{2}{5} = \frac{1}{3} + \frac{1}{15}$. In the case of $\frac{2}{7}$ one applies the halving procedure twice to $\frac{1}{7}$ to reach the result $\frac{2}{7} = \frac{1}{4} + \frac{1}{28}$; successive halving yields also the Ahmes decomposition $\frac{2}{13} = \frac{1}{8} + \frac{1}{52} + \frac{1}{104}$. The Egyptian obsession with halving and taking a third is seen in the last entry in the table $2/n$ for $n = 101$, for it is not at all clear to us why the decomposition $2/n = 1/n + 1/2n + 1/3n + 1/2 \cdot 3 \cdot n$ is better than $1/n + 1/n$. Perhaps one of the objects of the $2/n$ decomposition was to arrive at unit fractions smaller than $1/n$.

The $2/n$ table in the Ahmes Papyrus is followed by a short $n/10$ table for **5** n from 1 to 9, the fractions again being expressed in terms of the favorites— unit fractions and the fraction $\frac{2}{3}$. The fraction $\frac{9}{10}$, for example, is broken into $\frac{1}{30}$ and $\frac{1}{5}$ and $\frac{2}{3}$. Ahmes had begun his work with the assurance that it would provide a "complete and thorough study of all things ... and the knowledge of all secrets," and therefore the main portion of the material, following the $2/n$ and $n/10$ tables, consists of eighty-four widely assorted problems. The first six of these require the division of one or two or six or seven or eight or nine loaves of bread among ten men, and the scribe makes use of the $n/10$ table that he has just given. In the first problem the scribe goes to considerable trouble to show that it is correct to give to each of the ten men one tenth of a loaf! If one man receives $\frac{1}{10}$ loaf, two men will receive $\frac{2}{10}$ or $\frac{1}{5}$ and four men will receive $\frac{2}{5}$ of a loaf or $\frac{1}{3} + \frac{1}{15}$ of a loaf. Hence eight men will receive $\frac{2}{3} + \frac{2}{15}$ of a loaf or $\frac{2}{3} + \frac{1}{10} + \frac{1}{30}$ of a loaf, and eight men plus two men will receive $\frac{2}{3} + \frac{1}{5} + \frac{1}{10} + \frac{1}{30}$, or a whole loaf. Ahmes seems to have had a kind of equivalent to our least common multiple which enabled him to complete the proof. In the division of seven loaves among ten men, the scribe might have chosen $\frac{1}{2} + \frac{1}{5}$ of a loaf for each, but the predilection for $\frac{2}{3}$ led him instead to $\frac{2}{3}$ and $\frac{1}{30}$ of a loaf for each.[7]

The fundamental arithmetic operation in Egypt was addition, and our operations of multiplication and division were performed in Ahmes' day through successive doubling or "duplation." Our own word "multiplication" or manifold is, in fact, suggestive of the Egyptian process. A multiplication of, say, 69 by 19 would be performed by adding 69 to itself to obtain 138, then adding this to itself to reach 276, applying duplation again to get 552, and once more to obtain 1104, which is, of course, sixteen times 69. Inasmuch as $19 = 16 + 2 + 1$, the result of multiplying 69 by 19 is $1104 + 138 + 69$— that is, 1311. Occasionally a multiplication by ten also was used, for this was a natural concomitant of the decimal hieroglyphic notation. Multiplication of combinations of unit fractions was also a part of Egyptian arithmetic.

[7] For further details see R. J. Gillings, "Problems 1 to 6 of the Rhind Mathematical Papyrus," *The Mathematics Teacher*, **55** (1962), 61–69.

Problem 13 in the Ahmes Papyrus, for example, asks for the product of $\frac{1}{16} + \frac{1}{112}$ and $1 + \frac{1}{2} + \frac{1}{4}$; the result is correctly found to be $\frac{1}{8}$. For division the duplation process is reversed, and the *divisor* is successively doubled instead of the *multiplicand*. That the Egyptians had developed a high degree of artistry in applying the duplation process and the unit fraction concept is apparent from the calculations in the problems of Ahmes. Problem 70 calls for the quotient when 100 is divided by $7 + \frac{1}{2} + \frac{1}{4} + \frac{1}{8}$; the result, $12 + \frac{2}{3} + \frac{1}{42} + \frac{1}{126}$, is obtained as follows. Doubling the divisor successively, we first obtain $15 + \frac{1}{2} + \frac{1}{4}$, then $31 + \frac{1}{2}$, and finally 63, which is eight times the divisor. Moreover, two thirds of the divisor is known to be $5 + \frac{1}{4}$. Hence the divisor when multiplied by $8 + 4 + \frac{2}{3}$ will total $99\frac{3}{4}$, which is $\frac{1}{4}$ short of the product 100 that is desired. Here a clever adjustment was made. Inasmuch as eight times the divisor is 63, it follows that the divisor when multiplied by $\frac{2}{63}$ will produce $\frac{1}{4}$. From the $2/n$ table one knows that $\frac{2}{63}$ is $\frac{1}{42} + \frac{1}{126}$, hence the desired quotient is $12 + \frac{2}{3} + \frac{1}{42} + \frac{1}{126}$. Incidentally, this procedure makes use of a commutative principle in multiplication, with which the Egyptians evidently were familiar.

Many of Ahmes' problems show a knowledge of manipulations of proportions equivalent to the "rule of three." Problem 72 calls for the number of loaves of bread of "strength" 45 which are equivalent to 100 loaves of "strength" 10, and the solution is given as $100/10 \times 45$ or 450 loaves. In bread and beer problems the "strength" or *pesu* is the reciprocal of the grain density, being the quotient of the number of loaves or units of volume divided by the amount of grain. Bread and beer problems are numerous in the Ahmes Papyrus. Problem 63, for example, requires the division of 700 loaves of bread among four recipients if the amounts they are to receive are in the continued proportion $\frac{2}{3} : \frac{1}{2} : \frac{1}{3} : \frac{1}{4}$. The solution is found by taking the ratio of 700 to the sum of the fractions in the proportion. In this case the quotient of 700 divided by $1\frac{3}{4}$ is found by multiplying 700 by the reciprocal of the divisor, which is $\frac{1}{2} + \frac{1}{14}$. The result is 400; by taking $\frac{2}{3}$ and $\frac{1}{2}$ and $\frac{1}{3}$ and $\frac{1}{4}$ of this, the required shares of bread are found.

6   The Egyptian problems so far described are best classified as arithmetic, but there are others that fall into a class to which the term algebraic is appropriately applied. These do not concern specific concrete objects, such as bread and beer, nor do they call for operations on known numbers. Instead they require the equivalent of solutions of linear equations of the form $x + ax = b$ or $x + ax + bx = c$, where a and b and c are known and x is unknown. The unknown is referred to as "aha" or heap. Problem 24, for instance, calls for the value of heap if heap and a seventh of heap is 19. The solution given by Ahmes is not that of modern textbooks, but is characteristic of a procedure now known as the "method of false position" or the "rule of false." A specific value, most likely a false one, is assumed for heap, and the operations indicated on the left-hand side of the equality sign are performed on this assumed number. The result of these operations then is compared with the result desired, and by the use of proportions the correct answer is found. In problem 24 the tentative value of the unknown is taken as 7, so that $x + \frac{1}{7}x$ is 8, instead of the desired answer, which was 19. Inasmuch as $8(2 + \frac{1}{4} + \frac{1}{8}) = 19$, one must multiply 7 by $2 + \frac{1}{4} + \frac{1}{8}$ to obtain the correct heap; Ahmes found the answer to be $16 + \frac{1}{2} + \frac{1}{8}$. Ahmes then "checked" his result by showing that if to $16 + \frac{1}{2} + \frac{1}{8}$ one adds a seventh of this (which is $2 + \frac{1}{4} + \frac{1}{8}$), one does indeed obtain 19. Here we see another significant step in the development of mathematics, for the check is a simple instance of a proof. Although the method of false position was generally used by Ahmes, there is one problem (Problem 30) in which $x + \frac{2}{3}x + \frac{1}{2}x + \frac{1}{7}x = 37$ is solved by factoring the left-hand side of the equation and dividing 37 by $1 + \frac{2}{3} + \frac{1}{2} + \frac{1}{7}$, the result being $16 + \frac{1}{56} + \frac{1}{679} + \frac{1}{776}$.

Many of the "aha" calculations in the Rhind Papyrus evidently are practice exercises for young students. Although a large proportion of them are of a practical nature, in some places the scribe seems to have had puzzles or mathematical recreations in mind. Thus Problem 79 cites only "seven houses, 49 cats, 343 mice, 2401 ears of spelt, 16807 hekats." It is presumed that the scribe was dealing with a problem, perhaps quite well known, in which in each of seven houses there are seven cats each of which eats seven mice, each of which would have eaten seven ears of grain, each of which would have produced seven measures of grain. The problem evidently called not for the practical answer, which would be the number of measures of grain that were saved, but for the impractical sum of the numbers of houses, cats, mice, ears of spelt, and measures of grain. This bit of fun in the Ahmes Papyrus seems to be a forerunner of our familiar nursery rhyme:

> As I was going to St. Ives,
> I met a man with seven wives;
> Every wife had seven sacks,
> Every sack had seven cats,
> Every cat had seven kits.
> Kits, cats, sacks, and wives,
> How many were going to St. Ives?

The Greek historian Herodotus tells us that the obliteration of boundaries **7** in the overflow of the Nile emphasized the need for surveyors. The accomplishments of the "rope-stretchers" of Egypt evidently were admired by Democritus, an accomplished mathematician and one of the founders of an atomic theory, and today their achievements seem to be overvalued, in part as a result of the admirable accuracy of construction of the pyramids. It often is said that the ancient Egyptians were familiar with the Pythagorean theorem, but there is no hint of this in the papyri that have come down to us. There are nevertheless some geometrical problems in the Ahmes Papyrus. Problem 51 of Ahmes shows that the area of an isosceles triangle was found by taking half of what we would call the base and multiplying this by the altitude. Ahmes justified his method of finding the area by suggesting that the isosceles triangle can be thought of as two right triangles, one of which can be shifted in position, so that together the two triangles form a rectangle. The isosceles trapezoid is similarly handled in Problem 52, in which the larger base of a trapezoid is 6, the smaller base is 4, and the distance between them is 20. Taking half the sum of the bases, "so as to make a rectangle," Ahmes multiplied this by 20 to find the area. In transformations such as these, in which isosceles triangles and trapezoids are converted into rectangles, we see the beginnings of a theory of congruence and of the idea of proof in geometry, but the Egyptians did not carry their work further. A serious deficiency in their geometry was the lack of a clear-cut distinction between relationships that are exact and those that are approximations only. A surviving deed from Edfu, dating from a period some 1500 years after Ahmes, gives examples of triangles, trapezoids, rectangles, and more general quadrilaterals; the rule for finding the area of the general quadrilateral is to take the product of the arithmetic means of the opposite sides. Inaccurate though the rule is, the author of the deed deduced from it a corollary—that the area of a triangle is half the sum of two sides multiplied by half the third side. This is a striking instance of the search for relationships among geometric figures, as well as an early use of the zero concept as a replacement for a magnitude in geometry.

The Egyptian rule for finding the area of a circle has long been regarded as one of the outstanding achievements of the time. In Problem 50 the scribe Ahmes assumed that the area of a circular field with a diameter of nine units is the same as the area of a square with a side of eight units. If we compare this assumption with the modern formula $A = \pi r^2$, we find the Egyptian rule to be equivalent to giving π a value of about $3\frac{1}{6}$, a commendably close approximation; but here again we miss any hint that Ahmes was aware that the areas of his circle and square were not *exactly* equal. It is possible that Problem 48 gives a hint to the way in which the Egyptians were led to their area of the circle. In this problem the scribe formed an octagon from a square of side nine units by trisecting the sides and cutting off the four corner isosceles triangles, each having an area of $4\frac{1}{2}$ units. The area of the octagon, which does not differ greatly from that of a circle inscribed within the square, is sixty-three units, which is not far removed from the area of a square with eight units on a side. That the number $4(8/9)^2$ did indeed play a role comparable to our constant π seems to be confirmed by the Egyptian rule for the circumference of a circle, according to which the ratio of the area of a

circle to the circumference is the same as the ratio of the area of the circumscribed square to its perimeter. This observation represents a geometrical
relationship of far greater precision and mathematical significance than the
relatively good approximation for π. Degree of accuracy in approximation is,
after all, not a good measure of either mathematical or architectural achievement, and we should not overemphasize this aspect of Egyptian work.
Recognition by the Egyptians of interrelationships among geometrical
figures, on the other hand, has too often been overlooked, and yet it is here
that they came closest in attitude to their successors, the Greeks. No theorem
or formal proof is known in Egyptian mathematics, but some of the geometric
comparisons made in the Nile Valley, such as those on the perimeters and
areas of circles and squares, are among the first exact statements in history
concerning curvilinear figures.

Problem 56 of the Rhind Papyrus is of special interest in that it contains **8**
rudiments of trigonometry and a theory of similar triangles. In the construction of the pyramids it had been essential to maintain a uniform slope for the
faces, and it may have been this concern that led the Egyptians to introduce
a concept equivalent to the cotangent of an angle. In modern technology it is
customary to measure the steepness of a straight line through the ratio of the
"rise" to the "run." In Egypt it was customary to use the reciprocal of this
ratio. There the word "seqt" meant the horizontal departure of an oblique
line from the vertical axis for every unit change in the height. The seqt thus
corresponded, except for the units of measurement, to the *batter* used today
by architects to describe the inward slope of a masonry wall or pier. The
vertical unit of length was the cubit; but in measuring the horizontal distance,
the unit used was the "hand," of which there were seven in a cubit. Hence
the seqt of the face of a pyramid was the ratio of run to rise, the former
measured in hands, the latter in cubits. In Problem 56 one is asked to find
the seqt of a pyramid that is 250 ells or cubits high and has a square base
360 ells on a side. The scribe first divided 360 by 2 and then divided the result
by 250, obtaining $\frac{1}{2} + \frac{1}{5} + \frac{1}{50}$ in ells. Multiplying the result by 7, he gave
the seqt as $5\frac{1}{25}$ in hands per ell. In other pyramid problems in the Ahmes
Papyrus the seqt turns out to be $5\frac{1}{4}$, agreeing somewhat better with that of
the great Cheops Pyramid, 440 ells wide and 280 high, the seqt being $5\frac{1}{2}$ hands
per ell.

There are many stories about presumed geometrical relationships among
dimensions in the Great Pyramid, some of which are patently false. For
instance, the story that the perimeter of the base was intended to be precisely
equal to the circumference of a circle of which the radius is the height of the
pyramid is not in agreement with the work of Ahmes. The ratio of perimeter
to height is indeed very close to $\frac{44}{7}$, which is just twice the value of $\frac{22}{7}$ often

used today for π; but we must recall that the Ahmes value for π is about $3\frac{1}{6}$, not $3\frac{1}{7}$. That Ahmes' value was used also by other Egyptians is confirmed in a papyrus roll from the twelfth dynasty (the Kahun Papyrus, now in London) in which the volume of a cylinder is found by multiplying the height by the area of the base, the base being determined according to Ahmes' rule.

9   Much of our information about Egyptian mathematics has been derived from the Rhind or Ahmes Papyrus, the most extensive mathematical document from ancient Egypt; but there are other sources as well.[8] Besides the Kahun Papyrus, already mentioned, there is a Berlin Papyrus of the same period, two wooden tablets from Akhmim (Cairo) of about 2000 B.C., a leather roll containing lists of unit fractions and dating from the later Hyksos period, and an important papyrus, known as the Golenischev or Moscow Papyrus, purchased in Egypt in 1893. The Moscow Papyrus is about as long as the Rhind Papyrus—about 18 feet—but it is only one-fourth as wide, the width being about 3 inches. It was written, less carefully than the work of Ahmes, by an unknown scribe of the twelfth dynasty (ca. 1890 B.C.). It contains twenty-five examples, mostly from practical life and not differing greatly from those of Ahmes, except for two that have special significance. Associated with Problem 14 in the Moscow Papyrus is a figure that looks like an isosceles trapezoid (see Fig. 2.1), but the calculations associated with it indicate that

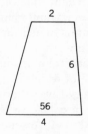

FIG. 2.1

a frustum of a square pyramid is intended. Above and below the figure are signs for two and four respectively, and within the figure are the hieratic symbols for six and fifty-six. The directions alongside make it clear that the problem calls for the volume of a frustum of a square pyramid six units high if the edges of the upper and lower bases are two and four units respectively. The scribe directs one to square the numbers two and four and to add to the sum of these squares the product of two and four, the result being twenty-eight. This result is then multiplied by a third of six; and the scribe concludes with the words, "See, it is 56; you have found it correctly." That is, the volume

[8] A good account of these appears in the work of Archibald cited in footnote 4.

of the frustum has been calculated in accordance with the modern formula $V = h(a^2 + ab + b^2)/3$, where h is the altitude and a and b are the sides of the square bases. Nowhere is this formula written out, but in substance it evidently was known to the Egyptians. If, as in the deed from Edfu, one takes $b = 0$, the formula reduces to the familiar formula, one-third the base times the altitude, for the volume of a pyramid. How these results were arrived at by the Egyptians is not known. An empirical origin for the rule on volume of a pyramid seems to be a possibility, but not for the volume of the frustum. For the latter a theoretical basis seems more likely; and it has been suggested that the Egyptians may have proceeded here as they did in the cases of the isosceles triangle and the isosceles trapezoid—they may in thought have broken the frustum into parallelepipeds, prisms, and pyramids.[9] Upon replacing the pyramids and prisms by equal rectangular blocks, a plausible grouping of the blocks leads to the Egyptian formula. One could, for example, have begun with a pyramid having a square base and with the vertex directly over one of the base vertices. An obvious decomposition of the frustum would be to break it into four parts as in Fig. 2.2—a rectangular parallelepiped

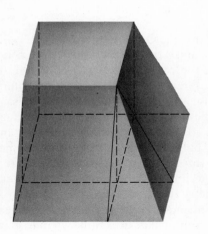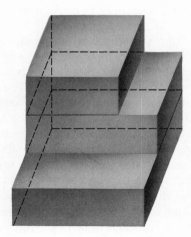

FIG. 2.2

having a volume b^2h, two triangular prisms, each with a volume of $b(a - b)h/2$, and a pyramid of volume $(a - b)^2h/3$. The prisms can be combined into a rectangular parallelepiped with dimensions b and $a - b$ and h; and the pyramid can be thought of as a rectangular parallelepiped with dimensions $a - b$ and $a - b$ and $h/3$. Upon cutting up the tallest parallelepipeds so that all altitudes are $h/3$, one can easily arrange the slabs so as to form three

[9] Van der Waerden, *Science Awakening*, p. 35. Cf. R. J. Gillings, "The Volume of a Truncated Pyramid in Ancient Egypt," *Mathematics Teacher*, **57** (1964), 552–555.

layers, each of altitude $h/3$, and having cross-sectional areas of a^2 and ab and b^2 respectively.

Problem 10 in the Moscow Papyrus presents a more difficult question of interpretation than does Problem 14. Here the scribe asks for the surface area of what looks like a basket with a diameter of $4\frac{1}{2}$. He proceeds as though he were using the equivalent of a formula $S = (1 - \frac{1}{9})^2(2x) \cdot x$, where x is $4\frac{1}{2}$, obtaining an answer of 32 units. Inasmuch as $(1 - \frac{1}{9})^2$ is the Egyptian approximation for $\pi/4$, the answer 32 would correspond to the surface of a hemisphere of diameter $4\frac{1}{2}$; and this was the interpretation given to the problem in 1930.[10] Such a result, antedating the oldest known calculation of a hemispherical surface by some 1500 years, would have been amazing, and it seems, in fact, to have been too good to be true. Later analysis[11] indicates that the "basket" may have been a roof—somewhat like that of a quonset hut in the shape of a half cylinder of diameter $4\frac{1}{2}$ and length $4\frac{1}{2}$. The calculation in this case calls for nothing beyond knowledge of the length of a semicircle; and the obscurity of the text makes it admissible to offer still more primitive interpretations, including the possibility that the calculation is only a rough estimate of the area of a domelike barn roof. In any case, we seem to have here an early calculation of a curvilinear surface area.

10   For many years it had been assumed that the Greeks had learned the rudiments of geometry from the Egyptians, and Aristotle argued that geometry had arisen in the Nile Valley because the priests there had the leisure to develop theoretical knowledge. That the Greeks did borrow some elementary mathematics from Egypt is probable, for the use of unit fractions persisted in Greece and Rome well into the Medieval period, but evidently they exaggerated the extent of their indebtedness. The knowledge indicated in extant Egyptian papyri is mostly of a practical nature, and calculation was the chief element in the questions. Where some theoretical elements appear to enter, the purpose may have been to facilitate technique rather than understanding. Even the once-vaunted Egyptian geometry turns out to have been mainly a branch of applied arithmetic. Where elementary congruence relations enter, the motive seems to be to provide mensurational devices rather than to gain insight. The rules of calculation seldom are motivated, and they concern specific concrete cases only. The Ahmes and Moscow papyri, our two chief sources of information, may have been only manuals intended for students,

---

[10] See W. W. Struve, "Mathematischer Papyrus des Staatlichen Museums der Schönen Künste in Moskau," *Quellen und Studien zur Geschichte der Mathematik*, Part A, *Quellen*, I (1930).

[11] See van der Waerden, *Science Awakening*, p. 34. Cf., however, R. J. Gillings, "The Area of the Curved Surface of a Hemisphere in Ancient Egypt," *The Australian Journal of Science*, **30** (1967), 113–116, in which the author concludes that the scribe of the Moscow Papyrus was indeed dealing correctly, in Problem 10, with the curved surface of a hemisphere.

but they nevertheless indicate the direction and tendencies in Egyptian mathematical instruction; further evidence provided by inscriptions on monuments, fragments of other mathematical papyri, and documents from related scientific fields serves to confirm the general impression. It is true that our two chief mathematical papyri are from a relatively early period, a thousand years before the rise of Greek mathematics, but Egyptian mathematics seems to have remained remarkably uniform throughout its long history. It was at all stages built around the operation of addition, a disadvantage that gave to Egyptian computation a peculiar primitivity combined with occasionally astonishing complexity. The fertile Nile Valley has been described as the world's largest oasis in the world's largest desert. Watered by one of the most gentlemanly of rivers and geographically shielded to a great extent from foreign invasion, it was a haven for peace-loving people who pursued, to a large extent, a calm and unchallenged way of life. Love of the beneficent gods, respect for tradition, and preoccupation with death and the needs of the dead, all encouraged a high degree of stagnation. Geometry may have been a gift of the Nile, as Herodotus believed, but the Egyptians did little with the gift. The mathematics of Ahmes was that of his ancestors and of his descendants. For more progressive mathematical achievements one must look to the more turbulent river valley known as Mesopotamia.

BIBLIOGRAPHY

Chace, A. B., L. S. Bull, H. P. Manning, and R. C. Archibald, eds., *The Rhind Mathematical Papyrus* (Oberlin, Ohio, 1927–1929, 2 vols.). This contains a comprehensive bibliography of works on Egyptian mathematics published in the interval from 1706 through 1927, as well as an extensive general account of Egyptian mathematics.

Gillings, R. J., "Problems 1 to 6 of the Rhind Mathematical Papyrus," *The Mathematics Teacher*, **55** (1962), 61–69. Continuations are found in later volumes of the journal.

Guggenbuhl, Laura, "Mathematics in Ancient Egypt: A Checklist," *The Mathematics Teacher*, **58** (1965), 630–634.

Neugebauer, O., *Die Grundlagen der ägyptischen Bruchrechnung* (Berlin: Springer, 1926).

Neugebauer, O., *The Exact Sciences in Antiquity*, 2nd ed. (Providence, R. I.: Brown University Press, 1957; paperback ed., New York; Harper Torchbook).

Parker, R. A., *The Calendars of Ancient Egypt* (Chicago: University of Chicago Press, 1950).

Struve, W. W., "Mathematischer Papyrus des Staatlichen Museums der Schönen Künste in Moskau," *Quellen and Studien zur Geschichte der Mathematik*, Part A, *Quellen*, **1** (1930).

Van der Waerden, B. L., "Die Entstehungsgeschichte der ägyptischen Bruchrechnung," *Quellen und Studien zur Geschichte der Mathematik*, Part B, *Studien*, IV (1937–1938), 359–382.

Van der Waerden, B. L., *Science Awakening*, trans. by Arnold Dresden (New York: Oxford University Press, 1961; paperback ed., New York: Wiley, 1963).

Vogel, Kurt., *Vorgriechische Mathematik*, Vol. I, *Vorgeschichte und Ägypten* (paperback ed., Hannover: Hermann Schroedel, ca. 1958).

Wheeler, Noel F., "Pyramids and Their Purpose," *Antiquity*, **9** (1935), 5–21, 161–189, 292–304.

EXERCISES

1. Describe the evidence on which our estimate of Egyptian mathematics is based. Do you think that this is likely to be altered by the discovery of new documents? Explain.

2. Do you think that astronomy was a more significant factor than surveying in the rise of Egyptian mathematics? Explain.

3. What does the word "geometry" mean etymologically? Is the use of the word justifiable in the light of the historical origin of the subject? Explain.

4. What do you regard as the three chief shortcomings in Egyptian mathematics? Explain why you regard these as the most significant.

5. What do you regard as the three chief contributions of Egypt to the development of mathematics? Explain why you regard them as important.

6. Write the number 7654 in Egyptian hieroglyphic form. How does this differ from the way in which Ahmes would write this number?

7. Express $\frac{2}{103}$ as a sum of two unequal unit fractions, and write these in Egyptian hieroglyphic notation. How does the hieratic form differ from this?

8. Solve by the method of false position the equation $x + \frac{1}{2}x = 16$. (This is Problem 25 in the Ahmes papyrus.)

9. Solve the following problem from the Ahmes Papyrus (Problem 40): Divide 100 loaves among five men so that the shares are in arithmetic progression and so that one seventh of the sum of the three largest shares is equal to the sum of the two smallest.

10. Solve in the Egyptian manner the simultaneous equations $x^2 + y^2 = 100$, $y = 3x/4$, taken from a Berlin papyrus from ancient Egypt. (Use the method of "double false," starting from an assumed value of x, finding the corresponding value of y from the second equation, and adjusting the values so that they satisfy the first equation.)

11. Through duplation and mediation (that is, successive doubling and halving) find $101 \div 16$, expressing the result in Egyptian hieroglyphic form.

12. Derive the Egyptian formula for volume of a frustum of a square pyramid algebraically from the known formula for volume of a pyramid, using proportions established in elementary geometry. Do you believe that the Egyptians could have derived their formula in this way? Explain.

13. To what extent is it fair to say that the Egyptians knew the area of the circle? Explain clearly.

14. Why do you think that the Egyptians preferred the decomposition $\frac{2}{15} = \frac{1}{10} + \frac{1}{30}$ to the alternative $\frac{2}{15} = \frac{1}{12} + \frac{1}{20}$?

15. Show that if n is a multiple of three, $2/n$ can be broken into the sum of two unit fractions, one of which is half of $1/n$.

16. Show that if n is a multiple of five, $2/n$ can be broken into the sum of two unit fractions, one of which is a third of $1/n$.

17. Justify the method of solution used by Ahmes in his Problem 63. (See text.)

18. Justify the assumption made by Ahmes that the ratio of the area of a circle to its circumference is the same as the ratio of the area of the circumscribed square to the perimeter of

this square.

19. If the seqt of a pyramid is 5 palms (or hands) and 1 finger per cubit, and if the side of its base is 140 cubits, what is its altitude? (This is Problem 57 in the Ahmes Papyrus.) There are five fingers in a palm.

*20. Using the Egyptian method of division, solve the following problem (Problem 31) from the Ahmes Papyrus: A quantity and its two thirds and its half and its one seventh together make 33. Find the quantity. [The answer given is $14 + \frac{1}{4} + \frac{1}{56} + \frac{1}{97} + \frac{1}{194} + \frac{1}{388} + \frac{1}{679} + \frac{1}{776}$.]

CHAPTER III

Mesopotamia

How much is one god beyond the other god?
An Old Babylonian astronomical text

1   The fourth millennium before our era was a period of remarkable cultural development, bringing with it the use of writing, of the wheel, and of metals. As in Egypt during the first dynasty, which began toward the end of this wonderful millennium, so also in the Mesopotamian valley there was at the time a high order of civilization. There the Sumerians had built homes and temples decorated with artistic pottery and mosaics in geometrical patterns. Powerful rulers united the local principates into an empire which completed vast public works, such as a system of canals to irrigate the land and to control flooding. The Biblical account of the Noachian flood had an earlier counterpart in the legend concerning the Sumerian hero Utnapischtum and the flooding of the region between the Tigris and Euphrates rivers, where the overflow of the rivers was not predictable, as was the inundation of the Nile Valley. The Bible tells us that Abraham came from the city of Ur, a Sumerian settlement where the Euphrates emptied into the Persian Gulf, for at that time the two rivers did not join, as they now do, before reaching the Gulf. The cuneiform pattern of writing that the Sumerians had developed during the fourth millennium, long before the days of Abraham, may have been the earliest form of written communication, for it probably antedates the Egyptian hieroglyphic, which may have been a derivative. Although they have nothing in common, it is an interesting coincidence that the origins of writing and of wheeled vehicles are roughly coeval.

The Mesopotamian civilizations of antiquity often are referred to as Babylonian, although such a designation is not strictly correct. The city of Babylon was not at first, nor was it always at later periods, the center of the culture associated with the two rivers, but convention has sanctioned the informal use of the name "Babylonian" for the region during the interval from about 2000 to roughly 600 B.C. When in 538 B.C. Babylon fell to Cyrus of Persia, the city was spared, but the Babylonian empire had come to an end. "Babylonian" mathematics, however, continued through the Seleucid period in Syria almost to the dawn of Christianity. Occasionally the area between the rivers is known also as Chaldea, because the Chaldeans (or Kaldis), originally from southern Mesopotamia, were for a time dominant, chiefly during the late seventh century B.C., throughout the region between the rivers. Then, as today, the Land of the Two Rivers was open to invasions from many directions, making of the Fertile Crescent a battlefield with frequently changing hegemony. One of the most significant of the invasions was that by the Semitic Akkadians under Sargon I (ca. 2276–2221 B.C.) or Sargon the Great. He established an empire that extended from the Persian Gulf in the south to the Black Sea in the north, and from the steppes of Persia on the east to the Mediterranean Sea on the west. Under Sargon there was begun a gradual absorption by the invaders of the indigenous Sumerian culture, including the cuneiform script. Later invasions and revolts brought varying racial strains—Ammorites, Kassites, Elamites, Hittites, Assyrians, Medes, Persians, and others—to political power at one time or another in the valley, but there remained in the area a sufficiently high degree of cultural unity to justify referring to the civilization simply as Mesopotamian. In particular, the use of cuneiform script formed a strong bond. Laws, tax accounts, stories, school lessons, personal letters—these and many other records were impressed on soft clay tablets with a stylus, and the tablets then were baked in the hot sun or in ovens. Such written documents, fortunately, were far less vulnerable to the ravages of time than were Egyptian papyri; hence there is available today a much larger body of evidence about Mesopotamian than about Nilotic mathematics. From one locality alone, the site of ancient Nippur, we have some 50,000 tablets. The university libraries at Columbia, Pennsylvania, and Yale, among others, have large collections of ancient tablets from Mesopotamia, some of them mathematical. Despite the availability of documents, however, it was the Egyptian hieroglyphic rather than the Babylonian cuneiform that first was deciphered in modern times. Some progress in the reading of Babylonian script had been made early in the nineteenth century by Grotefend, but it was only during the second quarter of the twentieth century that substantial accounts of Mesopotamian mathematics began to appear in histories of antiquity.[1]

The early use of writing in Mesopotamia is attested by hundreds of clay tablets found in Uruk and dating from about 5000 years ago. By this time picture writing had reached the point where conventionalized stylized forms were used for many things: ≈ for water, ⌓ for eye, and combinations of

---

[1] See especially O. Neugebauer, *The Exact Sciences in Antiquity* (1957) and B. L. van der Waerden, *Science Awakening* (1961). Cf. also O. Neugebauer and A. Sachs, *Mathematical Cuneiform Texts* (American Oriental Series, Vol. 29, 1945). A good secondary account and further references will be found in R. C. Archibald, *Outline of the History of Mathematics* (*American Mathematical Monthly*, **56** (1949), No. 1, supp.).

these to indicate weeping. Gradually the number of signs became smaller, so that of some 2000 Sumerian signs originally used only a third remained by the time of the Akkadian conquest. Primitive drawings gave way to combinations of wedges: water became ⑂ and eye ⊨⫪⊢. At first the scribe wrote from top to bottom in columns from right to left; later, for convenience, the table was rotated counter clockwise through 90°, and the scribe wrote from left to right in horizontal rows from top to bottom. The stylus, which formerly had been a triangular prism, was replaced by a right circular cylinder—or, rather, two cylinders of unequal radius. During the earlier days of the Sumerian civilization, the end of the stylus was pressed into the clay vertically to represent ten units and obliquely to represent a unit, using the smaller stylus; similarly, an oblique impression with the larger stylus indicated sixty units and a vertical impression indicated 3600 units. Combinations of these were used to represent intermediate numbers.

As the Akkadians adopted the Sumerian form of writing, lexicons were compiled giving equivalents in the two tongues, and forms of words and numerals became less varied. Thousands of tablets from about the time of the Hammurabi dynasty (ca. 1800–1600 B.C.) illustrate a number system that had become well established. The decimal system, common to most civilizations, both ancient and modern, had been submerged in Mesopotamia under a notation that made fundamental the base sixty. Much has been written about the motives behind this change; it has been suggested that astronomical considerations may have been instrumental or that the sexagesimal scheme may have been the natural combination of two earlier schemes, one decimal and the other using the base six. It appears more likely, however, that the base sixty was consciously adopted and legalized in the interests of metrology, for a magnitude of sixty units can be subdivided easily into halves, thirds, fourths, fifths, sixths, tenths, twelfths, fifteenths, twentieths, and thirtieths, thus affording ten possible subdivisions. Whatever the origin, the sexagesimal system of numeration has enjoyed a remarkably long life, for remnants survive, unfortunately for consistency, even to this day in units of time and angle measure, despite the fundamentally decimal form of our society.

2   Babylonian cuneiform numeration, for smaller whole numbers, proceeded along the same lines as did the Egyptian hieroglyphic, with repetitions of the symbols for units and tens. Where the Egyptian architect, carving on stone might write fifty-nine as ⸜⸜⸜|||, the Mesopotamian scribe could similarly represent the same number on a clay tablet through fourteen wedge-shaped marks—five broad sideways wedges or "angle-brackets," each representing ten units, and nine thin vertical wedges, each standing for a

unit, all juxtaposed in a neat group as ⟨𝖲𝖶𝖶. Beyond the number fifty-nine, however, the Egyptian and Babylonian systems differed markedly. Perhaps it was the inflexibility of the Mesopotamian writing materials, possibly it was a flash of imaginative insight that made the Babylonians aware that their two symbols for units and tens sufficed for the representation of any integer, however large, without excessive repetitiveness. This was made possible through their invention, some 4000 years ago, of the positional notation—the same principle that accounts for the effectiveness of our present numeral forms. That is, the ancient Babylonians saw that their symbols could do double, triple, quadruple, or any degree of duty simply by being assigned values that depend on their relative positions in the representation of a number. The wedges in the cuneiform symbol for fifty-nine are tightly grouped together so as to form almost the equivalent of a single cipher. Appropriate spacing between groups of wedges can establish positions, read from right to left, that correspond to ascending powers of the base; each group then has a "local value" that depends on its position. Our number 222 makes use of the same cipher three times, but with a different meaning each time. Once it represents two units, then it means two tens, and finally it stands for two hundreds (that is, twice the square of the base ten). In a precisely analogous way the Babylonians made multiple use of such a symbol as ᵣᵣ. When they wrote ᵣᵣ ᵣᵣ ᵣᵣ, clearly separating the three groups of two wedges each, they understood the right-hand group to mean two units, the next group to mean twice their base, sixty, and the left-hand group to signify twice the square of their base. This numeral therefore denoted $2(60)^2 + 2(60) + 2$ (or 7322 in our notation).

There is a wealth of primary material concerning Mesopotamian mathematics, but oddly enough most of it comes from two periods widely separated in time. There is an abundance of tablets from the first few hundred years of the second millennium B.C. (the Old Babylonian age), and there are many also from the last few centuries of the first millennium B.C. (the Seleucid period). Most of the important contributions to mathematics will be found to go back to the earlier period, but there is one contribution not in evidence until almost 300 B.C. The Babylonians seem at first to have had no clear way in which to indicate an "empty" position—that is, they did not have a zero symbol, although they sometimes left a space where a zero was intended. This meant that their forms for the numbers 122 and 7202 looked very much alike, for ᵣᵣ ᵣᵣ might mean either $2(60) + 2$ or $2(60)^2 + 2$. Context in many cases could be relied on to relieve some of the ambiguity; but the lack of a zero symbol, such as enables us to distinguish at a glance between 22 and 202, must have been quite inconvenient. By about the time of the conquest by Alexander the Great, however, a special sign, consisting of two small wedges placed obliquely, was invented to serve as a placeholder where a numeral was missing. From that time on, as long as cuneiform was used, the number ⴸⴸ, or $2(60)^2 + 0(60) + 2$, was readily distinguishable from ⴸⴸ, or $2(60) + 2$.

The Babylonian zero symbol apparently did not end all ambiguity, for the sign seems to have been used for intermediate empty positions only. There are no extant tablets in which the zero sign appears in a terminal position. This means that the Babylonians in antiquity never achieved an absolute positional system. Position was relative only; hence the symbol ⴸⴸ could represent $2(60) + 2$ or $2(60)^2 + 2(60)$ or $2(60)^3 + 2(60)^2$ or any one of indefinitely many other numbers in which two successive positions are involved.

3   Had Mesopotamian mathematics, like that of the Nile Valley, been based on the addition of integers and unit fractions, the invention of the positional notation would not have been of great significance at the time. It is not much more difficult to write 98,765 in hieroglyphic notation than in cuneiform, and the latter is definitely more difficult to write than the same number in hieratic script. The secret of the clear superiority of Babylonian mathematics over that of the Egyptians undoubtedly lies in the fact that those who lived "between the two rivers" took the most felicitous step of extending the principle of position to cover fractions as well as whole numbers. That is, the notation ⴸⴸ was used not only for $2(60) + 2$, but also for $2 + (60)^{-1}$ or for $2(60)^{-1} + 2(60)^{-2}$ or for other fractional forms involving two successive positions. This meant that the Babylonians had at their command the computational power that the modern decimal fractional notation affords us today. For the Babylonian scholar, as for the modern engineer, the addition or the multiplication of 23.45 and 9.876 was essentially no more difficult than was the addition or multiplication of the whole numbers 2345 and 9876; and the Mesopotamians were quick to exploit this important discovery. An Old Babylonian tablet from the Yale Collection (No. 7289) includes the calculation of the square root of two to three sexagesimal places, the answer being written ⴸⴸ. In modern characters this number can be appropriately written as $1;24,51,10$, where a semicolon is used to separate the integral and fractional parts and a comma is used as a separatrix for the sexagesimal positions. This form will generally be used throughout this chapter to designate numbers in sexagesimal notation. This Babylonian value for $\sqrt{2}$ is equal to approximately 1.414222, differing by about 0.000008 from the true value. Accuracy in approximations was relatively easy for the Babylonians to achieve with their fractional notation, the best that any civilization afforded until the time of the Renaissance.

4   The effectiveness of Babylonian computation did not result from their system of numeration alone. Mesopotamian mathematicians were skillful

in developing algorithmic procedures, among which was a square-root process often ascribed to later men. It sometimes is attributed to the Greek scholar Archytas (428–365 B.C.) or to Heron of Alexandria (ca. 100); occasionally one finds it called Newton's algorithm. This Babylonian procedure is as simple as it is effective. Let $x = \sqrt{a}$ be the root desired and let a_1 be a first approximation to this root; let a second approximation b_1 be found from the equation $b_1 = a/a_1$. If a_1 is too small, then b_1 is too large, and vice versa. Hence the arithmetic mean $a_2 = \frac{1}{2}(a_1 + b_1)$ is a plausible next approximation. Inasmuch as a_2 always is too large, the next approximation $b_2 = a/a_2$ will be too small, and one takes the arithmetic mean $a_3 = \frac{1}{2}(a_2 + b_2)$ to obtain a still better result; the procedure can be continued indefinitely. The value of $\sqrt{2}$ on Yale table 7289 will be found to be that of a_3, where $a_1 = 1;30$. In the Babylonian square-root algorithm one finds an iterative procedure that could have put the mathematicians of the time in touch with infinite processes, but scholars of the time did not pursue the implications of such problems.

The algorithm just described is equivalent to a two-term approximation to the binomial series, a case with which the Babylonians were familiar. If $\sqrt{a^2 + b}$ is desired, the approximation $a_1 = a$ leads to $b_1 = (a^2 + b)/a$ and $a_2 = (a_1 + b_1)/2 = a + b/(2a)$, which is in agreement with the first two terms in the expansion of $(a^2 + b)^{1/2}$ and provides an approximation found in Old Babylonian texts. Despite the efficacy of their rule for square roots, the Mesopotamian scribes seem to have imitated the modern applied mathematician in having frequent recourse to the ubiquitous tables that were available. In fact, a substantial proportion of the cuneiform tablets that have been unearthed are "table texts," including multiplication tables, tables of reciprocals, and tables of squares and cubes and of square and cube roots written, of course, in cuneiform sexagesimals. One of these, for example, carries the equivalents of the entries shown in the table below. The product

2	30
3	20
4	15
5	12
6	10
8	7,30
9	6,40
10	6
12	5

of elements in the same line is in all cases 60, the Babylonian number base, and the table apparently was thought of as a table of reciprocals. The sixth line, for example, denotes that the reciprocal of 8 is $7/60 + 30/(60)^2$. It will be noted that the reciprocals of 7 and 11 are missing from the table, because the reciprocals of such "irregular" numbers are nonterminating sexagesimals, just as in our decimal system the reciprocals of 3, 6, 7, and 9 are infinite when expanded decimally. Again the Babylonians were faced by the problem of infinity, but they did not consider it systematically. At one point, however, a Mesopotamian scribe seems to give upper and lower bounds for the reciprocal of the irregular number 7, placing it between $0;8,34,16,59$ and $0;8,34,18$. With their penchant for multipositional computations, it is tantalizing not to find among them a recognition of the simple three-place periodicity in the sexagesimal representation of $\frac{1}{7}$, a discovery that could have provoked considerations of infinite series.

It is clear that the fundamental arithmetic operations were handled by the Babylonians in a manner not unlike that which would be employed today, and with comparable facility. Division was not carried out by the clumsy duplication method of the Egyptians, but through an easy multiplication of the dividend by the reciprocal of the divisor, using the appropriate items in the table texts. Just as today the quotient of 34 divided by 5 is easily found by multiplying 34 by 2 and shifting the decimal point, so in antiquity the same division problem was carried out by finding the product of 34 by 12 and shifting one sexagesimal place to obtain $6\frac{48}{60}$. Tables of reciprocals in general furnished reciprocals of "regular" integers only—that is, those that can be written as products of twos, threes, and fives—although there are a few exceptions. One table text includes the approximations $\frac{1}{59} = \ ;1,1,1$ and $\frac{1}{61} = \ ;0,59,0,59$. Here we have sexagesimal analogues of our decimal expressions $\frac{1}{9} = .11\overline{1}$ and $\frac{1}{11} = .09\overline{09}$, unit fractions in which the denominator is one more or one less than the base; but it appears again that the Babylonians did not notice, or at least did not regard as significant, the infinite periodic expansions in this connection.[2]

One finds among the Old Babylonian tablets some table texts containing successive powers of a given number, analogous to our modern tables of logarithms, or, more properly speaking, of antilogarithms. Exponential (or logarithmic) tables have been found in which the first ten powers are listed for the bases 9 and 16 and 1,40 and 3,45 (all perfect squares). The question raised in a problem text, to what power must a certain number be raised in order to yield a given number, is equivalent to our question, what is the logarithm of the given number in a system with the certain number as base. The chief differences between the ancient tables and our own, apart from matters of language and notation, are that no single number was systematically used as a base in varied connections and that the gaps between entries in the ancient tables are far larger than in our tables. Then, too, their "logarithm tables" were not used for general purposes of calculation, but rather to solve certain very specific questions.

[2] In addition to the references cited in footnote 1, see also Kurt Vogel, *Vorgriechische Mathematik*, Vol. II, *Die Mathematik der Babylonier* (1959).

Despite the large gaps in their exponential tables, Babylonian mathematicians did not hesitate to interpolate by proportional parts to approximate intermediate values. Linear interpolation seems to have been a commonplace procedure in ancient Mesopotamia, and the positional notation lent itself conveniently to the rule of three. A clear instance of the practical use of interpolation within exponential tables is seen in a problem text that asks how long it will take money to double at 20 per cent annually; the answer given is 3;47,13,20. It seems to be quite clear that the scribe used linear interpolation between the values for $(1;12)^3$ and $(1;12)^4$, following the compound interest formula $a = P(1 + r)^n$, where r is 20 per cent or $\frac{12}{60}$, and reading values from an exponential table with powers of $1;12$.

One table for which the Babylonians found considerable use is not **5** generally included in handbooks of today. This is a tabulation of the values of $n^3 + n^2$ for integral values of n, a table essential in Babylonian algebra; this subject reached a considerably higher level in Mesopotamia than in Egypt. Many problem texts from the Old Babylonian period show that the solution of the complete three-term quadratic equation afforded the Babylonians no serious difficulty, for flexible algebraic operations had been developed. They could transpose terms in an equation by adding equals to equals, and they could multiply both sides by like quantities to remove fractions or to eliminate factors. By adding $4ab$ to $(a - b)^2$ they could obtain $(a + b)^2$, for they were familiar with many simple forms of factoring. They did not use letters for unknown quantities, for the alphabet had not yet been invented, but words such as "length," "breadth," "area," and "volume" served effectively in this capacity. That these words may well have been used in a very abstract sense is suggested by the fact that the Babylonians had no qualms about adding a "length" to an "area" or an "area" to a "volume." Such problems, if taken literally, could have had no practical basis in mensuration.

Egyptian algebra had been much concerned with linear equations, but the Babylonians evidently found these too elementary for much attention. In one problem the weight x of a stone is called for if $(x + x/7) + \frac{1}{11}(x + x/7)$ is one mina; the answer is simply given as 48;7,30 gin, where 60 gin make a mina. In another problem in an Old Babylonian text we find two simultaneous linear equations in two unknown quantities, called respectively the "first silver ring" and the "second silver ring." If we call these x and y in our notation, the equations are $x/7 + y/11 = 1$ and $6x/7 = 10y/11$. The answer is expressed laconically in terms of the rule

$$x/7 = \frac{11}{7 + 11} + \frac{1}{72} \quad \text{and} \quad \frac{y}{11} = \frac{7}{7 + 11} - \frac{1}{72}$$

In another pair of equations part of the method of solution is included in the text. Here $\frac{1}{4}$ width + length = 7 hands, and length + width = 10 hands. The solution is first found by replacing each "hand" by 5 "fingers" and then noticing that a width of 20 fingers and a length of 30 fingers will satisfy both equations. Following this, however, the solution is found by an alternative method equivalent to an elimination through combination. Expressing all dimensions in terms of hands, and letting the length and width be x and y respectively, the equations become $y + 4x = 28$ and $x + y = 10$. Subtracting the second equation from the first, one has the result $3x = 18$; hence $x = 6$ hands or 30 fingers and $y = 20$ fingers.

6   The solution of a three-term quadratic equation seems to have exceeded by far the algebraic capabilities of the Egyptians, but Neugebauer in 1930 disclosed that such equations had been handled effectively by the Babylonians in some of the oldest problem texts. For instance, one problem calls for the side of a square if the area less the side is 14,30. The solution of this problem, equivalent to solving $x^2 - x = 870$, is expressed as follows:

> Take half of 1, which is 0;30, and multiply 0;30 by 0;30, which is 0;15; add this to 14,30 to get 14,30;15. This is the square of 29;30. Now add 0;30 to 29;30, and the result is 30, the side of the square.

The Babylonian solution is, of course, exactly equivalent to the formula $x = \sqrt{(p/2)^2 + q} + p/2$ for a root of the equation $x^2 - px = q$—the quadratic formula that is familiar to schoolboys of today. In another text the equation $11x^2 + 7x = 6;15$ was reduced by the Babylonians to the standard type $x^2 + px = q$ by first multiplying through by 11 to obtain $(11x)^2 + 7(11x) = 1,8;15$. This is a quadratic in normal form in the unknown quantity $y = 11x$, and the solution for y is easily obtained by the familiar rule $y = \sqrt{(p/2)^2 + q} - p/2$, from which the value of x is then determined. This solution is remarkable as an instance of the use of algebraic transformations.

Until modern times there was no thought of solving a quadratic equation of the form $x^2 + px + q = 0$, where p and q are positive, for the equation has no positive root. Consequently, quadratic equations in ancient and Medieval times—and even in the early modern period—were classified under three types:

$$(1) \quad x^2 + px = q$$

$$(2) \quad x^2 = px + q$$

$$(3) \quad x^2 + q = px$$

All three types are found in Old Babylonian texts of some 4000 years ago.

The first two types are illustrated by the problems given above; the third type appears frequently in problem texts, where it is treated as equivalent to the simultaneous system $x + y = p$, $xy = q$. So numerous are problems in which one is asked to find two numbers when given their product and either their sum or their difference that these seem to have constituted for the ancients, both Babylonian and Greek, a sort of "normal form" to which quadratics were reduced. Then by transforming the simultaneous equations $xy = a$ and $x \pm y = b$ into the pair of linear equations $x \pm y = b$ and $x \mp y = \sqrt{b^2 \mp 4a}$, the values of x and y are found through an addition and a subtraction. A Yale cuneiform tablet, for example, asks for the solution of the system $x + y = 6;30$ and $xy = 7;30$. The instructions of the scribe are essentially as follows. First find

$$\frac{x + y}{2} = 3;15$$

and then find

$$\left(\frac{x + y}{2}\right)^2 = 10;33,45$$

Then

$$\left(\frac{x + y}{2}\right)^2 - xy = 3;3,45$$

and

$$\sqrt{\left(\frac{x + y}{2}\right)^2 - xy} = 1;45$$

Hence

$$\left(\frac{x + y}{2}\right) + \left(\frac{x - y}{2}\right) = 3;15 + 1;45$$

and

$$\left(\frac{x + y}{2}\right) - \left(\frac{x - y}{2}\right) = 3;15 - 1;45$$

From the last two equations it is obvious that $x = 5$ and $y = 1\frac{1}{2}$. Because the quantities x and y enter symmetrically in the given conditional equations, it is possible to interpret the values of x and y as the two roots of the quadratic equation $x^2 + 7;30 = 6;30x$. Another Babylonian text calls for a number which when added to its reciprocal becomes $2;0,0,33,20$. This leads to a quadratic of type 3, and again we have two solutions, $1;0,45$ and $0;59,15,33,20$.

7   The Babylonian reduction of a quadratic equation of the form $ax^2 + bx = c$ to the normal form $y^2 + by = ac$ through the substitution $y = ax$ shows the extraordinary degree of flexibility in Mesopotamian algebra. This facility, coupled with the place-value idea in computation, accounts in large measure for the superiority of the Babylonians in mathematics. There is no record in Egypt of the solution of a cubic equation, but among the Babylonians there are many instances of this. Pure cubics, such as $x^3 = 0;7,30$, were solved by direct reference to tables of cubes and cube roots, where the solution $x = 0;30$ was read off. Linear interpolation within the tables was used to find approximations for values not listed in the tables. Mixed cubics in the standard form $x^3 + x^2 = a$ were solved similarly by reference to the available tables which listed values of the combination $n^3 + n^2$ for integral values of n from 1 to 30. With the help of these tables they read off easily that the solution, for example, of $x^3 + x^2 = 4,12$ is approximately 6. For still more general cases of equations of third degree, such as $144x^3 + 12x^2 = 21$, the Babylonians used their method of substitution. Multiplying both sides by 12 and using $y = 12x$, the equation becomes $y^3 + y^2 = 4,12$, from which y is found to be approximately 6, hence x is about $\frac{1}{2}$ or $0;30$. Cubics of the form $ax^3 + bx^2 = c$ are reducible to the Babylonian normal form by multiplying through by a^2/b^3 to obtain $(ax/b)^3 + (ax/b)^2 = ca^2/b^3$, a cubic of standard type in the unknown quantity ax/b. Reading off from the tables the value of this unknown quantity, the value of x is determined. Whether or not the Babylonians were able to reduce the general four-term cubic, $ax^3 + bx^2 + cx = d$, to their normal form is not known. That it is not too unlikely that they could reduce it is indicated by the fact that a solution of a quadratic suffices to carry the four-term equation to the three-term form $px^3 + qx^2 = r$, from which, as we have seen, the normal form is readily obtained. There is, however, no evidence now available that would suggest that the Mesopotamian mathematicians actually carried out such a reduction of the general cubic equation.

The solution of quadratic and cubic equations in Mesopotamia is a remarkable achievement to be admired not so much for the high level of technical skill as for the maturity and flexibility of the algebraic concepts that are involved. With modern symbolism it is a simple matter to see that $(ax^3) + (ax)^2 = b$ is essentially the same type of equation as $y^3 + y^2 = b$; but to recognize this without our notation is an achievement of far greater significance for the development of mathematics than even the vaunted positional principle in arithmetic that we owe to the same civilization. Babylonian algebra had reached such an extraordinary level of abstraction that the equations $ax^4 + bx^2 = c$ and $ax^8 + bx^4 = c$ were recognized as nothing worse than quadratic equations in disguise—that is, quadratics in x^2 and x^4.

The algebraic achievements of the Babylonians are admirable, but the **8** motives behind this work are not easy to understand. It commonly has been supposed that virtually all pre-Hellenic science and mathematics were purely utilitarian; but what sort of real-life situation in ancient Babylon could possibly lead to problems involving the sum of a number and its reciprocal or a difference between an area and a length? If utility was the motive, then the cult of immediacy was less strong than it is now, for direct connections between purpose and practice in Babylonian mathematics are far from apparent. That there may well have been toleration for, if not encouragement of, mathematics for its own sake is suggested by a tablet (No. 322) in the Plimpton Collection at Columbia University.[3] The tablet dates from the Old Babylonian period (ca. 1900 to 1600 B.C.), and the tabulations it contains could easily be mistaken for a record of business accounts. Analysis, however, shows that it has deep mathematical significance in the theory of numbers and that it was perhaps related to a kind of prototrigonometry. Plimpton 322 was part of a larger tablet, as is illustrated by the break along the left-hand edge, and the remaining portion contains four columns of numbers arranged in fifteen horizontal rows. The right-hand column contains the digits from one to fifteen, and its purpose evidently was simply to identify in order the items in the other three columns, arranged as follows.

1,59,0,15	1,59	2,49	1
1,56,56,58,14,50,6,15	56,7	1,20,25	2
1,55,7,41,15,33,45	1,16,41	1,50,49	3
1,53,10,29,32,52,16	3,31,49	5,9,1	4
1,48,54,1,40	1,5	1,37	5
1,47,6,41,40	5,19	8,1	6
1,43,11,56,28,26,40	38,11	59,1	7
1,41,33,59,3,45	13,19	20,49	8
1,38,33,36,36	8,1	12,49	9
1,35,10,2,28,27,24,26,40	1,22,41	2,16,1	10
1,33,45	45,0	1,15,0	11
1,29,21,54,2,15	27,59	48,49	12
1,27,0,3,45	2,41	4,49	13
1,25,48,51,35,6,40	29,31	53,49	14
1,23,13,46,40	56	1,46	15

The tablet is not in such excellent condition that all the numbers can still be read, but the clearly discernible pattern of construction in the table made it possible to determine from context the few items that were missing because

---

[3] Further description of this table will be found in Neugebauer, *Exact Sciences in Antiquity*, pp. 36–40. A good account of it appears also in Howard Eves, *An Introduction to the History of Mathematics* (1964), pp. 35–37. A scholarly interpretation of the possible motivation behind the table text is given by D. J. de Solla Price, "The Babylonian 'Pythagorean Triangle' Tablet," *Centaurus*, **10** (1964), 219–231.

of small fractures. To understand what the entries in the table probably meant to the Babylonians, consider the right triangle *ABC* (Fig. 3.1). If the numbers in the second and third columns (from left to right) are thought of as the sides *a* and *c* respectively of the right triangle, then the first or left-hand column contains in each case the square of the ratio of *c* to *b*. The left-hand column therefore is a short table of values of $\sec^2 A$, but we must not assume that the Babylonians were familiar with our secant concept. Neither the Egyptians nor the Babylonians introduced a measure of angles in the modern sense. Nevertheless, the rows of numbers in Plimpton 322 are not arranged in haphazard fashion, as a superficial glance might imply. If the first comma

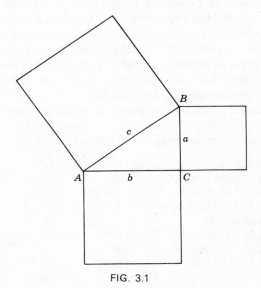

FIG. 3.1

in column one (on the left) is replaced by a semicolon, it is obvious that the numbers in this column decrease steadily from top to bottom. Moreover, the first number is quite close to $\sec^2 45°$, and the last number in the column is approximately $\sec^2 31°$, with the intervening numbers close to the values of $\sec^2 A$ as *A* decreases by degrees from 45° to 31°. This arrangement obviously is not the result of chance alone. Not only was the arrangement carefully thought out, but the dimensions of the triangle were also derived according to a rule. Those who constructed the table evidently began with two regular sexagesimal integers, which we shall call *p* and *q*, with $p > q$, and then formed the triple of numbers $p^2 - q^2$ and $2pq$ and $p^2 + q^2$. The three integers thus obtained are easily seen to form a Pythagorean triple in which the square of the largest is equal to the sum of the squares of the other two. Hence these numbers can be used as the dimensions of the right triangle ABC, with $a = p^2 - q^2$ and $b = 2pq$ and $c = p^2 + q^2$. Restricting themselves to values of p less than 60 and to corresponding values of q such that $1 < p/q < 1 + \sqrt{2}$—that is, to right triangles for which $a < b$—the Babylonians presumably found that there were just 38 possible pairs of values of p and q satisfying the conditions, and for these they apparently formed the 38 corresponding Pythagorean triples. Only the first 15, arranged in descending order for the ratio $(p^2 + q^2)/2pq$, are included in the table on the tablet, but it is likely that the scribe had intended to continue the table on the other side of the tablet. It has been suggested also[4] that the portion of Plimpton 322 that has been broken off from the left side contained four additional columns in which were tabulated the values of p and q and $2pq$ and what we should now call $\tan^2 A$.

The Plimpton tablet 322 might give the impression that it is an exercise in the theory of numbers, but it is likely that this aspect of the subject was merely ancillary to the problem of measuring the areas of squares on the sides of a right triangle. The Babylonians disliked working with the reciprocals of irregular numbers, for these could not be expressed exactly in finite

Plimpton 322.

[4] See the explanation given by Price (whose suggestion we have here been following) in the article cited in footnote 3.

sexagesimal fractions. Hence they were interested in values of p and q that should give them regular integers for the sides of right triangles of varying shape, from the isosceles right triangle down to one with a small value for the ratio a/b. For example, the numbers in the first row are found by starting with $p = 12$ and $q = 5$, with the corresponding values $a = 119$ and $b = 120$ and $c = 169$. The values of a and c are precisely those in the second and third positions from the left in the first row on the Plimpton tablet; the ratio $c^2/b^2 = 28561/14400$ is the number $1;59,0,15$ that appears in the first position in this row.[5] The same relationship is found in the other fourteen rows; the Babylonians carried out the work so accurately that the ratio c^2/b^2 in the tenth row is expressed as a fraction with eight sexagesimal places, equivalent to about fourteen decimal places in our notation.

So much of Babylonian mathematics is bound up with tables of reciprocals that it is not surprising to find that the items in Plimpton 322 are related to reciprocal relationships. If $a = 1$, then $1 = (c + b)(c - b)$, so that $c + b$ and $c - b$ are reciprocals. If one starts with $c + b = n$, where n is any regular sexagesimal, then $c - b = 1/n$; hence $a = 1$ and $b = \frac{1}{2}(n - 1/n)$ and and $c = \frac{1}{2}(n + 1/n)$ are a Pythagorean fraction triple which can easily be converted to a Pythagorean integer triple by multiplying each of the three by $2n$. All triples in the Plimpton tablet are easily calculated by this device.

The account of Babylonian algebra that we have given is representative of their work, but it is not intended to be exhaustive. There are in the Babylonian tablets many other things, although none so striking as those in the Plimpton tablet 322. For instance, in one tablet the geometrical progression $1 + 2 + 2^2 + \cdots + 2^9$ is summed and in another the sum of the series of squares $1^2 + 2^2 + 3^2 + \cdots + 10^2$ is found. One wonders if the Babylonians knew the general formulas for the sum of a geometrical progression and the sum of the first n perfect squares. It is quite possible that they did, and it has been conjectured that they were aware that the sum of the first n perfect cubes is equal to the square of the sum of the first n integers.[6] Nevertheless, it must be borne in mind that tablets from Mesopotamia resemble Egyptian papyri in that only specific cases are given, with no general formulations.

9   A few years ago it used to be held that the Babylonians were better in algebra than were the Egyptians, but that they had contributed less to geometry. The first half of this statement is clearly substantiated by what we have learned above; attempts to bolster the second half of the comparison

---

[5] Vogel, in *Vorgriechische Mathematik*, II, 37–41, interprets this number, and also the others in this column, as a^2/b^2 rather than as c^2/b^2—that is, as $\tan^2 A$ rather than $\sec^2 A$. The difference between these functions is always one, and the unit wedges in the left-hand column in Plimpton 322 have in most cases been broken away; but careful inspection of this edge seems to substantiate the interpretation of the column as squares of secants rather than of tangents.

[6] See Archibald, *Outline of the History of Mathematics*, p. 11.

generally are limited to the measure of the circle or to the volume of the frustrum of a pyramid.[7] In the Mesopotamian valley the area of a circle was generally found by taking three times the square of the radius, and in accuracy this falls considerably below the Egyptian measure. However, the counting of decimal places in the approximations for π is scarcely an appropriate measure of the geometrical stature of a civilization, and a recent discovery has effectively nullified even this weak argument. In 1936 a group of mathematical tables were unearthed at Susa, a couple of hundred miles from Babylon, and these include significant geometrical results. True to the Mesopotamian penchant for making tables and lists, one tablet in the Susa group compares the areas and the squares of the sides of the regular polygons of three, four, five, six, and seven sides. The ratio of the area of the pentagon, for example, to the square on the side of the pentagon is given as 1 ; 40, a value that is correct to two significant figures. For the hexagon and heptagon the ratios are expressed as 2 ; 37,30 and 3 ; 41 respectively. In the same tablet the scribe gives 0 ; 57,36 as the ratio of the perimeter of the regular hexagon to the circumference of the circumscribed circle; and from this we can readily conclude[8] that the Babylonian scribe had adopted 3 ; 7,30 or $3\frac{1}{8}$, as an approximation for π. This is at least as good as the value adopted in Egypt. Moreover, we see it in a more sophisticated context than in Egypt, for the tablet from Susa is a good example of the systematic comparison of geometric figures. One is almost tempted to see in it the genuine origin of geometry, but it is important to note that it was not so much the geometrical context that interested the Babylonians as the numerical approximations that they used in mensuration. Geometry for them was not a mathematical discipline in our sense, but a sort of applied algebra or arithmetic in which numbers are attached to figures.

There is some disagreement as to whether or not the Babylonians were familiar with the concept of similar figures, although this appears to be quite likely. The similarity of all circles seems to have been taken for granted in Mesopotamia, as it had been in Egypt, and the many problems on triangle measure in cuneiform tablets seem to imply a concept of similarity. A tablet in the Baghdad Museum has a right triangle ABC (Fig. 3.2) with sides $a = 60$ and $b = 45$ and $c = 75$, and it is subdivided into four smaller right triangles ACD, CDE, DEF, and EFB. The areas of these four triangles are then given as 8,6 and 5,11 ; 2,24 and 3,19 ; 3,56,9,36 and 5,53 ; 53,39,50,24 respectively. From these values the scribe computed the length of AD as 27, apparently using a sort of "similarity formula" equivalent to our theorem that areas of similar figures are to each other as squares on corresponding sides. The

---

[7] See, for example, George Sarton, *A History of Science*, Vol. I (Cambridge, Mass.: Harvard University Press, 1952), pp. 73–74.

[8] See Neugebauer, *Exact Sciences in Antiquity* (2), p. 47.

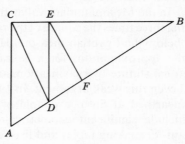

FIG. 3.2

lengths of CD and BD are found to be 36 and 48 respectively, and through an application of the "similarity formula" to triangles BCD and DCE the length of CE is found[9] to be 21 ; 36. The text breaks off in the middle of the calculation of DE.

10   Measurement was the keynote of algebraic geometry in the Mesopotamian valley, but a major flaw, as in Egyptian geometry, was that the distinction between exact and approximate measures was not made clear. The area of a quadrilateral was found by taking the product of the arithmetic means of the pairs of opposite sides, with no warning that this is in most cases only a crude approximation. Again, the volume of a frustum of a cone or pyramid sometimes was found by taking the arithmetic mean of the upper and lower bases and multiplying by the height; sometimes, for a frustum of a square pyramid with areas a^2 and b^2 for the lower and upper bases, the formula

$$V = \left(\frac{a + b}{2}\right)^2 h$$

was applied. However, for the latter the Babylonians used also a rule equivalent to

$$V = h\left[\left(\frac{a + b}{2}\right)^2 + \frac{1}{3}\left(\frac{a - b}{2}\right)^2\right]$$

a formula that is correct and reduces to the one known to the Egyptians.

It is not known whether Egyptian and Babylonian results were always independently discovered, but in any case the latter were definitely more extensive than the former, both in geometry and algebra. The Pythagorean theorem, for example, does not appear in any form in surviving documents from Egypt, but tablets even from the Old Babylonian period show that in Mesopotamia the theorem was widely used. A cuneiform text from the Yale

[9] See Vogel, *Vorgriechische Mathematik*, II, 78–79.

Collection, for example, contains a diagram of a square and its diagonals in which the number 30 is written along one side and the numbers 42;25,35 and 1;24,51,10 appear along a diagonal. The last number obviously is the ratio of the lengths of the diagonal and a side, and this is so accurately expressed that it agrees with $\sqrt{2}$ to within about a millionth. The accuracy of the result was made possible by knowledge of the Pythagorean theorem. Sometimes, in less precise computations, the Babylonians used 1;25 as a rough-and-ready approximation to this ratio. Of more significance than the precision of the values, however, is the implication that the diagonal of *any* square could be found by multiplying the side by $\sqrt{2}$. Thus there seems to have been some awareness of general principles, despite the fact that these are exclusively expressed in special cases.

Babylonian recognition of the Pythagorean theorem was by no means limited to the case of a right isosceles triangle. In one Old Babylonian problem text a ladder or beam of length 0;30 stands against a wall; the question is, how far will the lower end move out from the wall if the upper end slips down a distance of 0;6 units? The answer is correctly found by use of the Pythagorean theorem. Fifteen hundred years later similar problems, some with new twists, were still being solved in the Mesopotamian valley. A Seleucid tablet, for example, proposes the following problem. A reed stands against a wall. If the top slides down three units when the lower end slides away nine units, how long is the reed? The answer is given correctly as fifteen units.

Ancient cuneiform problem texts provide a wealth of exercises in what we might call geometry, but which the Babylonians probably thought of as applied arithmetic. A typical inheritance problem calls for the partition of a right triangular property among six brothers. The area is given as 11,22,30 and one of the sides is 6,30; the dividing lines are to be equidistant and parallel to the other side of the triangle. One is asked to find the difference in the allotments. Another text gives the bases of an isosceles trapezoid as 50 and 40 units and the length of the sides as 30; the altitude and area are required.[10]

The ancient Babylonians were aware of other important geometrical relationships. Like the Egyptians, they knew that the altitude in an isosceles triangle bisects the base. Hence, given the length of a chord in a circle of known radius, they were able to find the apothem. Unlike the Egyptians, they were familiar with the fact than an angle inscribed in a semicircle is a right angle, a proposition generally known as the Theorem of Thales, despite the fact that Thales lived well over a millennium after the Babylonians had begun to use it. This misnaming of a well-known theorem in geometry is symptomatic of the difficulty in assessing the influence of pre-Hellenic mathematics on later cultures. Cuneiform tablets had a permanence that

[10] These and other problems are found in van der Waerden, *Science Awakening*, pp. 76–77.

could not be matched by documents from other civilizations, for papyrus and parchment do not so easily survive the ravages of time. Moreover, cuneiform texts continued to be recorded down to the dawn of the Christian era; but were they read by neighboring civilizations, especially the Greeks? The center of mathematical development was shifting from the Mesopotamian valley to the Greek world half a dozen centuries before the beginning of our era, but reconstructions of early Greek mathematics are rendered hazardous by the fact that there are virtually no extant mathematical documents from the pre-Hellenistic period. It is important, therefore, to keep in mind the general characteristics of Egyptian and Babylonian mathematics so as to be able to make at least plausible conjectures concerning analogies that may be apparent between pre-Hellenic contributions and the activities and attitudes of later peoples.

11 A number of deficiencies in pre-Hellenic mathematics are quite obvious. Extant papyri and tablets contain specific cases and problems only, with no general formulations, and one may question whether these early civilizations really appreciated the unifying principles that are at the core of mathematics. Further study is somewhat reassuring, for the hundreds of problems of similar types in cuneiform tablets seem to be exercises that schoolboys were expected to work out in accordance with certain recognized methods or rules. That there are no surviving statements of these rules does not necessarily mean that the generality of the rules or principles was missing in ancient thought. Were a rule not there in essence, the similarity of the problems would be difficult to explain. Such large collections of similar problems could not have been the result of chance.

More serious, perhaps, than the lack of explicit statements of rules is the absence of clear-cut distinctions between exact and approximate results. The omission in the tables of cases involving irregular sexagesimals seems to imply some recognition of such distinctions, but neither the Egyptians nor the Babylonians seem to have raised the question of when the area of a quadrilateral (or of a circle) is found exactly and when only approximately. Questions about the solvability or unsolvability of a problem do not seem to have been raised; nor was there any investigation into the nature of proof. The word "proof" means various things at different levels and ages; hence it is hazardous to assert categorically that pre-Hellenic peoples had no concept of proof, nor any feeling of the need for proof. There are hints that these people occasionally were aware that certain area and volume methods could be justified through a reduction to simpler area and volume problems. Moreover, pre-Hellenic scribes not infrequently checked or "proved" their divisions by multiplication; occasionally they verified the procedure in a problem through a substitution that verified the correctness of the answer.

Nevertheless, there are no explicit statements from the pre-Hellenic period that would indicate a felt need for proofs or a concern for questions of logical principles. The lack of such statements often has led to judgments that pre-Hellenic civilizations had no true mathematics, despite the obviously high level of technical facility.

Critics also point to what they regard as an absence of abstraction in Egyptian and Babylonian mathematics. The language of the documents does seem always to remain close to concrete cases, as we have seen; but this, too, can be misleading. In Mesopotamian problems the words "length" and "width" should perhaps be interpreted much as we interpret the letters x and y, for the writers of cuneiform tablets may well have moved on from specific instances to general abstractions. How else does one explain the addition of a length to an area? In Egypt also the use of the word for quantity is not incompatible with an abstract interpretation such as we read into it today.

Evaluations of pre-Hellenic civilizations frequently point to the fact that there was no clearly discernible intellectual activity of a characteristically unified sort comparable to that which later carried the label "mathematics"; but here, too, it is easy to be excessively dogmatic. It may be true that geometry had not yet been crystallized out of a crude matrix of space experience that included all sorts of things that could be measured; but it is difficult not to see in Babylonian and Egyptian concern with number and its applications something very close to what usually, in ages since, has been known as algebra.

Pre-Hellenic cultures have been stigmatized also as entirely utilitarian, with little or no interest in mathematics for its own sake. Here, too, a matter of judgment, rather than of incontrovertible evidence, is involved. Then, as now, the vast majority of mankind were preoccupied with immediate problems of survival. Leisure was far scarcer than it is now, but even under this handicap there were in Egypt and Babylonia problems that have the earmarks of recreational mathematics. If a problem calls for a sum of cats and measures of grain, or of a length and an area, one cannot deny to the perpetrator either a modicum of levity or a feeling for abstraction. Of course much of pre-Hellenic mathematics was practical, but surely not all of it. The truth probably lies somewhere between extremes recently published by two historians of mathematics. One of them[11] claims that Babylonian mathematics was directed solely toward practical ends; the other has upheld the diametrically opposite view that "Sumerian mathematics was not used for the solution of problems in practical life, but only for enjoyment or for exultation of the spirit."[12] A cautious reader may safely assume that neither

[11] M. Cipolla, *Storia della matematica dai primordia a Leibniz* (Mozara: Società editrice siciliana, 1949), p. 23.

[12] Quoted from Ettore Bortolotti by Ettore Carruccio in his *Mathematics and Logic in History and in Contemporary Thought* (Chicago: Aldine, 1964), p. 15.

of these extreme positions can be held with impunity. In the practice of computation, which stretched over a couple of millennia, the schools of scribes used plenty of exercise material, often, perhaps, just as good clean fun.

BIBLIOGRAPHY

Archibald, R. C., *Outline of the History of Mathematics*, 6th ed. (Herbert Ellsworth Slaught Memorial Papers, No. 2, Buffalo, N.Y.: The Mathematical Association of America, 1949).

Bruins, E. M., and M. Rutten, *Textes mathématiques de Suse* (Paris, 1961).

Eves, Howard, *An Introduction to the History of Mathematics*, 2nd ed. (New York: Holt, 1964).

Kugler, F. X., *Sternkunst und Sterndienst in Babel* (Münster in Westphalia: Aschendoeff, 1907–1935, 2 vols. and 3 supps.).

Neugebauer, O., *The Exact Sciences in Antiquity*, 2nd ed. (Providence R. I.: Brown University Press, 1957; paperback ed., New York: Harper).

Neubebauer, O., *Mathematische Keilschift-Texte* (Berlin: Springer, 1935–1937, 3 vols.). This is Vol. II of *Quellen und Studien zur Geschichte der Mathematik, Astronomie und Physik*, Part A, *Quellen*. See also numerous articles by Neugebauer and others in *Quellen und Studien*, Part B, *Studien*, I–IV (1928–1938).

Neugebauer, O., *Vorgriechische Mathematik* (Berlin: Springer, 1934).

Neugebauer, O., and A. Sachs, *Mathematical Cuneiform Texts* (New Haven, Conn.: Yale University Press, 1945).

Thureau-Dangin, F., *Textes mathématiques Babyloniens* (Leiden: Brill, 1938).

Van der Waerden, B. L., *Science Awakening*, trans. by Arnold Dresden (New York: Oxford University Press, 1961; paperback ed., New York: Wiley, 1963).

Vogel, Kurt, *Vorgriechische Mathematik*, Vol. II, *Die Mathematik der Babylonier* (paperback ed., Hannover: Hermann Schroedel, ca. 1959).

EXERCISES

1. What do you regard as the four most significant contributions of the Mesopotamians to mathematics? Justify your answer.
2. What do you regard as the four chief weaknesses of Mesopotamian mathematics? Justify your answer.
3. Compare, as to significance and possible influence on later civilizations, the geometry and trigonometry of the Babylonians with that of the Egyptians.
4. Describe the relative advantages and disadvantages of the number notations of the Babylonians and the Egyptians.
5. Write the number 10,000 in Babylonian notation.
6. Write the number 0.0862 in Babylonian notation.
7. Use the Babylonian algorithm for square roots to find the square root of two to half a dozen decimal places and compare with the Babylonian value $1;24,51,10$.
8. Verify that if $(c/a)^2$ is $1;33,45$ and $b = 45$ and $c = 1,15$, then a, b, c form a Pythagorean triad.

9. Verify that the parameters $p = 9$ and $q = 4$ lead to the values in line 5 of the Plimpton Tablet 322.

10. Show that if p and q are positive numbers such that $p^2 - q^2 < 2pq$, then $1 < p/q < 1 + \sqrt{2}$.

11. How closely does the Babylonian approximation $3;41$ agree with the correct value for the ratio of the area of the regular heptagon to the square of a side?

12. The Babylonians estimated the ratio of the area of a regular hexagon to the square of one side as $2;37,30$. How does this compare with the correct ratio?

13. Solve the following Old Babylonian problem: The area of two squares together is 1000, and the side of one square is 10 less than two thirds of the side of the other square. Find the sides of the two squares.

14. Find as a sexagesimal fraction to the nearest minute the ratio of the area of a regular pentagon to the square of a side, and compare your answer with the value $1;40$ given by the Babylonians.

15. Solve the following Old Babylonian problem: One side of a right triangular property is 50 units long. Parallel to the other side and 20 units from the other side a line is drawn cutting off a right trapezoidal area of $5,20$ units. Find the lengths of the parallel sides of this trapezoid.

16. Verify the result of an Old Babylonian computation in which the area of an isosceles trapezoid whose sides are 30 units and whose bases are 14 and 50 is given as $12,48$.

17. Solve the following Old Babylonian problem: Ten brothers receive $1;40$ minas of silver, and brother over brother received a constant difference. If the eighth brother received 6 shekels, find how much each earned. (There are 60 shekels in a mina.)

18. Find the length of the ladder in the problem described in the text.

19. Solve the problem of the six brothers described in the text.

20. An Old Babylonian tablet unearthed at Susa asks for the radius of the circle circumscribing a triangle whose sides are 50, 50, and 60. Solve this problem.

21. Show that the sexagesimal representation of $\frac{1}{7}$ has a three-place periodicity. How many places are there in the periodicity of the decimal representation?

*22. Another tablet from Susa calls for the sides x and y of a rectangle if $xy = 20,0$ and $x^3 d = 14,48,53,20$, where d is the length of a diagonal. Solve this problem.

Chapter IV

Ionia and
the Pythagoreans

> To Thales . . . the primary question was not *What do we
> know*, but *How do we know it*.
>
> *Aristotle*

1 The intellectual activity of the Potamic civilizations in Egypt and Mesopotamia had lost its verve well before the Christian era; but as learning in the river valleys was declining, and as bronze was giving way to iron in weaponry, vigorous new cultures were springing up all along the shores of the Mediterranean Sea. To indicate this change in the centers of civilization, the interval from roughly 800 B.C. to A.D. 800 sometimes is known as the Thalassic Age (that is, the "sea" age). There was, of course, no sharp disruption to mark the transition in intellectual leadership from the valleys of the Nile, Tigris, and Euphrates rivers to the shores of the Mediterranean, for time and history flow continuously, and changing conditions are associated with antecedent causes. Egyptian and Babylonian scholars continued to produce papyrus and cuneiform texts for many centuries after 800 B.C.; but a new civilization meanwhile was rapidly preparing to take over scholarly hegemony, not only around the Mediterranean but, ultimately, in the chief river valleys as well. To indicate the source of the new inspiration, the first portion of the Thalassic Age is labeled the Hellenic era, so that the older cultures are consequently known as pre-Hellenic.

The Greeks of today still call themselves Hellenes, continuing a name used by their early forebears who settled along the coasts of the Mediterranean Sea. Greek history is traceable back into the second millennium B.C. when, as unlettered invaders, they pressed down relentlessly from the north. They brought with them no mathematical or literary tradition; they seem to have been very eager to learn, however, and it did not take them long to improve on what they were taught. For example, they took over, perhaps from the Phoenicians, an existing alphabet, consisting only of consonants, and to it they added vowels. The alphabet seems to have originated between the Babylonian and Egyptian worlds, possibly in the region of the Sinai Peninsula, through a process of drastic reduction in the number of cuneiform or hieratic symbols.

This alphabet found its way to the new colonies—Greek, Roman, and Carthaginian—through the activities of traders. It is presumed that some rudiments of computation traveled along the same routes, but the more esoteric portions of priestly mathematics may have remained undiffused. Before long, however, Greek traders, businessmen, and scholars made their way to the centers of learning in Egypt and Babylonia. There they made contact with pre-Hellenic mathematics; but they were not willing merely to receive the long-established traditions, for they made the subject so thoroughly their own that it shortly took a drastically different form.

The first Olympic Games were held in 776 B.C., and by that time a wonderful Greek literature already had developed, evidenced by the works of Homer and Hesiod. Of Greek mathematics at the time we know nothing. Presumably it lagged behind the development of literary forms, for the latter lend themselves more readily to continuity of oral transmission. It was to be almost another two centuries before there was any word, even indirectly, concerning Greek mathematics. Then, during the sixth century B.C., there appeared two men, Thales and Pythagoras, who seem to have played in mathematics a role similar to that of Homer and Hesiod in literature. Most of what is reported in this chapter centers on Thales and Pythagoras, but a note of warning is in order. Homer and Hesiod are somewhat shadowy figures, but at least we have a consistent tradition attributing to them certain literary masterpieces which, first transmitted orally from generation to generation, ultimately were copied down and preserved for posterity. Thales and Pythagoras also are somewhat indistinct figures, historically, although less so than Homer and Hesiod; but as far as their scholarly work is concerned, the parallel with Homer and Hesiod ceases. No mathematical masterpiece from either one has survived, nor is it even established that either Thales or Pythagoras ever composed such a work. What they may have done must be reconstructed on the basis of a none too trustworthy tradition that grew up around these two early mathematicians. Certain key phrases are attributed to them—such as "Know thyself," in the case of Thales, and "All is number," in the case of Pythagoras—but not much more of a specific nature. Nevertheless, the earliest Greek accounts of the history of mathematics, which no longer survive, ascribed to Thales and Pythagoras a number of very definite discoveries in mathematics. We outline these contributions in this chapter, but the reader should understand that it is largely persistent tradition, rather than any extant historical document, on which the account is based.

The Greek world for many centuries had its center between the Aegean and Ionian Seas, but Hellenic civilization was far from localized there. Greek settlements by about 600 B.C. were to be found scattered along the borders of most of the Black Sea and the Mediterranean Sea, and it was on these outskirts that a new surge in mathematics developed. In this respect the sea-bordering colonists, especially in Ionia, had two advantages: they had the bold and imaginative spirit typical of pioneers, and they were in closer proximity to the two chief river valleys from which knowledge could be derived. Thales of Miletus (ca. 624–548 B.C.) and Pythagoras of Samos (ca. 580–500 B.C.) had in addition a further advantage: they were in a position to travel to centers of ancient learning and there acquire firsthand information on astronomy and mathematics. In Egypt they are said to have learned geometry; in Babylon, under the enlightened Chaldean ruler Nebuchadnezzar, Thales probably came in touch with astronomical tables and instruments. Tradition has it that in 585 B.C. Thales amazed his countrymen by predicting the solar eclipse of that year. The historicity of this tradition is very much open to question, especially because an eclipse of the sun is visible over only a very small portion of the earth's surface, and it does not seem likely that there were in Babylon tables of solar eclipses that would have enabled Thales to make such a prediction. It is quite likely, on the other hand, that the gnomon or sundial entered Greece from Babylon, and perhaps the water clock came from Egypt. The Greeks were far from hesitant in taking over elements of foreign cultures, else they would never have learned so quickly how to advance beyond their predecessors; but everything they touched, they quickened.

2   What is really known about the life and work of Thales is very little indeed. His birth and death are estimated from the fact that the eclipse of 585 B.C. probably occurred when he was in his prime, say about forty, and that he was said to have been seventy-eight when he died. However, serious doubts about the authenticity of the eclipse story make such extrapolations hazardous, and they shake our confidence concerning the discoveries fathered upon Thales. Ancient opinion is unanimous in regarding Thales as an unusually clever man and the first philosopher—by general agreement the first of the Seven Wise Men. He was regarded as "a pupil of the Egyptians and the Chaldeans," an assumption that appears plausible. The proposition now known as the Theorem of Thales—that an angle inscribed in a semicircle is a right angle—may well have been learned by Thales during his travels to Babylon. However, tradition goes further and attributes to him some sort of demonstration of the theorem. For this reason Thales frequently has been hailed as the first true mathematician—as the originator of the deductive organization of geometry. This report—or legend—was embellished by adding to this theorem four others that Thales is said to have proved:

1. A circle is bisected by a diameter.
2. The base angles of an isosceles triangle are equal.
3. The pairs of vertical angles formed by two intersecting lines are equal.

4. If two triangles are such that two angles and a side of one are equal respectively to two angles and a side of the other, then the triangles are congruent.

There is no document from antiquity that can be pointed to as evidence of this achievement, and yet the tradition has been persistent. About the nearest one can come to reliable evidence on this point is derived from a source a thousand years after the time of Thales. A student of Aristotle by the name of Eudemus of Rhodes (fl. ca. 320 B.C.) wrote a history of mathematics. This has been lost, but before it disappeared, someone had summarized at least part of the history. The original of this summary also has been lost, but during the fifth century of our era information from the summary was incorporated by the Neoplatonic philosopher Proclus (410–485) in the early pages of his *Commentary on the First Book of Euclid's Elements*. Following introductory remarks on the origin of geometry in Egypt, the *Commentary* of Proclus reports that Thales

. . . first went to Egypt and thence introduced this study into Greece. He discovered many propositions himself, and instructed his successors in the principles underlying many others, his method of attack being in some cases more general, in others more empirical.[1]

It is largely upon this quotation at third hand that designations of Thales as the first mathematician hinge. Proclus later in his *Commentary*, again depending on Eudemus, attributes to Thales the four theorems mentioned above. There are other scattered references to Thales in ancient sources, but most of these describe his more practical activities. Diogenes Laertius, followed by Pliny and Plutarch, reported that he measured the heights of the pyramids in Egypt by observing the lengths of their shadows at the moment when the shadow of a vertical stick is equal to its height.[2] Herodotus, the historian, recounts the story of Thales' prediction of a solar eclipse; the philosopher Aristotle reports that Thales made a fortune by "cornering" the olive presses during a year in which the olive crop promised to be abundant. Still other legends picture Thales as a salt merchant, as a stargazer, as a defender of celibacy, or as a farsighted statesman. Such reports, however, provide no further evidence concerning the important question of whether or not Thales actually arranged a number of geometrical theorems in a deductive sequence. The tale that he calculated the distance of a ship at sea through the proportionality of sides of similar triangles is inconclusive, for the principles behind such a calculation had long been known in Egypt and Mesopotamia. Such stories do not establish the bold conjecture that Thales created

[1] The translation is taken from T. L. Heath, *History of Greek Mathematics* (1921), I, 128. Cf. Ivor Thomas, ed., *Selections Illustrating the History of Greek Mathematics* (1939–1941), I, 147.

[2] For a full account see Heath, *op. cit.*, I, 128–140.

demonstrative geometry; but in any case Thales is the first man in history to whom specific mathematical discoveries have been attributed.[3] We know now that a large body of mathematical material was familiar to the Babylonians a millennium before the time of Thales, and yet among the Greeks it was understood that Thales had made definite advances. It would appear reasonable to suppose, in the light of Proclus' statements, that Thales contributed something in the way of rational organization. That it was the Greeks who added the element of logical structure to geometry is virtually universally admitted today, but the big question remains whether this crucial step was taken by Thales or by others later—perhaps as much as two centuries later. On this point we must suspend final judgment until there is additional evidence on the development of Greek mathematics.

3 Pythagoras is scarcely less controversial a figure than Thales, for he has been more thoroughly enmeshed in legend and apotheosis. Thales had been a man of practical affairs, but Pythagoras was a prophet and a mystic, born at Samos, one of the Dodecanese islands not far from Miletus, the birthplace of Thales. Although some accounts picture Pythagoras as having studied under Thales, this is rendered unlikely by the half-century difference in their ages. Some similarity in their interests can readily be accounted for by the fact that Pythagoras also traveled to Egypt and Babylon—possibly even to India. During his peregrinations he evidently absorbed not only mathematical and astronomical information, but also much religious lore. Pythagoras was, incidentally, virtually a contemporary of Buddha, of Confucius, and of Lao-Tze, so that the century was a critical time in the development of religion as well as of mathematics. When he returned to the Greek world, Pythagoras settled at Croton on the southeastern coast of what is now Italy, but at that time was known as Magna Graecia. There he established a secret society which somewhat resembled an Orphic cult except for its mathematical and philosophical basis.

That Pythagoras remains a very obscure figure is due in part to the loss of documents from that age. Several biographies of Pythagoras were written in antiquity, including one by Aristotle, but these have not survived. A further difficulty in identifying clearly the figure of Pythagoras lies in the fact that the order he established was communal as well as secret. Knowledge and property were held in common, hence attribution of discoveries was not to be made to a specific member of the school. It is best, consequently, not to speak of the work of Pythagoras, but rather of the contributions of the Pythagoreans, although in antiquity it was customary to give all credit to the master.

[3] B. L. van der Waerden, in *Science Awakening*, p. 80, accepts the conjecture that Thales used deduction; O. Neugebauer, in *Exact Sciences in Antiquity*, pp. 142, 143, 148, rejects it.

The Pythagorean school of thought was politically conservative and with a strict code of conduct. Vegetarianism was enjoined upon the members, apparently because Pythagoreanism accepted the doctrine of metempsychosis, or the transmigration of souls, with the resulting concern lest an animal to be slaughtered might be the new abode of a friend who had died. Among other taboos of the school was the eating of beans (more properly lentils). Perhaps the most striking characteristic of the Pythagorean order was the confidence it maintained in the pursuit of philosophical and mathematical studies as a moral basis for the conduct of life. The very words "philosophy" (or "love of wisdom") and "mathematics" (or "that which is learned") are supposed to have been coined by Pythagoras himself to describe his intellectual activities. He is said to have given two categories of lectures, one for members of the school or order only, and the other for those in the larger community. It is presumed that it was in the lectures of the first category that Pythagoras presented whatever contributions to mathematics he may have made. Having described, in the quotation above, the work in geometry done by Thales, Proclus went on to say:

Pythagoras, who came after him, transformed this science into a liberal form of education, examining its principles from the beginning and probing the theorems in an immaterial and intellectual manner. He discovered the theory of proportionals and the construction of the cosmic figures.[4]

Even if we do not accept this statement at its face value, it is evident that the Pythagoreans played an important role—possibly the crucial role—in the history of mathematics. In Egypt and Mesopotamia the elements of arithmetic and geometry were primarily exercises in the application of numerical procedures to specific problems, whether concerned with beer or pyramids or the inheritance of land. There had been little in the way of intellectual structure, and perhaps nothing resembling philosophical discussion of principles. Thales is generally regarded as having made a beginning in this direction, although tradition supports the view of Eudemus and Proclus that the new emphasis in mathematics was due primarily to the Pythagoreans. With them mathematics was more closely related to a love of wisdom than to the exigencies of practical life; and it has had this tendency ever since. How far the Pythagoreans went in this direction is not at all clear, and at least one eminent scholar[5] regards all reports of important mathematical contributions by Pythagoras as unhistorical. It is indeed difficult to separate history and legend concerning the man, for he meant so many things to the populace—the philosopher, the astronomer, the mathematician, the abhorrer of beans, the

---

[4] See Ivor Thomas, *op. cit.*, I, 149. Cf. also Heath, *op. cit.*, I, 141, and van der Waerden, *op. cit.*, p. 90.

[5] See Neugebauer, *op. cit.*, p. 148.

saint, the prophet, the performer of miracles, the magician, the charlatan. That he was one of the most influential figures in history is difficult to deny, for his followers, whether deluded or inspired, spread their beliefs over most of the Greek world. The Pythagorean purification of the soul was accomplished in part through a strict physical regimen and in part through cultist rites reminiscent of worshippers of Orpheus and Dionysus; but the harmonies and mysteries of philosophy and mathematics also were essential parts in the rituals. Never before or since has mathematics played so large a role in life and religion as it did among the Pythagoreans. If, then, it is impossible to ascribe specific discoveries to Pythagoras himself, or even collectively to the Pythagoreans, it is nevertheless important to understand the type of activity with which, according to tradition, the school was associated.

4   The motto of the Pythagorean school is said to have been "All is number." Recalling that the Babylonians had attached numerical measures to things around them, from the motions of the heavens to the values of their slaves, we may perceive in the Pythagorean motto a strong Mesopotamian affinity. The very theorem to which the name of Pythagoras still clings quite likely was derived from the Babylonians. It has been suggested, as justification for calling it the Theorem of Pythagoras, that the Pythagoreans first provided a demonstration; but this conjecture cannot be verified. Legends that Pythagoras sacrificed an ox (a hundred oxen, according to some versions) upon discovering the theorem—or its proof—are implausible in view of the vegetarian rules of the school. Moreover, they are repeated, with equal incredibility, in connection with several other theorems. It is reasonable to assume that the earliest members of the Pythagorean school were familiar with geometrical properties known to the Babylonians; but when the Eudemus–Proclus summary ascribes to them the construction of the "cosmic figures" (that is, the regular solids), there is room for doubt. The cube, the octahedron, and the dodecahedron could perhaps have been observed in crystals, such as those of pyrite (iron disulphide); but a scholium in *Elements* XIII reports that the Pythagoreans knew only three of the regular polyhedra: the tetrahedron, the cube, and the dodecahedron. Familiarity with the last figure is rendered plausible by the discovery near Padua of an Etruscan dodecahedron of stone dating from before 500 B.C. It is not improbable, therefore, that even if the Pythagoreans did not know of the octahedron and the icosahedron, they knew of some of the properties of the regular pentagon. The figure of a five-pointed star (which is formed by drawing the five diagonals of a pentagonal face of a regular dodecahedron) is said to have been the special symbol of the Pythagorean school. The star pentagon had appeared earlier in Babylonian art, and it is possible that here, too, we find a connecting link between pre-Hellenic and Pythagorean mathematics.

One of the tantalizing questions in Pythagorean geometry concerns the construction of a pentagram or star pentagon. If we begin with a regular polygon *ABCDE* (Fig. 4.1) and draw the five diagonals, these diagonals intersect in points *A'B'C'D'E'* which form another regular pentagon. Noting that the triangle *BCD'*, for example, is similar to the isosceles triangle *BCE* and noting also the many pairs of congruent triangles in the diagram, it is not difficult to see that the diagonal points *A'B'C'D'E'* divide the diagonals in a striking manner. In each case a diagonal point divides a diagonal into two unequal segments such that the ratio of the whole diagonal is to the larger

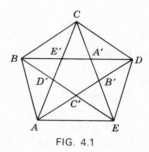

FIG. 4.1

segment as this segment is to the smaller segment. This subdivision of a diagonal is the well-known "golden section" of a line segment, but this name was not used until a couple of thousand years later—just about the time when Kepler wrote lyrically:

Geometry has two great treasures: one is the Theorem of Pythagoras; the other, the division of a line into extreme and mean ratio. The first we may compare to a measure of gold; the second we may name a precious jewel.

To the ancient Greeks this type of subdivision soon became so familiar that no need was felt for a special descriptive name; hence the longer designation "the division of a segment in mean and extreme ratio" generally was replaced by the simple words "the section."

One of the important properties of "the section" is that it is, so to speak, self-propagating. If a point P_1 divides a segment RS (Fig. 4.2) in mean and extreme ratio, with RP_1 the longer segment, and if on this larger segment we mark off a point P_2 such that $RP_2 = P_1S$, then segment RP_1 will in turn be subdivided in mean and extreme ratio at point P_2. Again, upon marking off

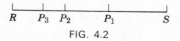

FIG. 4.2

on RP_2 point P_3 such that $RP_3 = P_2P_1$, segment RP_2 will be divided in mean and extreme ratio at P_3. This iterative procedure can be carried out as many times as desired, the result being an ever smaller segment RP_n divided in mean and extreme ratio by point P_{n+1}. Whether or not the earlier Pythagoreans noticed this unending process or drew significant conclusions from it is not known. Even the more fundamental question of whether or not the Pythagoreans of about 500 B.C. could divide a given segment into mean and extreme ratio cannot be answered with certainty, although the probability that they could and did seems to be high. The construction required is equivalent to the solution of a quadratic equation. To show this, let $RS = a$ and $RP_1 = x$ in Fig. 4.2. Then, by the property of the golden section, $a:x = x:(a - x)$, and upon multiplying means and extremes we have the equation $x^2 = a^2 - ax$. This is a quadratic equation of type 1 described in Chapter 3, and Pythagoras could have learned from the Babylonians how to solve this equation algebraically. However, if a is a rational number, then there is no rational number x satisfying the equation. Did Pythagoras realize this? It seems unlikely. Perhaps instead of the Babylonian algebraic type of solution, the Pythagoreans may have adopted a geometrical procedure similar to that found in Euclid's *Elements* II. 11 and VI. 30. To divide a line segment AB in mean and extreme ratio, Euclid first constructed on the segment AB the square $ABCD$ (Fig. 4.3). Then he bisected AC at point E, drew

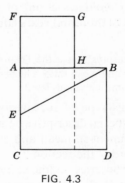

FIG. 4.3

line segment EB, and extended line CEA to F so that $EF = EB$. When the square $AFGH$ is completed, point H will be the point desired, for one can readily show that $AB:AH = AH:HB$. Knowing what solution, if any, the earlier Pythagoreans used for the golden section would go far toward clarifying the problem of the level and characteristics of pre-Socratic mathematics. If Pythagorean mathematics began under a Babylonian aegis, with strong faith that all is number, how (and when) did it happen that this gave way to the familiar emphasis on pure geometry that is so firmly enshrined in the classical treatises?

It has been customary to hold that most of the material in the first two books of the *Elements* was due to the Pythagoreans. This would presuppose a high level of achievement, implying a fairly rapid development of the subject after the days of Thales and Pythagoras. This view requires faith in what has been called the "Greek miracle," by which relatively unlettered newcomers on the Mediterranean scene mastered the material inherited from their neighbors and rapidly rose to new heights, establishing on the way the essential deductive pattern of theorems. In recent years serious doubt has been cast on the traditional view by those who call attention to relatively primitive concepts in Pythagorean arithmetic. If, for example, the leading Pythagorean mathematician of the early fourth century B.C., Archytas of Tarentum (428–365 B.C.), could assert that not geometry, but arithmetic alone, could provide satisfactory proofs,[6] there would appear to be little ground for placing the rise of the axiomatic method in geometry among the Pythagoreans of a century or two before this time. On the other hand, it may be argued that Archytas represented only one point of view, insisting on an orthodox Pythagorean numerology that others had abandoned or modified. Certainly there had been shifting attitudes in Pythagorean astronomy, and we can assume that there were comparable modifications in mathematics.

Number mysticism was not original with the Pythagoreans. The number seven, for example, had been singled out for special awe, presumably on account of the seven wandering stars or planets from which the week (hence our names for the days of the week) is derived. The Pythagoreans were not the only people who fancied that the odd numbers had male attributes and the even female—with the related (and not unprejudiced) assumption, found as late as Shakespeare, that "there is divinity in odd numbers." Many early civilizations shared various aspects of numerology, but the Pythagoreans carried number worship to its extreme, basing their philosophy and their way of life upon it. The number one, they argued, is the generator of numbers and the number of reason; the number two is the first even or female number, the number of opinion; three is the first true male number, the number of harmony, being composed of unity and diversity; four is the number of justice or retribution, indicating the squaring of accounts; five is the number of marriage, the union of the first true male and female numbers; and six is the number of creation. Each number in turn had its peculiar attributes. The holiest of all was the number ten or the *tetractys*, for it represented the

---

[6] Neugebauer, *Exact Sciences in Antiquity*, p. 148.

number of the universe, including the sum of all the possible geometric dimensions. A single point is the generator of dimensions, two points determine a line of dimension one, three points (not on a line) determine a triangle with area of dimension two, and four points (not in a plane) determine a tetrahedron with volume of dimension three; the sum of the numbers representing all dimensions therefore is the revered number ten. It is a tribute to the abstraction of Pythagorean mathematics that the veneration of the number ten evidently was not dictated by anatomy of the human hand or foot.

6   In Mesopotamia geometry had been not much more than number applied to spatial extension; it appears that at first it may have been much the same among the Pythagoreans—but with a modification. Number in Egypt had been the domain of the natural numbers and the unit fractions; among the Babylonians it had been the field of all rational fractions. In Greece the word number was used only for the integers. A fraction was not looked upon as a single entity, but as a ratio or relationship between two whole numbers. (Greek mathematics in its earlier stages frequently came closer to the "modern" mathematics of today than to the ordinary arithmetic of a generation ago.) As Euclid later expressed it (*Elements* V. 3), "A ratio is a kind of relation in respect of size of two magnitudes of the same kind." Such a view, focusing attention on the connection between pairs of numbers, tends to sharpen the theoretical or rational aspects of the number concept and to deemphasize the role of number as a tool in computation or approximation in mensuration. Arithmetic now could be thought of as an intellectual discipline as well as a technique, and a transition to such an outlook seems to have been nurtured in the Pythagorean school. If tradition is to be trusted, the Pythagoreans not only established arithmetic as a branch of philosophy; they seem to have made it the basis of a unification of all aspects of the world about them. Through patterns of points, or unextended units, they associated number with geometrical extension; this in turn led them to an arithmetic of the heavens. Philolaus (died ca. 390 B.C.), a later Pythagorean who shared the veneration of the *tetractys* or decad, wrote that it was "great, all-powerful and all-producing, the beginning and the guide of the divine as of the terrestrial life."[7] This view of the number ten as the perfect number, the symbol of health and harmony, seems to have provided the inspiration for the earliest nongeocentric astronomical system. Philolaus postulated that at the center of the universe there was a central fire about which the earth and the seven

[7] For an especially extensive account of Pythagoreanism see Eduard Zeller, *A History of Greek Philosophy from the Earliest Period to the Time of Socrates* (1881), I, 306–533. On the *tetractys* see especially pp. 428 ff. A longer description of the role of the tetractys is given on pp. 180–188 of Thomas Taylor, *The Theoretic Arithmetic of the Pythagoreans* (Los Angeles, 1934), but this book must be read with circumspection.

planets (including the sun and the moon) revolved uniformly. Inasmuch as this brought to only nine the number of heavenly bodies (other than the sphere of fixed stars), the Philolaic system assumed the existence of a tenth body—a "counterearth" collinear with the earth and the central fire—having the same period as the earth in its daily revolution about the central fire. The sun revolved about the fire once a year, and the fixed stars were stationary. The earth in its motion maintained the same uninhabited face toward the central fire, hence neither the fire nor the counterearth ever was seen. The postulate of uniform circular motion that the Pythagoreans adopted was to dominate astronomical thought for more than 2000 years. Copernicus, almost 2000 years later, accepted this assumption without question, and it was to the Pythagoreans that Copernicus referred to show that his doctrine of a moving earth was not so new or revolutionary.

The thoroughness with which the Pythagoreans wove number into their **7** thought is well illustrated by their concern for figurate numbers. Although no triangle can be formed by fewer than three points, it is possible to have triangles of a larger number of points, such as six, ten, or fifteen (see Fig. 4.4).

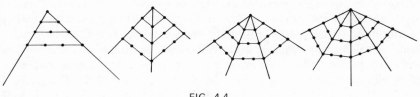

FIG. 4.4

Numbers such as three, six, ten, and fifteen or, in general, numbers given by the formula

$$N = 1 + 2 + 3 + \cdots + n = \frac{n(n + 1)}{2}$$

were called triangular; and the triangular pattern for the number ten, the holy tetractys, vied with the pentagon for veneration in Pythagorean number theory. There were, of course, indefinitely many other categories of privileged numbers. Successive square numbers are formed from the sequence $1 + 3 + 5 + 7 + \cdots + (2n - 1)$, where each odd number in turn was looked upon as a pattern of dots resembling a gnomon (the Babylonian shadow clock) placed around two sides of the preceding square pattern of dots (see Fig. 4.4). Hence the word gnomon (related to the word for knowing) came to be attached to the odd numbers themselves. The sequence of even numbers, $2 + 4 + 6 + \cdots + 2n = n(n + 1)$, produces what the Greeks called "oblong numbers," each of which is double a triangular number. Pentagonal patterns of points illustrated the pentagonal numbers given by the sequence

$$N = 1 + 4 + 7 + \cdots + (3n - 2) = \frac{n(3n - 1)}{2}$$

and hexagonal numbers were derived from the sequence

$$1 + 5 + 9 + \cdots + (4n - 3) = 2n^2 - n$$

In similar manner polygonal numbers of all orders are designated; the process, of course, is easily extended to three-dimensional space, where one deals with polyhedral numbers. Emboldened by such views, Philolaus is reported to have maintained that

All things which can be known have number; for it is not possible that without number anything can be either conceived or known.

The dictum of Philolaus seems to have been a tenet of the Pythagorean school, hence stories arose about the discovery by Pythagoras of some simple laws of music. Pythagoras is reputed to have noticed that when the lengths of vibrating strings are expressible as ratios of simple whole numbers, such as two to three (for the fifth) or as three to four (for the fourth), the tones will be harmonious. If, in other words, a string sounds the note C when plucked, then a similar string twice as long will sound the note C an octave below; and tones between these two notes are emitted by strings whose lengths are given by intermediate ratios: 16:9 for D, 8:5 for E, 3:2 for F, 4:3 for G, 6:5 for A, and 16:15 for B, in ascending order. Here we have perhaps the earliest quantitative laws of acoustics—possibly the oldest of all quantitative physical laws. So boldly imaginative were the early Pythagoreans that they extrapolated hastily to conclude that the heavenly bodies in their motions similarly emitted harmonious tones, the "harmony of the spheres." Pythagorean science, like Pythagorean mathematics, seems to have been an odd congeries of sober thought and fanciful speculation. The doctrine of a spherical earth often is ascribed to Pythagoras, but it is not known whether this conclusion[8] was based on observation (perhaps of new constellations as Pythagoras traveled southward) or on imagination. The very idea that the universe is a "cosmos," or a harmoniously ordered whole, seems to be a related Pythagorean contribution—one which at the time had little basis in direct observation but which has been enormously fruitful in the development

[8] The tradition that attributes the spherical-earth concept to the Pythagoreans has been questioned. See W. A. Heidel, *The Frame of the Ancient Greek Maps with a Discussion of the Sphericity of the Earth* (New York: Amer. Geog. Soc., 1937).

of astronomy. As we smile at ancient number fancies, we should at the same time be aware of the impulse these gave to the development of both mathematics and science. The Pythagoreans were among the earliest people to believe that the operations of nature could be understood through mathematics.

Proclus, quoting perhaps from Eudemus, ascribed to Pythagoras two **8** specific mathematical discoveries: (1) the construction of the regular solids and (2) the theory of proportionals. Although there is question about the extent to which this is to be taken literally, there is every likelihood that the statement correctly reflects the direction of Pythagorean thought. The theory of proportions clearly fits into the pattern of early Greek mathematical interests, and it is not difficult to find a likely source of inspiration. It is reported that Pythagoras learned in Mesopotamia of three means—the arithmetic, the geometric, and the subcontrary (later called the harmonic)— and of the "golden proportion" relating two of these: the first of two numbers is to their arithmetic mean as their harmonic mean is to the second of the numbers. This relationship is the essence of the Babylonian square-root algorithm, hence the report is at least plausible. At some stage, however, the Pythagoreans generalized this work by adding seven new means to make ten in all. If b is the mean of a and c, where $a < c$, then the three quantities are related according to one of the following ten equations:

$$(1)\quad \frac{b-a}{c-b} = \frac{a}{a} \qquad\qquad (6)\quad \frac{b-a}{c-b} = \frac{c}{b}$$

$$(2)\quad \frac{b-a}{c-b} = \frac{a}{b} \qquad\qquad (7)\quad \frac{c-a}{b-a} = \frac{c}{a}$$

$$(3)\quad \frac{b-a}{c-b} = \frac{a}{c} \qquad\qquad (8)\quad \frac{c-a}{c-b} = \frac{c}{a}$$

$$(4)\quad \frac{b-a}{c-b} = \frac{c}{a} \qquad\qquad (9)\quad \frac{c-a}{b-a} = \frac{b}{a}$$

$$(5)\quad \frac{b-a}{c-b} = \frac{b}{a} \qquad\qquad (10)\quad \frac{c-a}{c-b} = \frac{b}{a}$$

The first three equations are, of course, the equations for the arithmetic, the geometric, and the harmonic means respectively.

It is difficult to assign a date to the Pythagorean study of means, and similar problems arise with respect to the classification of numbers. The study of proportions or the equality of ratios presumably formed at first a part of Pythagorean arithmetic or theory of numbers. Later the quantities a, b, and c entering in such proportions were more likely to be regarded as geometrical magnitudes; but the period in which the change took place is not clear. In addition to the polygonal numbers mentioned above and the distinction between odd and even, the Pythagoreans at some stage spoke of odd-odd and even-odd numbers, according as the number in question was the product of two odd numbers or of an odd and an even number, so that sometimes the name even number was reserved for integral powers of two. By the time of Philolaus the distinction between prime and composite numbers seems to have become important. Speusippus, nephew of Plato and his successor as head of the Academy, asserted that ten was "perfect" for the Pythagoreans because, among other things, it is the smallest integer n for which there are just as many primes between one and n as nonprimes. (Occasionally prime numbers were called linear inasmuch as they usually are represented by dots in one dimension only.) Neo-Pythagoreans sometimes excluded two from the list of primes on the ground that one and two are not true numbers, but the generators of the odd and even numbers. The primacy of the odd numbers was assumed to be established by the fact that odd + odd is even, whereas even + even remains even.

To the Pythagoreans has been attributed the rule for Pythagorean triads given by $(m^2 - 1)/2, m, (m^2 + 1)/2$, where m is an odd integer; but inasmuch as this rule is so closely related to the Babylonian examples, it is perhaps not an independent discovery. Also ascribed to the Pythagoreans, with doubt as to the period in question, are the definitions of perfect, abundant, and deficient numbers according as the sum of the proper divisors of the number is equal to, greater than, or less than the number itself. According to this definition, six is the smallest perfect number, with twenty-eight next. That this view probably was a later development in Pythagorean thought is suggested by the early veneration of ten rather than six. Hence the related doctrine of "amicable" numbers also is likely to have been a later notion. Two integers a and b are said to be "amicable" if a is the sum of the proper divisors of b and if b is the sum of the proper divisors of a. The smallest such pair are the integers 220 and 284.

9   The picture of Pythagorean mathematics that has been presented is based largely on reports of commentators who lived many centuries later and who were, almost without exception, interested in philosophical aspects of thought. Although it appears plausible to assume, with the commentators, that it was the Pythagoreans who were largely responsible for the abstract and intellectual view that fashioned mathematics into a liberal discipline, the level of sophistication during the sixth and fifth centuries B.C. may not have been as high as that attributed to them by tradition. It must have been all too tempting to later devotees of a philosophical school, such as the

Pythagorean, to exaggerate the accomplishments of the founder and of the early members of the sect. It is highly probable that elements of primitivity were present during the early stages of Pythagoreanism, but went unreported. It is obvious also that the type of attitude toward mathematics represented by the Pythagoreans almost certainly was atypical of Greek thought as a whole. The Hellenes were celebrated as shrewd traders and businessmen, and there must have been a lower level of arithmetic or computation that satisfied the needs of the vast majority of Greek citizens. Number activities of this type would have been beneath the notice of philosophers, and recorded accounts of practical arithmetic were unlikely to find their way into libraries of scholars. If, then, there are not even fragments surviving of the more sophisticated Pythagorean works, it is clear that it would be unreasonable to expect manuals of trade mathematics to survive the ravages of time. Hence it is not possible to tell at this distance how the ordinary processes of arithmetic were carried out in Greece 2500 years ago. About the best one can do is to describe the systems of numeration that appear to have been in use.

In general there seem to have been two chief systems of numeration in Greece: one, probably the earlier, is known as the Attic (or Herodianic) notation; the other is called the Ionian (or alphabetic) system. Both systems are, for integers, based on the ten-scale, but the former is the more primitive, being based on a simple iterative scheme found in the earlier Egyptian hieroglyphic numeration and in the later Roman numerals. In the Attic system the numbers from one to four were represented by repeated vertical strokes. For the number five a new symbol—the first letter Π (or Γ) of the word for five, *pente*—was adopted. (Only capital letters were used at the time, both in literary works and in mathematics, lower-case letters being an invention of the later ancient or early Medieval period.) For numbers from six through nine, the Attic system combined the symbol Γ with unit strokes, so that eight, for example, was written as Γιιι. For positive integral powers of the base (ten), the initial letters of the corresponding number words were adopted—Δ for *deka* (ten), H for *hekaton* (hundred), x for *khilioi* (thousand), and M for *myrioi* (ten thousand). Except for the forms of the symbols, the Attic system is much like the Roman; but it had one advantage. Where the Latin world adopted distinctive symbols for 50 and 500, the Greeks wrote these numbers by combining letters for 5, 10, and 100, using Δ̵ (or 5 times 10) for 50, and H̵ (or 5 times 100) for 500. In the same way they wrote X̵ for 5000 and M̵ for 50000. In Attic script the number 45,678, for example, would appear as

$$\text{MMMM}\,\overline{X}\,\overline{H}\,H\,\overline{\Delta}\,\Delta\Delta\,\Gamma\text{ιιι}$$

The Attic system of notation (known also as Herodianic inasmuch as it **10** was described in a fragment attributed to Herodian, a grammarian of the second century) appears in inscriptions at various dates from 454 to 95 B.C.;[9] but by the early Alexandrian Age, at about the time of Ptolemy Philadelphus, it was being displaced by the Ionian or alphabetic numerals. Similar alphabetic schemes were used at one time or another by various Semitic peoples, including the Hebrews, Syrians, Arameans, and Arabs—as well as by other cultures, such as the Gothic—but these would seem to have been borrowed from the Greek notation. The Ionian system probably was used as early as the fifth century B.C. and perhaps as early as the eighth century B.C. One reason for placing the origin of the notation relatively early is that the scheme called for twenty-seven letters of the alphabet—nine for the integers less than 10, nine for multiples of 10 that are less than 100, and nine for multiples of 100 that are less than 1000. The classical Greek alphabet contains only twenty-four letters; hence use was made of an older alphabet that included three additional archaic letters— F (vau or digamma or stigma), ٩ (koppa), and Λ (sampi)—to establish the following association of letters and numbers:

A	B	Γ	Δ	E	F	Z	H	Θ	I	K	Λ	M	N
1	2	3	4	5	6	7	8	9	10	20	30	40	50

Ξ	O	Π	٩	P	Σ	T	Υ	Φ	X	Ψ	Ω	Λ
60	70	80	90	100	200	300	400	500	600	700	800	900

Since the three archaic letters occupy the positions in the numeral scheme that they held in the older alphabet, it has been suggested that the Ionian system was introduced before the abandonment of the three letters—say in the eighth century B.C.; this view becomes less convincing when we consider the long time interval between the presumed introduction and the ultimate triumph of the system in the third century B.C.[10] The obvious advantage in conciseness of the alphabetic system might have been expected to find a readier adoption for the system than the indicated delay of half a millennium. The cipherization in the Ionian notation bears to the Attic numeration essentially the same relationship as did the Egyptian hieratic to the more cumbersome hieroglyphic, where the superiority of the cursive script had been clear to scribes.

After the introduction of small letters in Greece, the association of letters and numbers appeared as follows:

α	β	γ	δ	ε	ς	ζ	η	θ	ι	κ	λ	μ	ν
1	2	3	4	5	6	7	8	9	10	20	30	40	50

ξ	o	π	ϙ	ρ	σ	τ	υ	φ	χ	ψ	ω	⅄
60	70	80	90	100	200	300	400	500	600	700	800	900

[9] See Heath, *op. cit.*, I. 30. See also James Gow, *A Short History of Greek Mathematics* (Cambridge, 1884).

[10] For further discussion and references see C. B. Boyer, "Fundamental Steps in the Development of Numeration," *Isis* **35** (1944), 153–168.

Since these forms are more familiar today, we shall use them here. For the first nine multiples of a thousand, the Ionian system adopted the first nine letters of the alphabet, a partial use of the positional principle; but for added clarity these letters were preceded by a stroke or accent:

$,\alpha$	$,\beta$	$,\gamma$	$,\delta$	$,\epsilon$	$,\varsigma$	$,\zeta$	$,\eta$	$,\theta$
1000	2000	3000	4000	5000	6000	7000	8000	9000

Within this system any number less than 10,000 was easily written with only four characters. The number 8888, for example, would appear as $,\eta\omega\pi\eta$ or as $\eta\omega\pi\eta$, the accent sometimes being omitted when the context was clear. The use of the same letters for thousands as for units should have suggested to the Greeks the full-fledged positional scheme in decimal arithmetic, but they do not seem to have appreciated the advantages of such a move. That they had such a principle more or less in mind is evident not only in the repeated use of the letters α through θ for units and thousands, but also in the fact that the symbols are arranged in order of magnitude, from the smallest on the right to the largest on the left. At 10,000, which for the Greeks was the beginning of a new count or category (much as we separate thousands from lower powers by a comma), the Ionian Greek notation adopted a multiplicative principle. A symbol for an integer from 1 to 9999, when placed above the letter M, or after it, separated from the rest of the number by a dot, indicated the product of the integer and the number 10,000—the Greek myriad. Thus the number 88888888 would appear as $M,\eta\omega\pi\eta \cdot \eta\omega\pi\eta$. Where still larger numbers are called for, the same principle could be applied to the double myriad, 100000000 or 10^8.

Early Greek notations for integers were not excessively awkward, and they served their purposes effectively. It was in the use of fractions that the systems were weak. Like the Egyptians, the Greeks were tempted to use unit fractions, and for these they had a simple representation. They wrote down the denominator and then simply followed this with a diacritical mark or accent to distinguish it from the corresponding integer. Thus $\frac{1}{34}$ would appear as $\lambda\delta'$. This could, of course, be confused with the number $30\frac{1}{4}$, but context or the use of words could be assumed to make the situation clear. In later centuries general common fractions and sexagesimal fractions were in use; these will be discussed later in connection with the work of Archimedes, Ptolemy, and Diophantus, for there are extant documents which, while not actually dating from the time of these men, are copies of works written by them—a situation strikingly different from that concerning mathematicians of the Hellenic period.

The history of mathematics during the time of Thales and the Pythagoreans **11** necessarily depends, to an undesirable degree, on conjecture and inference, since documents from the period are entirely missing. In this respect there is far more uncertainty about Greek mathematics from 600 to 450 B.C. than about Babylonian algebra or Egyptian geometry from about 1700 B.C. Not even mathematical artifacts have survived from the early days of Greece. It is evident that some form of counting board or abacus was used in calculation, but the nature and operation of the device must be inferred from the Roman abacus and from some casual references in Greek authors. Herodotus, writing in the early fifth century B.C., says that in counting with pebbles, as in writing, the Greek hand moved from left to right, the Egyptian from right to left. A vase from a somewhat later period pictures a collector of tribute with a counting board which was used not only for integral decimal multiples of the drachma, but for nondecimal fractional subdivisions. Beginning on the left, the columns designate myriads, thousands, hundreds, and tens of drachmas, respectively, the symbols being in Herodianic notation. Then, following the units column for drachmas, there are columns for obols (six obols = one drachma), for half the obol, and for the quarter obol. Here we see how ancient civilizations avoided an excessive use of fractions: they simply subdivided units of length, weight, and money so effectively that they could calculate in terms of integral multiples of the subdivisions. This undoubtedly is the explanation for the popularity in antiquity of duodecimal and sexagesimal subdivisions, for the decimal system here is at a severe disadvantage. Decimal fractions were rarely used, either by the Greeks or by other Western peoples, before the period of the Renaissance. The abacus can be readily adapted to any system of numeration or to any combination of systems; it is likely that the widespread use of the abacus accounts at least in part for the amazingly late development of a consistent positional system of notation for integers and fractions. In this respect the Pythagorean Age contributed little if anything. The point of view of the Pythagoreans seems to have been so overwhelmingly philosophical and abstract that technical details in computation were of little concern to them. Such techniques were relegated to a separate discipline, called logistic. This dealt with the numbering of things, rather than with the essence and properties of number as such, matters of concern in arithmetic. That is, the ancient Greeks made a clear distinction between mere calculation on the one hand and what today is known in America as theory of numbers (and in England as the higher arithmetic) on the other. Whether or not such a sharp distinction was a disadvantage to the historical development of mathematics may be a moot point, but it is not easy to deny to the early Ionian and Pythagorean mathematicians the primary role in establishing mathematics as a rational and liberal discipline. It is for this reason that Thales often is called the first mathematician and that Pythagoras is known as the father of mathematics. The extent to which we accept such ascriptions literally, in view of the absence of supporting documentary evidence, will depend on our confidence in tradition. It is obvious that tradition can be quite inaccurate, but it seldom is entirely misdirected.

BIBLIOGRAPHY

Allman, G. J., *Greek Geometry from Thales to Euclid* (Dublin: Dublin University Press, 1889).

Clagett, Marshall, *Greek Science in Antiquity* (New York: Abelard-Schuman, ca. 1955; 2nd paperback ed., New York: Collier, 1966).

Dantzig, Tobias, *The Bequest of the Greeks* (New York: Scribner's, 1955).

Freeman, Kathleen, *Ancilla to the Pre-Socratic Philosophers* (Cambridge, Mass.: Harvard University Press, 1948).

Gow, James, *A Short History of Greek Mathematics* (reprint, New York: Hafner, 1923).

Heath, T. L., *A History of Greek Mathematics* (Oxford: Clarendon, 1921, 2 vols.).

Heath, T. L., *Manual of Greek Mathematics* (New York: Oxford University Press, 1931; paperback ed., New York: Dover, 1963).

Loria, Gino, *Historie des sciences mathématiques dans l'antiquité hellénique* (Paris: Gauthier-Villars, 1929).

Michel, Paul-Henri, *De Pythagore à Euclide* (Paris, 1950).

Neugebauer, O., *The Exact Sciences in Antiquity*, 2nd ed. (Providence, R.I.: Brown University Press, 1957; paperback ed., New York: Harper).

Tannery, Paul, *La géométrie grecque, comment son histoire nous est parvenue et ce que nous en savons* (Paris, 1887).

Thomas, Ivor, ed., *Selections Illustrating the History of Greek Mathematics* (2 vols., Cambridge, Mass.: Harvard University Press, 1939–1941).

Van der Waerden, B. L., *Science Awakening*, trans. by Arnold Dresden (New York: Oxford University Press, 1961; paperback ed., New York: Wiley, 1963).

Zeller, Eduard: *A History of Greek Philosophy from the Earliest Period to the Time of Socrates*, trans. by S. F. Alleyne (London: Longmans, Green, 1881, 2 vols.).

EXERCISES

1. Prove two theorems attributed to Thales and tell, with reasons, whether or not you think he may have used similar reasoning.
2. Prove the Pythagorean theorem. Do you think Pythagoras used your method? Explain.
3. Theon of Smyrna, a Neoplatonist and Neo-Pythagorean of the second century, is said to have found that the sum of two consecutive triangular numbers is a square number. Prove this theorem.
4. What are the first four heptagonal numbers (corresponding to regular polygons of seven sides)?
5. Write the numbers 3456 and 4567 and their sum in the early Greek Attic notation and in the Ionian or alphabetic system.

6. Prove that if three numbers, a, b, c are in arithmetic progression in that order, and if A, B, C are their reciprocals respectively, B is the harmonic mean of A and C.

7. Philolaus called the cube a "geometrical harmony," because of the number of its faces, vertices, and edges. Justify his designation in the light of Pythagorean theory of proportions.

8. Show that 1184 and 1210 are amicable numbers.

9. Show, in the manner of the Pythagoreans, that an oblong number is the sum of two equal triangular numbers.

10. Prove carefully that the diagonals of a regular pentagon divide each other in mean and extreme ratio.

11. Using straightedge and compasses only, construct a regular pentagon, given the side of the pentagon.

12. Using straightedge and compasses only, construct a regular pentagon, given a diagonal of the pentagon.

13. In a given circle construct a regular pentagon, using straightedge and compasses only.

14. All polygonal numbers are of the form $P_m = an^2 + bn$, where m is the number of sides and n is the order. Using this fact, find a and b for octagonal numbers $(m = 8)$ and verify geometrically for $n = 3$.

15. Find the fifth pentagonal number and the sixth hexagonal number.

16. Is 4567 a heptagonal number? Justify your answer.

17. Show that, if $a > b > c$, the three equations

$$\frac{a-b}{b-c} = \frac{a}{a}, \qquad \frac{a-b}{b-c} = \frac{a}{b}, \qquad \frac{a-b}{b-c} = \frac{a}{c}$$

define b respectively as the arithmetic, the geometric, and the harmonic mean of a and c.

*18. All polyhedral numbers are of the form $P_m = an^3 + bn^2 + cn$, where m is the number of faces and n is the order. Use this fact to find a and b and c for tetrahedral numbers $(m = 4)$ and verify geometrically for $n = 4$.

*19. Polyhedral numbers are found by adding successive polygonal numbers of the same kind. Show how to generalize this procedure to define polytopal numbers in n-dimensional space and find three nontrivial polytopal numbers.

CHAPTER V

The Heroic Age

> I would rather discover one cause than gain the kingdom of Persia.
>
> *Democritus*

Accounts of the origins of Greek mathematics center on the so-called Ionian **1** and Pythagorean schools and the chief representative of each—Thales and Pythagoras—although reconstructions of their thought rest on fragmentary reports and traditions built up during later centuries. To a certain extent this situation prevails throughout the fifth century B.C. There are virtually no extant mathematical or scientific documents until the days of Plato in the fourth century B.C. Nevertheless, during the last half of the fifth century there circulated persistent and consistent reports concerning a handful of mathematicians who evidently were intensely concerned with problems that formed the basis for most of the later developments in geometry. We shall therefore refer to this period as the "Heroic Age of Mathematics," for seldom either before or since have men with so little to work with tackled mathematical problems of such fundamental significance. No longer was mathematical activity centered almost entirely in two regions nearly at opposite ends of the Greek world; it flourished all about the Mediterranean. In what is now southern Italy there were Archytas of Tarentum (born ca. 428 B.C.) and Hippasus of Metapontum (fl. ca. 400 B.C.); at Abdera in Thrace we find Democritus (born ca. 460 B.C.); nearer the center of the Greek world, on the Attic peninsula, there was Hippias of Ellis (born ca. 460 B.C.); and at nearby Athens there lived at various times during the critical last half of the fifth century B.C. three scholars from other regions: Hippocrates of Chios (fl. ca. 430 B.C.), Anaxagoras of Clazomenae (†428 B.C.), and Zeno of Elea (fl. ca. 450 B.C.). Through the work of these seven men we shall describe the fundamental changes in mathematics that took place a little before the year 400 B.C.

The fifth century B.C. was a crucial period in the history of Western **2** civilization, for it opened with the defeat of the Persian invaders and closed with the surrender of Athens to Sparta. Between these two events lay the great Age of Pericles, with its accomplishments in literature and art. The prosperity and intellectual atmosphere of Athens during the century attracted scholars from all parts of the Greek world, and a synthesis of diverse aspects was achieved. From Ionia came men, such as Anaxagoras, with a practical turn of mind; from southern Italy came others, such as Zeno, with stronger metaphysical inclinations. Democritus of Abdera espoused a materialistic view of the world, while Pythagoras in Italy held idealistic attitudes in science and philosophy. At Athens one found eager devotees of old and new branches of learning, from cosmology to ethics. There was a bold spirit of free inquiry that sometimes came into conflict with established mores. In particular, Anaxagoras was imprisoned at Athens for impiety in asserting that the sun was not a deity, but a huge red-hot stone as big as the whole Peloponnessus, and that the moon was an inhabited earth that borrowed its light from the sun. He well represents the spirit of rational inquiry, for he regarded as the aim of his life the study of the nature of the universe—a purposefulness that he derived from the Ionian tradition of which Thales had been a founder. The intellectual enthusiasm of Anaxagoras was shared with his countrymen through the first scientific best-seller—a book *On Nature* which could be bought in Athens for only a drachma. Anaxagoras was a teacher of Pericles, who saw to it that his mentor ultimately was released from prison. Socrates was at first attracted to the scientific ideas of Anaxagoras, but the gadfly of Athens found the naturalistic Ionian view less satisfying than the search for ethical verities.

Greek science had been rooted in a highly intellectual curiosity which often is contrasted with the utilitarian immediacy of pre-Hellenic thought; Anaxagoras clearly represented the typical Greek motive—the desire to know. In mathematics also the Greek attitude differed sharply from that of the earlier potamic cultures. The contrast was clear in the contributions generally attributed to Thales and Pythagoras, and it continues to show through in the more reliable reports on what went on in Athens during the Heroic Age. Anaxagoras was primarily a natural philosopher rather than a mathematician, but his inquiring mind led him to share in the pursuit of mathematical problems. We are told by Plutarch that while Anaxagoras was in prison he occupied himself in an attempt to square the circle. Here we have the first mention of a problem that was to fascinate mathematicians for more than 2000 years.[1] There are no further details concerning the origin of the problem or the rules governing it. At a later date it came to be understood that the required square, exactly equal in area to the circle, was to be

---

[1] See E. W. Hobson, *Squaring the Circle* (ca. 1913), p. 14. This work has been reprinted several times. The accuracy of Plutarch's statement in this connection has been questioned recently. On the work of Anaxagoras see D. E. Gershenson and D. A. Greenberg, *Anaxagoras and the Birth of Physics* (New York: Blaisdell, 1964).

constructed by the use of compasses and straightedge alone. Here we see a type of mathematics that is quite unlike that of the Egyptians and Babylonians. It is not the practical application of a science of number to a facet of life experience, but a theoretical question involving a nice distinction between accuracy in approximation and exactitude in thought. The mathematical problem that Anaxagoras here considered was no more the concern of the technologist than were those he raised in science concerning the ultimate structure of matter. In the Greek world mathematics was more closely related to philosophy than to practical affairs, and this kinship has persisted to the present day.

Anaxagoras died in 428 B.C., the year that Archytas was born, just one **3** year before Plato's birth and one year after Pericles' death. It is said that Pericles died of the plague that carried off perhaps a quarter of the Athenian population, and the deep impression that this catastrophe created is perhaps the origin of a second famous mathematical problem. It is reported that a delegation had been sent to the oracle of Apollo at Delos to inquire how the plague could be averted, and the oracle had replied that the cubical altar to Apollo must be doubled. The Athenians are said to have dutifully doubled the dimensions of the altar, but this was of no avail in curbing the plague. The altar had, of course, been increased eightfold in volume, rather than twofold. Here, according to the legend, was the origin of the "duplication of the cube" problem, one that henceforth was usually referred to as the "Delian problem"—given the edge of a cube, construct with compasses and straightedge alone the edge of a second cube having double the volume of the first. At about the same time there circulated in Athens still a third celebrated problem—given an arbitrary angle, construct by means of compasses and straightedge alone an angle one-third as large as the given angle. These three problems—the squaring of the circle, the duplication of the cube, and the trisection of the angle—have since been known as the "three famous (or classical) problems" of antiquity. More than 2200 years later it was to be proved that all three of the problems were unsolvable by means of straightedge and compasses alone. Nevertheless, the better part of Greek mathematics, and of much later mathematical thought, was suggested by efforts to achieve the impossible—or, failing this, to modify the rules. The Heroic Age failed in its immediate objective, under the rules, but the efforts were crowned with brilliant success in other respects.

Somewhat younger than Anaxagoras, and coming originally from about **4** the same part of the Greek world, was Hippocrates of Chios. He should not be confused with his still more celebrated contemporary, the physician Hippocrates of Cos. Both Cos and Chios are islands in the Dodecanese group; but Hippocrates of Chios in about 430 B.C. left his native land for

Athens in his capacity as a merchant. Aristotle reports that Hippocrates was less shrewd than Thales and that he lost his money in Byzantium through fraud; others say that he was beset by pirates. In any case, the incident was never regretted by the victim, for he counted this his good fortune in that as a consequence he turned to the study of geometry, in which he achieved remarkable success—a story typical of the Heroic Age. Proclus wrote that Hippocrates composed an "Elements of Geometry," anticipating by more than a century the better-known *Elements* of Euclid. However, the textbook of Hippocrates—as well as another reported to have been written by Leon, a later associate of the Platonic school—has been lost, although it was known to Aristotle. In fact, no mathematical treatise from the fifth century has survived; but we do have a fragment concerning Hippocrates which Simplicius (fl. ca. 520) claims to have copied literally from the *History of Mathematics* (now lost) by Eudemus. This brief statement, the nearest thing we have to an original source on the mathematics of the time, describes a portion of the work of Hippocrates dealing with the quadrature of lunes. A lune is a figure bounded by two circular arcs of unequal radii; the problem of the quadrature of lunes undoubtedly arose from that of squaring the circle. The Eudemian fragment attributes to Hippocrates the following theorem:

> Similar segments of circles are in the same ratio as the squares on their bases.

The Eudemian account reports that Hippocrates demonstrated this by first showing that the areas of two circles are to each other as the squares on their diameters. Here Hippocrates adopted the language and concept of proportion which played so large a role in Pythagorean thought. In fact, it is thought by some that Hippocrates became a Pythagorean. The Pythagorean school in Croton had been suppressed (possibly because of its secrecy, perhaps because of its conservative political tendencies), but the scattering of its adherents throughout the Greek world served only to broaden the influence of the school. This influence undoubtedly was felt, directly or indirectly, by Hippocrates.

The theorem of Hippocrates on the areas of circles seems to be the earliest precise statement on curvilinear mensuration in the Greek world. Eudemus believed that Hippocrates gave a proof of the theorem, but a rigorous demonstration at that time (say about 430 B.C.) would appear to be unlikely. The theory of proportions at that stage probably was established for commensurable magnitudes only. The proof as given in Euclid XII. 2 comes from Eudoxus, a man who lived halfway between Hippocrates and Euclid. However, just as much of the material in the first two books of Euclid seems to stem from the Pythagoreans, so it would appear reasonable to assume that the formulations, at least, of much of Books III and IV of the *Elements* came from the work of Hippocrates. Moreover, if Hippocrates did give a demonstration of his theorem on the areas of circles, he may have been responsible for the introduction into mathematics of the indirect method of proof. That is, the ratio of the areas of two circles is equal to the ratio of the squares on the diameters or it is not. By a *reductio ad absurdum* from the second of the two possibilities, the proof of the only alternative is established.

From his theorem on the areas of circles Hippocrates readily found the first rigorous quadrature of a curvilinear area in the history of mathematics. He began with a semicircle circumscribed about an isosceles right triangle, and on the base (hypotenuse) he constructed a segment similar to the circular segments on the sides of the right triangle (Fig. 5.1). Because the segments

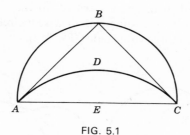

FIG. 5.1

are to each other as squares on their bases, and from the Pythagorean theorem as applied to the right triangle, the sum of the two small circular segments is equal to the larger circular segment. Hence the difference between the semicircle on AC and the segment $ADCE$ equals triangle ABC. Therefore the lune $ABCD$ is precisely equal to triangle ABC; and since triangle ABC is equal to the square on half of AC, the quadrature of the lune has been found.[2]

Eudemus describes also an Hippocratean lune quadrature based on an isosceles trapezoid $ABCD$ inscribed in a circle so that the square on the longest side (base) AD is equal to the sum of the squares on the three equal shorter sides AB and BC and CD (Fig. 5.2). Then if on side AD one constructs a circular segment $AEDF$ similar to those on the three equal sides, lune $ABCDE$ is equal to trapezoid $ABCDF$.

That we are on relatively firm ground historically in describing the quadrature of lunes by Hippocrates, is indicated by the fact that scholars other than Simplicius also refer to this work. Simplicius lived in the sixth century, but he depended not only on Eudemus (fl. ca. 320 B.C.) but also on Alexander of Aphrodisias (fl. ca. A.D. 200), one of the chief commentators on

---

[2] An excellent account of Hippocrates' quadratures is found in B. L. van der Waerden, *Science Awakening* (1961), pp. 131 ff.

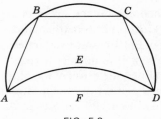

FIG. 5.2

Aristotle. Alexander describes two quadratures other than those given above. (1) If on the hypotenuse and sides of an isosceles right triangle one constructs semicircles (Fig. 5.3), then the lunes created on the smaller sides together equal the triangle. (2) If on a diameter of a semicircle one constructs an isosceles trapezoid with three equal sides (Fig. 5.4), and if on the three equal

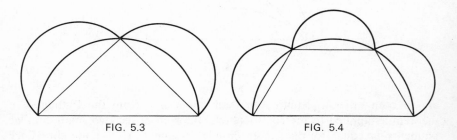

FIG. 5.3   FIG. 5.4

sides semicircles are constructed, then the trapezoid is equal in area to the sum of four curvilinear areas: the three equal lunes and a semicircle on one of the equal sides of the trapezoid. From the second of these quadratures it would follow that if the lunes can be squared, the semicircle—hence the circle—can also be squared. This conclusion seems to have encouraged Hippocrates, as well as his contemporaries and early successors, to hope that ultimately the circle would be squared.

5   The Hippocratean quadratures are significant not so much as attempts at circle-squaring as indications of the level of mathematics at the time. They show that Athenian mathematicians were adept at handling transformations of areas and proportions. In particular, there was evidently no difficulty in converting a rectangle of sides a and b into a square. This required finding the mean proportional or geometric mean between a and b. That is, if $a:x = x:b$, geometers of the day easily constructed the line x. It was natural, therefore, that geometers should seek to generalize the problem by inserting *two* means between two given magnitudes a and b. That is, given two line segments a and b, they hoped to construct two other segments x and y such that $a:x = x:y = y:b$. Hippocrates is said to have recognized that this problem is equivalent to that of duplicating the cube; for if $b = 2a$, the continued proportions, upon the elimination of y, lead to the conclusion that $x^3 = 2a^3$.

There are three views on what Hippocrates deduced from his quadrature of lunes. Some have accused him of believing that he could square all lunes, hence also the circle; others think that he knew the limitations of his work, concerned as it was with some types of lunes only. At least one scholar has held that Hippocrates knew he had not squared the circle but tried to deceive his countrymen into thinking that he had succeeded.[3] There are other questions, too, concerning Hippocrates' contributions, for to him has been ascribed, with some uncertainty, the first use of letters in geometric figures. It is interesting to note that whereas he advanced two of the three famous problems, he seems to have made no progress in the trisection of the angle, a problem studied somewhat later by Hippias of Ellis.

Toward the end of the fifth century B.C. there flourished at Athens a group **6** of professional teachers quite unlike the Pythagoreans. Disciples of Pythagoras had been forbidden to accept payment for sharing their knowledge with others. The Sophists, however, openly supported themselves by tutoring fellow citizens—not only in honest intellectual endeavor, but also in the art of "making the worse appear the better." To a certain extent the accusation of shallowness directed against the Sophists was warranted; but this should not conceal the fact that Sophists usually were very widely informed in many fields and that some of them made real contributions to learning. Among these was Hippias, a native of Ellis who was active at Athens in the second half of the fifth century B.C. He is one of the earliest mathematicians of whom we have firsthand information, for we learn much about him from Plato's dialogues. We read, for example, that Hippias boasted that he had made more money than any two other Sophists. He is said to have written much, from mathematics to oratory, but none of his work has survived. He had a remarkable memory, he boasted immense learning, and he was skilled in handicrafts. To this Hippias (there were many others in Greece who bore the same name) we apparently owe the introduction into mathematics of the first curve beyond the circle and the straight line. Proclus and other commentators ascribe to him the curve since known as the trisectrix or quadratix of Hippias.[4] This is drawn as follows: In the square $ABCD$ (Fig. 5.5) let side

---

[3] See Björnbo's article "Hippocrates" in Pauly-Wissowa, *Real-Enzyklopädie der klassischen Altertumswissenschaft*, Vol. VIII, p. 1796.

[4] An excellent account of this is found in Kathleen Freeman, *The Pre-Socratic Philosophers. A Companion to Diels, Fragmente der Vorsokratiker* (1949), pp. 381–391. See also the article on Hippias in Pauly-Wissowa, *op. cit.*, VIII, 1707 ff.

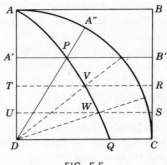

FIG. 5.5

AB move down uniformly from its present position until it coincides with *DC* and let this motion take place in exactly the same time that side *DA* rotates clockwise from its present position until it coincides with *DC*. If the positions of the two moving lines at any given time are given by *A′B′* and *DA″* respectively and if *P* is the point of intersection of *A′B′* and *DA″*, the locus of *P* during the motions will be the trisectrix of Hippias—curve *APQ* in the figure. Given this curve, the trisection of an angle is carried out with ease. For example, if *PDC* is the angle to be trisected, one simply trisects segments *B′C* and *A′D* at points *R*, *S*, *T*, and *U*. If lines *TR* and *US* cut the trisectrix in *V* and *W* respectively, lines *VD* and *WD* will, by the property of the trisectrix, divide angle *PDC* in three equal parts.

The curve of Hippias generally is known as the quadratrix, since it can be used to square the circle. Whether or not Hippias himself was aware of this application cannot now be determined. It has been conjectured that Hippias knew of this method of quadrature but that he was unable to justify it. Since the quadrature through Hippias' curve was specifically given later by Dinostratus, we shall describe this work in the next chapter.

Hippias lived at least as late as Socrates (†399 B.C.), and from the pen of Plato we have an unflattering account of him as a typical Sophist—vain, boastful, and acquisitive. Socrates is reported to have described Hippias as handsome and learned, but boastful and shallow. Plato's dialogue on *Hippias* satirizes his show of knowledge, and Xenophon's *Memorabilia* includes an unflattering account of Hippias as one who regarded himself an expert in everything from history and literature to handicrafts and science. In judging such accounts, however, we must remember that Plato and Xenophon were uncompromisingly opposed to the Sophists in general. It is well to bear in mind also that both Protagoras, the "founding father of the Sophists," and Socrates, the archopponent of the movement, were antagonistic to mathematics and the sciences. With respect to character, Plato contrasts Hippias with Socrates, but one can bring out much the same contrast by comparing Hippias with another contemporary—the Pythagorean mathematician Archytas of Tarentum.

Pythagoras is said to have retired to Metapontum toward the end of his **7** life and to have died there about 500 B.C. Tradition holds that he left no written works, but his ideas were carried on by a large number of eager disciples. The center at Croton was abandoned when a rival political group from Sybaris surprised and murdered many of the leaders, but those who escaped the massacre carried the doctrines of the school to other parts of the Greek world. Among those who received instruction from the refugees was Philolaus of Tarentum, and he is said to have written the first account of Pythagoreanism—permission having been granted, so the story goes, to repair his damaged fortunes. Apparently it was this book from which Plato derived his knowledge of the Pythagorean order. The number fanaticism that was so characteristic of the brotherhood evidently was shared by Philolaus, and it was from his account that much of the mystical lore concerning the *tetractys* was derived, as well as knowledge of the Pythagorean cosmology. The Philolaean cosmic scheme is said to have been modified by two later Pythagoreans, Ecphantus and Hicetas, who abandoned the central fire and counterearth and explained day and night by placing a rotating earth at the center of the universe. The extremes of Philolaean number worship also seem to have undergone some modification, more especially at the hands of Archytas, a student of Philolaus at Tarentum.

The Pythagorean sect had exerted a strong intellectual influence throughout Magna Graecia, with political overtones that may be described as a sort of "reactionary international," or perhaps better as a cross between Orphism and Freemasonry. At Croton political aspects were especially noticeable, but at outlying Pythagorean centers, such as Tarentum, the impact was primarily intellectual. Archytas believed firmly in the efficacy of number; his rule of the city, which allotted him autocratic powers, was just and restrained, for he regarded reason as a force working toward social amelioration. For many years in succession he was elected general, and he was never defeated; yet he was kind and a lover of children, for whom he is reported to have invented "Archytas' rattle." Possibly also the mechanical dove, which he is said to have fashioned of wood, was built to amuse the young folk.

Archytas continued the Pythagorean tradition in placing arithmetic above geometry, but his enthusiasm for number had less of the religious and mystical admixture found earlier in Philolaus. He wrote on the application of the arithmetic, geometric, and subcontrary means to music, and it was probably either Philolaus or Archytas who was responsible for changing the name of

the last one to "harmonic mean." Among his statements in this connection
was the observation that between two whole numbers in the ratio $n:(n + 1)$
there could be no integer that is a geometric mean. Archytas gave more
attention to music than had his predecessors, and he felt that this subject
should play a greater role than literature in the education of children. Among
his conjectures was one that attributed differences in pitch to varying rates
of motion resulting from the flow causing the sound. Archytas seems to
have paid considerable attention to the role of mathematics in the curriculum,
and to him has been ascribed the designation of the four branches in the
mathematical quadrivium—arithmetic (or numbers at rest), geometry (or
magnitudes at rest), music (or numbers in motion), and astronomy (or
magnitudes in motion). These subjects, together with the trivium consisting
of grammar, rhetoric, and dialectics (which Aristotle traced back to Zeno),
later constituted the seven liberal arts; hence the prominent role that
mathematics has played in education is in no small measure due to Archytas.

8   It is likely that Archytas had access to an earlier treatise on the elements
of mathematics, and the iterative square-root process often known by the
name of Archytas had been used long before in Mesopotamia. Nevertheless,
Archytas was himself a contributor of original mathematical results. The
most striking contribution was a three-dimensional solution of the Delian
problem which may be most easily described, somewhat anachronistically,
in the modern language of analytic geometry. Let a be the edge of the cube
to be doubled, and let the point $(a, 0, 0)$ be the center of three mutually
perpendicular circles of radius a and each lying in a plane perpendicular
to a coordinate axis. Through the circle perpendicular to the x-axis construct
a right circular cone with vertex $(0, 0, 0)$; through the circle in the xy-plane
pass a right circular cylinder; and let the circle in the yz-plane be revolved
about the z-axis to generate a torus. The equations of these three surfaces
are respectively $x^2 = y^2 + z^2$ and $2ax = x^2 + y^2$ and $(x^2 + y^2 + z^2)^2 = 4a^2(x^2 + y^2)$. These three surfaces intersect in a point whose x-coordinate is
$a\sqrt[3]{2}$; hence the length of this line segment is the edge of the cube desired.

The achievement of Archytas is the more impressive when we recall that
his solution was worked out synthetically without the aid of coordinates.
Nevertheless, the most important contribution of Archytas to mathematics
may have been his intervention with the tyrant Dionysius to save the life
of his friend, Plato. The latter remained to the end of his life deeply committed to the Pythagorean veneration of number and geometry, and the
supremacy of Athens in the mathematical world of the fourth century B.C.
resulted primarily from the enthusiasm of Plato, the "maker of mathematicians." However, before taking up the role of Plato it is necessary to discuss
the work of an earlier Pythagorean—an apostate by the name of Hippasus.

Hippasus of Metapontum (or Croton), roughly contemporaneous with Philolaus, is reported to have been originally a Pythagorean but to have been expelled from the brotherhood. One account has it that the Pythagoreans erected a tombstone to him, as though he were dead; another story reports that his apostasy was punished by death at sea in a shipwreck. The exact cause of the break is unknown, in part because of the rule of secrecy, but there are three suggested possibilities. According to one, Hippasus was expelled for political insubordination, having headed a democratic movement against the conservative Pythagorean rule. A second tradition attributes the expulsion to disclosures concerning the geometry of the pentagon or the dodecahedron—perhaps a construction of one of the figures. A third explanation holds that the expulsion was coupled with the disclosure of a mathematical discovery of devastating significance for Pythagorean philosophy—the existence of incommensurable magnitudes.

It had been a fundamental tenet of Pythagoreanism that the essence of **9** all things, in geometry as well as in the practical and theoretical affairs of man, are explainable in terms of *arithmos*, or intrinsic properties of whole numbers or their ratios. The dialogues of Plato show, however, that the Greek mathematical community had been stunned by a disclosure that virtually demolished the basis for the Pythagorean faith in whole numbers. This was the discovery that within geometry itself the whole numbers and their ratios are inadequate to account for even simple fundamental properties. They do not suffice, for example, to compare the diagonal of a square or a cube or a pentagon with its side. The line segments are incommensurable, no matter how small a unit of measure is chosen. Just when and how the discovery was made is not known, but much ink has been spilled in support of one hypothesis or another. Earlier arguments in favor of a Hindu origin of the discovery[5] lack foundation, and there seems to be little chance that Pythagoras himself was aware of the problem of incommensurability. The most plausible suggestion is that the discovery was made by the later Pythagoreans at some time before 410 B.C.[6] Some would attribute it specifically to Hippasus of Metapontum during the earlier portion of the last quarter of the fifth century B.C.,[7] while others place it about another half a century later.

---

[5] See Heinrich Vogt, "Haben die alten Inder den Pythagoreischen Lehrsatz und das Irrationale gekannt?," *Bibliotheca Mathematica* (3), **7** (1906–1907), 6–23; also Leopold von Schroeder, *Pythagoras und die Inder* (Leipzig, 1884).

[6] See especially Heinrich Vogt, "Die Entdeckungsgeschichte des Irrationalen nach Plato und anderen Quellen des 4. Jahrhunderts," *Bibliotheca Mathematica* (3), **10** (1910), 97–155, and the same author's paper, "Zur Entdeckungsgeschichte des Irrationalen," *Bibliotheca Mathematica* (3), **14** (1914), 9–29. Cf. Heath, *History of Greek Mathematics* (1921), 1, 157.

[7] See Kurt von Fritz, "The Discovery of Incommensurability by Hippasus of Metapontum," *Annals of Mathematics* (2), **46** (1945), 242–264.

The circumstances surrounding the earliest recognition of incommensurable line segments are as uncertain as is the time of the discovery. Ordinarily it is assumed that the recognition came in connection with the application of the Pythagorean theorem to the isosceles right triangle. Aristotle refers to a proof of the incommensurability of the diagonal of a square with respect to a side, indicating that it was based on the distinction between odd and even.[8] Such a proof is easy to construct. Let d and s be the diagonal and side of a square, and assume that they are commensurable—that is, that the radio d/s is rational and equal to p/q, where p and q are integers with no common factor. Now, from the Pythagorean theorem it is known that $d^2 = s^2 + s^2$; hence $(d/s)^2 = p^2/q^2 = 2$, or $p^2 = 2q^2$. Therefore p^2 must be even; hence p must be even. Consequently q must be odd. Letting $p = 2r$ and substituting in the equation $p^2 = 2q^2$, we have $4r^2 = 2q^2$, or $q^2 = 2r^2$. Then q^2 must be even; hence q must be even. However, q was shown above to be odd, and an integer cannot be both odd and even. It follows therefore, by the indirect method, that the assumption that d and s are commensurable must be false.

10 In this proof the degree of abstraction is so high that the possibility that it was the basis for the original discovery of incommensurability has been questioned. There are, however, other ways in which the discovery could have come about. Among these is the simple observation that when the five diagonals of a regular pentagon are drawn, these diagonals form a smaller regular pentagon (Fig. 5.6), and the diagonals of the second pentagon in

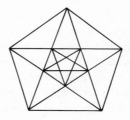

FIG. 5.6

turn form a third regular pentagon, which is still smaller. This process can be continued indefinitely, resulting in pentagons that are as small as desired and leading to the conclusion that the ratio of a diagonal to a side in a regular pentagon is not rational. The irrationality of this ratio is, in fact, a consequence

[8] See H. G. Zeuthen, "Sur l'origine historique de la connaissance des quantités irrationelles," *Oversigt over det Kongelige Danske Videnskabernes Selskabs. Forhandlinger*, 1915, pp. 333–362.

of the argument presented in connection with Fig. 4.2 in which the golden section was shown to repeat itself over and over again. Was it perhaps this property that led to the disclosure, possibly by Hippasus, of incommensurability? There is no surviving document to resolve the question, but the suggestion is at least a plausible one. In this case, it would not have been $\sqrt{2}$ but $\sqrt{5}$ that first disclosed the existence of incommensurable magnitudes, for the solution of the equation $a:x = x:(a - x)$ leads to $(\sqrt{5} - 1)/2$ as the ratio of the side of a regular pentagon to a diagonal. The ratio of the diagonal of a cube to an edge is $\sqrt{3}$, and here, too, the spectre of the incommensurable rears its ugly head.

A geometric proof somewhat analogous to that for the ratio of the diagonal of a pentagon to its side can be provided also for the ratio of the diagonal of a square to its side. If in the square $ABCD$ (Fig. 5.7) one lays off on the diagonal

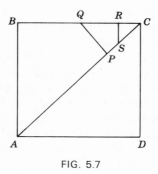

FIG. 5.7

AC the segment $AP = AB$ and at P erects the perpendicular PQ, the ratio of CQ to PC will be the same as the ratio of AC to AB. Again, if on CQ one lays off $QR = QP$ and constructs RS perpendicular to CR, the ratio of hypotenuse to side again will be what it was before. This process, too, can be continued indefinitely, thus affording a proof that no unit of length, however small, can be found so that the hypotenuse and a side will be commensurable.

The Pythagorean doctrine that "Numbers constitute the entire heaven" **11** was now faced with a very serious problem indeed; but it was not the only one, for the school was confronted also by arguments propounded by the neighboring Eleatics, a rival philosophical movement. Ionian philosophers of Asia Minor had sought to identify a first principle for all things. Thales had thought to find this in water, but others preferred to think of air or fire as the basic element. The Pythagoreans had taken a more abstract direction, postulating that number in all its plurality was the basic stuff behind phenomena; this numerical atomism, beautifully illustrated in the geometry of figurate numbers, had come under attack by the followers of Parmenides of Elea (fl. ca. 450 B.C.). The fundamental tenet of the Eleatics was the unity and permanence of being, a view that contrasted with the Pythagorean ideas of multiplicity and change. Of Parmenides' disciples the best known was Zeno the Eleatic (fl. ca. 450 B.C.) who propounded arguments to prove the inconsistency in the concepts of multiplicity and divisibility. The method Zeno adopted was dialectical, anticipating Socrates in this indirect mode of argument: starting from his opponent's premises, he reduced these to an absurdity.

The Pythagoreans had assumed that space and time can be thought of as consisting of points and instants; but space and time have also a property, more easily intuited than defined, known as "continuity." The ultimate elements making up a plurality were assumed on the one hand to have the characteristics of the geometrical unit—the point—and on the other to have certain characteristics of the numerical units or numbers. Aristotle described a Pythagorean point as "unity having position" or as "unity considered in space." It has been suggested[9] that it was against such a view that Zeno propounded his paradoxes, of which those on motion are cited most frequently. As they have come down to us, through Aristotle and others, four of them seem to have caused the most trouble: (1) the *Dichotomy*, (2) the *Achilles*, (3) the *Arrow*, and (4) the *Stade*. The first argues that before a moving object can travel a given distance, it must first travel half this distance; but before it can cover this, it must travel the first quarter of the distance; and before this, the first eighth, and so on through an infinite number of subdivisions. The runner wishing to get started, must make an infinite number of contacts in a finite time; but it is impossible to exhaust an infinite collection, hence the beginning of motion is impossible. The second of the paradoxes is similar to the first except that the infinite subdivision is progressive rather than regressive. Here Achilles is racing against a tortoise that has been given a headstart, and it is argued that Achilles, no matter how swiftly he may run, can never overtake the tortoise, no matter how slow it may be. By the time that Achilles will have reached the initial position of the tortoise, the latter will have advanced some short distance; and by the time that Achilles will have covered this distance, the tortoise will have advanced somewhat farther; and so the process continues indefinitely, with the result that the swift Achilles can never overtake the slow tortoise.

[9] See Paul Tannery, *La géométrie grecque* (1887), pp. 217–261. For a different view see B. L. van der Waerden, "Zenon und die Grundlagenkrise der griechischen Mathematik," *Mathematische Annalen*, **117** (1940), 141–161.

The *Dichotomy* and the *Achilles* argue that motion is impossible under the assumption of the infinite subdivisibility of space and time; the *Arrow* and the *Stade*, on the other hand, argue that motion is equally impossible if one makes the opposite assumption—that the subdivisibility of space and time terminates in indivisibles. In the *Arrow* Zeno argues that an object in flight always occupies a space equal to itself; but that which always occupies a space equal to itself is not in motion. Hence the flying arrow is at rest at all times, so that its motion is an illusion.

Most controversial of the paradoxes on motion, and most awkward to describe, is the *Stade* (or *Stadium*), but the argument can be phrased somewhat as follows. Let A_1, A_2, A_3, A_4 be bodies of equal size that are stationary; let B_1, B_2, B_3, B_4 be bodies, of the same size as the A's, that are moving to the right so that each B passes each A in an instant—the smallest possible interval of time. Let C_1, C_2, C_3, C_4 also be of equal size with the A's and B's and let them move uniformly to the left with respect to the A's so that each C passes each A in an instant of time. Let us assume that at a given time the bodies occupy the following relative positions:

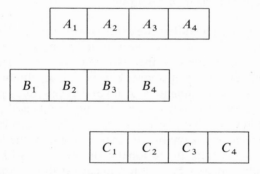

Then after the lapse of a single instant—that is, after an indivisible subdivision of time—the positions will be as follows:

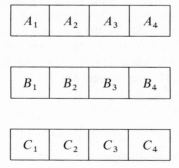

It is clear, then, that C_1 will have passed two of the B's; hence the instant cannot be the minimum time interval, for we can take as a new and smaller unit the time it takes C_1 to pass one of the B's.

The arguments of Zeno[10] seem to have had a profound influence on the development of Greek mathematics, comparable to that of the discovery of the incommensurable, with which it may have been related. Originally, in Pythagorean circles, magnitudes were represented by pebbles or calculi, from which our word calculation comes, but by the time of Euclid there is a complete change in point of view. Magnitudes are not in general associated with numbers or pebbles, but with line segments. In the *Elements* even the integers themselves are represented by segments of lines. The realm of number continued to have the property of discreteness, but the world of continuous magnitudes (and this included most of pre-Hellenic and Pythagorean mathematics) was a thing apart from number and had to be treated through geometrical method. It seemed to be geometry rather than number that ruled the world. This was perhaps the most far-reaching conclusion of the Heroic Age, and it is not unlikely that this was due in large measure to Zeno of Elea and Hippasus of Metapontum.

12   It has generally been held that the deductive element had been introduced into mathematics by Thales, but recently it has been argued against this thesis that the mathematics of the sixth and fifth centuries B.C. was too primitive to countenance such a contribution. Those who hold to this thesis sometimes refer to the arguments of Zeno and Hippasus as possible inspiration for the deductive approach. Certainly the doubts and problems raised in this connection would have been a fertile field for the growth of deduction; and it would not be unreasonable to regard the end of the fifth century B.C. as a *terminus ante quem* for the rational deductive form with which we have become so familiar. It may be well to indicate at this point, therefore, that there are several conjectures as to the causes leading to the conversion of the mathematical prescriptions of pre-Hellenic peoples into the deductive structure appearing in Greece. Some have suggested[11] that Thales in his travels had noted discrepancies in pre-Hellenic mathematics—such as the Egyptian and Babylonian rules for the area of a circle—and that he and his early successors therefore saw the need for a strict rational method. Others, more conservative, would place the deductive form much later—perhaps even as late as the early fourth century, following the discovery of the

[10] The bibliography on the paradoxes is enormous. Among the most informative historical treatments is that by Florian Cajori, "History of Zeno's Arguments on Motion," *American Mathematical Monthly*, **22** (1915), 1–6, 39–47, 77–82, 109–115, 145–149, 179–186, 215–220, 253–258, 292–297. For sources, see *Zeno of Elea* (text, translation, and notes by H. D. P. Lee; 1936).

[11] See van der Waerden, *Science Awakening* (1961), p. 89.

incommensurable.[12] Other suggestions find the cause outside mathematics. One, for example, sees in the sociopolitical development of the Greek city-states the rise of dialectrics and a consequent requirement of a rational basis for mathematics and other studies; another somewhat similar suggestion is that deduction may have come out of logic in attempts to convince an opponent of a conclusion by looking for premises from which the conclusion necessarily follows.[13]

Whether deduction came into mathematics in the sixth century B.C. or **13** the fourth and whether incommensurability was discovered before or after 400 B.C., there can be no doubt that Greek mathematics had undergone drastic changes by the time of Plato. The dichotomy between number and continuous magnitude required a new approach to the Babylonian algebra that the Pythagoreans had inherited. The old problems in which, given the sum and the product of the sides of a rectangle, the dimensions were required, had to be dealt with differently from the numerical algorithms of the Babylonians. A "geometrical algebra" had to take the place of the older "arithmetical algebra," and in this new algebra there could be no adding of lines to areas or of areas to volumes. From now on there had to be a strict homogeneity of terms in equations, and the Mesopotamian normal forms, $xy = A, x \pm y = b$, were to be interpreted geometrically. The obvious conclusion, which the reader can arrive at by eliminating y, is that one must construct on a given line b a rectangle whose unknown width x must be such that the area of the rectangle exceeds the given area A by the square x^2 or (in the case of the minus sign) falls short of the area A by the square x^2 (Fig. 5.8). In this way the

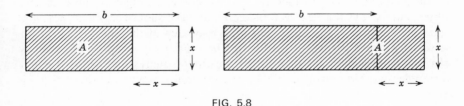

FIG. 5.8

Greeks built up the solution of quadratic equations by their process known as "the application of areas," a portion of geometrical algebra that is fully covered by Euclid's *Elements*. Moreover, the uneasiness resulting from incommensurable magnitudes led to an avoidance of ratios, insofar as

[12] Neugebauer, *The Exact Sciences in Antiquity*, pp. 148–149.

[13] See Arpád Szabó, "Anfänge des euklidischen Axiomensystems," *Archive for History of Exact Sciences*, **1** (1960), 37–106.

possible, in elementary mathematics. The linear equation $ax = bc$, for example, was looked upon as an equality of the areas ax and bc, rather than as a proportion—an equality between the two ratios $a:b$ and $c:x$. Consequently, in constructing the fourth proportion x in this case, it was usual to construct a rectangle $OCDB$ with sides $b = OB$ and $c = OC$ (Fig. 5.9) and

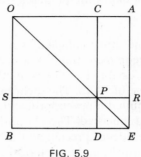

FIG. 5.9

then along OC to lay off $OA = a$. One completes rectangle $OAEB$ and draws the diagonal OE cutting CD in P. It is now clear that CP is the desired line x, for rectangle $OARS$ is equal in area to rectangle $OCDB$. Not until Book V of the *Elements* did Euclid take up the difficult matter of proportionality.

Greek geometrical algebra strikes the modern reader as excessively artificial and difficult; to those who used it and became adept at handling its operations, however, it probably appeared to be a convenient tool. The distributive law $a(b + c + d) = ab + ac + ad$ undoubtedly was far more obvious to a Greek scholar than to the beginning student of algebra today, for the former could easily picture the areas of the rectangles in this theorem, which simply says that the rectangle on a and the sum of segments b, c, d is equal to the sum of the rectangles on a and each of the lines b, c, d taken separately (Fig. 5.10). Again, the identity $(a + b)^2 = a^2 + 2ab + b^2$ becomes obvious from a diagram that shows the three squares and the two equal rectangles in the identity (Fig. 5.11); and a difference of two squares $a^2 - b^2 =$

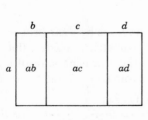

FIG. 5.10

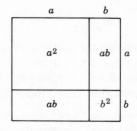

FIG. 5.11

$(a + b)(a - b)$ can be pictured in a similar fashion (Fig. 5.12). Sums, differences, products, and quotients of line segments can easily be constructed with straightedge and compasses. Square roots also afford no difficulty in geometric algebra. If one wishes to find a line x such that $x^2 = ab$, one simply follows the procedure found in elementary geometry textbooks today. One lays off on a straight line the segment ABC, where $AB = a$ and $BC = b$

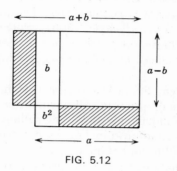

FIG. 5.12

(Fig. 5.13). With AC as diameter one constructs a semicircle (with center O) and at B erects the perpendicular BP, which is the segment x desired. It is interesting that here too the proof as given by Euclid, probably following the earlier avoidance of ratios, makes use of areas rather than proportions. If in our figure we let $PO = AO = CO = r$ and $BO = s$, Euclid would say essentially that $x^2 = r^2 - s^2 = (r - s)(r + s) = ab$.

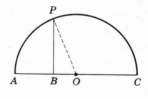

FIG. 5.13

The Heroic Age in mathematics produced half a dozen great figures, and among them must be included a man who is better known as a chemical philosopher. Democritus of Abdera (ca. 460 B.C.–ca. 370 B.C.) is today celebrated as a proponent of a materialistic atomic doctrine, but in his time he had acquired also a reputation as a geometer. He is reported to have traveled more widely than anyone of his day—to Athens, Egypt and Mesopotamia, and possibly India—acquiring what learning he could; but his own achievements in mathematics were such that he boasted that not even the

14

"rope-stretchers" in Egypt excelled him. He wrote a number of mathematical works, not one of which is extant today, but we have the titles of a few: On Numbers, On Geometry, On Tangencies, On Mappings, and On Irrationals. So great was his fame that in later centuries many treatises in chemistry and mathematics were unwarrantedly attributed to him. In particular, early alchemical works by a pseudo-Democritus are not to be ascribed to our Abderite; but other books, On the Pythagoreans, On the World Order, and On Ethics, may have been genuine. His scientific material was said to be clear, but clothed in a literary style; Cicero wrote of Democritus that he had rhythm that made him more poetical than the poets. Yet of the mass of writings thought to have been by Democritus, nothing beyond a few words has survived.

The key to the mathematics of Democritus is without doubt to be found in his physical doctrine of atomism. All phenomena were to be explained, he argued, in terms of indefinitely small and infinitely varied (in size and shape) impenetrably hard atoms moving about ceaselessly in empty space. The creation of our world—and of innumerable others also—was the result of an ordering or coagulation of atoms into groups having certain similarities. This was not a new theory, for it had been proposed earlier by Leucippus; therefore the opponents of Democritus (and there were many of these) accused him of plagiarism from others, including Anaxagoras and Pythagoras. The physical atomism of Leucippus and Democritus may indeed have been suggested by the geometrical atomism of the Pythagoreans, and it is not surprising that the mathematical problems with which Democritus was chiefly concerned were those that demand some sort of infinitesimal approach. The Egyptians, for example, were aware that the volume of a pyramid is one-third the product of the base and the altitude, but a proof of this fact almost certainly was beyond their capabilities, for it requires a point of view equivalent to the calculus. Archimedes later wrote that this result was due to Democritus, but that the latter did not prove it rigorously. This creates a puzzle, for if Democritus added anything to the Egyptian knowledge here, it must have been some sort of demonstration, albeit inadequate. Perhaps Democritus showed that a triangular prism can be divided into three triangular pyramids which, two by two, are equal in height and area of the base, and then deduced, from the assumption that pyramids of the same height and equal bases are equal, the familiar Egyptian theorem.

This assumption can be justified only by the application of infinitesimal techniques. If, for example, one thinks of two pyramids of equal bases and the same height as composed of indefinitely many infinitely thin equal cross sections in one-to-one correspondence (a device usually known as Cavalieri's principle in deference to the seventeenth-century geometer), the assumption appears to be justified. Such a fuzzy geometrical atomism might have been at the base of Democritus' thought, although this has not been established. In any case, following the paradoxes of Zeno and the awareness of incommensurables, such arguments based on an infinity of infinitesimals were not acceptable. Archimedes consequently could well hold that Democritus had not given a rigorous proof. The same judgment would be true with respect to the theorem, also attributed by Archimedes to Democritus, that the volume of a cone is one-third the volume of the circumscribing cylinder. This result probably was looked upon by Democritus as a corollary to the theorem on the pyramid, for the cone is essentially a pyramid whose base is a regular polygon of infinitely many sides.

Democritean geometrical atomism was immediately confronted by certain problems. If the pyramid or the cone, for example, is made up of infinitely many infinitely thin triangular or circular sections parallel to the base, a consideration of any two adjacent laminae creates a paradox. If the adjacent sections are equal in area, then, since all sections are equal, the totality will be a prism or a cylinder, and not a pyramid or a cone. If, on the other hand, adjacent sections are unequal, the totality will be a step pyramid or a step cone, and not the smooth-surfaced figure one has in mind. This problem is not unlike the difficulties with the incommensurable and with the paradoxes of motion. Perhaps, in his On the Irrational, Democritus analysed the difficulties here encountered, but there is no way of knowing what direction his attempts may have taken. His extreme unpopularity in the two dominant philosophical schools of the next century, those of Plato and Aristotle, may have encouraged the disregard of Democritean ideas. Nevertheless, the chief mathematical legacy of the Heroic Age can be summed up in six problems: the squaring of the circle, the duplication of the cube, the trisection of the angle, the ratio of incommensurable magnitudes, the paradoxes on motion, and the validity of infinitesimal methods. To some extent these can be associated, although not exclusively, with men considered in this chapter: Hippocrates, Archytas, Hippias, Hippasus, Zeno, and Democritus. Other ages were to produce a comparable array of talent, but perhaps never again was any age to make so bold an attack on so many fundamental mathematical problems with such inadequate methodological resources. It is for this reason that we have called the period from Anaxagoras to Archytas the Heroic Age.

BIBLIOGRAPHY

Allman, G. J., *Greek Geometry from Thales to Euclid* (Dublin: Dublin University Press, 1889).

Cajori, Florian, "History of Zeno's Arguments on Motion," *American Mathematical Monthly*, **22** (1915), 1–6, 39–47, 77–82, 109–115, 145–149, 179–186, 215–220, 253–258, 292–297.

Freeman, Kathleen, *The Pre-Socratic Philosophers*, 2nd ed. (Oxford: Blackwell, 1949).

Gow, James, *A Short History of Greek Mathematics* (reprint, New York: Hafner, 1923).

Heath, T. L., *History of Greek Mathematics* (New York: Oxford University Press, 1921, 2 vols.).

Hobson, E. W., *Squaring the Circle* (Cambridge, ca. 1913).

Lee, H. D. P., ed., *Zeno of Elea* (Cambridge: Cambridge University Press, 1936).

Michel, Paul-Henri, *De Pythagore à Euclide* (Paris: Société d'Edition "Les Belles Lettres," 1950).

Neugebauer, O., *The Exact Sciences in Antiquity*, 2nd ed., (Providence R.I.: Brown University Press, 1957; paperback, New York: Harper).

Szabó, Arpád, "The Transformation of Mathematics into Deductive Science and the Beginnings of its Foundation on Definitions and Axioms." *Scripta Mathematica*, **27** (1964), 27–48, 113–139.

Tannery, Paul, *La géométrie grecque, comment son histoire nous est parvenue et ce que nous en savons* (Paris, 1887).

Thomas, Ivor, ed., *Selections Illustrating the History of Greek Mathematics* (Cambridge, Mass.: Harvard University Press, 1939–1941, 2 vols).

Van der Waerden, B. L., *Science Awakening* (trans. by Arnold Dresden, New York: Oxford University Press, 1961; paperback ed., New York, Wiley, 1963).

Von Fritz, Kurt, "The Discovery of Incommensurability by Hippasus of Metapontum," *Annals of Mathematics* (2), **46** (1945), 242–264.

EXERCISES

1. Justify the two quadratures attributed by Alexander of Aphrodisias to Hippocrates.
2. Draw an angle of 60° and use Hippias' trisectrix to divide the angle into seven equal parts.
3. Prove carefully that the segments into which the diagonals of a regular pentagon divide each other are incommensurable with respect to the diagonals.
4. Which do you believe was discovered first, the irrationality of $\sqrt{2}$ or of $\sqrt{5}$? Justify your answer in terms of historical evidence.
5. Using ruler and compasses only, construct the line x if $ax = b^2$, where a and b are any given line segments.
6. Given line segments a and b and using compasses and straightedge only, construct x and y if $x + y = a$ and $xy = b^2$.
7. Given line segments a, b, and c, construct x and y if $x - y = a$ and $xy = bc$.
8. Solve the equation $x^2 + ax = b^2$ by constructing a line segment satisfying the given condition.
9. Given a unit line segment (of length 1), construct a line segment of length $\sqrt{3} + \frac{4}{5}$.
10. Show that in polar coordinates the equation of Hippias' trisectrix is $\pi r \sin \theta = 2a\theta$. Sketch the main branch of this curve for $-\pi/2 < \theta < 3\pi/2$ and tell why Hippias did not draw this complete branch.
11. Are the diagonals of a regular hexagon incommensurable with respect to a side? Explain fully, indicating whether your conclusion could have been reached in antiquity.
*12. Carry out the steps needed to show that Archytas' construction duplicates the cube.

CHAPTER VI

The Age of Plato and Aristotle

Willingly would I burn to death like Phaeton, were this
the price for reaching the sun and learning its shape, its
size, and its substance.

Eudoxus

The Heroic Age lay largely in the fifth century B.C., and from this period **1**
little remains in the way of direct evidence about mathematical developments.
The histories of Herodotus and Thucydides and the plays of Aeschylus,
Euripides, and Aristophanes have in some measure survived, but scarcely
a line is extant of what was written by mathematicians of the time. Firsthand
mathematical sources from the fourth century B.C. are almost as scarce, but
this inadequacy is made up for in large measure by accounts written by
philosophers who were *au courant* with the mathematics of their day. We
have most of what Plato wrote and about half of the work of Aristotle;
with the writings of these intellectual leaders of the fourth century B.C. as a
guide, we can give a far more dependable account of what happened in their
day than we could about the Heroic Age.

We included Archytas among the mathematicians of the Heroic Age, but
in a sense he really is a transition figure in mathematics during Plato's time.
Archytas was among the last of the Pythagoreans, both literally and figura-
tively. He could still believe that number was all-important in life and in
mathematics, but the wave of the future was to elevate geometry to the
ascendancy, largely because of the problem of incommensurability. On the
other hand, Archytas is reported to have established the quadrivium—
arithmetic, geometry, music, and astronomy—as the core of a liberal education, and here his views were to dominate much of pedagogical thought to our
day. The seven liberal arts, which remained a shiboleth for almost two
millennia, were made up of Archytas' quadrivium and the trivium of grammar, rhetoric, and Zeno's dialectic. Consequently, one may with some justice
hold that the mathematicians of the Heroic Age were responsible for much of

Plato and Aristotle in Raphael's "School of Athens."

the direction in Western educational traditions, especially as transmitted through the philosophers of the fourth century B.C.[1]

The fourth century B.C. had opened with the death of Socrates, a scholar **2** who adopted the dialectic method of Zeno and repudiated the Pythagoreanism of Archytas. Socrates admitted that in his youth he had been attracted by such questions as why the sum 2 + 2 was the same as the product 2 × 2, as well as by the natural philosophy of Anaxagoras; but upon realizing that neither mathematics nor science could satisfy his desire to know the essence of things, he gave himself up to his characteristic search for the good.

In the *Phaedo* of Plato, the dialogue in which the last hours of Socrates are so beautifully described, we see how deep metaphysical doubts precluded a Socratic concern with either mathematics or natural science.

> I cannot satisfy myself that, when one is added to one, the one to which the addition is made becomes two, or that the two units added together make two by reason of the addition. I cannot understand how when separated from the other, each of them was one and not two, and now, when they are brought together, the mere juxtaposition or meeting of them should be the cause of their becoming two.[2]

Hence the influence of Socrates in the development of mathematics was negligible, if not actually negative. This makes it all the more surprising that it was his student and admirer, Plato, who became the mathematical inspiration of the fourth century B.C. We shall concentrate in this chapter on the mathematical achievements of half a dozen men who lived between the death of Socrates in 399 B.C. and the death of Aristotle in 322 B.C. The six men whose work we shall describe (in addition to that of Plato and Aristotle) are Theodorus of Cyrene (fl. ca. 390 B.C.), Theaetetus (†368 B.C.), Eudoxus of Cnidus († ca. 355 B.C.), Menaechmus (fl. ca. 350 B.C.) and his brother Dinostratus (fl. ca. 350 B.C.), and Autolycus of Pitane (fl. ca. 330 B.C.).

The six mathematicians were not scattered throughout the Greek world, **3** as had been those in the fifth century B.C.; they were associated more or less closely with the Academy of Plato at Athens. Although Plato himself made no outstanding specific contribution to technical mathematical results, he was the center of the mathematical activity of the time and guided and inspired its development. Over the doors of his school was inscribed the motto, "Let no one ignorant of geometry enter here"; his enthusiasm for the

---

[1] The firm establishment of this particular group of seven liberal arts was, however, achieved only in the fourth century of our era, and the recognized division of these into the trivium and quadrivium became traditional only with the Carolingian renaissance. Marshall Clagett, in *Greek Science in Antiquity*, 2nd ed. New York: Collier, 1966, p. 185, writes that the use of the Latin term *quadrivium* seems to stem from Boethius (ca. 480–524).

[2] *Dialogues of Plato* (1875), I, 476–477.

subject led him to become known not as a mathematician, but as "the maker of mathematicians." It is clear that Plato's high regard for mathematics did not come from Socrates; in fact, the earlier Platonic dialogues seldom refer to mathematics. The one who converted Plato to a mathematical outlook undoubtedly was Archytas, a friend whom he visited in Sicily in 388 B.C. Perhaps it was there that he learned of the five regular solids, which were

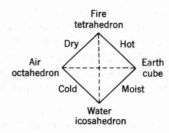

associated with the four elements of Empedocles in a cosmic scheme that fascinated men for centuries. Possibly it was the Pythagorean regard for the dodecahedron that led Plato to look on this, the fifth and last, regular solid as a symbol of the universe. Plato put his ideas on the regular solids into a dialogue entitled the *Timaeus*, presumably named for a Pythagorean who serves as the chief interlocutor. It is not known whether Timaeus of Locri really existed or whether Plato invented him as a character through whom to express the Pythagorean views that still were strong in what is now Southern Italy. The regular polyhedra have often been called "cosmic bodies" or "Platonic solids" because of the way in which Plato in the *Timaeus* applied them to the explanation of scientific phenomena. Although this dialogue, probably written when Plato was near seventy, provides the earliest definite evidence for the association of the four elements with the regular solids, much of this fantasy may be due to the Pythagoreans. Proclus attributes the construction of the cosmic figures to Pythagoras; but the scholiast Suidas reported that Plato's friend Theaetetus, born about 414 B.C. and the son of one of the richest patricians in Attica, first wrote on them. A scholium (of uncertain date) to Book XIII of Euclid's *Elements* reports that only three of the five solids were due to the Pythagoreans, and that it was through Theaetetus that the octahedron and icosahedron became known. It seems likely that in any case Theaetetus made one of the most extensive studies of the five regular solids, and to him probably is due the theorem that there are five and only five regular polyhedra. Perhaps he is responsible also for the calculations in the *Elements* of the ratios of the edges of the regular solids to the radius of the circumscribed sphere.

Theaetetus was a young Athenian who died in 369 B.C. from a combination of wounds received in battle and of dysentery, and the Platonic dialogue bearing his name was a commemorative tribute by Plato to his friend. In the dialogue, purporting to take place some thirty years earlier, Theaetetus discusses with Socrates and Theodorus the nature of incommensurable magnitudes. It has been assumed that this discussion took somewhat the form that we find in the opening of Book X of the *Elements*. Here distinctions are made not only between commensurable and incommensurable magnitudes, but also between those that while incommensurable in length are, or are not, commensurable in square. Surds such as $\sqrt{3}$ and $\sqrt{5}$ are incommensurable in length, but they are commensurable in square, for their squares have the ratio 3 to 5. The magnitudes $\sqrt{1 + \sqrt{3}}$ and $\sqrt{1 + \sqrt{5}}$, on the other hand, are incommensurable both in length and in square.

The dialogue that Plato composed in memory of his friend, Theaetetus, **4** contains information on another mathematician whom Plato admired and who contributed to the early development of the theory of incommensurable magnitudes. Reporting on the then recent discovery of what we call the irrationality of $\sqrt{2}$, Plato in the *Theaetetus* says that his teacher, Theodorus of Cyrene—of whom Theaetetus also was a pupil—was the first to prove the irrationality of the square roots of the nonsquare integers from 3 to 17 inclusive. It is not known how he did this, nor why he stopped with $\sqrt{17}$. The proof in any case could have been constructed along the lines of that for $\sqrt{2}$ as given by Aristotle and interpolated in later versions of Book X of the *Elements*. References in ancient historical works indicate that Theodorus made discoveries in elementary geometry that later were incorporated in Euclid's *Elements*; but the works of Theodorus are lost.

Plato is important in the history of mathematics largely for his role as **5** inspirer and director of others, and perhaps to him is due the sharp distinction in ancient Greece between arithmetic (in the sense of the theory of numbers) and logistic (the technique of computation). Plato regarded logistic as appropriate for the businessman and for the man of war, who "must learn the art of numbers or he will not know how to array his troops." The philosopher, on the other hand, must be an arithmetician "because he has to arise out of the sea of change and lay hold of true being." Moreover, Plato says in the *Republic*, "arithmetic has a very great and elevating effect, compelling the mind to reason about abstract number." So elevating are Plato's thoughts concerning number that they reach the realm of mysticism and apparent fantasy. In the last book of the *Republic* he refers to a number that he calls "the lord of better and worse births." There has been much speculation concerning this "Platonic number," and one theory is that it is the number $60^4 = 12,960,000$—important in Babylonian numerology and possibly transmitted to Plato through the Pythagoreans. In the *Laws* the number of citizens in the ideal state is given as 5040 (that is, $7 \cdot 6 \cdot 5 \cdot 4 \cdot 3 \cdot 2 \cdot 1$). This sometimes is referred to as the Platonic nuptial number, and various theories have been advanced to suggest what Plato had in mind.

As in arithmetic Plato saw a gulf separating the theoretical and computational aspects, so also in geometry he espoused the cause of pure mathematics as against the materialistic views of the artisan or technician. Plutarch, in his *Life of Marcellus*, speaks of Plato's indignation at the use of mechanical contrivances in geometry. Apparently Plato regarded such use as "the mere corruption and annihilation of the one good of geometry, which was thus shamefully turning its back upon the unembodied objects of pure intelligence." Plato may consequently have been largely responsible for the prevalent restriction in Greek geometrical constructions to those that can be effected by straightedge and compasses alone. The reason for the limitation is not likely to have been the simplicity of the instruments used in constructing lines and circles, but rather the symmetry of the configurations. Any one of the infinitely many diameters of a circle is a line of symmetry of the figure; any point on an infinitely extended straight line can be thought of as a center of symmetry, just as any line perpendicular to the given line is a line with respect to which the given line is symmetric. Platonic philosophy, with its apostatization of ideas, would quite naturally find a favored role for the line and the circle among geometrical figures. In a somewhat similar manner Plato glorified the triangle. The faces of the five regular solids in Plato's view were not simple triangles, squares, and pentagons. Each of the four faces of the tetrahedron, for example, is made up of six smaller right triangles formed by altitudes of the equilateral triangular faces. The regular tetrahedron he therefore thought of as made up of twenty-four scalene right triangles in which the hypotenuse is double one side; the regular octahedron contains 8×6 or 48 such triangles, and the icosahedron is made up of 20×6 or 120 triangles. In a similar way the hexahedron (or cube) is constructed of twenty-four isosceles right triangles, for each of the six square faces contains four right triangles when the diagonals of the squares are drawn.

To the dodecahedron Plato had assigned a special role as representative of the universe, cryptically saying that "God used it for the whole" (*Timaeus* 55C).[3] Plato looked upon the dodecahedron as composed of 360 scalene right triangles, for when the five diagonals and five medians are drawn in each of the pentagonal faces, each of the twelve faces will contain thirty right triangles. The association of the first four regular solids with the

---

[3] References here and elsewhere, unless otherwise noted, are to the dialogues of Plato and are from Plato, *Dialogues*, trans. by Benjamin Jowett (Oxford, 1871, 4 vols.).

traditional four universal elements provided Plato in the *Timaeus* with a beautifully unified theory of matter according to which everything was constructed of ideal right triangles. The whole of physiology, as well as the sciences of inert matter, is based in the *Timaeus* on these triangles. Normal growth of the body, for example, is explained as follows:

> When the frame of the whole creature is young and the triangles of its constituent bodies are still as it were fresh from the workshop, their joints are firmly locked together. ...Accordingly, since any triangles composing the meat and drink...are older and weaker than its own, with its new-made triangles, it gets the better of them and cuts them up, and so causes the animal to wax large.

In old age, on the other hand, the triangles of the body are so loosened by use that "they can no longer cut up into their own likeness the triangles of the nourishment as they enter, but are themselves easily divided by the intruders from without," and the creature wastes away.[4]

Pythagoras is reputed to have established mathematics as a liberal subject, **6** but Plato was influential in making the subject an essential part of the curriculum for the education of statesmen. Influenced perhaps by Archytas, Plato would add to the original subjects in the quadrivium a new subject, stereometry, for he believed that solid geometry had not been sufficiently emphasized. Plato also discussed the foundations of mathematics, clarified some of the definitions, and reorganized the assumptions. He emphasized that the reasoning used in geometry does not refer to the visible figures that are drawn but to the absolute ideas that they represent. The Pythagoreans had defined a point as "unity having position," but Plato would rather think of it as the beginning of a line. The definition of a line as "breadthless length" seems to have originated in the school of Plato, as well as the idea that a line "lies evenly with the points on it." In arithmetic Plato emphasized not only the distinction between odd and even numbers, but also the categories "even times even," "odd times even," and "odd times odd." Although we are told that Plato added to the axioms of mathematics, we do not have an account of his premises.

Few specific mathematical contributions are attributed to Plato. A formula for Pythagorean triples—$(2n)^2 + (n^2 - 1)^2 = (n^2 + 1)^2$, where n is any natural number—bears Plato's name, but this is merely a slightly modified version of a result known to the Babylonians and the Pythagoreans. Perhaps more genuinely significant is the ascription to Plato of the so-called analytic method. In demonstrative mathematics one begins with what is given, either generally in the axioms and postulates or more specifically in the problems at hand. Proceeding step by step, one then arrives at the statement that was

---

[4] *Timaeus* 81B–81D. Translation is from F. M. Cornford, *Plato's Cosmology* (1937), p. 329.

to have been proven. Plato seems to have pointed out that often it is pedagogically convenient, when a chain of reasoning from premises to conclusion is not obvious, to reverse the process. One might begin with the proposition that is to be proved and from it deduce a conclusion that is known to hold. If, then, one can reverse the steps in this chain of reasoning, the result is a legitimate proof of the proposition. It is unlikely that Plato was the first to note the efficacy in the analytic point of view, for any preliminary investigation of a problem is tantamount to this. What Plato is likely to have done is to formalize this procedure, or perhaps to give it a name.

The role of Plato in the history of mathematics is still bitterly disputed. Some[5] regarded him as an exceptionally profound and incisive thinker; others picture him as a mathematical pied piper who lured men away from problems concerning the world's work and encouraged them in idle speculation.[6] In any case, few would deny that Plato had a tremendous effect on the development of mathematics. The Platonic Academy in Athens became the mathematical center of the world, and it was from this school that the leading teachers and research workers came during the middle of the fourth century B.C. Of these the greatest was Eudoxus of Cnidus (408?–355? B.C.), a man who was at one time a pupil of Plato and who became the most renowned mathematician and astronomer of his day.

7   We sometimes read of the "Platonic reform" in mathematics, and although the phrase tends to exaggerate the changes taking place, the work of Eudoxus was so significant that the word "reform" is not inappropriate. In Plato's youth the discovery of the incommensurable had caused a veritable logical scandal, for it had raised havoc with theorems involving proportions. Two quantities, such as the diagonal and side of a square, are incommensurable when they do not have a ratio such as a (whole) number has to a (whole) number. How, then, is one to compare ratios of incommensurable magnitudes? If Hippocrates really did prove that the areas of circles are to each other as squares on their diameters, he must have had some way of handling proportions or the equality of ratios. We do not know how he proceeded, or whether to some extent he anticipated Eudoxus, who gave a new and generally accepted definition of equal ratios. Apparently the Greeks had made use of the idea that four quantities are in proportion, $a:b = c:d$, if the two ratios $a:b$ and $c:d$ have the same mutual subtraction. That is, the smaller in each ratio can be laid off on the larger the same integral number of times, and the remainder in each case can be laid off on the smaller the same integral number of times, and the new remainder can be laid off on the former remainder the

---

[5] See, for example, François Lasserre, *The Birth of Mathematics in the Age of Plato* (1964).
[6] Lancelot Hogben, *Science for the Citizen* (New York: 1938), p. 64. Cf. George Sarton, *A History of Science* (Cambridge, Mass.: Harvard University Press, 1952), Vol. I, pp. 431 ff.

same integral number of times, and so on. Such a definition would be awkward to use, and it was a brilliant achievement of Eudoxus to discover the theory of proportion used in Book V of Euclid's *Elements*. The word ratio denoted essentially an undefined concept in Greek mathematics, for Euclid's "definition" of ratio as a kind of relation in size between two magnitudes of the same type is quite inadequate. More significant is Euclid's statement that magnitudes are said to have a ratio to one another if a multiple of either can be found to exceed the other. This is essentially a statement of the so-called "Axiom of Archimedes"—a property that Archimedes himself attributed to Eudoxus. The Eudoxian concept of ratio consequently excludes zero and clarifies what is meant by magnitudes of the same kind. A line segment, for example, is not to be compared, in terms of ratio, with an area; nor is an area to be compared with a volume.

Following these preliminary remarks on ratios, Euclid gives in Definition 5 of Book V the celebrated formulation by Eudoxus:

Magnitudes are said to be in the same ratio, the first to the second and the third to the fourth, when, if any equimultiples whatever be taken of the first and the third, and any equimultiples whatever of the second and fourth, the former equimultiples alike exceed, are alike equal to, or are alike less than, the latter equimultiples taken in corresponding order.[7]

That is, $a/b = c/d$ if, and only if, given integers m and n, whenever $ma < nb$, then $mc < nd$; or if $ma = nb$, then $mc = nd$, or if $ma > nb$, then $mc > nd$.

The Eudoxian definition of equality of ratios is not unlike the process of cross-multiplication that is used today for fractions—$a/b = c/d$ according as $ad = bc$—a process equivalent to a reduction to a common denominator. To show that $\frac{3}{6}$ is equal to $\frac{4}{8}$, for example, we multiply 3 and 6 by 4, to obtain 12 and 24, and we multiply 4 and 8 by 3, obtaining the same pair of numbers 12 and 24. We could have used 7 and 13 as our two multipliers, obtaining the pair 21 and 42 in the first case and 52 and 104 in the second; and as 21 is less than 52, so is 42 less than 104. (We have here interchanged the second and third terms in Eudoxus' definition to conform to the common operations as usually used today, but similar relationships hold in either case.) Our arithmetical example does not do justice to the subtlety and efficacy of Eudoxus' thought, for the application here appears to be trivial. To gain a heightened appreciation of his definition it would be well to replace a, b, c, d by surds or, better still, to let a and b be spheres and c and d cubes on the radii of the spheres. Here a cross-multiplication becomes meaningless, and the applicability of Eudoxus' definition is far from obvious. In fact, it will be noted that, strictly speaking, the definition is not far removed from the

[7] *The Thirteen Books of Euclid's Elements*, ed. by T. L. Heath (Cambridge, 1908, 3 vols.), II, 114.

nineteenth-century definitions of real number, for it separates the class of rational numbers m/n into two categories, according as $ma \leq nb$ or $ma > nb$. Because there are infinitely many rational numbers, the Greeks by implication were faced by the concept that they wished to avoid—that of an infinite set—but at least it was now possible to give satisfactory proofs of theorems involving proportions.

8   A crisis resulting from the incommensurable had been successfully met, thanks to the imagination of Eudoxus; but there remained another unsolved problem—the comparison of curved and straight-line configurations. Here, too, it seems to have been Eudoxus who supplied the key. Earlier mathematicians seem to have suggested that one try inscribing and circumscribing rectilinear figures in and about the curved figure and continue to multiply indefinitely the number of sides; but they did not know how to clinch the argument, for the concept of a limit was unknown at the time. According to Archimedes, it was Eudoxus who provided the lemma that now bears Archimedes' name—sometimes known as the axiom of continuity—which served as the basis for the method of exhaustion, the Greek equivalent of the integral calculus. The lemma or axiom states that, given two magnitudes having a ratio (that is, neither being zero), one can find a multiple of either one which will exceed the other. This statement excluded a fuzzy argument about indivisible line segments, or fixed infinitesimals, that was sometimes maintained in Greek thought. It also excluded the comparison of the so-called angle of contingency or "horn angle" (formed by a curve C and its tangent T at a point P on C) with ordinary rectilinear angles. The horn angle seemed to be a magnitude different from zero, yet it does not satisfy the axiom of Eudoxus with respect to the measures of rectilinear angles.

From the axiom of Eudoxus (or Archimedes) it is an easy step, by a *reductio ad absurdum*, to prove a proposition that formed the basis of the Greek method of exhaustion:

> If from any magnitude there be subtracted a part not less than its half, and if from the remainder one again subtracts not less than its half, and if this process of subtraction is continued, ultimately there will remain a magnitude less than any preassigned magnitude of the same kind.[8]

This proposition, which we shall refer to as the "exhaustion property," is equivalent to the modern statement that if M is a given magnitude, ϵ is a preassigned magnitude of the same kind, and r is a ratio such that $\frac{1}{2} \leq r < 1$, then we can find a positive integer N such that $M(1 - r)^n < \epsilon$ for all positive integers $n > N$. That is, the exhaustion property is equivalent to the modern

---

[8] See *Elements of Euclid* (ed. by T. L. Heath, reprinted, New York: Dover, 3 vols., 1956), III, 14. The axiom is, of course, still legitimate if half is changed to third or quarter or other proper part.

statement that $\lim_{n \to \infty} M(1 - r)^n = 0$. Moreover, the Greeks made use of this property to prove theorems about the areas and volumes of curvilinear figures. In particular, Archimedes ascribed to Eudoxus the earliest satisfactory proof that the volume of the cone is one-third the volume of the cylinder having the same base and altitude, a statement that would seem to indicate that the method of exhaustion was derived by Eudoxus. If so, then it is to Eudoxus (rather than to Hippocrates) that we probably owe the Euclidean proofs of theorems concerning areas of circles and volumes of spheres. Facile earlier suggestions had been made that the area of a circle could be exhausted by inscribing in it a regular polygon and then increasing the number of sides indefinitely, but the Eudoxian method of exhaustion first made such a procedure rigorous. (It should be noted that the phrase "method of exhaustion" was not used by the ancient Greeks, being a modern invention; but the phrase has become so well established in the history of mathematics that we shall continue to make use of it.) As an illustration of the way in which Eudoxus probably carried out the method, we give here, in somewhat modernized notation, the proof that areas of circles are to each other as squares on their diameters. The proof, as it is given in Euclid, *Elements* XII. 2, is probably that of Eudoxus.

Let the circles be c and C, with diameters d and D and areas a and A. It is to be proven that $a/A = d^2/D^2$. The proof is complete if we proceed indirectly and disprove the only other possibilities, namely, $a/A < d^2/D^2$ and $a/A > d^2/D^2$. Hence we first assume that $a/A > d^2/D^2$. Then there is a magnitude $a' < a$ such that $a'/A = d^2/D^2$. Let $a - a'$ be a preassigned magnitude $\epsilon > 0$. Within the circles c and C inscribe regular polygons of areas p_n and P_n, having the same number of sides n, and consider the intermediate areas outside the polygons but inside the circles (Fig. 6.1). If the

FIG. 6.1

number of sides should be doubled, it is obvious that from these intermediate areas we would be subtracting more than the half. Consequently, by the exhaustion property, the intermediate areas can be reduced through successive doubling of the number of sides (that is, by letting n increase) until

$a - p_n < \epsilon$. Then, since $a - a' = \epsilon$, we have $p_n > a'$. Now, from earlier theorems it is known that $p_n/P_n = d^2/D^2$ and since it was assumed that $a'/A = d^2/D^2$, we have $p_n/P_n = a'/A$. Hence if $p_n > a'$, as we have shown, we must conclude that $P_n > A$. Inasmuch as P_n is the area of a polygon *inscribed within* the circle of area A, it is obvious that P_n cannot be greater than A. Since a false conclusion implies a false premise, we have disproved the possibility that $a/A > d^2/D^2$. In an analogous manner we can disprove the possibility that $a/A < d^2/D^2$, thereby establishing the theorem that areas of circles are to each other as squares on their diameters.

9   The property that we have just demonstrated appears to have been the first precise theorem concerning the magnitudes of curvilinear figures; it marks Eudoxus as the apparent originator of the integral calculus, the greatest contribution to mathematics made by associates of the Platonic Academy. Eudoxus, moreover, was by no means a mathematician only, and in the history of science he is known as the father of scientific astronomy. Plato is said to have proposed to his associates that they attempt to give a geometrical representation of the movements of the sun, the moon, and the five known planets. It evidently was tacitly assumed that the movements were to be compounded of uniform circular motions. Despite such a restriction, Eudoxus was able to give for each of the seven heavenly bodies a satisfactory representation through a composite of concentric spheres with centers at the earth and with varying radii, each sphere revolving uniformly about an axis fixed with respect to the surface of the next larger sphere. For each planet, then, Eudoxus gave a system known to his successors as "homocentric spheres"; these geometrical schemes were combined by Aristotle into the well-known Peripatetic cosmology of crystalline spheres that dominated thought for almost 2000 years.

Eudoxus was without doubt the most capable mathematician of the Hellenic Age, but all of his works have been lost.[9] It is possible that the Aristotelian estimate for the circumference of the earth—about 400,000 stades, or 40,000 miles—is due to Eudoxus, for Archimedes reported that Eudoxus had calculated that the diameter of the sun was nine times that of the earth. In his astronomical scheme Eudoxus had seen that by a combination of circular motions he could describe the motions of the planets in looped orbits along a curve known as the hippopede or horse fetter. This curve, resembling a figure eight on a sphere, is obtained as the intersection of a sphere and a cylinder tangent internally to the sphere—one of the few new curves that the Greeks recognized. At the time there were only two means of defining curves:

[9] For an extensive and authoritative account of what Eudoxus probably did, see O. Becker, "Eudoxus-Studien," *Quellen und Studien zur Geschichte der Mathematik*, Part B, II (1933), 311–333, 369–387; III (1936), 236–244, 370–410.

(1) through combinations of uniform motions and (2) as the intersections of familiar geometric surfaces. The hippopede of Eudoxus is a good example of a curve that is derivable in either of these two ways. Proclus, who wrote some 800 years after the time of Eudoxus, reported that Eudoxus had added many general theorems in geometry and had applied the Platonic method of analysis to the study of *the section* (probably the golden section); but the two chief claims to fame of Eudoxus remain the theory of proportions and the method of exhaustion.

Eudoxus is to be remembered in the history of mathematics not only for **10** his own work, but also through that of his pupils. In Greece there was a strong thread of continuity of tradition from teacher to student. Thus Plato learned from Archytas, Theodorus, and Theaetetus; the Platonic influence in turn was passed on through Eudoxus to the brothers Menaechmus and Dinostratus, both of whom achieved eminence in mathematics. We saw that Hippocrates of Chios had shown that the duplication of the cube could be achieved provided that one could find, and was permitted to use, curves with the properties expressed in the continued proportion $a/x = x/y = y/2a$; we noted also that the Greeks had only two approaches to the discovery of new curves. It was consequently a signal achievement on the part of Menaechmus when he disclosed that curves having the desired property were near at hand. In fact, there was a family of appropriate curves obtainable from a single source—the cutting of a right circular cone by a plane perpendicular to an element of the cone. That is, Menaechmus is reputed to have discovered the curves that were later known as the ellipse, the parabola, and the hyperbola.

Of all the curves, other than circles and straight lines, that are apparent to the eye in everyday experience, the ellipse should be the most obvious, for it is present by implication whenever a circle is viewed obliquely or whenever one saws diagonally through a cylindrical log. Yet the first discovery of the ellipse seems to have been made by Menaechmus as a mere by-product in a search in which it was the parabola and hyperbola which proffered the properties needed in the solution of the Delian problem. Beginning with a single-napped right circular cone having a right angle at the vertex (that is, a generating angle of 45°), Menaechmus found that when the cone is cut by a plane perpendicular to an element, the curve of intersection is such that, in terms of modern analytic geometry, its equation can be written in the form $y^2 = lx$, where l is a constant depending on the distance of the cutting plane from the vertex. We do not know how Menaechmus derived this property, but it depends only on theorems from elementary geometry. Let the cone be ABC and let it be cut in the curve EDG by a plane perpendicular to the element ADC of the cone (Fig. 6.2). Then through P, any point on the curve, pass a horizontal plane cutting the cone in the circle PVR, and let Q be the other

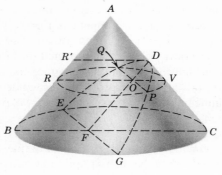

FIG. 6.2

point of intersection of the curve (parabola) and the circle. From the symmetries involved it follows that line $PQ \perp RV$ at O. Hence OP is the mean proportional between RO and OV. Moreover, from the similarity of triangles OVD and BCA it follows that $OV/DO = BC/AB$, and from the similarity of triangles $R'DA$ and ABC it follows that $R'D/AR' = BC/AB$. If $OP = y$ and $OD = x$ are coordinates of point P, we have $y^2 = RO \cdot OV$, or, on substituting equals,

$$y^2 = R'D \cdot OV = AR' \cdot \frac{BC}{AB} \cdot DO \cdot \frac{BC}{AB} = \frac{AR' \cdot BC^2}{AB^2} \cdot x$$

Inasmuch as segments AR', BC, and AB are the same for all points P on the curve $EQDPG$, we can write the equation of the curve, a "section of a right-angled cone," as $y^2 = lx$, where l is a constant, later to be known as the latus rectum of the curve. In an analogous way we can derive an equation of the form $y^2 = lx - b^2x^2/a^2$ for a "section of an acute-angled cone" and an equation of the form $y^2 = lx + b^2x^2/a^2$ for a "section of an obtuse-angled cone," where a and b are constants and the cutting plane is perpendicular to an element of the acute-angled or obtuse-angled right circular cone.

Menaechmus apparently derived these properties of the conic sections, and others as well. Since this material has a strong resemblance to the use of coordinates, as illustrated above, it has sometimes been maintained that Menaechmus had analytic geometry.[10] Such a judgment is warranted only in part, for certainly Menaechmus was unaware that any equation in two unknown quantities determines a curve. In fact, the general concept of an equation in unknown quantities was alien to Greek thought. It was short-

[10] See J. L. Coolidge, *A History of Geometrical Methods* (1940), pp. 117–119, and H. G. Zeuthen, "Sur l'usage des coordonnées dans l'antiquité," *Kongelige Danske Videnskabernes Selskabs, Forhandlinger. Oversigt*, 1888, pp. 127–144.

comings in algebraic notations that, more than anything else, operated against the Greek achievement of a full-fledged coordinate geometry.

Menaechmus had no way of foreseeing the hosts of beautiful properties **11** that the future was to disclose. He had hit upon the conics in a successful search for curves with the properties appropriate to the duplication of the cube. In terms of modern notation the solution is easily achieved. By shifting the cutting plane (Fig. 6.2), we can find a parabola with any latus rectum. If, then, we wish to duplicate a cube of edge a, we locate on a right-angled cone two parabolas, one with latus rectum a and another with latus rectum $2a$. If, then, we place these with vertices at the origin and with axes along the y- and x-axes respectively, the point of intersection of the two curves will have coordinates (x, y) satisfying the continued proportion $a/x = x/y = y/2a$ (Fig. 6.3)—that is, $x = a\sqrt[3]{2}$, $y = a\sqrt[3]{4}$. The x-coordinate therefore is the edge of the cube sought.

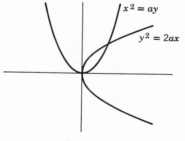

$$x^2 = ay$$
$$y^2 = 2ax$$

FIG. 6.3

It is probable that Menaechmus knew that the duplication could be achieved also by the use of a rectangular hyperbola and a parabola. If the parabola with equation $y^2 = (a/2)x$ and the hyperbola $xy = a^2$ are placed on a common coordinate system, the point of intersection will have coordinates $x = a\sqrt[3]{2}$, $y = a/\sqrt[3]{2}$, the x-coordinate being the side of the cube desired. Menaechmus probably was acquainted with many of the now familiar properties of the conic sections, including the asymptotes of the hyperbola which would have permitted him to operate with the equivalents of the modern equations that we used above. Proclus reported that Menaechmus was one of those who "made the whole of geometry more perfect"; but we know little concerning his actual work. We do know that Menaechmus taught Alexander the Great, and legend attributes to Menaechmus the celebrated comment, when his royal pupil asked for a shortcut to geometry: "O King, for travelling over the country there are royal roads and roads for common citizens; but in geometry there is one road for all." Among the chief authorities for attributing to Menaechmus the discovery of conic sections is a letter from Eratosthenes to King Ptolemy Euergetes, quoted some 700 years later by Eutocius, in which several duplications of the cube are mentioned. Among them is one by Archytas' unwieldy construction and another by "cutting the cone in the triads of Menaechmus."

12   Dinostratus, brother of Menaechmus, was also a mathematician, and where one of the brothers "solved" the duplication of the cube, the other "solved" the squaring of the circle. The quadrature became a simple matter once a striking property of the end point Q of the trisectrix of Hippias had been noted, apparently by Dinostratus. If the equation of the trisectrix (Fig. 6.4) is $\pi r \sin \theta = 2a\theta$, where a is the side of the square $ABCD$ associated with the curve, the limiting value of r as θ tends toward zero is $2a/\pi$. This is

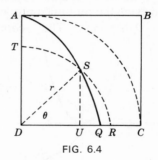

FIG. 6.4

obvious to one who has had calculus and recalls that $\lim_{\theta \to 0} \sin \theta/\theta = 1$ for radian measure. The proof as given by Pappus, and probably due to Dinostratus, is based only on considerations from elementary geometry. The theorem of Dinostratus states that side a is the mean proportional between the segment DQ and the arc of the quarter circle AC—that is, $\overset{\frown}{AC}/AB = AB/DQ$. Using a typically Greek indirect proof, we establish the theorem by demolishing the alternatives. Hence assume first that $\overset{\frown}{AC}/AB = AB/DR$ where $DR > DQ$. Then let the circle with center D and radius DR intersect the trisectrix at S and side AD of the square at T. From S drop the perpendicular SU to side CD. Inasmuch as it was known to Dinostratus that corresponding arcs of circles are to each other as the radii, we have $\overset{\frown}{AC}/AB = \overset{\frown}{TR}/DR$; and since by hypothesis $\overset{\frown}{AC}/AB = AB/DR$, it follows that $\overset{\frown}{TR} = AB$. But from the definitional property of the trisectrix it is known that $\overset{\frown}{TR}/\overset{\frown}{SR} = AB/SU$. Hence, since $\overset{\frown}{TR} = AB$, it must follow that $\overset{\frown}{SR} = SU$, which obviously is false, since the perpendicular is shorter than any other line or curve from point S to line DC. Hence the fourth term DR in the proportion $\overset{\frown}{AC}/AB = AB/DR$ cannot be greater than DQ. In a similar manner we can prove that this fourth proportional cannot be less than DQ; hence Dinostratus' theorem is established—that is, $\overset{\frown}{AC}/AB = AB/DQ$.

Given the intersection point Q of the trisectrix with DC, we then have a proportion involving three straight-line segments and the circular arc AC. Hence by a simple geometric construction of the fourth term in a proportion, a line segment b equal in length to AC can be easily drawn. Upon drawing a rectangle with $2b$ as one side and a as the other, we have a rectangle exactly equal in area to the area of the circle with radius a; a square equal to the rectangle is easily constructed by taking as the side of the square the geometric mean of the sides of the rectangle. Inasmuch as Dinostratus showed that the trisectrix of Hippias serves to square the circle, the curve more commonly came to be known as the quadratrix. It was, of course, always clear to the Greeks that the use of the curve in the trisection and quadrature problems violated the rules of the game—that circles and straight lines only were permitted. The "solutions" of Hippias and Dinostratus, as their authors realized, were sophistic; hence the search for further solutions, canonical or illegitimate, continued, with the result that several new curves were discovered by Greek geometers.

A few years after Dinostratus and Menaechmus there flourished a math- **13** ematician who has the distinction of having written the oldest surviving Greek mathematical treatise. We have described rather fully the work of earlier Hellenic mathematicians, but it must be borne in mind that the accounts have been based not on original works, but on later summaries, commentaries, or descriptions. Occasionally a commentator appears to be copying from an original work extant at the time, as when Simplicius in the sixth century of our era is describing the quadrature of lunes by Hippocrates. But not until we come to Autolycus of Pitane, a contemporary of Aristotle, do we find a Greek author one of whose works has survived. One reason for the survival of this little treatise, *On the Moving Sphere*, is that it formed part of a collection, known as the "Little Astronomy," widely used by ancient astronomers. *On the Moving Sphere* is not a profound, and probably not a very original work, for it includes little beyond elementary theorems on the geometry of the sphere that would be needed in astronomy. Its chief significance lies in the fact that it indicates that Greek geometry evidently had reached the form that we regard as typical of the classical age. Theorems are clearly enunciated and proved. Moreover, the author uses without proof or indication of source other theorems that he regards as well known; we conclude, therefore, that there was in Greece in his day, about 320 B.C., a thoroughly established textbook tradition in geometry.

Autolycus was a contemporary of Aristotle—the most widely learned **14** scholar of all times, whose death is usually taken to mark the end of the first great period, the Hellenic Age, in the history of Greek civilization.

Aristotle, like Eudoxus, was a student of Plato and, like Menaechmus, a tutor of Alexander the Great. Aristotle was primarily a philosopher and biologist, but he was thoroughly *au courant* with the activities of the mathematicians. He may have taken a role in one of the leading controversies of the day, for to him was ascribed a treatise *On Indivisible Lines*. Modern scholarship questions the authenticity of this work, but in any case it probably was the result of discussions carried on in the Aristotelian Lyceum. The thesis of the treatise is that the doctrine of indivisibles espoused by Xenocrates, a successor of Plato as head of the Academy, is untenable. The indivisible, or fixed infinitesimal of length or area or volume, has fascinated men of many ages; Xenocrates thought that this notion would resolve the paradoxes, such as those of Zeno, that plagued mathematical and philosophical thought. Aristotle, too, devoted much attention to the paradoxes of Zeno, but he sought to refute them on the basis of common sense. Inasmuch as he hesitated to follow Platonic mathematicians into the abstractions and technicalities of the day, Aristotle made no lasting contribution to the subject. He is said to have written a biography of Pythagoras, although this is lost; and Eudemus, one of his students, wrote a history of geometry, also lost. Moreover, through his foundation of logic and through his frequent allusion to mathematical concepts and theorems in his voluminous works,[11] Aristotle can be regarded as having contributed to the development of mathematics. The Aristotelian discussion of the potentially and actually infinite in arithmetic and geometry influenced many later writers on the foundations of mathematics; but Aristotle's statement that the mathematicians "do not need the infinite or use it" should be compared with the assertions of our day that the infinite is the mathematician's paradise. Of more positive significance are Aristotle's analysis of the roles of definitions and hypotheses in mathematics.

15   In 323 B.C. Alexander the Great suddenly died, and his empire fell apart. His generals divided the territory over which the young conqueror had ruled; Ptolemy took Egypt, Seleucus and Lysimachus vied for Syria and the East, and Antigonus and Cassander each for a while ruled Macedon. At Athens, where Aristotle had been regarded as a foreigner, the philosopher found himself unpopular, now that his powerful soldier-student was dead. He left Athens and died the following year. Throughout the Greek world the old order was changing, politically and culturally. Under Alexander there had been a gradual blending of Hellenic and Oriental customs and learning, so that it was more appropriate to speak of the newer civilization as Hellenistic, rather than Hellenic. Moreover, the new city of Alexandria, established by the

[11] See T. L. Heath, *Mathematics in Aristotle* (1949).

world conqueror, now took the place of Athens as the center of the mathematical world. In the history of civilization it is therefore customary to distinguish two periods in the Greek world, with the almost simultaneous deaths of Aristotle and Alexander (as well as that of Demosthenes) as a convenient dividing line. The earlier portion is known as the Hellenic Age, the later as the Hellenistic or Alexandrian Age; in the next few chapters we describe the mathematics of the first century of the new era, often known as the Golden Age of Greek mathematics.

BIBLIOGRAPHY

Becker, O., "Eudoxus-Studien," *Quellen und Studien zur Geschichte der Mathematik,* Part B, *Studien,* II (1933), 311–333, 369–387; III (1936), 236–244, 370–410.

Brumbaugh, R. S., *Plato's Mathematical Imagination* (Bloomington, Ind.: Indiana University Press, 1954).

Coolidge, J. L., *A History of the Conic Sections and the Quadric Surfaces* (Oxford: Clarendon, 1945).

Coolidge, J. L., *A History of Geometrical Methods* (Oxford: Clarendon, 1940; paperback ed., New York: Dover, 1963).

Cornford, F. M., *Plato's Cosmology,* The Timaeus of Plato translated with a running commentary (London: Routledge and Kegan Paul, 1937).

Görland, Albert, *Aristoteles und die Mathematik* (Marburg, 1899).

Heath, T. L., *History of Greek Mathematics* (Oxford, 1921, 2 vols.).

Heath, T. L., *Mathematics in Aristotle* (Oxford, 1949).

Heiberg, J. L., "Mathematisches zu Aristoteles," *Abhandlungen zur Geschichte der Mathematischen Wissenschaften,* **18** (1904), 1–49.

Lasserre, François, *The Birth of Mathematics in the Age of Plato,* trans. by Helen Mortimer (London: Hutchinson, 1964).

Loria, Gino, *Historie des sciences mathématiques dans l'antiquité hellénique* (Paris: Gauthier-Villars, 1929).

Michel, Paul-Henri, *De Pythagore à Euclide* (Paris: Société d'Edition "Les Belles Lettres," 1950).

Plato, *Dialogues,* trans. by Benjamin Jowett (Oxford, 1871, 4 vols.).

Solmsen, Friedrich, "Platos Einfluss auf die Bildung der mathematischen Methode," *Quellen und Studien zur Geschichte der Mathematik,* Part B, *Studien,* I (1929–1931), 93–107.

Toeplitz, Otto, "Das Verhältnis von Mathematik und Ideenlehre bei Plato," *Quellen und Studien zur Geschichte der Mathematik,* Part B, *Studien,* I (1931), 3–33.

Wedberg, Anders, *Plato's Philosophy of Mathematics* (Stockholm: Almquist & Wiksell, 1955).

Zeuthen, H. G., *Die Lehre von den Kegelschnitten im Altertum* (Copenhagen, 1886).

EXERCISES

1. It is believed that Theaetetus found for the regular solids the ratio of the edge to the radius of the circumscribed sphere. Do this for three of the regular solids. (See Book XIII of Euclid's *Elements*.)
2. Prove the theorem, probably due to Theaetetus, that there are not more than five regular solids. (See Book XIII of Euclid's *Elements*.)
3. Plato in the *Theaetetus* says that Theodorus proved $\sqrt{3}$ irrational. Give a careful proof of this theorem.
4. Find the angles of the 360 scalene right triangles that Plato indicated on the surface of the dodecahedron.
5. Complete the other half of the proof by exhaustion (see text) that the areas of circles are to each other as squares on their radii. (Use circumscribed polygons.)
6. Describe a method by which Eudoxus could have measured the circumference of the earth.
7. Using the method suggested in connection with the work of Menaechmus, prove that the section of a cylinder is an ellipse. This was proved by Serenus, who probably lived in the fourth century of our era.
8. In his theory of the rainbow Aristotle used a locus commonly attributed to Apollonius, a later mathematician: the locus of all points P such that the distances of P from two fixed points P_1 and P_2 are in a fixed ratio different from one. Identify the locus.
*9. Prove that a Menaechmean section (perpendicular to an element) of an acute-angled cone is an ellipse.
*10. Complete the proof by Dinostratus (see text) by showing that the assumption $DR < DQ$ leads to an absurdity.

CHAPTER VII

Euclid of Alexandria

> Ptolemy once asked Euclid whether there was any
> shorter way to a knowledge of geometry than by a
> study of the Elements, whereupon Euclid answered
> that there was no royal road to geometry.
>
> *Proclus Diadochus*

The death of Alexander the Great had led to internecine strife among the **1**
generals in the Greek army; but by 306 B.C. control of the Egyptian portion
of the empire was firmly in the hands of Ptolemy I, and this enlightened
ruler was able to turn his attention to constructive efforts. Among his early
acts was the establishment at Alexandria of a school or institute, known as
the Museum, second to none in its day. As teachers at the school he called a
band of leading scholars, among whom was the author of the most fabulously
successful mathematics textbook ever written—the *Elements* (Stoichia) of
Euclid. Considering the fame of the author and of his best seller, remarkably
little is known of Euclid's life. So obscure was his life that no birthplace is
associated with his name. Although editions of the *Elements* often bore the
identification of the author as Euclid of Megara and a portrait of Euclid of
Megara often appears in histories of mathematics, this is a case of mistaken
identity. The real Euclid of Megara was a student of Socrates and, although
concerned with logic, was no more attracted to mathematics than was his
teacher. Our Euclid, by contrast, is known as Euclid of Alexandria, for he was
called there to teach mathematics. From the nature of his work it is presumed
that he had studied with students of Plato, if not at the Academy itself.
Legends associated with Euclid picture him as a kindly and gentle old man.
The tale related above in connection with a request of Alexander the Great
for an easy introduction to geometry is repeated in the case of Ptolemy,
whom Euclid is reported to have assured that "there is no royal road to
geometry." Evidently Euclid did not stress the practical aspects of his subject,
for there is a tale told of him that when one of his students asked of what use
was the study of geometry, Euclid asked his slave to give the student three-
pence, "since he must needs make gain of what he learns."

Euclid and the *Elements* are often regarded as synonymous; in reality the man was the author of about a dozen treatises covering widely varying topics, from optics, astronomy, music, and mechanics to a book on the conic sections. With the exception of the *Sphere* of Autolycus, surviving works by Euclid are the oldest Greek mathematical treatises extant; yet of what Euclid wrote more than half has been lost, including some of his more important compositions, such as a treatise on conics. Euclid regarded Aristaeus, a contemporary geometer, as deserving great credit for having written an earlier treatise on *Solid Loci* (the Greek name for the conic sections, stemming presumably from the stereometric definition of the curves in the work of Menaechmus). The treatises on conics by Aristaeus and Euclid have both been lost, probably irretrievably, perhaps because they were soon superseded by the more extensive work on conics by Apollonius to be described below. Among Euclid's lost works are also one on *Surface Loci*, another on *Pseudaria* (or fallacies), and a third on *Porisms*. It is not even clear from ancient references what material these contained. The first one, for example, might have concerned the surfaces known to the ancients—the sphere, cone, cylinder, tore, ellipsoid of revolution, paraboloid of revolution, and hyperboloid of revolution of two sheets—or perhaps curves lying on these surfaces. As far as we know, the Greeks did not study any surface other than that of a solid of revolution.

The loss of the Euclidean *Porisms* is particularly tantalizing, for it may have represented an ancient approximation to an analytic geometry. Pappus later reported that a porism is intermediate between a theorem, in which something is proposed for demonstration, and a problem, in which something is proposed for construction. Others have described a porism as a proposition in which one determines a relationship between known and variable or undetermined quantities, perhaps the closest approach in antiquity to the concept of function. If a porism was, as has been thought, a sort of verbal equation of a curve, Euclid's book on *Porisms* may have differed from our analytic geometry largely in the lack of algebraic symbols and techniques. The nineteenth-century historian of geometry, Michel Chasles, suggested as a typical Euclidean porism the determination of the locus of a point for which the sum of the squares of its distances from two fixed points is a constant.

2   Five works by Euclid have survived to our day: the *Elements*, the *Data*, the *Division of Figures*, the *Phaenomena*, and the *Optics*. The last-mentioned is of interest as an early work on perspective, or the geometry of direct vision. The ancients had divided the study of optical phenomena into three parts: (1) optics (the geometry of direct vision), (2) catoptrics (the geometry of reflected rays), and (3) dioptrics (the geometry of refracted rays). A *Catoptrica* sometimes ascribed to Euclid is of doubtful authenticity, being perhaps by Theon of Alexandria who lived some six centuries later. Euclid's *Optics*[1] is noteworthy for its espousal of an "emission" theory of vision according to which the eye sends out rays that travel to the object, in contrast to a rival Aristotelian doctrine in which an activity in a medium travels in a straight line from the object to the eye. It should be noted that the mathematics of perspective (as opposed to the physical description) is the same no matter which of the two theories is adopted. Among the theorems found in Euclid's *Optics* is one widely used in antiquity—$\tan \alpha / \tan \beta < \alpha / \beta$ if $0 < \alpha < \beta < \pi/2$. One object of the *Optics* was to combat an Epicurean insistence that an object was just as large as it looked, with no allowance to be made for the foreshortening suggested by perspective.

Euclid's *Phaenomena* is much like the *Sphere* of Autolycus—that is, a work on spherical geometry of use to astronomers. A comparison of the two works indicates that both authors drew heavily on a textbook tradition that was well known to their generation. It is quite possible that much the same was true of Euclid's *Elements*, but in this case there is no contemporary work extant with which it can be compared.

The Euclidean *Division of Figures* is significant in that it is a work that would have been lost, had it not been for the learning of Arabic scholars. It has not survived in the original Greek; but before the disappearance of the Greek versions, an Arabic translation had been made (omitting some of the original proofs "because the demonstrations are easy"), which in turn was later translated into Latin, and ultimately into current modern languages.[2] This is not atypical of other ancient works. The *Division of Figures* includes a collection of thirty-six propositions concerning the division of plane configurations. For example, Proposition 1 calls for the construction of a straight line that shall be parallel to the base of a triangle and shall divide the triangle into two equal areas. Proposition 4 requires a bisection of a trapezoid *abqd* (Fig. 7.1) by a line parallel to the bases; the

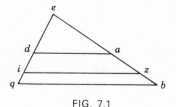

FIG. 7.1

[1] See M. R. Cohen and I. E. Drabkin: *A Source Book in Greek Science* (1948), pp. 257 ff.

[2] An English version entitled *Euclid's Book on Divisions of Figures* was edited by R. C. Archibald (1915).

required line zi is found by determining z such that $\overline{ze}^2 = \frac{1}{2}(\overline{eb}^2 + \overline{ea}^2)$. Other propositions call for the division of a parallelogram into two equal parts by a line drawn through a given point on one of the sides (Proposition 6) or through a given point outside the parallelogram (Proposition 10). The final proposition asks for the division of a quadrilateral in a given ratio by a line through a point on one of the sides of the quadrilateral. Somewhat similar in nature and purpose to the *Division of Figures* is Euclid's *Data*, a work that has come down to us through both the Greek and the Arabic. It seems to have been composed for use at the university of Alexandria, serving as a companion volume to the first six books of the *Elements* in much the way that a manual of tables supplements a textbook. It was to be useful as a guide to the analysis of problems in geometry in order to discover proofs. It opens with fifteen definitions concerning magnitudes and loci. The body of the text comprises ninety-five statements concerning the implications of conditions and magnitudes that may be given in a problem. The first two state that if two magnitudes a and b are given, their ratio is given, and that if one magnitude is given and also its ratio to a second, the second magnitude is given. There are about two dozen similar statements, serving as algebraic rules or formulas. Then follow simple geometrical rules concerning parallel lines and proportional magnitudes, reminding the student of the implications of the data given in a problem, such as the advice that when two line segments have a given ratio, then one knows the ratio of the areas of similar rectilinear figures constructed on these segments. Some of the statements are geometrical equivalents of the solution of quadratic equations. For example, we are told that if a given (rectangular) area AB is laid off along a line segment of given length AC (Fig. 7.2) and if the area BC by which the area AB falls short of

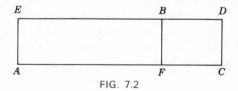

FIG. 7.2

the entire rectangle AD is given, the dimensions of the rectangle BC are known. The truth of this statement is easily demonstrated by modern algebra. Let the length of AC be a, the area of AB be b^2, and the ratio of FC to CD be $c:d$. Then if $FC = x$ and $CD = y$, we have $x/y = c/d$ and $(a - x)y = b^2$. Eliminating y we have $(a - x)dx = b^2 c$ or $dx^2 - adx + b^2 c = 0$, from which $x = a/2 \pm \sqrt{(a/2)^2 - b^2 c/d}$. The geometric solution given by Euclid is equivalent to this, except that the negative sign before the radical is used. Statements 84 and 85 in the *Data* are geometrical replacements of the familiar

Babylonian algebraic solutions of the systems $xy = a^2, x \pm y = b$, which again are the equivalents of solutions of simultaneous equations. The last few statements in the *Data* concern relationships between linear and angular measures in a given circle.

The university at Alexandria evidently was not unlike modern institutions **3** of higher learning. Some of the faculty probably excelled in research, others were better fitted to be administrators, and still others were noted for teaching ability. It would appear, from the reports we have, that Euclid very definitely fitted into the last category. There is no new discovery attributed to him, but he was noted for expository skill. This is the key to the success of his greatest work, the *Elements*. It was frankly a textbook and by no means the first one. We know of at least three earlier such elements, including that by Hippocrates of Chios; but there is no trace of these, nor of other potential rivals from ancient times. The *Elements* of Euclid so far outdistanced competitors that it alone survived. The *Elements* was not, as is sometimes thought, a compendium of all geometrical knowledge; it was instead an introductory textbook covering all *elementary* mathematics—that is, arithmetic (in the sense of the English "higher arithmetic" or the American "theory of numbers"), synthetic geometry (of points, lines, planes, circles, and spheres), and algebra (not in the modern symbolic sense, but an equivalent in geometrical garb). It will be noted that the art of calculation is not included, for this was not a part of university instruction; nor was the study of the conics or higher plane curves part of the book, for these formed a part of more advanced mathematics. Proclus described the *Elements* as bearing to the rest of mathematics the same sort of relation as that which the letters of the alphabet have in relation to language. Were the *Elements* intended as an exhaustive store of information, the author probably would have included references to other authors, statements of recent research, and informal explanations. As it is, the *Elements* is austerely limited to the business in hand—the exposition in logical order of the fundamentals of elementary mathematics. Occasionally, however, later writers interpolated into the text explanatory scholia, and such additions were copied by later scribes as part of the original text. Some of these appear in every one of the manuscripts now extant. Euclid himself made no claim to originality, and it is clear that he drew heavily from the works of his predecessors. It is believed that the arrangement is his own, and presumably some of the proofs were supplied by him; but beyond this it is difficult to estimate the degree of originality that is to be found in this, the most renowned mathematical work in history.

The *Elements* is divided into thirteen books or chapters, of which the first **4** half dozen are on elementary plane geometry, the next three on the theory of numbers, Book X on incommensurables, and the last three chiefly on solid geometry. There is no introduction or preamble to the work, and the first book opens abruptly with a list of twenty-three definitions. The weakness here is that some of the definitions do not define, inasmuch as there is no prior set of undefined elements in terms of which to define the others. Thus to say, as does Euclid, that "a point is that which has no part," or that "a line is breadthless length," or that "a surface is that which has length and breadth only," is scarcely to define these entities, for a definition must be expressed in terms of things that precede, and are better known than the things defined. Objections can easily be raised on the score of logical circularity to other so-called "definitions" of Euclid, such as "The extremities of a line are points," or "A straight line is a line which lies evenly with the points on itself," or "The extremities of a surface are lines," all of which may have been due to Plato. The Euclidean definition of a plane angle as "the inclination to one another of two lines in a plane which meet one another and do not lie in a straight line" is vitiated by the fact that "inclination" has not been previously defined and is not better known than the word "angle."

Following the definitions, Euclid lists five postulates and five common notions. Aristotle had made a sharp distinction between axioms (or common notions) and postulates; the former, he said, must be convincing in themselves—truths common to all studies—but the latter are less obvious and do not presuppose the assent of the learner, for they pertain only to the subject at hand. Some later writers distinguished between the two types of assumptions by applying the word axiom to something known or accepted as obvious, while the word postulate referred to something to be "demanded." We do not know whether Euclid subscribed to either of these views, or even whether he distinguished between two types of assumptions. Surviving manuscripts are not in agreement here, and in some cases the ten assumptions appear together in a single category. Modern mathematicians see no essential difference between an axiom and a postulate. In most manuscripts of the *Elements* we find the following ten assumptions:[3]

Postulates. Let the following be postulated:

1. To draw a straight line from any point to any point.
2. To produce a finite straight line continuously in a straight line.
3. To describe a circle with any center and radius.
4. That all right angles are equal.
5. That, if a straight line falling on two straight lines makes the interior angles on the same side less than two right angles, the two straight lines, if

[3] See *The Thirteen Books of Euclid's Elements*, translated and edited by T. L. Heath (1956, 3 vols.).

produced indefinitely, meet on that side on which the angles are less than the two right angles.

Common notions:

1. Things which are equal to the same thing are also equal to one another.
2. If equals be added to equals, the wholes are equal.
3. If equals be subtracted from equals, the remainders are equal.
4. Things which coincide with one another are equal to one another.
5. The whole is greater than the part.

Aristotle had written that "other things being equal, that proof is the better which proceeds from the fewer postulates," and Euclid evidently subscribed to this principle. For example, Postulate 3 is interpreted in the very limited literal sense, sometimes described as the use of Euclidean (collapsible) compasses, whose legs maintain a constant opening so long as the point stands on the paper, but fall back upon each other when they are lifted. That is, the postulate is not interpreted to permit the use of a pair of dividers to lay off a distance equal to one line segment upon a noncontiguous longer line segment, starting from an end point. It is proved in the first three propositions of Book I that the latter construction is always possible, even under the strict interpretation of Postulate 3. The first proposition justifies the construction of an equilateral triangle *ABC* on a given line segment *AB* by constructing through *B* a circle with a center at *A* and another circle through *A* with center at *B*, and letting *C* be the point of intersection of the two circles. (That they do intersect is tacitly assumed.) Proposition 2 then builds on Proposition 1 by showing that from any point *A* as extremity (Fig. 7.3) one

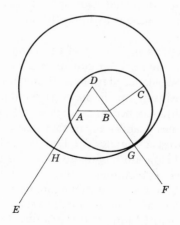

FIG. 7.3

can lay off a straight line segment equal to a given line segment *BC*. First Euclid draws *AB*, and on this he constructs the equilateral triangle *ABD*, extending the sides *DA* and *DB* to *E* and *F* respectively. With *B* as center describe the circle through *C*, intersecting *BF* in *G*; then with *D* as center draw a circle through *G*, intersecting *DE* in *H*. Line *AH* is then easily shown to be the line required. Finally, in Proposition 3, Euclid makes use of Proposition 2 to show that, given any two unequal straight lines, one can cut off from the greater a segment equal to the smaller.

5   In the first three propositions Euclid went to great pains to show that a very restricted interpretation of Postulate 3 nevertheless implies the free use of compasses as is usually done in laying off distances in elementary geometry. Nevertheless, by modern standards of rigor the Euclidean assumptions are woefully inadequate, and in his proofs Euclid often makes use of tacit postulates. In the first proposition of the *Elements*, for example, he assumes without proof that the two circles will intersect in a point. For this and similar situations it is necessary to add to the postulates one equivalent to a principle of continuity. Moreover, Postulates 1 and 2 as they were expressed by Euclid guarantee neither the uniqueness of the straight line through two non-coincident points nor even its infinitude; they simply assert that there is at least one and that it has no termini, yet in his proofs Euclid freely made use of the uniqueness and infinitude. It is, of course, easy to criticize the work of a man in the light of later developments and to forget that "sufficient unto the day is the rigor thereof." In its time the *Elements* evidently was the most tightly reasoned logical development of elementary mathematics that had ever been put together, and several thousand years were to pass before a more careful presentation occurred. During this long interval most mathematicians regarded the treatment as logically satisfying and pedagogically sound.

      Most of the propositions in Book I of the *Elements* are well known to anyone who has had a high school course in geometry. Included are the familiar theorems on congruence of triangles (but without an axiom justifying the method of superposition), on simple constructions by straightedge and compasses, on inequalities concerning angles and sides of a triangle, on properties of parallel lines (leading to the fact that the sum of the angles of a triangle is equal to two right angles), and on parallelograms (including the construction of a parallelogram having given angles and equal in area to a given triangle or to a given rectilinear figure). The book closes (in Propositions 47 and 48) with the proof of the Pythagorean theorem and its converse. The proof of the theorem as given by Euclid was not that usually given in textbooks of today, in which simple proportions are applied to the sides of similar triangles formed by dropping an altitude upon the hypotenuse. It has been suggested that Euclid avoided such a proof because of difficulties involved in commensurability. Only in Book V did Euclid turn to the well-founded theory of proportions, and up to that point the use of proportion-alities is avoided as far as possible. For the Pythagorean theorem Euclid used instead the beautiful proof with a figure sometimes described as a windmill or as the peacock's tail or as the bride's chair (Fig. 7.4). The proof is accomplished by showing that the square on AC is equal to twice the triangle FAB

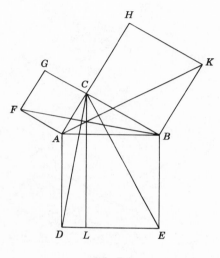

FIG. 7.4

or to twice the triangle CAD or to the rectangle AL, and that the square on BC is equal to twice the triangle ABK or to twice the triangle BCE or to the rectangle BL. Hence the sum of the squares is equal to the sum of the rect-angles, that is, to the square on AB. It has been assumed that this proof was original with Euclid, and many conjectures have been made as to the possible form of earlier proofs. Since the days of Euclid many alternative proofs have been proposed.

It is to Euclid's credit that the Pythagorean theorem is immediately followed by a proof of the converse: If in a triangle the square on one of the sides is equal to the sum of the squares on the other two sides, the angle between these other two sides is a right angle. Not infrequently in modern textbooks the exercises following the proof of the Pythagorean theorem are such that they require not the theorem itself, but the still unproved converse. There may be many a minor flaw in the *Elements*, but the book had all the major logical virtues.

6   Book II of the *Elements* is a short one, containing only fourteen propositions, not one of which plays any role in modern textbooks; yet in Euclid's
day this book was of great significance. This sharp discrepancy between
ancient and modern views is easily explained—today we have symbolic
algebra and trigonometry that have replaced the geometrical equivalents
from Greece. For instance, Proposition I of Book II states that "If there be

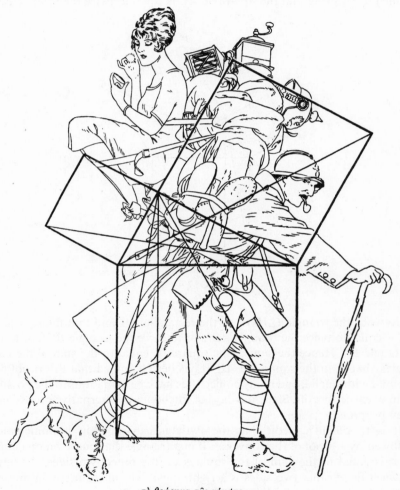

τὸ θεώρημα τῆς νύμφης.

(With apologies to *La Vie Parisienne*.)

The "Bride's Chair" diagram of Euclid's *Elements* I. 47 in a World War I setting. [*The
Mathematical Gazette*, **11** (1922–1923), 364.]

two straight lines, and one of them be cut into any number of segments whatever, the rectangle contained by the two straight lines is equal to the rectangles contained by the uncut straight line and each of the segments." This theorem, which asserts (Fig. 7.5) that $AD(AP + PR + RB) = AD \cdot AP +$

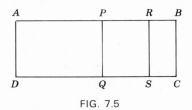

FIG. 7.5

$AD \cdot PR + AD \cdot RB$, is nothing more than a geometrical statement of one of the fundamental laws of arithmetic known today as the distributive law: $a(b + c + d) = ab + ac + ad$. In later books of the *Elements* (V and VII) we find demonstrations of the commutative and associative laws for multiplication. Whereas in our time magnitudes are represented by letters that are understood to be numbers (either known or unknown) on which we operate with the algorithmic rules of algebra, in Euclid's day magnitudes were pictured as line segments satisfying the axioms and theorems of geometry. It is sometimes asserted that the Greeks had no algebra, but this is patently false. They had Book II of the *Elements*, which is a geometrical algebra that served much the same purpose as does our symbolic algebra. There can be little doubt that modern algebra greatly facilitates the manipulation of relationships among magnitudes. But it is undoubtedly also true that a Greek geometer versed in the fourteen theorems of Euclid's "algebra" was far more adept in applying these theorems to practical mensuration than is an experienced geometer of today. Ancient geometrical algebra was not an ideal tool, but it was far from ineffective. Euclid's statement (Proposition 4), "If a straight line be cut at random, the square on the whole is equal to the squares on the segments and twice the rectangle contained by the segments," is a verbose way of saying that $(a + b)^2 = a^2 + 2ab + b^2$, but its visual appeal to an Alexandrian schoolboy must have been far more vivid than its modern algebraic counterpart can ever be. True, the proof in the *Elements* occupies about a page and a half; but how many high school students of today could give a careful proof of the algebraic rule they apply so unhesitatingly? The same holds true for *Elements* II. 5, which contains what we should regard as an impractical circumlocution for $(a - b)^2 = (a + b)(a - b)$:

> If a straight line be cut into equal and unequal segments, the rectangle contained by the unequal segments of the whole, together with the square on the straight line between the points of section, is equal to the square on the half.

The diagram that Euclid uses in this connection played a key role in Greek algebra; hence we reproduce it[4] with further explanation. If in the diagram (Fig. 7.6) we let $AC = CB = a$, and $CD = b$, the theorem asserts that

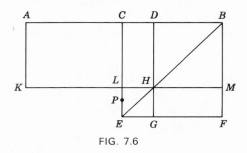

FIG. 7.6

$(a + b)(a - b) + b^2 = a^2$. The geometrical verification of this statement is not difficult. However, the significance of the diagram lies not so much in the proof of the theorem as in the use to which similar diagrams were put by Greek geometrical algebraists. The pride of the modern schoolboy in algebra is the solution of the quadratic equation (which he may or may not be able to justify), and a diagram similar to Fig. 7.6 was the Greek schoolboy's geometrical equivalent. If the Greek scholar were required to construct a line x having the property expressed by $ax - x^2 = b^2$, where a and b are given line segments, he would draw line $AB = a$ and bisect it at C. Then at C he would erect a perpendicular CP equal in length to b; with P as center and radius $a/2$ he would draw a circle cutting AB in point D. Then on AB he would construct rectangle $ABMK$ of width $BM = BD$ and complete the square $BDHM$. This square is the area x^2 having the property specified in the quadratic equation. As the Greeks expressed it, we have applied to the segment AB $(= a)$ a rectangle AH $(= ax - x^2)$ which is equal to a given square (b^2) and falls short (of AM) by a square DM. The demonstration of this is provided by the proposition cited above (II. 5) in which it is clear that the rectangle $ADHK$ equals the concave polygon $CBFGHL$—that is, it differs from $(a/2)^2$ by the square $LHGE$, the side of which by construction is $CD = \sqrt{(a/2)^2 - b^2}$.

In an exactly analogous manner the quadratic equation $ax + x^2 = b^2$ is solved through the use of II. 6:

> If a straight line be bisected and a straight line be added to it in a straight line, the rectangle contained by the whole (with the added straight line) and the added straight line together with the square on the half is equal to the square on the straight line made up of the half and the added straight line.

---

[4] Throughout this chapter the translations and most of the diagrams are from the *Thirteen Books of Euclid's Elements* as edited by T. L. Heath.

This time we "apply to a given straight line $(AB = a)$ a rectangle $(AM = ax + x^2)$ which shall be equal to a given square (b^2) and shall exceed (AH) by a square figure" (Fig. 7.7). In this case the distance $CD = \sqrt{(a/2)^2 + b^2}$; since from the proposition it is known that rectangle AM $(= ax + x^2)$ plus square LG $[=(a/2)^2]$ is equal to square CF $[=(a/2)^2 + b^2]$, it follows that the condition $ax + x^2 = b^2$ is satisfied.

The next few propositions of Book II are variations of the geometric algebra that we have illustrated, with II. 11 being an important special case of II. 6. Here Euclid solves the equation $ax + x^2 = a^2$ by drawing a square $ABCD$ with side a, bisecting side AD at E, drawing EB, extending side DA to F such that $EF = EB$, and completing the square $AFGH$ (Fig. 7.8). Then

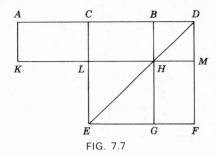

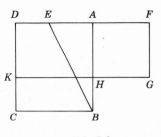

FIG. 7.7   FIG. 7.8

on extending GH to intersect DC in K, we shall have applied to segment AD a rectangle FK $(= ax + x^2)$ equal to a given square AC $(= a^2)$ and exceeding by a square (x^2).

The figure used by Euclid in *Elements* II. 11, and again in VI. 30 (our Fig. 7.8), is the basis for a diagram that appears today in many geometry books to illustrate the iterative property of the golden section. To the gnomon $BCDFGH$ (Fig. 7.8) we add point L to complete the rectangle $CDFL$ (Fig. 7.9), and within the smaller rectangle $LBGH$, which is similar to the larger rectangle $LCDF$, we construct, by making $GO = GL$, the gnomon $LBMNOG$ similar to gnomon $BCDFGH$. Now within the rectangle $BHOP$, which is similar to the larger rectangles $CDFL$ and $LBHG$, we construct the gnomon $PBHQRN$ similar to the gnomons $BCDFGH$ and $LBMNOG$. Continuing indefinitely in this manner, we have an unending sequence of nested similar rectangles tending toward a limiting point Z. It turns out that Z, which is easily seen to be the point of intersection of lines FB and DL, is also the pole of a logarithmic spiral tangent to the sides of the rectangles at points $C, A, G, P, M, Q, \ldots$. Other striking properties can be found in this fascinating diagram.[5]

[5] See, for example, H. S. M. Coxeter, "The Golden Section, Phyllotaxis, and Wythoff's Game," *Scripta Mathematica*, **19** (1953), 135–143.

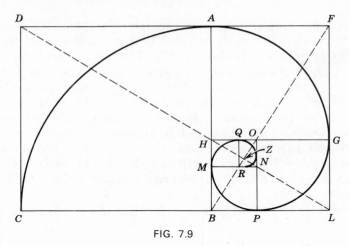

FIG. 7.9

Propositions 12 and 13 of Book II are of interest because they adumbrate the concern with trigonometry that was shortly to blossom in Greece. These propositions will be recognized by the reader as geometric formulations—first for the obtuse angle and then for the acute angle—of what later became known as the law of cosines for plane triangles:

Proposition 12

In obtuse-angled triangles the square on the side subtending the obtuse angle is greater than the squares on the sides containing the obtuse angle by twice the rectangle contained by one of the sides about the obtuse angle, namely that on which the perpendicular falls, and the straight line cut off outside by the perpendicular toward the obtuse angle.

Proposition 13

In acute-angled triangles the square on the side subtending the acute angle is less than the squares on the sides containing the acute angle by twice the rectangle contained by one of the sides about the acute angle, namely that on which the perpendicular falls, and the straight line cut off within by the perpendicular toward the acute angle.

The proofs of Propositions 12 and 13 are analogous to those used today in trigonometry through double application of the Pythagorean theorem.

7   It generally has been supposed that the contents of the first two books of the *Elements* are largely the work of the Pythagoreans. Books III and IV, on the other hand, deal with the geometry of the circle, and here the material is presumed to have been drawn largely from Hippocrates of Chios. The two books are not unlike the theorems on circles contained in textbooks of today.

The first proposition of Book III, for example, calls for the construction of the center of a circle; and the last, Proposition 37, is the familiar statement that if from a point outside a circle a tangent and a secant are drawn, the square on the tangent is equal to the rectangle on the whole secant and the external segment. Book IV contains sixteen propositions, largely familiar to modern students, concerning figures inscribed in, or circumscribed about, a circle. Theorems on the measure of angles are reserved until after a theory of proportions has been established.

Of the thirteen books of the *Elements* those most admired have been the   **8** fifth and the tenth—the one on the general theory of proportion and the other on the classification of incommensurables. The discovery of the incommensurable had threatened a logical crisis which cast doubt on proofs appealing to proportionality, but the crisis had been successfully averted through the principles enunciated by Eudoxus. Nevertheless, Greek mathematicians tended to avoid proportions. We have seen that Euclid put off their use as long as possible, and such a relationship among lengths as $x : a = b : c$ would be thought of as an equality of the areas $cx = ab$. Sooner or later, however, proportions are needed, and so Euclid tackled the problem in Book V of the *Elements*. Some commentators have gone so far as to suggest that the whole book, consisting of twenty-five propositions, was the work of Eudoxus, but this seems to be unlikely. Some of the definitions—such as that of a ratio—are so vague as to be useless. Definition 4, however, is essentially the axiom of Eudoxus and Archimedes: "Magnitudes are said to have a ratio to one another which are capable, when multiplied, of exceeding one another." Definition 5, the equality of ratios, is precisely that given earlier in connection with Eudoxus' definition of proportionality.

To the casual reader Book V might appear as superfluous as Book II, for both have now been displaced by corresponding rules in symbolic algebra. A more careful reader interested in axiomatics will see that Book V deals with topics of fundamental importance in all mathematics. It opens with propositions that are equivalent to such things as the left-hand and right-hand distributive laws for multiplication over addition, the left-hand distributive law for multiplication over subtraction, and the associative law for multiplication $(ab)c = a(bc)$. Then follow rules for "greater than" and "less than" and the well-known properties of proportions. It often is asserted that Greek geometrical algebra could not rise above the second degree in plane geometry, nor above the third degree in solid geometry, but this is not really the case. The general theory of proportions would permit work with products of any number of dimensions, for an equation of the form $x^4 = abcd$ is equivalent to one involving products of ratios of lines such as $x/a \cdot x/b = c/x \cdot d/x$.

Having developed the theory of proportions in Book V, Euclid exploited it in Book VI by proving theorems concerning ratios and proportions related to similar triangles, parallelograms, and other polygons. Noteworthy is Proposition 31, a generalization of the Pythagorean theorem: "In right-angled triangles the figure on the side subtending the right angle is equal to the similar and similarly described figures on the sides containing the right angle." Proclus credits this extension to Euclid himself. Book VI contains (in Propositions 28 and 29) also a generalization of the method of application of areas, for the sound basis for proportion given in Book V enabled the author now to make free use of the concept of similarity. The rectangles of Book II are now replaced by parallelograms, and it is required to apply to a given straight line a parallelogram equal to a given rectilinear figure and deficient (or exceeding) by a parallelogram similar to a given parallelogram. These constructions, like those of II. 5–6, are in reality solutions of the quadratic equations $bx = ac \pm x^2$, subject to the restriction (implied in IX. 27) that the discriminant is not negative.

9   The *Elements* of Euclid often is mistakenly thought of as restricted to geometry. We already have described two books (II and V) that are almost exclusively algebraic; three books (VII, VIII, and IX) are devoted to the theory of numbers. The word "number" to the Greeks always referred to what we call the natural numbers—the positive whole numbers or integers. Book VII opens with a list of twenty-two definitions distinguishing various types of number—odd and even, prime and composite, plane and solid (that is, those that are products of two or of three integers)—and finally defining a perfect number as "that which is equal to its own parts." The theorems in Books VII, VIII, and IX are likely to be familiar to the reader who has had an elementary course in the theory of numbers, but the language of the proofs will certainly be unfamiliar. Throughout these books each number is represented by a line segment, so that Euclid will speak of a number as *AB*. (The discovery of the incommensurable had shown that not all line segments could be associated with whole numbers; but the converse statement—that numbers can always be represented by line segments—obviously remains true.) Hence Euclid does not use the phrases "is a multiple of" or "is a factor of," for he replaces these by "is measured by" and "measures" respectively. That is, a number *n* is measured by another number *m* if there is a third number *k* such that $n = km$.

Book VII opens with two propositions that constitute a celebrated rule in the theory of numbers, which today is known as "Euclid's algorithm" for finding the greatest common divisor (measure) of two numbers. It is a scheme suggestive of a repeated inverse application of the axiom of Eudoxus. Given two unequal numbers, one subtracts the smaller *a* from the larger *b* repeatedly until a remainder r_1 less than the smaller is obtained; then one repeatedly subtracts this remainder r_1 from a until a remainder $r_2 < r_1$ results; then one repeatedly subtracts r_2 from r_1; and so on. Ultimately the process will lead to a remainder r_n which will measure r_{n-1}, hence all preceding remainders, as well as a and b; this number r_n will be the greatest common divisor of a and b. Among succeeding propositions we find equivalents of familiar theorems in arithmetic. Thus Proposition 8 states that if $an = bm$ and $cn = dm$, then $(a - c)n = (b - d)m$; Proposition 24 states that if a and b are prime to c, then ab is prime to c. The book closes with a rule (Proposition 39) for finding the least common multiple of several numbers.

Book VIII is one of the less rewarding of the thirteen books of the *Elements*. It opens with propositions on numbers in continued proportion (geometric progression) and then turns to some simple properties of squares and cubes, closing with Proposition 27: "Similar solid numbers have to one another the ratio which a cube number has to a cube number." This statement means simply that if we have a "solid number" $mg \cdot mb \cdot mc$ and a "similar solid number" $na \cdot nb \cdot nc$, then their ratio will be $m^3 : n^3$—that is, as a cube is to a cube.

Book IX, the last of the three books on theory of numbers, contains several **10** theorems that are of special interest. Of these the most celebrated is Proposition 20: "Prime numbers are more than any assigned multitude of prime numbers." That is, Euclid here gives the well-known elementary proof that the number of primes is infinite. The proof is indirect, for one shows that the assumption of a finite number of primes leads to a contradiction. Let P be the product of *all* the primes, assumed to be finite in number, and consider the number $N = P + 1$. Now, N cannot be prime, for this would contradict the assumption that P was the product of *all* primes. Hence N is composite and must be measured by some prime p. But p cannot be any of the prime factors in P, for then it would have to be a factor of 1. Hence p must be a prime different from all of those in the product P; therefore, the assumption that P was the product of *all* the primes must be false.

Proposition 35 of this book contains a formula for the sum of numbers in geometric progression, expressed in elegant but unusual terms:

> If as many numbers as we please be in continued proportion, and there be subtracted from the second and the last numbers equal to the first, then as the excess of the second is to the first, so will the excess of the last be to all those before it.

This statement is, of course, equivalent to the formula

$$\frac{a_{n+1} - a_1}{a_1 + a_2 + \cdots + a_n} = \frac{a_2 - a_1}{a_1}$$

which in turn is equivalent to

$$S_n = \frac{a - ar^n}{1 - r}$$

The following and last proposition in Book IX is the well-known formula for perfect numbers: "If as many numbers as we please, beginning from unity, be set out continuously in double proportion until the sum of all becomes prime, and if the sum is multiplied by the last, the product will be perfect." That is, in modern notation, if $S_n = 1 + 2 + 2^2 + \cdots + 2^{n-1} = 2^n - 1$ is prime, then $2^{n-1}(2^n - 1)$ is a perfect number. The proof is easily established in terms of the definition of perfect number given in Book VII. The ancient Greeks knew the first four perfect numbers: 6, 28, 496, and 8128. Euclid did not answer the converse question—whether or not his formula provides *all* perfect numbers. It is now known that all *even* perfect numbers are of Euclid's type, but the question of the existence of odd perfect numbers remains an unsolved problem.[6] Of the dozen and a half perfect numbers now known all are even, but to conclude by induction that all must be even would be hazardous.

In Propositions 21 through 36 of Book IX there is a unity which suggests that these theorems were at one time a self-contained mathematical system, possibly the oldest in the history of mathematics and stemming presumably from the middle or early fifth century B.C. It has even been suggested that Propositions 1 through 36 of Book IX were taken over by Euclid, without essential change, from a Pythagorean textbook.[7]

11   Book X of the *Elements* was, before the advent of early modern algebra, the most admired—and the most feared. It is concerned with a systematic classification of incommensurable line segments of the forms $a \pm \sqrt{b}$, $\sqrt{a} \pm \sqrt{b}$, $\sqrt{a \pm \sqrt{b}}$, and $\sqrt{\sqrt{a} \pm \sqrt{b}}$, where a and b, when of the same dimension, are commensurable. Today we would be inclined to think of this as a book on irrational *numbers* of the types above, where a and b are rational *numbers*; but Euclid regarded this book as a part of geometry, rather than of arithmetic. In fact, Propositions 2 and 3 of the book duplicate for geometrical magnitudes the first two propositions of Book VII, where the author had dealt with whole numbers. Here he proves that if to two unequal line segments one applies the process described above as Euclid's algorithm,

[6] For further details see L. E. Dickson, *History of the Theory of Numbers* (Washington, D.C., 1919–1923, 3 vols.), I, 3–33.

[7] See Arpád Szabó, "The Transformation of Mathematics into Deductive Science and the Beginnings of Its Foundations on Definitions and Axioms," *Scripta Mathematica*, **27** (1964), 27–48A.

and if the remainder never measures the one before it, the magnitudes are incommensurable. Proposition 3 shows that the algorithm, when applied to two commensurable magnitudes, will provide the greatest common measure of the segments.

Book X contains 115 propositions—more than any other— most of which contain geometrical equivalents of what we now know arithmetically as surds. Among the theorems are counterparts of rationalizing denominators of fractions of the form $a/(b \pm \sqrt{c})$ and $a/(\sqrt{b} \pm \sqrt{c})$. Line segments given by square roots, or by square roots of sums of square roots, are about as easily constructed by straightedge and compasses as are rational combinations. One reason that the Greeks turned to a geometrical rather than an arithmetical algebra was that, in view of the lack of the real-number concept, the former appeared to be more general than the latter. The roots of $ax - x^2 = b^2$, for example, can always be constructed (provided that $a > 2b$). Why, then, should Euclid have gone to great lengths to demonstrate, in Propositions 17 and 18 of Book X, the conditions under which the roots of this equation are commensurable with a? He showed that the roots are commensurable or incommensurable, with respect to a, according as $\sqrt{a^2 - 4b^2}$ and a are commensurable or incommensurable. It has been suggested[8] that such considerations indicate that the Greeks used their solutions of quadratic equations for *numerical* problems also, much as the Babylonians had in their system of equations $x + y = a$, $xy = b^2$. In such cases it would be advantageous to know whether the roots will or will not be expressible as quotients of integers. A close study of Greek mathematics seems to give evidence that beneath the geometrical veneer there was more concern for logistic and numerical approximations than the surviving classical treatises portray.

The material in Book XI, containing thirty-nine propositions on the geometry of three dimensions, will be largely familiar to one who has taken a course in the elements of solid geometry. Again the definitions are easily criticized, for Euclid defines a solid as "that which has length, breadth, and depth" and then tells us that "an extremity of a solid is a surface." The last four definitions are of four of the regular solids. The tetrahedron is not included, presumably because of an earlier definition of a pyramid as "a solid figure, contained by planes, which is constructed from one plane to any point." The eighteen propositions of Book XII are all related to the measurement of figures, using the method of exhaustion. The book opens with a careful proof of the theorem that areas of circles are to each other as squares on the diameters. Similar applications of the typical double *reductio ad*

---

[8] See Heath, *Elements of Euclid*, III, 43–45.

absurdum method then are applied to the volumetric mensuration of pyramids, cones, cylinders, and spheres. Archimedes ascribed the rigorous proofs of these theorems to Eudoxus, from whom Euclid probably adapted much of this material.

The last book is devoted entirely to properties of the five regular solids, a fact that has led some historians to say that the *Elements* was composed as a glorification of the cosmic or Platonic figures. Inasmuch as such a large proportion of the earlier material is far removed from anything relating to the regular polyhedra, such an assumption is quite gratuitous; but the closing theorems are a fitting climax to a remarkable treatise. Their object is to "comprehend" each of the regular solids in a sphere—that is, to find the ratio of an edge of the solid to the radius of the circumscribed sphere. Such computations are ascribed by Greek commentators to Theaetetus, to whom much of Book XIII is probably due. In preliminaries to these computations Euclid referred once more to the division of a line in mean and extreme ratio, showing that "the square on the greater segment added to half the whole is five times the square on the half"—as is easily verified by solving $a/x = x/(a - x)$—and citing other properties of the diagonals of a regular pentagon. Then in Proposition 10 Euclid proved the well-known theorem that a triangle whose sides are respectively sides of an equilateral pentagon, hexagon, and decagon inscribed in the same circle is a right triangle. Propositions 13 through 17 express the ratio of edge to diameter for each of the inscribed regular solids in turn: e/d is $\sqrt{\frac{2}{3}}$ for the tetrahedron, $\sqrt{\frac{1}{2}}$ for the octahedron, $\sqrt{\frac{1}{3}}$ for the cube or hexahedron, $\sqrt{(5 - \sqrt{5})/10}$ for the icosahedron, and $(\sqrt{5} - 1)/2\sqrt{3}$ for the dodecahedron. Finally, in Proposition 18, the last in the *Elements*, it is easily proved that there can be no regular polyhedron beyond these five. About 1900 years later the astronomer Kepler was so struck by this fact that he built a cosmology on the five regular solids, believing that they must have been the creator's key to the structure of the heavens.

13   In ancient times it was not uncommon to attribute to a celebrated author works that were not by him; thus some versions of Euclid's *Elements* include a fourteenth and even a fifteenth book, both shown by later scholars to be apocryphal. The so-called Book XIV continues Euclid's comparison of the regular solids inscribed in a sphere, the chief results being that the ratio of the surfaces of the dodecahedron and icosahedron inscribed in the same sphere is the same as the ratio of their volumes, the ratio being that of the edge of the cube to the edge of the icosahedron—that is, $\sqrt{10/[3(5 - \sqrt{5})]}$. It is thought that this book may have been composed by Hypsicles on the basis of a treatise (now lost) by Apollonius comparing the dodecahedron and icosahedron. (Hypsicles, who probably lived in the second half of the second century B.C., is thought to be the author of an astronomical work, *De ascensionibus*, from which the division of the circle into 360 parts may have been adopted.) That the same circle circumscribes both the pentagon of the dodecahedron and the triangle of the icosahedron (inscribed in the same sphere) was said to have been proved by Aristaeus, roughly contemporaneous with Euclid.

The spurious Book XV, which is inferior, is thought to have been (at least in part) the work of Isidore of Miletus (fl. ca. A.D. 532), architect of the cathedral of Holy Wisdom (Hagia Sophia) at Constantinople. This book also deals with the regular solids, showing how to inscribe certain of them within others, counting the number of edges and solid angles in the solids, and finding the measures of the dihedral angles of faces meeting at an edge. It is of interest to note that despite such enumerations, the ancients all missed the so-called polyhedral formula enunciated by Euler in the eighteenth century.

The *Elements* of Euclid not only was the earliest major Greek mathematical **14** work to come down to us, but also the most influential textbook of all times. It was composed in about 300 B.C. and was copied and recopied repeatedly after that. Errors and variations inevitably crept in, and some later editors, notably Theon of Alexandria in the late fourth century, sought to improve on the original. Nevertheless, it has been possible to obtain a good impression of the content of the Euclidean version through a comparison of more than half a dozen Greek manuscript copies dating mostly from the tenth to the twelfth century. Later accretions, generally appearing as scholia, add supplementary information, often of an historical nature, and in most cases they are readily distinguished from the original. Copies of the *Elements* have come down to us also through Arabic translations, later turned into Latin in the twelfth century, and finally, in the sixteenth century, into the vernacular. The first printed version of the *Elements* appeared at Venice in 1482, one of the very earliest of mathematical books to be set in type; it has been estimated that since then at least a thousand editions have been published. Perhaps no book other than the Bible can boast so many editions, and certainly no mathematical work has had an influence comparable with that of Euclid's *Elements*. How appropriate it was that Euclid's successors referred to him as "The Elementator!"

BIBLIOGRAPHY

Archibald, R. C., ed., *Euclid's Book on Divisions of Figures* (Cambridge: Cambridge University Press, 1915).

Cohen, M. R., and I. E. Drabkin, *A Source Book in Greek Science* (New York: McGraw-Hill, 1948; reprinted Cambridge, Mass.: Harvard University Press, 1958).

Frankland, W. B., *The First Book of Euclid's Elements, with a Commentary Based Principally upon that of Proclus Diadochus* (Cambridge: Cambridge University Press, 1905).

Heath, T. L., *History of Greek Mathematics* (Oxford: Clarendon, 1921, 2 vols.).

Heath, T. L., ed. *The Thirteen Books of Euclid's Elements* (Cambridge, 1908, 3 vols.; paperback ed., New York: Dover, 1956).

Hultsch, F. O., "Eukleides," in Pauly-Wissowa, *Real-Enzyclopädie der klassischen Altertumswissenschaft* (Stuttgart, 1909), Vol. VI, columns 1003–1052.

Loria, Gino, *Storia delle matematiche* (Turin: Sten, 1929–1933, 3 vols.).

Sarton, George, *Ancient Science and Modern Civilization* (Lincoln, Nebr.: University of Nebraska Press, 1954).

Szabó, Arpád, "Anfänge des euklidischen Axiomensystems," *Archive for History of Exact Sciences*, **1** (1960), 37–106.

Thomas, Ivor, ed., *Selections Illustrating the History of Greek Mathematics* (Cambridge, Mass.: Loeb Classical Library, 1939–1941, 2 vols.)

Thomas-Stanford, Charles, *Early Editions of Euclid's Elements* (London: Bibliographical Society, 1926).

Vogt, Heinrich, "Die Lebenzeit Euklids," *Bibliotheca Mathematica* (3), **13** (1913), 193–202.

EXERCISES

1. Describe the sources Euclid probably used in writing the *Elements*; justify your presumptions.
2. Which of the thirteen books of the *Elements* do you regard as the most important? Justify your answer.
3. Which of the thirteen books of the *Elements* do you regard as most dispensable? Justify your answer.
4. Given line segments a and b, construct, with straightedge and compasses alone, segments x and y satisfying the conditions $x + y = a$, $xy = b^2$.
5. Given line segments a and b, construct, with straightedge and compasses alone, a solution x of the equation $x^2 = ax + b^2$.
6. Use the Euclidean algorithm to find the greatest common divisor of 456 and 759.
7. Use the Euclidean algorithm to find the greatest common divisor of 567 and 839 and 432.
8. What is the greatest common measure of two line segments of lengths $\frac{2}{3}$ and $\frac{4}{7}$ respectively? Of two line segments of lengths a/b and c/d respectively, where a, b, c, d are relatively prime integers?
9. Given two unequal line segments a and b, prove that if a line segment c is obtained through the Euclidean algorithm, this is the greatest common measure of a and b.
10. Provide all the details of the proof of the Pythagorean theorem by the "windmill" method.
11. All even perfect numbers end in 6 or in 28 and, upon casting out nines, leave a remainder of 1 (except in the case of the first perfect number). Verify these statements for the first four perfect numbers.
12. Show how to construct a tangent to a circle from a point outside the circle.
13. Justify Euclid's formula for the sum of terms in a geometric progression.
14. The number $2^{13} - 1$ is a prime. Use this fact to find the fifth perfect number in order of magnitude.

15. Prove that there cannot be a regular convex solid other than those given by Euclid.
16. Prove the law of cosines for an acute-angled triangle, indicating just how far Euclid could go in expressing this relationship.
*17. In *Elements* IX. 14 it is proved that a number can be resolved into prime factors in only one way. Write out a proof of this proposition.
*18. Prove Euclid's formula for perfect numbers.
*19. Prove, by the method of exhaustion, that the volumes of spheres are to each other as cubes on their diameter (*Elements* XII. 18).
*20. Prove that if a pentagon, a hexagon, and a decagon are inscribed in the same circle, a triangle made up of a side of the pentagon and a side of the hexagon and a side of the decagon is a right triangle (*Elements* XIII. 10).
*21. Euclid's *Division of Figures* includes a construction of a line parallel to the bases of a trapezoid and dividing the trapezoid into two equal areas. Show how to carry out such a construction with straightedge and compasses alone.

CHAPTER VIII

Archimedes of Syracuse

> There was more imagination in the head of Archimedes
> than in that of Homer.
>
> *Voltaire*

1 Throughout the Hellenistic Age the center of mathematical activity remained at Alexandria, but the leading mathematician of that age—and of all antiquity—was not a native of the city. Archimedes may have studied for a while at Alexandria under the students of Euclid, and he maintained communication with mathematicians there, but he lived and died at Syracuse. Details of his life are scarce, but we have some information about him from Plutarch's account of the life of Marcellus, the Roman general. During the Second Punic War the city of Syracuse was caught in the power struggle between Rome and Carthage; having cast its lot with the latter, the city was besieged by the Romans during the years 214 to 212 B.C. We are told that throughout the siege Archimedes invented ingenious war machines to keep the enemy at bay—catapults to hurl stones; ropes, pulleys, and hooks to raise and smash the Roman ships; devices to set fire to the ships. Ultimately, however, Syracuse fell through a "fifth column"; in the sack of the city Archimedes was slain by a Roman soldier, despite orders from Marcellus that the life of the geometer be spared. Inasmuch as Archimedes at the time is reported to have been seventy-five years old, he was most likely born in 287 B.C. His father was an astronomer, and Archimedes also established a reputation in astronomy. Marcellus is said to have reserved for himself, as booty, ingenious planetaria that Archimedes had constructed to portray the motions of the heavenly bodies. Accounts of the life of Archimedes are in agreement, however, in depicting him as placing little value in his mechanical contrivances as compared with the products of his thought. Even when dealing with levers and other simple machines, he was far more concerned with general principles than with practical applications.

2 Archimedes was not, of course, the first to use the lever, nor even the first to formulate the general law. Aristotelian works contain the statement that

134

Mosaic representation of the death of Archimedes. Once thought to have been from the floor of a room in Pompeii, it is now believed to be a sixteenth-century copy (or falsification). (Municipal Art Institute, Frankfurt am Main.)

two weights on a lever balance when they are inversely proportional to their distances from the fulcrum; and the Peripatetics associated the law with their assumption that vertical rectilinear motion is the only natural terrestrial motion. They pointed out that the extremities of unequal arms of a lever will, in their displacement about the fulcrum, trace out circles rather than straight lines; the extremity of the longer arm will move in the circle that is larger, hence the path will approach more nearly to the natural vertical rectilinear motion than will the extremity of the shorter arm. Therefore the law of the lever is a natural consequence of this kinematic principle. Archimedes, on the other hand, deduced the law from a more plausible static postulate—that bilaterally symmetric bodies are in equilibrium. That is, let one assume that a weightless bar four units long and supporting three unit weights, one at either end and one in the middle (Fig. 8.1), is balanced by a

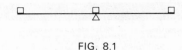

FIG. 8.1

fulcrum at the center. By the Archimedean axiom of symmetry the system is in equilibrium. But the principle of symmetry shows also that, considering only the right-hand half of the system, the balancing effect will remain the same if the two weights two units apart are brought together at the midpoint of the right-hand side. This means that a unit weight two units from the fulcrum will support on the other arm a weight of two units which is one unit from the fulcrum. Through a generalization of this procedure Archimedes established the law of the lever on static principles alone, without recourse to the Aristotelian kinematic argument. In the history of science during the Medieval period it will be found that a conjunction of static and kinematic views produced advances in both science and mathematics.

The work of Archimedes on the law of the lever is part of his treatise, in two books, *On the Equilibrium of Planes*. This is not the oldest extant book on what may be called physical science, for Aristotle about a century earlier had published an influential work, in eight books, entitled *Physics*; but whereas the Aristotelian approach was speculative and nonmathematical, the Archimedean development was similar to the geometry of Euclid. From a set of simple postulates Archimedes deduced some very abstruse conclusions, establishing the close relationship between mathematics and mechanics that was to become so significant for both physics and mathematics.[1] The first book in the *Equilibrium of Planes* is concerned with rectilinear figures and closes with the centers of gravity of the triangle and the trapezoid. Book II concentrates attention on the center of gravity of a parabolic segment and includes a proof of the fact that this center lies on the diameter of the segment and divides this diameter into segments in the ratio of 3 to 2. The procedure used is the familiar method of exhaustion, but a student familiar with the calculus and the principle of moments (or law of the lever) can easily verify the result.

3   Archimedes can well be called the father of mathematical physics, not only for his *On the Equilibrium of Planes*, but also for another treatise, in two books, *On Floating Bodies*. Again, beginning from a simple postulate about the nature of fluid pressure, he obtains some very deep results. Among the earlier propositions are two that formulate the well-known Archimedean hydrostatic principle:

[1] See E. J. Dijksterhuis, *Archimedes* (1957), pp. 286 ff., where attention is called to differences of opinion concerning the rigor of Archimedes' proofs.

Any solid lighter than a fluid will, if placed in a fluid, be so far immersed that the weight of the solid will be equal to the weight of the fluid displaced (I. 5).

A solid heavier than a fluid will, if placed in it, descend to the bottom of the fluid, and the solid will, when weighed in the fluid, be lighter than its true weight by the weight of the fluid displaced (I. 7).[2]

The mathematical derivation of this principle of buoyancy is undoubtedly the discovery that led the absentminded Archimedes to jump from his bath and run home naked, shouting "Eureka" ("I have found it"). It is also possible, although less likely, that the principle aided him in checking on the honesty of a goldsmith suspected of fraudulently substituting some silver for gold in a crown (or more likely a wreath) made for King Hiero of Syracuse, a friend (if not a relative) of Archimedes. Such fraud could easily have been detected by the simpler method of comparing the densities of gold, silver, and the crown by the simple device of measuring displacements of water when equal weights of each are in turn immersed in a vessel full of water. The later Roman architect, Vitruvius, attributed this method to Archimedes, whereas an anonymous Latin poetic account, *De ponderibus et mensuris*, written probably about A.D. 500, has Archimedes use the principle of buoyancy.

The Archimedean treatise *On Floating Bodies* contains much more than the simple fluid properties so far described. Virtually the whole of Book II, for example, is concerned with the position of equilibrium of segments of paraboloids when placed in fluids, showing that the position of rest depends on the relative specific gravities of the solid paraboloid and the fluid in which it floats. Typical of these is Proposition 4:

Given a right segment of a paraboloid of revolution whose axis a is greater than $\frac{3}{4}p$ (where p is the parameter), and whose specific gravity is less than that of a fluid but bears to it a ratio not less than $(a - \frac{3}{4}p)^2 : a^2$, if the segment of the paraboloid be placed in the fluid with its axis at any inclination to the vertical, but so that its base does not touch the surface of the fluid, it will not remain in that position but will return to the position in which its axis is vertical.

Still more complicated cases, with long proofs, follow. Archimedes could well have taught a theoretical course in naval architecture, although he probably would have preferred a graduate course in pure mathematics. No armchair scholar, he came to the rescue in mechanical emergencies. At one time, so it was reported, a ship had been built for King Hiero that was too heavy to be launched, but Archimedes, by a combination of levers and pulleys, accomplished the task. He is supposed to have boasted that if he were given a lever long enough, and a fulcrum on which to rest it, he could move the earth. It was probably at Alexandria that Archimedes became interested

---

[2] Translations in this chapter are taken from *The Works of Archimedes*, edited by T. L. Heath (1897).

in the technical problem of raising water from the Nile River to irrigate the arable portions of the valley; for this purpose he invented a device, now known as the Archimedean screw, made up of helical pipes or tubes fastened to an inclined axle with a handle by which it was rotated.

4   A clear distinction was made in Greek antiquity not only between theory and application, but also between routine mechanical computation and the theoretical study of the properties of number. The former, for which Greek scholars are said to have shown despite, was given the name logistic, while arithmetic, an honorable philosophical pursuit, was understood to be concerned solely with the latter. It has even been maintained that the classical attitude toward routine calculation mirrored the social structure of antiquity in which computations were relegated to slaves. Whatever truth there is in this view seems to have been exaggerated, for the Greeks took the trouble to replace their older Attic or Herodianic system of numeration by one distinctly more advantageous—the Ionian or alphabetic. Archimedes lived at about the time that the transition from Attic to Ionian numeration was made effective,[3] and this may account for the fact that he stooped to make a contribution to logistic. In a work entitled the *Psammites* ("Sand-Reckoner") Archimedes boasted that he could write down a number greater than the number of grains of sand required to fill the universe. In doing so he referred to one of the boldest astronomical speculations of antiquity—that in which Aristarchus of Samos, toward the middle of the third century B.C., proposed putting the earth in motion about the sun. Such an astronomical system would suggest that the relative positions of the fixed stars should change as the earth is displaced by many millions of miles while going around the sun. The absence of such parallactic displacement was the factor that led the greatest astronomers of antiquity (including, presumably, also Archimedes) to reject the heliocentric hypothesis; but Aristarchus asserted that the lack of parallax can be attributed to the enormity of the distance of the fixed stars from the earth. Now, to make good his boast, Archimedes had perforce to provide against all possible dimensions for the universe, and so he showed that he could enumerate the grains of sand needed to fill even Aristarchus' immense world. Archimedes began with certain estimates that had been made in his day concerning the sizes of the earth, the moon, and the sun and the distances of the moon, the sun, and the stars. An estimate of the earth's circumference in his day, he reported, had been given as 300,000 stades (about 30,000 miles, for the stade generally used was roughly a tenth of a mile); Archimedes allowed for an underestimate and assumed a circumference of 3,000,000 stades. Moreover, Aristarchus had estimated the diameter

[3] However, O. Neugebauer, in *Exact Sciences in Antiquity*, 2nd ed. (Providence, R.I.: Brown University Press, 1957), p. 11, believes that the alphabetic system was in use several centuries before the time of Archimedes.

of the sun as eighteen to twenty times that of the moon, which in turn is smaller than the earth. To play safe Archimedes took the diameter of the sun to be not more than thirty times that of the moon (or, a fortiori, of the earth). Next Archimedes assumed that the apparent size of the sun was greater than a thousandth part of a circle, for Aristarchus had estimated it to be about half a degree, a result confirmed by observation. Knowing an upper bound for the sun's actual size and a lower bound for its apparent size, an upper bound for its distance is easily established. Finally, Archimedes interpreted Aristarchus' universe to have a radius that is to the sun's distance as this distance is to the earth's radius.[4] From these assumptions Archimedes shows that the diameter of the ordinary universe as far as the sun is less than 10^{10} stades. Next he had to estimate the size of a grain of sand; remaining on the safe side, he assumed that 10,000 grains of sand are not smaller than a poppy seed, that the diameter of a poppy seed is not less than one fortieth of a finger breadth, and that a stadium in turn is less than 10,000 finger breadths. Putting together all these inequalities, Archimedes concluded that the number of grains of sand required to fill the sphere of the then generally accepted universe is less than a number that we should write as 10^{51}. For the universe of Aristarchus, which is to the ordinary universe as the latter is to the earth, Archimedes showed that not more than 10^{63} grains of sand are required. Archimedes did not use this notation, but instead described the number as ten million units of the eighth order of numbers (where the numbers of second order begin with a myriad-myriads and the numbers of eighth order begin with the seventh power of a myriad-myriads). To show that he could express numbers ever so much larger even than this, Archimedes extended his terminology to call all numbers of order less than a myriad-myriads those of the first period, the second period consequently beginning with the number $(10^8)^{10^8}$, one that would contain 800,000,000 ciphers. The periods of course continue through the 10^8th period. That is, his system would go up to a myriad-myriad units of the myriad-myriadth order of the myriad-myriadth period—a number that would be written as one followed by some eighty thousand million millions of ciphers. It was in connection with this work on huge numbers that Archimedes mentioned, all too incidentally, a principle that later led to the invention of logarithms—the addition of "orders" of numbers (the equivalent of their exponents when the base is 100,000,000) corresponds to finding the product of the numbers.

In his approximate evaluation of the ratio of the circumference to diameter **5** for a circle Archimedes again showed his skill in computation. Beginning

[4] The language in the *Psammites* is not clear at this point, but the interpretation adopted here seems to be appropriate. Erika and Rudolf von Erhardt, "Archimedes' Sand-Reckoner," *Isis*, **33** (1942), 578–602, question the authenticity of the *Psammites*, but this is defended by O. Neugebauer, "Archimedes and Aristarchus," *Isis*, **34** (1942), 4–6.

with the inscribed regular hexagon, he computed the perimeters of polygons obtained by successively doubling the number of sides until one reached ninety-six sides. His iterative procedure for these polygons was related to what is sometimes called the Archimedean algorithm. One sets out the sequence $P_n, p_n, P_{2n}, p_{2n}, P_{4n}, p_{4n} \ldots$, where P_n and p_n are the perimeters of the circumscribed and inscribed regular polygons of n sides. Beginning with the third term, one calculates any term from the two preceding terms by taking alternately their harmonic and geometric means. That is, $P_{2n} = 2p_nP_n/(p_n + P_n)$, $p_{2n} = \sqrt{p_nP_{2n}}$, etc. If one prefers, one can use instead the sequence $a_n, A_n, a_{2n}, A_{2n}, \ldots$, where a_n and A_n are the areas of the inscribed and circumscribed regular polygons of n sides. The third and successive terms are calculated by taking alternately the geometric and harmonic means, so that $a_{2n} = \sqrt{a_nA_n}$, $A_{2n} = 2A_na_{2n}/(A_n + a_{2n})$, etc. His method for computing square roots, in finding the perimeter of the circumscribed hexagon, and for the geometric means was similar to that used by the Babylonians. The result of the Archimedean computation on the circle was an approximation to the value of π expressed by the inequality $3\frac{10}{71} < \pi < 3\frac{10}{70}$, a better estimate than those of the Egyptians and the Babylonians. (It should be borne in mind that neither Archimedes nor any other Greek mathematician ever used our notation π for the ratio of circumference to diameter in a circle.) This result was given in Proposition 3 of the treatise *On the Measurement of the Circle*, one of the most popular of the Archimedean works during the Medieval period. This little work, probably incomplete as it has come down to us, includes only three propositions, of which one is the proof, by the method of exhaustion, that the area of the circle is the same as that of a right triangle having the circumference of the circle as one side and the radius of the circle as the other. It is unlikely that Archimedes was the discoverer of this theorem, for it is presupposed in the quadrature of the circle attributed to Dinostratus.

6   Archimedes, like his predecessors, was attracted by the three famous problems of geometry, and the well-known Archimedean spiral provided solutions to two of these (but not, of course, with straightedge and compasses alone). The spiral is defined as the plane locus of a point which, starting from the end point of a ray or half line, moves uniformly along this ray while the ray in turn rotates uniformly about its end point. In polar coordinates the equation of the spiral is $r = a\theta$. Given such a spiral, the trisection of an angle is easily accomplished. The angle is so placed that the vertex and initial side of the angle coincide with the initial point O of the spiral and the initial position OA of the rotating line. Segment OP, where P is the intersection of the terminal side of the angle with the spiral, is then trisected at points R and S (Fig. 8.2), and circles are drawn with O as center

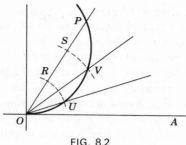

FIG. 8.2

and *OR* and *OS* as radii. If these circles intersect the spiral in points *U* and *V*, lines *OU* and *OV* will trisect the angle *AOP*.

Greek mathematics sometimes has been described as essentially static, with little regard for the notion of variability; but Archimedes, in his study of the spiral, seems to have found the tangent to a curve through kinematic considerations akin to the differential calculus. Thinking of a point on the spiral $r = a\theta$ as subjected to a double motion—a uniform radial motion away from the origin of coordinates and a circular motion about the origin—he seems to have found (through the parallelogram of velocities) the direction of motion (hence of the tangent to the curve) by noting the resultant of the two component motions. This seems to be the first instance in which a tangent was found to a curve other than a circle.

Archimedes' study of the spiral, a curve that he ascribed to his friend Conon of Alexandria, was part of the Greek search for solutions of the three famous problems. The curve lends itself so readily to angle multisections that it may well have been devised by Conon for this purpose. As in the case of the quadratrix, however, it can serve also to square the circle, as Archimedes showed. At point *P* let the tangent to the spiral *OPR* be drawn and let this tangent intersect in point *Q* the line through *O* that is perpendicular to *OP*. Then, Archimedes showed, the straight-line segment *OQ* (known as the polar subtangent for point *P*) is equal in length to the circular arc *PS* of the circle with center *O* and radius *OP* (Fig. 8.3) that is intercepted between

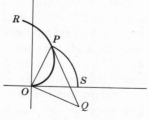

FIG. 8.3

the initial line (polar axis) and line OP (radius vector). This theorem, proved by Archimedes through a typical double *reductio ad absurdum* demonstration, can be verified by a student of the calculus who recalls that tan $\psi = r/r'$, where $r = f(\theta)$ is the polar equation of a curve, r' is the derivative of r with respect to θ, and ψ is the angle between the radius vector at a point P and the tangent line to the curve at the point P. A large part of the work of Archimedes is such that it would now be included in a calculus course, which is particularly true of the work *On Spirals*. If point P on the spiral is chosen as the inter-section of the spiral with the 90° line in polar coordinates, the polar sub-tangent OQ will be precisely equal to quarter of the circumference of the circle of radius OP. Hence the entire circumference is easily constructed as four times the segment OQ, and by Archimedes' theorem a triangle equal in area to the area of the circle is found. A simple geometrical transformation will then produce a square in place of the triangle, and the quadrature of the circle is effected.

Among the twenty-eight propositions in *On Spirals* are several concerning areas associated with the spiral. For example, it is shown in Proposition 24 that the area swept out by the radius vector in its first complete rotation is one third of the area of the "first circle"—that is, the circle with center at the pole and radius equal to the length of the radius vector following the first complete rotation. Archimedes used the method of exhaustion, but again a student today can easily verify the result if he recalls that this area is $\frac{1}{2}\int_0^{2\pi} r^2 \, d\theta$. Moreover, it can readily be shown by the calculus, as Archimedes did by the more difficult method of exhaustion, that on the next rotation the area of the additional ring R_2 (bounded by the first and second turns of the spiral and the portion of the polar axis between the two intercepts following the first and second rotations) is six times the region R_1 swept out in the first rotation. Areas of the *additional* rings added on successive rotations are given by the simple rule of succession $R_{n+1} = nR_n/(n - 1)$, as Archimedes showed.

7   The work *On Spirals* was much admired but little read, for it was generally regarded as the most difficult of all Archimedean works. Of the treatises concerned chiefly with the method of exhaustion (that is, the integral calculus), the most popular was *Quadrature of the Parabola*. The conic sections had been known for almost a century when Archimedes wrote, yet no progress had been made in finding their areas. It took the greatest mathematician of antiquity to square a conic section—a segment of the parabola—which he accomplished in Proposition 17 of the work in which the quadrature was the goal. The proof by the standard method of exhaustion is long and in-volved, but Archimedes rigorously proved that the area K of a parabolic segment $APBQC$ (Fig. 8.4) is four-thirds the area of a triangle T having the

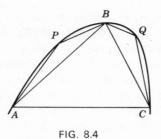

FIG. 8.4

same base and equal height. In the succeeding (and last) seven propositions Archimedes gave a second but different proof of the same theorem. He first showed that the area of the largest inscribed triangle, ABC, on the base AC is four times the sum of the corresponding inscribed triangles on each of the lines AB and BC as base. By continuing the process suggested by this relationship, it becomes clear that the area K of the parabolic segment ABC is given by the sum of the infinite series $T + T/4 + T/4^2 + \cdots + T/4^n + \cdots$, which, of course, is $\frac{4}{3}T$. Archimedes did not refer to the sum of the infinite series, for infinite processes were frowned on in his day; instead he proved by a double *reductio ad absurdum* that K can be neither more nor less than $\frac{4}{3}T$. (Archimedes, like his predecessors, did not use the name "parabola," but the word "orthotome," or "section of a right cone.")

In the preamble to the *Quadrature of the Parabola* we find the assumption or lemma that is usually known today as the axiom of Archimedes: "That the excess by which the greater of two unequal areas exceeds the less can, by being added to itself, be made to exceed any given finite area." This axiom in effect rules out the fixed infinitesimal or indivisible that had been much discussed in Plato's day. It is essentially the same as the axiom of exhaustion, and Archimedes freely admitted that

> The earlier geometers have also used this lemma, for it is by the use of this same lemma that they have shown that circles are to one another in the duplicate ratio of their diameters, and that spheres are to one another in the triplicate ratio of their diameters, and further that every pyramid is one third part of the prism which has the same base with the pyramid and equal height; also, that every cone is one third part of the cylinder having the same base as the cone and equal height they proved by assuming a certain lemma similar to that aforesaid.

The "earlier geometers" mentioned here presumably included Eudoxus and his successors.

Archimedes apparently was unable to find the area of a general segment **8** of an ellipse or hyperbola. Finding the area of a parabolic segment by modern integration involves nothing worse than polynomials, but the integrals arising in the quadrature of a segment of an ellipse or hyperbola (as well as the arcs of these curves or the parabola) require transcendental functions. Nevertheless, in his important treatise *On Conoids and Spheroids* Archimedes found the area of the *entire* ellipse: "The areas of ellipses are as the rectangles under their axes" (Proposition 6). This is, of course, the same as saying that the area of $x^2/a^2 + y^2/b^2 = 1$ is πab or that the area of an ellipse is the same as the area of a circle whose radius is the geometric mean of the semiaxes of the ellipse. Moreover, in the same treatise Archimedes showed how to find the volumes of segments cut from an ellipsoid or a paraboloid or a hyperboloid (of two sheets) of revolution about the principal axis. The process that he used is so nearly the same as that in modern integration that we shall describe it for one case. Let ABC be a paraboloidal segment (or paraboloidal "conoid") and let its axis be CD (Fig. 8.5); about the solid circumscribe the

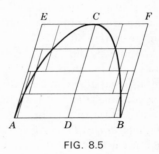

FIG. 8.5

circular cylinder $ABEF$, also having CD as axis. Divide the axis into n equal parts of length h, and through the points of division pass planes parallel to the base. On the circular sections that are cut from the paraboloid by these planes construct inscribed and circumscribed cylindrical fustra, as shown in the figure. It is then easy to establish, through the equation of the parabola and the sum of an arithmetic progression, the following proportions and inequalities:

$$\frac{\text{cylinder } ABEF}{\text{inscribed figure}} = \frac{n^2 h}{h + 2h + 3h + \cdots + (n-1)h} > \frac{n^2 h}{\frac{1}{2}n^2 h}$$

$$\frac{\text{cylinder } ABEF}{\text{circumscribed figure}} = \frac{n^2 h}{h + 2h + 3h + \cdots + nh} < \frac{n^2 h}{\frac{1}{2}n^2 h}$$

Archimedes had previously shown that the difference in volume between the circumscribed and inscribed figures was equal to the volume of the lowest slice of the circumscribed cylinder; by increasing the number n of subdivisions on the axis, thereby making each slice thinner, the difference between the circumscribed and inscribed figures can be made less than any preassigned magnitude. Hence the inequalities lead to the necessary conclusion that the volume of the cylinder is twice the volume of the conoidal segment. This work differs from the modern procedure in integral calculus chiefly in the lack of the concept of limit of a function—a concept that was so near at hand and yet was never formulated by the ancients, not even by Archimedes, the man who came closest to achieving it.

Archimedes composed many marvelous treatises, of which his successors **9** were inclined to admire most the one *On Spirals*. The author himself seems to have been partial to another, *On the Sphere and Cylinder*. Archimedes requested that on his tomb be carved a representation of a sphere inscribed in a right circular cylinder the height of which is equal to its diameter, for he had discovered, and proved, that the ratio of the volumes of cylinder and sphere is the same as the ratio of the areas—that is, three to two. This property, which Archimedes discovered subsequent to his *Quadrature of the Parabola*, remained unknown, he says, to geometers before him. It once had been thought[5] that the Egyptians knew how to find the area of a hemisphere; but Archimedes appears now as the first one to have known, and proved, that the area of a sphere is just four times the area of a great circle of the sphere. Moreover, Archimedes showed that "the surface of any segment of a sphere is equal to a circle whose radius is equal to the straight line drawn from the vertex of the segment to the circumference of the circle which is the base of the segment." This, of course, is equivalent to the more familiar statement that the surface area of any segment of a sphere is equal to that of the curved surface of a cylinder whose radius is the same as that of the sphere and whose height is the same as that of the segment. That is, the surface area of the segment does not depend on the distance from the center of the sphere, but only on the altitude (or thickness) of the segment. The crucial theorem on the surface of the sphere appears in Proposition 33, following a long series of preliminary theorems, including one that is equivalent to an integration of the sine function:

If a polygon be inscribed in a segment of a circle *LAL′* so that all its sides excluding the base are equal and their number even, as *LK* ... *A* ... *K′L′*, *A* being the middle point of the segment; and if the lines *BB′*, *CC′*, ... parallel to the base *LL′* and joining pairs of angular points be drawn, then (*BB′* + *CC′* + ··· + *LM*):*AM* = *A′B*:*BA*, where *M* is the middle point of *LL′* and *AA′* is the diameter through M (Fig. 8.6).

[5] See, E. G. Archibald, *Outline of the History of Mathematics*, 6th ed., (The American Mathematical Monthly, Slaught Memorial Papers, No. 2, January, 1949), pp. 15–16. Cf. footnotes 10 and 11 of Chapter 2.

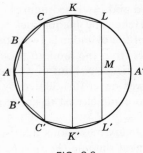

FIG. 8.6

This is the geometrical equivalent of the trigonometric equation

$$\sin\frac{\theta}{n} + \sin\frac{2\theta}{n} + \cdots + \sin\frac{n-1}{n}\theta + \tfrac{1}{2}\sin\frac{n\theta}{n} = \frac{1-\cos\theta}{2}\cot\frac{\theta}{2n}$$

From this theorem it is easy to derive the modern expression $\int_0^\phi \sin x \, dx = 1 - \cos\phi$ by multiplying both sides of the equation above by θ/n and taking limits as n increases indefinitely. The left-hand side becomes

$$\lim_{n\to\infty}\sum_{i=1}^{n}\sin x_i \Delta x_i$$

where $x_i = i\theta/n$ for $i = 1, 2, \ldots n$, $\Delta x_i = \theta/n$ for $i = 1, 2, \ldots n - 1$, and $\Delta x_n = \theta/2n$. The right-hand side becomes

$$(1 - \cos\theta)\lim_{n\to\infty}\frac{\theta}{2n}\cot\frac{\theta}{2n} = 1 - \cos\theta.$$

The equivalent of the special case $\int_0^\pi \sin x \, dx = 1 - \cos\pi = 2$ had been given by Archimedes in the preceding proposition.

The familiar formula for the volume of a sphere appears in *On the Sphere and Cylinder* I. 34:

> Any sphere is equal to four times the cone which has its base equal to the greatest circle in the sphere and its height equal to the radius of the sphere.

The theorem is proved by the usual method of exhaustion, and the Archimedean ratio for the volume and surface area of the sphere and circumscribed cylinder followed as an easy corollary. The sphere-in-a-cylinder diagram was indeed carved on the tomb of Archimedes, as we know from a report by Cicero. When he was quaestor in Sicily, the Roman orator found the neglected tomb with the engraving. He restored the tomb—almost the only contribution of a Roman to the history of mathematics—but all traces of it have since vanished.

An interesting light on Greek geometrical algebra is cast by a problem **10** in Book II of *On the Sphere and Cylinder*. In Proposition 2 Archimedes justified his formula for the volume of a segment of a given sphere; in Proposition 3 he showed that to cut a given sphere by a plane so that the *surfaces* of the segments are in a given ratio, one simply passes a plane perpendicular to a diameter through a point on the diameter which divides the diameter into two segments having the desired ratio. He then showed in Proposition 4 how to cut a given sphere so that the *volumes* of the two segments are in a given ratio—a far more difficult problem. In modern notation, Archimedes was led to the equation

$$\frac{4a^2}{x^2} = \frac{(3a - x)(m + n)}{ma},$$

where $m:n$ is the ratio of the segments. This is a cubic equation, and Archimedes attacked its solution as had his predecessors in solving the Delian problem—through intersecting conics. Interestingly, the Greek approach to the cubic was quite different from that to the quadratic equation. By analogy with the "application of areas" in the latter case, we would anticipate an "application of volumes," but this was not adopted. Through substitutions Archimedes reduced his cubic equation to the form $x^2(c - x) = db^2$ and promised to give separately a complete analysis of this cubic with respect to the number of positive roots. This analysis had apparently been lost for many centuries when Eutocius, an important commentator of the early sixth century, found a fragment that seems to contain the authentic Archimedean analysis. The solution was carried out by means of the intersection of the parabola $cx^2 = b^2y$ and the hyperbola $(c - x)y = cd$. Going further, he found a condition on the coefficients that determines the number of real roots satisfying the given requirements—a condition equivalent to finding the discriminant, $27b^2d - 4c^3$, of the cubic equation $b^2d = x^2(c - x)$. (This can easily be verified by the application of a little elementary calculus.) Inasmuch as all cubic equations can be transformed to the Archimedean type, we have here the essence of a complete analysis of the general cubic. Interest in the cubic equation disappeared shortly after Archimedes, to be revived for a while by Eutocius and then centuries later still by the Arabs.

Most of the Archimedean treatises that we have described are a part of **11** advanced mathematics, but the great Syracusan was not above proposing elementary problems. In his *Book of Lemmas*, for example, we find a study of the so-called *arbelos* or "shoemaker's knife." The shoemaker's knife is the region bounded by the three semicircles tangent in pairs in Fig. 8.7, the area in question being that which lies inside the largest semicircle and outside the two smallest. Archimedes showed in Proposition 4 that if *CD* is perpendicular to *AB*, the area of the circle with *CD* as diameter is equal to the area of the arbelos. In the next proposition it is shown that the two circles inscribed within the two regions into which *CD* divides the shoemaker's knife are equal.

The *Book of Lemmas* contains also a theorem (Proposition 14) on what Archimedes called the *salinon* or "salt cellar." Draw semicircles with the segments *AB*, *AD*, *DE*, and *EB* as diameters (Fig. 8.8), with *AD = EB*. Then

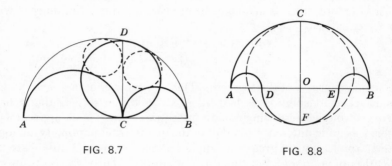

FIG. 8.7 FIG. 8.8

the total area bounded by the salinon (bounded entirely by semicircular arcs) is equal to the area of the circle having as its diameter the line of symmetry of the figure, *FOC*.

It is in the *Book of Lemmas* that we find also (as Proposition 8) the well-known Archimedean trisection of the angle. Let *ABC* be the angle to be trisected (Fig. 8.9). Then with *B* as center, draw a circle of any radius intersecting *AB* in *P* and *BC* in *Q*, with *BC* extended in *R*. Then draw a line *STP*

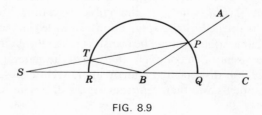

FIG. 8.9

such that *S* lies on *CQBR* extended and *T* lies on the circle and such that *ST = BQ = BP = BT*. It is then readily shown, since triangles *STB* and *TBP* are isosceles, that angle *BST* is precisely a third of angle *QBP*, the angle that was to have been trisected. Archimedes and his contemporaries were of course aware that this is not a canonical trisection in the Platonic sense, for it involves what they called a *neusis*—that is, an insertion of a given length, in this case $ST = BQ$, between two figures, here the line QR extended, and the circle.

The *Book of Lemmas* has not survived in the original Greek, but through Arabic translation that later was turned into Latin. (Hence it often is cited by its Latin title of *Liber assumptorum*.) In fact, the work as it has come down to us cannot be genuinely Archimedean, for his name is quoted several times within the text. However, even if the treatise is nothing more than a collection of miscellaneous theorems that were attributed by the Arabs to Archimedes, the work probably is substantially authentic. There is doubt also about the authenticity of the *Cattle-problem*, which is generally thought to be Archimedean, and certainly dates back to within a few decades of his death. The *Cattle-problem* is a challenge to mathematicians to solve a set of indeterminate simultaneous equations in eight unknown quantities—the number of bulls and cows of each of four colors. There is some ambiguity in the formulation of the problem, but according to one interpretation it would take a volume of more than 600 pages to give the values for the eight unknowns contained in one of the possible solutions! The problem, which involves the solution of $x^2 = 1 + 4729494y^2$, incidentally provides a first example of what later (see below) was to be known as a "Pell equation."

It is certain that not all of the works of Archimedes have survived, for **12** in a later commentary we learn (from Pappus) that Archimedes discovered all of the thirteen possible so-called semiregular solids. Whereas a regular solid or polyhedron has faces that are regular polygons of the same type, a semiregular solid is a convex polyhedron whose faces are regular polygons, but not all of the same type. For example, if from the eight corners of a cube a we cut off tetrahedra with edges $a(2 - \sqrt{2})/2$, the resulting figure will be a semiregular or Archimedean solid with surfaces made up of eight equilateral triangles and six regular hexagons.

That quite a number of Archimedean works have been lost is clear from many references. Arabic scholars inform us that the familiar area formula for a triangle in terms of its three sides, usually known as Heron's formula—$K = \sqrt{s(s - a)(s - b)(s - c)}$, where s is the semiperimeter—was known to Archimedes several centuries before Heron lived. Arabic scholars also attribute to Archimedes the "theorem on the broken chord"—if AB and BC make up any broken chord in a circle (with $AB \neq BC$) and if M is the midpoint of the arc ABC and F the foot of the perpendicular from M to the longer chord, F will be the midpoint of the broken chord ABC (Fig. 8.10). Archimedes is reported by the Arabs to have given several proofs of the theorem,

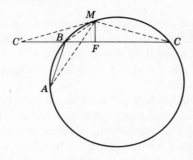

FIG. 8.10

one of which is carried out by drawing in the dotted lines in the figure, making $FC' = FC$, and proving that $\triangle MBC' \simeq \triangle MBA$. Hence $BC' = BA$, and it therefore follows that $C'F = AB + BF = FC$. We do not know whether Archimedes saw any trigonometric significance in the theorem, but it has been suggested[6] that it served for him as a formula analogous to our $\sin(x - y) = \sin x \cos y - \cos x \sin y$. To show the equivalence we let $\widehat{MC} = 2x$ and $\widehat{BM} = 2y$. Then $\widehat{AB} = 2x - 2y$. Now, the chords corresponding to these three arcs are respectively $MC = 2 \sin x$, $BM = 2 \sin y$, and $AB = 2 \sin (x - y)$. Moreover, the projections of MC and MB on BC are $FC = 2 \sin x \cos y$ and $FB = 2 \sin y \cos x$. If, finally, we write the broken-chord theorem in the form $AB = FC - FB$, and if for these three chords we substitute their trigonometric equivalents, the formula for $\sin(x - y)$ results. Other trigonometric identities can, of course, be derived from the same broken-chord theorem, indicating that Archimedes may have found it a useful tool in his astronomical calculations.

13   Unlike the *Elements* of Euclid, which have survived in many Greek and Arabic manuscripts, the treatises of Archimedes have reached us through a slender thread. Almost all copies are from a single Greek original which was in existence in the early sixteenth century and itself was copied from an original of about the ninth or tenth century. The *Elements* of Euclid has been familiar to mathematicians virtually without interruption since its composition, but Archimedean treatises have had a more checkered career. There have been times when few or even none of Archimedes' works were known. In the days of Eutocius, a first-rate scholar and skillful commentator of the sixth century, only three of the many Archimedean works were generally known—*On the Equilibrium of Planes*, the incomplete *Measurement of a*

---

[6] See Johannes Tropfke, "Archimedes und die Trigonometrie," *Archiv für die Geschichte der Mathematik,* **10** (1927–1928), 432–463.

Circle, and the admirable *On the Sphere and Cylinder*. Under the circumstances it is a wonder that such a large proportion of what Archimedes wrote has survived to this day. Among the amazing aspects of the provenance of Archimedean works is the discovery within the twentieth century of one of the most important treatises—one which Archimedes simply called *The Method* and which had been lost since the early centuries of our era until its rediscovery in 1906.

The Method of Archimedes is of particular significance because it discloses for us a facet of Archimedes' thought that is not found elsewhere. His other treatises are gems of logical precision, with little hint of the preliminary analysis that may have led to the definitive formulations. So thoroughly without motivation did his proofs appear to some writers of the seventeenth century that they suspected Archimedes of having concealed his method of approach in order that his work might be admired the more. How unwarranted such an ungenerous estimate of the great Syracusan was became clear in 1906 with the discovery of the manuscript containing *The Method*. Here Archimedes had published, for all the world to read, a description of the preliminary "mechanical" investigations that had led to many of his chief mathematical discoveries. He thought that his "method" in these cases lacked rigor, since it assumed an area, for example, to be a sum of line segments.

The Method, as we have it, contains most of the text of some fifteen propositions sent in the form of a letter to Eratosthenes, mathematician and librarian at the university at Alexandria. The author opened by saying that it is easier to supply a proof of a theorem if we first have some knowledge of what is involved; as an example he cites the proofs of Eudoxus on the cone and pyramid, which had been facilitated by the preliminary assertions, without proof, made by Democritus. Then Archimedes announced that he himself had a "mechanical" approach that paved the way for some of his proofs. The very first theorem that he discovered by this approach was the one on the area of a parabolic segment; in Proposition 1 of *The Method* the author describes how he arrived at this theorem by balancing lines as one balances weights in mechanics. He thought of the areas of the parabolic segment *ABC* and the triangle *AFC* (where *FC* is tangent to the parabola at *C*) as the totality of a set of lines parallel to the diameter *QB* of the parabola, such as *OP* (Fig. 8.11) for the parabola and *OM* for the triangle. If, now, one were to place at *H* (where *HK = KC*) a line segment equal to *OP*, this would just balance the line *OM* where it now is, *K* being the fulcrum. (This can be shown through the law of the lever and the property of the parabola.) Hence the area of the parabola, if placed with its center of gravity at *H*, will just balance the triangle, whose center of gravity is along *KC* and a third of the way from *K* to *C*. From this one easily sees that the area of the parabolic

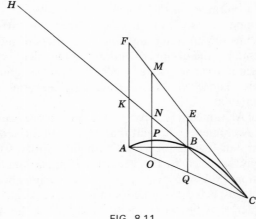

FIG. 8.11

segment is one-third the area of triangle AFC, or four-thirds the area of the inscribed triangle ABC.

14   The favorite theorem of Archimedes, represented on his tomb, was also suggested to him by his mechanical method. It is described in Proposition 2 of *The Method*:

> Any segment of a sphere has to the cone with the same base and height the ratio which the sum of the radius of the sphere and the height of the complementary segment has to the height of the complementary segment.

The theorem follows readily from a beautiful balancing property which Archimedes discovered (and which can be easily verified in terms of modern formulas). Let $AQDCP$ be a cross section of a sphere with center O and diameter AC (Fig. 8.12) and let AUV be a plane section of a right circular cone with axis AC and UV as diameter of the base. Let $IJUV$ be a right circular cylinder with axis AC and with $UV = IJ$ as diameter and let $AH = AC$. If a plane is passed through any point S on the axis AC and perpendicular to AC, the plane will cut the sphere, the cone, and the cylinder in circles of radii $r_1 = SR$, $r_2 = SP$, and $r_3 = SN$ respectively. If we call the areas of these circles A_1, A_2, and A_3, then, Archimedes found, A_1 and A_2, when placed with their centers at H, will just balance A_3 where it now is, with A as the fulcrum. Hence if we call the volumes of the sphere, the cone, and the cylinder V_1, V_2, V_3, it follows that $V_1 + V_2 = \frac{1}{2}V_3$; and since $V_2 = \frac{1}{3}V_3$, the sphere must be $\frac{1}{6}V_3$. Because the volume V_3 of the cylinder is known (from Democritus and Eudoxus), the volume of the sphere also is known—in modern notation, $V = \frac{4}{3}\pi r^3$. By applying the same balancing technique to the spherical segment with base diameter BD, to the cone with base diameter

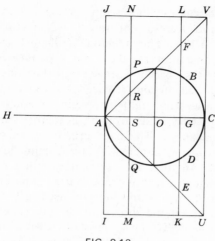

FIG. 8.12

EF, and to the cylinder with base diameter *KL*, the volume of the spherical segment is found in the same manner as for the whole sphere.

The method of equilibrium of circular sections about a vertex as fulcrum **15** was applied by Archimedes to discover the volumes of the segments of three solids of revolution—the ellipsoid, the paraboloid, and the hyperboloid, as well as the centers of gravity of the paraboloid (conoid), of any hemisphere, and of a semicircle. The *Method* closes with the determination of volumes of two solids that are favorites in modern calculus books—a wedge cut from a right circular cylinder by two planes (as in Fig. 8.13) and the volume common

FIG. 8.13

to two equal right circular cylinders intersecting at right angles. The work containing such marvelous results of more than 2000 years ago was recovered almost by accident in 1906. The indefatigable Danish scholar J. L. Heiberg had read that at Constantinople there was a palimpsest of mathematical content. (A palimpsest is a parchment the original writing on which has been only imperfectly washed off and replaced with a new and different text.) Close inspection showed him that the original manuscript had contained something by Archimedes, and through photographs he was able to read most of the Archimedean text. The manuscript consisted of 185 leaves, mostly of parchment but a few of paper, with the Archimedean text copied in a tenth-century hand. An attempt—fortunately, none too successful—had been made to expunge this text in order to use the parchment for a Euchologion (a collection of prayers and liturgies used in the Eastern Orthodox Church) written in about the thirteenth century. The mathematical text contained *On the Sphere and Cylinder*, most of the work *On Spirals*, part of the *Measurement of a Circle* and of *On the Equilibrium of Planes*, and *On Floating Bodies*, all of which have been preserved in other manuscripts; most important of all, the palimpsest gives us the only surviving copy of *The Method*. In a sense the palimpsest is symbolic of the contribution of the Medieval Age. Intense preoccupation with religious concerns very nearly wiped out one of the most important works of the greatest mathematician of antiquity; yet in the end it was Medieval scholarship that inadvertently preserved this, and much besides, which might otherwise have been lost.

BIBLIOGRAPHY

Bromwich, T. J., "The Methods Used by Archimedes for Approximating to Square Roots, *The Mathematical Gazette,* **14** (1928–1929), 253–257.

Clagett, Marshall, *Archimedes in the Middle Ages* (Madison, Wis.: University of Wisconsin Press, 1964–   , 2 vols.).

Cohen, M. R., and I. E. Drabkin, *A Source Book in Greek Science* (New York: McGraw-Hill, 1948; reprinted Cambridge, Mass.: Harvard University Press, 1958).

Davis, H. T., "Archimedes and Mathematics," *School Science and Mathematics,* **44** (1944), 136–145, 213–221.

Dijksterhuis, E. J., *Archimedes* (New York: Humanities Press, 1957).

Erhardt, Erika von, and Rudolf von Erhardt, "Archimedes' Sand-Reckoner," *Isis,* **33** (1942), 578–602.

Heath, T. L., *The Works of Archimedes* (Cambridge, 1897; paperback reprint, including Archimedes' *Method,* New York: Dover, n.d.).

Heiberg, J. L., *Quaestiones archimedae* (Copenhagen, 1879).

Heiberg, J. L., "Le rôle d'Archimède dans le développement des sciences exactes," *Scientia,* **20** (1916), 81–89.

Heiberg, J. L., ed., *Archimedes, Opera omnia* (Leipzig, 1880–1881, 3 vols.).

Heiberg, J. L., and H. G. Zeuthen, "Eine neue Schrift des Archimedes," *Bibliotheca Mathematica* (3), **7** (1906–1907), 321–363.

Hofmann, J. E., "Erklärungsversuche für Archimeds Berechnung von $\sqrt{3}$," *Archiv für die Geschichte der Mathematik*, **12** (1929), 386–408.

Hoppe, Edmund, "Die zweite Methode des Archimedes zur Berechnung von π," *Archiv für die Geschichte der Mathematik*, **9** (1920–1922), 104–107.

Midolo, P., *Archimede e il suo tempo* (Syracuse, 1912).

Neugebauer, O., "Archimedes and Aristarchus," *Isis*, **34** (1942), 4–6.

Smith, D. E., "A Newly Discovered Treatise of Archimedes," *Monist*, **19** (1909), 202–230.

Thomas, Ivor, *Selections Illustrating the History of Greek Mathematics* (Cambridge, Mass.: Leob Classical Library, 1939–1941, 2 vols.).

Tropfke, Johannes, "Archimedes und die Trigonometrie," *Archiv für die Geschichte der Mathematik*, **10** (1927–1928), 432–463.

Weissenborn, Hermann, "Die irrationalen Quadratwurzeln bei Archimedes und Heron," *Berliner Studien für Klassische Philologie und Archaeologie*, **1** (1884), 357–408.

EXERCISES

1. Archimedes is sometimes regarded as the inventor of the integral calculus. To what extent do you agree or disagree with this view?

2. Euclid depended heavily on the works of his predecessors. To what extent is this true also of Archimedes?

3. Aristotle knew the law of the lever before Archimedes was born. Why, then, is the law sometimes attributed to Archimedes? Explain.

4. Of the many treatises by Archimedes with which we are familiar, which do you regard as the most significant for the development of mathematics? Explain.

5. Archimedes generally is regarded as the greatest mathematician of antiquity. Explain fully the justification for such a view, comparing his work with that of at least two potential earlier rivals.

6. If a_i and A_i are respectively the areas of regular polygons of i sides inscribed in and circumscribed about a circle, prove the Archimedean recursion formulas $a_{2n} = \sqrt{a_n A_n}$ and $A_{2n} = 2A_n a_{2n}/(A_n + a_{2n})$.

7. If p_i and P_i are *perimeters* of regular polygons inscribed in and circumscribed about a circle, prove the Archimedean algorithm $P_{2n} = 2P_n p_n/(P_n + p_n)$ and $p_{2n} = \sqrt{p_n P_{2n}}$.

8. Beginning, as did Archimedes, with a regular hexagon inscribed in a circle, use an Archimedean recursion algorithm to find either p_{12} and P_{12} or a_{12} and A_{12}. What value of π would be implied by the arithmetic mean of your answers?

9. Find the area lying between the portions of the spiral $r = a\theta$ formed for $0 \le \theta \le 2\pi$ and for $2\pi \le \theta \le 4\pi$.

10. Show clearly how to divide the surface area of a sphere by two parallel planes into three numerically equal areas.

11. Prove the Archimedean theorem that the area of the "shoemaker's knife" is equal to the area of the circle with CD as diameter (Fig. 8.7).

12. Prove the Archimedean trisection method described in the text.

13. Either construct or draw diagrams of three Archimedean semiregular solids.

*14. Find, for the Archimedean spiral $r = a\theta$, the length of the polar subtangent for $\theta = 2\pi$ and show how this can be used to square the circle.

*15. Prove the Archimedean theorem on the broken chord.

*16. Using either the Archimedean balancing property or modern integration, give a proof of the formula for the volume of a segment of a sphere.

*17. Prove the Archimedean theorem on the salinon.

*18. In the diagram of the Archimedean theorem on the broken chord (Fig. 8.10), use the equation $BF + FC = BC$ to derive the familiar trigonometric identity for $\sin(x + y)$.

*19. Can you, either exactly or approximately, divide the unit sphere by two parallel planes into three segments equal in volume? Explain.

*20. Prove that the two circles inscribed in the two portions into which line CD divides the "shoemaker's knife" (Fig. 8.7) are equal.

CHAPTER IX

Apollonius of Perga

> It seems to me that all the evidence points to
> Apollonius as the founder of Greek mathematical
> astronomy.
>
> *Otto Neugebauer*

During the first century or so of the Hellenistic Age three mathematicians **1**
stood head and shoulders above all others of the time, as well as above most
of their predecessors and successors. These men were Euclid, Archimedes,
and Apollonius; it is their work that leads to the designation of the period
from about 300 to 200 B.C. as the "Golden Age" of Greek mathematics.
In a sense mathematics had lagged behind the arts and literature, for it was
the Age of Pericles, in the middle of the fifth century B.C., that in the broader
sense is known as the "Golden Age of Greece." Throughout the Hellenistic
period the city of Alexandria remained the mathematical focus of the
Western world, but Apollonius, like Archimedes, was not a native there.
He was born at Perga in Pamphilia (southern Asia Minor); but he may have
been educated at Alexandria, and he seems to have spent some time teaching
there at the university. For a while he was at Pergamum, where there was a
university and a library second only to that at Alexandria, through the
patronage of Alexander's general, Lysimachus, and his successors. Inasmuch
as the ancient world had many men named Apollonius (of these 129 with
biographies are listed in Pauly-Wissowa, *Real-Enzyclopädie der klassischen
Altertumswissenschaft*), our mathematician is distinguished from others by
use of the full name, Apollonius of Perga. We do not know the precise dates
of his life, but he is reported to have flourished during the reigns of Ptolemy
Euergetes and Ptolemy Philopater; one report makes him a treasurer-
general of Ptolemy Philadelphus, and it was said that he was twenty-five to
forty years younger than Archimedes. The years 262 to 190 B.C. have been
suggested for his life, about which little is known. He seems to have felt
himself to be a rival of Archimedes; he thus touched on several themes that
we discussed in the preceding chapter. He developed a scheme of "tetrads"
for expressing large numbers, using an equivalent of exponents of the single
myriad, whereas Archimedes had used the double myriad as a base. The

numerical scheme of Apollonius probably was the one of which part is described in the surviving last portion of Book II of the *Mathematical Collection* of Pappus. (All of Book I and the first part of Book II have been lost.) Here the number $5,462,360,064 \times 10^6$ is written as $\mu^{\gamma},\epsilon\upsilon\xi\beta\ \mu^{\beta},\gamma\chi\ \mu,^{\varkappa}\varsigma\upsilon$, where μ^{γ}, μ^{β}, and μ^{α} are the third, the second, and the first powers, respectively, of a myriad.

Apollonius wrote a work (now lost) entitled *Quick Delivery* which seems to have taught speedy methods of calculation. In it the author is said to have calculated a closer approximation to π than that given by Archimedes— probably the value we know as 3.1416. We do not know how this value, which appeared later in Ptolemy and also in India, was arrived at. In fact, there are more unanswered questions about Apollonius and his work than about Euclid or Archimedes, for more of his works have disappeared. We have the titles of many lost works, such as one on *Cutting-off of a Ratio*, another on *Cutting-off of an Area*, one *On Determinate Section*, another on *Tangencies* (or *Contacts*), one on *Vergings* (or *Inclinations*), and one on *Plane Loci*. In some cases we know what the treatise was about, for Pappus later gave brief descriptions of a few. Six of the works of Apollonius were included, together with a couple of Euclid's more advanced treatises (now lost), in a collection known as the "Treasury of Analysis." Pappus described this as a special body of doctrine for those who, after going through the usual elements, wish to obtain power to solve problems involving curves. The "Treasury," made up largely of works by Apollonius, consequently must have included much of what we now call analytic geometry; it was with good reason that Apollonius, rather than Euclid, was known in antiquity as "The Great Geometer."

2   From the descriptions given by Pappus and others, it is possible to obtain a good idea of the contents of some of the lost Greek works, and when in the seventeenth century the game of reconstructing lost geometrical books was at its height, the treatises of Apollonius were among the favorites.[1] From restorations of the *Plane Loci*, for example, we infer that the following were two of the loci considered: (1) The locus of points the difference of the squares of whose distances from two fixed points is constant is a straight line perpendicular to the line joining the points; (2) the locus of points the ratio of whose distances from two fixed points is constant (and not equal to one) is a circle. The latter locus is, in fact, now known as the "Circle of Apollonius," but this is a misnomer since it had been known to Aristotle who had used it to give a mathematical justification of the semicircular form of the rainbow.[2]

---

[1] For an account of these "restorations" see the article on "Apollonius" by T. L. Heath in the *Encyclopaedia Britannica*, 11th ed. (1910).

[2] See C. B. Boyer, *The Rainbow* (New York: Yoseloff, 1959), pp. 45–46.

The *Cutting-off of a Ratio* dealt with the various cases of a general problem—given two straight lines and a point on each, draw through a third given point a straight line that cuts off on the given lines segments (measured from the fixed points on them respectively) that are in a given ratio. This problem is equivalent to solving a quadratic equation of the type $ax - x^2 = bc$, that is, of applying to a line segment a rectangle equal to a rectangle and falling short by a square. In *Cutting-off of an Area* the problem is similar except that the intercepted segments are required to contain a given rectangle, rather than being in a given ratio. This problem leads to a quadratic of the form $ax + x^2 = bc$, so that one has to apply to a segment a a rectangle equal to a rectangle and exceeding by a square. The Apollonian treatise *On Determinate Section* dealt with what might be called an analytic geometry of one dimension. It considered the following general problem, using the typical Greek algebraic analysis in geometric form: Given four points A, B, C, D on a straight line, determine a fifth point P on it such that the rectangle on AP and CP is in a given ratio to the rectangle on BP and DP. Here, too, the problem reduces easily to the solution of a quadratic; and, as in other cases, Apollonius treated the question exhaustively, including the limits of possibility and the number of solutions.

The treatise on *Tangencies* is of a different sort from the three above, for as Pappus describes it we see the problem familiarly known today as the "Problem of Apollonius." Given three things, each of which may be a point, a line, or a circle, draw a circle that is tangent to each of the three given things (where tangency to a point is to be understood to mean that the circle passes through the point). This problem involves ten cases, from the two easiest (in which the three things are three points or three lines) to the most difficult of all (to draw a circle tangent to three circles). The two easiest had appeared in Euclid's *Elements* in connection with inscribed and circumscribed circles of a triangle; another six cases were handled in Book I of *Tangencies*, and the case covering two lines and a circle, as well as the case of three circles, occupied all of Book II. We do not have the solutions of Apollonius, but they can be reconstructed on the basis of information from Pappus. Nevertheless, scholars of the sixteenth and seventeenth centuries generally were under the impression that Apollonius had not solved the last case; hence they regarded this problem as a challenge to their abilities. Newton was among those who gave a solution, using straightedge and compasses alone.[3]

The trisection of the angle by Archimedes, in which a given length is inserted between a line and a circle along a straight line that is shifted so as to pass through a given point (point P in Fig. 8.9), is a typical example of a solution by means of a *neusis* (verging or inclination). Apollonius' treatise

---

[3] *Arithmetica universalis*, Problem XLVII.

on *Vergings* considered the class of *neusis* problems that can be solved by "plane" methods—that is, by the use of compasses and straightedge only. (The Archimedean trisection, of course, is not such a problem, for in modern times it has been proved that the general angle cannot be trisected by "plane" methods.) According to Pappus, one of the problems dealt with in *Vergings* is the insertion within a given circle of a chord of given length verging to a given point.

There were in antiquity allusions to still other works by Apollonius, including one on *Comparison of the Dodecahedron and the Icosahedron*. In this the author gave a proof of the theorem (known perhaps to Aristaeus) that the plane pentagonal faces of a dodecahedron are the same distance from the center of the circumscribing sphere as are the plane triangular faces of an icosahedron inscribed in the same sphere. The theorem in the spurious Book XIV of the *Elements*—that in this case the ratio of the areas of the icosahedron and the dodecahedron is equal to the ratio of their volumes—follows immediately from the Apollonian proposition; and it may be that the author of *Elements* XIV made use of the treatise of Apollonius.

4   Apollonius was also a celebrated astronomer; the favorite mathematical device in antiquity for the representation of the motions of the planets is apparently due to him. Whereas Eudoxus had used concentric spheres, Apollonius proposed instead two alternative systems, one made up of epicyclic motions and the other involving eccentric motions. In the first scheme a planet *P* was assumed to move uniformly about a small circle (epicycle), the center *C* of which in turn moved uniformly along the circumference of a larger circle (deferent) with center at the earth *E* (Fig. 9.1). In the eccentric scheme the planet *P* moves uniformly along the circumference of a large circle, the center *C'* of which in turn moves uniformly in a small circle

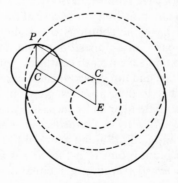

FIG. 9.1

with center at E. If $PC = C'E$, the two geometric schemes will be equivalent, as Apollonius evidently knew.[4] While the theory of homocentric spheres had become, through the work of Aristotle, the favorite astronomical scheme of those satisfied by a gross representation of the approximate motions, the theory of cycles and epicycles, or of eccentrics, became, through the work of Ptolemy, the choice of mathematical astronomers who wanted refinement of detail and predictive precision. For some 1800 years the two schemes—the one of Eudoxus and the other of Apollonius—were friendly rivals vying for the favor of scholars.

Despite his scholarly productivity, only two of the many treatises by Apollonius have in large part survived. All Greek versions of the *Cutting-off of a Ratio* were lost long ago, but not before an Arabic translation had been made. In 1706 Halley, Newton's friend, published a Latin translation of the work, and it has since appeared in vernacular tongues. Apart from this treatise, only one Apollonian work has substantially survived, which, however, was by all odds his chef-d'oeuvre—the *Conics*. Of this famous work only half—the first four of the original eight books—remains extant in Greek ; fortunately, an Arabic mathematician, Thabit ibn Qurra, had translated the next three books, and this version has survived. In 1710 Edmund Halley provided a Latin translation of the seven books, and editions in many languages have appeared since then.

The conic sections had been known for about a century and a half when Apollonius composed his celebrated treatise on these curves. At least twice in the interval general surveys had been written—by Aristaeus and by Euclid—but just as Euclid's *Elements* had displaced earlier elementary textbooks, so on the more advanced level of the conic sections the *Conics* of Apollonius superseded all rivals in its field, including the *Conics* of Euclid, and no attempt to improve on it seems to have been made in antiquity. If survival is a measure of quality, the *Elements* of Euclid and the *Conics* of Apollonius were clearly the best works in their fields.

Book I of the *Conics* opens with an account of the motivation for writing the work. While Apollonius was at Alexandria, he was visited by a geometer, named Naucrates, and it was at the latter's request that Apollonius wrote out a hasty draft of the *Conics* in eight books. Later at Pergamum the author took the time to polish the books one at a time, hence Books IV through VII open with greetings to Attalus, King of Pergamum. The first four books the author describes as forming an elementary introduction, and it has been assumed that much of this material had appeared in earlier treatises on conics. However, Apollonius expressly says that some of the theorems in Book III

---

[4] See O. Neugebauer, "Eccentric and Epicyclic Motion According to Apollonius," *Scripta Mathematica*, **24** (1959), 5–21.

were his own, for Euclid had not completed the loci there considered. The last four books he describes as extensions of the subject beyond the essentials, and we shall see that in them the theory is advanced in more specialized directions.[5]

Before the time of Apollonius the ellipse, parabola, and hyperbola were derived as sections of three distinctly different types of right circular cones, according as the vertex angle was acute, right, or obtuse. Apollonius, apparently for the first time, systematically showed that it is not necessary to take sections perpendicular to an element of the cone and that from a single cone one can obtain all three varieties of conic section simply by varying the inclination of the cutting plane. This was an important step in linking the three types of curve. A second important generalization was made when Apollonius demonstrated that the cone need not be a right cone—that is, one whose axis is perpendicular to the circular base—but can equally well be an oblique or scalene circular cone. If Eutocius, in commenting on the *Conics*, was well informed, we can infer that Apollonius was the first geometer to show that the properties of the curves are not different according as they are cut from oblique cones or from right cones. Finally, Apollonius brought the ancient curves closer to the modern point of view by replacing the single-napped cone (somewhat like a modern ice-cream cone) by a double-napped cone (resembling two oppositely oriented indefinitely long ice-cream cones placed so that the vertices coincide and the axes are in a straight line). Apollonius gave, in fact, the same definition of a circular cone as that used today:

> If a straight line, indefinite in length and passing always through a fixed point be made to move around the circumference of a circle which is not in the same plane with the point so as to pass successively through every point of that circumference, the moving straight line will trace out the surface of a double cone.

This change made the hyperbola the double-branched curve familiar to us today. Geometers often referred to the "two hyperbolas," rather than to the "two branches" of a single hyperbola, but in either case the duality of the curve was recognized.

6   Concepts are more important in the history of mathematics than is terminology, but there is more than ordinary significance in a change of name for the conic sections that was due to Apollonius. For about a century and a half the curves had had no more distinctive appellations than banal descriptions of the manner in which the curves had been discovered—sections

---

[5] See T. L. Heath, *Apollonius of Perga. Treatise on Conic Sections* (1896), pp. xxvi–xxvii. Here, and throughout this chapter, we depend on Heath's valuable volume, from which passages in translation have been taken.

of an acute-angled cone (oxytome), sections of a right-angled cone (orthotome), and sections of an obtuse-angled cone (amblytome). Archimedes had continued these names (although he is reported to have used also the word parabola as a synonym for section of a right-angled cone). It was Apollonius (possibly following up a suggestion of Archimedes) who introduced the names ellipse and hyperbola in connection with these curves. The words "ellipse," "parabola," and "hyperbola" were not newly coined for the occasion; they were adapted from an earlier use, perhaps by the Pythagoreans, in the solution of quadratic equations through the application of areas. *Ellipsis* (meaning a deficiency) had been used when a rectangle of given area was applied to a given line segment and fell short by a square (or other specified figure), and the word *hyperbola* (a throwing beyond) had been adopted when the area exceeded the line segment. The word *parabola* (a placing beside or comparison) had indicated neither excess nor deficiency. Apollonius now applied these words in a new context as names for the conic sections. The familiar modern equation of the parabola with vertex at the origin is $y^2 = lx$ (where l is the "latus rectum" or parameter, now often represented by $2p$, or occasionally by $4p$). That is, the parabola has the property that no matter what point on the curve one chooses, the square on the ordinate is precisely equal to the rectangle on the abscissa x and the parameter l. The equations of the ellipse and hyperbola, similarly referred to a vertex as origin, are $(x - a)^2/a^2 \pm y^2/b^2 = 1$, or $y^2 = lx \mp b^2x^2/a^2$ (where l again is the latus rectum or parameter $2b^2/a$). That is, for the ellipse $y^2 < lx$ and for the hyperbola $y^2 > lx$, and it is the properties of the curves that are represented by these inequalities that prompted the names given by Apollonius more than two millennia ago and still firmly attached to them.[6]

In deriving all conic sections from a single double-napped oblique circular cone, and in giving them eminently appropriate names, Apollonius made an important contribution to geometry; but he failed to go as far in generality as he might have. He could as well have begun with an elliptic cone—or with any quadric cone—and still have derived the same curves. That is, any plane section of Apollonius' "circular" cone could have served as the generating curve or "base" in his definition, and the designation "circular cone" is unnecessary. In fact, as Apollonius himself showed (Book I, Proposition 5), every oblique circular cone has not only an infinite number of circular sections parallel to the base, but also another infinite set of circular sections given by what he called subcontrary sections. Let *BFC* be the base of the

---

[6] The commentator Eutocius was responsible for an erroneous impression, still fairly widespread, that the words ellipse, parabola, and hyperbola were adopted by Apollonius to indicate that the cutting plane fell short of, or ran along with, or ran into the second nappe of the cone. This is not at all what Apollonius reported in the *Conics*.

oblique circular cone and let ABC be a triangular section of the cone (Fig. 9.2). Let P be any point on a circular section DPE parallel to BFC and let HPK be a section by a plane such that triangles AHK and ABC are similar but oppositely oriented. Apollonius then called the section HPK a subcontrary section and showed that it is a circle. The proof is easily established in terms of the similarity of triangles HMD and EMK, from which it follows that $HM \cdot MK = DM \cdot ME = PM^2$, the characteristic property of a circle. (In the language of analytic geometry, if we let $HM = x$, $HK = a$, and $PM = y$, then $y^2 = x(a - x)$ or $x^2 + y^2 = ax$, which is the equation of a circle.)

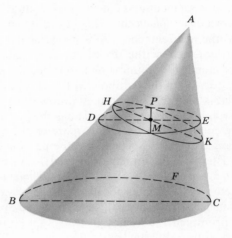

FIG. 9.2

8 Greek geometers divided curves into three categories. The first, known as "plane loci," consisted of all straight lines and circles; the second, known as "solid loci," was made up of all conic sections; the third category, known as "linear loci," lumped together all other curves. The name applied to the second category undoubtedly was suggested by the fact that the conics were not defined as loci in a plane which satisfy a certain condition, as is done today; they were described stereometrically as sections of a three-dimensional figure. Apollonius, like his predecessors, derived his curves from a cone in three-dimensional space, but he dispensed with the cone as promptly as possible. From the cone he derived a fundamental plane property or "symptome" for the section, and thereafter he proceeded with a purely planimetric study based on this property. This step, which we here illustrate for the ellipse (Book I, Proposition 13), probably was much the same as that used by his predecessors, including Menaechmus. Let ABC be a triangular section of an oblique circular cone (Fig. 9.3) and let P be any point on a section HPK cutting all elements of the cone. Extend HK to meet BC in G and through P pass a horizontal plane cutting the cone in the circle DPE and the plane HPK in the line PM. Draw DME, a diameter of the circle perpendicular to PM. Then from the similarity of triangles HDM and HBG we have $DM/HM = BG/HG$, and from the similarity of triangles MEK and KCG we have $ME/MK = CG/KG$. Now, from the property of the circle we have $PM^2 = DM \cdot ME$; hence $PM^2 = (HM \cdot BG/HG)(MK \cdot CG)/KG$. If $PM = y$, $HM = x$, and $HK = 2a$, the property in the preceding sentence is equivalent to the equation $y^2 = kx(2a - x)$, which we recognize as the equation of an ellipse with H as vertex and HK as major axis. In a similar manner Apollonius derived for the hyperbola the equivalent of the equation $y^2 = kx(x + 2a)$. These forms are easily reconciled with the "name" forms above by taking $k = b^2/a^2$ and $l = 2b^2/a$.

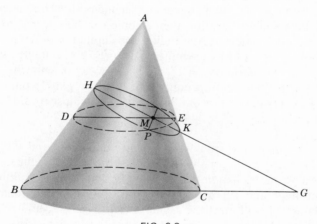

FIG. 9.3

After Apollonius had derived from a stereometric consideration of the **9** cone the basic relationship between what we should now call the plane coordinates of a point on the curve—given by the three equations $y^2 = lx - b^2x^2/a^2$, $y^2 = lx$, and $y^2 = lx + b^2x^2/a^2$—he derived further properties from the plane equations without reference to the cone. The author of the *Conics* reported that in Book I he had worked out the fundamental properties of the curves "more fully and generally than in the writings of other authors." The extent to which this statement holds true is suggested by the fact that here, in the very first book, the theory of conjugate diameters of a conic is developed. That is, Apollonius showed that the midpoints of a set of chords parallel to one diameter of an ellipse or hyperbola will constitute a second diameter, the two being called "conjugate diameters." In fact, whereas today we invariably refer a conic to a pair of mutually perpendicular lines as axes, Apollonius generally used a pair of conjugate diameters as equivalents of oblique coordinate axes. The system of conjugate diameters provided an exceptionally useful frame of reference for a conic, for Apollonius showed that if a line is drawn through an extremity of one diameter of an ellipse or hyperbola parallel to the conjugate diameter, the line "will touch the conic, and no other straight line can fall between it and the conic"—that is, the line will be tangent to the conic. Here we see clearly the Greek static concept of a tangent to a curve, in contrast to the Archimedean kinematic view. In fact, often in the *Conics* we find a diameter and a tangent at its extremity used as a coordinate frame of reference.

Among the theorems in Book I are several (Propositions 41 through 49) that are tantamount to a transformation of coordinates from a system based on the tangent and diameter through a point P on the conic to a new system determined by a tangent and diameter at a second point Q on the same curve, together with the demonstration that a conic can be referred to any such system as axes. In particular, Apollonius was familiar with the properties of the hyperbola referred to its asymptotes as axes, given, for the equilateral hyperbola, by the equation $xy = c^2$. He had no way of knowing, of course, that some day this relationship, equivalent to Boyle's law, would be fundamental in the study of gases or that his study of the ellipse would be essential to modern astronomy.

10   Book II continues the study of conjugate diameters and tangents. For example, if P is any point on any hyperbola, with center C, the tangent at P will cut the asymptotes in points L and L' (Fig. 9.4) that are equidistant from P (Propositions 8 and 10). Moreover (Propositions 11 and 16), any chord QQ' parallel to CP will meet the asymptotes in points K and K' such that $QK = Q'K'$ and $QK \cdot QK' = CP^2$. (These properties were verified synthetically, but the reader can convince himself of their validity by use of modern analytic methods.) Later propositions in Book II show how to draw tangents

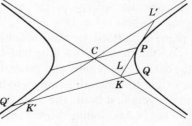

FIG. 9.4

to a conic by making use of the theory of harmonic division. In the case of the ellipse (Proposition 49), for example, if Q is a point on the curve (Fig. 9.5), Apollonius dropped a perpendicular QN from Q to the axis AA' and found the harmonic conjugate T of N with respect to A and A'. (That is, he found the point T on line AA' extended such that $AT/A'T = AN/NA'$; in other words,

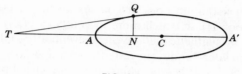

FIG. 9.5

he determined the point T that divides the segment AA' externally in the same ratio as N divides AA' internally.) The line through T and Q then will be tangent to the ellipse. The case in which Q does not lie on the curve can be reduced to this through familiar properties of harmonic division. (It can be proved that there are no plane curves other than the conic sections such that, given the curve and a point, a tangent can be drawn, with straightedge and compasses, from the point to the curve; but this was of course unknown to Apollonius.)

Apollonius apparently was especially proud of Book III, for in the General **11** Preface to the *Conics* he wrote:

The third book contains many remarkable theorems useful for the synthesis of solid loci and determinations of limits; the most and prettiest of these theorems are new and, when I had discovered them, I observed that Euclid had not worked out the synthesis of the locus with respect to three and four lines, but only a chance portion of it and that not successfully: for it was not possible that the synthesis could have been completed without my additional discoveries.

The three-and-four-line locus, to which reference is made, played an important role in mathematics from Euclid to Newton. Given three lines (or four lines) in a plane, find the locus of a point P that moves so that the square of the distance from P to one of these is proportional to the product of the distances to the other two (or, in the case of four lines, the product of the distances to two of them is proportional to the product of the distances to the other two), the distances being measured at given angles with respect to the lines. Through modern analytic methods, including the normal form of the straight line, it is easy to show that the locus is a conic section—real or imaginary, reducible or irreducible. If, for the three-line locus, equations of the given lines are $A_1x + B_1y + C_1 = 0$, $A_2x + B_2y + C_2 = 0$, and

$A_3x + B_3y + C_3 = 0$, and if the angles at which the distances are to be measured are θ_1, θ_2, and θ_3, then the locus of $P(x, y)$ is given by

$$\frac{(A_1x + B_1y + C_1)^2}{(A_1{}^2 + B_1{}^2)\sin^2 \theta_1} = \frac{K(A_2x + B_2y + C_2)}{\sqrt{A_2{}^2 + B_2{}^2}\sin \theta_2} \cdot \frac{(A_3x + B_3y + C_3)}{\sqrt{A_3{}^2 + B_3{}^2}\sin \theta_3}$$

This equation is, in general, of second degree in x and y; hence the locus is a conic section. Our solution does not do justice to the treatment given by Apollonius in Book III, in which more than fifty carefully worded propositions, all proved by synthetic methods, lead eventually to the required locus. Half a millennium later Pappus suggested a generalization of this theorem for n lines, where $n > 4$, and it was against this generalized problem that Descartes in 1637 tested his analytic geometry. Thus few problems have played as important a role in the history of mathematics as did the "locus to three and four lines."

12   Book IV of the *Conics* is described by its author as showing "in how many ways the sections of cones meet one another," and he is especially proud of theorems, "none of which has been discussed by earlier writers," concerning the number of points in which a section of a cone meets the "opposite branches of a hyperbola." The idea of the hyperbola as a double-branched curve was new with Apollonius, and he thoroughly enjoyed the discovery and proof of theorems concerning it. For example, he showed (IV. 42) that if one branch of a hyperbola meets both branches of another hyperbola, the opposite branch of the first hyperbola will not meet either branch of the second hyperbola in two points; or again (IV. 54), if a hyperbola is tangent to one of the branches of a second hyperbola with its concavity in the opposite direction, the opposite branch of the first will not meet the opposite branch of the second. It is in connection with the theorems in this book that Apollonius makes a statement implying that in his day, as in ours, there were narrow-minded opponents of pure mathematics who pejoratively inquired about the usefulness of such results. The author proudly asserted: "They are worthy of acceptance for the sake of the demonstrations themselves, in the same way as we accept many other things in mathematics for this and for no other reason."[7]

13   The preface to Book V, relating to maximum and minimum straight lines drawn to a conic, again argues that "the subject is one of those which seem worthy of study for their own sake." While one must admire the author for his lofty intellectual attitude, it may be pertinently pointed out that what in his day was beautiful theory, with no prospect of applicability to the science

---

[7] See Heath, *Apollonius of Perga. Treatise on Conic Sections,* p. lxxiv.

or engineering of his time, has since become fundamental in such fields as terrestrial dynamics and celestial mechanics. Apollonius' theorems on maxima and minima are in reality theorems on tangents and normals to conic sections. Without a knowledge of the properties of tangents to a parabola, an analysis of local trajectories would be impossible; and a study of the paths of the planets is unthinkable without reference to the tangents to an ellipse. It is clear, in other words, that it was the pure mathematics of Apollonius that made possible, some 1800 years later, the *Principia* of Newton; the latter, in turn, has given scientists of today the hope that some day a round-trip visit to the moon will be possible. Even in ancient Greece the Apollonian theorem that every oblique cone has two families of circular sections was applicable to cartography in the stereographic transformation, used by Ptolemy and possibly by Hipparchus, of a spherical region into a portion of a plane. It has often been true in the development of mathematics that topics that originally could be justified only as "worthy of study for their own sake" later became of inestimable value to the "practical man."

Greek mathematicians had no satisfactory definition of tangent to a curve *C* at a point *P*, thinking of it as a line *L* such that no other line could be drawn through *P* between *C* and *L*. Perhaps it was dissatisfaction with this definition that led Apollonius to avoid defining a normal to a curve *C* from a point *Q* as a line through *Q* which cuts the curve *C* in a point *P* and is perpendicular to the tangent to *C* at *P*. Instead he made use of the fact that the normal from *Q* to *C* is a line such that the distance from *Q* to *C* is a relative maximum or minimum. In *Conics* V. 8, for example, Apollonius proved a theorem concerning the normal to a parabola which today generally is part of a course in the calculus. In modern terminology the theorem states that the subnormal of the parabola $y^2 = 2px$ for any point *P* on the curve is constant and equal to *p*; in the language of Apollonius this property is expressed somewhat as follows:

> If *A* is the vertex of a parabola $y^2 = px$, and if *G* is a point on the axis such that *AG* > *p*, and, if *N* is a point between *A* and *G* such that *NG* = *p*, and if *NP* is drawn perpendicular to the axis meeting the parabola in *P* (Fig. 9.6), then *PG* is the minimum straight line from *G* to the curve and hence is normal to the parabola at *P*).

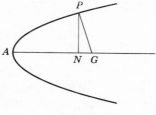

FIG. 9.6

The proof by Apollonius is of the typical indirect kind—it is shown that if P' is any other point on the parabola, $P'G$ increases as P' moves further from P in either direction. A proof of the corresponding, but more involved, theorem concerning the normal to an ellipse or hyperbola from a point on the axis is then given; and it is shown that if P is a point on a conic, only one normal can be drawn through P, whether the normal be regarded as a minimum or a maximum, and this normal is perpendicular to the tangent at P. Note that the perpendicularity that we take as a definition is here proved as a theorem, whereas the maximum-minimum property that we take as a theorem serves, for Apollonius, as a definition. Later propositions in Book V carry the topic of normals to a conic to such a point that the author gives criteria enabling one to tell how many normals can be drawn from a given point to a conic section. These criteria are tantamount to what we should describe as the equations of the evolutes to the conics. For the parabola $y^2 = 2px$ Apollonius showed in essence that points whose coordinates satisfy the cubic equation $27py^2 = 8(x - p)^3$ are limiting positions of the point of intersection of normals to the parabola at points P and P' as P' approaches P. That is, points on this cubic are the centers of curvature for points on the conic (that is, the centers of osculating circles for the parabola). In the case of the ellipse and the hyperbola, whose equations are respectively $x^2/a^2 \pm y^2/b^2 = 1$, the corresponding equations of the evolute are $(ax)^{\frac{2}{3}} \pm (by)^{\frac{2}{3}} = (a^2 \mp b^2)^{\frac{2}{3}}$.

After giving the conditions for the evolute of a conic, Apollonius showed how to construct a normal to a conic section from a point Q. In the case of the parabola $y^2 = 2px$, and for Q outside the parabola and not on the axis, one drops a perpendicular QM to the axis AK, measures off $MH = p$, and erects HR perpendicular to HA (Fig. 9.7). Then through Q one draws the rectangular

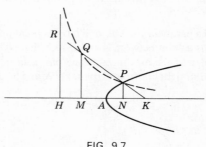

FIG. 9.7

hyperbola with asymptotes HA and HR, intersecting the parabola in a point P. Line QP is the normal required, as one can prove by showing that $NK = HM = p$. If point Q lies inside the parabola, the construction is

similar except that P lies between Q and R. Apollonius also gave constructions, likewise making use of an auxiliary hyperbola, for the normal from a
point to a given ellipse or hyperbola. It should be noted that the construction
of normals to the ellipse and hyperbola, unlike the construction of tangents,
requires more than straightedge and compasses. As the ancients described
the two problems, the drawing of a *tangent* to a conic is a "plane problem,"
for intersecting circles and straight lines suffice; by contrast, the drawing of a
normal from an arbitrary point in the plane to a given central conic is a
"solid problem," for it cannot be accomplished by use of lines and circles
alone, but can be done through the use of solid loci (in our case, a hyperbola).
Pappus later severely criticized Apollonius for his construction of a normal to
the parabola in that he treated it as a solid problem rather than a plane problem. That is, the hyperbola that Apollonius used could have been replaced
by a circle. Perhaps Apollonius felt that the line-and-circle fetish should give
way, in his construction of normals, to a desire for uniformity of approach
with respect to the three types of conic.

When Apollonius sent King Attalus the sixth book of the *Conics*, he **14**
described it as embracing propositions about "segments of conics equal and
unequal, similar and dissimilar, besides some other matters left out by those
who have preceded me. In particular, you will find in this book how, in a
given right cone, a section is to be cut equal to a given section." Two conics
are said to be similar if the ordinates, when drawn to the axis at proportional
distances from the vertex, are respectively proportional to the corresponding
abscissas. Among the easier of the propositions in Book VI are those demon-
strating that all parabolas are similar (VI. 11) and that a parabola cannot be
similar to an ellipse or hyperbola nor an ellipse to a hyperbola (VI. 14, 15).
Other propositions (VI. 26, 27) prove that if any cone is cut by two parallel
planes making hyperbolic or elliptic sections, the sections will be similar but
not equal.

Book VII returns to the subject of conjugate diameters and "many new
propositions concerning diameters of sections and the figures described
upon them." Among these are some that are found in modern textbooks, such
as the proof (VII. 12, 13, 29, 30) that

> In every ellipse the sum, and in every hyperbola the difference, of the squares
> on any two conjugate diameters is equal to the sum or difference respectively
> of the squares on the axes.

There is also the proof of the familiar theorem that if tangents are drawn at
the extremities of a pair of conjugate axes of an ellipse or hyperbola, the
parallelogram formed by these four tangents will be equal to the rectangle
on the axes. It has been conjectured that the lost Book VIII of the *Conics*

continued with similar problems, for in the preface to Book VII the author wrote that the theorems of Book VII were used in Book VIII to solve determinate conic problems, so that the last book "is by way of an appendix."

15   The *Conics* of Apollonius is a treatise of such extraordinary breadth and depth that we are startled to note the omission of some of the properties that to us appear so obviously fundamental. As the curves are now introduced in textbooks, the foci play a prominent role; yet Apollonius had no name for these points, and he referred to them only indirectly. It is presumed that he, and perhaps also Aristaeus and Euclid, was indeed familiar with the focus-directrix property of the curves, but this is not even mentioned in the *Conics*. There is no numerical concept in the ancient treatment of conics corresponding to what we call the eccentricity, and although the focus of the parabola by implication appears in many an Apollonian theorem, it is not clear that the author was aware of the now familiar role of the directrix. He seems to have known how to determine a conic through five points, but this topic, which later loomed large in the *Principia* of Newton, is omitted in the *Conics* of Apollonius. It is quite possible, of course, that some or all of such tantalizing omissions resulted from the fact they had been treated elsewhere, in works no longer extant, by Apollonius or other authors. So much of ancient mathematics has been lost that an argument *e silencio* is precarious indeed. Moreover, the words of Leibniz should serve as a warning that one should not underestimate ancient accomplishments: "He who understands Archimedes and Apollonius will admire less the achievements of the foremost men of later times."

16   The methods of Apollonius in the *Conics* in many respects are so similar to the modern approach that his work sometimes is judged to be an analytic geometry anticipating that of Descartes by 1800 years. The application of reference lines in general, and of a diameter and a tangent at its extremity in particular, is of course not essentially different from the use of a coordinate frame, whether rectangular or, more generally, oblique. Distances measured along the diameter from the point of tangency are the abscissas, and segments parallel to the tangent and intercepted between the axis and the curve are the ordinates. The Apollonian relationships between these abscissas and the corresponding ordinates are nothing more nor less than rhetorical forms of the equations of the curves. However, Greek geometrical algebra did not provide for negative magnitudes; moreover, the coordinate system was in every case superimposed a posteriori upon a given curve in order to study its properties. There appear to be no cases in ancient geometry in which a coordinate frame of reference was laid down a priori for purposes of graphical representation of an equation or relationship, whether symbolically or rhetorically expressed. Of Greek geometry we may say that equations are determined by curves, but not that curves were defined by equations. Coordinates, variables, and equations were subsidiary notions derived from a specific geometrical situation; and one gathers that in the Greek view it was not sufficient to define curves abstractly as loci satisfying given conditions on two coordinates. To guarantee that a locus was really a curve, the ancients felt it incumbent upon them to exhibit it stereometrically as a section of a solid or to describe a kinematic mode of construction.

The Greek definition and study of curves compare quite unfavorably with the flexibility and extent of the modern treatment. Indeed, the ancients overlooked almost entirely the part that curves of various sorts played in the world about them. Aesthetically one of the most gifted people of all times, the only curves that they found in the heavens and on the earth were combinations of circles and straight lines. They did not even effectively exploit the two means of definition for curves that they recognized. The kinematic approach and the use of plane sections of surfaces are capable of far-reaching generalization, yet scarcely a dozen curves were familiar to the ancients. Even the cycloid, generated by a point on a circle that rolls along a straight line, seems to have escaped their notice. That Apollonius, the greatest geometer of antiquity, failed to develop analytic geometry, was probably the result of a poverty of curves rather than of thought. General methods are not necessary when problems concern always one of a limited number of particular cases. Moreover, the early modern inventors of analytic geometry had all Renaissance algebra at their disposal, whereas Apollonius necessarily worked with the more rigorous but far more awkward tool of geometrical algebra.

BIBLIOGRAPHY

Apollonius of Perga, *Les coniques*, trans. by Paul Ver Eecke (Bruges: Desclée, de Brouwer, 1924).

Coolidge, J. L., *History of the Conic Sections and Quadric Surfaces* (Oxford: Clarendon, 1945).

Coolidge, J. L., *History of Geometrical Methods* (Oxford: Clarendon, 1940; paperback ed., New York: Dover, 1963).

Coxeter, H. S. M., "The Problem of Apollonius," *American Mathematical Monthly*, **75** (1968), 5–15.

Dingeldey, F., "Coniques," in *Encyclopédie des sciences mathématiques*, **3** (3), 1–256.

Fladt, K., *Geschichte und Theorie der Kegelschnitte und der Flächen zweiten Grades* (Stuttgart, 1965).

Heath, T. L., "Apollonius," in *Encyclopaedia Britannica*, 11th ed. (Cambridge, 1910), II, 186–188.

Heath, T. L., ed., *Apollonius of Perga. Treatise on Conic Sections* (Cambridge: Cambridge University Press, 1896; reprinted, New York: Barnes and Noble, 1961).

Neugebauer, O., "Apollonius-Studien," *Quellen und Studien zur Geschichte der Mathematik*, Part B, *Studien*, II (1932), 215–253.

Neugebauer, O., "Eccentric and Epicyclic Motion According to Apollonius," *Scripta Mathematica*, **24** (1959), 5–21.

Taylor, Charles, *An Introduction to the Ancient and Modern Geometry of Conics* (Cambridge, 1881).

Thomas, Ivor, *Selections Illustrating the History of Greek Mathematics* (Cambridge, Mass. : Loeb Classical Library, 1939–1941, 2 vols.).

Van der Waerden, B. L., *Science Awakening*, trans. by Arnold Dresden (New York : Oxford, 1961 ; paperback ed., New York : Wiley, 1963).

Zeuthen, H. G., *Die Lehre von den Kegelschnitten im Altertum* (Copenhagen, 1886 and 1902).

EXERCISES

1. The names of Aristotle, Euclid, Archimedes, and Apollonius are associated respectively with those of four powerful rulers—Alexander, Ptolemy, Hiero, and Attalus. Tell where these men ruled and in what connection their names are associated with those of the scholars.

2. Describe several respects in which the mathematics of Apollonius differs from that of Euclid and several respects in which their works are similar.

3. In what respects does the work of Apollonius resemble that of Archimedes and in what ways do their works differ?

4. Would you say that Apollonius used analytic geometry? Justify your answer, showing in what respects his methods resemble the modern subject and in what ways they differ.

5. Write the number 12,345,678,987,654,321 as Apollonius would have written it.

6. Prove the theorem of Apollonius that the locus of points the difference of the squares of whose distances from two fixed points is constant is a straight line perpendicular to the line joining the two fixed points.

7. Prove the theorem concerning the "circle of Apollonius"; that is, show that the locus of points whose distances from two fixed points are unequal, but are in a fixed ratio, is a circle.

8. Given the points $P_1(3, 0)$, $P_2(0, 4)$, and $P_3(1, 2)$, find the equation of a line through P_3 which intersects the x-axis in a point P_4 and the y-axis in a point P_5 such that (a) P_1P_4 is twice P_2P_5 and (b) $P_1P_4 \times P_2P_5$ is 10.

9. Solve the "Problem of Apollonius" for (a) the case of two points and a line and (b) the case of two lines and a point.

10. Beginning from the standard equations of the ellipse, the parabola, and the hyperbola with a vertex at the origin, complete the proof of the "name property" of Apollonius.

11. If one diameter of the ellipse $x^2/a^2 + y^2/b^2 = 1$ has slope m, find the slope of the conjugate diameter.

12. Find the slope of the system of parallel chords of $y^2 = 2px$ bisected by the "diameter" $y = a$.

13. Given a diameter of a hyperbola, show precisely how, with straightedge and compasses, you would construct the conjugate diameter.

14. Find equations of the tangents from the point $(-1, 2)$ to the parabola $y^2 = 2px$ and show how to construct the tangents with compasses and straightedge.

15. Find the coordinates of the feet of the four normals that can be drawn from the point $(1, 0)$ to the ellipse $x^2/25 + y^2/16 = 1$. How many normals can be drawn from $(2, 0)$ to this ellipse?

16. For what values of K can four normals be drawn from the point $(K, 0)$ to the ellipse $x^2/a^2 + y^2/b^2 = 1$?

17. Prove that the length of the subnormal to a parabola at a point P on the parabola is constant (hence independent of the position of the point P on the curve).

18. Apollonius knew that a tangent to an ellipse or hyperbola at a point P on the curve makes equal angles with the focal radii through P. Prove this theorem.

19. Prove the Apollonian theorem that the segment of a tangent to a hyperbola intercepted between the asymptotes is bisected by the point of tangency.

*20. Find an equation of the locus of points P such that the product of the perpendicular distances of P from the coordinate axes is equal to the product of the perpendicular distances of P from the lines $y = x$ and $y = 1 - x$.

*21. Find an equation of the polar of the point (a, b) with respect to the parabola $y^2 = 2px$.

*22. Prove, in the manner Apollonius used for the cone, that an oblique section of a circular cylinder is an ellipse.

*23. Prove that if AA' is the major axis of an ellipse, if the tangent to the ellipse at any point P intersects this axis (extended) in T, and if N is the projection of P on AA', then (AA', TN') form a conjugate set of points. (See Fig. 9.5.)

*24. How many normals can be drawn from the point $(1, 2)$ to the parabola $y^2 = 2x$? Justify your answer.

CHAPTER X

Greek Trigonometry and Mensuration

> When I trace at my pleasure the windings to and fro of the heavenly bodies, I no longer touch the earth with my feet: I stand in the presence of Zeus himself and take my fill of ambrosia, food of the gods.
>
> *Ptolemy*

1 Trigonometry, like other branches of mathematics, was not the work of any one man—or nation. Theorems on ratios of the sides of similar triangles had been known to, and used by, the ancient Egyptians and Babylonians. In view of the pre-Hellenic lack of the concept of angle measure, such a study might better be called "trilaterometry," or the measure of three-sided polygons (trilaterals), than "trigonometry," the measure of parts of a triangle. With the Greeks we first find a systematic study of relationships between angles (or arcs) in a circle and the lengths of chords subtending these. Properties of chords, as measures of central and inscribed angles in circles, were familiar to the Greeks of Hippocrates' day, and it is likely that Eudoxus had used ratios and angle measures in determining the size of the earth and the relative distances of the sun and the moon. In the works of Euclid there is no trigonometry in the strict sense of the word, but there are theorems equivalent to specific trigonometric laws or formulas. Propositions II. 12 and 13 of the *Elements*, for example, are the laws of cosines for obtuse and acute angles respectively, stated in geometric rather than trigonometric language and proved by a method similar to that used by Euclid in connection with the Pythagorean theorem. Theorems on the lengths of chords are essentially applications of the modern law of sines. We have seen that Archimedes' theorem on the broken chord can readily be translated into trigonometric language analogous to formulas for sines of sums and differences of angles. More and more the astronomers of the Alexandrian Age—notably Eratosthenes of Cyrene (ca. 276–ca. 194 B.C.) and Aristarchus of Samos (ca. 310–ca. 230 B.C.)—handled problems pointing to a need for more systematic relationsips between angles and chords.

176

Aristarchus, according to Archimedes and Plutarch, proposed a heliocentric system, anticipating Copernicus by more than a millennium and a
half;[1] but whatever he may have written on this scheme has been lost.
Instead we have an Aristarchan treatise, perhaps composed earlier (ca.
260 B.C.), *On the Sizes and Distances of the Sun and Moon*, which assumes a
geocentric universe.[2] In this work Aristarchus made the observation that
when the moon is just half-full, the angle between the lines of sight to the
sun and the moon is less than a right angle by one-thirtieth of a quadrant.
(The systematic introduction of the 360° circle came a little later.) In trigonometric language of today this would mean that the ratio of the distance of
the moon to that of the sun (the ratio ME to SE in Fig. 10.1) is sin 3°. Trigonometric tables not having been developed yet, Aristarchus fell back upon a

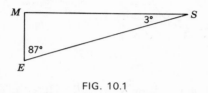

FIG. 10.1

well-known geometrical theorem of the time which now would be expressed
in the inequalities sin α/sin β < α/β < tan α/tan β, where 0° < β < α < 90°.
From these he derived the conclusion that $\frac{1}{20}$ < sin 3° < $\frac{1}{18}$, hence he asserted
that the sun is more than eighteen, but less than twenty, times as far from the
earth as is the moon. This is far from the modern value—somewhat less than
400—but it is better than the values nine and twelve that Archimedes ascribed
respectively to Eudoxus and to Phidias (Archimedes' father). Moreover, the
method used by Aristarchus was unimpeachable, the result being vitiated
only by the error of observation in measuring the angle MES as 87° (when
in actuality it should have been about 89° 50′).

Having determined the relative distances of the sun and moon, Aristarchus
knew also that the sizes of the sun and moon were in the same ratio. This
follows from the fact that the sun and moon have very nearly the same
apparent size—that is, they subtend about the same angle at the eye of an
observer on the earth. In the treatise in question, this angle is given as 2°, but
Archimedes attributed to Aristarchus the much better value of $\frac{1}{2}$°. From
this ratio Aristarchus was able to find an approximation for the sizes of the
sun and moon as compared with the size of the earth. From lunar eclipse

[1] The most complete account of Aristarchus and his place in astronomy is found in T. L.
Heath, *Aristarchus of Samos* (1913).
[2] It is possible that Aristarchus had been anticipated, in determining these distances, by
Eudoxus. See Paul Tannery, *Mémoires scientifiques*, I, 371.

observations he concluded that the breadth of the shadow cast by the earth at the distance of the moon was twice the width of the moon. Then if R_s, R_e, and R_m are the radii of the sun, earth, and moon respectively and if D_s and D_m are the distances of the sun and moon from the earth, then from the similarity of triangles BCD and ABE (Fig. 10.2), one has the proportion $(R_e - 2R_m)/(R_s - R_e) = D_m/D_s$. If in this equation one replaces D_s and R_s by the approximate values $19D_m$ and $19R_m$, one obtains the equation

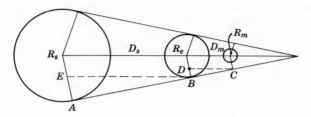

FIG. 10.2

$(R_e - 2R_m)/(19R_m - R_e) = \frac{1}{19}$ or $R_m = \frac{20}{57}R_e$. Here the actual computations of Aristarchus have been considerably simplified. His reasoning was in reality much more carefully carried out and led to the conclusion that

$$\frac{108}{43} < \frac{R_e}{R_m} < \frac{60}{19} \quad \text{and} \quad \frac{19}{3} < \frac{R_s}{R_e} < \frac{43}{6}$$

3   All that was needed to arrive at an estimate of the actual sizes of the sun and moon was a measure of the radius of the earth. Aristotle had mentioned a figure equivalent to about 40,000 miles for the circumference of the earth (a figure possibly due to Eudoxus), and Archimedes reported that some of his contemporaries estimated the perimeter to be about 30,000 miles.[3] A much better calculation, and by far the most celebrated, was one due to Eratosthenes, a younger contemporary of Archimedes and Aristarchus. Eratosthenes was a native of Cyrene who had spent much of his early life at Athens. He had achieved prominence in many fields—poetry, astronomy, history, mathematics, athletics—when, in middle life, he was called by Ptolemy III (Philopator) to Alexandria to tutor his son (later Ptolemy Philadelphus) and to serve as librarian of the university there. It was to Eratosthenes at Alexandria that Archimedes had sent the treatise on *Method*. Today Eratosthenes is best remembered for his measurement of the earth—not the first or last such estimate made in antiquity, but by all odds the most successful. Eratosthenes

[3] A. Diller, "The Ancient Measurements of the Earth," *Isis*, **40** (1949), 6–9.

observed that at noon on the day of the summer solstice the sun shone directly down a deep well at Syene. At the same time at Alexandria, taken to be on the same meridian and 5000 stades north of Syene, the sun was found to cast a shadow indicating that the sun's angular distance from the zenith was one fiftieth of a circle. From the equality of the corresponding angles $S'AZ$ and $S''OZ$ in Fig. 10.3 it is clear that the circumference of the earth must be fifty times the distance between Syene and Alexandria. This results in a perimeter of 250,000 stades, or, since a stade was about a tenth of a mile, of 25,000 miles. (Later accounts placed the figure at 252,000 stades, possibly in order to lead to the round figure of 700 stades per degree.)

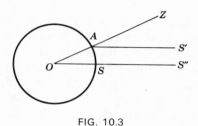

FIG. 10.3

A contributor to many fields of learning, Eratosthenes is well known in mathematics for the "sieve of Eratosthenes," a systematic procedure for isolating the prime numbers. With all the natural numbers arranged in order, one simply strikes out every second number following the number two, every third number (in the original sequence) following the number three, every fifth number following the number five, and continues in this manner to strike out every nth number following the number n. The remaining numbers, from two on, will of course be primes. Eratosthenes wrote also works on means and on loci, but these have been lost. Even his treatise *On the Measurement of the Earth* is no longer extant, although some details from it have been preserved by others, including Heron and Ptolemy of Alexandria.

For some two and a half centuries, from Hippocrates to Eratosthenes, **4** Greek mathematicians had studied relationships between lines and circles and had applied these in a variety of astronomical problems, but no systematic trigonometry had resulted. Then, presumably during the second half of the second century B.C., the first trigonometric table apparently was compiled by the astronomer Hipparchus of Nicaea (ca. 180–ca. 125 B.C.), who thus earned the right to be known as "the father of trigonometry." Aristarchus had known that in a given circle the ratio of arc to chord decreased as the angle decreases from 180° to 0°, tending toward a limit of 1. However, it appears that not until Hipparchus undertook the task had anyone tabulated corresponding values of arc and chord for a whole series of angles.[4] It has, however, been suggested that Apollonius may have anticipated Hipparchus in this respect, and that the contribution of the latter to trigonometry was simply the calculation of a better set of chords than had been drawn up by his predecessors. Hipparchus evidently drew up his tables for use in his astronomy, about the origin of which little is known.[5] Hipparchus was a transitional figure between Babylonian astronomy and the work of Ptolemy. Astronomy was flourishing in Mesopotamia when in about 270 B.C. Berossos, about the only Babylonian astronomer known by name, moved to the island of Cos, and it is not unlikely that the foundations of Near Eastern theory were transmitted to Greece by that time. The chief contributions attributed to Hipparchus in astronomy were his organization of the empirical data derived from the Babylonians, the drawing up of a star catalogue, improvement in important astronomical constants (such as the length of the month and year, the size of the moon, and the angle of obliquity of the ecliptic), and, finally, the discovery of the precession of the equinoxes. It generally has been assumed that he was largely responsible for the building of geometrical planetary systems, but this is uncertain because it is not clear to what extent Apollonius may have applied trigonometric methods to astronomy somewhat earlier.

It is not known just when the systematic use of the 360° circle came into mathematics, but it seems to be due largely to Hipparchus in connection with his table of chords. It is possible that he took over from Hypsicles, who earlier had divided the day into 360 parts, a subdivision that may have been suggested by Babylonian astronomy. Just how Hipparchus made up his table is not known, for his works are not extant (except for a commentary on a popular astronomical poem by Aratus). It is likely that his methods were similar to those of Ptolemy, to be described below, for Theon of Alexandria, commenting on Ptolemy's table of chords, reported that Hipparchus earlier had written a treatise in twelve books on chords in a circle.

5   Theon mentions also another treatise, in six books, by Menelaus of Alexandria (ca. 100) dealing with *Chords in a Circle*. Other mathematical and astronomical works by Menelaus are mentioned by later Greek and Arabic commentators, including an *Elements of Geometry*, but the only one that has survived—and only through the Arabic—is his *Sphaerica*. In Book I of this treatise Menelaus established a basis for spherical triangles

---

[4] See Paul Tannery, *Recherches sur l'histoire de l'astronomie ancienne* (Paris, 1893), pp. 66 ff.

[5] How little is known is made clear in O. Neugebauer, *The Exact Sciences in Antiquity*, 2nd ed. (Providence, R.I.; Brown University Press, 1957), especially pp. 167–168.

analogous to that of Euclid I for plane triangles. Included is a theorem without Euclidean analogue—that two spherical triangles are congruent if corresponding angles are equal (Menelaus did not distinguish between congruent and symmetric spherical triangles); and the theorem $A + B + C > 180°$ is established. The second book of the *Spherica* describes the application of spherical geometry to astronomical phenomena and is of little mathematical interest. Book III, the last, contains the well-known "theorem of Menelaus" as part of what is essentially spherical trigonometry in the typical Greek form—a geometry or trigonometry of chords in a circle. In the circle in Fig. 10.4 we should write that chord AB is twice the sine of half the central

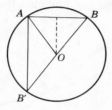

FIG. 10.4

angle AOB (multiplied by the radius of the circle). Menelaus and his Greek successors instead referred to AB simply as the chord corresponding to the arc AB. If BOB' is a diameter of the circle, then chord AB' is twice the cosine of half the angle AOB (multiplied by the radius of the circle). Hence the theorems of Thales and Pythagoras, which lead to the equation $AB^2 + AB'^2 = r^2$, are equivalent to the modern trigonometric identity $\sin^2 \theta + \cos^2 \theta = 1$. Menelaus, as also probably Hipparchus before him, was familiar with other identities, two of which he used as lemmas in proving his theorem on transversals. The first of these lemmas may be stated in modern terminology as follows. If a chord AB in a circle with center O (Fig. 10.5) is cut in point C

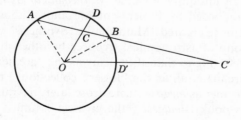

FIG. 10.5

by a radius OD, then $AC/CB = \sin \widehat{AD}/\sin \widehat{DB}$. The second lemma is similar: if the chord AB extended is cut in point C' by a radius OD' extended, then $AC'/BC' = \sin \widehat{AD'}/\sin \widehat{BD'}$. These lemmas were assumed by Menelaus without proof, presumably because they could be found in earlier works, possibly in Hipparchus' twelve books on chords. (The reader can prove the lemmas easily by drawing AO and BO, dropping perpendiculars from A and B to OD, and using similar triangles.[6]

It is probable that the "theorem of Menelaus" for the case of plane triangles had been known to Euclid, perhaps having appeared in the lost *Porisms*. The theorem in the plane states that if the sides AB, BC, CA of a triangle are cut by a transversal in points D, E, F respectively (Fig. 10.6), then $AD \cdot BE \cdot CF = BD \cdot CE \cdot AF$. In other words, any line cuts the sides of a triangle so that

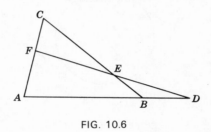

FIG. 10.6

the product of three nonadjacent segments equals the product of the other three, as can readily be proved by elementary geometry or through the application of simple trigonometric relationships. This theorem was assumed by Menelaus to be well known to his contemporaries, but he went on to extend it to spherical triangles in a form equivalent to $\sin AD \sin BE \sin GF = \sin BD \sin CE \sin AF$. If sensed segments are used rather than absolute magnitudes, the two products are equal in magnitude but differ in sign.

6   The theorem of Menelaus played a fundamental role in spherical trigonometry and astronomy, but by far the most influential and significant trigonometric work of all antiquity was the *Mathematical Syntaxis*, a work in thirteen books composed by Ptolemy of Alexandria about half a century after Menelaus. This celebrated "Mathematical Synthesis" was distinguished from another group of astronomical treatises by other authors (including Aristarchus) by referring to that of Ptolemy as the "greater" collection and to that of Aristarchus et al. as the "lesser" collection. From the frequent reference to the former as *megiste*, there arose later in Arabia the custom of calling Ptolemy's book *Almagest* ("the greatest"), and it is by this name that the work has since been known.

[6] See T. L. Heath, *History of Greek mathematics* (1921), II, 265–267.

Of the life of its author we are as little informed as we are of that of the author of the *Elements*. We do not know when or where Euclid and Ptolemy were born. We know that Ptolemy made observations at Alexandria from 127 to 151 and therefore assume that he was born at the end of the first century. Suidas, a writer who lived in the tenth century, reported that Ptolemy was still alive under Marcus Aurelius (emperor from 161 to 180).

Ptolemy's *Almagest* is presumed to be heavily indebted for its methods to the *Chords in a Circle* of Hipparchus, but the extent of the indebtedness cannot be reliably assessed. It is clear that in astronomy Ptolemy made use of the catalogue of star positions bequeathed by Hipparchus, but whether or not Ptolemy's trigonometric tables were derived in large part from his distinguished predecessor cannot be determined. Fortunately, Ptolemy's *Almagest* has survived the ravages of time; hence we have not only his trigonometric tables but also an account of the methods used in their construction. Central to the calculation of Ptolemy's chords was a geometrical proposition still known as "Ptolemy's theorem": If *ABCD* is a (convex) quadrilateral inscribed in a circle (Fig. 10.7), then $AB \cdot CD + BC \cdot DA = AC \cdot BD$; that is, the sum of the products of the opposite sides

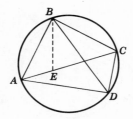

FIG. 10.7

of a cyclic quadrilateral is equal to the product of the diagonals. The proof of this is easily carried through by drawing *BE* so that angle *ABE* is equal to angle *DBC* and noting the similarity of the triangles *ABE* and *BCD*. A special case of Ptolemy's theorem had appeared in Euclid's *Data* (Proposition 93): If *ABC* is a triangle inscribed in a circle, and if *BD* is a chord bisecting angle *ABC*, then $(AB + BC)/BD = AC/AD$.

Another, and more useful, special case of the general theorem of Ptolemy is that in which one side, say *AD*, is a diameter of the circle (Fig. 10.8). Then if $AD = 2r$, we have $2r \cdot BC + AB \cdot CD = AC \cdot BD$. If we let arc $BD = 2\alpha$ and arc $CD = 2\beta$, then $BC = 2r \sin(\alpha - \beta)$, $AB = 2r \sin(90° - \alpha)$, $BD = 2r \sin \alpha$, $CD = 2r \sin \beta$, and $AC = 2r \sin(90° - \beta)$. Ptolemy's theorem therefore leads to the result $\sin(\alpha - \beta) = \sin \alpha \cos \beta - \cos \alpha \sin \beta$. Similar reasoning leads to the formula $\sin(\alpha + \beta) = \sin \alpha \cos \beta + \cos \alpha \sin \beta$, and to the

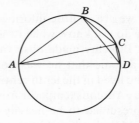

FIG. 10.8

analogous pair $\cos(\alpha \pm \beta) = \cos\alpha\cos\beta \mp \sin\alpha\sin\beta$. These four sum-and-difference formulas consequently are often known today as Ptolemy's formulas.

It was the formula for sine of the difference—or, more accurately, chord of the difference—that Ptolemy found especially useful in building up his tables. Another formula that served him effectively was the equivalent of our half-angle formula. Given the chord of an arc in a circle, Ptolemy found the chord of half the arc as follows. Let D be the midpoint of arc BC in a circle with diameter $AC = 2r$ (Fig. 10.9), let $AB = AE$, and let DF bisect EC

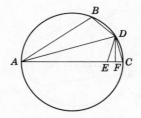

FIG. 10.9

(perpendicularly). Then it is not difficult to show that $FC = \frac{1}{2}(2r - AB)$. But from elementary geometry it is known that $DC^2 = AC \cdot FC$, from which it follows that $DC^2 = r(2r - AB)$. If we let arc $BC = 2\alpha$, then $DC = 2r\sin\alpha/2$ and $AB = 2r\cos\alpha$; hence we have the familiar modern formula $\sin\alpha/2 = \sqrt{(1 - \cos\alpha)/2}$. In other words, if the chord of any arc is known, the chord of half the arc is also known. Now Ptolemy was equipped to build up a table of chords as accurate as might be desired, for he had the equivalent of our fundamental formulas.

7   It should be recalled that from the days of Hipparchus until modern times there were no such things as trigonometric *ratios*. The Greeks, and after them the Hindus and the Arabs, used trigonometric *lines*. These at first took the form, as we have seen, of chords in a circle, and it became incumbent upon Ptolemy to associate numerical values (or approximations) with the chords. To do this two conventions were needed: (1) some scheme for subdividing the circumference of a circle and (2) some rule for subdividing the diameter. The division of a circumference into 360 degrees seems to have been in use in Greece since the days of Hipparchus, although it is not known just how the convention arose. It is not unlikely that the 360-degree measure was carried over from astronomy, where the zodiac had been divided into twelve "signs" or 36 "decans." A cycle of the seasons of roughly 360 days could readily be made to correspond to the system of zodiacal signs and decans by subdividing each sign into thirty parts and each decan into ten parts. Our common system of angle measure may stem from this correspondence. Moreover, since the Babylonian positional system for fractions was so obviously superior to the Egyptian unit fractions and the Greek common fractions, it was natural for Ptolemy to subdivide his degrees into sixty "partes minutae primae," each of these latter into sixty "partes minutae secundae," and so on. It is from the Latin phrases that translators used in this connection that our words "minute" and "second" have been derived. It undoubtedly was the sexagesimal system that led Ptolemy to subdivide the diameter of his trigonometric circle into 120 parts; each of these he further subdivided into sixty minutes and each minute of length into sixty seconds.

Our trigonometric identities are easily converted into the language of Ptolemaic chords through the simple relationships

$$\sin x = \frac{\text{chord } 2x}{120} \quad \text{and} \quad \cos x = \frac{\text{chord } (180° - 2x)}{120}$$

The formulas $\cos (x \pm y) = \cos x \cos y \mp \sin x \sin y$ become (chord is abbreviated to cd)

$$\text{cd } \overline{2x \pm 2y} = \frac{\text{cd } \overline{2x} \text{ cd } \overline{2y} \mp \text{cd } 2x \text{ cd } 2y}{120}$$

where a line over an arc (angle) indicates the supplementary arc. Note that not only angles and arcs, but also their chords were expressed sexagesimally. In fact, whenever scholars in antiquity wished an accurate system of approximation, they turned to the sixty-scale for the fractional portion; this led to the phrases "astronomers' fractions" and "physicists' fractions" to distinguish sexagesimal from common fractions.

Having decided upon his system of measurement, Ptolemy was ready to **8** compute the chords of angles within the system. For example, since the radius of the circle of reference contained sixty parts, the chord of an arc of sixty degrees also contained sixty linear parts. The chord of 120° will be $60\sqrt{3}$ or approximately 103 parts and 55 minutes and 33 seconds, or, in Ptolemy's Ionic or alphabetic notation, $\rho\gamma^p$ $\nu\epsilon'$ $\lambda\gamma''$. Ptolemy could now have used his half-angle formula to find the chord of 30°, then the chord of 15°, and so on for still smaller angles. However, he preferred to delay the application of this formula, and computed instead the chords of 36° and of 72°. He used a theorem from *Elements* XIII. 9 which shows that a side of a regular pentagon, a side of a regular hexagon, and a side of a regular decagon, all being inscribed within the same circle, constitute the sides of a right triangle. Incidentally, this theorem from Euclid provides the justification for Ptolemy's elegant construction of a regular pentagon inscribed in a circle. Let O be the center of a circle and AB a diameter (Fig. 10.10). Then if C is the midpoint of OB

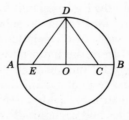

FIG. 10.10

and OD is perpendicular to AB, and if CE is taken equal to CD, the sides of the right triangle EDO are the sides of the regular inscribed pentagon, hexagon, and decagon. Then if the radius OB contains 60 parts, from the properties of the pentagon and the golden section it follows that OE, the chord of 36°, is $30(\sqrt{5}-1)$ or about 37.083 or 37^p $4'$ $55''$ or $\lambda\zeta^p$ δ' $\nu\epsilon''$. By the Pythagorean theorem the chord of 72° is $30\sqrt{10-2\sqrt{5}}$ or approximately 70.536 or 70^p $32'$ $3''$ or o^p $\lambda\beta'$ γ''.

Knowing the chord of an arc of s degrees in a circle, one can easily find the chord of the arc 180° $- s$ from the theorems of Thales and Pythagoras, for $\mathrm{cd}^2\,\bar{s} + \mathrm{cd}^2\,s = 120^2$. Hence Ptolemy knew the chords of the supplements of 36° and 72°. Moreover, from the chords of 72° and 60° he found chord 12° by means of his formula for the chord of the difference of two arcs. Then by successive applications of his half-angle formula he derived the chords of arcs of 6°, 3°, $1\frac{1}{2}$°, and $\frac{3}{4}$°, the last two being 1^p $34'$ $15''$ and 0^p $47'$ $8''$ respectively. Through a linear interpolation between these values Ptolemy arrived at 1^p $2'$ $50''$ as the chord of 1°. By using the half-angle formula—or, since the angle is very small, simply dividing by two—he found the value of 0^p $31'$ $25''$ for the chord of 30'. This is equivalent to saying that sin 15' is 0.00873, which is correct to almost half a dozen decimal places.

Ptolemy's value of the chord of $\frac{1}{2}°$ is, of course, the length of a side of a polygon of 720 sides inscribed in a circle of radius 60 units. Whereas Archimedes' polygon of 96 sides had led to 22/7 as an approximation to the value of π, Ptolemy's is equivalent to $6(0^p\ 31'\ 25'')$ or $3;8,30$. This approximation to π, used by Ptolemy in the *Almagest*, is the same as $\frac{377}{120}$, which leads to a decimal equivalent of about 3.1416, a value that may have been given earlier by Apollonius.

Armed with formulas for the chords of sums and differences and chords **9** of half an arc, and having a good value of chord $\frac{1}{2}°$, Ptolemy went on to build up his table, correct to the nearest second, of chords of arcs from $\frac{1}{2}°$ to $180°$ for every $\frac{1}{2}°$. This is virtually the same as a table of sines from $\frac{1}{4}°$ to $90°$, proceeding by steps of $\frac{1}{4}°$. The table formed an integral part of Book I of the *Almagest* and remained an indispensable tool of astronomers for more than a thousand years. The remaining twelve books of this celebrated treatise contain, among other things, the beautifully developed theory of cycles and epicycles for the planets known as the Ptolemaic system. Like Archimedes, Hipparchus, and most other great thinkers of antiquity, Ptolemy postulated an essentially geocentric universe, for a moving earth appeared to be faced with difficulties—such as lack of apparent stellar parallax and seeming inconsistency with the phenomena of terrestrial dynamics. In comparison with these problems, the implausibility of an immense speed required for the daily rotation of the sphere of the "fixed" stars seemed to shrink into insignificance. Besides appealing to common sense, the Ptolemaic system had the advantage of easy representation. Planetaria generally are constructed as though the universe were geocentric, for in this way the *apparent* motions are most easily reproduced.

Plato had set for Eudoxus the astronomical problems of "saving the phenomena"—that is, producing a mathematical device, such as a combination of uniform circular motions, which should serve as a model for the apparent motions of the planets. The Eudoxian system of homocentric spheres had been largely abandoned by mathematicians in favor of the system of cycles and epicycles of Apollonius and Hipparchus. Ptolemy in turn made an essential modification in the latter scheme. In the first place, he displaced the earth somewhat from the center of the deferent circle, so that he had eccentric orbits. Such changes had been made before him, but Ptolemy introduced a novelty so drastic in scientific implication that Copernicus later could not accept it, effective though the device, known as the equant, was in reproducing the planetary motions. Try as he would, Ptolemy had not been able to arrange a system of cycles, epicycles, and eccentrics in close agreement with the observed motions of the planets. His solution was to abandon the Greek insistence on uniformity of circular motions and to introduce instead a geometrical point, the equant E collinear with the earth G and the center C of the deferent circle, such that the *apparent* angular motion of the center Q of the epicycle in which a planet P revolves is uniform as seen from E (Fig. 10.11). In this way Ptolemy achieved accurate representations of planetary motions, but of course the device was kinematic only and made no effort to answer the questions in dynamics raised by nonuniform circular movements.

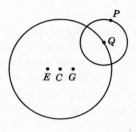

FIG. 10.11

10   Ptolemy's fame today is associated largely with a single book, the *Almagest*, but there are other Ptolemaic works as well. Among the more important was a *Geography*, in eight books, which was as much a bible to geographers of his day as the *Almagest* was to astronomers. The *Geography* of Ptolemy introduced the system of latitudes and longitudes as used today, described methods of cartographic projection, and catalogued some 8000 cities, rivers, and other important features of the earth. Unfortunately, there was at the time no satisfactory means of determining longitudes, hence substantial errors were inevitable. Even more significant was the fact that Ptolemy seems to have made a poor choice when it came to estimating the size of the earth. Instead of accepting the figure 252,000 stadia, given by Eratosthenes, he preferred the value 180,000 stadia proposed by Posidonius, a Stoic teacher of Pompey and Cicero. Hence Ptolemy thought that the known Eurasian world was a larger fraction of the circumference than it really is— more than 180° in longitude, instead of an actual figure of about 130°. This large error suggested to later navigators, including Columbus, that a voyage westward from Europe to India would not be nearly so far as it turned out to be. Had Columbus known how badly Ptolemy had underestimated the size of the earth, he might never have set sail.

Ptolemy's geographical methods were better in theory than in practice, for in separate monographs, which have survived only through Latin translations from the Arabic, Ptolemy described two types of map projection. Orthographic projection is explained in the *Analemma*, the earliest account we have of this method, although it may have been used by Hipparchus. In this transformation from a sphere to a plane, points on the spherical surface are projected orthogonally upon three mutually perpendicular planes. In the *Planisphaerium* Ptolemy described the stereographic projection in which points on the sphere are projected by lines from a pole onto a plane—in Ptolemy's case from the south pole to the plane of the equator. He knew that under such a transformation a circle not through the pole of projection went into a circle in the plane, and that a circle through the pole was projected into a straight line. Ptolemy was aware also of the important fact that such a transformation is conformal—that is, angles are preserved. The importance of Ptolemy for geography can be gauged from the fact that the earliest maps in the Middle Ages that have come down to us in manuscripts, none before the thirteenth century, had as prototypes the maps made by Ptolemy more than a thousand years before.[7]

Ptolemy wrote also an *Optics* which has survived, imperfectly, through a **11** Latin version of an Arabic translation. This deals with the physics and psychology of vision, with the geometry of mirrors, and with an early attempt at a law of refraction. From Ptolemy's table of angles of refraction from air to water (and also from air to glass and from water to glass) for angles of incidence from 10° to 80° at intervals of 10° we see that he assumed a law of the form $r = ai + bi^2$, for the second differences in his values of r are constant. For angles of incidence of 10° and 80° he assumed angles of refraction of 8° and 50° respectively, and the second differences are all equal to $\frac{1}{2}$°. The second differences in the old Pythagorean formulas for polygonal numbers also were constant, and perhaps Ptolemy was influenced by these to seek a quadratic rather than a trigonometric law for refraction. Trigonometry for the first millennium and a half of its existence was almost exclusively an adjunct of astronomy and geography, and only in the seventeenth century were trigonometric applications in refraction and other parts of physics discovered.

No account of Ptolemy's work would be complete without mention of his *Tetrabiblos* (or *Quadripartitum*), for it shows us a side of ancient scholarship that we are prone to overlook. Greek authors were not always the rational and clear-thinking men they are presumed to have been. The *Almagest* is indeed a model of good mathematics and accurate observational data put to work in building a sober scientific astronomy; but the *Tetrabiblos* (or work in four books) represents a kind of sidereal religion to which much of the ancient world had succumbed. With the end of the Golden Age, Greek mathematics and philosophy became allies of Chaldean arithmetic and astrology, and the resulting pseudoreligion filled the gap left by repudiation

---

[7] See George Sarton, *Ancient Science and Modern Civilization* (1954), pp. 53–54.

of the old mythology. Ptolemy seems to have shared the prejudices of his time; in the *Tetrabiblos* he argued that one should not, because of the possibility of error, discourage the astrologer any more than the physician. The further one reads in the work, the more dismayed one becomes, for the author showed no hesitation in accepting the superstitions of his day.

The *Tetrabiblos* differs from the *Almagest* not only as astrology differs from astronomy; the two works also make use of different types of mathematics. The latter is a sound and sophisticated work that makes good use of synthetic Greek geometry; the former is typical of the pseudoscience of the day in the adoption of primitive Babylonian arithmetic devices. From the classical works of Euclid, Archimedes, and Apollonius one might obtain the impression that Greek mathematics was exclusively occupied with the highest levels of logical geometrical reasoning; but Ptolemy's *Tetrabiblos* suggests that the populace in general were more concerned with arithmetical computation than with rational thought. At least from the days of Alexander the Great to the close of the classical world, there undoubtedly was much intercommunication between Greece and Mesopotamia, and it seems to be clear that the Babylonian arithmetic and algebraic geometry continued to exert considerable influence in the Hellenistic world. This aspect of mathematics, for example, appears so strongly in Heron of Alexandria (fl. ca. 100) that Heron once was thought to be Egyptian or Phoenician rather than Greek. Now it is thought that Heron portrays a type of mathematics that had long been present in Greece but does not find a representative among the greatest figures—except perhaps as betrayed by Ptolemy in the *Tetrabiblos*. Greek deductive geometry, on the other hand, seems not to have been welcomed in Mesopotamia until after the Arabic conquest.

12   Heron of Alexandria is best known in the history of mathematics for the formula, bearing his name, for the area of a triangle:

$$K = \sqrt{s(s-a)(s-b)(s-c)}$$

where *a*, *b*, *c* are the sides and *s* is half the sum of these sides, that is, the semiperimeter. The Arabs tell us that "Heron's formula" was known earlier to Archimedes, who undoubtedly had a proof of it, but the demonstration of it in Heron's *Metrica* is the earliest that we have. Although now the formula usually is derived trigonometrically, Heron's proof is conventionally geometric. The *Metrica*, like the *Method* of Archimedes, was long lost, until rediscovered at Constantinople in 1896 in a manuscript dating from about 1100. The word "geometry" originally meant "earth measure," but classical geometry, such as that found in Euclid's *Elements* and Apollonius' *Conics*, was far removed from mundane surveying. Heron's work, on the other hand, shows us that not all mathematics in Greece was of the "classical" type. There evidently were two levels in the study of configurations—comparable to the distinction made in numerical context between arithmetic (or theory of numbers) and logistic (or techniques of computation)—one of which, eminently rational, might be known as geometry and the other, crassly practical, might better be described as geodesy. The Babylonians lacked the former, but were strong in the latter, and it was essentially the Babylonian type of mathematics that is found in Heron. It is true that in the *Metrica* an occasional demonstration is included, but the body of the work is concerned with numerical examples in mensuration of lengths, areas, and volumes. There are strong resemblances between his results and those found in ancient Mesopotamian problem texts. For example, Heron gave a tabulation[8] of the areas A_n of regular polygons of n sides in terms of the square of one side s_n, beginning with $A_3 = \frac{13}{30}s_3{}^2$ and continuing to $A_{12} = \frac{45}{4}s_{12}{}^2$. As was the case in pre-Hellenic mathematics, Heron also made no distinction between results that are exact and those that are only approximations. For A_5, for example, Heron gave two formulas—$\frac{5}{3}s_5{}^2$ and $\frac{12}{7}s_5{}^2$—the first of which agrees with a value found in a Babylonian table,[9] but neither of which is precisely correct. For the hexagon Heron's ratio of A_6 to $s_6{}^2$ is $\frac{13}{5}$, the Babylonian is $2;37,30$, whereas the true value lies between these and is of course irrational. In such calculations we should have expected Heron to use trigonometric tables such as Hipparchus had drawn up a couple of hundred years before, but apparently trigonometry was at the time largely the handmaid of the astronomer rather than of the practical man.

The gap that separated classical geometry from Heronian mensuration is clearly illustrated by certain of the problems set and solved by Heron in another of his works, the *Geometrica*. One problem calls for the diameter, perimeter, and area of a circle, given the sum of these three magnitudes. The axiom of Eudoxus would rule out such a problem from theoretical consideration, for the three magnitudes are of unlike dimensions, but from an uncritical numerical point of view the problem makes sense. Moreover, Heron did not solve the problem in general terms but, taking a cue again from pre-Hellenic methods, chose the specific case in which the sum is 212; his solution is like the ancient recipes in which steps only, without reasons, are given. The diameter 14 is easily found by taking the Archimedean value for π and using the Babylonian method of completing the square to solve a quadratic equation. Heron simply gives the laconic instructions, "Multiply 212 by 154, add 841, take the square root and subtract 29, and divide by 11." This is scarcely the way to teach mathematics, but Heron's books were intended as manuals for the practitioner.

[8] See D. E. Smith, *History of Mathematics* (Boston: Ginn, 1923–1925, 2 vols.), II, 606.

[9] See Neugebauer: *Exact Sciences in Antiquity*, p. 47.

Heron paid as little attention to the uniqueness of his answer as he did to the dimensionality of his magnitudes. In one problem he called for the sides of a right triangle if the sum of the area and perimeter is 280. This is, of course, an indeterminate problem, but Heron gave only one solution, making use of the Archimedean formula for area of a triangle. In modern notation, if s is the semiperimeter of the triangle and r the radius of the inscribed circle, then $rs + 2s = s(r + 2) = 280$. Following his own cookbook rule, "Always look for the factors," he chose $r + 2 = 8$ and $s = 35$. Then the area rs is 210. But the triangle is a right triangle, hence the hypotenuse c is equal to $s - r$ or $35 - 6$ or 29; the sum of the two sides a and b is equal to $s + r$ or 41. The values of a and b are then easily found to be 20 and 21. Heron says nothing about other factorizations of 280, which of course would lead to other answers.

13   Heron was interested in mensuration in all its forms—in optics and mechanics, as well as in geodesy. The law of reflection for light had been known to Euclid and Aristotle (probably also to Plato); but it was Heron who showed by a simple geometrical argument, in a work on *Catoptrics* (or reflection), that the equality of the angles of incidence and reflection is a consequence of the Aristotelian principle that nature does nothing the hard way. That is, if light is to travel from a source S to a mirror MM' and then to the eye E of an observer (Fig. 10.12), the shortest possible path SPE is that in which the

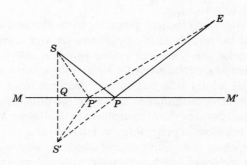

FIG. 10.12

angles SPM and EPM' are equal. That no other path $SP'E$ can be as short as SPE is apparent on drawing SQS' perpendicular to MM', with $SQ = QS'$, and comparing the path SPE with the path $SP'E$. Since paths SPE and $SP'E$ are equal in length to paths $S'PE$ and $S'P'E$ respectively, and inasmuch as $S'PE$ is a straight line (because angle $M'PE$ is equal to angle MPS), it follows that $S'PE$ is the shortest path.

Heron is remembered in the history of science as the inventor of a primitive type of steam engine, described in his *Pneumatics*, of a forerunner of the thermometer, and of various toys and mechanical contrivances based on the properties of fluids and on the laws of the simple machines. He suggested in the *Mechanics* a law (clever but incorrect) of the simple machine whose principle had eluded even Archimedes—the inclined plane. His name is attached also to "Heron's algorithm" for finding square roots, but this method of iteration was in reality due to the Babylonians of 2000 years before his day. Although Heron evidently learned much of Mesopotamian mathematics, he seems not to have appreciated the importance of the positional principle for fractions. Sexagesimal fractions had become the standard tool of scholars in astronomy and physics, but it is likely that they remained unfamiliar to the common man. Common fractions were used to some extent by the Greeks, at first with numerator placed below the denominator, later with the positions reversed (and without the bar separating the two), but Heron, writing for the practical man, seems to have preferred unit fractions. In dividing 25 by 13 he wrote the answer as $1 + \frac{1}{2} + \frac{1}{3} + \frac{1}{13} + \frac{1}{78}$. The old Egyptian addiction to unit fractions continued in Europe for at least a thousand years after the time of Heron.

The period from Hipparchus to Ptolemy, covering three centuries, was **14** one in which applied mathematics was in the ascendant, and Heron's books resemble notes taken by a student at the equivalent of an institute of technology at Alexandria. It sometimes is held[10] that mathematics develops most effectively when in close touch with the world's work; but the period we have been considering would argue for the opposite thesis. The loss of nerve in religion and philosophy, which led the Greeks to pursue cults and mysticism, was paralleled in mathematics by a movement toward applications which persisted for more than three centuries. From Hipparchus to Ptolemy there were advances in astronomy and geography, optics and mechanics, but no significant developments in mathematics. It is true that these centuries saw the development of trigonometry, but this subject, now an integral part of pure mathematics, was then at best a mensurational application of elementary geometry which met the needs of astronomy. Moreover, it is not even clear whether or not there was any significant advance in the trigonometry of Ptolemy in A.D. 150 over that of Hipparchus, in 150 B.C.—or even, perhaps, over that of Apollonius and Archimedes a century earlier still. It is evident that the rapid growth of mathematics from Eudoxus to Apollonius, when theoretical considerations were in the forefront, had come to an end. Perhaps the trend toward applications was the

[10] Especially by Lancelot Hogben in his many works on mathematics and its history, such as *Mathematics for the Million* (New York: W. W. Norton, ca. 1937).

result of the decline, rather than its cause, but in any case the two were con-comitant. Some[11] attribute the decline to the inadequacies and difficulties in Greek geometrical algebra, others[12] to the cold breath of Rome. In any case, the period during which trigonometry and mensuration came to the fore was characterized by lack of progress—if not actual decline; yet it was precisely these aspects of Greek mathematics that most attracted the Hindu and Arabic scholars who served as a bridge to the modern world. Before we turn to these peoples, however, we must look at the Indian summer of Greek mathematics, sometimes known as the "Silver Age."

BIBLIOGRAPHY

Aaboe, Asger, *Episodes from the Early History of Mathematics* (New York: Random House, 1964).

Braunmühl, Anton von, *Vorlesungen über Geschichte der Trigonometrie* (Leipzig, 1900–1903, 2 vols.).

Cohen, M. R., and I. E. Drabkin, *Source Book in Greek Science* (New York: McGraw-Hill, 1948; reprinted Cambridge, Mass.: Harvard University Press, 1958).

Dantzig, Tobias, *The Bequest of the Greeks* (New York: Scribner, 1955).

Heath, T. L., *Aristarchus of Samos* (Oxford: Clarendon, 1913).

Heath, T. L., *A History of Greek Mathematics* (Oxford: Clarendon, 1921, 2 vols.).

Lammert, Friedrich, "Klaudios Ptolemaios," in Pauly-Wissowa, *Real-Enzyclopädie der klassischen Altertumswissenschaft* (Stuttgart, 1959), Vol. XXIII, Part 2, columns 1788–1858.

Manitius, Karl, *Des Ptolemäus Handbuch der Astronomie* (Leipzig, 1912–1913), 2 vols.).

Peters, C. H. F., and E. B. Knobel, *Ptolemy's Catalogue of Stars; a Revision of the Almagest* (Washington, D.C.: Carnegie Institution, 1915).

Ptolemy, Claudius, *L'optique*, ed. by Albert Lejeune (Louvain, Belgium: Louvain University, 1956).

Ptolemy, Claudius, *Cosmographia*, ed. by R. A. Skelton (Amsterdam: Meridian, 1963).

Sarton, George, *Ancient Science and Modern Civilization* (Lincoln, Nebr.: University of Nebraska Press, 1954).

Stahl, W. H., *Ptolemy's Geography; a Select Bibliography* (New York: Bulletin of the New York Public Library, 1951–1952).

Tannery, Paul, *Mémoires scientifiques* (Toulouse, 1912, etc.), especially Vols. I and II.

Thomas, Ivor, *Selections Illustrating the History of Greek Mathematics* (Cambridge, Mass.: Loeb Classical Library, 1939–1941, 2 vols.).

Thomson, J. O., *History of Ancient Geography* (Cambridge, 1948).

Van der Waerden, B. L., *Science Awakening*, trans. by Arnold Dresden (New York: Oxford, 1961; paperback ed., New York: Wiley, 1963).

[11] For example, B. L. van der Waerden in *Science Awakening* (1961), pp. 265–266.
[12] E. T. Bell in *Development of Mathematics* (New York: McGraw-Hill, 1940).

EXERCISES

1. How can one account for the fact that the period of the rise of Greek trigonometry was a time of decline in Greek geometry?

2. Why did the ancients prefer a geocentric astronomical system to a heliocentric scheme? Explain clearly.

3. How far would Columbus have had to sail from Gibraltar to India, assuming the latter to be accessible from the east by water, if Ptolemy's ideas on the size of the earth had been correct?

4. What happens to circles on a sphere if projected orthogonally on a plane?

5. Using the information given in the text, find Ptolemy's law of refraction for rays going from air to water.

6. Prove, either geometrically or trigonometrically, Heron's formula for the area of a triangle.

7. Posidonius is said to have used observations of the stars to estimate the size of the earth. Show how this can be done.

8. Which of Heron's formulas for the ratio of A_5 to s_5^2 is the better approximation?

9. Heron gave the ratio of the area of a regular heptagon to the square of a side as $\frac{43}{12}$, and the Babylonians expressed this as $3:41$. Which is the better approximation?

10. Find to the nearest tenth of a per cent the error in Heron's value $\frac{45}{4}$ for the ratio $A_{12}:s_{12}^2$.

11. Complete the steps in Heron's solution of the problem of finding the diameter of a circle if the sum of the diameter and the perimeter and the area is 212.

12. Prove Aristarchus' inequality $\frac{1}{20} < \sin 3° < \frac{1}{18}$.

13. Hipparchus knew from eclipse observations that the lunar parallax (that is, the angle subtended by the earth at a point on the moon) is about 2°. What lunar distance does this imply?

14. Write in Greek notation the chord of 45°.

15. Find, without tables, sin 15° and from this write down in Greek alphabetic notation Ptolemy's value for chord 30°.

16. Write in Greek notation the chord of 150°.

17. If the Archimedean and Ptolemaic values of π are expressed as improper common fractions, and if a new fraction is formed by the difference of the two numerators over the difference of the two denominators, a better approximation, known to the Chinese, is found. How accurate is this new approximation?

18. Prove the theorem of Aristarchus that if $\beta < \alpha < 90°$, then $\sin \alpha/\sin \beta < \alpha/\beta$.

19. Prove the two lemmas of Menelaus.

20. Prove, either geometrically or trigonometrically, the theorem of Menelaus for plane triangles.

21. Complete the proof of Ptolemy's theorem.

22. Using the theorem of Ptolemy (with a diameter of the circle as one side of the quadrilateral), derive the formulas for $\sin(x + y)$ and $\cos(x \pm y)$.

23. Using Ptolemy's method for half angles, derive a formula for $\cos x/2$.

*24. Find exactly, in terms of radicals, the ratio of the area of a regular decagon to the square on a side. Is your value greater or less than the value $\frac{15}{2}$ given by Heron?

CHAPTER XI

Revival and Decline of Greek Mathematics

> Bees... by virtue of a certain geometrical forethought
> ... know that the hexagon is greater than the square
> and the triangle and will hold more honey for the same
> expenditure of material.
>
> *Pappus of Alexandria*

1 Today we use the conventional phrase "Greek mathematics" as though it indicated a homogeneous and well-defined body of doctrine. Such a view can be very misleading, however, for it implies that the sophisticated geometry of the Archimedean–Apollonian type was the only sort that the Hellenes knew. We must remember that mathematics in the Greek world spanned a time interval from at least 600 B.C. to at least A.D. 600 and that it traveled from Ionia to the toe of Italy, to Athens, to Alexandria, and to other parts of the civilized world. The intervals in time and space alone produced changes in the depth and extent of mathematical activity, for Greek science did not have the sameness, century after century, that is found in pre-Hellenic thought. Moreover, even at any given time and place in the Greek world (as in our civilization today) there were sharp differences in the level of mathematical interest and accomplishment. We have seen how even in the work of a single individual, such as Ptolemy, there can be two types of scholarship—the *Almagest* for the "tough-minded" rationalists and the *Tetrabiblos* for the "tender-minded" mystics. It is probable that there always were at least two levels of mathematical understanding, but that the paucity of surviving works, especially on the lower level, tends to obscure this fact. The phrase used as the title for this chapter must itself be accepted with some hesitation, for although it is justified in the light of what we know about the Greek world, our knowledge is far from complete. The period that we consider in this chapter, from Ptolemy to Proclus, covers almost four centuries (from the second to the sixth), but our account is based in large part on only two chief treatises, only portions of which are now extant, as well as on a number of works of lesser significance.

Heron and Ptolemy were Greek scholars, but they lived in a world dominated politically by Rome. The death of Archimedes by the hand of a Roman soldier may have been inadvertent, but it was truly portentous. Throughout its long history, ancient Rome contributed little to science or philosophy and less to mathematics. Whether during the Republic or in the days of the Empire, Romans were little attracted to speculative or logical investigation. The practical arts of medicine and agriculture were cultivated with some eagerness, and descriptive geography met with favor. Impressive engineering projects and architectural monuments were related to the simpler aspects of science, but Roman builders were satisfied with elementary rule-of-thumb procedures that called for little in the way of understanding of the great corpus of Greek thought. The extent of Roman acquaintance with science may be judged from the *De architectura* of Vitruvius, written during the middle part of the Augustine Age and dedicated to the emperor. At one point the author describes what to him appeared to be the three greatest mathematical discoveries: the incommensurability of the side and diagonal of a cube; the right triangle with sides 3, 4, and 5; and Archimedes' calculation on the composition of the king's crown. Marcus Vitruvius Pollio, the author, was especially interested in surveying instruments and in problems involving approximate mensurations. The perimeter of a wheel of diameter 4 feet is given by Vitruvius as $12\frac{1}{2}$ feet, implying a value of $3\frac{1}{8}$ for π. This is not so good an approximation as that of Archimedes, with whose works Vitruvius was probably only slightly acquainted, but it is of a respectable degree of accuracy for Roman purposes. It is sometimes claimed that impressive works of engineering, such as the Egyptian pyramids and the Roman aqueducts, imply a high level of mathematical achievement, but historical evidence does not bear this out. Just as earlier Egyptian mathematics had been on a lower plane than that in Babylon of the same period, so Roman mathematics was on a much lower level than that in Greece during the same years. The Romans were almost completely lacking in mathematical drive, so that their best efforts, such as those of Vitruvius, were not comparable to the poorer results in Greece, as exemplified by the work of Heron.[1]

We have seen that Greek mathematics was not uniformly on a high level, for the glorious period of the third century B.C. had been followed by a decline, perhaps to some extent arrested in the days of Ptolemy, but not effectively reversed until the century of the "Silver Age," about A.D. 250 to 350. At the beginning of this period, also known as the Later Alexandrian Age, we find the leading Greek algebraist, Diophantus of Alexandria, and toward its close there appeared the last significant Greek geometer, Pappus of Alexandria.

[1] A devastating comparison of Roman science with that of Greece is presented by W. H. Stahl, *Roman Science* (1962).

No other city has been the center of mathematical activity for so long a period as was Alexandria from the days of Euclid (ca. 300 B.C.) to the time of Hypatia (†415). It was a very cosmopolitan center, and the mathematics that resulted from Alexandrian scholarship was not all of the same type. The results of Heron were markedly different from those of Euclid or Apollonius or Archimedes, and again there is an abrupt departure from the classical Greek tradition in the extant work of Diophantus. Little is known of Diophantus' life beyond a tradition that is reported in a collection of problems dating from the fifth or sixth century, known as the "Greek Anthology" (described below):

> God granted him to be a boy for the sixth part of his life, and adding a twelfth part to this, He clothed his cheeks with down; He lit him the light of wedlock after a seventh part, and five years after his marriage He granted him a son. Alas! late-born wretched child; after attaining the measure of half his father's life, chill Fate took him. After consoling his grief by this science of numbers for four years he ended his life.[2]

If this conundrum is historically accurate, Diophantus lived to be eighty-four-years old. It should definitely not be taken as typical of the problems that interested Diophantus, for he paid little attention to equations of first degree.

3   Diophantus is often called the father of algebra, but we shall see that such a designation is not to be taken literally. His work is not at all the type of material forming the basis of modern elementary algebra; nor is it yet similar to the geometric algebra found in Euclid. The chief Diophantine work known to us is the *Arithmetica*, a treatise originally in thirteen books, only the first six of which have survived.[3] It should be recalled that in ancient Greece the word arithmetic meant theory of numbers, rather than computation. Often Greek arithmetic had more in common with philosophy than with what we think of as mathematics; hence the subject had played a large role in Neoplatonism during the Later Alexandrian Age. This had been particularly true of the *Introductio arithmeticae* of Nicomachus of Gerasa, a Neo-Pythagorean who lived not far from Jerusalem about the year 100. The author sometimes is held to be of Syrian background, but Greek philosophical tendencies certainly predominate in his work. The *Introductio* of Nicomachus, as we have it, contains only two books, and it is possible that this is only an abridged version of what originally was a more extensive treatise. At all events, the possible loss in this case is far less to be regretted than the loss of seven books of the *Arithmetica* of Diophantus, for there is a world of difference between

---

[2] Quoted from Cohen and Drabkin, *Source Book in Greek Science* (1958), p. 27. Uncertainty about the life of Diophantus is so great that we do not know definitely in which century he lived. Generally he is assumed to have flourished about 250, but dates a century or more earlier or later are sometimes suggested.

[3] For a full account see T. L. Heath, *Diophantus of Alexandria* (1910).

Four antique mathematicians who contributed also to music: Boethius, Pythagoras, Plato, Nicomachus; from a Boethius manuscript, Cambridge.

the two authors. Nicomachus had, so far as we can see, little mathematical competence and was concerned only with the most elementary properties of numbers. The level of the work may be judged from the fact that the author found it expedient to include a multiplication table up to ι times ι (that is, 10 times 10). If this is genuine and not just a later interpolation, it is the oldest surviving Greek instance of such a table, although many older Babylonian multiplication tables are extant.

The *Introductio* of Nicomachus opens with the anticipated Pythagorean classification of numbers into even and odd, then into evenly even (powers of two) and evenly odd ($2^n \cdot p$, where p is odd and $p > 1$ and $n > 1$) and oddly even ($2 \cdot p$, where p is odd and $p > 1$). Prime, composite, and perfect numbers are defined, including a description of the sieve of Eratosthenes and a list of the first four perfect numbers (6 and 28 and 496 and 8128). The work includes also a classification of ratios and combinations of ratios (for ratios of integers are essential in the Pythagorean theory of musical intervals), an extensive treatment of figurate numbers (which had loomed so large in Pythagorean arithmetic) in both two and three dimensions, and a comprehensive account of the various means (again a favorite topic in Pythagorean philosophy). As some other writers, Nicomachus regarded the number three as the first number in the strict sense of the word, for one and two were really only the generators of the number system. For Nicomachus, numbers were endowed with such qualities as better or worse, younger or older; and they could transmit characters, as parents to their progeny. Despite such arithmetical anthropomorphism as a background, the *Introductio* contains a moderately sophisticated theorem. Nicomachus noticed that if the odd integers are grouped in the pattern $1; 3 + 5; 7 + 9 + 11; 13 + 15 + 17 + 19; \ldots,$ the successive sums are the cubes of the integers. This observation, coupled with the early Pythagorean recognition that the sum of the first n odd numbers is n^2, leads to the conclusion that the sum of the first n perfect cubes is equal to the square of the sum of the first n integers.

The *Introductio* of Nicomachus[4] was neither a treatise on calculation nor one on algebra, but a handbook on those elements of mathematics that were essential to an understanding of Pythagorean and Platonic philosophy; as such it served as a model for later imitators and commentators. Among these the best known were Theon of Smyrna (fl. ca. 125), who wrote his *Expositio* in Greek, and Boethius (†524), who wrote his *Arithmetica*, long afterward, in Latin. These men, like Nicomachus, were far more concerned about the application of arithmetic to music and Platonic philosophy than in advancing

---

[4] For an English translation see Nicomachus of Gerasa, *Introduction to Arithmetic*, trans. by M. L. D'Ooge (1926). This very useful edition includes also an extensive introduction that places the work of Nicomachus in clear historical perspective. D'Ooge concluded from the evidence that Nicomachus was Greek rather than Syrian.

the subject itself. The full title of the *Expositio* indicates, in fact, that it is an exposition of mathematical matters useful to an understanding of Plato.[5] It explains, for example, that the tetractys consisting of the numbers 1, 2, 3, and 4 contains all the musical consonances inasmuch as it makes up the ratios 4:3, 3:2, 2:1, 3:1, and 4:1. The *Arithmetica* of Boethius is quite unoriginal, being almost a translation of the earlier work by Nicomachus.[6]

Quite different from the works of Nicomachus, Theon, and Boethius   **4** was the *Arithmetica* of Diophantus, a treatise characterized by a high degree of mathematical skill and ingenuity. In this respect the book can be compared with the great classics of the earlier Alexandrian Age; yet it has practically nothing in common with these or, in fact, with any traditional Greek mathematics. It represents essentially a new branch and makes use of a different approach. Being divorced from geometrical methods, it resembles Babylonian algebra to a large extent; but whereas Babylonian mathematicians had been concerned primarily with the *approximate* solution of *determinate* equations as far as the third degree, the *Arithmetica* of Diophantus (such as we have it) is almost entirely devoted to the *exact* solution of equations, both *determinate* and *indeterminate*. Because of the emphasis given in the *Arithmetica* to the solution of indeterminate problems, the subject dealing with this topic, sometimes known as indeterminate analysis, has since become known as Diophantine analysis. Since this type of work today is generally a part of courses in theory of numbers, rather than elementary algebra, it is not an appropriate basis for regarding Diophantus as the father of algebra. There is another respect, however, in which such a paternity is justified. Algebra now is based almost exclusively on symbolic forms of statement, rather than on the customary written language of ordinary communication in which earlier Greek mathematics, as well as Greek literature, had been expressed. It is generally held that three stages in the historical development of algebra can be recognized: (1) the rhetorical or early stage, in which everything is written out fully in words; (2) a syncopated or intermediate stage, in which some abbreviations are adopted; and (3) a symbolic or final stage. Such an arbitrary division of the development of algebra into three stages is of course a facile oversimplification; but it can serve effectively as a first approximation to what has happened, and within such a framework the *Arithmetica* of Diophantus is to be placed in the second category.

Throughout the six surviving books of the *Arithmetica* there is a systematic use of abbreviations for powers of numbers and for relationships and operations. An unknown number is represented by a symbol resembling the Greek

---

[5] There is an excerpt, in English translation, in Cohen and Drabkin, *Source Book in Greek Science*, pp. 294–298.
[6] Marshall Clagett, *Greek Science in Antiquity*, pp. 185–186.

letter ς (perhaps for the last letter of arithmos); the square of this appears as Δ^{γ}, the cube as K^{γ}, the fourth power, called square-square, as $\Delta^{\gamma}\Delta$, the fifth power or square-cube as ΔK^{γ}, and the sixth power or cube-cube as $K^{\gamma}K$. Diophantus was of course familiar with the rules of combination equivalent to our laws of exponents, and he had special names for the reciprocals of the first six powers of the unknowns, quantities equivalent to our negative powers. Numerical coefficients were written after the symbols for the powers with which they were associated; addition of terms was understood in the appropriate juxtaposition of the symbols for the terms, and subtraction was represented by a single letter-abbreviation placed before the terms to be subtracted. With such a notation Diophantus was in a position to write polynomials in a single unknown almost as concisely as we do today. The expression $2x^4 + 3x^3 - 4x^2 + 5x - 6$, for example, might appear in a form equivalent to $SS2\ C3\ x5\ M\ S4\ u6$, where the English letters S, C, x, M, and u have been used for "square," "cube," the "unknown," "minus," and "unit," and with our present numerals in place of the Greek alphabetic notation that was used in the days of Diophantus. Greek algebra now no longer was restricted to the first three powers or dimensions, and the identities $(a^2 + b^2)(c^2 + d^2) = (ac + bd)^2 + (ad - bc)^2 = (ac - bd)^2 + (ad + bc)^2$, which played important roles in Medieval algebra and modern trigonometry, appear in the work of Diophantus. The chief difference between the Diophantine syncopation and the modern algebraic notation is in the lack of special symbols for operations and relations, as well as of the exponential notation. These missing elements of notation were largely contributions of the period from the late fifteenth to the early seventeenth centuries in Europe.

5   If we think primarily of matters of notation, Diophantus has a good claim to be known as the father of algebra, but in terms of motivation and concepts the claim is less appropriate. The *Arithmetica* is not a systematic exposition of the algebraic operations or of algebraic functions or of the solution of algebraic equations. It is instead a collection of some 150 problems, all worked out in terms of specific numerical examples, although perhaps generality of method was intended. There is no postulational development, nor is an effort made to find all possible solutions. In the case of quadratic equations with two positive roots, only the larger is given, and negative roots are not recognized. No clear-cut distinction is made between determinate and indeterminate problems, and even for the latter, for which the number of solutions generally is unlimited, only a single answer is given. Diophantus solved problems involving several unknown numbers by skillfully expressing all unknown quantities, where possible, in terms of only one of them. Two problems from the *Arithmetica* will serve to illustrate the Diophantine

approach. In finding two numbers such that their sum is 20 and the sum of their squares is 208, the numbers are not designated as x and y, but as $10 + x$ and $10 - x$ (in terms of our modern notation). Then $(10 + x)^2 + (10 - x)^2 = 208$, hence $x = 2$; so the numbers sought are 8 and 12. Diophantus handled also the analogous problem in which the sum of the two numbers and the sum of the cubes of the numbers are given as 10 and 370 respectively.

In these problems he is dealing with a determinate equation, but Diophantus used much the same approach in indeterminate analysis. In one problem it is required to find two numbers such that either when added to the square of the other will yield a perfect square. This is a typical instance of Diophantine analysis in which only rational numbers are acceptable as answers. In solving the problem Diophantus did not call the numbers x and y, but rather x and $2x + 1$. Here the second, when added to the square of the first, will yield a perfect square no matter what value one chooses for x. Now, it is required also that $(2x + 1)^2 + x$ must be a perfect square. Here Diophantus does not point out the infinity of possible answers. He is satisfied to choose a particular case of a perfect square, in this instance the number $(2x - 2)^2$, such that when equated to $(2x + 1)^2 + x$ an equation that is linear in x results. Here the result is $x = \frac{3}{13}$, so that the other number, $2x + 1$, is $\frac{19}{13}$. One could, of course, have used $(2x - 3)^2$ or $(2x - 4)^2$, or expressions of similar form, instead of $(2x - 2)^2$, to arrive at other pairs of numbers having the desired property. Here we see an approach that comes close to a "method" in Diophantus' work: when two conditions are to be satisfied by two numbers, the two numbers are so chosen that one of the two conditions is satisfied; and then one turns to the problem of satisfying the second condition. That is, instead of handling *simultaneous* equations on two unknowns, Diophantus operates with *successive* conditions so that only a single unknown number appears in the work.

Among the indeterminate problems in the *Arithmetica* are some involving **6** equations such as $x^2 = 1 + 30y^2$ and $x^2 = 1 + 26y^2$, which are instances of the so-called "Pell equation" $x^2 = 1 + py^2$; again a single answer is thought to suffice.[7] In a sense it is not fair to criticize Diophantus for being satisfied with a single answer, for he was solving problems, not *equations*. In a sense the *Arithmetica* is not an algebra textbook, but a problem collection in the application of algebra. In this respect Diophantus is like the Babylonian algebraists; and his work sometimes is regarded as "the finest flowering of

[7] See D. J. Struik, *A Concise History of Mathematics*, 3rd ed. (New York: Dover, 1967), p. 62. For a full account of the work of Diophantus see T. L. Heath, *Diophantus of Alexandria*. Cf. also J. A. Sánchez Pérez: *La arithmética en Grecia* (1947) and the article on Diophantus by F. O. Hultsch in Pauly-Wissowa, *Real-Encyclopädie der klassischen Altertumswissenschaft*, Vol. V (Stuttgart: Metzler, 1905), columns 1051–1073.

Babylonian algebra."[8] To some extent such a characterization is unfair to Diophantus, for his numbers are entirely abstract and do not refer to measures of grain or dimensions of fields or monetary units, as was the case in Egyptian and Mesopotamian algebra. Moreover, he is interested only in *exact* rational solutions, whereas the Babylonians were computationally inclined and were willing to accept approximations to irrational solutions of equations. Hence cubic equations seldom enter in the work of Diophantus, whereas among the Babylonians attention had been given to the reduction of cubics to the standard form $n^3 + n^2 = a$ in order to solve approximately through interpolation in a table of values of $n^3 + n^2$.

We do not know how many of the problems in the *Arithmetica* were original or whether Diophantus had borrowed from other similar collections. Possibly some of the problems or methods are traceable back to Babylonian sources, for puzzles and exercises have a way of reappearing generation after generation. To us today the *Arithmetica* of Diophantus looks strikingly original, but possibly this impression results from the loss of rival problem collections. Our view of Greek mathematics is derived from a relatively small number of surviving works, and conclusions derived from these necessarily are precarious. Indications that Diophantus may have been less isolated a figure than has been supposed are found in a collection of problems from about the early second century of our era (hence presumably antedating the *Arithmetica*) in which some Diophantine symbols appear.[9] Nevertheless, Diophantus has had a greater influence on modern number theory than any other nongeometric Greek algebraist. In particular, Fermat was led to his celebrated "great" or "last" theorem (see below) when he sought to generalize a problem that he had read in the *Arithmetica* of Diophantus (II. 8): to divide a given square into two squares.[10]

7 The *Arithmetica* of Diophantus is a brilliant work worthy of the period of revival in which it was written, but it is, in motivation and content, far removed from the beautifully logical treatises of the great geometrical triumvirate of the earlier Alexandrian Age. Algebra seemed to be more appropriate for problem-solving than for deductive exposition, and the great work of Diophantus remained outside the mainstream of Greek mathematics. A minor work on polygonal numbers by Diophantus comes closer to the earlier Greek interests, but even this cannot be regarded as approaching the Greek logical ideal. Classical geometry had found no ardent supporter, with the

---

[8] See J. D. Swift, "Diophantus of Alexandria," *American Mathematical Monthly,* **43** (1956), 163–170.

[9] See F. E. Robbins, "P. Mich. 620: A Series of Arithmetical Problems," *Classical Philology,* **24** (1929), 321–329, and Kurt Vogel, "Die algebraischen Probleme des P. Mich. 620," *Classical Philology,* **25** (1930), 373–375.

[10] See Heath, *Diophantus of Alexandria,* pp. 144–145.

possible exception of Menelaus, since the death of Apollonius some four hundred and more years before. But during the reign of Diocletian (284–305) there lived again at Alexandria a scholar who was moved by the spirit that had possessed Euclid, Archimedes, and Apollonius. Pappus of Alexandria in about 320 composed a work with the title *Collection* (Synagoge) which is important for several reasons. In the first place it provides a most valuable historical record of parts of Greek mathematics that otherwise would be unknown to us. For instance, it is in Book V of the *Collection* that we learn of Archimedes' discovery of the thirteen semiregular polyhedra or "Archimedian solids." Then, too, the *Collection* includes alternative proofs and supplementary lemmas for propositions in Euclid, Archimedes, Apollonius, and Ptolemy. Finally, the treatise includes new discoveries and generalizations not found in any earlier work. The *Collection*, Pappus' most important treatise, contained eight books, but the first book and the first part of the second book are now lost. In this case the loss is less to be regretted than is that of the last books of Diophantus' *Arithmetica*, for it appears that the first two books of the *Collection* were chiefly concerned with the principles of Apollonius' system of tetrads in Greek numeration. Since we have, in the *Sand-Reckoner*, the corresponding system of octads from Archimedes, we can judge quite well what material has been lost from the exposition of Pappus.

Book III of the *Collection* shows that Pappus shared thoroughly the **8** classical Greek appreciation of the niceties of logical precision in geometry. Here he distinguishes sharply between "plane," "solid," and "linear" problems—the first being constructible with circles and straight lines only, the second being solvable through the use of conic sections, and the last requiring curves other than lines, circles, and conics. Then Pappus describes some solutions of the three famous problems of antiquity, the duplication and trisection being problems in the second or solid category and the squaring of the circle being a linear problem. Pappus virtually here asserts the fact that the classical problems are impossible of solution under the Platonic conditions, for they do not belong among the plane problems; but rigorous proofs were not given until the nineteenth century.

In Book IV Pappus again is insistent that one should give for a problem a construction appropriate to it. That is, one should not use linear loci in the solution of a solid problem, nor solid or linear loci in the solution of a plane problem. Asserting that the trisection of an angle is a solid problem, he therefore suggests methods that make use of conic sections, whereas Archimedes in one case had used a *neusis* or sliding-ruler type of construction and in another the spiral, which is a linear locus. One of the Pappus trisections is as follows. Let the given angle *AOB* be placed in a circle with center

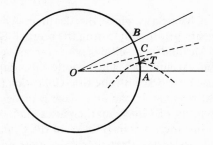

FIG. 11.1

O (Fig. 11.1) and let *OC* be the angle bisector. Draw the hyperbola having *A* as one focus, *OC* as the corresponding directrix, and with an eccentricity equal to 2. Then one branch of this hyperbola will cut the circumference of the circle in a point *T* such that ∠ *AOT* is one-third ∠ *AOB*.

A second trisection construction proposed by Pappus makes use of an equilateral hyperbola as follows. Let the side *OB* of the given angle *AOB* be a diagonal of a rectangle *ABCO* and through *A* draw the equilateral hyperbola having *BC* and *OC* (extended) as asymptotes (Fig. 11.2). With *A* as center

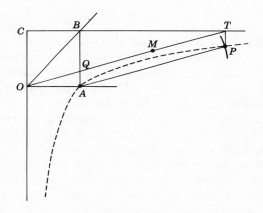

FIG. 11.2

and with radius twice *OB* draw a circle intersecting the hyperbola in *P* and from *P* drop the perpendicular *PT* to the line *CB* extended. Then it is readily proved, from the properties of the hyperbola, that the straight line through *O* and *T* is parallel to *AP* and that ∠ *AOT* is one-third ∠ *AOB*. Pappus gives no source for his trisections, and we cannot help but wonder if this trisection was known to Archimedes. If we draw the semicircle passing through *B*, having

QT as diameter and M as center, we have essentially the Archimedean *neusis* construction, for $OB = QM = MT = MB$.

In Book III Pappus describes also the theory of means and gives an attractive construction that includes the arithmetic, the geometric, and the harmonic mean within a single semicircle. Pappus shows that if in the semicircle ADC with center O (Fig. 11.3) one has $DB \perp AC$ and $BF \perp OD$, then DO is the arithmetic mean, DB the geometric mean, and DF the harmonic mean of the magnitudes AB and BC. Here Pappus claims for himself only the proof, attributing the diagram to an unnamed geometer. Even when Pappus names his source, it sometimes is not otherwise known to us, indicating how inadequate is our information on mathematicians of his day.

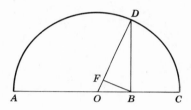

FIG. 11.3

The *Collection* of Pappus is replete with bits of interesting information and **9** significant new results. In many cases the novelties take the form of generalizations of earlier theorems, and a couple of these instances appear in Book IV. Here we find an elementary generalization of the Pythagorean theorem. If ABC is *any* triangle (Fig. 11.4) and if $ABDE$ and $CBGF$ are *any* parallelograms constructed on two of the sides, then Pappus constructs on side AC a third parallelogram $ACKL$ equal to the sum of the other two. This is easily accomplished by extending sides FG and ED to meet in H, then drawing HB and extending it to meet side AC in J, and finally drawing AL and CK parallel

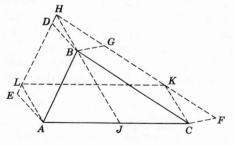

FIG. 11.4

to *HBJ*. It is not known whether or not this generalization, usually bearing the name of Pappus was original with Pappus, and it has been suggested that possibly it was known earlier to Heron.

Another instance of generalization in Book IV, also bearing Pappus' name, extends theorems of Archimedes on the shoemaker's knife. It asserts that if circles $C_1, C_2, C_3, C_4, \ldots, C_n, \ldots$ are inscribed successively as in Fig. 11.5, all being tangent to the semicircles on *AB* and on *AC*, and successively to each other, the perpendicular distance from the center of the *n*th circle to the base line *ABC* is *n* times the diameter of the *n*th circle.[11]

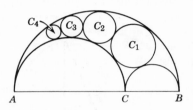

FIG. 11.5

10   Book V of the *Collection* was a favorite with later commentators, for it raised the question of the sagacity of bees. Inasmuch as Pappus showed that of two regular polygons having equal perimeters the one with the greater number of sides has the greater area, he concluded that bees demonstrated some degree of mathematical understanding in constructing their cells as hexagonal, rather than square or triangular, prisms. The book goes into other problems of isoperimetry, including a demonstration that the circle has a greater area, for a given perimeter, than does any regular polygon. Here Pappus seems to have been following closely a work *On Isometric Figures* written almost half a millennium earlier by Zenodorus (ca. 180 B.C.), some fragments of which were preserved by later commentators. Among the propositions in Zenodorus' treatise was one asserting that of all solid figures the surfaces of which are equal the sphere has the greatest volume, but only an incomplete justification was given.[12]

Books VI and VIII of the *Collection* are chiefly on applications of mathematics to astronomy, optics, and mechanics (including an unsuccessful

[11] An indicated proof of the theorem will be found in R. A. Johnson, *Modern Geometry* (New York: Houghton Mifflin, 1929), p. 117.

[12] See Heath: *History of Greek Mathematics* (1921), II, 207 ff. A fascinating account of such matters is found in D'Arcy Wentworth Thompson: *On Growth and Form*, 2nd ed. (Cambridge University Press, 1942).

attempt at finding the law of the inclined plane). Of far more significance in the history of mathematics is Book VII, in which, through his penchant for generalization, Pappus came close to the fundamental principle of analytic geometry. The only means recognized by the ancients for defining plane curves were (1) kinematic definitions in which a point moves subject to two superimposed motions and (2) the section by a plane of a geometrical surface, such as a cone or sphere or cylinder. Among the latter curves were certain quartics known as spiric sections, described by Perseus (ca. 150 B.C.), obtained by cutting the anchor ring or torus by a plane. Occasionally a twisted curve caught the attention of the Greeks, including the cylindrical helix and an analogue of the Archimedean spiral described on a spherical surface, both of which were known to Pappus; but Greek geometry was primarily restricted to the study of plane curves, in fact, to a very limited number of plane curves. It is significant to note, therefore, that in Book VII of the *Collection* Pappus proposed a generalized problem that implied infinitely many new types of curves. This problem, even in its simplest form, usually is known as the "Pappus problem," but the original statement, involving three or four lines, seems to go back to the days of Euclid. As first considered, the problem is referred to as "the locus to three or four lines," described above in connection with the work of Apollonius. Euclid evidently had identified the locus for certain special cases only, but it appears that Apollonius, in a work now lost, had given a complete solution. Pappus nevertheless gave the impression that geometers had failed in attempts at a general solution and implied that it was he who had first shown the locus in all cases to be a conic section.

More importantly, Pappus then went on to consider the analogous problem for more than four lines. For six lines in a plane he recognized that a curve is determined by the condition that the product of the distances from three of the lines shall be in a fixed ratio to the product of the distances to the other three lines. In this case a curve is defined by the fact that a solid is in a fixed ratio to another solid. Pappus hesitated to go on to cases involving more than six lines inasmuch as "there is not anything contained by more than three dimensions." But, he continued, "men a little before our time have allowed themselves to interpret such things, signifying nothing at all comprehensible, speaking of the product of the content of such and such lines by the square of this or the content of those. These things might however be stated and shown generally by means of compounded proportions." The unnamed predecessors evidently were prepared to take a highly important step in the direction of an analytic geometry that should include curves of degree higher than three, just as Diophantus had used the expressions square-square and cube-cube for higher powers of numbers. Had Pappus pursued the suggestion further, he might have anticipated Descartes in a general classification and theory of curves far beyond the classical distinction between plane, solid, and linear loci. His recognition that, no matter what the number of lines in the Pappus problem, a specific curve is determined, is the most general observation on loci in all of ancient geometry, and the algebraic syncopations that Diophantus had developed would have been adequate to have disclosed some of the properties of the curves. But Pappus was at heart a geometer only, as Diophantus had been an algebraist only; hence Pappus merely remarked with surprise that no one had made a synthesis of this problem for any case beyond that of four lines. Pappus himself made no deeper study of these loci, "of which one has no further knowledge and which are simply called curves."[13] What was needed for the next step in this connection was the appearance of a mathematician equally concerned for algebra and geometry; it is significant to note that when such a figure appeared in the person of Descartes, it was this very problem of Pappus that served as the point of departure in the invention of analytic geometry.

11   There are other important topics in Book VII of the *Collection*, apart from the Pappus problem. For one thing, there is a full description of what was called the method of analysis and of a collection of works known as the *Treasury of Analysis*. Pappus describes analysis as "a method of taking that which is sought as though it were admitted and passing from it through its consequences in order to something which is admitted as a result of synthesis." That is, he recognized analysis as a "reverse solution," the steps of which must be retraced in opposite order to constitute a valid demonstration. If analysis leads to something admitted to be impossible, the problem also will be impossible, for a false conclusion implies a false premise. Pappus explains that the method of analysis and synthesis is used by the authors whose works constitute the *Treasury of Analysis*: "This is a body of doctrine furnished for the use of those who, after going through the usual elements, wish to obtain power to solve problems set to them involving curves"; and Pappus lists among the works in the *Treasury of Analysis* the treatises on conics by Aristaeus, Euclid, and Apollonius. It is from Pappus' description that we learn that Apollonius' *Conics* contained 487 theorems. Since the seven books now extant comprise 382 propositions, we can conclude that the lost eighth book had 105 propositions. About half of the works listed by Pappus in the *Treasury of Analysis* are now lost, including Apollonius' *Cutting-off of a Ratio*, Eratosthenes' *On Means*, and Euclid's *Porisms*. It has been suggested that a porism was an antique equivalent of our equation of a curve or locus, indicating that Euclid and Pappus may not have been as far removed from what we call "analytic geometry" as generally is supposed.

[13] There is no English translation of the *Collection* of Pappus, but extensive accounts of it will be found in Heath, *History of Greek Mathematics*, and in I. Thomas, *Selections Illustrating the History of Greek Mathematics*. There is a convenient French translation of the *Collection*, made by Paul Ver Eecke, (Paris: Desclée de Brouwer, 1933, 2 vols.).

Book VII of the *Collection* contains the first statement on record of the **12** focus-directrix property of the three conic sections. It appears that Apollonius knew of the focal properties for central conics, but it is possible that the focus-directrix property for the parabola was not known before Pappus. Another theorem in Book VII that appears for the first time is one usually named for Paul Guldin, a seventeenth-century mathematician: If a closed plane curve is revolved about a line not passing through the curve, the volume of the solid generated is found by taking the product of the area bounded by the curve and the distance traversed during the revolution by the center of gravity of the area. Pappus was rightfully proud of this very general theorem, for it included "a large number of theorems of all sorts about curves, surfaces and solids, all of which are proved simultaneously by one demonstration." It is indeed the most general theorem involving the calculus to be found in antiquity. Pappus gave also the analogous theorem that the surface area generated by the revolution of a curve about a line not cutting the curve is equal to the product of the length of the curve and the distance traversed by the centroid of the curve during the revolution.[14]

The *Collection* of Pappus is the last truly significant ancient mathematical treatise, for the attempt of the author to revive geometry was not successful. Mathematical works continued to be written in Greek for about another thousand years, continuing an influence that had begun almost a millennium before, but authors following Pappus never again rose to his level. Their works are almost exclusively in the form of commentary on earlier treatises. Pappus himself is in part responsible for the ubiquitous commentaries that ensued, for he had composed commentaries on the *Elements* of Euclid and on the *Almagest* of Ptolemy, among others, only fragments of which survive. Later commentaries, such as those of Theon of Alexandria (fl. 365), are more useful for historical information than for mathematical results. Theon was responsible also for an important edition of the *Elements* that has survived; he is remembered also as the father of Hypatia, a learned young lady who wrote commentaries on Diophantus, Ptolemy, and Apollonius. An ardent devotee of pagan learning, Hypatia incurred the enmity of a fanatical Christian mob at whose hands she suffered a cruel death in 415. The dramatic impact of her death in Alexandria has caused that year to be taken by some to mark the end of ancient mathematics, but a more appropriate close is found another century later.

Alexandria produced in Proclus (410–485) a young mathematical scholar **13** who went to Athens, where be became the head of the Neoplatonic school.

---

[14] There is a possibility that the "Guldin theorem" represents an interpolation in the manuscript of the *Collection*. (See the Ver Eecke translation cited in footnote 12.) In any case, the theorem represents a striking advance by someone during or following the long period of decline.

Proclus was more the philosopher than the mathematician, but his remarks are often critical for the history of early Greek geometry. Of great significance is his *Commentary on Book I of the Elements* of Euclid, for, while writing this, Proclus undoubtedly had at hand a copy of the *History of Geometry* by Eudemus, now lost, as well as Pappus' *Commentary on the Elements*, largely lost. For our information on the history of geometry before Euclid we are heavily indebted to Proclus, who included in his *Commentary* a summary or substantial extract from Eudemus' *History*. This passage, which has come to be known as the *Eudemian Summary*, may be taken as Proclus' chief contribution to mathematics, although to him is ascribed the theorem that if a line segment of fixed length moves with its end points on two intersecting lines, a point on the segment will describe a portion of an ellipse.

14   During the years when Proclus was writing in Athens, the Roman Empire in the West was gradually collapsing. The end of the empire usually is placed at 476, for in this year the incumbent Roman emperor was displaced by Odoacer, a Goth. Some of the old Roman senatorial pride remained, but the senatorial party had lost political control. In this situation Boethius (ca. 480–524) found his position difficult, for he came of an old distinguished patrician family. He was not only a philosopher and mathematician, but also a statesman, and he probably viewed with distaste the rising Ostrogothic power. Although Boethius may have been the foremost mathematician produced by ancient Rome, the level of his work is a far cry from that characteristic of Greek writers. He was the author of textbooks for each of the four mathematical branches in the liberal arts, but these were jejune and exceedingly elementary abbreviations of earlier classics—an *Arithmetic* that was only an abridgement of the *Introductio* of Nicomachus; a *Geometry* based on Euclid and including statements only, without proof, of some of the simpler portions of the first four books of the *Elements*; an *Astronomy* derived from Ptolemy's *Almagest*; and a *Music* that is indebted to the earlier works of Euclid, Nicomachus, and Ptolemy. In some cases these primers, used extensively in medieval monastic schools, may have suffered later interpolations, hence it is difficult to determine precisely what is genuinely due to Boethius himself. It is nevertheless clear that the author was concerned primarily with two aspects of mathematics: its relationship to philosophy and its applicability to simple problems of mensuration. Of mathematics as a logical structure there is little trace.

Boethius seems to have been a statesman of high purpose and unquestioned integrity. He and his sons in turn served as consuls, and Boethius was among the chief advisers of Theodoric; but for some reason, whether political or religious, the philosopher incurred the displeasure of the emperor. It has been suggested that Boethius was a Christian (as perhaps Pappus was also) and that he espoused Trinitarian views that alienated the Arian emperor. It is possible also that Boethius was too closely associated with political elements that looked to the Eastern Empire for help in restoring the old Roman order in the West.[15] In any case, Boethius was executed in 524 or 525, following a long imprisonment. (Theodoric, incidentally, died only about a year later, in 526.) It was while in prison that he wrote his most celebrated work, *De consolatione philosophiae*. This essay, written in prose and verse while he faced death, discusses moral responsibility in the light of Aristotelian and Platonic philosophy.

The death of Boethius may be taken to mark the end of ancient math-   **15** ematics in the Western Roman Empire, as the death of Hypatia had marked the close of Alexandria as a mathematical center; but work continued for a few years longer at Athens. There one found no great original mathematician, but the Peripatetic commentator Simplicius (fl. 520) was sufficiently concerned about Greek geometry to have preserved for us what may be the oldest fragment extant. Aristotle in the *Physica* had referred to the quadrature of the circle or of a segment, and Simplicius took this opportunity to quote "word for word" what Eudemus had written on the subject of the quadrature of lunes by Hippocrates. The account, several pages long, gives full details on the quadratures of lunes, quoted by Simplicius from Eudemus, who in turn is presumed to have given at least part of the proofs in Hippocrates' own words, especially where certain archaic forms of expression are used. This source is the closest we can come to direct contact with Greek mathematics before the days of Plato.

Simplicius was primarily a philosopher, but in his day there circulated a   **16** work usually described as the *Greek Anthology*, the mathematical portions of which remind us strongly of the problems in the Ahmes Papyrus of more than two millennia earlier. The *Anthology* contained some six thousand epigrams; of these more than forty are mathematical problems, collected presumably by Metrodorus, a grammarian of perhaps the fifth or sixth century. Most of them, including the epigram above on the age of Diophantus, lead to simple linear equations. For example, one is asked to find how many apples are in a collection if they are to be distributed among six persons so that the first person receives one third of the applies, the second receives one fourth, the third person receives one fifth, the fourth person receives one eighth, the fifth person receives ten apples, and there is one apple left for the last person. Another problem is typical of elementary algebra texts of our

[15] See Helen M. Barrett, *Boethius. Some Aspects of His Times and Work* (Cambridge University Press, 1940). Brief extracts from works of Boethius are included in Cohen and Drabkin, *Source Book in Greek Science*, pp. 291–294, 298–299.

day: If one pipe can fill a cistern in one day, a second in two days, a third in three days, and a fourth in four days, how long will it take all four running together to fill it? The problems presumably were not original with Metrodorus, but were collected from various sources. Some probably go back before the days of Plato, reminding us that not all Greek mathematics was of the type that we think of as classical.

17   Simplicius and Metrodorus were not the outstanding mathematicians of their day, for there were contemporary commentators with training adequate for an understanding of the works of Archimedes and Apollonius. Among these was Eutocius (born ca. 480), who commented on several Archimedean treatises and on the Apollonian *Conics*. It is to Eutocius that we owe the Archimedean solution of a cubic through intersecting conics, referred to in *The Sphere and Cylinder* but not otherwise extant except through the commentary of Eutocius. The commentary by Eutocius on the *Conics* of Apollonius was dedicated to Anthemius of Tralles (†534), an able mathematician and architect of St. Sophia of Constantinople, who described the string construction of the ellipse and wrote a work *On Burning-mirrors* in which the focal properties of the parabola are described. His colleague and successor in the building of St. Sophia, Isidore of Miletus (fl. 520), also was a mathematician of some ability. It was Isidore who made known the commentaries of Eutocius and spurred a revival of interest in the works of Archimedes and Apollonius. To him perhaps we owe the familiar T-square and string construction of the parabola—and possibly also the apocryphal Book XV of Euclid's *Elements*. It may be in large measure due to the activities of the Constantinople group—Eutocius, Isidore, and Anthemius—that Greek versions of Archimedean works and of the first four books of Apollonius' *Conics* have survived to this day.

Isidore of Miletus was one of the last directors of the Platonic Academy at Athens. The school had of course undergone many changes throughout its existence of more than 900 years, and during the days of Proclus it had become a center of Neoplatonic learning. When in 527 Justinian became emperor in the East, he evidently felt that the pagan learning of the Academy and other philosophical schools at Athens was a threat to orthodox Christianity; hence in 1529 the philosophical schools were closed and the scholars dispersed. Rome at the time was scarcely a very hospitable home for scholars, and Simplicius and some of the other philosophers looked to the East for a haven. This they found in Persia, where under King Chosroes they established what might be called the "Athenian Academy in Exile."[16] The date 529 may therefore be taken to mark the close of European mathematical

[16] See George Sarton, *The History of Science* (Cambridge, Mass.: Harvard University Press, 1952–1959, 2 vols.), I, 400.

development in antiquity. Henceforth the seeds of Greek science were to develop in Near and Far Eastern countries until, some 600 years later, the Latin world was in a more receptive mood. The date 529 has another significance that may be taken as symptomatic of a change in values—in this year the venerable monastery of Monte Cassino was established. Mathematics did not of course entirely disappear from Europe in 529, for undistinguished commentaries continued to be written in Greek in the Byzantine Empire and versions of the jejune Latin texts of Boethius continued in use in Western schools. The spirit of mathematics languished, however, while men argued less about the value of geometry and more about the way to salvation. For the next steps in mathematical development we must therefore turn our backs on Europe and look toward the East.

BIBLIOGRAPHY

Clagett, Marshall, *Greek Science in Antiquity* (New York: Abelard Schuman, 1955; paperback ed., Collier Books, 1963).

Cohen, M. R., and I. E. Drabkin, *Source Book in Greek Science* (New York: McGraw-Hill, 1948; reprinted, Cambridge, Mass.: Harvard University Press, 1958).

Chasles, Michel, *Les trois livres de porismes d'Euclide, rétablis... d'après la notice... de Pappus* (Paris: Mallet-Bachelier, 1860).

Heath, T. L., *Diophantus of Alexandria: A Study in the History of Greek Algebra*, 2nd ed. (New York: Cambridge University Press, 1910; paperback ed., New York: Dover, 1964).

Heath, T. L., *History of Greek Mathematics* (Oxford: Clarendon, 1921, 2 vols.).

Nesselmann, G. H. F., *Die Algebra der Griechen* (Berlin, 1842).

Nicomachus of Gerasa, *Introduction to Arithmetic*, trans. by M. L. D'Ooge, with studies in Greek arithmetic by F. E. Robbins and L. C. Karpinski (New York: Macmillan, 1926).

Pappus of Alexandria, *Collectionis quae supersunt*, ed. by F. Hultsch (Berlin, 1876–1878, 3 vols.).

Pappus of Alexandria, *La collection mathématique*, trans. by Paul Ver Eecke (Paris, 1933, 2 vols.).

Proclus Diadochus, *Les commentaires sur le premier livre des Éléments d'Euclide*, trans. by Paul Ver Eecke (Bruges: Desclee de Brouwer, 1948).

Sánchez Pérez, José Augusto, *La aritmética en Grecia* (Madrid: Instituto Jorge Juan, 1947).

Sánchez Pérez, José Augusto, *La aritmética en Roma, en India y en Arabia* (Madrid: Instituto Miguel Asín, 1949).

Stahl, W. H., *Roman Science* (Madison, Wis.: University of Wisconsin Press, 1962).

Swift, J. D., "Diophantus of Alexandria," *American Mathematical Monthly*, **43** (1956), 163–170.

Thomas, Ivor, ed., *Selections Illustrating the History of Greek Mathematics* (Cambridge, Mass.: Loeb Classical Library, 1939–1941, 2 vols.).

Van der Waerden, B. L., *Science Awakening*, trans. by Arnold Dresden (New York: Oxford, 1961; paperback ed., New York: Wiley, 1963).

Ziegler, Konrat, "Pappos," in Pauly-Wissowa, *Real-Enzyclopädie der klassischen Wissenschaft* (Stuttgart, 1949), Vol. XVIII, Part 3, columns 1084–1106.

EXERCISES

1. Do you think the conditions in Alexandria were more or less favorable for the development of mathematics in the days of Pappus than at the time of Ptolemy? Explain.

2. How did the intellectual conditions at Alexandria compare with those at Rome in the days of Diophantus and Pappus?

3. Would the development of mathematics have been essentially modified if Rome had not fallen in 476? Give reasons for your answer.

4. If you were a mathematician living in the year 500, would you have chosen Alexandria, Rome, Athens, or Constantinople as your home? Give reasons for your answer.

5. Show that the epigram concerning the age of Diophantus leads to the conclusion that he died at the age of eighty-four.

6. Verify that the four numbers listed by Nicomachus as perfect are indeed perfect numbers.

7. Solve the problem of Diophantus in which it is required to find two numbers such that their sum is 10 and the sum of their cubes is 370.

8. Find two rational fractions, other than $\frac{3}{13}$ and $\frac{9}{13}$ satisfying Diophantus' condition that either one when added to the other will produce a perfect square.

9. Prove that the lines OC, BD, and DF in Fig. 11.3 are indeed the arithmetic, the geometric, and the harmonic means, respectively, of AB and BC, as Pappus asserted.

10. Prove Pappus' generalization of the Pythagorean theorem illustrated in Fig. 11.4.

11. Draw carefully a diagram similar to Fig. 11.5 in which AB is 3 inches and BC is 2 inches and find approximately, by measurements, the diameter of the circle C_3 and the distance of its center from the line AC, thus verifying roughly the assertion of Pappus.

12. Solve the problem of the distribution of apples described in the text.

13. Solve the problem of the three pipes described in the text.

14. Show analytically that the Pappus problem for six lines leads to a locus the equation of which is not higher than third degree.

*15. Prove the first Pappus trisection given in the text.

*16. Prove that OT is parallel to AP in Fig. 11.2.

*17. Using the result in Exercise 16, complete the proof of the second Pappus trisection given in the text.

*18. Justify the Pappus theorem on solids of revolution.

*19. Prove the theorem of Proclus on the generation of an ellipse for the case in which the intersecting lines are mutually perpendicular.

CHAPTER XII

China and India

> A mixture of pearl shells and sour dates... or of costly crystal and common pebbles.
>
> *Al-Biruni's* India

The civilizations of China and India are of far greater antiquity than those of Greece and Rome, although not older than those in the Nile and Mesopotamian valleys. They go back to the Potamic Age, whereas the cultures of Greece and Rome were of the Thalassic Age. Civilizations along the Yangtze and Yellow rivers are comparable in age with those along the Nile or between the Tigris and Euphrates; but chronological accounts in the case of China are less dependable than those for Egypt and Babylonia. Claims that the Chinese made astronomical observations of importance, or described the twelve signs of the zodiac, by the fifteenth millennium B.C. are certainly unfounded, but a tradition that places the first Chinese empire about 2750 B.C. is not unreasonable. More conservative views place the early civilizations of China nearer 1000 B.C. The dating of mathematical documents from China is far from easy, and estimates concerning the *Chou Pei Suan Ching*, generally considered to be the oldest of the mathematical classics, differ by almost a thousand years. The problem of its date is complicated by the fact that it may well have been the work of several men of differing periods. Some consider the *Chou Pei* to be a good record of Chinese mathematics of about 1200 B.C. but others place the work in the first century before our era. A date of about 300 B.C. would appear reasonable, thus placing it in close competition with another treatise, the *Chiu-chang suan-shu*, composed about 250 BC.,[1] that is, shortly before the Han dynasty (202 B.C.). The words "Chou Pei" seem to refer to the use of the gnomon in studying the circular paths of the heavens, and the book of this title is concerned with astronomical calculations, although it includes an introduction on the properties of the right triangle and some work on the use of fractions. The work is cast in the form

1

---

[1] Histories of mathematics generally devote little space to Chinese contributions. Exceptional in this respect are D. E. Smith, *History of Mathematics* (1923–1925), and J. E. Hofmann, *Geschichte der Mathematik*, 2nd ed. (Berlin, 1963), Vol. I. An unusually thorough and up-to-date account in the Near and Far East is given in A. P. Juschkewitsch, *Geschichte der Mathematik im Mittelalter* (1964).

of a dialogue between a prince and his minister concerning the calendar; the minister tells his ruler that the art of numbers is derived from the circle and the square, the square pertaining to the earth and the circle belonging to the heavens. The *Chou Pei* indicates that in China, as Herodotus held in Egypt, geometry arose from mensuration; and, as in Babylonia, Chinese geometry was essentially only an exercise in arithmetic or algebra. There seem to be some indications in the *Chou Pei* of the Pythagorean theorem, a theorem treated algebraically by the Chinese.

2   Almost as old as the *Chou Pei*, and perhaps the most influential of all Chinese mathematical books,[2] was the *Chui-chang suan-shu*, or *Nine Chapters on the Mathematical Art*. This book includes 246 problems on surveying, agriculture, partnerships, engineering, taxation, calculation, the solution of equations, and the properties of right triangles. Whereas the Greeks of this period were composing logically ordered and systematically expository treatises, the Chinese were repeating the old custom of the Babylonians and Egyptians of compiling sets of specific problems. The *Nine Chapters* resembles Egyptian mathematics also in its use of the method of "false position," but the invention of this scheme, like the origin of Chinese mathematics in general, seems to have been independent of Western influence.

In Chinese works, as in Egyptian, one is struck by the juxtaposition of accurate and inaccurate, primitive and sophisticated results. Correct rules are used for the areas of triangles, rectangles, and trapezoids. The area of the circle was found by taking three fourths the square on the diameter or one-twelfth the square of the circumference—a correct result if the value three is adopted for π—but for the area of a segment of a circle the *Nine Chapters* uses the approximate results $s(s + c)/2$, where s is the sagitta (that is, the radius minus the apothem) and c the chord or base of the segment. There are problems that are solved by the rule of three; in others square and cube roots are found. Chapter eight of the *Nine Chapters* is significant for its solution of problems in simultaneous linear equations, using both positive and negative numbers. The last problem in the chapter involves four equations in five unknowns, and the topic of indeterminate equations was to remain a favorite among Oriental peoples. The ninth and last chapter includes problems on right-angled triangles, some of which later reappeared in India and Europe. One of these asks for the depth of a pond 10 feet square if a reed growing in the center and extending 1 foot above the water just reaches the surface if drawn to the edge of the pond. Another of these well-known problems is that of the Broken bamboo: There is a bamboo 10 feet high, the

[2] See Joseph Needham, *Science and Civilization in China* (1959), Vol. III, pp. 24–25. For recent mathematical works see Tung-Li Yuan, *Bibliography of Chinese Mathematics 1918–1960* (Washington, D.C., published by the author. 1963).

upper end of which being broken reaches the ground 3 feet from the stem. Find the height of the break.[3]

  The Chinese were especially fond of patterns; hence it is not surprising **3** that the first record (of ancient but unknown origin) of a magic square appeared there. The square

$$
\begin{array}{ccc}
4 & 9 & 2 \\
2 & 5 & 7 \\
8 & 1 & 6
\end{array}
$$

was supposedly brought to man by a turtle from the River Lo in the days of the legendary Emperor Yii, reputed to be a hydraulic engineer.[4] The concern for such patterns led the author of the *Nine Chapters* to solve the system of simultaneous linear equations

$$3x + 2y + z = 39$$
$$2x + 3y + z = 34$$
$$x + 2y + 3z = 26$$

by performing column operations on the matrix

$$
\begin{array}{ccc}
1 & 2 & 3 \\
2 & 3 & 2 \\
3 & 1 & 1 \\
26 & 34 & 39
\end{array}
\quad \text{to reduce it to} \quad
\begin{array}{ccc}
0 & 0 & 3 \\
0 & 5 & 2 \\
36 & 1 & 1 \\
99 & 24 & 39
\end{array}
$$

The second form represented the equations $36z = 99$, $5y + z = 24$, and $3x + 2y + z = 39$, from which the values of z, y, and x are successively found with ease.

  Had Chinese mathematics enjoyed uninterrupted continuity of tradition, **4** some of the striking anticipations of modern methods might have significantly modified the development of mathematics, but Chinese culture was seriously hampered by abrupt breaks. In 213 B.C., for example, the Chinese

---

[3] See Yoshio Mikami, *The Development of Mathematics in China and Japan* (1913), p.23.

[4] See D. J. Struik, "On Ancient Chinese Mathematics," *The Mathematics Teacher*, **56** (1963), 424–432.

emperor ordered the burning of books. Some works must obviously have survived, either through the persistence of copies or through oral transmission; and learning did indeed persist, with mathematical emphasis on problems of commerce and the calendar.

There seems to have been contact between India and China, as well as between China and the West, but scholars differ on the extent and direction of borrowing. The temptation to see Babylonian or Greek influence in China, for example, is faced with the problem that the Chinese did not make use of sexagesimal fractions. Chinese numeration remained essentially decimal, with notations rather strikingly different from those in other lands. In China, from early times, two schemes of notation were in use. In one the multiplicative principle predominated, in the other a form of positional notation was used. In the first of these there were distinct ciphers for the digits from one to ten and additional ciphers for the powers of ten, and in the written forms the digits in odd positions (from left to right or from bottom to top) were multiplied by their successor. Thus the number 678 would be written as a six followed by the symbol for one hundred, then a seven followed by the symbol for ten, and finally the symbol for eight.

In the system of "rod numerals" the digits from one to nine appeared as Ⅰ Ⅱ Ⅲ Ⅲⅰ ⅢⅠⅠ Т ТТ ⅢТ ⅢⅢ and the first nine multiples of ten as — = ≡ ≣ ≣ ⊥ ⊥ ⊥ ⩵. By the use of these eighteen symbols alternately in positions from right to left, numbers as large as desired could be represented. The number 56,789, for instance, would appear as ⅢⅠⅠ⊥ ТТ⩵ⅢТ. As in Babylonia, a symbol for an empty position appeared only relatively late. In a work of 1247 the number 1,405,536 is written with a round zero symbol as Ⅰ ≡ Ο ≡ ⅢⅠⅠ ≡ Т. (Occasionally, as in the fourteenth-century form of the arithmetic triangle, the vertical and horizontal rods or strokes were interchanged.)

The precise age of the original rod numerals cannot be determined, but they were certainly in use several hundred years before our era, that is, long before the positional notation had been adopted in India. The use of a centesimal, rather than a decimal, positional system in China was convenient for adaptation to computations with the counting board. Distinctive notations for neighboring powers of ten enabled the Chinese to use, without confusion, a counting board with unmarked vertical columns. Before the eighth century the place in which a zero was required was simply left blank. Although in texts older than A.D. 300 the numbers and multiplication tables were written out in words, calculations actually were made with rod numerals on a counting board.

5   The rod numerals of about 300 B.C. were not merely a notation for the written result of a computation. Actual bamboo, ivory, or iron rods were carried about in a bag by administrators and used as a calculating device. Counting rods were manipulated with such dexterity that an eleventh-century writer described them as "flying so quickly that the eye could not follow their movement." Cancellations probably were more rapidly carried out with rods on a counting board than in written calculations. So effective, in fact, was the use of the rods on a counting board that the abacus or rigid counting frame with movable markers on wires was not used so early as has been generally supposed. First clear descriptions of the modern forms, known in China as the *suan phan* and in Japan as the *soroban*, are of the sixteenth century; but anticipations would appear to have been in use perhaps a thousand years earlier. The word *abacus* probably is derived from the Semitic word *abq* or dust, indicating that in other lands, as well as in China, the device grew out of a dust or sand tray used as a counting board. It is possible, but by no means certain, that the use of the counting board in China antedates the European, but clear-cut and reliable dates are not available. In the National Museum in Athens there is a marble slab, dating probably from the fourth century B.C., which appears to be a counting board; and when a century earlier Herodotus wrote, "The Egyptians move their hand from right to left in calculation, while the Greeks move it from left to right," he probably was referring to the use of some sort of counting board. Just when such devices gave way to the abacus proper is difficult to determine; nor can we tell whether or not the appearances of the abacus in China, Arabia, and Europe were independent inventions. The Arabic abacus had ten balls on each wire and no center bar, whereas the Chinese had five lower and two upper counters on each wire, separated by a bar. Each of the upper counters on a wire of the Chinese abacus is equivalent to five on the lower wire; a number is registered by sliding the appropriate counters against the separating bar. (See the accompanying illustration of an abacus.)

No description of Chinese numeration would be complete without reference to the use of fractions. The Chinese were familiar with operations on common fractions, in connection with which they found lowest common denominators. As in other contexts, they saw analogies with the differences in the sexes, referring to the numerator as the "son" and to the denominator as the "mother." Emphasis on *yin* and *yang* (opposites, especially in sex) made it easier to follow the rules for the manipulation of fractions. More important than these, however, was the tendency in China toward decimalization of fractions. As in Mesopotamia a sexagesimal metrology led to sexagesimal numeration, so also in China adherence to the decimal idea in weights and measures resulted in a decimal habit in the treatment of fractions that, it is said, can be traced back as far as the fourteenth century B.C.[5]

[5] See Needham, *op. cit.*, III, 89.

Marble counting board, probably from the fourth century B.C., found on the island of Salamis and now in the National Museum in Athens.

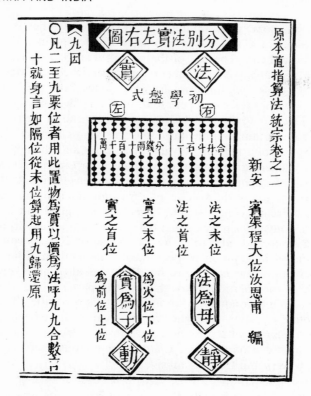

An early printed picture of the abacus, from the *Suan Fa Thung Tsung*, 1593. (Reproduced from Joseph Needham, *Science and Civilization in China*, III, 76.)

Decimal devices in computation sometimes were adopted to lighten manipulations of fractions. In a first-century commentary on the *Nine Chapters*, for example, we find the use of the now familiar rules for square and cube roots, equivalent to $\sqrt{a} = \sqrt{100a}/10$ and $\sqrt[3]{a} = \sqrt[3]{1000a}/10$, which facilitate the decimalization of root extractions.

The idea of negative numbers seems not to have occasioned much difficulty for the Chinese since they were accustomed to calculating with two sets of rods—a red set for positive coefficients or numbers and a black set for negatives. Nevertheless, they did not accept the notion that a negative number might be a solution of an equation.

The earliest Chinese mathematics is so different from that of comparable periods in other parts of the world that the assumption of independent development would appear to be justified. At all events, it seems safe to say that if there was some intercommunication before 400, then more mathematics came out of China than went in. For later periods the question becomes more difficult. The use of the value three for π in early Chinese mathematics is scarcely an argument for dependence on Mesopotamia, especially since the search for more accurate values, from the first centuries of the Christian era, was more persistent in China than elsewhere. Values such as 3.1547, $\sqrt{10}$, 92/29, and 142/45 are found; and in the third century Liu Hui, an important commentator on the *Nine Chapters*, derived the figure 3.14 by use of a regular polygon of 96 sides and the approximation 3.14159 by considering a polygon of 3072 sides. In Liu Hui's reworking of the *Nine Chapters* there are many problems in mensuration, including the correct determination of the volume of a frustum of a square pyramid. For a frustum of a circular cone a similar formula was applied, but with a value of three for π. Unusual is the rule that the volume of a tetrahedron with two opposite edges perpendicular to each other is one-sixth the product of these two edges and their common perpendicular. The method of false position is used in solving linear equations, but there are also more sophisticated results, such as the solution, through a matrix pattern, of a Diophantine problem involving four equations in five unknown quantities. The approximate solution of equations of higher degree seems to have been carried out by a device similar to what we know as "Horner's method." Liu Hui also included, in his work on the *Nine Chapters*, numerous problems involving inaccessible towers and trees on hillsides.[6]

The Chinese fascination with the value of π reached its high point in the work of Tsu Ch'ung-chih (430–501). One of his values was the familiar Archimedean 22/7, described by Tsu Ch'ung-chih as "inexact"; his "accurate" value was 355/113. If one persists in seeking possible Western influence, one can explain away this remarkably good approximation, not equaled anywhere until the fifteenth century, by subtracting the numerator and denominator, respectively, of the Archimedean value from the numerator and denominator of the Ptolemaic value 377/120. However, Tsu Ch'ung-chih went even further in his calculations, for he gave 3.1415927 as an "excess" value and 3.1415926 as a "deficit value."[7] The calculations by which he arrived at these bounds, apparently aided by his son Tsu Cheng-chih, were probably contained in one of his books, since lost. In any case, his results were remarkable for that age, and it is fitting that today a landmark on the moon bears his name.

We should bear in mind that accuracy in the value of π is more a matter

---

[6] See the excellent article on Liu Hui, written by Ho Peng-Yoke, to appear in the forthcoming volumes of the *Dictionary of Scientific Biography*.

[7] See the article cited in footnote 6. There seems to be some confusion in the citation of this value by Mikami, *op. cit.*, p. 50, by Smith, *op. cit.*, II, 309, and Hofmann, *op. cit.*, I, 76.

of computational stamina than of theoretical insight. The Pythagorean theorem alone suffices to give as accurate an approximation as may be desired. Starting with the known perimeter of a regular polygon of n sides inscribed in a circle, the perimeter of the inscribed regular polygon of $2n$ sides can be calculated by two applications of the Pythagorean theorem. Let C be a circle with center O and radius r (Fig. 12.1) and let $PQ = s$ be a

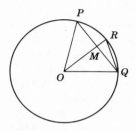

FIG. 12.1

side of a regular inscribed polygon of n sides having a known perimeter. Then the apothem $OM = u$ is given by $u = \sqrt{r^2 - (s/2)^2}$; hence the sagitta $MR = v = r - u$ is known. Then the side $RQ = w$ of the inscribed regular polygon of $2n$ sides is found from $w = \sqrt{v^2 + (s/2)^2}$; hence the perimeter of this polygon is known. The calculation, as Liu Hui saw, can be shortened by noting that $w^2 = 2rv$. An iteration of the procedure will result in an ever closer approximation to the perimeter of the circle, in terms of which π is defined.

Chinese mathematical problems often appear to be more picturesque **7** than practical, and yet Chinese civilization was responsible for a surprising number of technological innovations. The use of printing and gunpowder (eighth century) and of paper and the mariner's compass (eleventh century) was earlier in China than elsewhere, and earlier also than the high-water mark in Chinese mathematics that occurred in the thirteenth century, during the latter part of the Sung period. At that time there were mathematicians working in various parts of China; but relations between them seem to have been remote, and, as in the case of Greek mathematics, we evidently have relatively few of the treatises that once were available. The last and greatest of the Sung mathematicians was Chu Shih-chieh (fl. 1280–1303), yet we know little about him—not even when he was born or when he died. He was a resident of Yen-shan, near modern Peking, but he seems to have spent some twenty years as a wandering scholar who earned his living by teaching mathematics, even though he had the opportunity to write two treatises. The first of these, written in 1299, was the *Suan-hsüeh ch'i-meng* ("Introduction to Mathematical Studies"), a relatively elementary work that strongly influenced Korea and Japan, although in China it was lost until it reappeared in the nineteenth century.[8] Of greater historical and mathematical interest is the *Ssu-yüan yü-chien* ("Precious Mirror of the Four Elements") of 1303. In the eighteenth century this too disappeared in China, only to be rediscovered in the next century. The four elements, called heaven, earth, man, and matter, are the representations of four unknown quantities in the same equation. The book marks the peak in the development of Chinese algebra, for it deals with simultaneous equations and with equations of degrees as high as fourteen. In it the author describes a transformation method that he calls *fan fa*, the elements of which seem to have arisen long before in China, but which generally bears the name of Horner, who lived half a millennium later. In solving the equation $x^2 + 252x - 529 = 0$, for example, Chu Shih-chieh first obtained $x = 19$ as an approximation (a root lies between $x = 19$ and $x = 20$) and then used the *fan-fa*, in this case the transformation $y = x - 19$, to obtain the equation $y^2 + 290y - 143 = 0$ (with a root between $y = 0$ and $y = 1$). He then gave the root of the latter as (approximately) $y = 143/(1 + 290)$; hence the corresponding value of x is $19\frac{143}{291}$. For the equation $x^3 - 574 = 0$ he used $y = x - 8$ to obtain $y^3 + 24y^2 + 192y - 62 = 0$, and he gave the root as $x = 8 + 62/(1 + 24 + 192)$ or $x = 8\frac{2}{7}$. In some cases he found decimal approximations.

8   That the so-called Horner method was a commonplace in China is indicated by the fact that at least three other mathematicians of the later Sung period made use of similar devices. One of these was Li Chih (or Li Yeh, 1192–1279), a mathematician of Peking who was offered a government post by Khublai Khan in 1260, but politely found an excuse to decline it. His *Ts'e-yuan hai-ching* ("Sea-Mirror of the Circle Measurements") includes 170 problems dealing with circles inscribed within, or escribed without, a right triangle and with determining the relationships between the sides and the radii, some of the problems leading to equations of fourth degree. Although he did not describe his method of solution of equations, including some of sixth degree, it appears that it was not very different from that used by Chu Shih-chieh and Horner.[9] Others who used the Horner method were Ch'in Chiu-shao (ca. 1202–ca. 1261) and Yang Hui (fl. ca. 1261–1275). The former was an unprincipled governor and minister who acquired immense wealth within a hundred days of assuming office. His *Shu-shu chiu-chang* ("Mathematical Treatise in Nine Sections") marks the high point in Chinese indeter-

[8] See the extensive forthcoming article on Chu Shih-chieh by Ho Peng-Yoke to appear in the *Dictionary of Scientific Biography*. See also Needham, *op. cit.*, III, 38–53.

[9] See the article on Li Chih by Ho Peng-Yoke to appear in *Dictionary of Scientific Biography*.

minate analysis, with the invention of routines for solving simultaneous congruences. In this work also he found the square root of 71,824 by steps paralleling those in the Horner method. With 200 as the first approximation to a root of $x^2 - 71,824 = 0$, he diminished the roots of this by 200 to obtain $y^2 + 400y - 31,824 = 0$. For the latter equation he found 60 as an approximation, and diminished the roots by 60, arriving at a third equation, $z^2 + 520z - 4224 = 0$, of which 8 is a root. Hence the value of x is 268. In a similar way he solved cubic and quartic equations. The same "Horner" device was used by Yang Hui, about whose life almost nothing is known and whose work has survived only in part. Among his contributions that are extant are the earliest Chinese magic squares of order greater than three, including two each of orders four through eight and one each of orders nine and ten.[10]

Yang Hui's works included also results in the summation of series and the **9** so-called Pascal triangle, things that were published and better known through the *Precious Mirror* of Chu Shih-chieh, with which the Golden Age of Chinese mathematics closed. A few of the many summations of series found in the *Precious Mirror* are the following:

$$1^2 + 2^2 + 3^2 + \cdots + n^2 = n(n + 1)(2n + 1)/3!$$

$$1 + 8 + 30 + 80 + \cdots + n^2(n + 1)(n + 2)/3! = n(n + 1)(n + 2)(n + 3)$$

$$\times (4n + 1)/5!$$

However, no proofs are given, nor does the topic seem to have been continued again in China until about the nineteenth century. Chu Shih-chieh handled his summations through the method of finite differences, some elements of which seem to date in China from the seventh century; but shortly after his work the method disappeared for many centuries.

The *Precious Mirror* opens with a diagram of the arithmetic triangle, inappropriately known in the West as "Pascal's triangle." In Chu's arrangement we have the coefficients of binomial expansions through the eighth power, clearly given in rod numerals and a round zero symbol. Chu disclaims credit for the triangle, referring to it as a "diagram of the old method for finding eighth and lower powers." A similar arrangement of coefficients through the sixth power had appeared in the work of Yang Hui, but without the round zero symbol. There are references in Chinese works of about 1100 to tabulation systems for binomial coefficients, and it is likely that the arithmetic triangle originated in China by about that date. It is interesting to note that the Chinese discovery of the binomial theorem for integral

[10] Excellent articles, including much more on the work of Ch'in Chiu-shao and Yang Hui, written by Ho Peng-Yoke, will appear in the forthcoming *Dictionary of Scientific Biography*.

powers was associated in its origin with root extractions, rather than with powers. The equivalent of the theorem apparently was known to Omar Khayyam at about the time that it was being used in China, but the earliest extant Arabic work containing it is by Al-Kashi in the fifteenth century. By that time Chinese mathematics had failed to match achievements in Europe

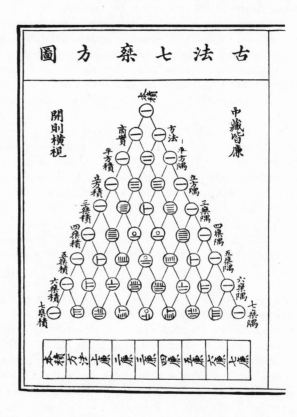

The "Pascal" Triangle as depicted in 1303 at the front of Chu Shih-Chieh's *Ssu Yuan Yii Chien*. It is entitled "The Old Method Chart of the Seven Multiplying Squares" and tabulates the binomial coefficients up to the eighth power. (Reproduced from Joseph Needham, *Science and Civilization in China*, III, 135.)

and the Near East, and it is likely that by then more mathematics went into China than came out. Still to be answered is the thorny problem of determining the relative influences of China and India on each other during the first millennium of our era.

Archeological excavations at Mohenjo Daro give evidence of an old and **10** highly cultured civilization in India during the era of the Egyptian pyramid builders, but we have no Indian mathematical documents from that age. Later the country was occupied by Aryan invaders who introduced the caste system and developed the Sanskrit literature. The great religious teacher, Buddha, was active in India at about the time that Pythagoras is said to have visited there, and it sometimes is suggested that Pythagoras learned his theorem from the Hindus. Recent studies make this highly unlikely in view of Babylonian familiarity with the theorem at least a thousand years earlier.

The fall of the Western Roman Empire traditionally is placed in the year 476; it was in this year that Aryabhata, author of one of the oldest Indian mathematical texts, was born. It is clear, however, that there had been mathematical activity in India long before this time—probably even before the mythical founding of Rome in 753 B.C. India, like Egypt, had its "rope-stretchers"; and the primitive geometrical lore acquired in connection with the laying out of temples and the measurement and construction of altars took the form of a body of knowledge known as the *Sulvasūtras* or "rules of the cord." *Sulva* (or *sulba*) refers to cords used for measurements, and *sūtra* means a book of rules or aphorisms relating to a ritual or science. The stretching of ropes is strikingly reminiscent of the origin of Egyptian geometry, and its association with temple functions reminds one of the possible ritual origin of mathematics. However, the difficulty of dating the rules is matched also by doubt concerning the influence they had on later Hindu mathematicians. Even more so than in the case of China, there is a striking lack of continuity of tradition in the mathematics of India; significant contributions are episodic events separated by intervals without achievement.[11]

Three versions, all in verse, of the work referred to as the *Sulvasūtras* are **11** extant, the best-known being that bearing the name of Apastamba. In this primitive account, dating back perhaps as far as the time of Pythagoras, we find rules for the construction of right angles by means of triples of cords the lengths of which form Pythagorean triads, such as 3, 4, and 5, or 5, 12, and 13, or 8, 15, and 17, or 12, 35, and 37. However, all of these triads are easily derived from the old Babylonian rule; hence Mesopotamian influence in the *Sulvasūtras* is not unlikely. Apastamba knew that the square on the diagonal of a rectangle is equal to the sum of the squares on the two adjacent

[11] The reader should be forewarned that there are a number of books in which the contributions from India are grossly overrated. One such instance is B. K. Sarkar, *Hindu Achievements in Exact Science* (New York, 1918). The two-volume *History of Hindu Mathematics* by B. Datta and A. N. Singh (1935–1938) is much more reliable, but even this must be qualified along the lines indicated by Otto Neugebauer when he reviewed Volume I in *Isis*, **25** (1936), 478–488.

sides, but this form of the Pythagorean theorem also may have been derived from Mesopotamia. Less easily explained is another rule given by Apastamba —one that strongly resembles some of the geometrical algebra in Book II of Euclid's *Elements*. To construct a square equal in area to the rectangle *ABCD* (Fig. 12.2), lay off the shorter sides on the longer, so that $AF = AB = BE = CD$, and draw *HG* bisecting segments *CE* and *DF*; extend *EF* to *K*, *GH* to *L*, and *AB* to *M* so that $FK = HL = FH = AM$, and draw *LKM*.

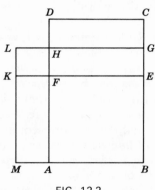

FIG. 12.2

Now construct a rectangle with diagonal equal to *LG* and with shorter side *HF*. Then the longer side of this rectangle is the side of the square desired.

So conjectural are the origin and period of the *Sulvasūtras* that we cannot tell whether or not the rules are related to early Egyptian surveying or to the later Greek problem of altar doubling. They are variously dated within an interval of almost a thousand years stretching from the eighth century B.C. to the second century of our era. Chronology in ancient cultures of the Far East is scarcely reliable when orthodox Hindu tradition boasts of important astronomical work more than 2,000,000 years ago[12] and when calculations lead to billions of days from the beginning of the life of Brahman to about A.D. 400.[13] References to arithmetic and geometric series in Vedic literature that purport to go back to 2000 B.C.[14] may be more reliable, but there are no contemporary documents from India to confirm this. It has been claimed also that the first recognition of incommensurables is to be found in India during the *Sulvasūtra* period,[15] but such claims are not well substantiated.

[12] G. R. Kaye, "Indian Mathematics," *Isis*, **2** (1914), 326–356.

[13] *Alberuni's India*, ed. by E. C. Sachan (London, 1960, 2 vols.), II, 32 f.

[14] A. N. Singh, "On the use of Series in Hindu Mathematics," *Osiris*, **1** (1936), 606–628.

[15] A. N. Singh, "A Review of Hindu Mathematics up to the XIIth Century," *Archeion* **18** (1936), 43–62; Saradakanta Ganguli, "On the Indian Discovery of the Irrational at the Time of the Sulvasutras," *Scripta Mathematica*, **1** (1932), 135–141.

The case for early Hindu awareness of incommensurable magnitudes is rendered most unlikely by the failure of Indian mathematicians to come to grips with fundamental concepts.

The period of the *Sulvasūtras*, which closed in about the second century, **12** was followed by the age of the *Siddhāntas*, or systems (of astronomy). The establishment of the dynasty of King Gupta (290) marked the beginning of a renaissance in Sanskrit culture, and the *Siddhāntas* seem to have been an outcome of this revival. Five different versions of the *Siddhāntas* are known by name, *Pauliśha Siddhānta, Sūrya Siddhānta, Vasisishta Siddhānta, Paitamaha Siddhānta*, and *Romanka Siddhānta*. Of these, the *Sūrya Siddhānta* ("System of the Sun"), written about 400, is the only one that seems to be completely extant. According to the text, written in epic stanzas, it is the work of Sūrya, the Sun God.[16] The main astronomical doctrines evidently are Greek, but with the retention of considerable old Hindu folklore. The *Pauliśha Siddhānta*, which dates from about 380, was summarized by the Hindu mathematician Varahamihira (fl. 505) and was referred to frequently by the Arabic scholar Al-Biruni, who suggested a Greek origin or influence. Later writers report that the *Siddhāntas* were in substantial agreement on substance, only the phraseology varying; hence we can assume that the others, like the *Sūrya Siddhānta*, were compendia of astronomy comprising cryptic rules in Sanskrit verse with little explanation and without proof.

It is generally agreed that the *Siddhāntas* stem from the late fourth or the early fifth century, but there is sharp disagreement about the origin of the knowledge that they contain. Hindu scholars insist on the originality and independence of the authors, whereas Western writers are inclined to see definite signs of Greek influence. It is not unlikely, for example, that the *Pauliśha Siddhānta* was derived in considerable measure from the work of the astrologer Paul who lived at Alexandria shortly before the presumed date of composition of the *Siddhāntas*. (Al-Biruni, in fact, explicitly attributes this *Siddhānta* to Paul of Alexandria.) This would account in a simple manner for the obvious similarities between portions of the *Siddhāntas* and the trigonometry and astronomy of Ptolemy. The *Pauliśha Siddhānta*, for example, uses the value 3 177/1250 for π, which is in essential agreement with the Ptolemaic sexagesimal value 3;8,30.

Even if the Hindus did acquire their knowledge of trigonometry from the cosmopolitan Hellenism at Alexandria, the material in their hands took on a significantly new form. Whereas the trigonometry of Ptolemy had been based on the functional relationship between the chords of a circle and the

---

[16] An English translation by Burgess and Whitney, together with extensive notes, was published in *Journal of the American Oriental Society*, **6** (1860), 141–498. See also George Sarton, *An Introduction to the History of Science* (1927), pp. 386–388.

central angles they subtend, the writers of the *Siddhāntas* converted this to a study of the correspondence between *half* of a chord of a circle and *half* of the angle subtended at the center by the whole chord. Thus was born, apparently in India, the predecessor of the modern trigonometric function known as the sine of an angle; and the introduction of the sine function represents the chief contribution of the *Siddhāntas* to the history of mathematics. Although it is generally assumed that the change from the whole chord to the half chord took place in India, it has been suggested by Paul Tannery, the leading historian of science at the turn of this century, that this transformation of trigonometry may have occurred at Alexandria during the post-Ptolemaic period. Whether or not this suggestion has merit, there is no doubt that it was through the Hindus, and not the Greeks, that our use of the half chord has been derived; and our word "sine", through misadventure in translation (see below), has descended from the Hindu name, *jiva*.

13   During the sixth century, shortly after the composition of the *Siddhāntas*, there lived two Hindu mathematicians who are known to have written books on the same type of material. The older, and more important, of the two was Aryabhata, whose best known work, written in 499 and entitled *Aryabhatiya*, is a slim volume, written in verse, covering astronomy and mathematics. The names of several Hindu mathematicians before this time are known, but nothing of their work has been preserved beyond a few fragments. In this respect, then, the position of the *Aryabhatiya* of Aryabhata in India is somewhat akin to that of the *Elements* of Euclid in Greece some eight centuries before. Both are summaries of earlier developments, compiled by a single author. There are, however, more striking differences than similarities between the two works. The *Elements* is a well-ordered synthesis of pure mathematics with a high degree of abstraction, a clear logical structure, and an obvious pedagogical inclination; the *Aryabhatiya* is a brief descriptive work, in 123 metrical stanzas, intended to supplement rules of calculation used in astronomy and mensurational mathematics, with no feeling for logic or deductive methodology. About a third of the work is on *ganitapada* or mathematics. This section opens with the names of the powers of ten up to the tenth place and then proceeds to give instructions for square and cube roots of integers. Rules of mensuration follow, about half of which are erroneous. The area of a triangle is correctly given as half the product of the base and altitude, but the volume of a pyramid also is taken to be half the product of the base and altitude.[17] The area of a circle is found correctly as the product of the circumference and half the diameter, but the volume of a sphere is incorrectly stated to be the product of the area of a great circle

---

[17] *The Aryabhatiya of Aryabhata*, trans. by W. E. Clark (1930), p. 26.

and the square root of this area. Again, in the calculation of areas of quadrilaterals, correct and incorrect rules appear side by side. The area of a trapezoid is expressed as half the sum of the parallel sides multiplied by the perpendicular between them; but then follows the incomprehensible assertion that the area of any plane figure is found by determining two sides and multiplying them. One statement in the *Aryabhatiya* to which Hindu scholars have pointed with pride is as follows:[18]

> Add 4 to 100, multiply by 8, and add 62,000. The result is approximately the circumference of a circle of which the diameter is 20,000.

Here we see the equivalent of 3.1416 for π, but it should be recalled that this is essentially the value Ptolemy had used. The likelihood that Aryabhata here was influenced by Greek predecessors is strengthened by his adoption of the myriad, 10,000, as the number of units in the radius.

A typical portion of the *Aryabhatiya* is that involving arithmetic progressions, which contains arbitrary rules for finding the sum of the terms in a progression and for determining the number of terms in a progression when given the first term, the common difference, and the sum of the terms. The first rule had long been known by earlier writers. The second is a curiously complicated bit of exposition:

> Multiply the sum of the progression by eight times the common difference, add the square of the difference between twice the first term, and the common difference, take the square root of this, subtract twice the first term, divide by the common difference, add one, divide by two. The result will be the number of terms.

Here, as elsewhere in the *Aryabhatiya*, no motivation or justification is given for the rule. It was probably arrived at through a solution of a quadratic equation, knowledge of which might have come from Mesopotamia or Greece. Following some complicated problems on compound interest (that is, geometrical progressions), the author turns, in flowery language, to the very elementary problem of finding the fourth term in a simple proportion:

> In the rule of three multiply the fruit by the desire and divide by the measure. The result will be the fruit of the desire.

This, of course, is the familiar rule that if $a/b = c/x$, then $x = bc/a$, where a is the "measure," b the "fruit," c the "desire," and x the "fruit of the desire." The work of Aryabhata is indeed a potpourri of the simple and the complex, the correct and the incorrect. The Arabic scholar Alberuni, half a millennium later, characterized Hindu mathematics as a mixture of common pebbles and costly crystals, a description quite appropriate to *Aryabhatiya*.

[18] *Aryabhatiya*, p. 28. Translations, here and below, are from the Clark edition cited in footnote 17.

14   The second half of the *Aryabhatiya* is on the reckoning of time and on spherical trigonometry; here we note an element that was to leave a permanent impress on the mathematics of later generations—the decimal place-value numeration. It is not known just how Aryabhata carried out his calculations, but his phrase "from place to place each is ten times the preceding" is an indication that the application of the principle of position was in his mind. "Local value" had been an essential part of Babylonian numeration, and perhaps the Hindus were becoming aware of its applicability to the decimal notation for integers in use in India. The development of numerical notations in India seems to have followed about the same pattern found in Greece. Inscriptions from the earliest period at Mohenjo-Daro show at first simple vertical strokes, arranged into groups, but by the time of Asoka (third century B.C.) a system resembling the Herodianic was in use. In the newer scheme the repetitive principle was continued, but new symbols of higher order were adopted for four, ten, twenty, and one hundred. This so-called Karosthi script then gradually gave way to another notation, known as the Brahmi characters, which resembled the alphabetic cipherization in the Greek Ionian system; one wonders if it was only a coincidence that the change in India took place shortly after the period when in Greece the Herodianic numerals were displaced by the Ionian.

From the Brahmi ciphered numerals to our present-day notation for integers two short steps are needed. The first is a recognition that, through the use of the positional principle, the ciphers for the first nine units can serve also as the ciphers for the corresponding multiples of ten, or equally well as ciphers for the corresponding multiples of any power of ten. This recognition would make superfluous all of the Brahmi ciphers beyond the first nine. It is not known when the reduction to nine ciphers occurred, and it is likely that the transition to the more economical notation was made only gradually. It appears from extant evidence that the change took place in India, but the source of the inspiration for the change is uncertain. Possibly the so-called Hindu numerals were the result of internal development alone; perhaps they developed first along the western interface between India and Persia, where remembrance of the Babylonian positional notation may have led to modification of the Brahmi system. It is possible that the newer system arose along the eastern interface with China where the pseudopositional rod numerals may have suggested the reduction to nine ciphers. There is also a theory that this reduction may first have been made at Alexandria within the Greek alphabetic system and that subsequently the idea spread to India.[19] During the later Alexandrian period the earlier Greek habit of writing common fractions with the numerator beneath the denominator was

[19] See Harriet P. Lattin, "The Origin of Our Present System of Notation According to the Theories of Nicholas Bubnov," *Isis*, **19** (1933), 181–194.

reversed, and it is this form that was adopted by the Hindus, without the bar between the two. Unfortunately, the Hindus did not apply the new numeration for integers to the realm of decimal fractions; hence the chief potential advantage of the change from Ionian notation was lost.

The earliest specific reference to the Hindu numerals is found in 662 in the writings of Severus Sebokt, a Syrian bishop. After Justinian closed the Athenian philosophical schools some of the scholars moved to Syria, where they established centers of Greek learning. Sebokt evidently felt piqued by the disdain for non-Greek learning expressed by some associates; hence he found it expedient to remind those who spoke Greek that "there are also others who know something." To illustrate his point he called attention to the Hindus and their "subtle discoveries in astronomy," especially "their valuable methods of calculation, and their computing that surpasses description. I wish only to say that this computation is done by means of nine signs."[20] That the numerals had been in use for some time is indicated by the fact that the first Indian occurrence is on a plate of the year 595, where the date 346 is written in decimal place-value notation.[21]

It should be remarked that the reference to *nine* symbols, rather than *ten*, implies that the Hindus evidently had not yet taken the second step in the transition to the modern system of numeration—the introduction of a notation for a missing position, that is, a zero symbol. The history of mathematics holds many anomalies, and not the least of these is the fact that "the earliest undoubted occurrence of a zero in India is in an inscription of 876"[22] —that is, more than two centuries after the first reference to the other nine numerals. It is not even established that the number zero (as distinct from a symbol for an empty position) arose in conjunction with the other nine Hindu numerals. It is quite possible that zero originated in the Greek world, perhaps at Alexandria, and that it was transmitted to India after the decimal positional system had been established there.[23]

The history of the zero placeholder in positional notation is further complicated by the fact that the concept appeared independently, well before the days of Columbus, in the western, as well as the eastern hemisphere. The Mayas of Yucatan, in their representation of time intervals between dates in their calendar, used a place-value numeration, generally with twenty as the primary base and with five as an auxiliary (corresponding to the Babylonian use of sixty and ten respectively). Units were represented by dots and fives by horizontal bars, so that the number seventeen, for example,

[20] Quoted from D. E. Smith. *History of Mathematics*. I, 167.

[21] See D. J. Struik, *A Concise History of Mathematics*, 3rd ed. (New York: Dover, 1967), p. 71.

[22] Smith, *History of Mathematics*, II, 69.

[23] See, for example, B. L. van der Waerden, *Science Awakening* (1961), pp. 56–58.

would appear as ☰ [that is, as 3(5) + 2]. A vertical positional arrangement was used, with the larger units of time above; hence the notation ☰ denoted 352 [that is 17(20) + 12]. Because the system was primarily for counting days within a calendar having 360 days in a year, the third position usually did not represent multiples of (20)(20), as in a pure vigesimal system, but (18)(20). However, beyond this point the base twenty again prevailed. Within this positional notation the Mayas indicated missing positions

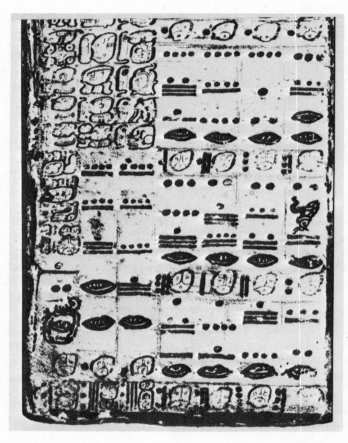

From the Dresden Codex, of the Maya, displaying numbers. The second column on the left, from above down, displays the numbers 9, 9, 16, 0, 0, which stand for 9 × 144,000 + 9 × 7200 + 16 × 360 + 0 + 0 = 1,366,560. In the third column are the numerals 9, 9, 9, 16, 0, representing 1,364,360. The original appears in black and red colors. (Taken from Morley, *An Introduction to the Study of the Maya Hieroglyphs*, p. 266.)

through the use of a symbol, appearing in variant forms, somewhat resembling a half-open eye. In their scheme, then, the notation ⚇ denoted
$17(20 \cdot 18 \cdot 20) + 0(18 \cdot 20) + 13(20) + 0$.

With the introduction, in the Hindu notation, of the tenth numeral, a round goose egg for zero, the modern system of numeration for integers was completed. Although the Medieval Hindu forms of the ten numerals differ considerably from those in use today, the principles of the system were established. The new numeration, which we generally call the Hindu system, is merely a new combination of three basic principles, all of ancient origin: (1) a decimal base; (2) a positional notation; and (3) a ciphered form for each of the ten numerals. Not one of these three was due originally to the Hindus, but it presumably is due to them that the three were first linked to form the modern system of numeration.

It may be well to say a word about the form of the Hindu symbol for zero—which is also ours. It once was assumed that the round form stemmed originally from the Greek letter omicron, initial letter in the word "ouden" or empty, but recent investigations seem to belie such an origin. Although the symbol for an empty position in some of the extant versions of Ptolemy's tables of chords does seem to resemble an omicron, the early zero symbols in Greek sexagesimal fractions are round forms variously embellished and differing markedly from a simple goose egg. Moreover, when in the fifteenth century in the Byzantine Empire a decimal positional system was fashioned out of the old alphabetic numerals by dropping the last eighteen letters and adding a zero symbol to the first nine letters, the zero sign took forms quite unlike an omicron.[24] Sometimes it resembled an inverted form of our small letter h, sometimes it appeared as a dot.

The development of our system of notation for integers was one of the **16** two most influential contributions of India to the history of mathematics. The other was the introduction of an equivalent of the sine function in trigonometry to replace the Greek tables of chords. The earliest tables of the sine relationship that have survived are those in the *Siddhāntas* and the *Aryabhatiya*. Here the sines of angles up to 90° are given for twenty-four equal intervals of $3\frac{3}{4}°$ each. In order to express arc length and sine length in terms of the same unit, the radius was taken as 3438 and the circumference as $360 \cdot 60 = 21{,}600$. This implies a value of π agreeing to four significant figures with that of Ptolemy. In another connection Aryabhata used the value $\sqrt{10}$ for π, which appeared so frequently in India that it sometimes is known as the Hindu value.

---

[24] See O. Neugebauer, *The Exact Sciences in Antiquity*, 2nd ed. (Providence, R.I.: Brown University Press, 1957), p. 14.

For the sine of $3\frac{3}{4}°$ the *Siddhāntas* and the *Aryabhatiya* took the number of units in the arc—that is, $60 \times 3\frac{3}{4}$ or 225. In modern language, the sine of a small angle is very nearly equal to the radian measure of the angle (which is virtually what the Hindus were using). For further items in the sine table the Hindus used a recursion formula which may be expressed as follows. If the nth sine in the sequence from $n = 1$ to $n = 24$ is designated as s_n, and if the sum of the first n sines is S_n, then $s_{n+1} = s_n + s_1 - S_n/s_1$. From this rule one easily deduces that $\sin 7\frac{1}{2}° = 449$, $\sin 11\frac{1}{4}° = 671$, $\sin 15° = 890$, and so on up to $\sin 90° = 3438$—the values listed in the table in the *Siddhāntas* and the *Aryabhatiya*. Moreover, the table also includes values for what we call the versed sine of the angle—that is, $1 - \cos\theta$ in modern trigonometry or $3438\,(1 - \cos\theta)$ in Hindu trigonometry—from vers $3\frac{3}{4}° = 7$ to vers $90° = 3438$. If we divide the items in the table by 3438, the results are found to be in close agreement with the corresponding values in modern trigonometric tables.[25]

17   Hindu trigonometry evidently was a useful and accurate tool in astronomy. How the Hindus arrived at results such as the recursion formula is uncertain, but it has been suggested[26] that an intuitive approach to difference equations and interpolation may have prompted such rules. Indian mathematics frequently is described as "intuitive," in contrast to the stern rationalism of Greek geometry. Although in Hindu trigonometry there is evidence of Greek influence, the Indians seem to have had no occasion to borrow Greek geometry, concerned as they were with simple mensurational rules. Of the classical geometrical problems, or the study of curves other than the circle, there is little evidence in India, and even the conic sections seem to have been overlooked by the Hindus, as by the Chinese. Hindu mathematicians were fascinated instead by work with numbers, whether it involved the ordinary arithmetic operations or the solution of determinate or indeterminate equations. Addition and multiplication were carried out in India much as they are by us today, except that they seem at first to have preferred to write numbers with the smaller units on the left, hence to work from left to right, using small blackboards with white removable paint or a board covered with sand or flour. Among the devices used for multiplications was one that is known under various names: lattice multiplication, *gelosia* multiplication, or cell or grating or quadrilateral multiplication. The scheme behind this is readily recognized in two examples. In the first example (Fig. 12.3) the number 456 is multiplied by 34. The multiplicand has been written above the lattice and the multiplier appears to the left, with the partial products

---

[25] The table from the *Sūrya Siddhānta* is reproduced in Smith, *History of Mathematics*, II.

[26] E. S. Kennedy in the article "Trigonometry," to appear in the *Yearbook on History of Mathematics* of the National Council of Teachers of Mathematics.

occupying the square cells. Digits in the diagonal rows are added, and the product 15,504 is read off at the bottom and the right. To indicate that other arrangements are possible, a second example is given in Fig. 12.4, in which

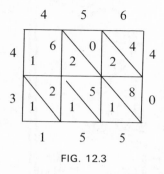

FIG. 12.3

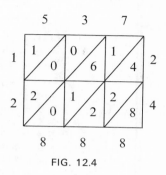

FIG. 12.4

the multiplicand 537 is placed at the top and the multiplier 24 is on the right, the product 12,888 appearing to the left and along the bottom. Still other modifications are easily devised. In fundamental principle *gelosia* multiplication is of course the same as our own, the cell arrangement being merely a convenient device for relieving the mental concentration called for in "carrying over" from place to place the tens arising in the partial products. The only "carrying" required in lattice multiplication is in the final additions along the diagonals.

It is not known when or where *gelosia* multiplication arose, but India seems to be the most likely source. It was used there at least by the twelfth century, and from India it seems to have been carried to China and Arabia. From the Arabs it passed over to Italy in the fourteenth and fifteenth centuries, where the name *gelosia* was attached to it because of the resemblance to gratings placed before windows in Venice and elsewhere. (The current word jalousie seems to stem from the Italian *gelosia* and is used for Venetian blinds in France, Germany, Holland, and Russia.) The Arabs (and through them the later Europeans) appear to have adopted most of their arithmetic devices from the Hindus, and so it is likely that the pattern of long division known as the "scratch method" or the "galley method" (from its resemblance to a boat) came also from India. To illustrate the method, let it be required to divide 44,977 by 382. In Fig. 12.5 we give the modern method, in Fig. 12.6 the galley method.[27] The latter parallels the former closely except that the

---

[27] For further description of the innumerable computational devices that have been used, see F. A. Yeldham, *The Story of Reckoning in the Middle Ages* (1926).

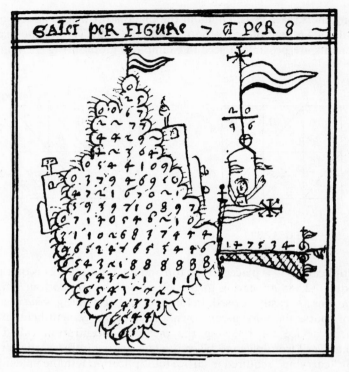

Galley division, sixteenth century. From an unpublished manuscript of a Venetian monk. The title of the work is "Opus Arithmeticá D. Honorati veneti monachj coenobij S. Lauretig." From Mr. Plimpton's library.

dividend appears in the middle, for subtractions are performed by canceling digits and placing differences *above* rather than *below* the minuends. Hence the remainder, 283, appears above and to the right, rather than below.

```
       117
382)44977
    382
    ----
     677
     382
    ----
    2957
    2674
    ----
     283
```

FIG. 12.5

			2			
		~~2~~	~~3~~			
		~~3~~	~~9~~	8		
	~~1~~	~~6~~	~~7~~	~~5~~	3	
382	~~4~~	~~4~~	~~9~~	~~7~~	~~7~~	117
	~~3~~	~~8~~	~~2~~	~~2~~	4	
		~~3~~	~~8~~	~~7~~		
		~~2~~	~~6~~			

FIG. 12.6

The process in Fig. 12.6 is easy to follow if we note that the digits in a given subtrahend, such as 2674, or in a given difference, such as 2957, are not necessarily all in the same row and that subtrahends are written below the middle and differences above the middle. Position in a column is significant, but not position in a row. The determination of roots of numbers probably followed a somewhat similar "galley" pattern, coupled in the later years with the binomial theorem in "Pascal triangle" form; but Hindu writers did not provide explanations for their calculations or proofs for their statements. It is possible that Babylonian and Chinese influences played a role in the problem of evolution or root extraction. It is often said that the "proof by nines," or the "casting out of nines," is a Hindu invention, but it appears that the Greeks knew earlier of this property, without using it extensively, and that the method came into common use only with the Arabs of the eleventh century.

The last few paragraphs may leave the unwarranted impression that there **19** was a uniformity in Hindu mathematics, for frequently we have localized developments as merely "of Indian origin," without specifying the period. The trouble is that there is a high degree of uncertainty in Hindu chronology. Material in the important Bakshali manuscript, containing an anonymous arithmetic, is supposed by some to date from the third or fourth century, by others from the sixth century, by others from the eighth or ninth century or later; and there is a suggestion that it may not even be of Hindu origin.[28] We have placed the work of Aryabhata around the year 500, but the date is doubtful since there were two mathematicians named Aryabhata and we cannot with certainty ascribe results to our Aryabhata, the elder. Hindu mathematics presents more historical problems than does Greek mathematics, for Indian authors referred to predecessors infrequently, and they exhibited surprising independence in mathematical approach. Thus it is that Brahmagupta (fl. 628), who lived in Central India somewhat more than a century after Aryabhata, has little in common with his predecessor, who had lived in eastern India. Brahmagupta mentions two values of π—the "practical value" 3 and the "neat value" $\sqrt{10}$—but not the more accurate value of Aryabhata; in the trigonometry of his best-known work, the *Brahmasphuta Siddhānta*, he adopted a radius of 3270 instead of Aryabhata's 3438. In one respect he does resemble his predecessor—in the juxtaposition of good and bad results. He found the "gross" area of an isosceles triangle by multiplying half the base by one of the equal sides; for the scalene triangle with base fourteen and sides thirteen and fifteen he found the "gross area" by multiplying half the base by the arithmetic mean of the other sides. In finding the

[28] See Florian Cajori, *A History of Mathematics* (1919), pp. 84-85; Smith, *History of Mathematics*, I, 164; Hofmann, *Geschichte der Mathematik*, I, 59.

"exact" area he utilized the Archimedean-Heronian formula. For the radius of the circle circumscribed about a triangle he gave the equivalent of the correct trigonometric result $2R = a/\sin A = b/\sin B = c/\sin C$, but this of course is only a reformulation of a result known to Ptolemy in the language of chords. Perhaps the most beautiful result in Brahmagupta's work is the generalization of "Heron's" formula in finding the area of a quadrilateral. This formula— $K = \sqrt{(s - a)(s - b)(s - c)(s - d)}$, where a, b, c, d are the sides and s is the semiperimeter—still bears his name; but the glory of his achievement is dimmed by failure to remark that the formula is correct only in the case of a *cyclic* quadrilateral.[29] (The correct formula for an arbitrary quadrilateral is $\sqrt{(s - a)(s - b)(s - c)(s - d)} - abcd \cos^2 \alpha$, where α is half the sum of two opposite angles.) As a rule for the "gross" area of a quadrilateral Brahmagupta gave the pre-Hellenic formula, the product of the arithmetic means of the opposite sides. For the quadrilateral with sides $a = 25$, $b = 25$, $c = 25$, $d = 39$, for example, he found a "gross" area of 800.

20   Brahmagupta's contributions to algebra are of a higher order than are his rules of mensuration, for here we find general solutions of quadratic equations, including two roots even in cases in which one of them is negative. The systematized arithmetic of negative numbers and zero is, in fact, first found in his work. The equivalents of rules on negative magnitudes were known through the Greek geometrical theorems on subtraction, such as $(a - b)(c - d) = ac + bd - ad - bc$, but the Hindus converted these into numerical rules on positive and negative numbers. Moreover, although the Greeks had a concept of nothingness, they never interpreted this as a number, as did the Hindus. However, here again Brahmagupta spoiled matters somewhat by asserting that $0 \div 0 = 0$, and on the touchy matter of $a \div 0$, for $a \neq 0$, he did not commit himself:

> Positive divided by positive, or negative by negative, is affirmative. Cipher divided by cipher is naught. Positive divided by negative is negative. Negative divided by affirmative is negative. Positive or negative divided by cipher is a fraction with that for denominator.[30]

It should be mentioned also that the Hindus, unlike the Greeks, regarded irrational roots of numbers as numbers. This was of enormous help in algebra, and Indian mathematicians have been much praised for taking this step; but one must remember that the Hindu contribution in this case was the result of logical innocence rather than of mathematical insight. We have

[29] A proof of the formula can be found in R. A. Johnson, *Modern Geometry* (New York: Houghton Mifflin, 1929), pp. 81–82.

[30] See H. T. Colebrooke, *Algebra, with Arithmetic and Mensuration, from the Sanscrit of Brahmagupta and Bhaskara* (1817).

seen the lack of nice distinction on the part of Hindu mathematicians between exact and inexact results, and it was only natural that they should not have taken seriously the difference between commensurable and incommensurable magnitudes. For them there was no impediment to the acceptance of irrational numbers, and later generations followed their lead uncritically until in the nineteenth century mathematicians established the real number system on a sound basis.

Indian mathematics was, as we have said, a mixture of good and bad. But some of the good was superlatively good, and here Brahmagupta deserves high praise. Hindu algebra is especially noteworthy in its development of indeterminate analysis, to which Brahmagupta made several contributions. For one thing, in his work we find a rule for the formation of Pythagorean triads expressed in the form $m, \frac{1}{2}(m^2/n - n), \frac{1}{2}(m^2/n + n)$; but this is only a modified form of the old Babylonian rule, with which he may have become familiar. Brahmagupta's area formula for a quadrilateral, mentioned above, was used by him in conjunction with the formulas

$$\sqrt{(ab + cd)(ac + bd)/(ad + bc)} \quad \text{and} \quad \sqrt{(ac + bd)(ad + bc)/(ab + cd)}$$

for the diagonals,[31] to find quadrilaterals whose sides, diagonals, and areas are all rational. Among them was the quadrilateral with sides $a = 52, b = 25$, $c = 39, d = 60$, and diagonals 63 and 56. Brahmagupta gave the "gross" area as $1933\frac{3}{4}$, despite the fact that his formula provides the exact area, 1764, in this case.

Like many of his countrymen, Brahmagupta evidently loved mathematics **21** for its own sake, for no practical-minded engineer would raise questions such as those Brahmagupta asked about quadrilaterals. One admires his mathematical attitude even more when one finds that apparently he was the first one to give a *general* solution of the linear Diophantine equation $ax + by = c$, where a, b, and c are integers. For this equation to have integral solutions, the greatest common divisor of a and b must divide c; and Brahmagupta knew that if a and b are relatively prime, all solutions of the equation are given by $x = p + mb, y = q - ma$, where m is an arbitrary integer. He suggested also the Diophantine quadratic equation $x^2 = 1 + py^2$, named mistakenly for John Pell (1611–1685), but first appearing in the Archimedean cattle problem. The Pell equation was solved for some cases by Brahmagupta's countryman, Bhaskara (1114–ca. 1185).

It is greatly to the credit of Brahmagupta that he gave *all* integral solutions of the linear Diophantine equation, whereas Diophantus himself had been

[31] For indications of a proof of these formulas see Howard Eves, *An Introduction to the History of Mathematics* (1964), pp. 202–203.

satisfied to give one particular solution of an indeterminate equation. Inasmuch as Brahmagupta used some of the same examples as Diophantus, we see again the likelihood of Greek influence in India—or the possibility that they both made use of a common source, possibly from Babylonia. It is interesting to note also that the algebra of Brahmagupta, like that of Diophantus, was syncopated. Addition was indicated by juxtaposition, subtraction by placing a dot over the subtrahend, and division by placing the divisor below the dividend, as in our fractional notation but without the bar. The operations of multiplication and evolution (the taking of roots), as well as unknown quantities, were represented by abbreviations of appropriate words.

22   India produced a number of later Medieval mathematicians, but we shall describe the work of only one of these—Bhaskara (1114–ca. 1185), the leading mathematician of the twelfth century. It was he who filled some of the gaps in Brahmagupta's work, as by giving a general solution of the Pell equation and by considering the problem of division by zero. Aristotle once had remarked that there is no ratio by which a number such as four exceeds the number zero;[32] but the arithmetic of zero had not been part of Greek mathematics, and Brahmagupta had been noncommittal on the division of a number other than zero by the number zero. It is therefore in Bhaskara's *Vija-Ganita* that we find the first statement that such a quotient is infinite.

> Statement: Dividend 3. Divisor 0. Quotient the fraction 3/0. This fraction of which the denominator is cipher, is termed an infinite quantity. In this quantity consisting of that which has cipher for a divisor, there is no alteration, though many be inserted or extracted; as no change takes place in the infinite and immutable God.

This statement sounds promising, but lack of clear understanding of the situation is suggested by Bhaskara's further assertion that $a/0 \cdot 0 = a$.

Bhaskara was the last significant Medieval mathematician from India, and his work represents the culmination of earlier Hindu contributions. In his best known treatise, the *Lilavati*, he compiled problems from Brahmagupta and others, adding new observations of his own. The very title of this book may be taken to indicate the uneven quality of Hindu thought, for the name in the title is that of Bhaskara's daughter who, according to legend, lost the opportunity to marry because of her father's confidence in his astrological predictions. Bhaskara had calculated that his daughter might propitiously marry only at one particular hour on a given day. On what was to have been her wedding day the eager girl was bending over the water clock, as the hour

[32] See C. B. Boyer, "An Early Reference to Division by Zero," *American Mathematical Monthly*, **50** (1943), 487–491.

for the marriage approached, when a pearl from her headdress fell, quite unnoticed, and stopped the outflow of water. Before the mishap was noted, the propitious hour had passed. To console the unhappy girl, the father gave her name to the book we are describing.

The *Lilavati*, like the *Vija-Ganita*, contains numerous problems dealing **23** with favorite Hindu topics: linear and quadratic equations, both determinate and indeterminate, simple mensuration, arithmetic and geometric progressions, surds, Pythagorean triads, and others. The "broken bamboo" problem, popular in China (and included also by Brahmagupta), appears in the following form: If a bamboo 32 cubits high is broken by the wind so that the tip meets the ground 16 cubits from the base, at what height above the ground was it broken? Also making use of the Pythagorean theorem is the following problem: A peacock is perched atop a pillar at the base of which is a snake's hole. Seeing the snake at a distance from the pillar which is three times the height of the pillar, the peacock pounced upon the snake in a straight line before it could reach its hole. If the peacock and the snake had gone equal distances, how many cubits from the hole did they meet?

These two problems illustrate well the heterogeneous nature of the *Lilavati*, for despite their apparent similarity and the fact that only a single answer is required, one of the problems is determinate and the other is indeterminate. In treating of the circle and the sphere the *Lilavati* fails also to distinguish between exact and approximate statements. The area of the circle is correctly given as one-quarter the circumference multiplied by the diameter and the volume of the sphere as one-sixth the product of the surface area and the diameter, but for the ratio of circumference to diameter in a circle Bhaskara suggests either 3927 to 1250 or the "gross" value 22/7. The former is equivalent to the ratio mentioned, but not used, by Aryabhata. There is no hint in Bhaskara or other Hindu writers that they were aware that all ratios that had been proposed were approximations only. However, Bhaskara severely condemns his predecessors for using the formulas of Brahmagupta for the area and diagonals of a general quadrilateral, because he saw that a quadrilateral is not uniquely determined by its sides. Evidently he did not realize that the formulas are indeed exact for all cyclic quadrilaterals.

Many of Bhaskara's problems in the *Lilavati* and the *Vija-Ganita* evidently were derived from earlier Hindu sources; hence it is no surprise to note that the author is at his best in dealing with indeterminate analysis. In connection with the Pell equation, $x^2 = 1 + py^2$, proposed earlier by Brahmagupta, Bhaskara gave particular solutions for the five cases $p = 8, 11, 32, 61,$ and 67. For $x^2 = 1 + 61y^2$, for example, he gave the solution $x = 1,776,319,049$ and $y = 22,615,390$. This is an impressive feat in calculation, and its verification alone will tax the efforts of the reader unless he is equipped with a modern computer. Bhaskara's books are replete with other instances of Diophantine problems.[33]

24 Bhaskara died toward the end of the twelfth century, and for several hundred years there were few mathematicians in India of comparable stature. It is of interest to note, nevertheless, that Srinivasa Ramanujan (1887–1920), the twentieth-century Hindu genius, had the same uncanny manipulative ability in arithmetic and algebra that is found in Bhaskara. The British mathematician G. H. Hardy once visited Ramanujan in a hospital at Putney and mentioned to his ill friend that he had arrived in a taxi with the dull number 1729, whereupon Ramanujan without hesitation pointed out that this number is indeed interesting, for it is the least integer that can be represented in two different ways as the sum of two cubes—$1^3 + 12^3 = 1729 = 9^3 + 10^3$. In Ramanujan's work we note also the disorganized character, the strength of intuitive reasoning, and the disregard for geometry that stood out so clearly in his predecessors. Although in Ramanujan these characteristics had perhaps developed largely because he was self-taught, we cannot help but see how strikingly different the development of mathematics in India has been from that in Greece. Even when the Hindus borrowed from their neighbors, they fashioned the material in their own peculiar manner. Although in attitude and interests they had more in common with the Chinese, they did not share the latter's fascination with accurate approximations, such as led to Horner's method. And although they shared with the Mesopotamians a preponderately algebraic view, they tended to avoid sexagesimal numeration. In short, the eclectic Hindu mathematicians adopted and developed only such aspects as appealed to them. In one respect it was unfortunate that their first love should have been theory of numbers in general, and indeterminate analysis in particular, for it was not from these aspects that later developments in mathematics grew. Analytic geometry and calculus had Greek rather than Indian roots, and European algebra came from the Islamic countries rather than India. Nevertheless, in modern mathematics there are at least two reminders that mathematics owes its development to India, as well as to many other lands. The trigonometry of the sine function came presumably from India; our own system of numeration for integers is appropriately called the Hindu-Arabic system to indicate its probable origin in India and its transmission through Arabia.

BIBLIOGRAPHY

Boyer, C. B., "Fundamental Steps in the Development of Numeration," *Isis*, **35** (1944), 153–168.

Cajori, Florian, *A History of Mathematics*, 2nd ed. (New York: Macmillan, 1919).

[33] An exceptionally full account of Bhaskara's work is found in J. F. Scott, *A History of Mathematics* (1958). See also Colebrooke, *op. cit.*

Clark, W. E., ed., *The Aryabhatiya of Aryabhata* (Chicago: Open Court, 1930).

Colebrooke, H. T., *Algebra, with Arithmetic and Mensuration, from the Sanscrit of Brahmagupta and Bhaskara* (London, 1817).

Datta, B., and A. N. Singh, *History of Hindu Mathematics* (Lahore, 1935–1938, 2 vols.; Bombay: Asia Publishing House, 1962).

Eves, Howard, *An Introduction to the History of Mathematics*, 2nd ed. (New York: Holt, 1964).

Goldschmidt, Victor, *Die Entstehung unserer Ziffern* (Heidelberg: C. Winter, 1932).

Hill, G. F., *The Development of Arabic Numerals in Europe* (Oxford, 1915).

Ho Peng-Yoke, articles on Liu Hui, Chu Shih-chieh, Ch'in Chiu-shao, Li Chih, and Yang Hui, in *Dictionary of Scientific Biography* (New York: Scribner's) in press.

Juschkewitsch, A. P., *Geschichte der Mathematik im Mittelalter* (Leipzig: Teubner, 1964).

Kaye, G. R., "Indian Mathematics," *Isis*, **2** (1914), 326–356.

Lattin, Harriet P., "The Origin of Our Present System of Notation According to the Theories of Nicholas Bubnov," *Isis*, **19** (1933), 181–194.

Loeffler, Eugen, *Ziffern und Ziffernsysteme* (Leipzig and Berlin: Teubner, 1912).

Menninger, Karl, *Zahlwort und Ziffer* (2nd ed., Göttingen: Vandenhoeck and Ruprecht, 1957–1958, 2 vols.).

Mikami, Yoshio, *The Development of Mathematics in China and Japan* (1913; reprinted, New York: Chelsea, n.d.).

Morley, S. G., *An Introduction to the Study of Maya Hieroglyphics* (Washington: Carnegie Institution, 1915).

Needham, Joseph, *Science and Civilization in China* (Cambridge: Cambridge University Press, 1959), Vol. III.

Rajagopal, C. T., and T. V. Vedamurthi Aiyar, "On the Hindu Proof of Gregory's Series," *Scripta Mathematica*, XVII (1951) 65–74. See also XV (1949) 201–209 and XVIII (1952) 25–30.

Sarton, George, *An Introduction to the History of Science* (Baltimore: Carnegie Institution of Washington, 1927–1948, 3 vols. in 5).

Scott, J. F., *A History of Mathematics* (London: Taylor & Francis, 1958).

Smith, D. E., *History of Mathematics* (Boston: Ginn, 1923–1925, 2 vols., paperback reprint, New York: Dover, 1958).

Smith, D. E. and L. C. Karpinski: *The Hindu-Arabic Numerals* (Boston: Ginn, 1911).

Struik, D. J., "On Ancient Chinese Mathematics," *The Mathematics Teacher*, **56** (1963), 424–432.

Winter, H. J. J., *Eastern Science* (London: John Murray, 1952).

Yeldham, F. A., *The Story of Reckoning in the Middle Ages* (London: G. G. Harrap, 1926).

EXERCISES

1. Compare Hindu and Chinese mathematics with respect to favorite topics and level of achievement.
2. Which had the greater influence on modern thought, Chinese or Hindu mathematics? Explain clearly.

3. What evidences are there of Greek influence in Hindu mathematics? Are there evidences of Hindu influence in Greece? Explain.
4. Is it likely that the ancient Chinese and Babylonian mathematicians borrowed from each other? Explain.
5. How can one account for the Chinese and the Hindu indifference toward conic sections?
6. Describe some respects in which Hindu algebra differed markedly from Greek algebra.
7. Solve the system

$$4x + y + z = 40$$

$$2x + 3y + z = 30$$

$$x + y + 2z = 20$$

by the Chinese matrix method.
8. Write the number 7,834,679 in Chinese rod numerals and in Mayan positional notation.
9. Using the method of Ch'in Chiu-shao, find the square root of 29,584.
10. Write in the notation of Chu Shih-chieh the coefficients in the expansion of the ninth power of a binomial.
11. Justify Aryabhata's rule for finding the number of terms in an arithmetic progression, given the first term, the common difference, and the sum of the terms.
12. Find sin 15° by the *Siddhānta* recursion formula and compare this with the value found in modern tables.
13. Use a *gelosia* pattern to find the product of 345 and 256.
14. Divide 56,789 by 273, using the "galley" method.
15. Check the multiplication in Exercise 13 by casting out nines in the multiplicand, the multiplier, and the product.
16. From Brahmagupta's formula for area deduce Heron's formula as a special case.
17. Show that $21x + 14y = 3$ has no solution in integers.
18. From Brahmagupta's formulas for the diagonals of a (cyclic) quadrilateral deduce Ptolemy's theorem.
19. Solve Bhaskara's broken-bamboo problem.
20. Solve Bhaskara's peacock-and-the-snake problem.
*21. Verify that Brahmagupta's quadrilateral with sides $a = 52$, $b = 25$, $c = 39$, $d = 60$ and diagonals $e = 56$ and $f = 63$ is a cyclic quadrilateral.
*22. Is it possible for Brahmagupta's quadrilateral with sides $a = 25$, $b = 25$, $c = 25$, $d = 39$ to be cyclic? Explain.
*23. Show that the formula of Liu Hui holds for the volume of the tetrahedron $(0, 0, 0)$, $(0, 0, a)$, $(b, 0, 0)$, $(c, d, 0)$. Is the formula valid for all tetrahedra with a pair of opposite edges orthogonal? Explain.

CHAPTER XIII

The Arabic Hegemony

Ah, but my Computations, People say, Have squared
the Year to human Compass, eh? If so, by striking from
the Calendar Unborn To-morrow, and dead Yesterday.
Omar Khayyam (Rubaiyat *in the FitzGerald version*)

At the time that Brahmagupta was writing, the Sabean Empire of Arabia   **1**
Felix had fallen and the peninsula was in a severe crisis. It was inhabited
largely by desert nomads, known as Bedouins, who could neither read nor
write; among them was the prophet Mohammed, born at Mecca in about
570. During his journeys Mohammed came in contact with Jews and Christians, and the amalgam of religious feelings that were raised in his mind led
him to regard himself as the apostle of God sent to lead his people. For some
ten years he preached at Mecca, but in 622, faced by a plot on his life, he
accepted an invitation to Medina. This "flight," known as the Hejira, marked
the beginning of the Mohammedan era—one that was to exert a strong
influence on the development of mathematics. Mohammed now became a
military, as well as a religious leader. Ten years later he had established a
Mohammedan state, with center at Mecca, within which Jews and Christians,
being also monotheistic, were afforded protection and freedom of worship.
In 632, while planning to move against the Byzantine Empire, Mohammed
died at Medina. His sudden death in no way impeded the expansion of the
Islamic state, for his followers overran neighboring territories with astonishing rapidity. Within a few years Damascus and Jerusalem and much of the
Mesopotamian Valley fell to the conquerors; by 641 Alexandria, which for
many years had been the mathematical center of the world, was captured.
There is a legend that the leader of the victorious troops, having asked what
was to be done with the books in the library, was told to burn them; for if
they were in agreement with the Koran they were superfluous, if they were in
disagreement they were worse than superfluous. However, stories that the
baths were long heated by the fires of burning books undoubtedly are
exaggerated. Following depredations by earlier military and religious
fanatics, and long ages of sheer neglect, there probably were relatively few
books in the library that once had been the greatest in the world.

For more than a century the Arab conquerors fought among themselves and with their enemies, until by about 750 the warlike spirit subsided. By this time a schism had arisen between the western Arabs in Morocco and the eastern Arabs who, under the caliph al-Mansur, had established a new capital at Baghdad, a city that was shortly to become the new center for mathematics. However, the caliph at Baghdad could not command the allegiance even of all Moslems in the eastern half of his empire, although his name appeared on coins of the realm and was included in the prayers of his "subjects." The unity of the Arab world, in other words, was more economic and religious than it was political. Arabic was not necessarily the common language, although it was a kind of lingua franca for intellectuals. Hence it might be more appropriate to speak of the culture as Islamic, rather than Arabic, although we shall use the terms more or less interchangeably.

During the first century of the Arabic conquests there had been political and intellectual confusion, and possibly this accounts for the difficulty in localizing the origin of the modern system of numeration. The Arabs were at first without intellectual interest, and they had little culture, beyond a language, to impose on the peoples they conquered. In this respect we see a repetition of the situation when Rome conquered Greece, of which it was said that, in a cultural sense, captive Greece took captive the captor Rome. By about 750 the Arabs were ready to have history repeat itself, for the con-querors became eager to absorb the learning of the civilizations they had overrun. By 766 we learn that an astronomical-mathematical work, known to the Arabs as the *Sindhind* was brought to Baghdad from India. It is generally thought that this was the *Brahmasphuta Siddhānta*, although it may have been the *Surya Siddhānta*. A few years later, perhaps about 775, this *Siddhānta* was translated into Arabic, and it was not long afterward (ca. 780) that Ptolemy's astrological *Tetrabiblos* was translated into Arabic from the Greek. Alchemy and astrology were among the first studies to appeal to the dawning intellectual interests of the conquerors. The "Arabic miracle" lies not so much in the rapidity with which the political empire rose as in the alacrity with which, their tastes once aroused, the Arabs absorbed the learn-ing of their neighbors.

2   The first century of the Muslim empire had been devoid of scientific achievement. This period (from about 650 to 750) had been, in fact, perhaps the nadir in the development of mathematics, for the Arabs had not yet achieved intellectual drive, and concern for learning in other parts of the world had pretty much faded. Had it not been for the sudden cultural awakening in Islam during the second half of the eighth century, consid-erably more of ancient science and mathematics undoubtedly would have been lost. To Baghdad at that time were called scholars from Syria, Iran, and

Mesopotamia, including Jews and Nestorian Christians; under three great Abbasid patrons of learning—al-Mansur, Haroun al-Raschid, and al-Mamun—the city became a new Alexandria. During the reign of the second of these caliphs, familiar to us today through the *Arabian Nights*, part of Euclid was translated. It was during the caliphate of al-Mamun (809–833), however, that the Arabs fully indulged their passion for translation. The caliph is said to have had a dream in which Aristotle appeared, and as a consequence al-Mamun determined to have Arabic versions made of all the Greek works he could lay his hands on, including Ptolemy's *Almagest* and a complete version of Euclid's *Elements*. From the Byzantine Empire, with which the Arabs maintained an uneasy peace, Greek manuscripts were obtained through treaties.

Al-Mamun established at Baghdad a "House of Wisdom" (Bait al-hikma) comparable to the ancient Museum at Alexandria. Among the faculty members was a mathematician and astronomer, Mohammed ibn-Musa al-Khowarizmi, whose name, like that of Euclid, later was to become a household word in Western Europe.[1] This scholar, who died sometime before 850, wrote more than half a dozen astronomical and mathematical works, of which the earliest were probably based on the *Sindhind* derived from India. Besides astronomical tables, and treatises on the astrolabe and the sundial, al-Khowarizmi wrote two books on arithmetic and algebra which played very important roles in the history of mathematics. One of these survives only in a unique copy of a Latin translation with the title *De numero indorum* ("Concerning the Hindu Art of Reckoning"), the original Arabic version having since been lost. In this work, based presumably on an Arabic translation of Brahmagupta, al-Khowarizmi gave so full an account of the Hindu numerals that he probably is responsible for the widespread but false impression that our system of numeration is Arabic in origin. Al-Khowarizmi made no claim to originality in connection with the system, the Hindu source of which he assumed as a matter of course; but when subsequently Latin translations of his work appeared in Europe, careless readers began to attribute not only the book, but also the numeration, to the author. The new notation came to be known as that of al-Khowarizmi, or more carelessly, algorismi; ultimately the scheme of numeration making use of the Hindu numerals came to be called simply algorism or algorithm, a word that, originally derived from the name al-Khowarizmi, now means, more generally, any peculiar rule of procedure or operation—such as the Euclidean method for finding the greatest common divisor.

Through his arithmetic, al-Khowarizmi's name has become a common **3** English word; through the title of his most important book, *Al-jabr wa'l*

[1] For two recent studies on the science of al-Khowarizmi, see *Isis*, **54** (1963), 97–119.

muqābalah, he has supplied us with an even more popular household term. From this title has come the word *algebra,* for it is from this book that Europe later learned the branch of mathematics bearing this name. Diophantus sometimes is called "the father of algebra," but this title more appropriately belongs to al-Khowarizmi. It is true that in two respects the work of al-Khowarizmi represented a retrogression from that of Diophantus. First, it is on a far more elementary level than that found in the Diophantine problems and, second, the algebra of al-Khowarizmi is thoroughly rhetorical, with none of the syncopation found in the Greek *Arithmetica* or in Brahmagupta's work. Even numbers were written out in words rather than symbols! It is quite unlikely that al-Khowarizmi knew of the work of Diophantus, but he must have been familiar with at least the astronomical and computational portions of Brahmagupta; yet neither al-Khowarizmi nor other Arabic scholars made use of syncopation or of negative numbers. Nevertheless, the *Al-jabr* comes closer to the elementary algebra of today than the works of either Diophantus or Brahmagupta, for the book is not concerned with difficult problems in indeterminate analysis but with a straightforward and elementary exposition of the solution of equations, especially of second degree. The Arabs in general loved a good clear argument from premise to conclusion, as well as systematic organization—respects in which neither Diophantus nor the Hindus excelled. The Hindus were strong in association and analogy, in intuition and an aesthetic and imaginative flair, whereas the Arabs were more practical-minded and down-to-earth in their approach to mathematics.

The *Al-jabr* has come down to us in two versions, Latin and Arabic, but in the Latin translation, *Liber algebrae et almucabola,* a considerable portion of the Arabic draft is missing. The Latin, for example, has no preface, perhaps because the author's preface in Arabic gave fulsome praise to Mohammed, the prophet, and to al-Mamun, "the Commander of the Faithful." Al-Khowarizmi wrote that the latter had encouraged him to

... compose a short work on Calculating by (the rules of) Completion and Reduction, confining it to what is easiest and most useful in arithmetic, such as men constantly require in cases of inheritance, legacies, partitions, law-suits, and trade, and in all their dealings with one another, or where the measuring of lands, the digging of canals, geometrical computation, and other objects of various sorts and kinds are concerned.[2]

It is not certain just what the terms "al-jabr" and "muqābalah" mean, but the usual interpretation is similar to that implied in the translation above. The word "al-jabr" presumably meant something like "restoration" or "completion" and seems to refer to the transposition of subtracted terms to

---

[2] See *Robert of Chester's Latin Translation of the Algebra of al-Khowarizmi,* ed. by L. C. Karpinski (1915), p. 46. Translations used by us are taken from this edition.

the other side of an equation; the word "muqābalah" is said to refer to "reduction" or "balancing"—that is, the cancellation of like terms on opposite sides of the equation.[3] Arabic influence in Spain long after the time of al-Khowarizmi is found in *Don Quixote*, where the word *algebrista* is used for a bone-setter, that is, a "restorer."

The Latin translation of al-Khowarizmi's *Algebra* opens with a brief **4** introductory statement of the positional principle for numbers and then proceeds to the solution, in six short chapters, of the six types of equations made up of the three kinds of quantities: roots, squares, and numbers (that is, x, x^2, and numbers). As abu-Kamil Shoja ben Aslam, a slightly later textbook writer, expressed the situation,

The first thing which is necessary for students in this science [algebra] is to understand the three species which are noted by Mohammed ibn Musa al-Khowarizmi in his book. These are *roots, squares* and *numbers.*[4]

Chapter I, in three short paragraphs, covers the case of squares equal to roots, expressed in modern notation as $x^2 = 5x$, $x^2/3 = 4x$, and $5x^2 = 10x$, giving the answers $x = 5$, $x = 12$, and $x = 2$ respectively. (The root $x = 0$ was not recognized.) Chapter II covers the case of squares equal to numbers, and Chapter III solves the case of roots equal to numbers, again with three illustrations per chapter to cover the cases in which the coefficient of the variable term is equal to, more than, or less than one. Chapters IV, V, and VI are more interesting, for they cover in turn the three classical cases of three-term quadratic equations: (1) squares and roots equal to numbers, (2) squares and numbers equal to roots, and (3) roots and numbers equal to squares. The solutions are "cookbook" rules for "completing the square" applied to specific instances. Chapter IV, for example, includes the three illustrations $x^2 + 10x = 39$, $2x^2 + 10x = 48$, and $\frac{1}{2}x^2 + 5x = 28$. In each case only the positive answer is given. In Chapter V only a single example— $x^2 + 21 = 10x$—is used, but both roots, 3 and 7, are given, corresponding to the rule $x = 5 \mp \sqrt{25 - 21}$. Al-Khowarizmi here calls attention to the fact that what we designate as the discriminant must be positive:

You ought to understand also that when you take the half of the roots in this form of equation and then multiply the half by itself; if that which proceeds or results from the multiplication is less than the units above-mentioned as accompanying the square, you have an equation.

---

[3] It should be noted, however, that this interpretation has been questioned by Solomon Gandz, "The Origin of the Term 'Algebra'," *American Mathematical Monthly*, **33** (1926), 437–440. Gandz thinks that "jabr" was an Assyrian word for equation and that "al-muqābalah" is simply the Arabic translation of "al-jabr."

[4] L. C. Karpinski, "The Algebra of Abu Kamil," *American Mathematical Monthly*, **21** (1914), 37–48.

In Chapter VI the author again uses only a single example—$3x + 4 = x^2$—for whenever the coefficient of x^2 is not unity, the author reminds one to divide first by this coefficient (as in Chapter IV). Once more the steps in completing the square are meticulously indicated, without justification, the procedure being equivalent to the solution $x = 1\frac{1}{2} + \sqrt{(1\frac{1}{2})^2 + 4}$. Again only one root is given, for the other is negative.

5   The six cases of equations given above exhaust all possibilities for linear and quadratic equations having a positive root. So systematic and exhaustive was al-Khowarizmi's exposition that his readers must have had little difficulty in mastering the solutions. In this sense, then, al-Khowarizmi is entitled to be known as "the father of algebra." However, no branch of mathematics springs up fully grown, and we cannot help but ask where the inspiration for Arabic algebra came from. To this question no categorical answer can be given; but the arbitrariness of the rules and the strictly numerical form of the six chapters remind us of ancient Babylonian and medieval Indian mathematics. The exclusion of indeterminate analysis, a favorite Hindu topic, and the avoidance of any syncopation, such as is found in Brahmagupta, might suggest Mesopotamia as more likely a source than India. As we read beyond the sixth chapter, however, an entirely new light is thrown on the question. Al-Khowarizmi continued:

> We have said enough so far as numbers are concerned, about the six types of equations. Now, however, it is necessary that we should demonstrate geometrically the truth of the same problems which we have explained in numbers.

The ring in this passage is obviously Greek rather than Babylonian or Indian. There are, therefore, three main schools of thought on the origin of Arabic algebra; one emphasizes Hindu influences, another stresses the Mesopotamian, or Syriac-Persian, tradition, and the third points to Greek inspiration.[5] The truth is probably approached if we combine the three theories. The philosophers of Islam admired Aristotle to the point of aping him, but eclectic Mohammedan mathematicians seem to have chosen appropriate elements from various sources.

6   The *Algebra* of al-Khowarizmi betrays unmistakable Hellenic elements, but the first geometrical demonstrations have little in common with classical Greek mathematics. For the equation $x^2 + 10x = 39$ al-Khowarizmi drew a square ab to represent x^2, and on the four sides of this square he placed rectangles $c, d, e,$ and f, each $2\frac{1}{2}$ units wide. To complete the larger square one

---

[5] See Solomon Gandz, "The Sources of al-Khowarizmi's Algebra," *Osiris*, **1** (1936), 263–277; also H. J. J. Winter, "Formative Influences in Islamic Science," *Archives Internationales d'Histoire des Sciences*, **6** (1953), 171–192.

must add the four small corner squares (dotted in Fig. 13.1), each of which has an area of $6\frac{1}{4}$ units. Hence to "complete the square" we add 4 times $6\frac{1}{4}$ units or 25 units, thus obtaining a square of total area $39 + 25 = 64$ units

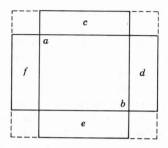

FIG. 13.1

(as is clear from the right-hand side of the given equation). The side of the large square must therefore be 8 units, from which we subtract 2 times $2\frac{1}{2}$ or 5 units to find that $x = 3$, thus proving that the answer found in Chapter IV is correct.

The geometrical proofs for Chapters V and VI are somewhat more involved. For the equation $x^2 + 21 = 10x$ the author draws the square ab to represent x^2 and the rectangle bg to represent 21 units. Then the large rectangle, comprising the square and the rectangle bg, must have an area equal to $10x$, so that the side ag or hd must be 10 units. If, then, one bisects hd at e, draws et perpendicular to hd, extends te to c so that $tc = tg$, and completes the squares $tclg$ and $cmne$ (Fig. 13.2), the area tb is equal to area md. But square tl is 25, and the gnomon $tenmlg$ is 21 (since the gnomon is equal to the rectangle bg). Hence the square nc is 4, and its side ec is 2. Inasmuch as $ec = be$, and since $he = 5$, we see that $x = hb = 5 - 2$ or 3, which proves that the arithmetic solution given in Chapter V is correct. A modified diagram

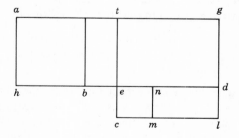

FIG. 13.2

is given for the root $x = 5 + 2 = 7$, and an analogous type of figure is used to justify geometrically the result found algebraically in Chapter VI.

7   A comparison of Fig. 13.2, taken from al-Khowarizmi's *Algebra*, with diagrams found in the *Elements* of Euclid in connection with Greek geometrical algebra (such as our Fig. 7.7) leads to the inevitable conclusion that Arabic algebra had much in common with Greek geometry; yet the first, or arithmetical part, of al-Khowarizmi's *Algebra* obviously is alien to Greek thought. What apparently happened in Baghdad was just what one would expect in a cosmopolitan intellectual center. Arabic scholars had great admiration for Greek astronomy, mathematics, medicine, and philosophy—subjects that they mastered as best they could. However, they could scarcely help but notice that, as the Nestorian Bishop Sebokt had observed when in 662 he first called attention to the nine marvelous digits of the Hindus, "there are also others who know something." It is probable that al-Khowarizmi typified the Arabic electicism that will so frequently be observed in other cases. His system of numeration most likely came from India, his systematic algebraic solution of equations may have been a development from Mesopotamia, and the logical geometric framework for his solutions palpably was derived from Greece.

The *Algebra* of al-Khowarizmi contains more than the solution of equations, material that occupies about the first half. There are, for example, rules for operations on binomial expressions, including products such as $(10 + 2)(10 - 1)$ and $(10 + x)(10 - x)$. Although the Arabs rejected negative roots and absolute negative magnitudes, they were familiar with the rules governing what are now known as signed numbers. There are also alternative geometrical proofs of some of the author's six cases of equations. Finally, the *Algebra* includes a wide variety of problems illustrating the six chapters or cases. As an illustration of the fifth chapter, for example, al-Khowarizmi asks for the division of ten into two parts in such a way that "the sum of the products obtained by multiplying each part by itself is equal to fifty eight." The extant Arabic version, unlike the Latin, includes also an extended discussion of inheritance problems, such as the following:

> A man dies, leaving two sons behind him, and bequeathing one-third of his capital to a stranger. He leaves ten dirhems of property and a claim of ten dirhems upon one of the sons.

The answer is not what one would expect, for the stranger gets only 5 dirhems. According to Arabic law, a son who owes to the estate of his father an amount greater than the son's portion of the estate retains the whole sum that he owes, one part being regarded as his share of the estate and the remainder as a gift from his father. To some extent it seems to have been the complicated nature of laws governing inheritance that encouraged the study of algebra in Arabia.

A few of al-Khowarizmi's problems give rather clear evidence of Arabic **8** dependence on the Babylonian-Heronian stream of mathematics. One of them presumably was taken directly from Heron, for the figure and dimensions are the same. Within an isosceles triangle having sides 10 yards and base 12 yards (Fig. 13.3) a square is to be inscribed, and the side of this square

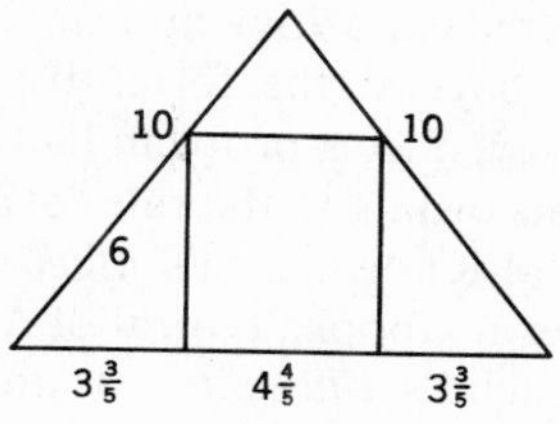

FIG. 13.3

is called for. The author of the *Algebra* first finds through the Pythagorean theorem that the altitude of the triangle is 8 yards, so that the area of the triangle is 48 square yards. Calling the side of the square the "thing," he notes that the square of the "thing" will be found by taking from the area of the large triangle the areas of the three small triangles lying outside the square but inside the large triangle. The sum of the areas of the two lower small triangles he knows to be the product of the "thing" by six less half the "thing"; and the area of the upper small triangle is the product of eight less the "thing" by half the "thing." Hence he is led to the obvious conclusion that the "thing" is $4\frac{4}{5}$ yards—the side of the square. The chief difference between the form of this problem in Heron and that of al-Khowarizmi is that Heron had expressed the answer in terms of unit fractions as $4\frac{1}{2}\frac{1}{5}\frac{1}{10}$. The similarities are so much more pronounced than the differences that we may take this case as confirmation of the general axiom that continuity in the history of mathematics is the rule rather than the exception. Where a discontinuity seems to arise, we should first consider the possibility that the apparent saltus may be explained by the loss of intervening documents.

The *Algebra* of al-Khowarizmi usually is regarded as the first work on **9** the subject, but a recent publication in Turkey raises some question about this. A manuscript of a work by abd-al-Hamid ibn-Turk, entitled "Logical Necessities in Mixed Equations," was part of a book on *Al-jabr wa'l muqāba-lah* which was evidently very much the same as that by al-Khowarizmi and was published at about the same time—possibly even earlier. The surviving chapters on "Logical Necessities" give precisely the same type of geometrical demonstration as al-Khowarizmi's *Algebra* and in one case the same illustrative example $x^2 + 21 = 10x$. In one respect abd-al-Hamid's exposition is more thorough than that of al-Khowarizmi for he gives geometrical figures to prove that if the discriminant is negative, a quadratic equation has no solution. Similarities in the work of the two men and the systematic organization found in them seem to indicate that algebra in their day was not so recent a development as has usually been assumed.[6] When textbooks with a conventional and well-ordered exposition appear simultaneously, a subject is likely to be considerably beyond the formative stage. Successors of al-Khowarizmi were able to say, once a problem had been reduced to the form of an equation, "Operate according to the rules of algebra and almucabala." In any case, the survival of al-Khowarizmi's *Algebra* can be taken to indicate that it was one of the better textbooks typical of Arabic algebra of the time. It was to algebra what Euclid's *Elements* was to geometry—the best elementary exposition available until modern times—but al-Khowarizmi's work had a serious deficiency that had to be removed before it could serve its purpose effectively in the modern world: a symbolic notation had to be developed to replace the rhetorical form. This step the Arabs never took, except for the replacement of number words by number signs.

10   The ninth century was a glorious one in Arabic mathematics, for it produced not only al-Khowarizmi in the first half of the century, but also Thabit ibn-Qurra (826–901) in the second half. If al-Khowarizmi resembled Euclid as an "elementator," then Thabit is the Arabic equivalent of Pappus, the commentator on higher mathematics. Thabit was the founder of a school of translators, especially from Greek and Syriac, and to him we owe an immense debt for translations into Arabic of works by Euclid, Archimedes, Apollonius, Ptolemy, and Eutocius. (Note the omission of Diophantus and Pappus, authors who evidently were not at first known in Arabia, although the Diophantine *Arithmetica* became familiar before the end of the tenth century.) Had it not been for his efforts, the number of Greek mathematical works extant today would be smaller. For example, we should have only the first four, rather than the first seven, books of Apollonius' *Conics*. Moreover, Thabit had so thoroughly mastered the content of the classics he translated that he suggested modifications and generalizations. To him is due a remarkable formula for amicable numbers: If p, q, and r are prime numbers, and if they are of the form $p = 3 \cdot 2^n - 1$, $q = 3 \cdot 2^{n-1} - 1$, and $r = 9 \cdot 2^{2n-1} - 1$, then $2^n pq$ and $2^n r$ are amicable numbers, for each is equal to the sum of the

[6] See Aydin Sayili, *Logical Necessities in Mixed Equations by 'Abd al Hamid ibn Turk and the Algebra of His Time* (1962).

proper divisors of the other. Like Pappus, he also gave a generalization of the Pythagorean theorem that is applicable to all triangles, whether right or scalene. If from vertex A of any triangle ABC one draws lines intersecting BC in points B' and C' such that angles $AB'B$ and $AC'C$ are each equal to angle A (Fig. 13.4), then $\overline{AB}^2 + \overline{AC}^2 = \overline{BC}(\overline{BB'} + \overline{CC'})$. Thabit gave no

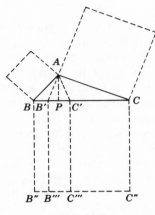

FIG. 13.4

proof of the theorem, but this is easily supplied through theorems on similar triangles. In fact, the theorem provides a beautiful generalization of the pinwheel diagram used by Euclid in the proof of the Pythagorean theorem. If, for example, angle A is obtuse, then the square on side AB is equal to the rectangle $BB'B''B'''$, and the square on AC is equal to the rectangle $CC'C''C'''$, where $BB'' = CC'' = BC = B''C''$. That is, the sum of the squares on AB and AC is the square on BC less the rectangle $B'C'B'''C'''$. If angle A is acute, then the positions of B' and C' are reversed with respect to AP, where P is the projection of A on BC, and in this case the sum of the squares on AB and AC is equal to the square on BC *increased* by the rectangle $B'C'B'''C'''$. If A is a right angle, then B' and C' coincide with P, and for this case Thabit's theorem becomes the Pythagorean theorem. (Thabit[7] did not draw the dotted lines that are shown in Fig. 13.4, but he did consider the several cases.)

Alternative proofs of the Pythagorean theorem, works on parabolic and paraboloidal segments, a discussion of magic squares, angle trisections, and new astronomical theories are among Thabit's further contributions to scholarship. The Arabs sometimes are described as servile imitators of the Greeks in science and philosophy, but such accusations are exaggerated. Thabit, for instance, boldly added a ninth sphere to the eight previously

[7] See Aydin Sayili, "Thabit ibn Qurra's Generalization of the Pythagorean Theorem," *Isis*, **51** (1960), 35–37. See also *Isis*, **55** (1964), 68–70, and **57** (1966), 56–66.

assumed in simplified versions of Aristotelian–Ptolemaic astronomy; and instead of the Hipparchan precession of the equinoxes in one direction or sense only, Thabit proposed a "trepidation of the equinoxes" in a reciprocating type of motion. Such questioning of points in Greek astronomy may well have been a factor in paving the way for the revolution in astronomy initiated by Copernicus.

11   We have mentioned several times that the Arabs were quick to absorb learning from the neighbors they conquered; it should be noted also that within the confines of the Arabic empire lived peoples of very varied ethnic backgrounds: Syrian, Greek, Egyptian, Persian, Turkish, and many others. Most of them shared a common faith, Islam, although Christians and Jews were tolerated; very many shared a common language, Arabic, although Greek and Hebrew were sometimes used. Nevertheless, we should not expect a high degree of uniformity in learning. There was considerable factionalism at all times, and it sometimes erupted into conflict. Thabit himself lived in a pro-Greek community, which opposed him for his pro-Arabic sympathies. In Arabic mathematics such cultural differences occasionally became quite apparent, as in the works of the tenth- and eleventh-century scholars Abu'l-Wefa (940–998) and al-Karkhi (or al-Karagi, ca. 1029). In some of their works they used the Hindu numerals, which had reached Arabia through the astronomical *Sindhind*; at other times they adopted the Greek alphabetic pattern of numeration (with, of course, Arabic equivalents for the Greek letters). Ultimately the superior Hindu numerals won out, but even within the circle of those who used the Indian numeration, the forms of the numerals differed considerably. Variations had obviously been prevalent in India, but in Arabia variants were so striking that there are theories suggesting entirely different origins for forms used in the eastern and western halves of the Arabic world. Perhaps the numerals of the Saracens in the east came directly from India, while the numerals of the Moors in the west were derived from Greek or Roman forms. More likely the variants were the result of gradual changes taking place in space and time, for the Arabic numerals of today are strikingly different from the modern Devanagari (or "divine") numerals still in use in India. After all, it is the principles within the system of numeration that are important, and not the specific forms of the numerals. Our numerals often are known as Arabic, despite the fact that they bear little resemblance to those now in use in Egypt, Iraq, Syria, Arabia, Iran, and other lands within the Islamic culture—that is, the forms ١٢٣٤٥٦٧٨٩٠. We call our numerals Arabic because the principles in the two systems are the same and because our forms may have been derived from the Arabic. However, the principles behind the Arabic numerals presumably were derived from India; hence it is better to call ours the Hindu or the Hindu-Arabic system.

As in numeration there was competition between systems of Greek and **12** Indian origin, so also in astronomical calculations there were at first in Arabia two types of trigonometry—the Greek geometry of chords, as found in the *Almagest*, and the Hindu tables of sines, as derived through the *Sindhind*. Here, too, the conflict resulted in triumph for the Hindu aspect,

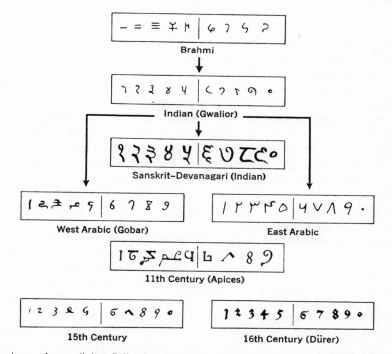

Brahmi

Indian (Gwalior)

Sanskrit–Devanagari (Indian)

West Arabic (Gobar)

East Arabic

11th Century (Apices)

15th Century

16th Century (Dürer)

Genealogy of our digits. Following Karl Menninger, *Zahlwort und Ziffer* (Göttingen: Vanderhoeck & Ruprecht, 1957–1958, 2 vols.), II, 233.

and most Arabic trigonometry ultimately was built on the sine function. It was, in fact, again through the Arabs, rather than directly from the Hindus, that this trigonometry of the sine reached Europe. The astronomy of al-Battani (ca. 850–929), known in Europe as Albategnius, served as the primary vehicle of transmission, although Thabit ibn Qurra seems to have used sines somewhat earlier. In a book entitled *On the Motion of the Stars* Albategnius gave formulas, such as $b = [a \sin(90° - A)]/\sin A$ (see Fig. 13.5), in which the sine and versed sine functions appear. By the time of Abu'l-Wefa, a century later, the tangent function was fairly well known, so that one could express the above relationship more simply as $a = b \tan A$. Here one is in more immediate touch with modern trigonometry, for the Arabic tangent function,

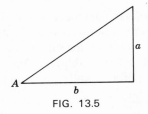

FIG. 13.5

unlike the Hindu sine function, generally was given for a unit circle. Moreover, with Abu'l-Wefa trigonometry assumes a more systematic form in which such theorems as double and half-angle formulas are proved. Although the Hindu sine function had displaced the Greek chord, it was nevertheless the *Almagest* of Ptolemy that motivated the logical arrangement of trigonometric results. The law of sines had been known to Ptolemy in essence and is implied in the work of Brahmagupta, but it frequently is attributed to Abu'l-Wefa because of his clear-cut formulation of the law for spherical triangles. Abu'l-Wefa also made up a new sine table for angles differing by $\frac{1}{4}°$, using the equivalent of eight decimal places. He contributed also a table of tangents and made use of all six of the common trigonometric functions, together with relations among them, but his use of the new functions seems not to have been followed widely in the medieval period.

Sometimes attempts are made to attribute the functions tangent, cotangent, secant, and cosecant to specific times and even to specific individuals, but this cannot be done with any assurance. In India and Arabia there had been a general theory of shadow lengths, as related to a unit of length or gnomon, for varying solar altitudes. There was no one standard unit of length for the staff or gnomon used, although a handspan or a man's height was frequently adopted. The horizontal shadow, for a vertical gnomon of given length, was what we call the cotangent of the angle of elevation of the sun. The "reverse shadow"—that is, the shadow cast on a vertical wall by a stick or gnomon projecting horizontally from the wall—was what we know as the tangent of the solar elevation. The "hypotenuse of the shadow"—that is, the distance from the tip of the gnomon to the tip of the shadow—was the equivalent of our cosecant function; and the "hypotenuse of the reverse shadow" played the role of our secant. This shadow tradition seems to have been well established in Asia by the time of Thabit ibn Qurra,[8] but values of the hypotenuse (secant or cosecant) were seldom tabulated.

13   Abu'l-Wefa was a capable algebraist as well as a trigonometer. He commented on al-Khowarizmi's *Algebra* and translated from the Greek one of the last great classics—the *Arithmetica* of Diophantus. His successor

[8] See E. S. Kennedy, "Overview on Trigonometry," to appear in the *Yearbook on History of Mathematics* of the National Council of Teachers of Mathematics.

al-Karkhi evidently used this translation to become an Arabic disciple of Diophantus—but without Diophantine analysis! That is, al-Karkhi was concerned with the algebra of al-Khowarizmi rather than the indeterminate analysis of the Hindus; but like Diophantus (and unlike al-Khowarizmi) he did not limit himself to quadratic equations—despite the fact that he followed the Arabic custom of giving geometric proofs for quadratics. In particular, to al-Karkhi is attributed the first numerical solutions of equations of the form $ax^{2n} + bx^n = c$ (only equations with positive roots were considered), where the Diophantine restriction to rational numbers was abandoned. It was in just this direction, toward the algebraic solution (in terms of radicals) of equations of more than second degree, that the early developments in mathematics in the Renaissance were destined to take place.

The time of al-Karkhi—the early eleventh century—was a brilliant era **14** in the history of Arabic learning, and a number of his contemporaries deserve brief mention—brief not because they were less capable, but because they were not primarily mathematicians. Ibn-Sina (980–1037), better known to the West as Avicenna, was the foremost scholar and scientist in Islam, but in his encyclopedic interests mathematics played a smaller role than medicine and philosophy. He made a translation of Euclid and explained the casting-out of nines (which consequently is sometimes unwarrantedly attributed to him), but he is better remembered for his application of mathematics to astronomy and physics. As Avicenna reconciled Greek learning with Muslim thought, so his contemporary al-Biruni (973–1048) made the Arabs—hence us—familiar with Hindu mathematics and culture through his well-known book entitled *India*. An indefatigable traveler and a critical thinker, he gave a sympathetic but candid account, including full descriptions of the *Siddhāntas* and the positional principle of numeration. It is he who tells us that Archimedes was familiar with Heron's formula and gives a proof of this and of Brahmagupta's formula, correctly insisting that the latter applies only to a cyclic quadrilateral. In inscribing a nonagon in a circle al-Biruni reduced the problem, through the trigonometric formula for $\cos 3\theta$, to solving the equation $x^3 = 1 + 3x$, and for this he gave the approximate solution in sexagesimal fractions as 1;52,15,17,13—equivalent to more than six-place accuracy.[9] Al-Biruni also gave us, in a chapter on gnomon lengths, an account of the Hindu shadow reckoning. The boldness of his thought is illustrated by his discussion of whether or not the earth rotates on its axis, a question to which he did not give an answer. (Aryabhata seems earlier to have suggested a rotating earth at the center of space.) Al-Biruni contributed also to physics, especially through studies in specific gravity and the causes

[9] See Pierre Dedron and Jean Itard, *Mathématiques et mathématiciens* (1959), p. 126.

of artesian wells; but as a physicist and mathematician he was excelled by ibn-al-Haitham (ca. 965–1039), known to the West as Alhazen. The most important treatise written by Alhazen was the *Treasury of Optics*, a book which was inspired by work of Ptolemy on reflection and refraction and which in turn inspired scientists of medieval and early modern Europe. Among the questions that Alhazen considered were the structure of the eye, the apparent increase in the size of the moon when near the horizon, and an estimate, from the observation that twilight lasts until the sun is 19° below the horizon, of the height of the atmosphere. The problem of finding the point on a spherical mirror at which light from a source will be reflected to the eye of an observer is known to this day as "Alhazen's problem." It is a "solid problem" in the old Greek sense, solvable by conic sections, a subject with which Alhazen was quite familiar. He extended Archimedes' results on conoids by finding the volume generated by revolving about the tangent at the vertex the area bounded by a parabolic arc and the axis and an ordinate of the parabola.

15   Arabic mathematics can with some propriety be divided into four parts: (1) an arithmetic derived presumably from India and based on the principle of position; (2) an algebra which, although from Greek, Hindu, and Babylonian sources, nevertheless in Muslim hands assumed a characteristically new and systematic form; (3) a trigonometry the substance of which came chiefly from Greece but to which the Arabs applied the Hindu form and added new functions and formulas; (4) a geometry which came from Greece but to which the Arabs contributed generalizations here and there. In connection with (3) it should be noted that ibn-Yunus (†1008), Alhazen's contemporary and fellow countryman (they both lived in Egypt), introduced the formula $2 \cos x \cos y = \cos(x + y) + \cos(x - y)$. This is one of the four "product to sum" formulas that in sixteenth-century Europe served, before the invention of logarithms, to convert products to sums by the method known as "prosthaphaeresis" (Greek for addition and subtraction). In connection with (4) there was a significant contribution about a century after Alhazen by a man who in the East is known as a scientist but whom the West recalls as one of the greatest Persian poets. Omar Khayyam (ca. 1050–1122), the "tentmaker," wrote an *Algebra*[10] that went beyond that of al-Khowarizmi to include equations of third degree. Like his Arabic predecessors, Omar Khayyam provided for quadratic equations both arithmetic and geometric solutions; for general cubic equations, he believed (mistakenly, as the sixteenth century later showed), arithmetic solutions were impossible; hence

---

[10] See *The Algebra of Omar Khayyam*, ed. by D. S. Kasir (1931); also D. J. Struik, "Omar Khayyam, Mathematician," *The Mathematics Teacher*, **51** (1958), 280–285.

he gave only geometric solutions. The scheme of using intersecting conics to solve cubics had been used earlier by Menaechmus, Archimedes, and Alhazen, but Omar Khayyam took the praiseworthy step of generalizing the method to cover all third-degree equations (having positive roots). When in an earlier work he came across a cubic equation, he specifically remarked: "This cannot be solved by plane geometry [i.e., using straightedge and compasses only] since it has a cube in it. For the solution we need conic sections."[11]

For equations of higher degree than three Omar Khayyam evidently did not envision similar geometric methods, for space does not contain more than three dimensions, "what is called square-square by algebraists in continuous magnitude is a theoretical fact. It does not exist in reality in any way." The procedure that Omar Khayyam so tortuously—and so proudly—applied to cubic equations can be stated with far greater succinctness in modern notation and concepts as follows. Let the cubic be $x^3 + ax^2 + b^2x + c^3 = 0$. Then if for x^2 in this equation we substitute $2py$, we obtain (recalling that $x^3 = x^2 \cdot x$) the result $2pxy + 2apy + b^2x + c^3 = 0$. Since the resulting equation represents an hyperbola, and the equality $x^2 = 2py$ used in the substitution represents a parabola, it is clear that if the hyperbola and the parabola are sketched on the same set of coordinate axes, then the abscissas of the points of intersection of the two curves will be the roots of the cubic equation. Obviously many other pairs of conic sections can be used in a similar way to solve the cubic.

Our exposition of Omar Khayyam's work does not do justice to his genius, for, lacking the concept of negative coefficients, he had to break the problem into many separate cases according as the parameters a, b, c are positive, negative, or zero. Moreover, he had to specifically identify his conic sections for each case, for the concept of a general parameter was not at hand in his day. Not all roots of a given cubic equation were given, for he did not accept the appropriateness of negative roots and did not note all intersections of the conic sections. It should be remarked also that in the earlier Greek geometric solutions of cubic equations the coefficients had been line segments, whereas in the work of Omar Khayyam they were specific numbers. One of the most fruitful contributions of Arabic eclecticism was the tendency to close the gap between numerical and geometrical algebra. The decisive step in this direction came much later with Descartes, but Omar Khayyam was moving in this direction when he wrote, "Whoever thinks algebra is a trick in obtaining unknowns has thought it in vain. No attention should be paid to the fact that algebra and geometry are different in appearance. Algebras are geometric facts which are proved."[12] In replacing Euclid's theory of proportions by a

[11] A. R. Amir-Moez, "A Paper of Omar Khayyam," *Scripta Mathematica*, **26** (1963), 323–337, p. 328.

[12] Amir-Moez, *op. cit.*, p. 329.

numerical approach, he came close to a definition of the irrational and struggled with the concept of real number in general.[13]

16   In his *Algebra* Omar Khayyam wrote that he had set forth elsewhere a rule that he had discovered for finding fourth, fifth, sixth, and higher powers of a binomial, but such a work is not extant. It is presumed that he is referring to the Pascal triangle arrangement, one that seems to have appeared in China at about the same time. Such a coincidence is not easy to explain, but until further evidence is available, independence of discovery is to be assumed. Intercommunication between Arabia and China was not extensive at that time; but there was a silk route connecting China with Persia, and information might have trickled along it.

The Arabs were clearly more attracted to algebra and trigonometry than to geometry, but one aspect of geometry held a special fascination for them— the proof of Euclid's fifth postulate. Even among the Greeks the attempt to prove the postulate had become virtually a "fourth famous problem of geometry," and several Muslim mathematicians continued the effort. Alhazen had begun with a trirectangular quadrilateral (sometimes known as "Lambert's quadrangle" in recognition of efforts in the eighteenth century) and thought that he had proved that the fourth angle must also be a right angle. From this "theorem" on the quadrilateral the fifth postulate can easily be shown to follow. In his "proof" Alhazen had assumed that the locus of a point that moves so as to remain equidistant from a given line is necessarily a line parallel to the given line—an assumption shown in modern times to be equivalent to Euclid's postulate. Omar Khayyam criticized Alhazen's proof on the ground that Aristotle had condemned the use of motion in geometry. Omar Khayyam then began with a quadrilateral the two sides of which are equal and are both perpendicular to the base (usually known as a "Saccheri quadrilateral," again in recognition of eighteenth-century efforts), and he asked about the other (upper) angles of the quadrilateral, which necessarily are equal to each other. There are of course, three possibilities. The angles may be (1) acute, (2) right, or (3) obtuse. The first and third possibilities Omar Khayyam ruled out on the basis of a principle, which he attributed to Aristotle, that two converging lines must intersect—again an assumption equivalent to Euclid's parallel postulate.

17   When Omar Khayyam died in 1123, Arabic science was in a state of decline. Excesses of political and religious factionalism—a condition that is well illustrated by the origin of our word "assassin"—would seem to be

---

[13] See D. J. Struik, "Omar Khayyam, Mathematician," *The Mathematics Teacher*, **51** (1958), 280–285.

among the causes of the decline. Islam never again was to reach the scholarly level of the glorious age of Avicenna and al-Karkhi, but Muslim contributions did not come to a sudden stop after Omar Khayyam. Both in the thirteenth century and again in the fifteenth century we find an Arabic mathematician of note. At Maragha, for example, Nasir Eddin al-Tusi (or at-Tusi, 1201–1274), astronomer to Hulagu Khan, grandson of the conqueror Genghis Khan and brother of Kublai Khan, continued efforts to prove the parallel postulate, starting from the usual three hypotheses on a Saccheri quadrilateral. His "proof" depends on the following hypothesis, again equivalent to Euclid's:

> If a line u is perpendicular to a line w at A, and if line v if oblique to w at B, then the perpendiculars drawn from u upon v are less than AB on the side on which v makes an acute angle with w and greater on the side on which v makes an obtuse angle with w.[14]

The views of Nasir Eddin, the last in the sequence of three Arabic precursors of non-Euclidean geometry, were translated and published by Wallis in the seventeenth century; it appears that this work was the starting point for the developments by Saccheri in the first third of the eighteenth century.

Nasir Eddin followed characteristic Arabic interests; hence he made contributions also to trigonometry and astronomy. Continuing the work of Abu'l-Wefa, he was responsible for the first systematic treatise on plane and spherical trigonometry, treating the material as an independent subject in its own right and not simply as the handmaid of astronomy, as had been the case in Greece and India. The six usual trigonometric functions are used, and rules for solving the various cases of plane and spherical triangles are given. Unfortunately, the work of Nasir Eddin had limited influence inasmuch as it did not become well known in Europe. In astronomy, however, Nasir Eddin made a contribution that may have come to the attention of Copernicus. The Arabs had adopted theories of both Aristotle and Ptolemy for the heavens; noticing elements of conflict between the cosmologies, they sought to reconcile them and to refine them. In this connection Nasir Eddin observed that a combination of two uniform circular motions in the usual epicyclic construction can produce a reciprocating rectilinear motion. That is, if a point moves with uniform circular motion clockwise around the epicycle while the center of the epicycle moves counterclockwise with half this speed along an equal deferent circle, the point will describe a straight-line segment. (In other words, if a circle rolls without slipping along the inside of a circle whose diameter is twice as great, the locus of a point on the circumference of the smaller circle will be a diameter of the larger circle.) This "theorem of

---

[14] See Roberto Bonola, *Non-Euclidean Geometry* (New York: Dover reprint, 1955), p. 10. See also D. E. Smith, "Euclid, Omar Khayyam, and Saccheri," *Scripta Mathematica*, **3** (1935), 5–10.

Nasir Eddin" became known to, or was rediscovered by, Copernicus and Cardan in the sixteenth century.[15]

18   Arabic mathematics continued to decline, following Nasir Eddin, but our account of the Muslim contribution would not be adequate without reference to the work of a figure in the early fifteenth century. Al-Kashi († ca. 1436) found a patron in the prince Ulugh Beg, grandson of the Mongol conqueror Tamerlane. At Samarkand, where he held his court, Ulugh Beg had built an observatory, and al-Kashi joined the group of scientists gathered there. In numerous works, written in Persian and Arabic, al-Kashi contributed to mathematics and astronomy. Noteworthy is the accuracy of his computations, especially in connection with the solution of equations by Horner's method, derived perhaps from the Chinese. From China, too, al-Kashi may have taken the practice of using decimal fractions. Al-Kashi is an important figure in the history of decimal fractions, and he realized the significance of his contribution in this respect, regarding himself as the inventor of decimal fractions.[16] Although to some extent he had had precursors, he is perhaps the first user of sexagesimal fractions to suggest that decimals are just as convenient for problems requiring many-place accuracy. Nevertheless, in his systematic computations of roots he continued to make use of sexagesimals. In illustrating his method for finding the nth root of a number, he took the sixth root of the sexagesimal

$$34,59,1,7,14,54,23,3,47,37\,;40$$

This was a prodigious feat of computation, using the steps that we follow in Horner's method—locating the root, diminishing the roots, and stretching or multiplying the roots—and using a pattern similar to our synthetic division.

Al-Kashi evidently delighted in long calculations, and he was justifiably proud of his approximation for π, which was more accurate than any of the values given by his predecessors. True to the penchant of the Arabs for alternative notations, he expressed his value of 2π in *both* sexagesimal and decimal forms. The former—6 ;16,59,28,34,51,46,15,50—is more reminiscent of the past and the latter—6.2831853071795865—in a sense presaged the future use of decimal fractions. No mathematician approached the accuracy in this tour de force of computation until the late sixteenth century. (The

[15] See C. B. Boyer, "Note on Epicycles and the Ellipse from Copernicus to Lahire," *Isis*, **38** (1947).

[16] See Abdul-Kader Kakhel, *Al-Kashi on Root Extraction* (1960), p. 2. An unusually extensive account of some of the work of al-Kashi is found in P. Luckey, "Die Ausziehung der n-ten Wurzel und der binomische Lehrsatz in der islamischen Mathematik," *Mathematische Annalen*, **120** (1948), 217–274. Very recently it has been pointed out that use of decimal fractions in Arabia is found in a work by abu-al-Hasan, Ahmad ibn-Ibrahim al-Uqlidisi dating from 952–953. See A. S. Saidan, "The Earliest Extant Arabic Arithmetic," *Isis*, **57** (1966), 475–490.

following mnemonic device will aid in memorizing a good approximation to π: "How I want a drink, alcoholic of course, after the heavy lectures involving quantum mechanics." The number of letters in the words will provide the values for the successive digits in 3.14159265358979, and these will be found to be in full agreement with al-Kashi's value for 2π.) In al-Kashi the binomial theorem in "Pascal triangle" form again appears, just about a century after its publication in China and about a century before it was printed in European books.

With the death of al-Kashi in about 1436 we can close the account of Arabic mathematics, for the cultural collapse of the Muslim world was more complete than the political disintegration of the empire. The number of significant Arabic contributors to mathematics before al-Kashi was considerably larger than our exposition would suggest, for we have concentrated only on major figures;[17] but following him the number is negligible. It was very fortunate indeed that when Arabic learning began to decline, scholarship in Europe was on the upgrade and was prepared to accept the intellectual legacy bequeathed by earlier ages. It is sometimes held that the Arabs had done little more than to put Greek science into "cold storage" until Europe was ready to accept it. But the account in this chapter has shown that at least in the case of mathematics the tradition handed over to the Latin world in the twelfth and thirteenth centuries was richer than that with which the unlettered Arabic conquerors had come into contact in the seventh century.

BIBLIOGRAPHY

Amir-Moez, A. R., "A Paper of Omar Khayyam," *Scripta Mathematica*, **26** (1963), 323–337.

Cajori, Florian, *History of Mathematics*, 2nd ed. (New York: Macmillan, 1919).

Dedron, Pierre, and Jean Itard, *Mathématiques et mathématiciens* (Paris: Magnard, 1959).

Gandz, Solomon, "The Sources of al-Khowarizmi's Algebra," *Osiris*, **1** (1936), 263–277.

Hill, G. F., *The Development of Arabic Numerals in Europe* (Oxford: Clarendon, 1915).

Kakhel, Abdul-Kader, *Al-Kashi on Root Extraction* (Lebanon, 1960).

Kasir, D. S., ed., *The Algebra of Omar Khayyam* (New York: Columbia Teachers College, 1931).

Karpinski, L. C., "The Algebra of Abu Kamil," *American Mathematical Monthly*, **21** (1914), 37–48.

Karpinski, L. C., ed., *Robert of Chester's Latin Translation of the Algebra of al-Khowarizmi* (New York: Macmillan, 1915).

Kennedy, E. S., "Overview on Trigonometry," *Yearbook on History of Mathematics*, The National Council of Teachers of Mathematics (Washington, D.C.), in press.

[17] See Heinrich Suter, *Die Mathematiker und Astronomer der Araber und ihre Werke* (1900), for an account of more than 500 scholars.

Levey, Martin, ed., *The Algebra of Abu Kamil* (Madison, Wis.: University of Wisconsin Press, 1966).

Luckey, P., "Die Ausziehung der *n*-ten Wurzel und der binomische Lehrsatz in der islamischen Mathematik," *Mathematische Annalen*, **120** (1948), 217–274.

Rosenfeld, B. A., and A. P. Youschkevitch, *Omar Khayyam* (in Russian, Moscow: Izdatelestvo "Nauka," 1965).

Saidan, A. S., "The Earliest Extant Arabic Arithmetic," *Isis*, **57** (1966), 475–490.

Sánchez Pérez, José, *La arithmética en Roma, en India y en Arabia* (Madrid: Instituto Miguel Asín, 1949).

Sarton, George, *Introduction to the History of Science* (Baltimore: Carnegie Institution of Washington, 1927–1948, 3 vols. in 5).

Sayili, Aydin, *Logical Necessities in Mixed Equations by 'Abd al Hamid ibn Turk and the Algebra of His Time* (Ankara, 1962).

Sayili, Aydin, "Thabit ibn Qurra's Generalization of the Pythagorean Theorem," *Isis*, **51** (1960), 35–37.

Smith, D. E., *History of Mathematics* (Boston: Ginn, 1923–1925, 2 vols.; paperback reprint, New York: Dover, 1958).

Smith, D. E., and L. C. Karpinski, *The Hindu-Arabic Numerals* (Boston, 1911).

Struik, D. J., "Omar Khayyam, Mathematician," *The Mathematics Teacher*, **51** (1958), 280–285.

Suter, Heinrich, *Die Mathematiker und Astronomer der Araber und ihre Werke* (Leipzig, 1900).

Vogel, Kurt, ed., *Mohammed ibn Musa Alchwarizmis Algorismus* (Aalen: O. Zeller, 1963).

Winter, H. J. J., "Formative Influences in Islamic Science," *Archives Internationales d'Histoire des Sciences*, **6** (1953), 171–192.

EXERCISES

1. Compare, in its effect on learning, the Arabic conquest of neighboring lands with the earlier conquests of Alexander the Great and with the conquests of the Romans.
2. Explain why al-Khowarizmi's *Algebra* contains no quadratic equation of the case squares and roots and numbers equal zero.
3. Which of the numerals used in modern Arabia most closely resemble our own? Are there any advantages or disadvantages in the Arabic forms?
4. Was it fortunate or unfortunate for the future of mathematics that Charles Martel turned back the Arabs at Tours in 732? Give reasons for your answer.
5. How would you account for the fact that after 1500 the Arabs made virtually no further contribution to mathematics?
6. Mention some parts of Greek mathematics that would be lost except for Arabic assistance.
7. Compare Arabic and Hindu mathematics with respect to form, content, level, and influence.
8. Compare the roles of logic and philosophy in Greek and Arabic mathematics.
9. Using a geometrical diagram like that of al-Khowarizmi, solve $x^2 + 12x = 64$.
10. Verify the answer given by al-Khowarizmi and Heron for the dimensions of a square inscribed in a triangle of sides 10, 10, and 12.
11. Verify the theorem of Thabit ibn-Qurra on amicable numbers.
12. Prove Thabit ibn-Qurra's generalization of the Pythagorean theorem.
13. Solve al-Biruni's cubic $x^3 = 1 + 3x$ for the positive root, correct to the nearest hundredth, and verify that to this extent your answer agrees with his.

14. Prove the formula of ibn-Yunus $2 \cos x \cos y = \cos (x + y) + \cos (x - y)$.
15. Use this formula to convert the product of 0.4567 and 0.5678 to a sum.
16. Solve the equation $x^3 = x^2 + 20$ geometrically in the manner of Omar Khayyam.
17. Solve the equation $x^3 + x = 20$ geometrically in the manner of Omar Khayyam.
18. Using Alhazen's estimate for the length of twilight and taking the radius of the earth as 4000 miles, find approximately the height of the atmosphere. (Twilight is caused by the reflection of the sun's rays in particles in the atmosphere.)
19. Find the volume obtained by revolving about the y-axis the area bounded by $y^2 = 2px$ and the line $x = a$. Which of the Greeks and the Arabs were able to handle this problem?
20. Show that the first three sexagesimals of al-Kashi's value of 2π are in agreement with the first five places in his decimal form.
21. Nasir Eddin showed that the sum of two odd squares cannot be a square. Prove this theorem, making use of properties of squares of odd and even numbers.
*22. As a special case of Alhazen's problem, consider a spherical mirror with circular section given by the equation $x^2 + y^2 = 1$, let a source of light be at the point $(0, 3)$, and let the eye be at the point $(4, 0)$. Show that the point at which the light will be reflected by the mirror can be found through the intersection of the circle and a hyperbola.

CHAPTER XIV

Europe in
the Middle Ages

> Neglect of mathematics works injury to all knowledge,
> since he who is ignorant of it cannot know the other
> sciences or the things of this world.
>
> *Roger Bacon*

1 Time and history are, of course, seamless wholes, like the continuum of mathematics, and any subdivision into periods is man's handiwork; but just as a coordinate framework is useful in geometry, so also the subdivision of events into periods or eras is convenient in history. For purposes of political history it has been customary to designate the fall of Rome in 476 as the beginning of the Middle Ages and the fall of Constantinople to the Turks in 1453 as the end. Disregarding politics, it might be better to close the ancient period with the year 524, which is both the year of Boethius' death and the approximate time when the Roman abbot Dionysius Exiguus proposed the chronology based on the Christian era that has since come into common use. For the history of mathematics we indicated in Chapter II a preference for the year 529 as a marker for the beginning of the medieval period, and we shall somewhat arbitrarily designate the year 1436 as the close.

The date 1436 is the probable year of death of al-Kashi, a very capable mathematician whom we already have described as somewhat Janus-faced —looking back on the old and in some respects anticipating the new. The year 1436 marks also the birth of another eminent mathematician, Johann Müller (1436–1476), better known under the name Regiomontanus, a Latinized form of his place of birth in Königsberg. The year 1436, in other words, symbolizes the fact that during the Middle Ages those who excelled in mathematics wrote in Arabic and lived in Islamic Africa and Asia, whereas during the new age that was dawning the leading mathematicians wrote in Latin and lived in Christian Europe.

An oversimplified view of the Middle Ages often results from a predominantly Europe-centered historical account; hence we remind readers that five great civilizations, writing in five different tongues, make up the bulk of the history of medieval mathematics. In the two preceding chapters we described contributions from China, India, and Arabia, three of the five leading medieval cultures. In this chapter we look at the mathematics of the other two: (1) the Eastern or Byzantine Empire, with center at Constantinople (or Byzantium), in which Greek was the official language; and (2) the Western or Roman Empire, which had no one center and no single spoken language, but in which Latin was the lingua franca of scholars.

When Justinian in 529 closed the pagan philosophical schools at Athens, **2** the scholars were dispersed, and some of them made permanent homes in Syria, Persia, and elsewhere. Nonetheless, some of the scholars remained, and others returned some years later, with the result that there was no serious hiatus in Greek learning in the Byzantine world. We have mentioned briefly the work of several Greek scholars of the sixth century: Eutocius, Simplicius, Isidore of Miletus, and Anthemius of Tralles. It was Justinian himself who put the building of Hagia Sophia in charge of the last two. To the list of Byzantine scholars should also be added the name of John Philoponus, who flourished at Alexandria in the early sixth century and was the leading physicist of his age anywhere in the world. Philoponus argued against the Aristotelian laws of motion and the impossibility of a vacuum, and he suggested the operation of a kind of inertia principle under which bodies in motion continued to move. Like Galileo later, he denied that the speed acquired by a freely falling body is proportional to its weight:

> If you let fall from the same height two weights of which one is many times as heavy as the other, you will see that the ratio of the times required for the motion does not depend on the ratio of the weights, but that the difference in time is a very small one.[1]

Philoponus was a Christian scientist (as were also perhaps Eutocius and Anthemius) who was making use of ancient pagan sources and whose ideas influenced later Islamic thinkers, thus indicating the continuity of the scientific tradition despite religious and political differences.

Philoponus was not primarily a mathematician, but some of his work, such as his treatise on the astrolabe, can be thought of as applied mathematics. Most Byzantine contributions to mathematics were on an elementary level and consisted chiefly of commentaries on ancient classics. Byzantine mathematics, far more than Arabic, was a sort of holding action to preserve as much of antiquity as possible until the West was ready to carry on. Philoponus aided in this work through his commentary on the *Introduction to Arithmetic* of Nicomachus. Neoplatonic thought continued to exert a strong influence in the Eastern Empire, which accounts for the popularity of Nicomachus' treatise. Again in the eleventh century it was the subject of a

[1] Quoted from Marshall Clagett, *The Science of Mechanics in the Middle Ages* (1959), p. 546.

commentary, this time by Michael Constantine Psellus (1018–1080?), a philosopher of Athens and Constantinople who counted among his pupils the Emperor Michael VII. Another of Psellus' works, a very elementary compendium on the quadrivium, enjoyed quite a vogue in the West during the sixteenth-century Renaissance period. Two centuries later we note another Greek summary of the mathematical quadrivium, this time by Georgios Pachymeres (1242–1316). Such compendia were significant only in showing that a thin thread of the old Greek tradition continued in the Eastern Empire to the very end of the medieval period.

Pachymeres wrote also a commentary on the *Arithmetic* of Diophantus, as did his contemporary, Maximos Planudes (1255?–1310). The latter, a Greek monk, was ambassador to Venice of the Emperor Andronicus II, indicating that there was some scholarly contact between the East and the West. Planudes wrote also a work on the Hindu system of numeration, which had finally reached the Greek world. In Byzantium, as might have been anticipated, the alphabetic numerals were not wholly abandoned, for they have continued to our own day in Greece in legal, administrative, and ecclesiastical documents. Section LXXVIII of a document, for example is *οη* (that is, omicron eta) as in Alexandrian days. Moreover, even within the new Hindu system the Byzantine scholars of the fourteenth century retained the first nine letters of the old alphabetic scheme, adding to these a zero symbol, like an inverted h. The number 7890, for example, would be written as *ζηθч*, a form every bit as convenient as our own. Manuel Moschopoulos (fl. 1300), a disciple of Planudes, wrote on magic squares, and the account of Planudes on numeration was commented on by the arithmetician and geometer Nicholas Rhabdas (†1350). The latter composed also a work on finger reckoning; but Byzantine mathematics, never very strong, by this time had become negligible. By the fourteenth century the Greek world had been clearly surpassed by the Latin world in the West, to which we now turn.

3   Chapter II included reference to the Latin treatises of Boethius at the end of the ancient period, with an indication of their very elementary level. Even from that level it was possible for mathematics to deteriorate still further, as we see in the trivial compendium on the liberal arts composed by Cassiodorus (ca. 480–ca. 575), a disciple of Boethius who spent his last years in a monastery that he had established. The primitive works of Cassiodorus served as textbooks in church schools in the early Middle Ages and sometimes also as the source for still lower-level books, such as the *Origines* or *Etymologies* of Isidore of Seville (570–636), one book of the twenty being a brief summary of the arithmetic of Boethius. When we consider that his contemporaries regarded Isidore as the most learned man of his time, we can well appreciate the lament of his day that "the study of letters is dead in

our midst." These were truly the "Dark Ages" of science, but we should not make the mistake of assuming that this was true of the Middle Ages as a whole. For the next two centuries the gloom continued to such an extent that it has been said that nothing scholarly could be heard in Europe but the scratching of the pen of the Venerable Bede (ca. 673–735) writing in England about the mathematics needed for the ecclesiastical calendar, or about the representation of numbers by means of the fingers.

Alcuin of York (ca. 735–804) was born the year that Bede died; he was **4** called by Charlemagne to revitalize education in France, and sufficient improvement was apparent to lead some historians to speak of a Carolingian Renaissance. Alcuin explained that the act of creation had taken six days because six was a perfect number; but beyond some arithmetic, geometry, and astronomy that Alcuin is reputed to have written for beginners, there was little mathematics in France or England for another two centuries. In Germany Hrabanus Maurus (784–856) continued the slight mathematical and astronomical efforts of Bede, especially in connection with the computation of the date of Easter. But not for another century and a half was there any notable change in the mathematical climate in Western Europe, and then it came in the person of one who rose ultimately to become Pope Sylvester II.

Gerbert (ca. 940–1003) was born in France and educated in Spain and Italy, and then served in Germany as tutor and later adviser to the Holy Roman Emperor, Otto III. Having served as archbishop, first at Reims and later at Ravenna, Gerbert in 999 was elevated to the papacy, taking the name Sylvester—possibly in recollection of an earlier pope who had been noted for scholarship, but more probably because Sylvester I, pope during the days of Constantine, symbolized the unity of papacy and empire. Gerbert was active in politics, both lay and ecclesiastical, but he had time also for educational matters. He wrote on both arithmetic and geometry, depending probably on the Boethian tradition, which had dominated the teaching in Western church schools and had not improved! More interesting than these expository works, however, is the fact that Gerbert was perhaps the first one in Europe to have taught the use of the Hindu-Arabic numerals. It is not clear how he came in contact with these. A possible explanation is that when he went to Spain in 967 he came in touch, perhaps at Barcelona, with Moorish learning, including Arabic numeration with the western or Gobar (dust) forms of the numerals, although there is little evidence of Arabic influence in extant documents. A Spanish copy of the *Origines* of Isidore, dating from 992, contains the numerals, without the zero, and Gerbert probably never knew of this last part of the Hindu-Arabic system. In certain manuscripts of Boethius, however, similar numeral forms, or apices, appear as counters for use on a computing board or abacus; and perhaps it was from these that Gerbert first learned of the new system. The Boethian apices, on the other hand, may themselves have been later interpolations. The situation with respect to the introduction of the numerals into Europe is about as confused as is that surrounding the invention of the system perhaps half a millennium earlier. Moreover, it is not clear that there was any continued use of the new numerals in Europe during the two centuries following Gerbert. Not until the thirteenth century was the Hindu-Arabic system definitively introduced into Europe, and then the achievement was not the work of one man, but of several.[2]

5   Europe, before and during the time of Gerbert, was not yet ready for developments in mathematics. The Christian attitude, expressed by Tertullian, had at first been somewhat the same as that of early Islam, cited with respect to the library at Alexandria. Scientific research, Tertullian wrote, had become superfluous since the gospel of Jesus Christ had been received. The time of Gerbert was the high point of Muslim learning, but contemporary Latin scholars could scarcely have appreciated Arabic treatises if they had learned about them. By the early twelfth century the situation began to change in a direction reminiscent of the ninth century in Arabia. One cannot absorb the wisdom of one's neighbors if one cannot understand their language. The Moslems had broken down the language barrier to Greek culture in the ninth century, and the Latin Europeans overcame the language barrier to Arabic learning in the twelfth century. At the beginning of the twelfth century no European could expect to be a mathematician or an astronomer, in any real sense, without a good knowledge of Arabic; and Europe, during the earlier part of the twelfth century, could not boast of a mathematician who was not a Moor, a Jew, or a Greek. By the end of the century the leading and most original mathematician in the whole world came from Christian Italy. So obviously was the period one of transition from an older to a newer point of view that C. H. Haskins entitled his work *The Renaissance of the Twelfth Century*.[3] The revival of which he wrote began of necessity with a spate of translations. At first these were almost exclusively from Arabic into Latin, but by the thirteenth century there were many variants—Arabic to Spanish, Arabic to Hebrew, Greek to Latin, or combinations such as Arabic to Hebrew to Latin. The *Elements* of Euclid was among the earliest of the mathematical classics to appear in Latin translation from the Arabic, the version being produced in 1142 by Adelard of Bath (ca. 1075–1160). It is not clear how the Englishman had come into contact with Muslim learning. There were at the time three chief bridges between Islam

[2] See G. F. Hill, *The Development of Arabic Numerals in Europe* (1915), and D. E. Smith and L. C. Karpinski, *The Hindu-Arabic Numerals* (1911).

[3] A paperback edition (New York: Meridian Books, 1957) is readily available.

and the Christian world—Spain, Sicily, and the Eastern Empire—and of these the first was the most important. Adelard, however, seems not to have been one of the many who made use of the Spanish intellectual bridge. It is not easy to tell whether the religious crusades had a positive influence on the transmission of learning, but it is likely that they disrupted channels of communication more than they facilitated them. At all events, the channels through Spain and Sicily were the most important in the twelfth century, and these were largely undisturbed by the marauding armies of the crusaders from 1096 to 1272. The revival of learning in Latin Europe took place *during* the crusades, but probably *in spite of* the crusades.

Adelard's translation of the *Elements* did not become very influential for another century, but it was far from an isolated event. Adelard earlier (1126) had translated al-Khowarizmi's astronomical tables from Arabic into Latin, and later (ca. 1155) Ptolemy's *Almagest* from Greek into Latin. Among the early translators, however, Adelard was an exception in that he was not one of the large group working in Spain. There, especially at Toledo, where the archbishop encouraged such work, a veritable school of translation was developing. The city, once a Visigothic capital and later in the hands of the Moors for several centuries before falling to the Christians, was an ideal spot for the transfer of learning. In Toledo libraries there was a wealth of Muslim manuscripts; and of the populace, including Christians, Moham- medans, and Jews, many spoke Arabic, facilitating the interlingual flow of information. The cosmopolitanism of the translators in Spain is evident from some of the names: Robert of Chester, Hermann the Dalmatian, Plato of Tivoli, Rudolph of Bruges, Gerard of Cremona, and John of Seville, the last a converted Jew. These are but a small portion of the men associated in the translation projects in Spain.[4]

Of the translators in Spain, perhaps the greatest was Gerard of Cremona (1114–1187). He had gone to Spain to learn Arabic in order to understand Ptolemy, but he devoted the rest of his life to translations from the Arabic. Among these was the translation into Latin of a revised version of Thabit ibn Qurra's Arabic of Euclid's *Elements*, a better piece of work than that of Adelard. In 1175 Gerard translated the *Almagest*, and it was chiefly through this work that Ptolemy came to be known in the West. Translations of more than eighty-five works are ascribed to Gerard of Cremona, but only the translation of Ptolemy is dated. Among the works of Gerard was a Latin adaptation of the *Algebra* of al-Khowarizmi, but an earlier and more popular translation of the *Algebra* had been made in 1145 by Robert of Chester. This, the first translation of al-Khowarizmi's treatise (as Robert's translation of the Koran, a few years before, had marked another "first"), may be taken as marking the beginning of European algebra.

[4] For others see George Sarton, *Introduction to the History of Science*, II (1), 113 ff, 338 ff.

Robert of Chester returned to England in 1150, but the Spanish work of translation continued unabated through Gerard and others. The works of al-Khowarizmi evidently were among the more popular subjects of the time, and the names of Plato of Tivoli and John of Seville are attached to still other adaptations of the *Algebra*. Western Europe suddenly took far more favorably to Arabic mathematics than it ever had to Greek geometry. Perhaps part of the reason for this is that Arabic arithmetic and algebra were on a more elementary level than Greek geometry had been during the days of the Roman republic and empire. However, the Romans had never displayed much interest in Greek trigonometry, relatively useful and elementary though it was; yet Latin scholars of the twelfth century devoured Arabic trigonometry as it appeared in astronomical works. It was Robert of Chester's translation from the Arabic that resulted in our word "sine." The Hindus had given the name *jiva* to the half chord in trigonometry, and the Arabs had taken this over as *jiba*. In the Arabic language there is also a word *jaib* meaning "bay" or "inlet." When Robert of Chester came to translate the technical word *jiba*, he seems to have confused this with the word *jaib* (perhaps because vowels were omitted); hence he used the word *sinus*, the Latin word for "bay" or "inlet." Sometimes the more specific phrase *sinus rectus*, or "vertical sine," was used; hence the phrase *sinus versus*, or our "versed sine," was applied to the "sagitta," or the "sine turned on its side."

It was during the twelfth-century period of translation and the following century that the confusion arose concerning the name al-Khowarizmi and led to the word "algorithm," as explained in the preceding chapter. The Hindu numerals had been explained to Latin readers by Adelard of Bath and John of Seville at about the same time that an analogous scheme was introduced to the Jews by Abraham ibn Ezra (ca. 1090–1167), author of books on astrology, philosophy, and mathematics. As in the Byzantine culture the first nine Greek alphabetic numerals, supplemented by a special zero symbol, took the place of the Hindu numerals, so Ibn Ezra used the first nine Hebraic alphabetic numerals, and a circle for zero, in the decimal positional system for integers. Despite the numerous accounts of the Hindu-Arabic numerals, the transition from the Roman number scheme was surprisingly slow. Perhaps this was because computation with the abacus was quite common, and in this case the advantages of the new scheme are not nearly so apparent as in calculation with pen and paper only. For several centuries there was keen competition between the "abacists" and the "algorists," and the latter triumphed definitively only in the sixteenth century.

6   It is sometimes claimed that in the later Middle Ages there were two classes of mathematicians—those in the church or university schools and those concerned with trade and commerce—and that rivalries are found between the two. There seems to be little basis for such a thesis; certainly in the spread of the Hindu-Arabic numerals both groups shared in the dissemination. Thirteenth-century authors from many walks of life helped to popularize "algorism," but we shall mention three in particular. One of them, Alexandre de Villedieu (fl. ca. 1225), was a French Franciscan; another,

A woodcut from Gregor Reisch, *Margarita Philosophica* (Freiburg, 1503). Arithmetic is instructing the algorist and the abacist, here inaccurately represented by Boethius and Pythagoras.

John of Halifax (ca. 1200–1256), known also as Sacrobosco, was an English schoolman; and the third was Leonardo of Pisa (ca. 1180–1250), better known as Fibonacci, or "son of Bonaccio," an Italian merchant. The *Carmen de algorismo* of Alexandre is a poem in which the fundamental operations on integers are fully described, using the Hindu-Arabic numerals and treating zero as a number. The *Algorismus vulgaris* of Sacrobosco was a practical account of reckoning that rivaled in popularity his *Sphaera*, an elementary tract on astronomy used in the schools throughout the later Middle Ages. The book in which Fibonacci described the new algorism is a celebrated classic, completed in 1202, but it bears a misleading title—*Liber abaci* (or book of the abacus). It is *not* on the abacus; it is a very thorough treatise on algebraic methods and problems in which the use of the Hindu-Arabic numerals is strongly advocated.

Leonardo's father was a Pisan engaged in business in northern Africa, and the son studied under a Muslim teacher and traveled in Egypt, Syria, and Greece. It therefore was natural that Fibonacci should have been steeped in Arabic algebraic methods, including, fortunately, the Hindu-Arabic numerals and, unfortunately, the rhetorical form of expression. The *Liber abaci* opens with an idea that sounds almost modern, but which was characteristic of both Islamic and Christian medieval thought—that arithmetic and geometry are connected and support each other. This view is, of course, reminiscent of al-Khowarizmi's *Algebra*, but it was equally accepted in the Latin Boethian tradition. The *Liber abaci*, nevertheless, is much more concerned with number than with geometry. It first describes "the nine Indian figures," together with the sign 0, "which is called zephirum in Arabic." Incidentally, it is from *zephirum* and its variants that our words "cipher" and "zero" are derived. Fibonacci's account of Hindu-Arabic numeration was important in the process of transmission; but it was not, as we have seen, the first such exposition, nor did it achieve the popularity of the later but more elementary descriptions by Sacrobosco and Villedieu. The horizontal bar in fractions, for example, was used regularly by Fibonacci (and was known before in Arabia), but it was only in the sixteenth century that it came into general use. (The slanted solidus was suggested in 1845 by De Morgan.)

7   The *Liber abaci*[5] is not a rewarding book for the modern reader, for after explanation of the usual algoristic or arithmetic processes, including the extraction of roots, it stresses problems in commercial transactions, using a complicated system of fractions in computing exchanges of currency. It is one of the ironies of history that the chief advantage of positional notation— its applicability to fractions—almost entirely escaped the users of the Hindu-Arabic numerals for the first thousand years of their existence. In this respect Fibonacci was as much to blame as anyone, for he used three types of fractions—common, sexagesimal, and unit—but not decimal fractions. In the *Liber abaci*, in fact, the two worst of these systems—unit fractions and

---

[5] There is no English translation of this important work, nor even a readily accessible Latin version. It is included in the *Bullettino di Bibliografia e di Storia delle Scienze Matematiche e Fisiche* of Baldassare Boncompagni (Rome, 1868–1887, 20 vols.). For notations used by Fibonacci and others, see Florian Cajori, *A History of Mathematical Notations* (Chicago, 1928–1929, 2 vols.).

common fractions—are extensively used. Moreover, problems of the following dull type abound: If 1 solidus imperial, which is 12 deniers imperial, is
sold for 31 deniers Pisan, how many deniers Pisan should one obtain for
11 deniers imperial? In a recipe type of exposition the answer is found
laboriously to be $\frac{5}{12}$ 28 (or, as we should write it, $28\frac{5}{12}$). Fibonacci customarily
placed the fractional part or parts of a mixed number before the integral part.
Instead of writing $11\frac{5}{6}$, for example, he wrote $\frac{1}{3}\frac{1}{2}$ 11, with juxtaposition of
unit fractions and integers implying addition.

Fibonacci evidently was fond of unit fractions—or he thought his readers
were—for the *Liber abaci* includes tables of conversion from common
fractions to unit fractions. The fraction $\frac{98}{100}$, for instance, is broken into
$\frac{1}{100}\frac{1}{50}\frac{1}{5}\frac{1}{4}\frac{1}{2}$, and $\frac{99}{100}$ appears as $\frac{1}{25}\frac{1}{5}\frac{1}{4}\frac{1}{2}$. An unusual quirk in his notation led
him to express the sum of $\frac{1}{5}\frac{3}{4}$ and $\frac{1}{10}\frac{2}{9}$ as $\frac{1}{2}\frac{6}{9}\frac{2}{10}$ 1, the notation $\frac{1}{2}\frac{6}{9}\frac{2}{10}$ meaning
in this case

$$\frac{1}{2 \cdot 9 \cdot 10} + \frac{6}{9 \cdot 10} + \frac{2}{10}$$

Analogously in another of the many problems on monetary conversion in the
Liber abaci we read that if $\frac{1}{4}\frac{2}{3}$ of a rotulus is worth $\frac{1}{7}\frac{1}{6}\frac{2}{5}$ of a bizantium, then
$\frac{1}{8}\frac{4}{9}\frac{7}{10}$ of a bizantium is worth $\frac{3}{4}\frac{8}{10}\frac{83}{149}\frac{11}{12}$ of a rotulus. Pity the poor medieval
businessman who had to operate with such a system!

Much of the *Liber abaci* makes dull reading, but some of the problems   **8**
were so lively that they were used by later writers. Among these is a hardy
perennial which may have been suggested by a similar problem in the Ahmes
papyrus. As expressed by Fibonacci, it read:

> Seven old women went to Rome; each woman had seven mules; each mule
> carried seven sacks, each sack contained seven loaves; and with each loaf
> were seven knives; each knife was put up in seven sheaths.

Without doubt the problem in the *Liber abaci* that has most inspired
future mathematicians was the following:

> How many pairs of rabbits will be produced in a year, beginning with a single
> pair, if every month each pair bears a new pair which becomes productive
> from the second month on?

This celebrated problem gives rise to the "Fibonacci sequence" 1, 1, 2, 3,
5, 8, 13, 21, ..., u_n, ..., where $u_n = u_{n-1} + u_{n-2}$, that is, where each term
after the first two is the sum of the two terms immediately preceding it. This
sequence has been found to have many beautiful and significant properties.
For instance, it can be proved that any two successive terms are relatively

prime and that $\lim_{n \to \infty} u_{n-1}/u_n$ is the golden section ratio $(\sqrt{5} - 1)/2$. The sequence is applicable also to questions in phyllotaxy and organic growth.[6]

9   The *Liber abaci* was Fibonacci's best known book, appearing in another edition in 1228, but it evidently was not appreciated widely in the schools, and it did not appear in print until the nineteenth century. Leonardo of Pisa was without doubt the most original and most capable mathematician of the medieval Christian world, but much of his work was too advanced to be understood by his contemporaries. His treatises other than the *Liber abaci* also contain many good things. In the *Flos*, dating from 1225, there are indeterminate problems reminiscent of Diophantus and determinate problems reminiscent of Euclid, the Arabs, and the Chinese.

Fibonacci evidently drew from many and varied sources. Especially interesting for its interplay of algorithm and logic is Fibonacci's treatment of the cubic equation $x^3 + 2x^2 + 10x = 20$. The author showed an attitude close to that of the modern period in first proving the impossibility of a root in the Euclidean sense, such as a ratio of integers, or a number of the form $a + \sqrt{b}$, where a and b are rational. As of that time, this meant that the equation could not be solved exactly by algebraic means. Fibonacci then went on to express the positive root approximately as a sexagesimal fraction to half a dozen places—1;22,7,42,33,4,40. This was a remarkable achievement, but we do not know how he did it. Perhaps through the Arabs he had learned what we call "Horner's method," a device known before this time in China. This is the most accurate European approximation to an irrational root of an algebraic equation up to that time—or anywhere in Europe for another 300 years and more. It is characteristic of the time that Fibonacci should have used sexagesimal fractions in theoretical mathematical work but not in mercantile affairs. Perhaps this explains why the Hindu-Arabic numerals were not promptly used in astronomical tables, such as the Alfonsine Tables of the thirteenth century. Where the "Physicists'" (sexagesimal) fractions were in use, there was less urgency in displacing them than there was in connection with the common and unit fractions in commerce.

10   In 1225 Leonardo of Pisa published not only the *Flos*, but also the *Liber quadratorum*, a brilliant work on indeterminate analysis. This, like *Flos*, contains a variety of problems, some of which stemmed from the mathematical contests held at the court of the emperor Frederick II, to which Fibonacci

---

[6] For some further mathematical properties see N. N. Vorob'ev, *Fibonacci Numbers*, trans. by H. Mors (New York: Blaisdell, 1961); S. M. Plotnick, "The Sum of Terms of the Fibonacci Series," *Scripta Mathematica*, **9** (1943), 197. For the relevance of the sequence in biology, see D. W. Thompson, *On Growth and Form*, 2nd ed. (Cambridge University Press, 1952). See also issues of *The Fibonacci Quarterly*. Interesting applications, and further references, are given in H. S. M. Coxeter, "The Golden Section, Phyllotaxis, and Wythoff's Game," *Scripta Mathematica*, **19** (1953), 135–143.

had been invited. One of the problems proposed strikingly resembles the type in which Diophantus had delighted—to find a rational number such that if five is added to, or subtracted from, the square of the number, the result will be the square of a rational number. Both the problem and a solution, $3\frac{5}{12}$, are given in *Liber quadratorum*. The book makes frequent use of the identities

$$(a^2 + b^2)(c^2 + d^2) = (ac + bd)^2 + (bc - ad)^2$$

$$= (ad + bc)^2 + (ac - bd)^2$$

which had appeared in Diophantus and had been widely used by the Arabs. Fibonacci, in some of his problems and methods, seems to follow the Arabs closely.[7]

Fibonacci was primarily an algebraist, but he wrote also, in 1220, a book entitled *Practica geometriae*. This seems to be based on an Arabic version of Euclid's *Division of Figures* (now lost) and on Heron's works on mensuration. It contains among other things a proof that the medians of a triangle divide each other in the ratio 2 to 1, and a three-dimensional analogue of the Pythagorean theorem. Continuing a Babylonian and Arabic tendency, he used algebra to solve geometrical problems.

It will be clear from the few illustrations we have given that Leonardo of Pisa was an unusually capable mathematician. It is true that he was without a worthy rival during the 900 years of medieval European culture, but he was not quite the isolated figure he is sometimes held to be. He had an able though less gifted younger contemporary in Jordanus Nemorarius (date uncertain). Some[8] identify this man with Jordanus Teutonicus or Jordanus of Saxony, leader of the Dominican Order, who died in 1237. In any case, our Jordanus Nemorarius, or Jordanus de Nemore, represents a more Aristotelian aspect of science than others we have met in the thirteenth century, and he became the founder of what sometimes is known as the medieval school of mechanics. To him we owe the first correct formulation of the law of the inclined plane, a law that the ancients had sought in vain: the force along an oblique path is inversely proportional to the obliquity, where obliquity is measured by the ratio of a given segment of the oblique path to the amount of the vertical intercepted by that path[9]—that is, the "run" over the "rise." In the language of trigonometry this means that $F : W = 1/\csc\theta$,

11

---

[7] See L. C. Karpinski, "The Algebra of Abu Kamil," *American Mathematical Monthly*, **21** (1914), 37–48.

[8] See, for example, D. E. Smith, *History of Mathematics*, I, 226, and George Sarton, *Introduction to the History of Science*, II (2), 613 f. The identification is denied by Joseph Hoffmann, *Geschichte der Mathematik*, 2nd ed. (Berlin, 1963), I, 96.

[9] See Clagett, *The Science of Mechanics in the Middle Ages*, p. 74.

which is equivalent of the modern formulation $F = W \sin \theta$, where W is weight, F is force, and θ is the angle of inclination.

Jordanus was the author of books on arithmetic, geometry, and astronomy, as well as mechanics. His *Arithmetica* in particular was the basis of popular commentaries at the University of Paris as late as the sixteenth century; this was not a book on computation, but a quasi-philosophical work in the tradition of Nicomachus and Boethius. It contains such theoretical results as the theorem that any multiple of a perfect or abundant number is abundant and that a divisor of a perfect number is deficient. The *Arithmetica* is significant especially for the use of letters instead of numerals as numbers, thus making possible the statement of general algebraic theorems. In the arithmetical theorems in Euclid's *Elements* VII–IX, numbers had been represented by line segments to which letters had been attached, and the geometrical proofs in al-Khowarizmi's *Algebra* made use of lettered diagrams; but all coefficients in the equations used in the *Algebra* are specific numbers, whether represented by numerals or written out in words. The idea of generality is implied in al-Khowarizmi's exposition, but he had no scheme for expressing algebraically the general propositions that are so readily available in geometry. In the *Arithmetica* the use of letters suggests the concept of "parameter"; but Jordanus' successors generally overlooked his scheme of letters. They seem to have been more interested in the Arabic aspects of algebra found in another Jordanian work, *De numeris datis*, a collection of algebraic rules for finding, from a given number, other numbers related to it according to certain conditions, or for showing that a number satisfying specific restrictions is determined. A typical instance is the following: If a given number is divided into two parts such that the product of one part by the other is given, then each of the two parts is necessarily determined. The rule is expressed awkwardly by Jordanus as follows:

> Let the given number be *abc* and let it be divided into two parts *ab* and *c*, and let *d* be the given product of the parts *ab* and *c*. Let the square of *ab* be *e* and let four times *d* be *f*, and let *g* be the result of taking *f* from *e*. Then *g* is the square of the difference between *ab* and *c*. Let *h* be the square root of *g*. Then *h* is the difference between *ab* and *c*. Since *h* is known, *c* and *ab* are determined.[10]

Note that Jordanus' use of letters is somewhat confusing, for, like Euclid, he sometimes uses two letters for a number and sometimes only a single letter. He evidently followed Euclid in picturing the given number as a line segment *ac* and the two parts into which it is subdivided as *ab* and *bc*; but he uses both end-point letters to designate the first part or number, and only the single letter *c* to represent the number of line segment *bc*. It is greatly to his credit, however, that he first stated the rule, equivalent to the solution of a

[10] For an extensive account of many aspects of the work of Jordanus see Moritz Cantor, *Vorlesungen über Geschichte der Mathematik* (1880–1908), II, 49–79.

quadratic equation, completely in general form. Only later did he provide a specific example of it, expressed in Roman numerals : to divide the number X into two parts the product of which is to be XXI, Jordanus follows through the steps indicated above to find that the parts are III and VII.

To Jordanus is attributed also an *Algorismus* (or *Algorithmus*) *demonstra-* **12** *tus*, an exposition of arithmetic rules that was popular for three centuries. The *Algorismus demonstratus* again shows Boethian and Euclidean inspiration, as well as Arabic algebraic characteristics. Still greater preponderance of Euclidean influence is seen in the work of Johannes Campanus of Novara (fl. ca. 1260), chaplain to Pope Urban IV. To him the late medieval period owed the authoritative translation of Euclid from Arabic into Latin, the one that first appeared in printed form in 1482. In making the translation Campanus used various Arabic sources, as well as the earlier Latin version by Adelard. Both Jordanus and Campanus discussed the angle of contact, or horn angle, a topic that produced lively discussion in the later medieval period when mathematics took on a more philosophical and speculative aspect. Companus noticed that if one compared the angle of contact—that is, the angle formed by an arc of a circle and the tangent at an end point—with the angle between two straight lines, there appears to be an inconsistency with Euclid's *Elements* X. 1, the fundamental proposition of the method of exhaustion. The rectilineal angle is obviously greater than the horn angle. Then if from the larger angle we take away more than half, and if from the remainder we take away more than half, and if we continue in this way, each time taking away more than half, ultimately we should reach a rectilineal angle less than the horn angle ; but this obviously is not true. Campanus correctly concluded that the proposition applies to magnitudes of the same kind, and horn angles are different from rectilineal angles.

Similarity in the interests of Jordanus and Campanus is seen in the fact that Campanus, at the end of Book IV of his translation of the *Elements*, describes an angle trisection which is exactly the same as that which had appeared in Jordanus' *De triangulis*. The only difference is that the lettering of the Campanus diagram is Latin, whereas that of Jordanus is Greco-Arabic. The trisection, unlike those in antiquity, is essentially as follows. Let the angle *AOB* that is to be trisected be placed with its vertex at the center of a circle of any radius $OA = OB$ (Fig. 14.1). From O draw a radius $OC \perp OB$, and through A place a straight line *AED* in such a way that $DE = OA$. Finally, through O draw line *OF* parallel to *AED*. Then $\angle FOB$ is one-third $\angle AOB$, as required.[11]

[11] Marshall Clagett, *Archimedes in the Middle Ages* (1964), I, 681. See also Moritz Cantor, *Vorlesungen über Geschichte der Mathematik*, II, 75 f, 94. A more sophisticated trisection, using the limaçon, is attributed to Jordanus (see Clagett, *The Science of Mechanics in the Middle Ages*, pp. 666–677).

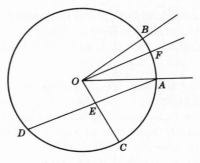

FIG. 14.1

13 The thirteenth century presents such a striking advance over the earlier Middle Ages that it has occasionally been viewed, none too impartially, as "the greatest of centuries."[12] We have seen how, in the work of Leonardo of Pisa, Western Europe had come to rival other civilizations in the level of its mathematical achievement; but this was only a small part of what was taking place in Latin culture as a whole. Many of the famous universities—Bologna, Paris, Oxford, and Cambridge—were established in the late twelfth and early thirteenth centuries, and this was the period in which great Gothic cathedrals—Chartres, Notre Dame, Westminster, Reims—were built. Aristotelian philosophy and science had been recovered and were taught in the universities and church schools. The thirteenth century is the period of great scholars and churchmen, such as Albertus Magnus, Robert Grosseteste, Thomas Aquinas, and Roger Bacon. Incidentally, two of these in particular, Grosseteste and Bacon, made strong pleas for the importance of mathematics in the curriculum, although neither was himself much of a mathematician. It was during the thirteenth century that many practical inventions became known in Europe—gunpowder and the compass, both perhaps from China, and spectacles from Italy, with mechanical clocks appearing only a little later.

The twelfth century had seen the great tide of translation from Arabic into Latin, but there now were other crosscurrents of translations. Most of the works of Archimedes, for example, had been virtually unknown to the medieval West; but in 1269 William of Moerbeke (ca. 1215–1286) published a translation (the original manuscript of which was discovered in 1884 in the Vatican) from Greek into Latin of the chief Archimedean scientific and mathematical treatises. Moerbeke, who came from Flanders and was named Archbishop of Corinth, knew little mathematics; hence his excessively literal translation (helpful now in reconstructing the original Greek text)

---

[12] J. J. Walsh, *The Thirteenth, Greatest of Centuries* (New York, 1909).

was of limited usefulness, but from this time on most of the works of Archimedes were at least accessible. In fact, the Moerbeke translation included parts of Archimedes with which the Arabs evidently were not familiar, such as the treatise *On Spirals*, the *Quadrature of the Parabola*, and *Conoids and Spheroids*. Nevertheless, the Muslims had been able to make more progress in understanding the mathematics of Archimedes than did the Europeans during the medieval period.

During the twelfth century the works of Archimedes had not completely escaped the attention of the indefatigable Gerard of Cremona, who had converted into Latin an Arabic version of the short work on *Measurement of the Circle*, which was used in Europe for several centuries. There had circulated also, before 1269, a portion of the Archimedean *Sphere and Cylinder*. These two examples could provide only a very inadequate idea of what Archimedes had done, and therefore the translation by Moerbeke was of the greatest importance, including as it did a number of major treatises. It is true that the version was only occasionally used during the next two centuries, but it at least remained extant. It was this translation that became known to Leonardo da Vinci and other Renaissance scholars, and it was Moerbeke's version that was first printed in the sixteenth century.[13]

The history of mathematics has not been a record of smooth and continuous development; hence it should come as no surprise that the upward surge during the thirteenth century should have lost some of its momentum. There was no Latin equivalent of Pappus to stimulate a revival of classical higher geometry. The works of Pappus were not available in Latin or Arabic. Even Apollonius' *Conics* was little known, beyond some of the simplest properties of the parabola that arose in connection with the ubiquitous treatises on optics, a branch of science that fascinated the scholastic philosophers. The science of mechanics, too, appealed to the scholars of the thirteenth and fourteenth centuries, for now they had at hand both the statics of Archimedes and the kinematics of Aristotle.

We noted earlier that the Aristotelian conclusions on motion had not gone unchallenged and modifications had been suggested, especially by Philoponus. During the fourteenth century the study of change in general, and of motion in particular, was a favorite topic in the universities, especially at Oxford and Paris. In Merton College at Oxford the scholastic philosophers had deduced a formulation for uniform rate of change which today generally is known as the Merton rule. Expressed in terms of distance and time, the rule says essentially that if a body moves with uniformly accelerated motion, then the distance covered will be that which another body would have

[13] For further details see Marshall Clagett, "The Impact of Archimedes on Medieval Science," *Isis*, **1** (1959), 419–429. See also Clagett's definitive work, *Archimedes in the Middle Ages*.

covered had it been moving uniformly for the same length of time with a speed equal to that of the first body at the midpoint of the time interval. As we should formulate it, the average velocity is the arithmetic mean of the initial and terminal velocities. Meanwhile, at the University of Paris there was developed a more specific and clear-cut doctrine of impetus, in which we can recognize a concept akin to our inertia, than that proposed by Philoponus.

15   The late medieval physicists comprised a large group of university teachers and churchmen, but we call attention to only two, for these were also prominent mathematicians. The first is Thomas Bradwardine (1290?–1349), a philosopher, theologian, and mathematician who rose to the position of Archbishop of Canterbury; the second is Nicole Oresme (1323?–1382), a Parisian scholar who became Bishop of Lisieux. To these two men was due a broadened view of proportionality.[14] The *Elements* of Euclid had included a logically sound theory of proportion, or the equality of ratios, and this had been applied by ancient and medieval scholars to scientific questions. For a given time, the distance covered in uniform motion is proportional to the speed; and for a given distance, the time is universely proportional to the speed. Aristotle had thought, none too correctly, that the speed of an object subject to a moving force acting in a resisting medium is proportional to the force and inversely proportional to the resistance. In some respects this formulation seemed to later scholars to contradict common sense. When force F is equal to or less than resistance, a velocity V will be imparted accordingly to the law $V = KF/R$, where K is a nonzero constant of proportionality; but when resistance balances or exceeds force, one should expect no velocity to be acquired. To avoid this absurdity Bradwardine made use of a generalized theory of proportions. In his *Tractatus de proportionibus* of 1328, Bradwardine developed the Boethian theory of double or triple or, more generally, what we would call "n-tuple" proportion. His arguments are expressed in words, but in modern notation we would say that in these cases quantities vary as the second or third or nth power. In the same way the theory of proportions included subduple or subtriple or sub-n-tuple proportion, in which quantities vary as the second or third or nth root. Now Bradwardine was ready to propose an alternative to the Aristotelian law of motion. To double a velocity that arises from some ratio or proportion F/R, he said, it was necessary to square the ratio F/R; to triple the velocity, one must cube the "proportio" or ratio F/R; to increase the velocity n-fold, one must take the nth power of the ratio F/R. This is tantamount to asserting that velocity

[14] See especially Nicole Oresme, *De proportionibus proportionum* and *Ad pauca respicientes*, ed. and trans. by Edward Grant (Madison, Wis.: University of Wisconsin Press, 1966). Cf. Edward Grant, "Part I of Nicole Oresme's Algorismus proportionum," *Isis*, **56** (1965), 327–341.

is given, in our notation, by the relationship $V = K \log F/R$, for $\log(F/R)^n = n \log F/R$. That is, if $V_0 = \log F_0/R_0$, then $V_n = \log(F_0/R_0)^n = n \log F_0/R_0 = nV_0$. Bradwardine himself evidently never sought experimental confirmation of his law, and it seems not to have been widely accepted.

Bradwardine wrote also several other mathematical works, all pretty much in the spirit of the times. His *Arithmetic* and his *Geometry* show the influence of Boethius, Aristotle, Euclid, and Campanus. Bradwardine, known in his day as "Doctor profundus," was attracted also to topics such as the angle of contact and star polygons, both of which occur in Campanus and earlier works. Star polygons, which include regular polygons as special cases, go back to ancient times. A star polygon is formed by connecting with straight lines every mth point, starting from a given one, of the n points that divide the circumference of a circle into n equal parts, where $n > 2$ and m is prime to n. There is in the *Geometry* even a touch of Archimedes' *Measurement of the Circle*. The philosophical bent in all of Bradwardine's works is seen most clearly in the *Geometrica speculativa* and the *Tractatus de continuo*, in which he argued[15] that continuous magnitudes, although including an infinite number of indivisibles, are not made up of such mathematical atoms, but are composed instead of an infinite number of continua of the same kind. His views sometimes are said to resemble those of the modern intuitionists; at any rate, medieval speculations on the continuum, popular among Scholastic thinkers like St. Thomas Aquinas, later influenced the Cantorian infinite of the nineteenth century.

Nicole Oresme lived later than Bradwardine, and in the work of the former we see extensions of ideas of the latter. In *De proportionibus proportionum*, composed about 1360, Oresme generalized Bradwardine's proportion theory to include any rational fractional power and to give rules for combining proportions that are the equivalents of our laws of exponents, now expressed in the notations $x^m \cdot x^n = x^{m+n}$ and $(x^m)^n = x^{mn}$. For each rule specific instances are given; and the latter part of another work, the *Algorismus proportionum*, applies the rules in geometrical and physical problems. Oresme suggested also the use of special notations for fractional powers, for in his *Algorismus proportionum* there are expressions such as

p	1
1	2

---

[15] See Edward Stamm, "Tractatus de continuo von Thomas Bradwardina. Eine Handschrift aus dem XIV. Jahrhundert," *Isis*, **26** (1936), 13–32.

to denote the "one and one-half proportion"—that is, the cube of the principal square root—and forms such as

$$\frac{1 \cdot p \cdot 1}{4 \cdot 2 \cdot 2}$$

for $\sqrt[4]{2\frac{1}{2}}$. We now take for granted our symbolic notations for powers and roots, with little thought for the slowness with which these developed in the history of mathematics. Even more imaginative than Oresme's notations was his suggestion that irrational proportions are possible. Here he was striving toward what we should write as $x^{\sqrt{2}}$, for example, which is perhaps the first hint in the history of mathematics of a higher transcendental function; but lack of adequate terminology and notation prevented him from effectively developing his notion of irrational powers.[16]

17   The notion of irrational powers may have been Oresme's most brilliant idea, but it was not in this direction that he was most influential. For almost a century before his time Scholastic philosophers had been discussing the quantification of variable "forms," a concept of Aristotle roughly equivalent to qualities. Among these forms were such things as the velocity of a moving object and the variation in temperature from point to point in an object with nonuniform temperature. The discussions were interminably prolix, for the available tools of analysis were inappropriate. Despite this handicap the logicians at Merton College had reached, as we saw, an important theorem concerning the mean value of a "uniformly difform" form—that is, one in which the rate of change of the rate of change is constant. Oresme was well aware of this result, and to him occurred, some time before 1361, a brilliant thought—why not draw a picture or graph of the way in which things vary?[17] Here we see, of course, an early suggestion of what we now describe as the graphical representation of functions. Everything measurable, Oresme wrote, is imaginable in the manner of continuous quantity; hence he drew a velocity-time graph for a body moving with uniform acceleration. Along a horizontal line he marked points representing instants of time (or longitudes), and for each instant he drew perpendicular to the line of longitudes a line

[16] For an admirable account of this work see Edward Grant, "Nichole Oresme and his *De proportionibus protortionum*," *Isis*, **51** (1960), 293–314. Cf. Edward Grant, "Bradwardine and Galileo: Equality of Velocities in the Void," *Archive for History of Exact Sciences*, **2** (1965), 344–364. See also references in footnote 14.

[17] We here imply, for simplicity of exposition, that Oresme was the first one to have this idea, but this is not necessarily the case. Marshall Clagett has found what looks like an earlier graph, drawn by Giovanni di Cosali, in which the line of longitude is placed in a vertical position. See Marshall Clagett, *Science of Mechanics in the Middle Ages*, pp. 332–333, 414. In any event, the exposition of Oresme surpasses that of Cosali in clarity and influence, and so our account does not do any real violence to history.

segment (latitude) the length of which represented the velocity. The end points of these segments, he saw, lie along a straight line; and if the uniformly accelerated motion starts from rest, the totality of velocity lines (which we call ordinates) will make up the area of a right triangle (see Fig. 14.2.)

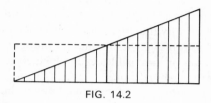

FIG. 14.2

Inasmuch as this area represents the distance covered, Oresme has provided a geometrical verification of the Merton rule, for the velocity at the midpoint of the time interval is half the terminal velocity. Moreover, the diagram leads obviously to the law of motion generally ascribed to Galileo in the seventeenth century. It is clear from the geometrical diagram that the area in the first half of the time is to that in the second half in the ratio 1 to 3. If we subdivide the time into three equal parts, the distances covered (given by the areas) are in the ratio 1:3:5. For four equal subdivisions the distances are in the ratio 1:3:5:7. In general, as Galileo later observed, the distances are to each other as the odd numbers; and since the sum of the first n consecutive odd numbers is the square of the last, the total distance covered varies as the square of the time, the familiar Galilean law for falling bodies.

The terms latitude and longitude that Oresme used are in a general sense equivalent to our ordinate and abscissa, and his graphical representation is akin to our analytic geometry. His use of coordinates was not, of course, new, for Apollonius, and others before him, had used coordinate systems, but his graphical representation of a variable quantity was novel. He seems to have grasped the essential principle that a function of one unknown can be represented as a curve, but he was unable to make any effective use of this observation except in the case of the linear function. Moreover, Oresme was chiefly interested in the area under the curve; hence it is not very likely that he saw the other half of the fundamental principle of analytic geometry—that every plane curve can be represented, with respect to a coordinate system, as a function of one variable. Where we say that the velocity graph in uniformly accelerated motion is a straight line, Oresme wrote, "Any uniformly difform quality terminating in zero intensity is imagined as a right triangle." That is, Oresme was more concerned with the calculus aspects of the situation : (1) the way in which the function varies (that is, the differential equation of the curve), and (2) the way in which the area under the curve varies (that is, the integral of the function). He pointed out the constant-slope property for his graph of uniformly accelerated motion—an observation equivalent to the modern two-point equation of the line in analytic geometry and leading to the concept of the differential triangle. Moreover, in finding the distance function, the area, Oresme obviously is performing geometrically a simple integration that results in the Merton rule. He did not explain why the area under a velocity-time curve represents the distance covered, but it is probable that he thought of the area as made up of many vertical lines or indivisibles each of which represented a velocity that continued for a very short time.

The graphical representation of functions, known then as the latitude of forms, remained a popular topic from the time of Oresme to that of Galileo. The *Tractatus de latitudinibus formarum*, written perhaps by a student of Oresme, if not by Oresme himself, appeared in numerous manuscript forms and was printed at least four times between 1482 and 1515; but this was only a précis of a larger work by Oresme entitled *Tractatus de figuratione potentiarum et mensurarum*.[18] Here Oresme went so far as to suggest a three-dimensional extension of his "latitude of forms" in which a function of two independent variables was pictured as a volume made up of all the ordinates erected according to a given rule at points in a portion of the reference plane. We even find a hint of a geometry of four dimensions when Oresme speaks of representing the intensity of a form for each point in a reference body or volume. What he really needed here was, of course, an *algebraic* geometry rather than a pictorial representation such as he had in mind; but weakness in technique hampered Europe throughout the medieval period.

18   Mathematicians of the Western world during the fourteenth century had imagination and precision of thought, but they were lacking in algebraic and geometrical facility; hence their contributions lay not in extensions of classical work, but in new points of view. Among these was an occupation with infinite series, an essentially novel topic anticipated only by some ancient iterative algorithms and Archimedes' summation of an infinite geometrical progression. Where the Greeks had a *horror infiniti*, the late medieval Scholastic philosophers referred frequently to the infinite, both as a potentiality and as an actuality (or something "completed"). In England in the fourteenth century a logician by the name of Richard Suiseth (fl. ca.

[18] See especially two articles by Heinrich Wieleitner, "Der 'Tractatus de latitudinibus formarum' des Oresme," *Bibliotheca Mathematica* (3), **13** (1913), 113–145, and "Ueber den Funktionsbegriff und die graphische Darstellung bei Oresme," *Bibliotheca Mathematica* (3), **14** (1914), 193–243. See also Marshall Clagett, *Science of Mechanics in the Middle Ages*.

1350), but better known as Calculator, solved the following problem in the latitude of forms:

> If throughout the first half of a given time interval a variation continues at a certain intensity, throughout the next quarter of the interval at double this intensity, throughout the following eighth at triple the intensity and so ad infinitum; then the average intensity for the whole interval will be the intensity of the variation during the second subinterval (or double the initial intensity).

This is equivalent to saying that the sum of the infinite series

$$\frac{1}{2} + \frac{2}{4} + \frac{3}{8} + \cdots + n/2^n + \cdots$$

is 2. Calculator gave a long and tedious verbal proof, for he did not know about graphical representation, but Oresme used his graphical procedure to prove the theorem more easily. Oresme handled also other cases, such as

$$\frac{1 \cdot 3}{4} + \frac{2 \cdot 3}{16} + \frac{3 \cdot 3}{64} + \cdots + \frac{n \cdot 3}{4^n} + \cdots$$

in which the sum is $\frac{4}{3}$. Problems similar to these continued to occupy scholars during the next century and a half.[19]

Among other contributions of Oresme to infinite series was his proof, evidently the first in the history of mathematics, that the harmonic series is divergent. He grouped the successive terms in the series

$$\frac{1}{2} + \frac{1}{3} + \frac{1}{4} + \frac{1}{5} + \frac{1}{6} + \frac{1}{7} + \frac{1}{8} + \cdots + \frac{1}{n} + \cdots$$

placing the first term in the first group, the next two terms in the second group, the next four terms in the third group, and so on, the mth group containing 2^{m-1} terms. Then it is obvious that we have infinitely many groups and that the sum of the terms within each group is at least $\frac{1}{2}$. Hence by adding together enough terms in order, we can exceed any given number.[20]

We have traced the history of mathematics in Europe through the Dark **19** Ages of the early medieval centuries to the high point in the time of the Scholastics. From the nadir in the seventh century to the work of Fibonacci and Oresme in the thirteenth and fourteenth centuries the improvement had been striking; but the combined efforts of all medieval civilizations were in no sense comparable to the mathematical achievements in Ancient Greece. The progress of mathematics had not been steadily upward in any part of the world—Babylonia, Greece, China, India, Arabia, or the Roman world— and it should come as no surprise that in Western Europe a decline set in

[19] For more details see C. B. Boyer, *History of the Calculus* (1959), pp. 86–87, and H. Busard, "Über unendliche Reihen im Mittelalter," *L'Enseignement Mathématique*, **8**, Nos. 3–4 (1962).

[20] See John Murdoch, "Oresme's Commentary on Euclid," *Scripta Mathematica*, **27** (1964), 67–91.

following the work of Bradwardine and Oresme. In 1349 Thomas Bradwardine had succumbed to the Black Death, the worst scourge ever to strike Europe. Estimates of the number of those who died of the plague within the short space of a year or two run between a third and a half of the population. This catastrophe inevitably caused severe dislocations and loss of morale. If we note that England and France, the nations that had seized the lead in mathematics in the fourteenth century, were further devastated in the fifteenth century by the Hundred Year's War and the Wars of the Roses, the decline in learning will be understandable. Italian, German, and Polish universities during the fifteenth century took over the lead in mathematics from the waning Scholasticism of Oxford and Paris, and it is primarily to representatives from these lands that we now turn.

BIBLIOGRAPHY

Boncompagni, Baldassare, ed., *Bullettino di bibliografia e di storia delle scienze mathematische e fisiche* (Rome, 1868–1887, 20 vols; reprint, New York: Johnson Reprint).

Boyer, C. B., *History of the Calculus* (paperback ed., New York: Dover, 1959).

Busard, H., "Über unendliche Reihen im Mittelalter," *L'Enseignement Mathématique,* **8**, Nos. 3–4 (1962).

Cantor, Moritz, *Vorlesungen über Geschichte der Mathematik* (Leipzig: Teubner, 1900–1908, 4 vols.).

Clagett, Marshall, *The Science of Mechanics in the Middle Ages* (Madison, Wis.: University of Wisconsin Press, 1959).

Clagett, Marshall, *Archimedes in the Middle Ages* (Madison, Wis.: University of Wisconsin Press, 1964–   , 2 vols.).

Duhem, Pierre, *Les origines de la statique* (Paris, 1905–1906, 2 vols.).

Ginsburg, Benjamin, "Duhem and Jordanus Nemorarius," *Isis,* **25** (1936), 340–362.

Grant, Edward, "Bradwardine and Galileo: Equality of Velocities in the Void," *Archive for History of Exact Sciences,* **2** (1965), 344–364.

Grant, Edward, "Nicole Oresme and his *De proportionibus proportionum,*" *Isis,* **51** (1960), 293–314.

Grant, Edward, "Part I of Nicole Oresme's Algorismus proportionum," *Isis,* **56** (1965), 327–341.

Grant, Edward, ed., *Nicole Oresme: De proportionibus proportionum and Ad pauca respicientes* (Madison, Wis.: University of Wisconsin Press, 1966).

Hill, G. F., *The Development of Arabic Numerals in Europe* (Oxford: Clarendon, 1915).

Murdoch, John, "Oresme's Commentary on Euclid," *Scripta Mathematica,* **27** (1964), 67–91.

Sarton, George, *Introduction to the History of Science* (Baltimore: Carnegie Institution of Washington, 1927–1948, 3 vols in 5).

Smith, D. E., *History of Mathematics* (Boston: Ginn, 1923–1925, 2 vols.; paperback reprint, New York: Dover, 1958).

Smith, D. E., and L. C. Karpinski, *The Hindu-Arabic Numerals* (Boston: Ginn, 1911).

Sullivan, J. W. N., *The History of Mathematics in Europe from the Fall of Greek Science to the Rise of the Conception of Mathematical Rigour* (New York: Oxford University Press, 1925).

Wieleitner, Heinrich, "Der 'Tractatus de latitudinibus formarum' des Oresme," *Bibliotheca Mathematica* (3), **13** (1913), 113–145.

Wieletner, Heinrich, "Ueber den Funktionsbegriff und die graphische Darstellung bei Oresme," *Bibliotheca Mathematica* (3), **14** (1914), 193–243.

Wieleitner, Heinrich, "Zur Geschichte der unendlichen Reihen im christlichen Mittelalter," *Bibliotheca Mathematica* (3), **14** (1914), 150–168.

Youschkevitch, A. P., *Geschichte der Mathematik im Mittelalter* (Leipzig: Teubner, 1964).

EXERCISES

1. Compare the mathematical work of one representative, living in about the year 500, from each of the following civilizations: China, India, Rome, Greece.
2. In what ways were the crusades likely to help or hinder the transmission of mathematics from Islam to the Christian world?
3. Was Western Europe in 1150 in closer touch with the Arabic or the Greek world? Which had relatively more to offer in mathematics? Give reasons for your answers.
4. Which three of the following—Euclid, Archimedes, Apollonius, Diophantus, Boethius, al-Khowarizmi—would you think were the most influential mathematical authors in Europe in 1250? Give reasons.
5. Compare the sources of support for mathematicians in medieval Europe with those in medieval Arabia.
6. Write the number 980,765 in the notation of Planudes.
7. For a unit circle express the versed sine of an angle in terms of the sine of the same angle. Explain how the names sine and versed sine arose.
8. Verify the answer given by Fibonacci in the problem (see text) of converting from a fractional part of a bizantium to a fractional part of a rotulus.
9. Find the ratio of u_{12} to u_{13} in the Fibonacci sequence. To how many significant figures is this in agreement with the golden-section ratio?
10. Prove that Fibonacci's cubic, $x^3 + 2x^2 + 10x = 20$, has no rational root.
11. Prove that the equation in Exercise 10 has no root of the form $a + \sqrt{b}$, where a and b are rational.
12. Find to the nearest hundredth a root of the cubic in Exercise 10 and show that to this extent Fibonacci's answer and yours are in agreement.
13. Verify Jordanus' rule (see text) for dividing a "given number *abc*."
14. Prove the Jordanus-Campanus trisection construction.
15. Using Bradwardine's law, and assuming that a force of 10 lb produces in a body a velocity of 20 ft/sec against a resistance of 2 lb, what velocity will be produced in the body against the same resistance by a force of 40 lb?
16. Draw Bradwardine's star polygon for eleven points on a circle if we connect in order every seventh point.
17. Prove for three equal subdivisions of the time interval that Oresme's ratio $1:3:5$ for the distances covered is correct.

*18. Verify Calculator's summation of the series

$$\sum_{n=1}^{\infty} \frac{n}{2^n}.$$

*19. Verify Oresme's summation of the series

$$\sum_{n=1}^{\infty} \frac{3n}{4^n}.$$

*20. Prove, using Oresme's method, that the series

$$\frac{1}{1} + \frac{1}{3} + \frac{1}{5} + \frac{1}{7} + \cdots + \frac{1}{2n-1} + \cdots$$

is divergent.

CHAPTER XV

The Renaissance

> I will sette as I doe often in woorke use, a paire of
> paralleles, or Gemowe [twin] lines of one lengthe, thus:
> =====, bicause noe 2. thynges, can be moare equalle.
>
> *Robert Recorde*

The fall of Constantinople in 1453 signaled the collapse of the Byzantine **1**
Empire, and in this respect it serves a convenient chronological placeholder
in the history of political events. The significance of the date for the history
of mathematics however, is, a moot point. It is frequently asserted that at that
time refugees fled to Italy with treasured manuscripts of ancient Greek
treatises, thereby putting the Western European world in touch with the
works of antiquity. It is as likely, though, that the fall of the city had just
the opposite effect: that now the West no longer could count on what had
been a dependable source of manuscript material for ancient classics, both
literary and mathematical. Whatever the ultimate decision may be on this
matter, there can be no question that mathematical activity was again rising
during the middle years of the fifteenth century. Europe was recovering from
the physical and spiritual shock of the Black Death, and the then-recent
invention of printing with movable type made it possible for learned works
to become much more widely available than ever before. The earliest printed
book from Western Europe is dated 1447, and by the end of the century
over 30,000 editions of various works were available. Of these, few were
mathematical; but the few, coupled with existing manuscripts, provided a
base for expansion. The recovery of unfamiliar Greek geometrical classics
was at first less significant than the printing of medieval Latin translations
of Arabic algebraic and arithmetic treatises, for few men of the fifteenth century either read Greek or were sufficiently proficient in mathematics to profit
from the works of the better Greek geometers. A substantial portion of the
treatises of Archimedes had, in fact, been accessible in Latin through the
translation of William of Moerbeke, but to little avail, for there were few to
appreciate classical mathematics. In this respect mathematics differed from
literature, and even from the natural sciences. As Humanists of the fifteenth

297

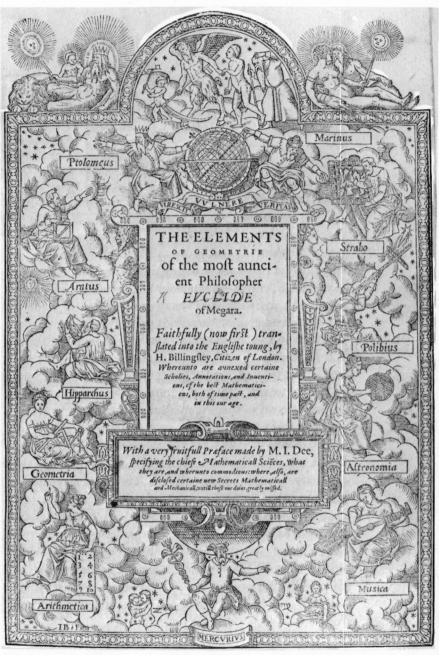

Title page of the first English version of Euclid's *Elements* (London, 1570). The translation purports to be by Sir Henry Billingsley, later Lord Mayor of London, but part or all of it may be by John Dee, writer of the preface.

and sixteenth centuries fell ever more deeply in love with the newly rediscovered Greek treasures in science and the arts, their estimate of the immediately preceding Latin and Arabic achievements declined. Classical
mathematics, except for the most elementary portions of Euclid, was an
intensely esoteric discipline, accessible only to those with a high degree of
preliminary training; hence the disclosure of Greek treatises in this field
did not at first seriously impinge on the continuing medieval mathematical
tradition. Medieval Latin studies in elementary geometry and the theory of
proportions, as well as Arabic contributions to arithmetic operations and
algebraic methods, did not present difficulties comparable to those associated
with the works of Archimedes and Apollonius. It was the more elementary
branches that were to attract notice and to appear in printed works.

Oresme had argued that everything measurable can be represented by a   **2**
line (latitude); and a mathematics of mensuration, both from a theoretical
and a practical standpoint, flourished during the early Renaissance period.
A similar view was adopted by Nicholas of Cusa (1401–1464), a man who well
represents the weaknesses of the age, for he was on the border line between
medieval and modern times. (Cusa was a Latin place-name for a city on the
Mosel.) Nicholas saw that a scholastic weakness in science had been a failure
to measure; *mens*, he thought, was etymologically related to *mensura*, so
that knowledge must be based on measurement. Cusa (or Cusanus, the Latin
form) also was influenced by the Humanist concern for antiquity and
espoused Neoplatonic views. Moreover, he had access to a translation of
some of Archimedes work made in 1450 by Jacob of Cremona. But, alas,
Nicholas of Cusa was better as an ecclesiastic than as a mathematician.[1]
In the Church he rose to the rank of cardinal, but in the field of mathematics
he is known as a misguided circle-squarer. His philosophical doctrine of the
"concordance of contraries" led him to believe that maxima and minima
are related, hence that the circle (a polygon with the greatest possible number
of sides) must be reconcilable with the triangle (the polygon with the smallest
number of sides). He believed that through an ingenuous averaging of
inscribed and circumscribed polygons he had arrived at a quadrature. That he
was wrong was of less significance than that he was one of the first modern
Europeans to attempt a problem that had fascinated the best minds of antiquity, and that his effort stimulated contemporaries to criticism of his work.

Among those who pointed out the error in Cusa's reasoning was Regio-   **3**
montanus (1436–1476), probably the most influential mathematician of the

[1] For an overappreciative account of his work see Max Simon, *Cusanus als Mathematiker*
(Strassburg, 1911, in *Festschrift Heinrich Weber*, Leipzig and Berlin: Teubner, 1912, pp. 298–337).
For a modern edition (in German) of the works of Nicholas of Cusa see his *Mathematische
Schriften*, ed. by J. E. Hofmann, (Hamburg: F. Meiner, ca. 1952, 1950).

Title page of Gregor Reisch, *Margarita philosophica* (1503). Around the three-headed figure in the center are grouped the seven liberal arts, with arithmetic seated in the middle and holding a counting board.

fifteenth century, and one whose birth date might be taken to mark the beginning of the new age. Having studied at the universities of Leipzig and Vienna, where he developed a love for mathematics and astronomy, Regiomontanus accompanied Cardinal Bessarion to Rome, where he acquired a proficiency in Greek and became acquainted with the crosscurrents of scientific and philosophical thought. Bessarion, once Archbishop of Nicaea, had won a cardinal's hat from Pope Eugenius IV in Rome (1439) for efforts to unite the Greek and Latin churches. He thus became a link between the classical learning preserved at Constantinople and the young Renaissance movement in the West. It probably was his association with the cardinal that inspired in Regiomontanus the ambition to acquire, translate, and publish the scientific legacy of antiquity. After travel and study in Italy, Regiomontanus returned to Germany, where he set up a printing press and an observatory at Nuremberg in order to advance the interests of science and literature. He hoped to print translations of Archimedes, Apollonius, Heron, Ptolemy, and other scientists, but his tragic death at the early age of forty cut short his ambitious project. In 1475 he had been invited to Rome by Pope Sixtus IV to share in one of the perennial attempts to reform the calendar, but he died there (some said he was poisoned by enemies) shortly after he had arrived. The trade list of books he planned to print survives,[2] and this indicates that the development of mathematics undoubtedly would have been accelerated had he survived. He was, in his wide and varied interests, a typical "Renaissance man," as his adopted name indicates. He was born "Johann Müller of Königsberg," but like others of his day he preferred to be known by the Latin form of his birthplace, the Germanic Königsberg ("king's mountain") becoming Regiomontanus.

Regiomontanus had become familiar, during his stay in Italy, with some of the leading figures of his day, and he entered into correspondence with others on current questions. His interests were broad, but he seems to have had little sympathy with the speculative thought of Nicholas of Cusa, which he criticized severely. In astronomy his chief contribution was the completion of a new Latin version, begun by his teacher at Vienna, Georg Peuerbach (1423–1469), of Ptolemy's *Almagest*. Peuerbach's *Theoricae novae planetarum*, a new textbook of astronomy, which was published in Regiomontanus' shop in 1472, was an improvement on the ubiquitous copies of the *Sphere* of Sacrobosco; but Humanists felt the need for a better Latin edition of the *Almagest* than the medieval version that had been derived from the Arabic. (The Humanists insisted on elegance and purity in their classical languages; hence they abhorred the barbarous medieval Latin, as well as the Arabic from which it often was derived.) Peuerbach had planned to make a trip to Italy with Regiomontanus to seek a good manuscript copy, but he died prematurely and the completion of the plan devolved upon his student. Regiomontanus' translation project resulted also in textbooks of his own. His *Epitome of Ptolemy's Almagest* is noteworthy for its emphasis on the mathematical portions that had often been omitted in commentaries dealing with elementary descriptive astronomy. Of greater significance for mathematics, however, was his *De triangulis omnimodis*, a systematic account of the methods for solving triangles which marked the rebirth of trigonometry.

[2] See George Sarton, "The Scientific Literature Transmitted Through the Incunabula," *Osiris*, **5** (1938), 41–247.

New works on astronomy invariably had been accompanied by tables of trigonometric functions, and Peuerbach's works had included a new table of sines. In these cases, however, trigonometry was serving merely as the handmaid of astronomy. In India, where the sine function evidently had its birth, there had been little interest in this function apart from its role in the astronomical systems or *Siddhāntas*. Even among the Arabs, for whom trigonometry was second only to algebra in mathematical appeal, the subject had had no independent existence, except in the *Treatise on the Quadrilateral* of Nasir Eddin, a work that owed more to the Greeks than to the Hindus. The twelfth-century age of translation in Europe had included some Arabic trigonometry, but for several centuries Latin contributions were only pale imitations of the Arabic. The *Practica geometriae* of Fibonacci and the works of Bradwardine had contained some fundamentals of trigonometry gleaned from Muslim sources, but it was not until Regiomontanus began writing his *De triangulis* that Europe gained preeminence in this field. It appears that Regiomontanus was acquainted with the work of Nasir Eddin, and this may have been the source of his desire to organize trigonometry as a discipline independent of astronomy.

The first book of *De triangulis*, composed in about 1464, opens with fundamental notions, derived largely from Euclid, on magnitudes and ratios; then there are more than fifty propositions on the solution of triangles, using the properties of right triangles. Book II begins with a clear statement and proof of the law of sines, and then includes problems on determining sides, angles, and areas of plane triangles when given determinate conditions. Among the problems, for example, is the following: If the base of a triangle and the angle opposite are known, and if either the altitude to the base or the area is given, then the sides can be found. Book III contains theorems of the sort found in ancient Greek texts on "spherics" before the use of trigonometry; Book IV is on spherical trigonometry, including the spherical law of sines.

The use of area "formulas," written out in words, was among the novelties in Regiomontanus' *De triangulis*, but in the avoidance of the tangent function the work falls short of Nasir Eddin's treatment. The tangent function nevertheless was included in another trigonometric treatise by Regiomontanus— *Tabulae directionum*. Revisions of Ptolemy had suggested the need for new tables, and these were supplied by a number of fifteenth-century astronomers, of whom Regiomontanus was one. In order to avoid fractions it was customary to adopt a large value for the radius of the circle, or the *sinus totus*. For one of his sine tables Regiomontanus followed his immediate predecessors in using a radius of 600,000; for others he adopted 10,000,000 or 600,000,000. For his tangent table in *Tabulae directionum* he chose 100,000. He does not call the function "tangent," but uses only the word "numerus" for the entries, degree by degree, in a tabulation headed "Tabula fecunda" ("Productive Table"). The entry for 89° is 5,729,796, and for 90° simply *infinite*.

The sudden death of Regiomontanus occurred before his two trigonometric works were published, and this considerably delayed their effect. The *Tabulae directionum* was published in 1490, but the more important treatise, *De triangulis*, appeared in print only in 1533 (and again in 1561). Nevertheless, the works were known in manuscript form to the circle of mathematicians at Nuremberg, where Regiomontanus was working, and it is very likely that they influenced work of the early sixteenth century.[3] For a hundred years after the fall of Constantinople, cities in central Europe, notably Vienna, Cracow, Prague, and Nuremberg, were leaders in astronomy and mathematics. The last of these became a center for the printing of books (as well as for learning, art, and invention), and some of the greatest scientific classics were published there toward the middle of the sixteenth century.

A general study of triangles led Regiomontanus to a consideration of **4** problems of geometrical construction somewhat reminiscent of Euclid's *Division of Figures*. For example, one is asked to construct a triangle given one side, the altitude to this side, and the ratio of the other two sides. Here, however, we find a striking departure from ancient customs: whereas Euclid's problems invariably had been given in terms of general quantities, Regiomontanus gave his lines specific numerical values, even where he intended that his methods should be general. This enabled him to make use of the algorithmic methods developed by Arabic algebraists and transmitted to Europe in twelfth-century translations. In the construction problem above, one of the unknown sides can be expressed as a root of a quadratic equation with known numerical coefficients, and this root is constructible by devices familiar from Euclid's *Elements*, or Al-Khowarizmi's *Algebra*. (As Regiomontanus expressed it, he let one part be the "thing" and then solved by the rule of "thing" and "square"—that is, through quadratic equations.) Another problem in which Regiomontanus called for the construction of a cyclic quadrilateral, given the four sides, can be handled similarly.

The algebra of Regiomontanus, like that of the Arabs, was rhetorical. The *Arithmetica* of Diophantus, in which some syncopation had been adopted, was known in Greek to Regiomontanus, who hoped ultimately to translate it; but it was from al-Khowarizmi that Europe learned the routine algebraic procedures. The *Arithmetica* was, after all, concerned primarily with the more recondite aspects of number theory. Moreover, Regiomontanus did not get

---

[3] An extensive account of his work and influence is included in Sister Mary Claudia Zeller, *The Development of Trigonometry from Regiomontanus to Pitiscus* (1944). There is an English translation of *De triangulis* under the title *Regiomontanus On Triangles*, ed. by Barnabas Hughes (1967).

around to publishing it, and few Latin scholars were aware of its contents for another century, until 1575 when it appeared in Latin. In fact, the influence of Regiomontanus in algebra was restricted not only by his adherence to the rhetorical form of expression and by his early death. His manuscripts, on his death, came into the hands of a Nuremberg patron who failed to make the work effectively accessible to posterity. Europe learned its algebra painfully and slowly from the thin Greek, Arabic, and Latin tradition that trickled down through the universities, the church scribes, the rising mercantile activities, and scholars from other fields.

5   Regiomontanus stood at a critical juncture in the history of science, and he had the tastes and the abilities to make the most of this. His love of classical learning was shared by the Humanists, but unlike them he was strongly inclined toward the sciences. Moreover, he did not indulge in the Humanist contempt for Scholastic and Arabic learning, and he was a Renaissance man in his concern for the practical arts as well as for scholarship. What better combination could a modern scientist have had than a good library, an observatory, a printing press, and a love of knowledge? Regiomontanus was aware, through his contact with Averroists in the Italian universities, that the Arabic astronomers had been worried about inconsistencies between the schemes of Aristotle and Ptolemy; and he undoubtedly knew also that Oresme and Cusa had seriously raised the possibility of the earth's moving. It is reported that he planned to reform astronomy; had he lived, he might have anticipated Copernicus. His premature death cut short all such schemes, and astronomy and mathematics had to look to others for the next steps, including in particular an isolated French figure outside of the mainstream of development.

6   It was Germany and Italy that provided most of the early Renaissance mathematicians, but in France in 1484 a manuscript was composed which in level and significance was perhaps the most outstanding since the *Liber abaci* of Fibonacci, almost three centuries before and which, like the *Liber abaci*, was not printed until the nineteenth century. This work, entitled *Triparty en la sciences des nombres*, was by Nicolas Chuquet († ca. 1500), about whom we know virtually nothing except that he was born at Paris, took his bachelor's degree in medicine, and practiced at Lyons. The *Triparty* does not closely resemble any earlier work in arithmetic or algebra, and the only writers the author mentions are Boethius and Campanus. There is evidence of Italian influence, which possibly resulted from acquaintance with Fibonacci's *Liber abaci*.

The first of the "Three Parts" concerns the rational arithmetic operations on numbers, including an explanation of the Hindu-Arabic numerals. Of these Chuquet says that "the tenth figure does not have or signify a value, and hence it is called cipher or nothing or figure of no value." The work is essentially rhetorical, the four fundamental operations being indicated by the words and phrases *plus*, *moins*, *multiplier par*, and *partyr par*, the first two sometimes abbreviated in the medieval manner as $\bar{p}$ and $\bar{m}$. In connection with the computation of averages, Chuquet gave a *regle des nombres moyens* according to which $(a + c)/(b + d)$ lies between a/b and c/d if a, b, c, d are positive numbers. In the second part, concerning roots of numbers, there is some syncopation, so that the modern expression $\sqrt{14 - \sqrt{180}}$ appears in the not very dissimilar form $R)^2 \cdot 14 \cdot \bar{m} \cdot R)^2 180$.

The last and by far the most important part of the *Triparty* concerns the "Regle des premiers"—that is, the rule of the unknown, or what we should call algebra. During the fifteenth and sixteenth centuries there were various names for the unknown thing, such as *res* (in Latin), or *chose* (in French) or *cosa* (in Italian) or *coss* (in German); Chuquet's word *premier* is unusual in this connection. The second power he called *champs* (whereas the Latin had been *census*), the third *cubiez*, and the fourth *champs de champ*. For multiples of these Chuquet invented an exponential notation of great significance. The *denominacion* or power of the unknown quantity was indicated by an exponent associated with the coefficient of the term, so that our modern expressions $5x$ and $6x^2$ and $10x^3$ appeared in the *Triparty* as $.5.^1$ and $.6.^2$ and $.10.^3$. Moreover, zero and negative exponents take their place along with the positive integral powers, so that our $9x^0$ became $.9.^0$, and $9x^{-2}$ was written as $.9.^{2.m.}$, meaning $.9.$ *seconds moins*. Such a notation laid bare the laws of exponents, with which Chuquet may have become familiar through the work of Oresme on proportions. Chuquet wrote, for example, that $.72.^1$ divided by $.8.^3$ is $.9.^{2.m.}$—that is, $72x \div 8x^3 = 9x^{-2}$. Related to these laws is his observation of the relationships between the powers of the number two, and the indices of these powers set out in a table from 0 to 20, in which sums of the indices correspond to products of the powers. Except for the magnitude of the gaps between entries, this constituted a miniature table of logarithms to the base two. Observations similar to those of Chuquet were to be repeated several times during the next century, and these undoubtedly played a role in the ultimate invention of logarithms.

The second half of the last part of the *Triparty* is devoted to the solution of equations. Here are many of the problems that had appeared among his predecessors, but there is also at least one significant novelty. In writing $.4.^1$ egaulx a $\bar{m}.2.^0$—that is, $4x = -2$—Chuquet was for the first time expressing an isolated negative number in an algebraic equation. Generally he rejected zero as a root of an equation, but on one occasion he remarked that the number sought was 0. In considering equations of the form $ax^m + bx^{m+n} = cx^{m+2n}$ (where the coefficients and exponents are specific positive integers), he found that some implied imaginary solutions; in these cases he simply added, "Tel nombre est ineperible."[4]

The *Triparty* of Chuquet, like the *Collectio* of Pappus, is a book in which the extent of the author's originality cannot be determined. Each undoubtedly was indebted to his immediate predecessors, but we are unable to identify any of them. Moreover, in the case of Chuquet we cannot determine his influence on later writers. The *Triparty* was not printed until 1880, and probably was known to few mathematicians; but one of those into whose hands it fell used so much of the material that he can be charged with plagiarism, even though he mentioned Chuquet's name. The *Larismethique nouvellement composee*, published at Lyons by Etienne de la Roche in 1520, and again in 1538, depended heavily, as we now know, on Chuquet; hence it is safe to say that the *Triparty* was not without effect.

7   The earliest Renaissance algebra, that of Chuquet, was the product of a Frenchman, but the best known algebra of that period was published ten years later in Italy. In fact, the *Summa de arithmetica, geometrica, proportioni et proportionalita* of the friar Luca Pacioli (1445–1514) overshadowed the *Triparty* so thoroughly that older historical accounts of algebra leap directly from the *Liber abaci* of 1202 to the *Summa* of 1494 without mentioning the work of Chuquet or other intermediaries. The way for the *Summa*, however, had been prepared by a generation of algebraists, for the *Algebra* of al-Khowarizmi was translated into Italian at least by 1464, the date of a manuscript copy in the Plimpton Collection in New York; the writer of this manuscript stated that he based his work on numerous predecessors in this field, naming some from the earlier fourteenth century. The Renaissance in science often is assumed to have been sparked by the recovery of ancient Greek works; but the Renaissance in mathematics was characterized especially by the rise of algebra, and in this respect it was but a continuation of the medieval tradition. Regiomontanus had been well versed in Greek; but he had not shared the Humanists' apotheosis of Hellenism, and he had been ready to recognize the importance of medieval Arabic and Latin algebra. He obviously had been familiar with the works of al-Khowarizmi and Fibonacci and had planned to print the *De numeris datis* of Jordanus Nemorarius. Had Regiomontanus achieved his plans for publication, the *Summa* of Pacioli (or Paciuolo) would certainly not today be regarded as the first printed work on algebra.

[4] Good accounts of this work are found in Ch. Lambo, S. J., "Une algèbre française de 1484. Nicolas Chuquet," *Revue des Questions Scientifiques*, (3), **2** (1902), 442–472, and in Aristide Marre, "Notice sur Nicolas Chuquet et son Triparty en la science des nombres," *Bullettino di Bibliografia e di Storia delle Scienze Matematiche e Fisiche*, **13** (1880), 555–659, 693–814; **14** (1881), 413–460.

The *Summa*, the writing of which had been completed by 1487, was more influential than it was original. It is an impressive compilation (with sources of information not generally indicated) of material in four fields: arithmetic, algebra, very elementary Euclidean geometry, and double-entry bookkeeping. Pacioli (also known as Luca di Borgo) for a time had been tutor to the sons of a wealthy merchant at Venice, and he undoubtedly was familiar with the rising importance in Italy of commercial arithmetic. The earliest printed arithmetic, appearing anonymously at Treviso in 1478, had featured the fundamental operations, the rules of two and three, and business applications. Several more technical commercial arithmetics appeared shortly thereafter, and Pacioli borrowed freely from them. One of these, the *Compendio de lo abaco* of Francesco Pellos (fl. 1450–1500), which was published at Torino in the year Columbus discovered America, made use of a dot to denote the division of an integer by a power of ten, thus adumbrating our decimal point.

The *Summa*, which like the *Triparty* was written in the vernacular, was a summing up of unpublished works that the author had composed earlier, as well as of general knowledge at the time. The portion on arithmetic is much concerned with devices for multiplication and for finding square roots; the section on algebra includes the standard solution of linear and quadratic equations. Although it lacks the exponential notation of Chuquet, there is increased use of syncopation through abbreviations. The letters p and m were by this time widely used in Italy for addition and subtraction, and Pacioli used co, ce, and ae for *cosa* (the unknown), *censo* (the square of the unknown), and *aequalis* respectively. For the fourth power of the unknown he naturally used *cece* (for square-square). Echoing a sentiment of Omar Khayyam, he believed that cubic equations could not be solved algebraically.

Pacioli's work in geometry in the *Summa* was not significant, although some of his geometrical problems remind one of the algebraic geometry of Regiomontanus, specific numerical cases being employed. For example, it is required to find the sides of a triangle if the radius of the inscribed circle is four and the segments into which one side is divided by the point of contact are six and eight. Although Pacioli's geometry did not attract much attention, so popular did the commercial aspect of the book become that the author generally is regarded as the father of double-entry bookkeeping.

Pacioli, the first mathematician of whom we have an authentic portrait, **8** in 1509 tried his hand twice more at geometry, publishing an undistinguished edition of Euclid and a work with the impressive title *De divina proportione*. The latter concerns regular polygons and solids and the ratio later known as "the golden section." It is noteworthy for the excellence of the figures,[5] which

---

[5] For a further description of this work and of contemporary activity, see R. Emmett Taylor, *No Royal Road. Luca Pacioli and His Times* (1942).

have been attributed to Leonardo da Vinci (1452–1519). Leonardo frequently is thought of as a mathematician, but his restless mind did not dwell on arithmetic or algebra or geometry long enough to make a significant contribution. In his notebooks we find quadratures of lunes, constructions of regular polygons, and thoughts on centers of gravity and on curves of double curvature; but he is best known for his application of mathematics to science and the theory of perspective. Da Vinci is pictured as the typical all-round Renaissance man; and in fields other than mathematics there is much to support such a view. Leonardo was a genius of bold and original thought, a man of action as well as contemplation, at once an artist and an engineer; but he appears not to have been in close touch with the chief mathematical trend of the time—the development of algebra. Few subjects depend as heavily on a continuous bookish tradition and long-continued concentration as does mathematics, and Leonardo was not one to maintain concentrated library research or even to pursue his own imaginative ideas to their conclusions. Ultimately, hundreds of years later, Renaissance notions on mathematical perspective were to blossom into a new branch of geometry, but these developments were not perceptibly influenced by the thoughts that the left-handed Leonardo entrusted to his notebooks in the form of mirror-written entries.

9   The word Renaissance inevitably brings to mind Italian literary, artistic, and scientific treasures, for renewed interest in art and learning became apparent in Italy earlier than in the other parts of Europe. There, in a rough-and-tumble conflict of ideas, men learned to put greater trust in independent observations of nature and judgments of the mind. Moreover, Italy had been one of the two chief avenues along which Arabic learning, including algorism and algebra, had entered Europe. Nevertheless, other parts of Europe did not remain far behind, as the work of Regiomontanus and Chuquet shows. In Germany, for example, books on algebra became so numerous that for a time the Germanic word *coss* for the unknown triumphed in other parts of Europe, and the subject became known as the "cossic art." Moreover, the Germanic symbols for addition and subtraction ultimately displaced the Italian p and m. In 1489, before the publication of Pacioli's *Summa*, a German lecturer at Leipzig, Johann Widman (born ca. 1460), had published a commercial arithmetic, *Rechenung auff allen Kauffmanschafft*, the oldest book in which our familiar + and − signs appear in print. At first used to indicate excess and deficiency in warehouse measures, they later became symbols of the familiar arithmetic operations.[6] Widman, incidentally, possessed a manuscript

---

[6] See J. W. L. Glaisher, "On the Early History of the Signs + and − and on the Early German Arithmeticians," *Messenger of Mathematics*, **51** (1921–1922), 1–148.

copy of the *Algebra* of al-Khowarizmi, a work well known to other German mathematicians.

Among the numerous Germanic algebras was *Die Coss*, written in 1524 by Germany's celebrated Rechenmeister, Adam Riese (1492–1559). The author was the most influential German writer in the move to replace the old computation (in terms of counters and Roman numerals) by the newer method (using the pen and Hindu-Arabic numerals); so effective were his numerous arithmetic books that the phrase "nach Adam Riese" still survives in Germany as a tribute to accuracy in arithmetic processes. Riese, in his *Coss*, mentions the *Algebra* of al-Khowarizmi and refers to a number of Germanic predecessors in the field.

The first half of the sixteenth century saw a flurry of German algebras, among the most important of which were the *Coss* (1525) of Christoph Rudolff (ca. 1500—ca. 1545), the *Rechnung* (1527) of Peter Apian (1495–1552), and the *Arithmetica integra* (1544) of Michael Stifel (ca. 1487–1567). The first is especially significant as one of the earliest printed works to make use of decimal fractions, as well as of the modern symbol for roots; the second is

Title page of an edition (1529) of one of the *Rechenbücher* of Adam Riese, the celebrated *Rechenmeister*. It depicts a contest between an algorist and an abacist.

worth recalling for the fact that here, in a commercial arithmetic, the so-called "Pascal triangle" was printed on the title page, almost a century before Pascal was born. The third work, Stifel's *Arithmetica integra*, was the most important of all the sixteenth-century German algebras. It, too, includes the Pascal triangle, but it is more significant for its treatment of negative numbers, radicals, and powers. Through the use of negative coefficients in equations, Stifel was able to reduce the multiplicity of cases of quadratic equations to what appeared to be a single form; but he had to explain, under a special rule, when to use + and when −. Moreover, even he failed to admit negative numbers as roots of an equation. Stifel, a onetime monk turned itinerant Lutheran preacher, and for a time Professor of Mathematics at Jena, was one of the many writers who popularized the "German" symbols + and − at the expense of the "Italian" p and m notation. He was thoroughly familiar with the properties of negative numbers, despite the fact that he called them "numeri absurdi." About irrational numbers he was somewhat hesitant, saying that they are "hidden under some sort of cloud of infinitude." Again calling attention to the relations between arithmetic and geometric progressions, as had Chuquet for powers of two from 0 to 20, Stifel extended the table to include $2^{-1} = \frac{1}{2}$ and $2^{-2} = \frac{1}{4}$ and $2^{-3} = \frac{1}{8}$ (without, however, using exponential notation). For powers of the unknown quantity in algebra Stifel in *Arithmetica integra* used abbreviations for the German words coss, zensus, cubus, and zenzizensus; but in a later treatise, *De algorithmi numerorum cossicorum*, he proposed using a single letter for the unknown and repeating the letter for higher powers of the unknown, a scheme later employed by Harriot.[7]

10   The *Arithmetica integra* was a thorough treatment of algebra as generally known up to 1544, but by the following year it was in a sense quite outmoded. Stifel gave many examples leading to quadratic equations, but none of his problems lead to mixed cubic equations, for the simple reason that he knew no more about the algebraic solution of the cubic than did Pacioli or Omar Khayyam. In 1545, however, the solution not only of the cubic but of the quartic as well became common knowledge through the publication of the *Ars magna* of Geronimo Cardano (1501–1576). Such a striking and unanticipated development made so strong an impact on algebraists that the year 1545 frequently is taken to mark the beginning of the modern period in mathematics. It must be pointed out immediately, however, that Cardano (or Cardan) was not the original discoverer of the solution of either the cubic or the quartic. He himself candidly admitted this in his book. The hint for solving

[7] For accounts of early books in arithmetic and algebra see especially D. E. Smith, *Rara arithmetica* (1908). For lists of early arithmetics see J. E. Hofmann: *Geschichte der Mathematik* (1963), Vol. I, pp. 142–145.

the cubic, he averred, he had obtained from Niccolo Tartaglia (ca. 1500–1557); the solution of the quartic was first discovered by Cardan's *quondam amanuensis*, Ludovico Ferrari (1522–1565). What Cardan failed to mention in *Ars magna* is the solemn oath he had sworn to Tartaglia that he would not disclose the secret, for the latter intended to make his reputation by publishing the solution of the cubic as the crowning part of his treatise on algebra.

Lest one feel undue sympathy for Tartaglia, it may be noted that he had published an Archimedean translation (1543), derived from Moerbeke, leaving the impression that it was his own, and in his *Quesiti et inventioni diverse* (Venice, 1546) he gave the law of the inclined plane, presumably derived from Jordanus Nemorarius, without proper credit. It is, in fact, possible that Tartaglia himself had received a hint concerning the solution of the cubic from an earlier source. Whatever may be the truth in a rather complicated and sordid controversy between proponents of Cardan and Tartaglia, it is clear that neither of the principals was first to make the discovery. The hero in the case evidently was one whose name is scarcely remembered today—Scipione del Ferro (ca. 1465–1526), professor of mathematics at Bologna, one of the oldest of the medieval universities and a school with a strong mathematical tradition. How or when del Ferro made his wonderful discovery is not known. He did not publish the solution, but before his death he had disclosed it to a student, Antonio Maria Fior (or Floridus in Latin), a mediocre mathematician.

Word of the existence of an algebraic solution of the cubic seems to have gotten around, and Tartaglia tells us that knowledge of the possibility of solving the equation inspired him to devote himself to finding the method for himself. Whether independently or on the basis of a hint, Tartaglia did indeed learn, by 1541, how to solve cubic equations. When news of this spread, a mathematical contest between Fior and Tartaglia was arranged. Each contestant proposed thirty questions for the other to solve within a stated time interval. When the day for decision had arrived, Tartaglia had solved all questions posed by Fior, whereas the latter had not solved a single one set by his opponent. The explanation is relatively simple. Today we think of cubic equations as all essentially of one type and as amenable to a single unified method of solution. At that time, however, when negative coefficients were virtually unused, there were as many types of cubics as there are possibilities in positive or negative signs for coefficients. Fior was able to solve only equations of the type in which cubes and roots equal a number—that is, those of the type $x^3 + px = q$, although at that time only specific numerical (positive) coefficients were used. Tartaglia meanwhile had learned how to solve also equations of the form where cubes and squares equal a number. It is likely that Tartaglia had learned how to reduce this case to Fior's by removing the squared term, for it became known by this time that if the leading coefficient is unity, then the coefficient of the squared term, when it appears on the other side of the equality sign, is the sum of the roots.

News of Tartaglia's triumph reached Cardan, who promptly invited the winner to his home, with a hint that he would arrange to have him meet a prospective patron. Tartaglia had been without a substantial source of support, partly perhaps because of his speech impediment. As a child he had received a sabre cut in the fall of Brescia to the French in 1512, which impaired his speech. This earned him the nickname Tartaglia, or stammerer, a name that he thereafter used instead of the name Niccolo Fontana that had been given him at birth. Cardan, in contrast to Tartaglia, had achieved worldly success as a physician. So great was his fame that he was once called to Scotland to diagnose an ailment of the Archbishop of St. Andrews (evidently a case of asthma). By birth illegitimate, and by habit an astrologer, gambler, and heretic, Cardan nevertheless was a respected professor at Bologna and Milan, and ultimately he was granted a pension by the pope. One of his sons poisoned his own wife, the other son was a scoundrel, and Cardan's secretary Ferrari probably died of poison at the hands of his own sister. Despite such distractions, Cardan was a prolific writer on topics ranging from his own life and praise of gout to science and mathematics.

In his chief scientific work, a ponderous volume with the title *De subtilitate*, Cardan is clearly a child of his age, discussing interminably the Aristotelian physics handed down through Scholastic philosophy, while at the same time he waxed enthusiastic about the new discoveries of the then-recent times. Much the same can be said of his mathematics, for this too was typical of the day. He knew little of Archimedes and less of Apollonius, but he was thoroughly familiar with algebra and trigonometry. He already had published a *Practica arithmetice* in 1539, which included among other things the rationalization of denominators containing cube roots. By the time he published the *Ars magna*, half a dozen years later, he probably was the ablest algebraist in Europe. Nevertheless, the *Ars magna* makes dull reading today. Case after case of the cubic equation is laboriously worked out in detail according as terms of the various degrees appear on the same or on opposite sides of the equality, for coefficients were necessarily positive. Despite the fact that he is dealing with equations on numbers, he followed al-Khowarizmi in thinking geometrically, so that we might refer to his method as "completing the cube." There are, of course, certain advantages in such an approach. For instance, since x^3 is a volume, $6x$, in Cardan's equation below, must also be thought of as a volume. Hence the number 6 must have the dimensionality of an area, suggesting the type of substitution that Cardan used, as we shall shortly see.

11   Cardan used little syncopation, being a true disciple of al-Khowarizmi, and, like the Arabs, he thought of his equations with specific numerical coefficients as representative of general categories. For example, when he wrote, "Let the cube and six times the side be equal to 20" (or $x^3 + 6x = 20$), he obviously was thinking of this equation as typical of *all* those having "a cube and thing equal to a number"—that is, of the form $x^3 + px = q$. The solution of this equation covers a couple of pages of rhetoric that we should now put in symbols as follows: Substitute $u - v$ for x and let u and v be related

Jerome Cardan.

so that their product (thought of as an area) is one-third the x coefficient in the cubic equation—that is, $uv = 2$. Upon substitution in the equation, the result is $u^3 - v^3 = 20$; and, on eliminating v, we have $u^6 = 20u^3 + 8$, a quadratic in u^3. Hence u^3 is known to be $\sqrt{108} + 10$. From the relationship $u^3 - v^3 = 20$, we see that $v^3 = \sqrt{108} - 10$; hence, from $x = u - v$, we have $x = \sqrt[3]{\sqrt{108} + 10} - \sqrt[3]{\sqrt{108} - 10}$. Having carried through the method for this specific case, Cardan closes with a verbal formulation of the rule equivalent to our modern solution of $x^3 + px = q$ as

$$x = \sqrt[3]{\sqrt{(p/3)^3 + (q/2)^2} + q/2} - \sqrt[3]{\sqrt{(p/3)^3 + (q/2)^2} - q/2}$$

Cardan then went on to other cases, such as "cube equal to thing and number." Here one makes the substitution $x = u + v$ instead of $x = u - v$, the rest of the method remaining essentially the same. In this case, however, there is a difficulty. When the rule is applied to $x^3 = 15x + 4$, for example, the result is $x = \sqrt[3]{2 + \sqrt{-121}} + \sqrt[3]{2 - \sqrt{-121}}$. Cardan knew that there was no square root of a negative number, and yet he knew $x = 4$ to be a root. He was unable to understand how his rule could make sense in this situation. He had toyed with square roots of negative numbers in another connection when he asked that one divide 10 into two parts such that the product of the parts is 40. The usual rules of algebra lead to the answers $5 + \sqrt{-15}$ and $5 - \sqrt{-15}$ (or, in Cardan's notation, 5p:R̃m:15 and 5m:R̃m:15). Cardan referred to these square roots of negative numbers as "sophistic" and concluded that his result in this case was "as subtile as it is useless." Later writers were to show that such manipulations were indeed subtle but far from useless. It is to Cardan's credit that at least he paid some attention to this puzzling situation.[8]

12 Of the rule for solving quartic equations Cardan in the *Ars magna* wrote that it "is due to Luigi Ferrari, who invented it at my request." Again separate cases, twenty in all, are considered in turn, but for the modern reader one case will suffice. Let square-square and square and number be equal to side. (Cardan knew how to eliminate the cubic term by increasing or diminishing the roots by one-fourth the coefficient in the cubic term.) Then the steps in the solution of $x^4 + 6x^2 + 36 = 60x$ are expressed by Cardan essentially as follows:

1. First add enough squares and numbers to both sides to make the left-hand side a perfect square, in this case $x^4 + 12x^2 + 36$ or $(x^2 + 6)^2$.

[8] There is no published English translation of the whole of the *Ars magna*, but a selection from it appears in D. E. Smith, *A Source Book in Mathematics* (1929). In a recent communication D. J. Struik informed me that there exists in manuscript an English translation of the *Ars magna* by J. R. Witner in Washington.

2. Now add to both sides of the equation terms involving a new unknown y such that the left-hand side remains a perfect square, such as $(x^2 + 6 + y)^2$. The equation now becomes

$$(x^2 + 6 + y)^2 = 6x^2 + 60x + y^2 + 12y + 2yx^2$$
$$= (2y + 6)x^2 + 60x + (y^2 + 12y)$$

3. The next, and crucial, step is to choose y so that the trinomial on the right-hand side will be a perfect square. This is done, of course, by setting the discriminant equal to zero—an ancient and well-known rule equivalent in this case to $60^2 - 4(2y + 6)(y^2 + 12y) = 0$.

4. The result of step 3 is a cubic equation in y—$y^3 + 15y^2 + 36y = 450$—today known as the "resolvent cubic" for the given quartic equation. This is now solved for y by the rules previously given for the solution of cubic equations, the result being

$$y = \sqrt[3]{287\tfrac{1}{2} + \sqrt{80449\tfrac{1}{4}}} + \sqrt[3]{287\tfrac{1}{2} - \sqrt{80449\tfrac{1}{4}}} - 5$$

5. Substitute a value of y from step 4 into the equation for x in step 2 and take the square root of both sides.

6. The result of step 5 is a quadratic equation, which must now be solved in order to find the value of x desired.

The solution of cubic and quartic equations was perhaps the greatest **13** contribution to algebra since the Babylonians, almost four millennia earlier, had learned how to complete the square for quadratic equations. No other discoveries had had quite the stimulus to algebraic development as did those disclosed in the *Ars magna*. The solutions of the cubic and quartic were in no sense the result of practical considerations, nor were they of any value to engineers or mathematical practitioners. Approximate solutions of some cubic equations had been known in antiquity, and al-Kashi a century before Cardan could have solved to any desired degree of accuracy any cubic equation resulting from a practical problem. The Tartaglia-Cardan formula is of great logical significance, but it is not nearly so useful to practical men as are methods of successive approximation.

The most important outcome of the discoveries published in the *Ars magna* was the tremendous stimulus they gave to algebraic research in various directions. It was natural that study should be generalized to include polynomial equations of any order and that in particular a solution should be sought for the quintic. Here mathematicians of the next couple of centuries were faced with an unsolvable algebraic problem comparable to the classical geometrical problems of antiquity. Much good mathematics, but only a negative conclusion, was the outcome. Another immediate result of the solution of the cubic was the first significant glance at a new kind of number.

Irrational numbers had been accepted by the time of Cardan, even though they were not soundly based, for they are readily approximated by rational numbers. Negative numbers afforded more difficulty because they are not readily approximated by positive numbers, but the notion of sense (or direction on a line) made them plausible. Cardan used them even while calling them "numeri ficti." If an algebraist wished to deny the existence of irrational or negative numbers, he would simply say, as had the ancient Greeks, that the equations $x^2 = 2$ and $x + 2 = 0$ are not solvable. In a similar way algebraists had been able to avoid imaginaries simply by saying that an equation such as $x^2 + 1 = 0$ is not solvable. There was no need for square roots of negative numbers. With the solution of the cubic equation, however, the situation became markedly different. Whenever the three roots of a cubic equation are real and different from zero, the Cardan-Tartaglia formula leads inevitably to square roots of negative numbers. The goal was known to be a real number, but it could not be reached without understanding something about imaginary numbers. The imaginary now had to be reckoned with even if one did agree to restrict oneself to real roots.

At this stage another important Italian algebraist, Rafael Bombelli (ca. 1526–1573), had what he called "a wild thought," for the whole matter "seemed to rest on sophistry." The two radicands of the cube roots resulting from the usual formula differ only in one sign. We have seen that the solution by formula of $x^3 = 15x + 4$ leads to $x = \sqrt[3]{2 + \sqrt{-121}} + \sqrt[3]{2 - \sqrt{-121}}$, whereas it is known by direct substitution that $x = 4$ is the only positive root of the equation. (Cardan had noted that when all terms on one side of the equality sign are of higher degree than the terms on the other side, the equation has one and only one positive root—an anticipation, in a small way, of part of Descartes' rule of signs.) Bombelli had the happy thought that the radicals themselves might be related in much the way that the radicands are related—that, as we should now say, they are conjugate imaginaries that lead to the real number 4. It is obvious that if the sum of the real parts is 4, then the real part of each is 2; and if a number of the form $2 + b\sqrt{-1}$ is to be a cube root of $2 + 11\sqrt{-1}$, then it is easy to see that b must be 1. Hence $x = 2 + 1\sqrt{-1} + 2 - 1\sqrt{-1}$, or 4.

Through his ingenious reasoning Bombelli had shown the important role that conjugate imaginary numbers were to play in the future; but at that time the observation was of no help in the actual work of solving cubic equations, for Bombelli had had to know beforehand what one of the roots is. In this case the equation is already solved, and no formula is needed; without such foreknowledge, Bombelli's approach fails. Any attempt to find algebraically the cube roots of the imaginary numbers in the Cardan-Tartaglia rule leads to the very cubic in the solution of which the cube roots arose in the first place, so that one is back where he started from. Because this impasse arises whenever all three roots are real, this is known as the "irreducible case." Here an expression for the unknown is indeed provided by the formula, but the form in which this appears is useless for most purposes.

Bombelli composed his *Algebra*[9] in about 1560, but it was not printed until 1572, about a year before he died, and then only in part. One of the significant things about this book is that it contains symbolisms reminiscent of those of Chuquet. Bombelli sometimes wrote 1 Z p.5Rm.4 (that is, 1 zenus plus 5 res minus 4) for $x^2 + 5x - 4$. But he used also another form of expression—$1\overset{2}{\smile}$p · $5\overset{1}{\smile}$m · 4—in which the power of the unknown quantity is represented simply as an Arabic numeral above a short circular arc, so that x, x^2, x^3 appear as $\overset{1}{\smile}, \overset{2}{\smile}, \overset{3}{\smile}$, for example, influenced perhaps by de la Roche's *Larismethique*. Bombelli's *Algebra* of course uses the standard Italian symbols p and m for addition and subtraction, but he still had no symbol for equality. Our standard equality sign had been published before Bombelli wrote his book, but the symbol had appeared in a distant part of Europe—in England in 1557 in the *Whetstone of Witte* of Robert Recorde (1510–1558).

Mathematics had not prospered in England during the period of almost **14** two centuries since the death of Bradwardine, and what little work was done there in the early sixteenth century depended much on Italian writers such as Pacioli. Recorde was, in fact, just about the only mathematician of any stature in England throughout the century. He was born in Wales and studied and taught mathematics at both Oxford and Cambridge. In 1545 he received his medical degree at Cambridge, and thereafter he became physician to Edward VI and Queen Mary. One of the remarkable things about the period was the surprisingly large number of physicians who contributed outstandingly to mathematics, Chuquet, Cardan, and Recorde being three of the best known. It is likely that Recorde was the most influential of these three within his own country, for he virtually established the English mathematical school. Like Chuquet and Pacioli before him, and Galileo after him, he wrote in the vernacular; this may have limited his effect on the Continent, although the easy dialogue form that he adopted was used also, some time later, by Galileo. Recorde's first extant mathematical work was the *Grounde of Artes* (1541), a popular arithmetic containing computation by abacus and algorism, with commercial applications. The level and style of this book, dedicated to Edward VI and appearing in more than two dozen editions, may be judged from the following problem:

> Then what say you to this equation? If I sold unto you an horse having 4 shoes, and in every shoe 6 nayles, with this condition, that you shall pay for

[9] There is no convenient edition. On Bombelli's life see articles by S. A. Jayawardine in *Isis*, **54** (1963), 391–395; **56** (1965), 298–306.

the first nayle one ob: for the second nayle two ob: for the third nayle foure ob: and so forth, doubling untill the end of all the nayles, now I ask you, how much would the price of the horse come unto?[10]

His *Castle of Knowledge*, an astronomy in which the Copernican system is cited with approval, and his *Pathewaie to Knowledge*, an abridgement of the *Elements* and the first geometry to appear in English, both appeared in 1551. The work of Recorde that is most often cited is *The Whetstone of Witte*, published in 1557, only a year before he died in prison. (Whether he was jailed for political or religious reasons or because of difficulties related to his position, from 1551 on, as Surveyor of the Mines and Monies of Ireland, is not known.[11]) The title *Whetstone* evidently was a play on the word "coss," for cos is the Latin for whetstone, and the book is devoted to "the cossike practise" (that is, algebra). It did for England what Stifel had done for Germany—with one addition. The well-known equality sign first appeared in it, explained by Recorde in the quotation at the beginning of this chapter. However, it was to be a century or more before the sign triumphed over rival notations.

15   Recorde died in 1558, the year in which Queen Mary also died, and no comparable English mathematical author appeared during the long reign of Elizabeth I. It was France, rather than England, Germany, or Italy, that produced the outstanding mathematician of the Elizabethan Age; but before we turn to his work in the next chapter, there are certain aspects of the earlier sixteenth century that should be clarified. The direction of greatest progress in mathematics during the sixteenth century was obviously in algebra, but developments in trigonometry were not far behind, although they were not nearly so spectacular. The construction of trigonometric tables is a dull task, but they are of great usefulness to astronomers and mathematicians; here early sixteenth-century Poland and Germany were very helpful indeed. Most of us today think of Nicholas Copernicus (1473–1543) as an astronomer who revolutionized the world view by successfully putting the earth in motion about the sun (where Aristarchus had tried and failed); but an astronomer is almost inevitably a trigonometer as well, and we owe to Copernicus a mathematical obligation as well as an astronomical debt.

During the lifetime of Regiomontanus, Poland had enjoyed a "Golden Age" of learning, and the University of Cracow, where Copernicus enrolled in 1491, enjoyed great prestige in mathematics and astronomy. After further

[10] See E. R. Ebert, "A Few Observations on Robert Recorde and his 'Grounde of Artes'," *The Mathematics Teacher*, **30** (1937), 110–121. See also Joy B. Easton, "A Tudor Euclid," *Scripta Mathematica*, **27** (1966), 339–355; F. R. Johnson and S. V. Larkey, "Robert Recorde's Mathematical Teaching and the Anti-Aristotelian Movement," *Huntington Library Bulletin*, **7** (1935), 59–87.

[11] See F. M. Clarke, "New Light on Robert Recorde," *Isis*, **7** (1926), 50–70.

The Arte

as their woꝛkes doe extende) to diftincte it onely into
twoo partes. Whereof the firfte, *when one number is
equalle vnto one other* And the feconde i. *when one nomber is compared as equalle vnto* ..*other nombers.*

Alwaies willyng you to remeber, that you reduce
your nombers, to their leafte denominations, and
fmallefte foꝛmes, befoꝛe you pꝛocede any farther.

And again, if your *equation* be foche, that the greatefte denomination *Cofike*, be ioined to any parte of a
compounde nomber, you ſhall tourne it fo, that the
nomber of the greatefte figne alone, maie ftande as
equalle to the refte.

And this is all that neadeth to be taughte, concer-
nyng this wooꝛke.

Howbeit, foꝛ eafie alteratiō of *equations.* I will pꝛo-
pounde a fewe eꝛaples, bicaufe the extraction of their
rootes, maie the moꝛe aptly bee wꝛoughte. And to a-
uoide the tedioufe repetition of thefe wooꝛdes: is e-
qualle to: I will fette as I doe often in wooꝛke bfe, a
paire of paralleles, oꝛ Ꝼemowe lines of one lengthe,
thus: ========, bicaufe noe. 2. thynges, can be moare
equalle. And now marke thefe nombers.

1. $14.\mathcal{z}e. \text{---} .15.\text{\textit{q}} == 71.\text{\textit{q}}.$

2. $20.\mathcal{z}e. \text{------} .18.\text{\textit{q}} === .102.\text{\textit{q}}.$

3. $26.\mathcal{z} \text{---} 10\mathcal{z}e === 9.\mathcal{z} \text{---} 10\mathcal{z}e \text{---} 213.\text{\textit{q}}.$

4. $19.\mathcal{z}e \text{---} 192.\text{\textit{q}} === 10\mathcal{z} \text{---} 108\text{\textit{q}} \text{---} 19\mathcal{z}e.$

5. $18.\mathcal{z}e \text{---} 24.\text{\textit{q}}. === 8.\mathcal{z}. \text{---} 2.\mathcal{z}e.$

6. $34\mathcal{z} \text{------} 12\mathcal{z}e === 40\mathcal{z}e \text{---} 480\text{\textit{q}} \text{---} 9.\mathcal{z}.$

1.   In the firfte there appeareth. 2. nombers, that is
                                          $14.\mathcal{z}e.$

A page from Robert Recorde's *Whetstone of Witte* (1557). Note that his symbols for
equality are much longer than ours.

studies in law, medicine, and astronomy at Bologna, Padua, and Ferrara, and after some teaching at Rome, Copernicus returned to Poland in 1510 to become Canon of Frauenburg. Despite multitudinous administrative obligations, including currency reform and the curbing of the Teutonic Order, Copernicus completed the celebrated treatise, *De revolutionibus orbium coelestium*, published in 1543, the year he died. This contains substantial sections on trigonometry that had been separately published in the previous year under the title *De lateribus et angulis triangulorum*. The trigonometric material is similar to that in Regiomontanus' *De triangulis*, published at Nuremberg only a decade earlier; but Copernicus' trigonometric ideas seem to date from before 1533, at which time he probably did not know of the work of Regiomontanus. It is quite likely, nevertheless, that the final form of Copernicus' trigonometry was in part derived from Regiomontanus, for in 1539 he received as a student the Prussian mathematician Georg Joachim Rheticus (or Rhaeticus, 1514–1576), a mathematician of Wittenberg who evidently had been in touch with Nuremberg mathematics. Rheticus worked with Copernicus for some three years, and it was he who, with his teacher's approval, published the first short account of Copernican astronomy in a work entitled *Narratio prima* (1540) and who made the first arrangements, completed by Andreas Osiander, for the printing of the celebrated *De revolutionibus*. It is likely, therefore, that the trigonometry in the classic work of Copernicus is closely related, through Rheticus, to that of Regiomontanus.

We see the thorough trigonometric capabilities of Copernicus not only in the theorems included in *De revolutionibus*, but also in a proposition originally included by the author in an earlier manuscript version of the book, but not in the printed work. The deleted proposition is a generalization of the theorem of Nasir Eddin (which does appear in the book) on the rectilinear motion resulting from the compounding of two circular motions. The theorem of Copernicus is as follows: If a smaller circle rolls without slipping along the inside of a larger circle with diameter twice as great, then the locus of a point which is not on the circumference of the smaller circle, but which is fixed with respect to this smaller circle, is an ellipse. Cardan, incidentally, knew of the Nasir Eddin theorem, but not of the Copernican locus, a theorem rediscovered in the seventeenth century.[12]

16   Through the trigonometric theorems in *De revolutionibus* Copernicus spread the influence of Regiomontanus, but his student Rheticus went further. He combined the ideas of Regiomontanus and Copernicus, together with views of his own, in the most elaborate treatise composed up to that

[12] See C. B. Boyer, "Note on Epicycles and the Ellipse from Copernicus to Lahire," *Isis*, **38** (1947), 54–56.

time—the two-volume *Opus palatinum de triangulis.* Here trigonometry really came of age. The author discarded the traditional consideration of the functions with respect to the arc of a circle and focused instead on the lines in a right triangle. Moreover, all six trigonometric functions now came into full use, for Rheticus calculated elaborate tables of all of them. Decimal fractions still had not come into common use; hence for the sine and cosine functions he used a hypotenuse (radius) of 10,000,000 and for the other four functions a base (or adjacent side or radius) of 10,000,000 parts, for intervals in the angle of 10″. He began tables of tangents and secants with a base of 10^{15} parts; but he did not live to finish them, and the treatise was completed and edited, with additions, by his pupil Valentin Otho (ca. 1550–1605) in 1596.[13]

The work of Rheticus, who like Copernicus, Chuquet, Cardan, and Recorde **17** had also studied medicine, was much admired by Pierre de la Ramée or Ramus (1515–1572), a man who contributed to mathematics in a pedagogical sense. At the Collège de Navarre he had in 1536 defended, for his master's degree, the audacious thesis that everything Aristotle had said was wrong—at a time when Peripateticism was the same as orthodoxy. In his intellectual criticism and pedagogical interests he may be compared with Recorde in England. Ramus was at odds with his age in many ways, and while his Humanist contemporaries had little use for mathematics, he had almost a blind faith in the subject. He proposed revisions in the university curricula so that logic and mathematics should receive more attention; his logic enjoyed considerable popularity in Protestant countries, in part because he died a martyr in the St. Bartholomew massacre. Not satisfied even with the *Elements* of Euclid, Ramus edited this with revisions. However, his competence in geometry was very limited, and his suggested changes in mathematics were in the opposite direction from those in our day. Ramus had more confidence in practical elementary mathematics than in speculative higher algebra and geometry; looking back on his age we see that the mathematics of that time seems already to have been excessively concerned with practical problems in arithmetic, while weakness in geometry was quite conspicuous.

Pappus in about 320 had wished to initiate a geometrical revival, but he **18** found no really capable successor in pure geometry in Greece. In China and India there never had been any real concern for geometry beyond problems in mensuration, but the Arabs, who appreciated demonstrative reasoning, used geometrical arguments in their algebra. In medieval Europe, as we have seen, there was a two-way tendency to relate algebra and geometry. In the

---

[13] See J. D. Bond, "The Development of Trigonometric Methods Down to the Close of the XVth Century," *Isis*, **4** (1921–1922), 295–323; also Sister Mary Claudia Zeller, *The Development of Trigonometry from Regiomontanus to Pitiscus* (1946).

medieval tradition, Books IV and VI of Bombelli's *Algebra* were full of problems in geometry that are solved algebraically—somewhat in the manner of Regiomontanus, but making use of new symbolisms. For example, Bombelli asked for the side of a square inscribed in a triangle with sides $ac = 13$, $cf = 14$, $fa = 15$, so that one side lies on cf (Fig. 15.1), which he solved as follows: Let $bg = 14^{\perp}$ (that is, $14x$). Then $ag = 15^{\perp}$ and $ab = 13^{\perp}$. Now $ah = 12^{\perp}$ and $hi = 14^{\perp}$. Since $ai = 12$, we have $26^{\perp} = 12$; then "cosa"

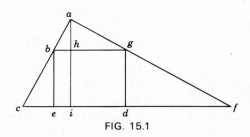

FIG. 15.1

or x is $\frac{6}{13}$, so that hi, or the side of the square, must be 14 times $\frac{6}{13}$ or $6\frac{6}{13}$. Here a highly symbolic algebra has come to the aid of geometry; but Bombelli worked in the other direction, too. In the *Algebra*, the algebraic solution of cubic equations is accompanied by geometric demonstrations in terms of the subdivision of the cube. Unfortunately for the future of geometry—and of mathematics in general—the last books of Bombelli's *Algebra* were not included in the publication of 1572, but remained in manuscript until 1929.[14]

19   Pure geometry in the sixteenth century was not entirely without representatives, for unspectacular contributions were made in Germany by Johannes Werner (1468–1528) and Albrecht Dürer (1471–1528), and in Italy by Francesco Maurolico (1494–1575) and Pacioli. Once more we note the preeminence of these two countries in contributions to mathematics during the Renaissance. Werner had aided in preserving the trigonometry of Regiomontanus, but of more geometrical significance was his Latin work, in twenty-two books, on the *Elements of Conics*, printed at Nuremberg in 1522. This cannot be compared favorably with the *Conics* of Apollonius, almost entirely unknown in Werner's day, but it marks the renewal of interest in the curves for almost the first time since Pappus. Because the author was concerned primarily with the duplication of the cube, he concentrated on the parabola and the hyperbola, deriving the standard plane equations stereometrically from the cone, as had his predecessors in Greece; but there seems

[14] See *L'algebra. Opera di Rafael Bombelli da Bologna* Books IV and V, comprising "La parte geometrica", ed. by Ettore Bortolotti (1929).

to be an element of originality in his plane method for plotting points on a parabola with compasses and straightedge. One first draws a pencil of circles tangent to each other and intersecting the common normal in points c, d, e. f. g. . . . (Fig. 15.2). Then along the common normal one marks off a distance

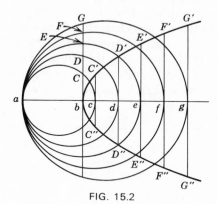

FIG. 15.2

ab equal to a desired parameter. At b one erects the line bG perpendicular to ab and cutting the circles in points C, D, E, F, G, . . . respectively. Then at c one erects line segments cC' and cC'' perpendicular to ab and equal to bC; at d one erects perpendicular segments dD' and dD'' equal to bD; at e one erects segments eE' and eE'' equal to bE, and so on. Then C', C'', D', D'', E', E'', . . . will all lie on the parabola with vertex b, axis along ab, and having ab as the magnitude of the parameter—as is readily seen from the relationships $(cC')^2 = ab \cdot bc$, $(dD')^2 = ab \cdot bd$, and so on.[15]

Werner's work is closely related to ancient studies of conics; but meanwhile **20** in Italy and Germany a relatively novel relationship between mathematics and art was developing. One important respect in which Renaissance art differed from art in the Middle Ages was in the use of perspective in the plane representation of objects in three-dimensional space. The Florentine architect Filippo Brunelleschi (1377–1446) is said to have given much attention to this problem, but the first formal account of some of the problems was given by Leon Battista Alberti (1404–1472) in a treatise of 1435 (printed in 1511) entitled *Della pictura*. Alberti opens with a general discussion of the principles of foreshortening and then describes a method he had invented for representing in a vertical "picture plane" a set of squares in a horizontal "ground

[15] See J. L. Coolidge, *A History of the Conic Sections and Quadratic Surfaces* (Oxford: Clarendon, 1945), pp. 26–28.

plane." Let the eye be at a "station point" S that is h units above the ground plane and k units in front of the picture plane. The intersection of the ground plane and the picture plane is called the "groundline," the foot V of the perpendicular from S to the picture plane is called the "center of vision" (or the principal vanishing point), the line through V parallel to the ground-line is known as the "vanishing line" (or horizon line), and the points P and Q on this line which are k units from V are called the "distance points." If we take points A, B, C, D, E, F, G marking off equal distances along the groundline RT (Fig. 15.3), where D is the intersection of this line with the vertical plane through S and V, and if we draw lines connecting these points with V, then the projection of these last lines, with S as a center, upon the

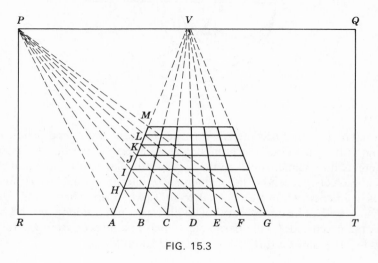

FIG. 15.3

ground plane will be a set of parallel and equidistant lines. If P (or Q) is con-nected with the points B, C, D, E, F, G to form another set of lines intersecting AV in points H, I, J, K, L, M, and if through the latter points parallels are drawn to the groundline RT, then the set of trapezoids in the picture plane will correspond to a set of squares in the ground plane.[16]

A further step in the development of perspective was taken by the Italian painter of frescoes, Piero della Francesca (1410?–1492), in *De prospectiva pingendi* (ca. 1478). Where Alberti had concentrated on representing on the picture plane figures in the ground plane, Piero handled the more complicated problem of depicting on the picture plane objects in three dimensions as seen

[16] Further details, as well as solid accounts of other work by Piero della Francesca, Leonardo da Vinci, and Albrecht Dürer, will be found in J. L. Coolidge, *The Mathematics of Great Amateurs* (Oxford: Clarendon, 1949), pp. 30–70.

from a given station point. He wrote also a *De corporibus regularibus* where he noted the "divine proportion" in which diagonals of a regular pentagon cut each other and where he found the volume common to two equal circular cylinders whose axes cut each other at right angles (unaware of Archimedes' *Method*, which was unknown at the time). The connection between art and

Albrecht Dürer's "Melancholia" (The British Museum). Note the four-celled magic square in the upper right-hand corner.

mathematics was strong also in the work of Leonardo da Vinci. He wrote a work, now lost, on perspective; his *Trattato della pittura* opens with the admonition. "Let no one who is not a mathematician read my works."[17] The same combination of mathematical and artistic interests is seen in Albrecht Dürer, a contemporary of Leonardo and a fellow townsman of Werner at Nuremberg. In Dürer's work we see also the influence of Pacioli, especially in the celebrated engraving of 1514 entitled *Melancholia*. Here the magic square figures prominently. This often is regarded as the first use of a

16	3	2	13
5	10	11	8
9	6	7	12
4	15	14	1

magic square in the West, but Pacioli had left an unpublished manuscript, *De viribus quantitatis*, in which interest in such squares is indicated. Dürer's interests in mathematics, however, were far more geometrical than arithmetic, as the title of his most important book indicates; "Investigation of the measurement with circles and straight lines of plane and solid figures." This work, which appeared in several German and Latin editions from 1525 to 1538, contains some striking novelties, of which the most important were his new curves. This is one direction in which the Renaissance could easily have improved on the work of the ancients, who had studied only a handful of types of curves. Dürer took a fixed point on a circle and then allowed the circle to roll along the circumference of another circle, generating an epicycloid; but, not having the necessary algebraic tools, he did not study this analytically. The same was true of other plane curves that he obtained by projecting helical space curves onto a plane to form spirals. Too often those working in perspective were not familiar with the foundations of mathematics and failed to distinguish between exact and approximate results. In Dürer's work we find the Ptolemaic construction of the regular pentagon, which is exact, as well as another original construction that is only an approximation. For the heptagon and nonagon he also gave ingenious, but of course inexact, constructions. Dürer's construction of an approximately regular nonagon is as follows: Let *O* be the center of a circle *ABC* in which *A*, *B*, and *C* are

[17] Morris Kline, *Mathematics in Western Culture* (New York: Oxford University Press, 1953). This book contains an eminently readable account of art as related to mathematics.

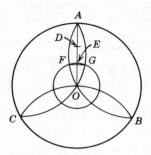

FIG. 15.4

vertices of the inscribed equilateral triangle (Fig. 15.4). Through A, O, and C draw a circular arc, draw similar arcs through B, O, and C and through B, O, and A. Let AO be trisected at points D and E, and through E draw a circle with center at O and cutting arcs AFO and AGO in points F and G respectively. Then the straight line segment FG will be very nearly equal to the side of the regular nonagon inscribed in this smaller circle, the angle FOG differing from 40° by less than 1°.[18] The relation of art and geometry might have been very productive indeed, had it gained the attention of professionally minded mathematicians, but in this respect it failed for more than a century after Dürer's time.

Dürer's contemporaries in pure mathematics failed to appreciate the **21** future of geometric transformations, but projections of various sorts are essential to cartographers. Geographical explorations had widened horizons and created a need for better maps, but Scholasticism and Humanism were of little help here since new discoveries had outmoded medieval and ancient maps. One of the most important of the innovators was a German mathematician and astronomer Peter Apian (or Bienewitz, 1495–1552). In 1520 he published perhaps the earliest map of the Old World and the New World in which the name "America" was used; in 1527 he issued a business arithmetic in which, on the title page, the arithmetic or "Pascal" triangle appeared in print for the first time. The maps of Apian were well done, but they followed Ptolemy closely wherever possible. For that novelty which is thought to be so characteristic of the Renaissance it is better to look instead to a Flemish geographer, Gerard Mercator (or Gerhard Kremer, 1512–1594), who was for a time associated with the court of Charles V at Brussels. Mercator may be said to have broken with Ptolemy in geography as Copernicus had revolted against Ptolemaic astronomy.

[18] See Moritz Cantor, *Vorlesungen über Geschichte der Mathematik* (Leipzig: Teubner, 1900–1908, 4 vols.), II, 425.

Pascal Triangle as first printed, 1527. Title page of the arithmetic of Petrus Apianus, Ingolstadt, 1527, more than a century before Pascal investigated the properties of the triangle.

For the first half of his life Mercator depended heavily on Ptolemy, but by 1554 he had emancipated himself sufficiently to cut down the Ptolemaic estimate of the width of the Mediterranean from 62° to 53°. (Actually it is close to 40°.) More importantly, in 1569 he published the first map, *Nova et aucta orbis terrae descriptio*, drawn up on a new principle. Maps in common use in Mercator's day were usually based upon a rectangular grid made up of two sets of equidistant parallel lines, one set for latitudes, the other for longitudes. The length of a degree of longitude, however, varies with the parallel of latitude along which it is measured, an inequality disregarded in common practice and resulting in distortion of shape and in errors of direction on the part of navigators who based a course upon the straight line drawn between two points on the map. The Ptolemaic stereographic projection preserved shapes, but it did not use the common grid of lines. In order to bring theory and practice into some accord, Mercator introduced the projection that bears his name and, with later improvement, has been basic in cartography ever since. The first step in the Mercator projection is to think of a spherical earth inscribed within an indefinitely long right circular cylinder touching the earth along the equator (or some other great circle), and to project, from the center of the earth, points on the surface of the earth onto the cylinder. If the cylinder then is cut along an element and flattened out, the meridians and parallels on the earth will have been transformed into a rectangular network of lines. Distances between successive meridian lines will be equal, but not distances between successive lines of latitude. In fact, the latter distances increase so rapidly, as one moves away from the equator, that distortions of shape and direction occur; but Mercator found that through an empirically determined modification of these distances preservation of direction and shape (although not of size) was possible.[19] In 1599 Edward Wright (1558–1615), a fellow at Cambridge, tutor to Henry, Prince of Wales, and a good sailor, developed the theoretical basis of the Mercator projection by computing the functional relationship $D = a \ln \tan(\phi/2 + 45°)$ between map distance D from the equator and latitude ϕ.

Mathematics during the Renaissance had been widely applied—to bookkeeping, mechanics, surveying, art, cartography, optics—and there were numerous books devoted to the practical arts. Nevertheless, interest in the classical works of antiquity continued strong, as we see in the case of Maurolico, a priest of Greek parentage who was born, lived, and died in Sicily. Maurolico was a scholarly geometer who did much to revive interest in the more advanced of the antique works.[20] Geometry in the first half of the sixteenth century had been far too heavily dependent on the elementary properties found in Euclid. Werner had been an exception to this rule, but few others were really familiar with the geometry of Archimedes, Apollonius, and Pappus. The reason for this was simple—Latin translations of these did not become generally available until the middle of the century. In this process of translation Maurolico was joined by an ardent Italian scholar, Federigo

---

[19] A compendious historical account of this and other projections is provided in the article "Map" by E. G. Ravenstein et al. in *Encyclopaedia Britannica*, 11th ed., **17**, 629–663.

[20] Some idea of the extent of his writings and the difficulty of dating them can be gained from Edward Rosen, "The Editions of Maurolico's Mathematical Works," *Scripta Mathematica*, **24** (1959), 59–76.

Commandino, who died in the same year—1575. We have mentioned Tartaglia's borrowed translation of Archimedes printed in 1543; this was followed by a Greek edition of 1544 and a Latin translation by Commandino at Venice in 1558.

Four books of the *Conics* of Apollonius had survived in Greek, and these had been translated into Latin and printed at Venice in 1537. Maurolico's translation, completed in 1548, was not published for more than a century, appearing in 1654, but another translation by Commandino was printed at Bologna in 1566. The *Mathematical Collection* of Pappus had been virtually unknown to the Arabs and the medieval Europeans, but this, too, was translated by the indefatigable Commandino, although it was not printed until 1588. Maurolico was acquainted with the vast treasures of ancient geometry that were becoming available, for he read Greek as well as Latin. In fact, from some indications in Pappus of Apollonius' work on maxima and minima—that is, on normals to the conic sections—Maurolico tried his hand at a reconstruction of the then-lost Book V of the *Conics*. In this respect he represented a vogue that was to be one of the chief stimuli to geometry before Descartes—the reconstruction of lost works in general and of the last four books of the *Conics* in particular. During the interval from Maurolico's death in 1575 to the publication of *La géométrie* by Descartes in 1637, geometry was marking time until developments in algebra had reached a level making algebraic geometry possible. The Renaissance could well have developed pure geometry in the direction suggested by art and perspective, but the possibility went unheeded until almost precisely the same time that algebraic geometry was created. Between Maurolico and Descartes, meanwhile, mathematics developed in several nongeometrical directions, and it is to these that we now turn.

BIBLIOGRAPHY

Bond, J. D., "The Development of Trigonometric Methods Down to the Close of the XVth Century," *Isis*, **4** (1921–1922), 295–323.

Bortolotti, Ettore, *Studi e ricerche sulla storia della matematica in Italia nei secoli XVI e XVII* (Bologna, 1928).

Bortolotti, Ettore, ed., *L'algebra. Opera di Rafael Bombelli da Bologna* (Bologna, 1929).

Cardan, Jerome, *The Book of My Life*, trans. by Jean Stoner (paperback ed., New York: Dover, 1963).

Clarke, F. M., "New Light on Robert Recorde," *Isis*, **7** (1926), 50–70.

Easton, Joy B, "A Tudor Euclid," *Scripta Mathematica*, **27** (1966), 339–355.

Ebert, E. R., "A Few Observations on Robert Recorde and his 'Grounde of Artes'," *The Mathematics Teacher*, **30** (1937), 110–121.

Glaisher, J. W. L., "On the Early History of the Signs + and − and on the Early German Arithmeticians," *Messenger of Mathematics*, **51** (1921–1922), 1–148.

Hofmann, J. E., *Geschichte der Mathematik*, 2nd ed. (Berlin: Walter de Gruyter, 1963), Vol. I.

Hughes, Barnabas, ed., *Regiomontanus on Triangles* (Madison, Wis.: University of Wisconsin Press, 1967).

Lambo, Ch., S. J., "Une algèbre française de 1484. Nicolas Chuquet," *Revue des Questions Scientifiques* (3), **2** (1902), 442–472.

Marre, Aristide, "Notice sur Nicolas Chuquet et son Triparty en la science des nombres," *Bullettino di Bibliografia e di Storia delle Scienze Matematiche e Fisiche*, **13** (1880), 555–659, 693–814; **14** (1881), 413–460.

Ore, Oystein, *Cardano, the Gambling Scholar* (Princeton, N. J.: Princeton University Press, 1953).

Sarton, George, "The Scientific Literature Transmitted Through the Incunabula," *Osiris*, **5** (1938), 41–247.

Simon, Max, *Cusanus als Mathematiker* (Strassburg, 1911, in *Festschrift Heinrich Weber*, Leipzig and Berlin: Teubner, 1912, pp. 298–337).

Smith, D. E., *Rara arithmetica* (Boston: Ginn, 1908).

Smith, D. E., ed., *A Source Book in Mathematics* (New York: McGraw-Hill, 1929; paperback ed., New York: Dover, 1959, 2 vols.).

Sullivan, J. W. N., *The History of Mathematics in Europe, from the Fall of Greek Science to the Rise of the Conception of Mathematical Rigour* (New York: Oxford University Press, 1925).

Taylor, R. Emmett, *No Royal Road. Luca Pacioli and His Times* (Chapel Hill, N.C.: University of North Carolina Press, 1942).

Waters, W. G., *Jerome Cardan, a Biographical Study* (London, 1898).

Zeller, Sister Mary Claudia, *The Development of Trigonometry from Regiomontanus to Pitiscus* (Ann Arbor, Mich.: University of Michigan, Ph. D. thesis, 1944).

EXERCISES

1. Which of the following factors were important in the development of Renaissance mathematics: (*a*) the fall of Constantinople, (*b*) the Protestant Reformation, (*c*) the rise of Humanism, (*d*) the invention of printing, (*e*) the rising mercantile class? Explain.

2. How do you account for the fact that algebra and trigonometry developed more rapidly than geometry during the Renaissance?

3. Why was the solution of the cubic so important for the development of imaginary numbers?

4. How would you account for the fact that many of the leading mathematicians in the sixteenth century were physicians?

5. Which countries took the lead, during the Renaissance, in the development of (*a*) algebra, (*b*) trigonometry, (*c*) geometry? Mention specific contributions in each case.

6. How does Regiomontanus' value for tan 89° compare with that in modern tables? How might he have found his value?

7. Construct, with compasses and straightedge, a triangle in which one side is 5, the altitude to this side is 3, and the ratio of the other two sides is $\sqrt{2}:1$. (*Suggestion*: apply the algebraic approach of Regiomontanus and Bombelli.)

8. Solve Pacioli's problem in which it is required to find the sides of a triangle if the radius of the inscribed circle is 4, and if the segments into which one side is divided by the point of contact are 6 and 8.

9. Derive a solution of Bombelli's equation $x^3 = 15x + 4$ as a sum or difference of cube roots of imaginary numbers.

10. Reduce the solution of Ferrari's quartic $x^4 + 6x^2 + 36 = 60x$ to the solution of a cubic equation.

11. Verify Bombelli's statement that $4 + \sqrt{-1}$ is a cube root of $52 + \sqrt{-2209}$.

12. Recorde's *Grounde of Artes* contains the following "simplified" scheme for multiplying two one-digit numbers (both more than 5): First subtract each number from 10. The product of these differences is the units digit in the product of the original numbers, and if either difference is subtracted from the *other* original number this will be the tens digit in the product of the original numbers. Prove this rule.

13. Form the cubic equation with roots $1 \pm \sqrt{3}$ and -3 and then apply the method of Cardan and Tartaglia to solve this cubic.

14. Solve Cardan's problem in dividing 10 into two parts the product of which shall be 40. Verify your answer.

*15. Justify Werner's construction of the parabola, indicating where the directrix lies.

*16. How do you know that Dürer's construction of a regular nonagon is not exact? Find to the nearest minute the arc FG.

*17. Show how to construct the cyclic quadrilateral with successive sides $a = 25, b = 33, c = 60, d = 16$.

*18. Prove the Copernican theorem on the epicyclic generation of the ellipse.

*19. Justify Alberti's method for representing in the picture plane a set of squares in the ground plane.

CHAPTER XVI

Prelude
to Modern Mathematics

> In mathematics I can report no deficiency, except it
> be that men do not sufficiently understand the excellent
> use of the Pure Mathematics.
>
> *Francis Bacon*

When in 1575 Maurolico and Commandino died, Western Europe had **1**
recovered most of the major mathematical works of antiquity now extant.
Arabic algebra had been thoroughly mastered and improved upon, both
through the solution of the cubic and quartic and through a partial use of
symbolism; and trigonometry had become an independent discipline. The
time was almost ripe for rapid strides beyond ancient, medieval, and Renaissance contributions—but not quite. There is in the history of mathematics
a high degree of continuity from one age to the next; the transition from the
Renaissance to the modern world was also made through a large number of
intermediate figures, a few of the more important of whom we shall now
consider. Two of these men, Galileo Galilei (1564–1642) and Bonaventura
Cavalieri (1598–1647), came from Italy; several more, such as Henry Briggs
(1561–1639), Thomas Harriot (1560–1621), and William Oughtred (1574–
1660), were English; two of them, Simon Stevin (1548–1620) and Albert
Girard (1590–1633), were Flemish; others came from varied lands—John
Napier (1550–1617) from Scotland, Jobst Bürgi (1552–1632) from Switzerland,
and Johann Kepler (1571–1630) from Germany. Most of Western Europe
now was involved in the advance of mathematics, but the central and most
magnificent figure in the transition was a Frenchman, François Viète (1540–
1603)—or, in Latin, Franciscus Vieta.

Viète was not a mathematician by vocation. As a young man he studied
and practiced law, becoming a member of the Bretagne parlement; later
he became a member of the king's council, serving first under Henry III
and later under Henry IV. It was during his service with the latter, Henry of
Navarre, that he became so successful in deciphering cryptic enemy messages
that the Spanish accused him of being in league with the devil. Only Viète's
leisure time was devoted to mathematics, yet he made contributions to

arithmetic, algebra, trigonometry, and geometry. There was a period of almost half a dozen years, before the accession of Henry IV, during which Viète was out of favor, and these years he spent largely on mathematical studies. In arithmetic he should be remembered for his plea for the use of decimal, rather than sexagesimal, fractions. In one of his earliest works, the *Canon-mathematicus* of 1579, he wrote:

> Sexagesimals and sixties are to be used sparingly or never in mathematics, and thousandths and thousands, hundredths and hundreds, tenths and tens, and similar progressions, ascending and descending, are to be used frequently or exclusively.[1]

In the tables and computations he adhered to his word and used decimal fractions. The sides of the squares inscribed in and circumscribed about a circle of diameter 200,000 he wrote as $141,421,^{356,24}$ and $200,000,^{000,00}$, and their mean as $177,245,^{385,09}$. A few pages further on he wrote the semi-circumference as $314,159,\frac{265,35}{1,000,00}$, and still later this figure appeared as **314,159,**265,36, with the integral portion in boldface type. Occasionally he used a vertical stroke to separate the integral and fractional portions, as when writing the apothem of the 96-sided regular polygon, in a circle of diameter 200,000, as about **99,946**|458,75.

The use of a decimal point separatrix generally is attributed either to G. A. Magini (1555–1617), a map-making friend of Kepler and rival of Galileo for a chair at Bologna, in his *De planis triangulis* of 1592, or to Christoph Clavius (1537–1612), a Jesuit friend of Kepler, in a table of sines of 1593. But the decimal point did not become popular until Napier used it more than twenty years later.

2 Without doubt it was in algebra that Viète made his most estimable contributions, for it was here that he came closest to modern views. Mathematics is a form of reasoning, and not a bag of tricks, such as Diophantus had possessed; yet algebra, during the Arabic and early modern periods, had not gone far in freeing itself from the treatment of special cases. There could be little advance in algebraic theory so long as the chief preoccupation was with finding "the thing" in an equation with specific numerical coefficients. Symbols and abbreviations for an unknown, and for powers of the unknown, as well as for operations and for the relationship of equality, had been developed. Stifel had gone so far as to write $AAAA$ for the fourth power of an unknown quantity; yet he had no scheme for writing an equation that might represent any one of a whole class of equations—of all quadratics, say, or of all cubics. A geometer, by means of a diagram, could let ABC represent all triangles, but an algebraist had no counterpart for writing down all equations

---

[1] For further details see C. B. Boyer, "Viète's Use of Decimal Fractions," *The Mathematics Teacher*, **55** (1962), 123–127.

of second degree. Letters had indeed been used to represent magnitudes, known or unknown, since the days of Euclid, and Jordanus had done this freely; but there had been no way of distinguishing magnitudes assumed to be known from those unknown quantities that are to be found. Here Viète introduced a convention as simple as it was fruitful. He used a vowel to represent the quantity in algebra that was assumed to be unknown or un-determined and a consonant to represent a magnitude or number assumed to be known or given. Here we find for the first time in algebra a clear-cut distinction between the important concept of a parameter and the idea of an unknown quantity.

Had Viète adopted other symbolisms extant in his day, he might have **3** written *all* quadratic equations in the single form $BA^2 + CA + D = 0$, where A is the unknown and B, C, and D are parameters; but unfortunately he was modern only in some ways and ancient and medieval in others. His algebra is fundamentally syncopated rather than symbolic, for although he wisely adopted the Germanic symbols for addition and subtraction and, still more wisely, used differing symbols for parameters and unknowns, the remainder of his algebra consisted of words and abbreviations. The third power of the unknown quantity was not A^3, or even AAA, but *A cubus*, and the second power was *A quadratus*. Multiplication was signified by the Latin word *in*, division was indicated by the fraction line, and for equality Viète used an abbreviation for the Latin *aequalis*. It is not given for one man to make the whole of a given change; it must come in steps.

One of the steps beyond the work of Viète was taken by Harriot when he revived the idea Stifel had had of writing the cube of the unknown as AAA. This notation was used systematically by Harriot in his posthumous book entitled *Artis analyticae praxis* and printed in 1631. Its title had been sug-gested by the earlier work of Viète, who had disliked the Arabic name algebra. In looking for a substitute Viète noted that in problems involving the "cosa" or unknown quantity, one generally proceeds in a manner that Pappus and the ancients had described as analysis. That is, instead of reasoning from what is known to what was to be demonstrated, algebraists invariably reasoned from the assumption that the unknown was given and deduced a necessary conclusion from which the unknown can be determined. In modern symbols, if we wish to solve $x^2 - 3x + 2 = 0$, for example, we proceed on the premise that there is a value of x satisfying this equation; from this assumption we draw the necessary conclusion that $(x - 2)(x - 1) = 0$, so that either $x - 2 = 0$ or $x - 1 = 0$ (or both) is satisfied, hence that x necessarily is 2 or 1. However, this does not mean that one or both of these numbers will satisfy the equation unless we can reverse the steps in the reasoning process. That is, the analysis must be followed by the synthetic demonstration.

In view of the type of reasoning so frequently used in algebra, Viète called the subject "the analytic art." Moreover, he had a clear awareness of the broad scope of the subject, realizing that the unknown quantity need not be either a number or a geometrical line. Algebra reasons about "types" or species, hence Viète contrasted *logistica speciosa* with *logistica numerosa*. His algebra was presented in the *Isagoge* (or *Introduction*), printed in 1591, but his several other algebraic works did not appear until many years after his death. In all of these he maintained a principle of homogeneity in equations, so that in an equation such as $x^3 + 3ax = b$ the a is designated as *planum* and the b as *solidum*. This suggests a certain inflexibility, which Descartes removed a generation later; but homogeneity also has certain advantages, as Viète undoubtedly saw.

4   The algebra of Viète is noteworthy for the generality of its expression, but there are also other novel aspects. For one thing, Viète suggested a new approach to the solution of the cubic. Having reduced it to the standard form equivalent to $x^3 + 3ax = b$, he introduced a new unknown quantity y that was related to x through the equation $y^2 + xy = a$. This changes the cubic in x into a quadratic equation in y^3, for which the solution is readily obtained. Moreover, Viète was aware of some of the relations between roots and coefficients of an equation, although here he was hampered by his failure to allow the coefficients and roots to be negative. He realized, for example, that if $x^3 + b = 3ax$ has two positive roots, x_1 and x_2, then $3a = x_1{}^2 + x_1x_2 + x_2{}^2$ and $b = x_1x_2{}^2 + x_2x_1{}^2$. This is, of course, a special case of our theorem that the coefficient of the term in x, in a cubic with leading coefficient unity, is the sum of the products of the roots taken two at a time, and the constant term is the negative of the product of the roots. Viète, in other words, was close to the subject of symmetric functions of the roots in the theory of equations. It remained for Girard in 1629, in *Invention nouvelle en l'algèbre*, to state clearly the relations between roots and coefficients, for he allowed for negative and imaginary roots, whereas Viète had recognized only the positive roots. In a general way Girard realized that negative roots are directed in a sense opposite to that for positive numbers, thus anticipating the idea of the number line. "The negative in geometry indicates a retrogression," he said, "where the positive is an advance." To him also seems to be largely due the realization that an equation can have as many roots as is indicated by the degree of the equation. Girard retained imaginary roots of equations because they show the general principles in the formation of an equation from its roots.

5   Discoveries much like those of Girard had been made even earlier by Thomas Harriot, but these did not appear in print until ten years after Harriot had died of cancer in 1621. Harriot had been hampered in publication by conflicting political currents during the closing years of the reign of Queen Elizabeth I. He had been sent by Sir Walter Raleigh as a surveyor on the latter's expeditions to the New World in 1585, becoming thus the first substantial mathematician to set foot in North America. (Brother Juan Diaz, a young chaplain with some mathematical training, had earlier joined Cortes on an expedition to Yucatan in 1518.) On his return he published *A Briefe and True Report of the New-found Land of Virginia* (1586). When his patron lost favor with the queen and was executed, Harriot was granted a pension of £300 a year by Henry, Earl of Northumberland; but in 1606 the earl was committed to the Tower by Elizabeth's successor, James I. Harriot continued to meet with Henry in the Tower, and distractions and poor health contributed to his failure to publish results.

Harriot knew of relationships between roots and coefficients and between roots and factors, but like Viète he was hampered by failure to take note of negative and imaginary roots. In notations, however, he advanced the use of symbolism, being responsible for the signs $>$ and $<$ for "greater than" and "less than."[2] It was partly also his use of Recorde's equality sign that led to the ultimate adoption of this sign. Harriot showed much more moderation in the use of new notations than did his younger contemporary, William Oughtred. The latter published his *Clavis mathematicae* in the same year, 1631, in which Harriot's *Praxis* was printed. In the *Clavis* the notation for powers was a step back toward Viète, for where Harriot had written $AAAAAAA$, for example, Oughted used $Aqqc$ (that is, A squared squared cubed). Of all Oughtred's new notations, only one is now widely used—the cross $\times$ for multiplication.[3]

The homogeneous form of his equations shows that Viète's thought was always close to geometry, but his geometry was not on the elementary level of so many of his predecessors; it was on the higher level of Apollonius and Pappus. Interpreting the fundamental algebraic operations geometrically, Viète realized that straightedge and compasses suffice up through square roots. However, if one permits the interpolation of two geometric means between two magnitudes, one can construct cube roots, or, a fortiori, solve geometrically any cubic equation. In this case one can, Viète showed,

[2] See J. A. Lohne, "Thomas Harriot als Mathematiker," *Centaurus,* **11** (1965), 19–45. For the life of Harriot see the article by Agnes M. Clerke in *Dictionary of National Biography,* XXIV (1890), 437–439. See also R. C. H. Tanner, "On the Role of Equality and Inequality in the History of Mathematics," *British Journal of the History of Science,* **1** (1962), 159–169; Robert Kargon, "Thomas Harriot, the Northumberland Circle and Early Atomism in England," *Journal of the History of Ideas,* **27** (1966), 128–136.

[3] For a life of Oughtred and further references, see Florian Cajori, *William Oughtred, a Great Seventeenth-Century Teacher of Mathematics* (1916), and the article on Oughtred by J. B. Mullinger in *Dictionary of National Biography,* XLII (1895), 356–358. On matters of symbolism one should be sure to consult Florian Cajori, *A History of Mathematical Notations* (1929).

construct the regular heptagon, for this construction leads to a cubic of the form $x^3 = ax + a$. In fact, every cubic or quartic equation is solvable by angle trisections and the insertion of two geometric means between two magnitudes. Here we see clearly a very significant trend—the association of the new higher algebra with the ancient higher geometry. Analytic geometry could not, then, be far away, and Viète might have discovered this branch had he not avoided the geometrical study of indeterminate equations. The mathematical interests of Viète were unusually broad, hence he had read Diophantus' *Arithmetica*; but when a geometrical problem led Viète to a final equation in two unknown quantities, he dismissed it with the casual observation that the problem is indeterminate. One wishes that, with his general point of view, he had inquired into the geometrical properties of the indeterminacy.

6   In many respects the work of Viète is greatly undervalued, but in one case it is possible that he has been given undue credit for a method known long before in China. In one of his later works, the *De numerosa potestatum ... resolutione* (1600), he gave a method for the approximate solution of equations which is virtually that known today as Horner's method. To solve $x^2 + 7x = 60{,}750$, for example, Viète found as a first lower approximation for x the value $x_1 = 200$. Then upon substituting $x = 200 + x_2$ in the original equation (or, as we should say, reducing the roots by 200), he found $x_2^2 + 407x_2 = 19{,}350$. This equation now leads to a second approximation $x_2 = 40$. Now substituting $x_2 = 40 + x_3$, the equation $x_3^2 + 487x_3 = 1470$ results, and the positive root of this is $x_3 = 3$. Hence $x_2 = 43$ and $x = 243$. This illustrative equation taken from Viète (but written in modern notation) could of course have been solved by completing the square; but the author solved in the same manner other cases in which no simple alternative was at hand, finding, for example, a solution of $x^6 + 6000x = 191{,}246{,}976$. One of the beauties of the method is that it is applicable to any polynomial equation with real coefficients and a real root.

7   The trigonometry of Viète, like his algebra, was characterized by a heightened emphasis on generality and breadth of view. As Viète was the effective founder of a literal algebra, so he may with some justification be called the father of a generalized analytic approach to trigonometry that sometimes is known as goniometry. Here too, of course, Viète started from the work of his predecessors, notably, Regiomontanus and Rheticus. Like the former, he thought of trigonometry as an independent branch of mathematics; like the latter, he generally worked without direct reference to half chords in a circle. Viète in the *Canon mathematicus* (1579) prepared extensive tables of all six functions for angles to the nearest minute. We have seen that he had urged the use of decimal, rather than sexagesimal, fractions; but to avoid all fractions as much as possible, Viète chose a "sinus totus" or hypotenuse of 100,000 parts for the sine and cosine table and a "basis" or "perpendiculum" of 100,000 parts for the tangent, cotangent, secant, and cosecant tables. (Except for the sine function, he did not, however, use these names.)

In solving oblique triangles, Viète in the *Canon mathematicus* broke them down into right triangles, but in another work a few years later, *Variorum de rebus mathematicis* (1593), there is a statement equivalent to our law of tangents:

$$\frac{\dfrac{(a+b)}{2}}{\dfrac{(a-b)}{2}} = \frac{\tan\dfrac{A+B}{2}}{\tan\dfrac{A-B}{2}}$$

Though Viète may have been the first to use this formula, it was first published by a more obscure figure, Thomas Finck, in 1583 in *Geometria rotundi*.

Trigonometric identities of various sorts were appearing about this time in all parts of Europe, resulting in reduced emphasis on computation in the solution of triangles and more on analytic functional relationships. Among these were a group of formulas known as the prosthaphaeretic rules—that is, formulas that would convert a product of functions into a sum or difference (hence the name *prosthaphaeresis*, a Greek word meaning addition and subtraction). From the following type of diagram, for example, Viète derived the formula $\sin x + \sin y = 2\sin\dfrac{x+y}{2}\cos\dfrac{x-y}{2}$. Let $\sin x = AB$ (Fig. 16.1) and $\sin y = CD$. Then $\sin x + \sin y = AB + CD = AE = AC\cos\dfrac{x-y}{2} = 2\sin\dfrac{x+y}{2}\cos\dfrac{x-y}{2}$. On making the substitutions $(x+y)/2 = A$ and $(x-y)/2 = B$, we have the more useful form $\sin(A+B) + \sin(A-B) = 2\sin A\cos B$. In a similar manner one derives $\sin(A+B) - \sin(A-B) = 2\cos A\sin B$ by placing the angles x and y on the same side of the radius OD.

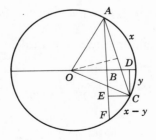

FIG. 16.1

The formulas $2 \cos A \cos B = \cos(A + B) + \cos(A - B)$ and $2 \sin A \sin B = \cos(A - B) - \cos(A + B)$ are somewhat similarly derived.

The rules above sometimes bear the name "formulas of Werner," for they seem to have been used by Werner to simplify astronomical calculations. At least one of these, that converting a product of cosines to a sum of cosines, had been known to the Arabs in the time of ibn-Yunus, but it was only in the sixteenth century, and more particularly near the end of the century, that the method of prosthaphaeresis came to be widely used. If, for example, one wished to multiply 98,436 by 79,253, one could let $\cos A = 49,218$ (that is, 98,436/2) and $\cos B = 79,253$. (In modern notation we would place a decimal point, temporarily, before each of the numbers and adjust the decimal point in the answer.) Then from the table of trigonometric functions one reads off angles A and B, and from the table one looks up $\cos(A + B)$ and $\cos(A - B)$, the sum of these being the product desired. Note that the product is found without any multiplication having been performed. In our example of prosthaphaeretic multiplication there is not a great saving of time and energy; but when we recall that at that time trigonometric tables of a dozen or fifteen significant figures were not uncommon, the laborsaving possibilities of prosthaphaeresis become more pronounced. The device was adopted at major astronomical observatories, including that of Tycho Brahe (1546–1601) in Denmark, from where word of it was carried to Napier in Scotland. Quotients are handled in the same manner by using a table of secants and cosecants.

Perhaps nowhere is Viète's generalization of trigonometry into goniometry more pronounced than in connection with his multiple-angle formulas. The double-angle formulas for the sine and cosine had of course been known to Ptolemy, and the triple-angle formulas are then easily derived from Ptolemy's formulas for the sine and cosine of the sum of two angles. By continuing to use the Ptolemy formulas recursively, a formula for $\sin nx$ or $\cos nx$ can be derived, but only with great effort. Viète used an ingenious manipulation of right triangles and the well-known identity

$$(a^2 + b^2)(c^2 + d^2) = (ad + bc)^2 + (bd - ac)^2 = (ad - bc)^2 + (bd + ac)^2$$

to arrive at formulas for multiple angles equivalent to what we should now write as

$$\cos nx = \cos^n x - \frac{n(n-1)}{1 \cdot 2} \cos^{n-2} x \sin^2 x$$

$$+ \frac{n(n-1)(n-2)(n-3)}{1 \cdot 2 \cdot 3 \cdot 4} \cos^{n-4} x \sin^4 x - \cdots$$

and

$$\sin nx = n \cos^{n-1} x \sin x - \frac{n(n-1)(n-2)}{1 \cdot 2 \cdot 3} \cos^{n-3} x \sin^3 x + \cdots$$

where the signs alternate and the coefficients are in magnitude the alternate numbers in the appropriate line of the arithmetic triangle. Here we see a striking link between trigonometry and the theory of numbers.

Viète noted also an important link between his formulas and the solution **8** of the cubic equation. Trigonometry could serve as a handmaid to algebra where the latter had run up against a stone wall—in the irreducible case of the cubic. This evidently occurred to Viète when he noticed that the angle trisection problem led to a cubic equation. If in the equation $x^3 + 3px + q = 0$ one substitutes $mx = y$ (to obtain a degree of freedom in the later selection of a value for m), the result is $y^3 + 3m^2 py + m^3 q = 0$. Comparing this with the formula $\cos^3 \theta - \frac{3}{4} \cos \theta - \frac{1}{4} \cos 3\theta = 0$, one notes that if $y = \cos \theta$, and if $3m^2 p = -\frac{3}{4}$, then $-\frac{1}{4} \cos 3\theta = m^3 q$. Since p is given, m is now known (and will be real whenever the three roots are real). Hence 3θ is readily determined, since q is known; hence $\cos \theta$ is known. Therefore y, and from it x, will be known. Moreover, by considering all possible angles satisfying the conditions, all three real roots will be found. This trigonometric solution of the irreducible case of the cubic, suggested by Viète, was carried out in detail later by Girard in 1629 in *Invention nouvelle en l'algèbre*.

Viète in 1593 found an unusual opportunity to use his multiple-angle formulas. A Belgian mathematician, Adriaen van Roomen (1561–1615) or Romanus, had issued a public challenge to anyone to solve an equation of forty-fifth degree:

$$x^{45} - 45x^{43} + 945x^{41} - \cdots - 3795x^3 + 45x = K$$

The ambassador from the Low Countries to the court of Henry IV boasted that France had no mathematician capable of solving the problem proposed by his countryman. Viète, called upon to defend the honor of his countrymen, noted that the proposed equation was one that arises in expressing $k = \sin 45\theta$ in terms of $x = \sin \theta$, and he promptly found the positive roots. The achievement so impressed van Roomen that he paid Viète a special visit.

In applying trigonometry to arithmetic and algebraic problems, Viète was broadening the scope of the subject.[4] Moreover, his multiple-angle formulas

---

[4] There is no generally accessible edition of the works of Viète, nor even a good general account in English of his life and work. His *Opera mathematica*, ed. by Fr. van Schooten (Leiden, 1646) is rare, as are most of Viète's published books. Useful is Frédéric Ritter, *François Viète* (1895). A recent communication from Professor D. J. Struik informs me that there exists an English translation (typeset) of Viète's *Isagoge* by J. Winfree Smith, St. John's College, Annapolis, Md., 1955.

should have disclosed the periodicity of the goniometric functions, but it probably was his hesitancy with respect to negative numbers that prevented him—or his contemporaries—from going as far as this. There was considerable enthusiasm for trigonometry in the late sixteenth and early seventeenth centuries, but this took the form primarily of synthesis and textbooks. It was during this period that the name "trigonometry" came to be attached to the subject. It was used as the title of an exposition by Bartholomaeus Pitiscus (1561–1613), which was first published in 1595 as a supplement to a book on spherics and again independently in 1600, 1606, and 1612.[5] Coincidentally the development of logarithms, ever since a close ally of trigonometry, was also taking place during these years.

9   John Napier (or Neper), like Viète, was not a professional mathematician. He was a Scottish laird, the Baron of Murchiston, who managed his large estates and wrote on varied topics. In a commentary on the Book of Revelations, for example, he argued that the pope at Rome was the anti-Christ. He was interested in certain aspects of mathematics only, chiefly those relating to computation and trigonometry. "Napier's rods" or "bones" were sticks on which items of the multiplication tables were carved in a form ready to be applied to lattice multiplication; "Napier's analogies" and "Napier's rule of circular parts" were devices to aid the memory in connection with spherical trigonometry.

Napier tells us that he had been working on his invention of logarithms for twenty years before he published his results, a statement that would place the origin of his ideas about 1594. He evidently had been thinking of the sequences, which had been published now and then, of successive powers of a given number—as in Stifel's *Arithmetica integra* fifty years before and as in the works of Archimedes. In such sequences it was obvious that sums and differences of indices of the powers corresponded to products and quotients of the powers themselves; but a sequence of integral powers of a base, such as two, could not be used for computational purposes because the large gaps between successive terms made interpolation too inaccurate. While Napier was pondering the matter, Dr. John Craig, physician to James VI of Scotland, called on him and told him of the use in Denmark of prosthaphaeresis. Craig presumably had been in the party when James VI of Scotland in 1590 had sailed with a delegation for Denmark to meet his bride-to-be, Anne of Denmark. The party had been forced by storms to land on the shore not far away from the observatory of Tycho Brahe, where, while awaiting more favorable weather, they were entertained by the astronomer. Reference apparently was made to the marvelous device of prosthaphaeresis, freely

[5] For a full account of this work see Sister Mary Claudia Zeller, *The Development of Trigonometry from Regiomontanus to Pitiscus* (1944).

used in the computations at the observatory; and word of this encouraged Napier to redouble his efforts and ultimately to publish in 1614 the *Mirifici logarithmorum canonis descriptio* ("A Description of the Marvelous Rule of Logarithms").

The key to Napier's work can be explained very simply. To keep the terms **10** in a geometrical progression of integral powers of a given number close together, it is necessary to take as the given number something quite close to one. Napier therefore chose to use $1 - 10^{-7}$ (or .9999999) as his given number. Now the terms in the progression of increasing powers are indeed close together—too close, in fact. To achieve a balance and to avoid decimals Napier multiplied each power by 10^7. That is, if $N = 10^7(1 - 1/10^7)^L$, then L is Napier's "logarithm" of the number N. Thus his logarithm of 10^7 is 0, his logarithm of $10^7(1 - 1/10^7) = 9999999$ is 1, and so on. If his numbers and his logarithms were to be divided by 10^7, one would have virtually a system of logarithms to the base $1/e$, for $(1 - 1/10^7)^{10^7}$ is close to $\lim_{n \to \infty} (1 - 1/n)^n = 1/e$. It must be remembered, however, that Napier had no concept of a base for a system of logarithms, for his definition was different from ours. The principles of his work were explained in geometrical terms as follows. Let a line segment AB and a half line or ray CDE ... be given (Fig. 16.2). Let a point P start from A and move along AB with variable speed decreasing in proportion to its distance from B; during the same time let a point Q start from C and move along CDE... with uniform speed equal to the rate with which point P began its motion. Napier called this variable distance CQ the logarithm of the distance PB.

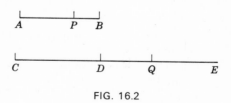

FIG. 16.2

Napier's geometrical definition is of course in agreement with the numerical description given above. To show this, let $PB = x$ and $CQ = y$. If AB is taken as 10^7, and if the initial speed of P is also taken as 10^7, then in modern calculus notations we have $dx/dt = -x$ and $dy/dt = 10^7$, $x_0 = 10^7$, $y_0 = 0$. Then $dy/dx = -10^7/x$, or $y = -10^7 \ln cx$, where c is found from the initial boundary conditions to be 10^{-7}. Hence $y = -10^7 \ln (x/10^7)$ or $y/10^7 = \log_{1/e}(x/10^7)$. That is, if the distances PB and CQ were divided by 10^7, the definition of Napier would lead precisely to a system of logarithms to the base $1/e$, as mentioned earlier. Needless to say, Napier built up his tables numerically rather than geometrically, as the word "logarithm," which he coined, implies. At first he called his power indices "artificial numbers," but later he made up the compound of the two Greek words *Logos* (or ratio) and *arithmos* (or number).

Napier did not think of a base for his system, but his tables nevertheless were compiled through repeated multiplications, equivalent to powers of .9999999. Obviously the power (or number) decreases as the index (or logarithm) increases. This is to be expected, because he was essentially using a base $1/e$ which is less than one. A more striking difference between his logarithms and ours lies in the fact that his logarithm of a product (or quotient) generally was not equal to the sum (or difference) of the logarithms. If $L_1 = \text{Log } N_1$ and $L_2 = \text{Log } N_2$, then $N_1 = 10^7(1 - 10^{-7})^{L_1}$ and $N_2 = 10^7(1 - 10^{-7})^{L_2}$, whence $N_1 N_2/10^7 = 10^7(1 - 10^{-7})^{L_1 + L_2}$, so that the sum of Napier's logarithms will be the logarithm not of $N_1 N_2$ but of $N_1 N_2/10^7$. Similar modifications hold, of course, for logarithms of quotients, powers, and roots. If $L = \text{Log } N$, for instance, then $nL = \text{Log } N^n/10^{7(n-1)}$. These differences are not too significant, for they merely involve shifting a decimal point. That Napier was thoroughly familiar with rules for products and powers is seen in his remark that all numbers (he called them "sines") in the ratio of 2 to 1 have differences of 6,931,469.22 in logarithms, and all those in the proportion of 10 to 1 have differences of 23,025,842.34 in logarithms. In these differences we see, if we shift the decimal point, the natural logarithms of the numbers two and ten. Hence it is not unreasonable to use the name "Napierian" for natural logarithms, even though these logarithms are not strictly the ones that Napier had in mind.

The concept of the logarithmic function is implied in Napier's definition and in all of his work with logarithms, but this relationship was not uppermost in his mind. He had laboriously built up his system for one purpose— the simplification of computations, especially of products and quotients. Moreover, that he had trigonometric computations in view is made clear by the fact that what we for simplification of exposition referred to as Napier's logarithm of a number, he actually called the logarithm of a sine. In Fig. 16.2, the line CQ was called the logarithm of the sine PB. This makes no real difference either in theory or in practice.

11   The publication in 1614 of the system of logarithms was greeted with prompt recognition, and among the most enthusiastic admirers was Henry Briggs, the first Savilian professor of geometry at Oxford. In 1615 he visited Napier at his home in Scotland, and there they discussed possible modifications in the method of logarithms. Briggs proposed that powers of ten should be used, and Napier said he had thought of this and was in agreement.

Napier at one time had proposed a table using $\log 1 = 0$ and $\log 10 = 10^{10}$ (to avoid fractions). The two men finally concluded that the logarithm of one should be zero and that the logarithm of ten should be one. Napier, however, no longer had the energy to put their ideas into practice. He died in 1617, the year in which his *Rhabdologia*, with its description of his rods, appeared. The second of his classic treatises on logarithms, the *Mirifici logarithmorum canonis constructio*, in which he gave a full account of the methods he used in building up his tables, appeared posthumously in 1619.[6] To Briggs, therefore, fell the task of making up the first table of common, or Briggsian, logarithms. Instead of taking *powers* of a number close to one, as had Napier, Briggs began with $\log 10 = 1$ and then found other logarithms by taking successive *roots*. By finding $\sqrt{10} = 3.162277$, for example, Briggs had $\log 3.162277 = .5000000$, and from $10^{\frac{3}{4}} = \sqrt{31.62277} = 5.623413$, he had $\log 5.623413 = .7500000$. Continuing in this manner, he computed other common logarithms. In the year of Napier's death, 1617, Briggs published his *Logarithmorum chilias prima*—that is, the logarithms of numbers from 1 to 1000, each carried out to fourteen places. In 1624, in *Arithmetica logarithmica*, Briggs extended the table to include common logarithms of numbers from 1 to 20,000 and from 90,000 to 100,000, again to fourteen places. Work with logarithms now could be carried out just as it is today, for all the usual laws of logarithms applied in connection with Briggs' tables. Incidentally, it is from Briggs' book of 1624 that our words "mantissa" and "characteristic" are derived. While Briggs was working out tables of common logarithms, a contemporary, John Speidell, drew up natural logarithms of trigonometric functions, publishing these in his *New Logarithmes* of 1619. A few natural logarithms had, in fact, appeared earlier in 1616 in an English translation by Edward Wright (1559–1615) of Napier's first work on logarithms. Seldom has a new discovery "caught on" so rapidly as did the invention of logarithms, and the result was the prompt appearance of tables of logarithms which were more than adequate for that time.

It has been implied, up to this point, that the invention of logarithms was **12** the work of one man alone, but such an impression must not be permitted to remain. Napier was indeed the first one to publish a work on logarithms,

---

[6] There are many good accounts of Napier's work. Among the best is the article on "Logarithms" by J. W. L. Glaisher in the *Encyclopaedia Britannica*, 11th ed., Vol. 16, pp. 868–877. Also excellent is the article by Florian Cajori, "History of the Exponential and Logarithmic Concepts," *American Mathematical Monthly*, **20** (1913), 5–14, 35–47, 75–84, 107–117, 148–151, 173–182, 205–210, as well as the article by Glaisher, "On Early Tables of Logarithms and Early History of Logarithms," *Quarterly Journal of Pure and Applied Mathematics*, **48** (1920), 151–192. See also E. W. Hobson, *John Napier and the Invention of Logarithms* (1914), and the *Napier Tercentary Memorial Volume*, ed. by C. G. Knott (1915). The latter, however, gives Napier more credit than is his due.

but very similar ideas were developed independently in Switzerland by Jobst Bürgi at about the same time. In fact, it is possible that the idea of logarithms had occurred to Bürgi[7] as early as 1588, which would be half a dozen years before Napier began work in the same direction. However, Bürgi printed his results only in 1620, half a dozen years after Napier had published his *Descriptio*. Bürgi's work appeared at Prague in a book entitled *Arithmetische und geometrische Progress-Tabulen*, and this indicates that the influences leading to his work were similar to those operating in the case of Napier. Both men proceeded from the properties of arithmetic and geometric sequences, spurred, probably, by the method of prosthaphaeresis. The differences between the work of the two men lie chiefly in their terminology and in the numerical values they used; the fundamental principles were the same. Instead of proceeding from a number a little *less* than one (as had Napier, who used $1 - 10^{-7}$), Bürgi chose a number a little *greater* than one— the number $1 + 10^{-4}$; and instead of multiplying powers of this number by 10^7, Bürgi multiplied by 10^8. There was one other minor difference: Bürgi multiplied all of his power indices by ten in his tabulation. That is, if $N = 10^8(1 + 10^{-4})^L$, Bürgi called $10L$ the "red" number corresponding to the "black" number N. If in this scheme we were to divide all the black numbers by 10^8 and all red numbers by 10^5, we should have virtually a system of natural logarithms. For instance, Bürgi gave for the black number 1,000,000,000 the red number 230,270.022, which, on shifting decimal points, is equivalent to saying that $\ln 10 = 2.30270022$. This is not a bad approxima- tion to the modern value, especially when we recall that $(1 + 10^{-4})^{10^4}$ is not quite the same as $\lim_{n \to \infty} (1 + 1/n)^n$, although the values agree to four significant figures.

In publishing his tables, Bürgi placed his red numbers on the side of the page and his black numbers in the body of the table, hence he had what we should describe as an antilogarithmic table; but this is a minor matter. The essence of the principle of logarithms is there, and Bürgi must be regarded as an independent discoverer who lost credit for the invention because of Napier's priority in publication. In one respect his logarithms come closer to ours than do Napier's, for as Bürgi's black numbers increase, so do the red numbers; but the two systems share the disadvantage that the logarithm of a product or quotient is not the sum or difference of the logarithms.

13 The invention of logarithms ultimately had a tremendous impact on the structure of mathematics, but at that time it could not be compared in theoretical significance with the work, say, of Viète. Logarithms were hailed gladly by Kepler not as a contribution to thought, but because they vastly

[7] See J. E. Hofmann: *Geschichte der Mathematik* (1963), p. 167.

increased the computational power of the astronomer. Viète was not exactly a "voice crying in the wilderness"; it is nevertheless true that most of his contemporaries were primarily concerned with the practical aspects of mathematics. Bürgi was a clockmaker, Galileo was a physicist and astronomer, and Stevin was an engineer. It was inevitable that these men should have preferred parts of mathematics that gave promise of applicability to their fields. Bürgi and Stevin, for example, aided in the development of decimal fractions, and Bürgi and Galileo were rivals in the manufacture and sale of a practical computing device known as the proportional compass. The so-called Renaissance in science, illustrated by the work of such men as Leonardo da Vinci and Copernicus, had been a ferment that to a large extent grew out of contact between old ideas and new and between the views of artisans and those of scholars.

In mathematics of the sixteenth century there were diverse and conflicting tendencies; but we can perceive there, as well as in science, the results of a confrontation of established ideas by new concepts and of theoretical views by the exigencies of practical problems. We have seen that the work of Viète grew out of two factors in particular: (1) the recovery of ancient Greek classics and (2) the relatively new developments in medieval and early modern algebra. Throughout the sixteenth century both professional and amateur theoretical mathematicians showed concern for the practical techniques of computation, which contrasted strongly with the dichotomy emphasized two millennia earlier by Plato. Viète, the outstanding mathematician in France, in 1579 had urged the replacement of sexagesimal fractions by decimal fractions. In 1585 an even stronger plea for the use of the ten-scale for fractions, as well as for integers, was made by the leading mathematician in the Low Countries, Simon Stevin of Bruges.

Stevin, a supporter of the Protestant faction under William of Orange in the struggle against Catholic Spain, was tolerant, if not indifferent, in religion. Under Prince Maurice of Nassau he served as quartermaster and as commissioner of public works, and for a time he tutored the prince in mathematics. Although Stevin was a great admirer of the theoretical treatises of Archimedes, there runs through the works of the Flemish engineer a strain of practicality that is more characteristic of the Renaissance period than of classical antiquity. Thus Stevin was largely responsible for the introduction into the Low Countries of double-entry bookkeeping fashioned after that of Pacioli in Italy almost a century earlier.[8] Of far more widespread influence in economic practice, in engineering, and in mathematical notations was Stevin's little book with the Flemish title *De thiende* ("The Tenth"), published at Leyden in 1585. A French version entitled *La disme* appeared in the same

[8] For an account of Stevin's life and work see *The Principal Works of Simon Stevin* (edited by E. J. Dijksterhuis, D. J. Struik, and others), Amsterdam, 1955–1958.

year, and the popularity of the book was such that its place in the development of mathematics has been often misunderstood.

It is clear that Stevin was in no sense the *inventor* of decimal fractions, nor was he the first systematic user of them. More than incidental use of decimal fractions is found in ancient China, in medieval Arabia, and in Renaissance Europe; by the time of Viète's forthright advocacy of decimal fractions in 1579 they were generally accepted by mathematicians on the frontiers of research. Among the common people, however, and even among mathematical practitioners, decimal fractions became widely known only when Stevin undertook to explain the system in full and elementary detail. He wished to teach everyone "how to perform with an ease, unheard of, all computations necessary between men by integers without fractions." That is, oddly enough Stevin was concentrating on his tenths, hundredths, thousandths, and so on, as integral numerators, much as we do in the common measure of time in minutes and seconds. How many of us think of 3 minutes and 4 seconds, say, as a fraction? We are far more likely to think of 3 minutes as an integer than as 3/60 of an hour; and this was precisely Stevin's view. For this reason he did not write his decimal expressions with denominators, as Viète had; instead, in a circle above or after each digit he wrote the power of ten assumed as a divisor. Thus the value of π, approximately, appeared as

$$3⓪\ 1①\ 4②\ 1③\ 6④ \qquad \text{or} \qquad \begin{array}{ccccc} ⓪ & ① & ② & ③ & ④ \\ 3 & 1 & 4 & 1 & 6 \end{array}$$

Instead of the words "tenth," "hundredth," and so on, he used "prime," "second," and so on, somewhat as we still designate the places in sexagesimal fractions.[9]

Stevin obviously had the right idea about decimal fractions, but his Bombelli-inspired notation for places was more appropriate for algebra than for arithmetic. Fortunately, the modern notation was not long delayed. In the 1616 English translation of Napier's *Descriptio* decimal fractions appear as today, with a decimal point separating the integral and fractional portions. In 1617 in the *Rhabdologia*, in which he described computation using his rods, Napier referred to Stevin's decimal arithmetic and proposed a point or a comma as the decimal separatrix. In the Napierian *Constructio* of 1619 the decimal point became standard in England, but many European countries continue to this day to use the decimal comma. Stevin urged also

[9] See D. J. Struik, "Simon Stevin and the Decimal Fractions," *The Mathematics Teacher*, **52** (1959), 474–478; also George Sarton, "Simon Stevin of Bruges (1548–1620)," *Isis*, **21** (1934), 241–303, and "The First Explanation of Decimal Fractions and Measures (1585)," *Isis*, **23** (1935), 153–244. See also D. E. Smith, "The Invention of the Decimal Fraction," *Teachers College Bulletin*, **5** (1910), 11–21.

THIENDE. 13

HET ANDER DEEL
DER THIENDE VANDE
WERCKINCHE.

I. VOORSTEL VANDE
VERGADERINGHE.

Wefende ghegeven Thiendetalen te ver-
gaderen: hare Somme te vinden.

T'GHEGHEVEN. Het fijn drie oirdens van
Thiendetalen, welcker eerfte 27 ⓪ 8 ① 4 ②
7 ③ , de tweede, 37. ⓪ 6 ① 7 ② 5 ③ , de derde,
875 ⓪ 7 ① 8 ② 2 ③ , T'BEGHEERDE. Wy
moeten haer Somme vinden . WERCKING.
Men fal de ghegheven ghe-
talen in oirden ftellen als
hier neven, die vergaderen-
de naer de ghemeene manie
re der vergaderinghe van
heelegetalen aldus:

⓪	①	②	③
2 7	8	4	7
3 7	6	7	5
8 7 5	7	8	2
9 4 1	3	0	4

Comt in Somme (door het 1 . probleme onfer
Franfcher Arith.) 9 4 1 3 0 4 dat fijn (t'welck de
teeckenen boven de ghetalen ftaende, anwijfen)
9 4 1 ⓪ 3 ① 0 ② 4 ③. Ick fegghe de felve te wefen
de ware begheerde Somme. BEWYS. De ghegeven 27 ⓪ 8 ① 4 ② 7 ③, doen (door de 3ᵉ. bepa-
ling) $27\frac{8}{10}, \frac{4}{100}, \frac{7}{1000}$, maecké t'famen $27\frac{847}{1000}$.
Ende door de felve reden fullen de 37 ⓪ 6 ① 7 ②
5 ③ weerdich fijn $37\frac{675}{1000}$; Ende de 875 ⓪ 7 ①
8 ③

A page from Stevin's work (1634 edition) showing Stevin's notations for decimal fractions.

a decimal system of weights and measures, but this part of his work has not
yet triumphed in England and America.

In the history of science, as well as in mathematics, Stevin is an important
figure. He and a friend dropped two spheres of lead, one ten times the weight
of the other, from a height of 30 feet onto a board and found the sounds of
their striking the board to be almost simultaneous. But Stevin's published
report (in Flemish in 1586) of the experiment has received far less notice than
the similar and later experiment attributed, on very doubtful evidence,

to Galileo. On the other hand, Stevin usually receives credit for the discovery of the law of the inclined plane, justified by his familiar "wreath of spheres" diagram, whereas this law had been given earlier by Jordanus Nemorarius.[10]

14   Stevin was a practical-minded mathematician who saw little point in the more speculative aspects of the subject. Of imaginary numbers he wrote: "There are enough legitimate things to work on without need to get busy on uncertain matter." Nevertheless, he was not narrow-minded, and his reading of Diophantus impressed him with the importance of appropriate notations as an aid to thought. Although he followed the custom of Viète and other contemporaries in writing out some words, such as that for equality, he preferred a purely symbolic notation for powers. Carrying over to algebra his positional notation for decimal fractions, he wrote ② instead of Q (or square), ③ for C (or cube), ④ for QQ (or square-square), and so on. This notation may well have been suggested by Bombelli's *Algebra*. It also paralleled a notation of Bürgi who indicated powers of an unknown by placing Roman numerals over the coefficients. Thus $x^4 + 3x^2 - 7x$, for example, would be written by Bürgi as

$$
\begin{array}{ccc}
\text{iv} & \text{ii} & \text{i} \\
1 \; + \; 3 & - & 7
\end{array}
$$

and by Stevin as

$$
\begin{array}{ccc}
④ & ② & ① \\
1 \; + \; 3 & - & 7
\end{array}
$$

Stevin went further than Bombelli or Bürgi in proposing that such notations be extended to fractional powers. (It is interesting to note that although Oresme had used both fractional-power indices and coordinate methods in geometry, these seem to have had only a very indirect influence, if any, on the progress of mathematics in the Low Countries and in France in the early seventeenth century.) Even though Stevin had no occasion to use the fractional index notation, he clearly stated that $\frac{1}{2}$ in a circle would mean square root and that $\frac{3}{2}$ in a circle would indicate the square root of the cube. A little later Girard, editor of Stevin's works, adopted the circled-numerical notation for powers, and he, too, indicated that this could be used for roots instead of such symbols as $\sqrt{}$ and $\sqrt[3]{}$. Symbolic algebra was developing apace, and it reached its maturity, only eight years after Girard's *Invention nouvelle*, in Descartes' *La géométrie*.

[10] See Marshall Clagett, *The Science of Mechanics in the Middle Ages* (Madison, Wis.: University of Wisconsin Press, 1959).

Simon Stevin was a typical mathematician of his day in that he enjoyed **15** the elementary applications of the subject; in this respect he was like Galileo. Galileo had originally intended to take a degree in medicine, but a taste for Euclid and Archimedes led him instead to become a professor of mathematics, first at Pisa and later at Padua. This does not mean, however, that he taught on the level of the authors he admired. Little mathematics was included in university curricula of the time, and a large proportion of what was taught in Galileo's courses would now be classified as physics or astronomy or engineering applications. Moreover, Galileo was not a "mathematician's mathematician," as was Viète; he came close to being what we should call a mathematical practitioner. This we see in his interest in computational techniques that led him in 1597 to construct and market a device that he called his "geometric and military compasses."

In a pamphlet of 1606 with the title *Le operazioni del compasso geometrico et militare*, he described in detail the way in which the instrument could be used to perform a variety of computations quickly without pen or paper or an abacus. The theory behind this was extremely elementary, and the degree of accuracy was very limited, but the financial success of Galileo's device shows that military engineers and other practitioners found a need for such an aid in calculation. Bürgi had constructed a similar device, but Galileo had a better entrepreneurial sense, one that gave him an advantage. The Galilean compasses consisted of two arms pivoted as in the ordinary compasses of today, but each of the arms was engraved with graduated scales of varying types. Fig. 16.3 shows only the simple equispaced markings up to 250, and only the simplest of the many possible computations, the first one explained by Galileo, is described here. If, for instance, one wishes to divide a given line segment into five equal parts, one opens a pair of ordinary compasses (or divider) to the length of the line segment. Then one opens the geometric compasses so that the distance between the points of the divider just spans the distance between two markings, one on each arm of the

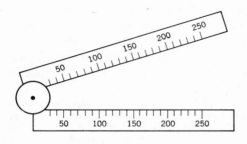

FIG. 16.3

geometric compasses, which are simple integral multiples of five—say, the number 200 on each scale. Then, if one holds the opening of the geometric compasses fixed and places the ends of the divider on the mark for 40 on each scale, the distance between the divider points will be the desired fifth of the length of the original line segment. The instructions Galileo provided with his compasses included many other operations, from changing the scale of a drawing to computing amounts of money under compound interest.[11]

16   Elementary though it was, Galileo's 1606 pamphlet on the geometric compasses, published when he was over forty years old, was his only strictly mathematical treatise. Nevertheless, it was far from his only contribution to the field. More significant are the many appeals in his astronomical and physical works to mathematical reasoning, and here he was frequently close to developments leading to the calculus. Much the same can be said also of Stevin and Kepler. Physics and astronomy had reached the point where there was increasing need for arguments concerning the infinitely large and small—the subject now known as analysis. Viète had been one of the first to use the word "analysis" as a synonym for algebra, but he was one of the earliest analysts also in the more modern sense of one who studies infinite processes.

Before the time of Viète there had been many good and bad approximations for the ratio of circumference to diameter in a circle,[12] such as that of V. Otho and A. Anthonisz who, evidently independently, rediscovered (about 1573) the approximation $\pi \approx 355/113$ by subtracting numerators and denominators of the Ptolemaic and Archimedean values, 377/120 and 22/7 respectively. Viète worked out π correctly to ten significant figures, apparently unaware of al-Kashi's still better approximation. The most impressive achievement of this type was by Ludolph van Ceulen (1540–1610). First he published in 1596 a twenty-place value obtained by starting with a polygon of fifteen sides and doubling the number of sides thirty-seven times. Using a still larger number of sides, he ultimately achieved a thirty-five place approximation, which his widow had engraved on his tombstone. This feat of computation so impressed his successors that π frequently has been known as the "Ludolphine constant." Such tours de force, however, have no theoretical significance. An exact expression was far more to be desired; and it is in this respect that Viète gave the first theoretically precise numerical

[11] A brief excerpt, in English translation, from *Le operazioni del compasso geometrico et militare* is included in D. E. Smith, *Source Book in Mathematics* (1929), pp. 186–191. The Italian original appears in Galileo's *Opere* (Florence, 1890–1909, 20 vols.), II, 335–424.

[12] An extensive list of very many of these values is given by H. C. Schepler, "The Chronology of Pi," *Mathematics Magazine*, **23** (1949–1950), 162–170, 216–228, 279–283.

expression for π—an infinite product that can be written as

$$2/\pi = \sqrt{\tfrac{1}{2}} \sqrt{\tfrac{1}{2} + \sqrt{\tfrac{1}{2}}} \sqrt{\tfrac{1}{2} + \sqrt{\tfrac{1}{2} + \sqrt{\tfrac{1}{2}}}} \cdots$$

In a sense Viète's approach is not novel. His product is easily derived by inscribing a square within a circle, then applying the recursive trigonometric formula $a_{2n} = a_n \sec \pi/n$, where a_n is the area of the inscribed regular polygon of n sides, and finally allowing n to increase indefinitely. Moreover, the same infinite product is readily derived by calculating radius vectors of points on the quadratrix of Hippias, $r \sin \theta = 2\theta$, for successive bisections of the angle, beginning with $\theta = \pi/2$ and noting that $r_n/r_{n-1} = \cos \pi/2^n$ and that $\lim\limits_{n \to \infty} r_n = 2/\pi$. Nevertheless, it was Viète who first expressed π analytically, a significant result because arithmetic, algebraic, and trigonometric notations were more and more invading the realm of the infinitely large and the infinitely small, a field once almost exclusively dominated by geometry.

Viète's last years were embittered by a controversy largely of his own making. Christopher Clavius (1537–1612), a well-known contemporary mathematician, had advised Pope Gregory XIII on the reform of the calendar, and Viète attacked the accuracy of this. The bitterness of Viète's statement may have resulted from resentment that his opponents failed to evaluate correctly the significance of the new "logistica speciosa." Viète had a few ardent disciples, one of whom, Alexander Anderson (1582–ca. 1620) of Scotland, published some of his work in 1615, but it was not until the 1630s that the "Analytic Art" began to receive the attention it deserved. This delay is in sharp contrast to the rapidity with which logarithms became widely known.

Viète was primarily an analyst, but he contributed also to pure geometry. **17** Here his work centered chiefly on problems raised in the works of Apollonius. Regiomontanus had doubted that the celebrated Apollonian problem (proposed in the lost book *On Tangencies*) of constructing a circle tangent to three circles could be solved with ruler and straightedge; van Rooman therefore solved it by means of two intersecting hyperbolas. Viète knew through a reference in Pappus' *Collection* that an elementary construction was indeed possible, and in his *Varia responsa* of 1600 he published his solution. In a reconstruction of what he thought Apollonius' book may have contained, Viète proceeded through the simpler cases, in which one or more of the three circles are replaced by points or lines, until he had reached the tenth and most difficult case—that of three circles. This construction was one of Viète's most beautiful contributions to mathematics. Such problems in geometry later had a significant attraction for Descartes, but Viète's immediate successors were far less attracted to the theoretical results of Apollonius than to the applicability of Archimedes' work.

18 Stevin, Kepler, and Galileo all had need for Archimedean methods, being practical men, but they wished to avoid the logical niceties of the method of exhaustion. It was largely the resulting modifications of the ancient infinitesimal methods that ultimately led to the calculus, and Stevin was one of the first to suggest changes. In his *Statics* of 1586, almost exactly a century before Newton and Leibniz published their calculus, the engineer of Bruges demonstrated as follows that the center of gravity of a triangle lies on its median. In the triangle *ABC* inscribe a number of parallograms of equal height whose sides are pairwise parallel to one side and to the median drawn to this side (Fig. 16.4). The center of gravity of the inscribed figures will lie

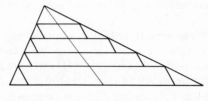

FIG. 16.4

on the median, by the Archimedean principle that bilaterally symmetrical figures are in equilibrium. However, we may inscribe in the triangle an infinite number of such parallelograms, and the greater the number of parallelograms, the smaller will be the difference between the inscribed figure and the triangle. Inasmuch as the difference can be made as small as one pleases, the center of gravity of the triangle also lies on the median. In some of the propositions on fluid pressure Stevin supplemented this geometrical approach by a "demonstration by numbers" in which a sequence of numbers tended to a limiting value; but the "Dutch Archimedes" had more confidence in a geometrical proof than an arithmetic one.[13]

19 Whereas Stevin was interested in physical applications of infinitely many infinitely small elements, Kepler had need for astronomical applications, especially in connection with his elliptical orbits of 1609. As early as 1604 Kepler had become involved with conic sections through work in optics and the properties of parabolic mirrors. Whereas Apollonius had been inclined to think of the conics as three distinct types of curves—ellipses,

[13] For further details see C. B. Boyer, *The Concepts of the Calculus* (1939), pp. 99–104.

Johann Kepler.

parabolas, and hyperbolas—Kepler preferred to think of five species of conics, all belonging to a single family or genus. With a strong imagination and a Pythagorean feeling for mathematical harmony, Kepler developed for conics in 1604 (in his *Ad Vitellionem paralipomena*, that is, "Introduction to

Vitello's Optics") what we call the principle of continuity. From the conic section made up simply of two intersecting lines, in which the two foci coincide at the point of intersection, we pass gradually through infinitely many hyperbolas as one focus moves farther and farther from the other. When the one focus is infinitely far away, we no longer have the double-branched hyperbola, but the parabola. As the moving focus passes beyond infinity and approaches again from the other side, we pass through infinitely many ellipses until, when the foci again coincide, we reach the circle.

The idea that a parabola has two foci, one at infinity, is due to Kepler, as is also the word "focus" (Latin for "hearthside"); we find this bold and fruitful speculation on "points at infinity" extended a generation later in the geometry of Desargues. Meanwhile, Kepler found a useful approach to the problem of the infinitely small in astronomy. In his *Astronomia nova* of 1609 he announced his first two laws of astronomy: (1) the planets move about the sun in elliptical orbits with the sun at one focus, and (2) the radius vector joining a planet to the sun sweeps out equal areas in equal times. In handling problems of areas such as these, Kepler thought of the area as made up of infinitely small triangles with one vertex at the sun and the other two vertices at points infinitely close together along the orbit. In this way he was able to use a crude type of integral calculus resembling that of Oresme. The area of a circle, for example, is found in this way by noting that the altitudes of the infinitely thin triangles (Fig. 16.5) are equal to the radius. If we call the infinitely small bases, lying along the circumference, $b_1, b_2,$ $\ldots, b_n, \ldots$, then the area of the circle—that is, the sum of the areas of the triangles—will be $\frac{1}{2}b_1 r + \frac{1}{2}b_2 r + \cdots + \frac{1}{2}b_n r + \cdots$ or $\frac{1}{2}r(b_1 + b_2 + \cdots + b_n + \cdots)$. Inasmuch as the sum of the b's is the circumference C, the area A will be given by $A = \frac{1}{2}rC$, the well-known ancient theorem which Archimedes had proved more carefully.

By analogous reasoning Kepler knew the area of the ellipse—a result of Archimedes not then extant. The ellipse can be obtained from a circle of radius a through a transformation under which the ordinate of the circle at

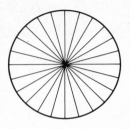

FIG. 16.5

each point is shortened according to a given ratio, say $b:a$. Then, following Oresme, one can think of the area of the ellipse and the area of the circle as made up of all the ordinates for points on the curves (Fig. 16.6); but inasmuch as the ratio of the components of the areas are in the ratio $b:a$, the areas themselves must have the same ratio. However, the area of the circle is known to be πa^2; hence the area of the ellipse $x^2/a^2 + y^2/b^2 = 1$ must be πab. This result is correct; but the best that Kepler could do for the circumference of the ellipse was to give the approximate formula $\pi(a + b)$. Lengths of curves in general, and of the ellipse in particular, were to elude mathematicians for another half a century.

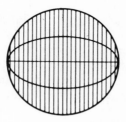

FIG. 16.6

Kepler had worked with Tycho Brahe, first in Denmark and later at Prague, where, following Brahe's death, Kepler became mathematician to the Emperor Rudolph II. One of his duties was the casting of horoscopes; mathematicians, whether for emperors or at universities, found various applications for their talents, as Kepler discovered while he was at Linz, in Austria. The year 1612 had been a very good one for wine, and Kepler began to meditate at this time on the crude methods then in use for estimating the volumes of wine casks. He compared these with the methods of Archimedes on the volumes of conoids and spheroids, and then he proceeded to find the volumes of various solids of revolution not previously considered by Archimedes. For example, he revolved a segment of a circle about its chord, calling the result a citron if the segment was less than a semicircle and an apple if the segment exceeded a semicircle. His volumetric method consisted in regarding the solids as composed of infinitely many infinitesimal elements, and he proceeded much as we have indicated above for areas. He dispensed with the Archimedean double *reductio ad absurdum*, and in this he was followed by most mathematicians from that time to the present.[14]

---

[4] See D. J. Struik, "Kepler as a Mathematician," in *Johann Kepler, 1571–1630. A Tercentenary Commemoration of His Life and Works*, ed. by F. E. Brasch (1931).

20   Kepler collected his volumetric thoughts in a book that appeared in 1615 under the title *Stereometria doliorum* ("Volume-measurement of Barrels"). For a score of years it seemed to have excited no great interest, but in 1635 the Keplerian ideas were systematically expanded in a celebrated book entitled *Geometria indivisibilibus*, written by Cavalieri, a disciple of Galileo. While Kepler had been studying wine barrels, Galileo had been scanning the heavens with his telescope and rolling balls down inclined planes. The results of Galileo's efforts were two famous treatises, one astronomical and the other physical. They were both written in Italian, but we shall refer to them in English as *The Two Chief Systems* (1632) and *The Two New Sciences* (1638). The first was a dialogue concerning the relative merits of the Ptolemaic and Copernican views of the universe, carried on by three men: Salviati (a scientifically informed scholar), Sagredo (an intelligent layman), and Simplicio (an obtuse Aristotelian). In the dialogue Galileo left little doubt about where his preferences lay, and the consequences were his trial and imprisonment. During the years of his detention he nevertheless prepared *The Two New Sciences*, a dialogue concerning dynamics and the strength of materials, carried out by the same three characters. Although neither of the two great Galilean treatises was in a strict sense mathematical, there are in both of them many points at which appeal is made to mathematics, frequently to the properties of the infinitely large and the infinitely small.

The infinitely small was of more immediate relevance to Galileo than the infinitely large, for he found it essential in his dynamics. Galileo gave the impression that dynamics was a totally new science created by him, and all too many writers since have agreed with this claim. It is virtually certain, however, that he was thoroughly familiar with the work of Oresme on the latitude of forms, and several times in the *Two New Sciences* Galileo had occasion to use a diagram of velocities similar to the triangle graph of Oresme. Nevertheless, Galileo organized the ideas of Oresme and gave them a mathematical precision that had been lacking. Among the new contributions to dynamics was Galileo's analysis of projectile motion into a uniform horizontal component and a uniformly accelerated vertical component. As a result he was able to show that the path of a projectile, disregarding air resistance, is a parabola. It is a striking fact that the conic sections had been studied for almost 2000 years before two of them almost simultaneously found applicability in science—the ellipse in astronomy and the parabola in physics. Galileo mistakenly thought he had found a further application of the parabola in the curve of suspension of a flexible rope or wire or chain (*catena*); but mathematicians later in the century proved that this curve, the catenary, not only is not a parabola, it is not even algebraic.

Galileo resembled Dürer in that they both were quick to notice new curves, but neither was mathematically equipped to analyze them. Galileo had

Galileo Galilei.

noted the curve now known as the cycloid, traced out by a point on the rim of a wheel as it rolls along a horizontal path, and he tried to find the area under one arch of it. The best he could do was to trace the curve on paper, cut out an arch, and weigh it, concluding that the area was a little less than three times the area of the generating circle. (French and Italian mathematicians later showed that the area of the arch is precisely three times the area of the circle.) Galileo abandoned study of the curve, suggesting only that the cycloid would make an attractive arch for a bridge; many years later his disciple Torricelli took up the study of the curve with great success.

A more important contribution to mathematics was made by Galileo in the *Two Chief Systems* of 1632 at a point on the "third day" when Salviati **21** adumbrated the idea of an infinitesimal of higher order. Simplicio had argued that an object on a rotating earth should be thrown off tangentially by the motion; but Salviati argued that the distance QR through which an object has to fall to remain on the earth, while the latter rotates through a small angle θ (Fig. 16.7), is infinitely small compared with the tangential distance PQ through which the object travels horizontally. Hence even a very small downward tendency, as compared with the forward impetus, will be sufficient to hold the object on the earth.[15] Galileo's argument here is equivalent to saying that $PS = \text{vers } \theta$ is an infinitesimal of higher order with respect to lines PQ or RS or arc PR.

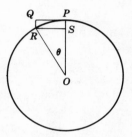

FIG. 16.7

A similar bit of reasoning arises also in Galileo's *Two New Sciences* of 1638, a very influential treatise on dynamics and the strength of materials. Here the author used the infinitely small sometimes to the point of whimsy, as when Salviati assures Simplicio that it is as easy to resolve a line segment into an infinite number of parts as it is to divide the line into finite parts. First he gets Simplicio to admit that one need not separate the parts, but merely to mark the points of division. If, for example, a line segment is bent into the form of a square or a regular octagon, one has resolved it into four or eight equal parts. Salviati then concluded that by bending the line segment into the shape of a circle, he has "reduced to actuality that infinite number of parts into which you claimed, while it was straight, were contained in it only potentially," for the circle is a polygon of an infinite number of sides. On another occasion, however, Galileo has Salviati assert that infinities and indivisibles "transcend our finite understanding, the former on account of their magnitude, the latter because of their smallness; Imagine what they are when combined."

From the infinite in geometry Salviati led Simplicio to the infinite in arithmetic, pointing out that a one-to-one correspondence can be set up

[15] There are two excellent English editions of *The Two Chief Systems*, one edited by Stillman Drake (1953), the other by Giorgio de Santillana (1953).

between all the integers and the perfect squares, despite the fact that the further one proceeds in the sequence of integers, the scarcer the perfect squares become. Through the simple expedient of counting the perfect squares, a one-to-one correspondence is established in which each integer inevitably is matched against a perfect square, and vice versa. Even though there are many whole numbers that are not perfect squares (and the proportion of this increases as we consider larger and larger numbers), "we must say that there are as many squares as there are numbers." Galileo here was face-to-face with the fundamental property of an infinite set—that a part of the set can be equal to the whole set—but Galileo did not draw this conclusion. While Salviati correctly concluded that the number of perfect squares is not less than the number of integers, he could not bring himself to make the statement that they are equal. Instead, he simply concluded that "the attributes 'equal,' 'greater,' and 'less' are not applicable to infinite, but only to finite quantities." He even asserted (incorrectly, we now know) that one cannot say that one infinite number is greater than another infinite number, or even that an infinite number is greater than a finite number. Galileo, like Moses, came within sight of the promised land, but he could not enter it.[16]

Galileo had intended to write a treatise on the infinite in mathematics, **22** but it has not been found. Meanwhile his disciple Cavalieri was spurred by Kepler's *Stereometria*, as well as by ancient and medieval views and by Galileo's encouragement, to organize his thoughts on infinitesimals in the form of a book. Cavalieri was a member of a religious order (a Jesuate, not a Jesuit as is frequently but incorrectly stated) who lived at Milan and Rome before becoming professor of mathematics at Bologna in 1629. Characteristically for that time, he wrote on many aspects of pure and applied mathematics—geometry, trigonometry, astronomy, and optics—and he was the first Italian writer to appreciate the value of logarithms. In his *Directorium universale uranometricum* of 1632 he published tables of sines, tangents, secants, and versed sines, together with their logarithms, to eight places; but the world remembers him instead for one of the most influential books of the early modern period, the *Geometria indivisibilibus continuorum*, published in 1635.

The argument on which the book is based is essentially that implied by Oresme, Kepler, and Galileo—that an area can be thought of as made up of lines or "indivisibles" and that a solid volume can be regarded similarly as composed of areas that are indivisible or quasi-atomic volumes. Although Cavalieri at the time could scarcely have realized it, he was following in very

---

[16] Galileo's *Dialogue Concerning Two New Sciences* (in English translation) is readily available in a paperback edition from Dover Publications, New York (no date).

respectable footsteps indeed, for this is precisely the type of reasoning that Archimedes had used in the *Method*, then lost. But Cavalieri, unlike Archimedes, felt no compunction about the logical deficiencies behind such procedures.

The general principle that in an equation involving infinitesimals those of higher order are to be discarded because they have no effect on the final result is frequently erroneously attributed to Cavalieri's *Geometria indivisibilibus*. The author undoubtedly was familiar with such an idea, for it is implied in some of the work of Galileo, and it appeared more specifically in results of contemporary French mathematicians; but Cavalieri assumed almost the opposite of this principle. There was in Cavalieri's method no process of continued approximation, nor any omission of terms, for he used a strict one-to-one pairing of the elements in two configurations. No elements are discarded, no matter what the dimension. The general approach and the specious plausibility of the method of indivisibles is well illustrated by the proposition still known in many solid geometry books as "the theorem of Cavalieri":

> If two solids have equal altitudes, and if sections made by planes parallel to the bases and at equal distances from them are always in a given ratio, then the volumes of the solids also are in this ratio.[17]

Cavalieri evidently had developed his method by 1626, for in that year he wrote to Galileo that he was going to publish a book on the subject. Galileo himself had once planned to write a book on the infinite, and perhaps Cavalieri delayed publishing his own work in deference to Galileo. However, Galileo's book undoubtedly would have been more philosophical and speculative, with emphasis on the nature of the infinitely large and small, a theme that Cavalieri avoided. Instead, Cavalieri concentrated on an extremely useful geometrical theorem equivalent to the modern statement in the calculus

$$\int_0^a x^n \, dx = \frac{a^{n+1}}{n+1}$$

The statement and the proof of the theorem are very different from those with which a modern reader is familiar, for Cavalieri compared powers of the lines in a parallelogram parallel to the base with the corresponding powers of lines in either of the two triangles into which a diagonal divides the parallelogram. Let the parallelogram $AFDC$ be divided into two triangles by the diagonal CF (Fig. 16.8) and let HE be an indivisible of triangle CDF which is parallel to the base CD. Then upon taking $BC = FE$ and drawing

---

[17] D. E. Smith, *Source Book in Mathematics*, pp. 605–609.

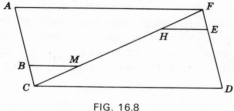

FIG. 16.8

BM parallel to *CD*, it is easy to show that the indivisible *BM* in triangle *ACF* will be equal to *HE*. Hence one can pair all of the indivisibles of triangle *CDF* with equal indivisibles in triangle *ACF*, and therefore the two triangles are equal. Inasmuch as the parallelogram is the sum of the indivisibles in the two triangles, it is clear that the sum of the first powers of the lines in one of the constituent triangles is half the sum of the first powers of the lines in the parallelogram; in other words,

$$\int_0^a x\,dx = \frac{a^2}{2}$$

Through a similar but considerably more involved argument Cavalieri showed that the sum of the squares of the lines in the triangle is one-third the sum of the squares of the lines in the parallelogram.[18] For the cubes of the lines he found the ratio to be 1/4. Later he carried the proof to higher powers, finally asserting, in *Exercitationes geometricae sex* (that is, "Six Geometrical Exercises") of 1647, the important generalization that for the *n*th powers the ratio will be $1/(n+1)$. This was known at the same time to French mathematicians, but Cavalieri was first to publish this theorem—one that was to open the way to many algorithms in the calculus. *Geometrica indivisibilibus*, which so greatly facilitated the problem of quadratures, appeared again in a second edition in 1553, but by that time mathematicians had achieved remarkable results in new directions that outmoded Cavalieri's laborious geometric approach.

The most significant theorem by far in Cavalieri's work was his equivalent **23** of

$$\int_0^a x^n\,dx = \frac{a^{n+1}}{n+1}$$

but another contribution was also to lead to important results. The spiral

[18] For further details see C. B. Boyer, "Cavalieri, Limits and Discarded Infinitesimals," *Scripta Mathematica*, **8** (1941), 79–91.

$r = a\theta$ and the parabola $x^2 = ay$ had been known since antiquity without anyone's having previously noted a relationship between them, until Cavalieri thought of comparing straight-line indivisibles with curvilinear indivisibles. If, for example, one were to twist the parabola $x^2 = ay$ (Fig. 16.9) around like a watch spring so that vertex O remains fixed while point P becomes point P',

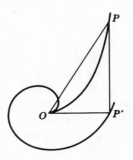

FIG. 16.9

then the ordinates of the parabola can be thought of as transformed into radius vectors through the relationships $x = r$ and $y = r\theta$ between what we now call rectangular and polar coordinates. The points on the Apollonian parabola $x^2 = ay$ then will lie on the Archimedian spiral $r = a\theta$. Cavalieri noted further that if PP' is taken equal to the circumference of the circle of radius OP', the area within the first turn of the spiral is exactly equal to the area between the parabolic arc OP and the radius vector OP. Here we see work that amounts to analytic geometry and the calculus; yet Cavalieri was writing before either of these subjects had been formally invented. As in other parts of the history of mathematics, we see that great milestones do not appear suddenly, but are merely the more clear-cut formulations along the thorny path of uneven development.

BIBLIOGRAPHY

Boyer, C. B., *The Concepts of the Calculus* (New York, 1939; paperback ed., New York: Dover, 1959).

Brasch, F. E., ed., *Johann Kepler, 1571–1630. A Tercentenary Commemoration of His Life and Works* (Baltimore: Williams and Wilkins, 1931).

Braunmühl, Anton von, *Vorlesungen über Geschichte der Trigonometrie* (Leipzig: Teubner, 1900, 2 vols.).

Cajori, Florian, *A History of the Logarithmic Slide Rule and Allied Instruments* (New York: McGraw-Hill, 1909).

Cajori, Florian, "History of the Exponential and Logarithmic Concepts," *American Mathematical Monthly,* **20** (1913), 5–14, 35–47, 75–84, 107–117.

Cajori, Florian, *William Oughtred, a Great Seventeenth-Century Teacher of Mathematics* (Chicago: Open Court, 1916).

Cajori, Florian, *A History of Mathematical Notations* (Chicago: Open Court, 1929, 2 vols.).

Caspar, Max, *Kepler,* trans. by C. Doris Hellman (New York: Abelard-Schuman, 1959).

Dedron, Pierre, and Jean Itard, *Mathématiques et mathématiciens* (Paris: Magnard, 1959).

Dijksterhuis, E. J., and D. J. Struik, eds., *The Principal Works of Simon Stevin* (Amsterdam: Swets and Zeitlinger, 1955–1958).

Galilei, Galileo, *Dialogue Concerning Two New Sciences,* ed. by Henry Crew and Alfonso de Salvio (paperback ed., New York: Dover, no date).

Galilei, Galileo, *Discourses on the Two Chief Systems,* ed. by Stillman Drake (Berkeley, Calif.: University of California Press, 1953).

Galilei, Galileo, *Discourses on the Two Chief Systems,* ed. by Giorgio de Santillana (Chicago: University of Chicago Press, 1953).

Glaisher, J. W. L., "Logarithms," *Encyclopaedia Britannica,* 11th ed., (1910–1911), Vol. XVI, pp. 868–877.

Glaisher, J. W. L., "On Early Tables of Logarithms and Early History of Logarithms," *Quarterly Journal of Pure and Applied Mathematics,* **48** (1920), 151–192.

Hobson, E. W., *John Napier and the Invention of Logarithms* (Cambridge, 1914).

Hofmann, J. E., *Geschichte der Mathematik,* 2nd ed. (Berlin: Walter de Gruyter, 1963).

Kepler, Johann, *Gesammelte Werke,* ed. by Walther von Dyck and Max Caspar (München: C. H. Beck, 1937–   ).

Knott, C. G., *Napier Tercentenary Memorial Volume* (London: Longmans, Green, 1915).

Ritter, Frederic, *François Viète* (Paris, 1895).

Sarton, George, "Simon Stevin of Bruges (1548–1620)," *Isis,* **21** (1934), 241–303.

Sarton, George, "The First Explanation of Decimal Fractions and Measures (1585)," *Isis,* **23** (1935), 153–244.

Smith, D. E., *A Source Book in Mathematics* (New York: McGraw-Hill, 1929; paperback ed., New York: Dover, 1959, 2 vols.).

Tropfke, Johannes, *Geschichte der Elementar-Mathematik* (Berlin: De Gruyter, 1923), Vol. V.

Turnbull, H. W., *The Great Mathematicians* (New York: New York University Press, 1961).

Zeller, Sister Mary Claudia, *The Development of Trigonometry from Regiomontanus to Pitiscus* (Ann Arbor, Mich.: University of Michigan, Ph.D. thesis, 1944).

EXERCISES

1. Compare the contributions to mathematics of Stevin with those of Bürgi.
2. Why is Viète sometimes called the first really modern mathematician? Explain clearly.
3. What were the first two curves, other than straight-line and circle, or combinations of these, to find application in science? Explain how they came to be applied.

4. What advantages do decimal fractions have over sexagesimal fractions? What reasons can you give for the late appearance of the former in Europe?

5. What is a parameter? Can you find instances of parameters before Viète? Explain.

6. Compare Viète's use of analytic method with that of Euclid.

7. What are the relative advantages and disadvantages of the algebraic notations of Viète and Harriot?

8. Prove Viète's observation that if x_1 and x_2 are positive roots of $x^3 + b = 3ax$, then $3a = x_1^2 + x_1 x_2 + x_2^2$ and $b = x_1 x_2^2 + x_2 x_1^2$.

9. Using Viète's method, solve $x^3 = 232x^2 + 465x + 702$ for the positive root (which lies between 200 and 300).

10. Prove Viète's form of the law of tangents.

11. Using Viète's method, prove that

$$\sin x - \sin y = 2 \cos \frac{x + y}{2} \sin \frac{x - y}{2}$$

12. Using Viète's method, prove that

$$\cos x + \cos y = 2 \cos \frac{x + y}{2} \cos \frac{x - y}{2}$$

13. Multiply 8743 by 5692 prosthaphaeretically.

14. Divide 8743 by 5692 prosthaphaeretically.

15. Write $\sin 10x$ and $\cos 10x$ in terms of powers of $\sin x$ and $\cos x$.

16. Using Napier's system of logarithms, what is the relationship between Log x, Log y, and Log x/y? Justify your answer.

17. Find approximately the number of which Napier's Log is 3.

18. Find approximately the number of which Bürgi's Log is 4.

19. What is the difference between Napier's logarithms of two numbers in the ratio of 3 to 1?

20. Using Briggs' method, find antilog 0.2500 to four decimal places.

*21. Using Bürgi's logarithms, what is the relationship between log x, log y, log z, and log xy/z?

22. Use Kepler's type of reasoning to prove that the volume of a sphere is one-third the surface area times the radius.

*23. Verify Galileo's observation that

$$\lim_{\theta \to 0} \frac{\text{vers } \theta}{\theta} = 0$$

*24. Verify Cavalieri's comparison of the areas of the spiral and the parabola.

*25. Prove Viète's infinite product for π by starting with an inscribed polygon of four sides and doubling the number of sides successively.

*26. Use the trigonometric method of Viète and Girard to solve the equation $x^3 - 9x^2 + 15x + 7 = 0$ for one root correct to the nearest thousandth.

CHAPTER XVII

The Time
of Fermat and Descartes

> Fermat, the true inventor of the differential calculus.
> *Laplace*

The year 1647 in which Cavalieri died marked the death also of another **1**
disciple of Galileo, the young Evangelista Torricelli (1608–1647); but in
many ways Torricelli represented the new generation of mathematicians who
were building rapidly on the infinitesimal foundation that Cavalieri had
sketched all too vaguely. Had Torricelli not died so prematurely, Italy might
have continued to share the lead in new developments; as it turned out,
France was the undisputed mathematical center during the second third of
the seventeenth century. The leading figures were René Descartes (1596–
1650) and Pierre de Fermat (1601–1665), but three other contemporary
Frenchmen also made important contributions, in addition to Torricelli—
Gilles Persone de Roberval (1602–1675), Girard Desargues (1591–1661), and
Blaise Pascal (1623–1662). This chapter, covering one of the most critical
periods in the history of mathematics, focuses attention on these six men,
not only as individuals, but also collectively, for not since the days of Plato
had there been such mathematical intercommunication as during the
seventeenth century.

No professional mathematical organizations yet existed, but in Italy,
France, and England there were loosely organized scientific groups: the
Accademia dei Lincei (to which Galileo belonged) and the Accademia del
Cimento in Italy, the Cabinet DuPuy in France, and the Invisible College in
England. There was in addition an individual who, during the period we are
now considering, served through correspondence as a clearing house for
mathematical information. This was the Minimite friar, Marin Mersenne
(1588–1648), a close friend of Descartes and Fermat, as of many another
mathematician of the time. Had Mersenne lived a century earlier the delay
in information concerning the solution of the cubic might not have occurred,
for when Mersenne knew something, the whole of the "Republic of Letters"
was shortly informed about it. From the seventeenth century on, therefore,

René Descartes.

mathematics developed more in terms of inner logic than through economic, social, or technological forces, as is apparent particularly in the work of Descartes, the best-known mathematician of the period.

2   Descartes was born of a good family and received a thorough education at the Jesuit college at La Flèche, where the textbooks of Clavius were fundamental. Later he took a degree at Poitier, where he had studied law,

without much enthusiasm. For a number of years he traveled about in conjunction with varied military campaigns, first in Holland with Maurice, Prince of Nassau, then with Duke Maximillian I of Bavaria, and later still with the French army at the siege of LaRochelle. Descartes was not really a professional soldier, and his brief periods of service in connection with campaigns were separated by intervals of independent travel and study during which he met some of the leading scholars in various parts of Europe—Faulhaber in Germany and Desargues in France, for example. At Paris he met Mersenne and a circle of scientists who freely discussed criticisms of Peripatetic thought; from such stimulation Descartes went on to become the "father of modern philosophy", to present a changed scientific world view, and to establish a new branch of mathematics. In his most celebrated treatise, the *Discours de la méthode pour bien conduire sa raison et chercher la vérité dans les sciences* ("Discourse on the Method of Reasoning Well and Seeking Truth in the Sciences") of 1637, he announced his program for philosophical research. In this he hoped, through systematic doubt, to reach clear and distinct ideas from which it would then be possible to deduce innumerably many valid conclusions. This approach to science led him to assume that everything was explainable in terms of matter (or extension) and motion. The entire universe, he postulated, was made up of matter in ceaseless motion in vortices, and all phenomena were to be explained mechanically in terms of forces exerted by contiguous matter. Cartesian science enjoyed great popularity for almost a century, but it then necessarily gave way to the mathematical reasoning of Newton. Ironically, it was in large part the mathematics of Descartes that later made possible the defeat of Cartesian science.

The philosophy and science of Descartes were almost revolutionary in **3** their break with the past; his mathematics, by contrast, was linked with earlier traditions. To some extent this may have resulted from the commonly accepted humanistic heritage—a belief that there had been a Golden Age in the past, a "reign of Saturn," the great ideas of which remained to be rediscovered. Probably in larger measure it was the natural result of the fact that the growth of mathematics is more cumulatively progressive than is the development of other branches of learning. Mathematics grows by accretions, with very little need to slough off irrelevancies, whereas science grows largely through substitutions when better replacements are found. It should come as no surprise, therefore, to see that Descartes' chief contribution to mathematics, the foundation of analytic geometry, was motivated by an attempt to return to the past.

Descartes had become seriously interested in mathematics by the time he spent the cold winter of 1619 with the Bavarian army, where he lay abed until ten in the morning, thinking out problems. It was during this early period in his life that he discovered the polyhedral formula usually named for Euler—$v + f = e + 2$, where v, f, and e are the number of vertices, faces, and edges, respectively, of a simple polyhedron. Nine years later Descartes wrote to a friend in Holland that he had made such strides in arithmetic and geometry that he had no more to wish for. Just what the strides were is not known, for Descartes had published nothing; but the direction of his thoughts is indicated in a letter of 1628 to his Dutch friend where he gave a rule for the construction of the roots of any cubic or quartic equation by means of a parabola. This is, of course, essentially the type of thing that Menaechmus had done for the duplication of the cube some 2000 years earlier and that Omar Khayyam had carried out for cubics in general around the year 1100.

Whether or not Descartes by 1628 was in full possession of his analytic geometry is not clear, but the effective date for the invention of Cartesian geometry cannot be much later than that. At this time Descartes left France for Holland, where he spent the next twenty years. Three or four years after settling down there, his attention was called by another Dutch friend, a classicist, to the three-and-four-line problem of Pappus. Under the mistaken impression that the ancients had been unable to solve this problem, Descartes applied his new methods to it and succeeded without difficulty. This made him aware of the power and generality of his point of view, and he consequently wrote the well-known work, *La géométrie*, which made analytic geometry known to his contemporaries.

4   *La géométrie* was not presented to the world as a separate treatise, but as one of three appendices to the *Discours de la méthode* in which he thought to give illustrations of his general philosophical method. The other two appendices were *La dioptrique*, containing the first publication of the law of refraction (discovered earlier by Snell), and *Les météores*, including, among other things, the first generally satisfactory quantitative explanation of the rainbow. Descartes' successors had difficulty seeing just how the three appendices were related to his general method, and in subsequent editions of the *Discours* they frequently were omitted. The original edition of the *Discours* was published without the name of the author, but the authorship of the work was generally known.

Cartesian geometry now is synonymous with analytic geometry, but the fundamental purpose of Descartes was far removed from that of modern textbooks. The theme is set by the opening sentence:

Any problem in geometry can easily be reduced to such terms that a knowledge of the lengths of certain lines is sufficient for its construction.

As this statement indicates, the goal is generally a geometric construction, and not necessarily the reduction of geometry to algebra. The work of Descartes far too often is described simply as the application of algebra to geometry, whereas actually it could be characterized equally well as the translation of the algebraic operations into the language of geometry. The very first section of *La géométrie* is entitled "How the calculations of arithmetic are related to the operations of geometry"; the second section describes "How multiplication, division, and the extraction of square roots are performed geometrically." Here Descartes was doing what had to some extent been done from al-Khowarizmi to Oughtred—furnishing a geometric background for the algebraic operations. The five arithmetic operations are shown to correspond to simple constructions with straightedge and compasses, thus justifying the introduction of arithmetical terms in geometry.

Descartes was more thorough in his symbolic algebra, and in the geometric interpretation of algebra, than any of his predecessors. Formal algebra had been advancing steadily since the Renaissance, and it found its culmination in Descartes' *La géométrie*, the earliest mathematical text that a present-day student of algebra can follow without encountering difficulties in notation. About the only archaic symbol in the book is the use of ∞ instead of $=$ for equality. The Cartesian use of letters near the beginning of the alphabet for parameters and those near the end as unknown quantities, the adaptation of exponential notation to these, and the use of the Germanic symbols $+$ and $-$, all combined to make Descartes' algebraic notation look like ours, for of course we took ours from him. There was, nevertheless, an important difference in view, for where we think of the parameters and unknowns as numbers, Descartes thought of them as line segments. In one essential respect he broke from Greek tradition, for instead of considering x^2 and x^3, for example, as an area and a volume, he interpreted them also as lines. This permitted him to abandon the principle of homogeneity, at least explicitly, and yet retain geometrical meaning. Descartes could write an expression such as $a^2b^2 - b$, for, as he expressed it, one "must consider the quantity a^2b^2 divided once by unity (that is, the unit line segment), and the quantity b multiplied twice by unity." It is clear that Descartes substituted homogeneity in thought for homogeneity in form, a step that made his geometric algebra more flexible—so flexible indeed that today we read xx as "x-squared" without ever seeing a square in our mind's eye.

Book I includes detailed instructions on the solution of quadratic equa- **5** tions, not in the algebraic sense of the ancient Babylonians, but geometrically, somewhat in the manner of the ancient Greeks. To solve the equation $z^2 = az + b^2$, for example, Descartes proceeded as follows. Draw a line segment LM of length b (Fig. 17.1) and at L erect a segment NL equal to $a/2$ and perpendicular to LM. With center N construct a circle of radius $a/2$

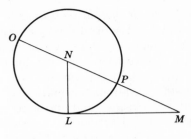

FIG. 17.1

and draw the line through M and N intersecting the circle at O and P. Then $z = OM$ is the line desired. (Descartes ignored the root PM of the equation because it is "false," that is negative.) Similar constructions are given for $z^2 = az - b^2$ and for $z^2 + az = b^2$, the only other quadratic equations with positive roots.

Having shown how algebraic operations, including the solution of quadratics, are interpreted geometrically, Descartes turned to the application of algebra to determinate geometrical problems, formulating far more clearly than the Renaissance cossists the general approach:

> If, then, we wish to solve any problem, we first suppose the solution already affected, and give names to all the lines that seem needful for its construction—to those that are unknown as well as to those that are known. Then, making no distinction between known and unknown lines, we must unravel the difficulty in any way that shows most naturally the relations between these lines, until we find it possible to express a single quantity in two ways. This will constitute an equation (in a single unknown), since the terms of the one of these two expressions are together equal to the terms of the other.[1]

Throughout Books I and III of *La géométrie* Descartes is concerned primarily with this type of geometrical problem, in which the final algebraic equation can contain only one unknown quantity. Descartes was well aware that it was the degree of this resulting algebraic equation that determined the geometrical means by which the required geometric construction can be carried out.

> If it can be solved by ordinary geometry, that is, by the use of straight lines and circles traced on a plane surface, when the last equation shall have been entirely solved there will remain at most only the square of an unknown quantity, equal to the product of its root by some known quantity, increased or diminished by some other quantity also known.

[1] Translations of passages from *La géométrie* here and elsewhere are from *The Geometry of René Descartes*, trans. by D. E. Smith and Marcia L. Latham (New York: Dover reprint, 1954). See pp. 7–9 for the above passage.

Here we see a clear-cut statement that what the Greeks had called "plane problems" lead to nothing worse than a quadratic equation. Since Viète already had shown that the duplication of the cube and the trisection of the angle lead to cubic equations, Descartes was able to state categorically that these cannot be solved with straightedge and compasses. Of the three ancient problems, therefore, only the squaring of the circle remained open to question.

The title *La géométrie* should not mislead one into thinking that the treatise is primarily geometrical. Already in the *Discourse*, to which the *Geometry* had been appended, Descartes had discussed the relative merits of algebra and geometry, without being partial to either. He charged the latter with relying too heavily on diagrams that unnecessarily fatigue the imagination, and he stigmatized the former as a confused and obscure art that embarrasses the mind. The aim of his method, therefore, was twofold: (1) through algebraic procedure to free geometry from the use of diagrams and (2) to give meaning to the operations of algebra through geometric interpretation. Descartes was convinced that all mathematical sciences proceed from the same basic principles, and he decided to use the best of each branch. His procedure in *La géométrie*, then, was to begin with a geometrical problem, to convert it to the language of an algebraic equation, and then, having simplified the equation as far as possible, to solve this equation geometrically, in a manner similar to that which he had used for the quadratics. Following Pappus, Descartes insisted that one should use in the geometric solution of an equation only the simplest means appropriate to the degree of the equation. For quadratic equations, lines and circles suffice; for cubics and quartics, conic sections are adequate. Now Descartes was ready to move beyond the point at which the Greeks had stopped.

Descartes was much impressed by the power of his method in handling **6** the three-and-four-line locus, and so he moved on to generalizations of this problem—a problem that runs like a thread of Ariadne through the three books of *La géométrie*. He knew that Pappus had been unable to tell anything about the loci when the number of lines was increased to six or eight or more; so Descartes proceeded to study such cases. He was aware that for five or six lines, the locus is a cubic, for seven or eight, it is a quartic, and so on. But Descartes showed no real interest in the shape of these loci, for he was obsessed with the question of the means needed to construct geometrically the ordinates corresponding to given abscissas. For five lines, for example, he remarked triumphantly that if they are not all parallel, then the locus is elementary in the sense that, given a value for x, the line representing y is constructible by ruler and compasses alone. If four of the lines are parallel and equal distances a apart and the fifth is perpendicular to the others (Fig. 17.2), and if the constant of proportionality in the Pappus problem is taken

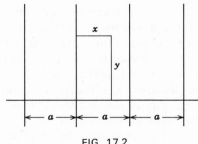

FIG. 17.2

as this same constant a, then the locus is given by $(a + x)(a - x)(2a - x) = axy$, a cubic that Newton later called the Cartesian parabola or trident— $x^3 - 2ax^2 - a^2x + 2a^3 = axy$. This curve comes up repeatedly in *La géométrie*, yet Descartes at no point gave a complete sketch of it. His interest in the curve was threefold: (1) deriving its equation as a Pappus locus, (2) showing its generation through the motion of curves of lower degree, and (3) using it in turn to construct the roots of equations of higher degree.

Descartes considered the trident constructible by plane means alone inasmuch as, for each point x on the axis of abscissas, the ordinate y can be drawn with ruler and compasses alone. This is not in general possible for five or more lines taken at random in the Pappus problem. In the case of not more than eight lines, the locus is a polynomial in x and y such that, for a given point on the x-axis, the construction of the corresponding ordinate y requires the geometric solution of a cubic or quartic equation which, as we have seen, usually calls for the use of conic sections. For not more than twelve lines in the Pappus problem, the locus is a polynomial in x and y of not more than sixth degree, and the construction in general requires curves beyond the conic sections. Here Descartes made an important advance beyond the Greeks in problems of geometric constructibility. The ancients had never really legitimized constructions that made use of curves other than straight lines or circles, although they sometimes reluctantly recognized, as Pappus did, the classes that they called solid problems and linear problems. The second category in particular was a catchall class of problems with no real standing.

Descartes now took the step of specifying an orthodox classification of determinate geometrical problems. Those that lead to quadratic equations, and can therefore be constructed by lines and circles, he placed in class one; those leading to cubic and quartic equations, the roots of which can be constructed by means of conic sections, he placed in class two; those leading to equations of degree five or six can be constructed by introducing a cubic curve, such as the trident or the higher parabola $y = x^3$, and these he placed in class three. Descartes continued in this manner, grouping geometric problems and algebraic equations into classes, assuming that the construction of the roots of an equation of degree $2n$ or $2n - 1$ was a problem of class n.

The Cartesian classification by pairs of degrees seemed to be confirmed by algebraic considerations. It was known that the solution of the quartic was reducible to that of the resolvent cubic, and Descartes extrapolated prematurely to assume that the solution of an equation of degree $2n$ can be reduced to that of a resolvent equation of degree $2n - 1$. Many years later it was shown that Descartes' tempting generalization does not hold. A number of his contemporaries were only too eager to point out a more serious error made by Descartes, for it is clear from the theory of algebraic elimination that curves of degree n suffice to solve equations not up to degree $2n$ only, but up to n^2. His classification, therefore, lost validity, but his work did have the salutary effect of encouraging the relaxation of the rules on constructibility so that higher plane curves might be used.

It will be noted that the Cartesian classification of geometric problems **7** included some, but not all, of those that Pappus had lumped together as "linear." In introducing the new curves that he needed for geometric constructions beyond the fourth degree, Descartes added to the usual axioms of geometry one more axiom:

Two or more lines (or curves) can be moved, one upon the other, determining by their intersection other curves.

This in itself is not unlike what the Greeks had actually done in their kinematic generation of curves such as the quadratrix, the cissoid, the conchoid, and the spiral; but whereas the ancients had lumped these together, Descartes now carefully distinguished between those, such as the cissoid and the conchoid, that we should call algebraic, and others, such as the quadratrix and the spiral, that are now known as transcendental. To the first type Descartes gave full-fledged geometrical status, along with the line, the circle, and the conics, calling all of these the "geometrical curves"; the second type he ruled out of geometry entirely, stigmatizing them as "mechanical curves." The basis upon which Descartes made this decision was "exactness of reasoning." Mechanical curves, he said, "must be conceived of as described by two separate movements whose relation does not admit of exact determination" —such as the ratio of circumference to diameter of a circle in the case of the motions describing the quadratrix and the spiral. In other words, Descartes thought of algebraic curves as *exactly* described and of transcendental curves as *inexactly* described, for the latter generally are defined in terms of arc lengths. On this matter he wrote, in *La géométrie*:

Geometry should not include lines (or curves) that are like strings, in that they are sometimes straight and sometimes curved, since the ratios between straight and curved lines are not known, and I believe cannot be discovered by human minds, and therefore no conclusion based upon such ratios can be accepted as rigorous and exact.

Descartes here is simply reiterating the dogma, suggested by Aristotle and affirmed by Averroës, that no algebraic curve can be exactly rectified. Interestingly enough, in 1638, the year after the publication of *La géométrie*, Descartes ran across a "mechanical" curve that turned out to be rectifiable. Through Mersenne, Galileo's representative in France, the question, raised in the *Two New Sciences*, of the path of fall of an object on a rotating earth (assuming the earth permeable) was widely discussed, and this led Descartes to the equiangular or logarithmic spiral $r = ae^{b\theta}$ as the possible path.[2] Had Descartes not been so firm in his rejection of such nongeometrical curves, he might have anticipated Torricelli in discovering, in 1645, the first modern rectification of a curve. Torricelli showed, by infinitesimal methods that he had learned from Archimedes, Galileo, and Cavalieri, that the total length of the logarithmic spiral from $\theta = 0$ as it winds backward about the pole O is exactly equal to the length of the polar tangent PT (Fig. 17.3) at the point for which $\theta = 0$. This striking result did not, of course, disprove the Cartesian doctrine of the nonrectifiability of *algebraic* curves. In fact, Descartes could have asserted not only that the curve was not exactly determined, being mechanical, but also that the arc of the curve has an asymptotic point at the pole, which it never reaches.

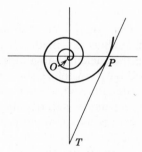

FIG. 17.3

8 Virtually the whole of *La géométrie* is devoted to a thoroughgoing application of algebra to geometry and of geometry to algebra; but there is little in the treatise that resembles what usually is thought of today as analytic geometry. There is nothing systematic about rectangular coordinates, for oblique ordinates usually are taken for granted; hence there are no formulas

[2] See *Oeuvres de Descartes*, ed. by Charles Adam and Paul Tannery (1897–1913), II, 222–245.

for distance, slope, point of division, angle between two lines, or other similar introductory material. Moreover, in the whole of the work there is not a single new curve plotted directly from its equation, and the author took so little interest in curve sketching that he never fully understood the meaning of negative coordinates. He knew in a general sort of way that negative ordinates are directed in a sense opposite to that taken as positive, but he never made use of negative abscissas. Moreover, the fundamental principle of analytic geometry—the discovery that indeterminate equations in two unknowns correspond to loci—does not appear until the second book, and then only somewhat incidentally.

> The solution of any one of these problems of loci is nothing more than the finding of a point for whose complete determination one condition is wanting. . . . In every such case an equation can be obtained containing two unknown quantities.

In one case only did Descartes examine a locus in detail, and this was in connection with the three-and-four-line locus problem of Pappus for which Descartes derived the equation $y^2 = ay - bxy + cx - dx^2$. This is the general equation of a conic passing through the origin; even though the literal coefficients are understood to be positive, this is by far the most comprehensive approach ever made to the analysis of the family of conic sections. Descartes indicated conditions on the coefficients for which the conic is a straight line, a parabola, an ellipse, or a hyperbola, the analysis being in a sense equivalent to a recognition of the characteristic of the equation of the conic. The author knew that by a proper choice of the origin and axes the simplest form of the equation is obtained, but he did not give any of the canonical forms. The omission of much of the elementary detail made the work exceedingly difficult for his contemporaries to follow. In concluding remarks Descartes sought to justify inadequacy of exposition by the incongruous assertion that he had left much unsaid in order not to rob the reader of the joy of discovery. A genius himself, he could not appreciate the difficulty that others were to have in understanding his new and profound thoughts. It is small wonder that the number of editions of *La géométrie*, apart from those with considerable amplification, was small in the seventeenth century and has been still smaller since then.

Inadequate though the exposition is, it is Book II of *La géométrie* that comes closest to modern views of analytic geometry. There is even a statement of a fundamental principle of solid analytic geometry:

> If two conditions for the determination of a point are lacking, the locus of the point is a surface.

However, Descartes did not give any illustrations of such equations or expand the brief hint of analytic geometry of three dimensions.

9   Descartes was so fully aware of the significance of his work that he regarded it as bearing to ancient geometry somewhat the same relationship as the rhetoric of Cicero bears to the a, b, c's of children. His mistake, from our point of view, was in emphasizing determinate equations rather than indeterminate. He realized that all the properties of a curve, such as the magnitude of its area, or the direction of its tangent, are fully determined when an equation in two unknowns is given, but he did not take full advantage of this recognition. He wrote:

> I shall have given here a sufficient introduction to the study of curves when I shall have given a general method of drawing a straight line making right angles with a curve at an arbitrarily chosen point upon it. And I dare say that this is not only the most useful and most general problem in geometry that I know, but even that I have ever desired to know.

Descartes was quite right that the problem of finding the normal (or the tangent) to a curve was of great importance, but the method that he published in *La géométrie* was less expeditious than that which Fermat had developed at about the same time. Descartes suggested that to find the normal to an algebraic curve at a fixed point P on the curve, one should take a second variable point Q on the curve, then find the equation of the circle with center on the coordinate axis (for he used only an axis of abscissas) and passing through P and Q. Now, by setting equal to zero the discriminant of the equation that determines the intersections of the circle with the curve, one finds the center of the circle where Q coincides with P. The center being known, the tangent and normal to the curve at P are then easily found.

Book II of *La géométrie* contains also much material on the "ovals of Descartes," which are very useful in optics and are obtained by generalizing the "gardner's method" for constructing an ellipse by means of strings. If D_1 and D_2 are the distances of a variable point P from two fixed points F_1 and F_2 respectively, and if m and n are positive integers and K is any positive constant, then the locus of P such that $mD_1 + nD_2 = K$ is now known as an oval of Descartes; but the author did not use the equations of the curves. Descartes realized that his methods can be extended to "all those curves which can be conceived of as generated by the regular movement of the points of a body in three-dimensional space," but he did not carry out any details. The sentence with which Book II concludes, "And so I think I have omitted nothing essential to an understanding of curved lines," is presumptuous indeed.

The third and last book of *La géométrie* resumes the topic of Book I—the construction of the roots of determinate equations. Here the author warned that in such constructions "We should always choose with care the simplest curve that can be used in the solution of a problem." This means, of course, that one must be fully aware of the nature of the roots of the equation under consideration, and in particular one must know whether or not the equation is reducible. For this reason, Book III is virtually a course in the elementary theory of equations. It tells how to discover rational roots, if any, how to depress the degree of the equation when a root is known, how to increase and decrease the roots of an equation by any amount, or to multiply or divide them by a number, how to eliminate the second term, how to determine the number of possible "true" and "false" roots (that is, positive and negative roots) through the well-known "Descartes' rule of signs," and how to find the algebraic solution of cubic and quartic equations. In closing, the author reminds the reader that he has given the simplest constructions possible for problems in the various classes mentioned earlier. In particular, the trisection of the angle and the duplication of the cube are in class two, requiring more than circles and lines for their construction.

Our account of Descartes' analytic geometry should make clear how far **10** removed the author's thought was from the practical considerations that are now so often associated with the use of coordinates. He did not lay down a coordinate frame in order to locate points as a surveyor or geographer might do, nor were his coordinates thought of as number pairs. In this respect the phrase "Cartesian product," so often used today, is an anachronism. *La géométrie* was in its day just as much a triumph of impractical theory as was the *Conics* of Apollonius in antiquity, despite the inordinately useful role that both were ultimately destined to play. Moreover, the use of oblique coordinates was much the same in both cases, thus confirming that the origin of modern analytic geometry lies in antiquity rather than in the medieval latitude of forms. The coordinates of Oresme, which influenced Galileo, are closer, both in motive and in appearance, to the modern point of view than are those of Apollonius and Descartes. Even if Descartes was familiar with Oresme's graphical representation of functions, and this is not evident, there is nothing in Cartesian thought to indicate that he would have seen any similarity between the purpose of the latitude of forms and his own classification of geometric constructions. The theory of functions ultimately profited greatly from the work of Descartes, but the notion of a form or function played no apparent role in leading to Cartesian geometry.

In terms of mathematical ability Descartes probably was the most able thinker of his day, but he was at heart not really a mathematician. His geometry was only an episode in a life devoted to science and philosophy, and although occasionally in later years he contributed to mathematics through correspondence, he left no other great work in this field. In 1649 he accepted an invitation from Queen Christina of Sweden to instruct her in philosophy and to establish an academy of sciences at Stockholm. Descartes had never enjoyed robust health, and the rigors of a Scandinavian winter were too much for him; he died early in 1650.

11   If Descartes had a rival in mathematical ability, it was Fermat, but the latter was in no sense a professional mathematician. Fermat studied law at Toulouse, where he then served in the local parlement, first as a lawyer and later as councillor. This meant that he was a busy man; yet he seems to have had time to enjoy as an avocation a taste for classical literature, including science and mathematics. The result was that by 1629 he began to make discoveries of capital importance in mathematics. In this year he joined in one of the favorite sports of the time—the "restoration" of lost works of antiquity on the basis of information found in extant classical treatises. Fermat undertook to reconstruct the *Plane Loci* of Apollonius, depending on allusions contained in the *Mathematical Collection* of Pappus. A by-product of this effort was the discovery, at least by 1636, of the fundamental principle of analytic geometry:

> Whenever in a final equation two unknown quantities are found, we have a locus, the extremity of one of these describing a line, straight or curved.

This profound statement, written a year before the appearance of Descartes' *Geometry*, seems to have grown out of Fermat's application of the analysis of Viète to the study of loci in Apollonius. In this case, as also in that of Descartes, the use of coordinates did not arise from practical considerations, nor from the medieval graphical representation of functions. It came about through the application of Renaissance algebra to problems from ancient geometry. However, Fermat's point of view was not entirely in conformity with that of Descartes, for Fermat emphasized the sketching of solutions of *indeterminate* equations, instead of the geometrical construction of the roots of *determinate* algebraic equations. Moreover, where Descartes had built his *Geometry* around the difficult Pappus problem, Fermat limited his exposition, in the short treatise entitled *Ad locus planos et solidos isagoge* ("Introduction to Plane and Solid Loci"), to the simplest loci only. Where Descartes had begun with the three-and-four-line locus, using one of the lines as an axis of abscissas, Fermat began with the linear equation and chose an arbitrary coordinate system upon which to sketch it.

Using the notation of Viète, Fermat sketched first the simplest case of a linear equation—given in Latin as "*D in A aequetur B in E*" (that is, $Dx = By$ in modern symbolism). The graph, is of course, a straight line through the origin of coordinates—or rather a half line with the origin as end point, for Fermat, like Descartes, did not use negative abscissas. The more general linear equation $ax + by = c^2$ (for Fermat retained Viète's homogeneity) he sketched as a line segment in the first quadrant terminated by the coordinate axes. Next, to show the power of his method for handling loci, Fermat announced the following problem that he had discovered by the new approach:

> Given any number of fixed lines, in a plane, the locus of a point such that the sum of any multiples of the segments drawn at given angles from the point to the given lines is constant, is a straight line.

That is, of course, a simple corollary of the fact that the segments are linear functions of the coordinates, and of Fermat's proposition that every equation of first degree represents a straight line.[3]

Fermat next showed that $xy = k^2$ is a hyperbola and that an equation of the form $xy + a^2 = bx + cy$ can be reduced to one of the form $xy = k^2$ (by a translation of axes). The equation $x^2 = y^2$ he considered as a single straight line (or ray), for he operated only in the first quadrant, and he reduced other homogeneous equations of second degree to this form. Then he showed that $a^2 \pm x^2 = by$ is a parabola, that $x^2 + y^2 + 2ax + 2by = c^2$ is a circle, that $a^2 - x^2 = ky^2$ is an ellipse, and that $a^2 + x^2 = ky^2$ is a hyperbola (for which he gave both branches). To more general quadratic equations, in which the several second-degree terms appear, Fermat applied a rotation of axes to reduce them to the earlier forms. As the "crowning point" of his treatise, Fermat considered the following proposition:

> Given any number of fixed lines, the locus of a point such that the sum of the squares of the segments drawn at given angles from the point to the lines is constant, is a solid locus.

This proposition is obvious in terms of Fermat's exhaustive analysis of the various cases of quadratic equations in two unknowns. As an appendix to the *Introduction to Loci*, Fermat added "The Solution of Solid Problems by Means of Loci," pointing out that determinate cubic and quartic equations can be solved by conics, the theme that had loomed so large in the geometry of Descartes.

Fermat's *Introduction to Loci* was not published during the author's **12** lifetime; hence analytic geometry in the minds of many was regarded as the unique invention of Descartes. It is now clear that Fermat had discovered essentially the same method well before the appearance of *La géométrie* and that his work circulated in manuscript form until its publication in 1679 in *Varia opera mathematica*. It is a pity that Fermat published almost nothing during his lifetime, for his exposition was much more systematic and didactic than that of Descartes. Moreover, his analytic geometry was somewhat

---

[3] For this and other aspects of Fermat's work see his *Oeuvres*, ed. by Paul Tannery and Charles Henry (1891–1922).

closer to ours in that ordinates usually are taken at right angles to the line of abscissas. Like Descartes, Fermat was aware of an analytic geometry of more than two dimensions, for in another connection he wrote:

> There are certain problems which involve only one unknown, and which can be called *determinate*, to distinguish them from the problems of loci. There are certain others which involve two unknowns and which can never be reduced to a single one; these are the problems of loci. In the first problems we seek a unique point, in the latter a curve. But if the proposed problem involves three unknowns, one has to find, to satisfy the equation, not only a point or a curve, but an entire surface. In this way surface loci arise, etc.[4]

Here in the final "etc." there is a hint of geometry of more than three dimensions, but if Fermat really had this in mind, he did not carry it further. Even the geometry of three dimensions had to wait until the eighteenth century for its effective development.

13   It is possible that Fermat was in possession of his analytic geometry as early as 1629, for about this time he made two significant discoveries that are closely related to his work on loci. The more important of these was described a few years later in a treatise, again unpublished in his lifetime, entitled *Method of Finding Maxima and Minima*. Fermat had been considering loci given (in modern notation) by equations of the form $y = x^n$; hence today they are often known as "parabolas of Fermat" if n is positive or "hyperbolas of Fermat" if n is negative. Here we have an analytic geometry of higher plane curves; but Fermat went further. For polynomial curves of the form $y = f(x)$ he noted a very ingenious method of finding points at which the function takes on a maximum or a minimum value. He compared the value of $f(x)$ at a point with the value $f(x + E)$ at a neighboring point. Ordinarily these values will be distinctly different, but at the top or bottom of a smooth curve the change will be almost imperceptible. Hence to find maximum and minimum points Fermat equated $f(x)$ and $f(x + E)$, realizing that the values, although not exactly the same, are almost equal. The smaller the interval E between the two points, the nearer the pseudoequality comes to being a true equation; so Fermat, after dividing through by E, set $E = 0$. The results gave him the abscissas of the maximum and minimum point of the polynomial. Here in essence is the process now called differentiation, for the method of Fermat is equivalent to finding

$$\lim_{E \to 0} \frac{f(x + E) - f(x)}{E}$$

and setting this equal to zero. Hence it is appropriate to follow Laplace in

[4] See Fermat, *Oeuvres*, I, 186–187.

acclaiming Fermat as the discoverer of the differential calculus, as well as a codiscoverer of analytic geometry. Obviously Fermat was not in possession of the limit concept, but otherwise his method of maxima and minima parallels that used in the calculus today, except that now the symbol h or Δx is customarily used in place of Fermat's E. Fermat's process of changing the variable slightly and considering neighboring values has ever since been the essence of infinitesimal analysis.

During the very years in which Fermat was developing his analytic geometry, he discovered also how to apply his neighborhood process to find the tangent to an algebraic curve of the form $y = f(x)$. If P is a point on the curve $y = f(x)$ at which the tangent is desired, and if the coordinates of P are (a, b), then a neighboring point on the curve with coordinates $x = a + E$, $y = f(a + E)$ will lie so close to the tangent that one can think of it as approximately on the tangent as well as on the curve. If, therefore, the subtangent at the point P is $TQ = C$ (Fig. 17.4), the triangles TPQ and $TP'Q'$ can be taken as being virtually similar. Hence one has the proportion

$$\frac{b}{c} = \frac{f(a + E)}{c + E}$$

Upon cross-multiplying, canceling like terms, recalling that $b = f(a)$, then dividing through by E, and finally setting $E = 0$, the subtangent c is readily found.

Fermat's procedure amounts to saying that

$$\lim_{E \to 0} \frac{f(a + E) - f(a)}{E}$$

is the slope of the curve at $x = a$; but Fermat did not explain his procedure satisfactorily, saying simply that it was similar to his method of maxima and minima. Descartes in particular, when the method was reported to him in 1638 by Mersenne, attacked it as not generally valid. He proposed as a challenge the curve ever since known as the "folium of Descartes"— $x^3 + y^3 = 3axy$. That mathematicians of the time were quite unfamiliar

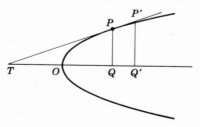

FIG. 17.4

with negative coordinates is apparent in that the curve was drawn as but a single *folium* or "leaf" in the first quadrant—or sometimes as a four-leaf clover, with one leaf in each quadrant! Ultimately Descartes grudgingly conceded the validity of Fermat's tangent method, but Fermat never was accorded the esteem to which he was entitled.

14   Mersenne, through correspondence and in his own printed works, made some of Fermat's results known in France and Italy, but it would have been ever so much better if Fermat had published his marvelous discoveries. Fermat not only had a method for finding the tangent to curves of the form $y = x^m$, he also, some time after 1629, hit upon a theorem on the area under these curves—the theorem that Cavalieri published in 1635 and 1647. In finding the area Fermat at first seems to have used formulas for the sums of powers of the integers, or inequalities of the form

$$1^m + 2^m + 3^m + \cdots + n^m > \frac{n^{m+1}}{m+1} > 1^m + 2^m + 3^m + \cdots + (n-1)^m$$

to establish the result for all positive integral values of m. This in itself was an advance over the work of Cavalieri, who limited himself to the cases from $m = 1$ to $m = 9$; but later Fermat developed a better method for handling the problem,[5] which was applicable to fractional as well as integral values of m. Let the curve be $y = x^n$, and let the area under the curve from $x = 0$ to $x = a$ be desired. Then Fermat subdivided the interval from $x = 0$ to $x = a$ into infinitely many subintervals by taking the points with abscissas a, aE, aE^2, $aE^3, \ldots$, where E is a quantity less than one. At these points he erected ordinates to the curve and then approximated to the area under the curve by means of rectangles (as indicated in Fig. 17.5). The areas of the successive

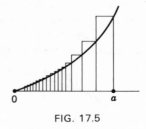

0   a

FIG. 17.5

approximating circumscribed rectangles, beginning with the largest, are given by the terms in geometric progression $a^n(a - aE)$, $a^n E^n(aE - aE^2)$,

---

[5] See Fermat, *Oeuvres*, I, 255–288; III, 216–240.

$a^n E^{2n}(aE^2 - aE^3), \ldots$. The sum to infinity of these terms is

$$\frac{a^{n+1}(1 - E)}{1 - E^{n+1}} \quad \text{or} \quad \frac{a^{n+1}}{1 + E + E^2 + \cdots + E^n}$$

As E tends toward one—that is, as the rectangles become narrower—the sum of the areas of the rectangles approaches the area under the curve. Upon letting $E = 1$ in the formula above for the sum of the rectangles, we obtain $(a^{n+1})/(n + 1)$, the desired area under the curve $y = x^n$ from $x = 0$ to $x = a$. To show that this holds also for rational fractional values, p/q, let $n = p/q$. The sum of the geometric progression then is

$$a^{(p+q)/q}\left(\frac{1 - E^q}{1 - E^{p+q}}\right) = a^{(p+q)/q}\left(\frac{1 + E + E + \ldots + E^{q-1}}{1 + E + E^2 + \ldots + E^{p+q-1}}\right)$$

and when $E = 1$, this becomes

$$\frac{q}{p + q}a^{(p+q)/q}$$

If, in modern notation, we wish to obtain $\int_a^b x^n \, dx$, it is only necessary to observe that this is $\int_0^b x^n \, dx - \int_0^a x^n \, dx$.

For negative values of n (except $n = -1$) Fermat used a similar procedure, except that E is taken as greater than one and tends toward one from above, the area found being that beneath the curve from $x = a$ to infinity. To find $\int_a^b x^{-n} \, dx$, then, it was only necessary to note that this is $\int_a^\infty x^{-n} \, dx - \int_b^\infty x^{-n} \, dx$.

For $n = -1$ the procedure fails; but Fermat's older contemporary, **15** Gregory of St. Vincent (1584–1667) disposed of this case in his *Opus geometricum quadraturae circuli et sectionum coni* ("Geometrical Work on the Squaring of the Circle and of Conic Sections"). Much of this work had been completed before the time that Fermat was working on tangents and areas, perhaps as early as 1622–1625, although it was not published until 1647. Gregory of St. Vincent, born at Ghent, was a Jesuit teacher at Rome and Prague and later became a tutor at the court of Philip IV of Spain. Through his travels he became separated from his papers, with the result that the appearance of the *Opus geometricum* was long delayed. In this treatise Gregory had shown that if along the x-axis one marks off from $x = a$ points the intervals between which are increasing in continued geometric proportion, and if at these points ordinates are erected to the hyperbola $xy = 1$, then the areas under the curve intercepted between successive ordinates are equal. That is, as the abscissa increases geometrically, the area under the curve increases arithmetically. Hence the equivalent of $\int_a^b x^{-1} \, dx = \ln b - \ln a$ was known to Gregory and his contemporaries. Unfortunately, a faulty application of the method of indivisibles had led Gregory of St. Vincent to believe that he had squared the circle, an error that damaged his reputation.

Fermat had been concerned with many aspects of infinitesimal analysis—tangents, quadratures, volumes, lengths of curves, centers of gravity. He could scarcely have failed to notice that in finding tangents to $y = kx^n$ one multiplies the coefficient by the exponent and lowers the exponent by one, whereas in finding areas one raises the exponent and divides by the new exponent. Could the inverse nature of these two problems have escaped him? Although this seems unlikely, it nevertheless appears that he nowhere called attention to the relationship now known as the fundamental theorem of the calculus. Perhaps he recognized the inverse nature of the problems but saw no great significance in this. Integration of x^n, about the only function that he really considered, was, after all, almost as easy for him as differentiation—and chronologically, at least for positive integral values of n, the former may have preceded the latter in Fermat's work. Thus also in the work of Gregory of St. Vincent the integral calculus came before the differential calculus for the logarithmic function.

The inverse relationship between area and tangent problems should have been apparent from a comparison of Gregory of St. Vincent's area under the hyperbola and Descartes' analysis of inverse tangent problems proposed through Mersenne in 1638. The problems had been set by Florimond Debeaune (1601–1652), a jurist at Blois who was also an accomplished mathematician, for whom even Descartes expressed admiration. One of the problems called for the determination of a curve whose tangent had the property now expressed by the differential equation $a\ dy/dx = x - y$. Descartes recognized the solution as nonalgebraic, but he evidently just missed seeing that logarithms were involved.[6]

16   Fermat's contributions to analytic geometry and to infinitesimal analysis were but two aspects of his work—and probably not his favorite topics. In 1621 the *Arithmetica* of Diophantus had come to life again through the Greek and Latin edition by Claude Gaspard de Bachet (1591–1639), a member of an informal group of scientists in Paris. Diophantus' *Arithmetica* had not been unknown, for Regiomontanus had thought of printing it; several translations had appeared in the sixteenth century, with little result for the theory of numbers. Perhaps the work of Diophantus was too impractical for the practitioners and too algorithmic for the speculatively inclined; but it appealed strongly to Fermat, who became the founder of the modern theory of numbers. Many aspects of the subject caught his fancy, including perfect

    [6] See C. J. Scriba, "Zur Lösung des 2. Debeauneschen Problems durch Descartes," *Archive for History of Exact Sciences*, **1** (1961), 406–419.

and amicable numbers, figurate numbers, magic squares, Pythagorean triads, divisibility, and, above all, prime numbers. Some of his theorems he proved by a method that he called his "infinite descent"—a sort of inverted mathematical induction, a process that Fermat was among the first to use. As an illustration of his process of infinite descent, let us apply it to an old and familiar problem—the proof that $\sqrt{3}$ is not rational. Let us assume that $\sqrt{3} = a_1/b_1$, where a_1 and b_1 are positive integers with $a_1 > b_1$. Since

$$\frac{1}{\sqrt{3} - 1} = \frac{\sqrt{3} + 1}{2}$$

upon replacing the first $\sqrt{3}$ by its equal a_1/b_1, we have

$$\sqrt{3} = \frac{3b_1 - a_1}{a_1 - b_1}$$

In view of the inequality $\frac{3}{2} < a_1/b_1 < 2$, it is clear that $3b_1 - a_1$ and $a_1 - b_1$ are positive integers, a_2 and b_2, each less than a_1 and b_1 respectively, and such that $\sqrt{3} = a_2/b_2$. This reasoning can be repeated indefinitely, leading to an infinite descent in which a_n and b_n are ever smaller integers such that $\sqrt{3} = a_n/b_n$. This implies the false conclusion that there is no smallest positive integer. Hence the premise that $\sqrt{3}$ is a quotient of integers must be false.

Using his method of infinite descent, Fermat was able to prove Girard's assertion that every prime number of the form $4n + 1$ can be written in one and only one way as the sum of two squares. He showed that if $4n + 1$ is not the sum of two squares, there always is a smaller integer of this form that is not the sum of two squares. Using this recursive relationship backward leads to the false conclusion that the smallest integer of this type, 5, is not the sum of two squares (whereas $5 = 1^2 + 2^2$). Hence the general theorem is proved to be true. Since it is easy to show that no integer of the form $4n - 1$ can be the sum of two squares and since all primes except 2 are of the form $4n + 1$ or $4n - 1$, by Fermat's theorem one can easily classify prime numbers into those that are and those that are not the sum of two squares. The prime 23, for example, cannot be so divided, whereas the prime 29 can be written as $2^2 + 5^2$. Fermat knew that a prime of either form can be expressed as the difference of two squares in one and only one way.

Fermat used his method of infinite descent to prove that there is no cube **17** that is divisible into two cubes—that is, that there are no positive integers x, y, and z such that $x^3 + y^3 = z^3$. Going further, Fermat stated the general proposition that for n an integer greater than two, there are no positive integral values x, y, and z such that $x^n + y^n = z^n$. He wrote in the margin of his copy of Bachet's *Diophantus* that he had a truly marvelous proof of this celebrated theorem, which since has become known as Fermat's "last" or "great" theorem. Fermat, most unfortunately, did not give his proof, but described it only as one "which this margin is too narrow to contain." If Fermat did indeed have such a proof, it has remained lost to this day. Despite all efforts to find a proof, once stimulated by a pre-World War I prize offer of 100,000 marks for a solution, the problem remains unsolved. However, the search for solutions has led to even more good mathematics than that which in antiquity resulted from efforts to solve the three classical and unsolvable geometrical problems. Like Horace Walpole's three princes of Serendip, mathematicians seem to have had the gift of finding along the way agreeable things not sought for.

Whether or not Fermat was correct in stating his "great" theorem is not yet known, but decisions have been reached on two of his other conjectures in the theory of numbers. Perhaps two millennia before his day there had been a "Chinese hypothesis" which held that n is prime if and only if $2^n - 2$ is divisible by n, where n is an integer greater than one. Half of this conjecture now is known to be false, for $2^{341} - 2$ is divisible by 341, and $341 = 11 \cdot 31$ is composite; but the other half is indeed valid, and Fermat's "lesser" theorem is a generalization of this. A consideration of many cases of numbers of the form $a^{p-1} - 1$, including $2^{36} - 1$, suggested that whenever p is prime and a is prime to p, then $a^{p-1} - 1$ is divisible by p. On the basis of an induction from only five cases ($n = 0, 1, 2, 3$, and 4), Fermat formulated a second conjecture—that integers of the form $2^{2^n} + 1$, now known as "Fermat numbers," always are prime. Euler a century later showed this conjecture to be false, for $2^{2^5} + 1$ is composite. In fact, it is known now that $2^{2^n} + 1$ is *not* prime for n between five and sixteen inclusive, and we begin to wonder if there is even one more prime Fermat number beyond those that Fermat knew.[7]

Fermat's lesser theorem fared better than his conjecture on prime Fermat numbers. A proof of the theorem was left in manuscript by Leibniz, and another elegant and elementary demonstration was published by Euler in 1736. The proof by Euler makes ingenious use of mathematical induction, a device with which Fermat, as well as Pascal, was quite familiar. In fact, mathematical induction, or reasoning by recurrence, sometimes is referred to as "Fermatian induction," to distinguish it from scientific or "Baconian" induction. (Today the former sometimes is known also as "complete induction," the latter as "incomplete induction.")

18   Fermat was truly "the prince of amateurs" in mathematics. No professional mathematician of his day made greater discoveries or contributed more to

[7] W. Sierpinski, "L'induction incomplète dans la théorie des nombres," *Scripta Mathematica.* **28** (1967), 5–13.

the subject; yet Fermat was so modest that he published virtually nothing. He was content to write of his thoughts to Mersenne (whose name, incidentally, is preserved in connection with the "Mersenne numbers," that is, primes of the form $2^p - 1$) and thus lost priority credit for much of his work. In this respect he shared the fate of one of his most capable friends and contemporaries—the unamiable professor Roberval, a member of the "Mersenne group" and the only truly professional mathematician among the Frenchmen whom we discuss in this chapter. Appointment to the chair of Ramus at the Collège Royal, which Roberval held for some forty years, was determined every three years on the basis of a competitive examination, the questions for which were set by the incumbent. In 1634 Roberval won the contest, probably because he had developed a method of indivisibles similar to that of Cavalieri; by not disclosing his method to others, he successfully retained his position in the chair until his death in 1675. This meant, however, that he lost credit for most of his discoveries and that he became embroiled in numerous quarrels with respect to priority. The bitterest of these controversies concerned the cycloid, to which the phrase "the Helen of geometers" came to be applied because of the frequency with which it provoked quarrels during the seventeenth century. Mersenne in 1615 had called the attention of mathematicians to the cycloid, perhaps having heard of the curve through Galileo; in 1628, when Roberval arrived in Paris, Mersenne proposed to the young man that he study the curve. By 1634, Roberval was able to show that the area under one arch of the curve is exactly three times the area of the generating circle. By 1638 he had found how to draw the tangent to the curve at any point (a problem solved at about the same time also by Fermat and Descartes) and had found the volumes generated when the area under an arch is revolved about the base line. Later still he found the volumes generated by revolving the area about the axis of symmetry or about the tangent at the vertex.[8]

Roberval did not publish his discoveries concerning the cycloid (which **19** he named the "trochoid," from the Greek word for wheel), for he may have wished to set similar questions for prospective candidates for his chair. Meanwhile Torricelli became interested in the cycloid, possibly on the suggestion of Mersenne, perhaps through Galileo, whom Torricelli, like Mersenne, greatly admired. In 1643 Torricelli sent Mersenne the quadrature of the cycloid, and in 1644 he published a work with the title *De parabole* to which he appended both the quadrature of the cycloid and the construction of the tangent. Torricelli made no mention of the fact that Roberval had

[8] An excellent account of all of this work and of the place of Roberval in the mathematics of the time is found in Evelyn Walker, *A Study of the Traité des Indivisibles of Gilles Persone de Roberval* (1932).

arrived at these results before him, and so in 1646 Roberval wrote a letter accusing Torricelli of plagiarism from him and from Fermat (on maxima and minima). It is clear now that priority of discovery belongs to Roberval, but priority in publication goes to Torricelli, who probably rediscovered the area and tangent independently. Roberval had used the method of indivisibles for the area problem; Torricelli gave two quadratures, one making use of Cavalieri's method of indivisibles and the other of the ancient method of exhaustion. For finding the tangent to the curve both men employed a composition of motions reminiscent of Archimedes' tangent to his spiral. Roberval thought of a point P on the cycloid as subject to two equal notions, one a motion of translation, the other a rotary motion. As the generating circle rolls along the base line AB (Fig. 17.6), P is carried horizontally, at the same time rotating about O, the center of the circle. Through P one therefore draws a horizontal line PS, for the motion of translation, and a line PR tangent to the generating circle, for the rotary component. Inasmuch as the motion of translation is equal to that of rotation, the bisector PT of the angle SPR is the required tangent to the cycloid.

The idea of the composition of movements was not original with Roberval, for Archimedes, Galileo, Descartes, and others had used it. Torricelli might have derived the idea from any one of these men; hence his application of the principle to the cycloid need not have been plagiarism from Roberval. Both Torricelli and Roberval applied the kinematic method to other curves as well. A point on the parabola, for example, moves away from the focus at the same rate at which it moves away from the directrix, hence the tangent will be the bisector of the angle between lines in these two directions. A similar argument holds for the ellipse, in which the motion away from one focus is equal to the motion toward the other focus. Torricelli made use also of Fermat's method of tangents for the higher parabolas, knowledge of which is known to have reached Italy.[9]

20   The works of Roberval and Torricelli include many excellent results, only a few of which can be mentioned here. Among the contributions of Roberval was the first sketch, in 1635, of half an arch of a sine curve. This was important as an indication that trigonometry gradually was moving away from the computational emphasis, which had dominated thought in that branch, toward a functional approach. By means of his method of indivisibles, Roberval was able to show the equivalent of $\int_a^b \sin x \, dx = \cos a - \cos b$, again indicating that area problems tended at that time to be easier to handle than tangent questions. Roberval and Torricelli, working

---

[9] There is no good account in English of the work of Torricelli, but certain aspects of it, especially tangents, are well treated in Evelyn Walker, *A Study of the Traité des Indivisibles of Gilles Persone de Roberval*. For other aspects see Torricelli's *Opere* (1919–1944).

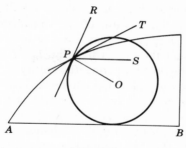

FIG. 17.6

independently but along remarkably similar lines, extended Cavalieri's comparison of the parabola and the spiral by considering arc length as well as area. In the 1640s they showed that the length of the first rotation of the spiral $r = a\theta$ is equal to the length of the parabola $x^2 = 2ay$ from $x = 0$ to $x = 2\pi a$. Interest in the spiral at the time may have arisen from correspondence between Galileo and Mersenne concerning the path of a freely falling object on a moving earth, but the discussion soon greatly broadened. Fermat, ever one to seek generalizations, introduced the higher spirals $r^n = a\theta$ and compared the arcs of these with the lengths of his higher parabolas $x^{n-1} = 2ay$. Torricelli studied spirals of various kinds, discovering the rectification of the logarithmic spiral, as we have seen. There was a remarkable unity in the mathematical interests of the period from about 1630 to 1650, attributable in part to the intercommunication through Mersenne. Problems involving infinitesimals were by far the most popular at the time, and Torricelli in particular delighted in these. In the *De dimensione parabolae*, for instance, Torricelli gave twenty-one different proofs of the quadrature of the parabola, using approaches about evenly divided between the use of indivisibles and the method of exhaustion. One in the first category is almost identical with the mechanical quadrature given by Archimedes in his *Method*, presumably not then extant; as might be anticipated, one in the second category is virtually that given in Archimedes' treatise *On the Quadrature of the Parabola*, extant and well-known at the time. Had Torricelli arithmetized his procedures in this connection, he would have been very close to the modern limit concept, but he remained under the heavily geometrical influence of Cavalieri. Nevertheless, Torricelli far outdid his master in the flexible use of indivisibles to achieve new discoveries.

One novel result of 1641 that greatly pleased Torricelli was his proof that if an infinite area, such as that bounded by the hyperbola $xy = a^2$, an ordinate $x = b$, and the axis of abscissas, is revolved about the x-axis, the volume of the solid generated may be finite. Torricelli believed that he was first to discover that a figure with infinite dimensions can have a finite magnitude; but in this respect he may have been anticipated by Fermat's work on the areas under the higher hyperbolas, or possibly by Roberval, and certainly by Oresme in the fourteenth century.

Among the problems that Torricelli handled just before his premature death in 1647 was one in which he sketched the curve whose equation we should write as $x = \log y$—perhaps the first graph of a logarithmic function, thirty years after the death of the discoverer of logarithms as a computational device. Torricelli found the area bounded by the curve, its asymptote, and an ordinate, as well as the volume of the solid obtained upon revolving the area about the x-axis.

Torricelli was one of the most promising mathematicians of the seventeenth century—often referred to as the century of genius. Mersenne had made the work of Fermat, Descartes, and Roberval known in Italy, both through correspondence with Galileo dating from 1635 and during a pilgrimage to Rome in 1644; Torricelli promptly mastered the new methods, although he always favored the geometric approach over the algebraic. Torricelli's brief association with the blind and aged Galileo in 1614–1642 had aroused in the young student an interest in physical science also, and today he is probably better recalled as the inventor of the barometer than as a mathematician. He studied the parabolic paths of projectiles fired from a point with fixed initial speeds but with varying angles of elevation, finding that the envelope of the parabolas is another parabola. In going from an equation for distance in terms of time to that for speed as a function of time, and inversely, Torricelli saw the inverse character of quadrature and tangent problems. Had he enjoyed the normal span of years, it is possible that he would have become the inventor of the calculus; but a cruel fate cut short his life in Florence only a few days after his thirty-ninth birthday.

21   The great developments in mathematics during the days of Descartes and Fermat were in analytic geometry and infinitesimal analysis. It is likely that it was the very success in these branches that made men of the time relatively oblivious to other aspects of mathematics. We already have seen that Fermat found no one to share his fascination with the theory of numbers; pure geometry likewise suffered a wholly undeserved neglect in the same period. The *Conics* of Apollonius once had been among Fermat's favorite works, but analytic methods redirected his views. Meanwhile, the *Conics* had attracted the attention of a practical man with a very impractical imagination—Girard Desargues, an architect and military engineer of Lyons. For some years Desargues had been at Paris, where he was part of the group of mathematicians that we have been considering; but his very unorthodox views on the role of perspective in architecture and geometry met with little favor, and he returned to Lyons to work out his new type of mathematics largely by himself. The result was one of the most unsuccessful great books ever produced. Even the ponderous title was repulsive—*Brouillon projet d'une atteinte aux événemens des rencontres d'un cone avec un plan* (Paris, 1639). This may be translated as "Rough Draft of an Attempt to Deal with the Outcome of a Meeting of a Cone with a Plane," the barbarity of which stands in sharp contrast to the brevity and simplicity of Apollonius' title, *Conics*. The thought on which Desargues' work is based nevertheless is simplicity itself—a thought derived from perspective in Renaissance art and from Kepler's principle of continuity. Everyone knows that a circle, when viewed obliquely, looks like an ellipse, or that the outline of the shadow of a lampshade will be a circle or a hyperbola according as it is projected upon the ceiling or a wall. Shapes and sizes change according to the plane of incidence that cuts the cone of visual rays or of light rays; but certain properties remain the same throughout such changes, and it is these properties that Desargues studied. For one thing, a conic section remains a conic section no matter how many times it undergoes a projection. The conics form a single close-knit family, as Kepler had suggested for somewhat different reasons. But in accepting this view Desargues had to assume, with Kepler, that the parabola has a focus "at infinity" and that parallel lines meet at "a point at infinity." The theory of perspective makes such ideas plausible, for light from the sun ordinarily is considered to be made up of rays that are parallel—comprising a cylinder or a parallel pencil of rays—whereas rays from a terrestrial light source are treated as a cone or a point pencil. The cylinder is merely a cone the vertex of which is infinitely distant, and a parallel pencil of lines is simply a family of lines all of which go through the same point at infinity. Desargues similarly studied a sheaf or bundle of planes through a point, finite or infinite.

Desargues' treatment of the conics is beautiful, although his language is **22** unconventional. He calls a conic section a "coup de rouleau" (that is, incidence with a rolling pin). About the only one of his many new terms that has survived is the word "involution"—that is, pairs of points on a line the product of whose distances from a fixed point is a given constant. He called points in harmonic division a four-point involution, and he showed that this configuration is projectively invariant, a result known, under a different point of view, to Pappus. Because of its harmonic properties, the complete quadrangle played a large role in Desargues' treatment, for he knew that when such a quadrangle (as *ABCD* in Fig. 17.7) is inscribed in a conic, the line through two of the diagonal points (*E*, *F*, and *G* in Fig. 17.7) is the polar line, with respect to the conic, of the third diagonal point. He knew, of

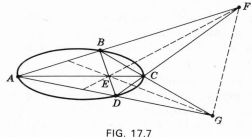

FIG. 17.7

course, that the intersections with the conic of the polar of a point with respect to the conic were the points of contact of the tangents from the point to the conic; and instead of defining a diameter metrically, Desargues introduced it as the polar of a point at infinity. There is a pleasing unity in Desargues' treatment of the conics through projective methods, but it was too thorough a break with the past to meet with acceptance.

The projective geometry of Desargues had a tremendous advantage in generality over the metric geometry of Apollonius, Descartes, and Fermat, for many special cases of a theorem blend into one all-inclusive statement. Yet mathematicians of the time not only failed to accept the methods of the new geometry, they actively opposed them as dangerous and unsound. So rare were copies of Desargues' *Brouillon projet* that by the end of the century all copies had disappeared, for Desargues published his works not to sell but to distribute to friends. The work was completely lost until in 1847 a handwritten copy made by Philippe de la Hire, one of Desargues' few admirers, was found in a Paris library. Part of the responsibility for the neglect of projective geometry falls on Desargues himself, for he wrote in a difficult and unconventional manner. He was not writing for professional scholars, who might have followed his imaginative flights, but for mechanics and practical mathematicians, who did not understand the meaning of his work. Moreover, he used a bizarre new vocabulary full of terms borrowed from botany, a terminology that repelled scholars and practitioners alike. Then, too, the projective approach was not in tune with the times, which had just celebrated triumphs in algebra and analysis. Descartes, who had known Desargues in Paris in 1626 and was with him in 1628 at the siege of LaRochelle, always had a high regard for his nonconformist friend; but even Descartes, when he heard that the *Brouillon projet* would treat of conic sections without the use of algebra, was dismayed. It did not seem possible to say anything about conics that could not more easily be expressed with algebra than without.[10] The commitment to algebra was so strong that for nearly two

---

[10] See W. M. Ivins, Jr., "A Note on Girard Desargues," *Scripta Mathematica*, **9** (1943), 33–48. Although Desargues' works have not been translated into English, accounts are readily available. See, for example, J. L. Coolidge, *A History of Geometrical Methods* (1940).

centuries the beauties of projective geometry went almost unnoticed. Even today the name of Desargues is familiar not as that of the author of *Brouillon projet* but for a proposition that does not appear in the book, the famous theorem of Desargues:

> If two triangles are so situated that lines joining pairs of corresponding vertices are concurrent, then the points of intersection of pairs of corresponding sides are collinear, and conversely.

This theorem,[11] which holds for either two or three dimensions, was first published in 1648 by Desargues' devoted friend and follower Abraham Bosse (1611–1678), an engraver. It appears in a book with the title *Manière universelle de S. Desargues, pour pratiquer la perspective.* The theorem, which Bosse explicitly attributes to Desargues, became, in the nineteenth century, one of the fundamental propositions of projective geometry. It is interesting to note that whereas in three dimensions the theorem is an easy consequence of incidence axioms, the proof for two dimensions requires an additional assumption.

Desargues was the prophet of projective geometry, but he went without honor in his day largely because his most promising disciple, Blaise Pascal, abandoned mathematics for theology. Pascal was a mathematical prodigy. His father, too, was mathematically inclined, and the "limaçon of Pascal" is named for the father, Etienne, rather than for the son, Blaise. The limaçon $r = a + b \cos \theta$ had been known to Jordanus Nemorarius, and possibly to the ancients, as "the conchoid of the circle," but Etienne Pascal studied the curve so thoroughly that, on the suggestion of Roberval, it has ever since borne his name. We are told that he at first kept mathematics books from his son Blaise in order to encourage the youngster to develop other interests, but at the age of twelve the youth showed such geometrical talent that thereafter his mathematical bent was encouraged.

When he was fourteen Blaise joined with his father in the informal meetings of the "Mersenne Academy" at Paris. Here he became familiar with the ideas of Desargues; two years later, in 1640, the young Pascal, then sixteen years old, published an *Essay pour les coniques.* This consisted of only a single printed page—but one of the most fruitful pages in history. It contained the proposition, described by the author as *mysterium hexagrammicum,* which has ever since been known as Pascal's theorem. This states, in essence, that the opposite sides of a hexagon inscribed in a conic intersect in three collinear points. Pascal did not state the theorem in this way, for it is not true unless, as in the case of a regular hexagon inscribed in a circle, one

23

[11] See N. A. Court, "Desargues and his Strange Theorem," *Scripta Mathematica,* **20** (1954), 5–13, 155–164.

resorts to the ideal points and line of projective geometry. Instead he followed the special language of Desargues, saying that if *A*, *B*, *C*, *D*, *E*, and *F* are successive vertices of a hexagon in a conic, and if *P* is the intersection point of *AB* and *DE* and *Q* is the point of intersection of *BC* and *EF* (Fig. 17.8), then *PQ* and *CD* and *FA* are lines "of the same order" (or, as we should say,

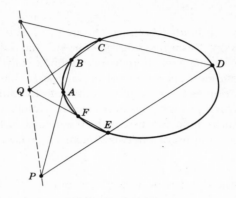

FIG. 17.8

the lines are members of a pencil, whether a point pencil or a parallel pencil). The young Pascal went on to say that he had deduced many corollaries from this theorem, including the construction of the tangent to a conic at a point on the conic. (The construction of the tangent at a point *P* on the conic is easy if we recall that the tangent is a line through two "consecutive points" and apply the Pascal theorem to these and any four other points on the conic.) The inspiration for the little *Essay* was candidly admitted, for after citing a theorem of Desargues the young author wrote, "I should like to say that I owe the little that I have found on this subject to his writings."

The *Essay* was an auspicious opening for a mathematical career, but Pascal's mathematical interests were chameleonlike. He next turned, when he was about eighteen, to plans for a calculating machine, and within a few years he had built and sold some fifty machines. Then in 1648 Pascal became interested in hydrostatics, and the results were the celebrated Puy-de-Dôme experiment confirming the weight of the air and the experiments on fluid pressure that clarified the hydrostatic paradox. In 1654 he returned again to mathematics and worked on two unrelated projects. One of these was to be a "Complete Work on Conics," evidently a continuation of the little *Essay* he had published when sixteen; but this larger work on conics was never printed and is not now extant. Leibniz saw a manuscript copy, and the notes

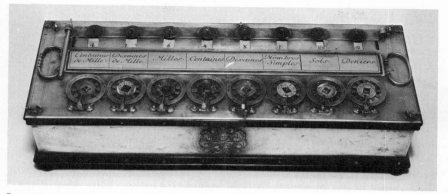

Pascal's calculating machine (from an original model in the collection of Arts and Sciences Department of IBM).

that he took are now all that we have of Pascal's larger work on conics.[12] (Only two copies of the smaller work have survived.) According to Leibniz' notes, the "Complete Work on Conics" contained a section on the familiar three-and-four-line locus and a section on the *magna problema*—to place a given conic on a given cone of revolution. The treatise made use of synthetic methods, for Pascal for some reason never developed a facility in symbolic algebra or saw the role that good notations play in mathematical discovery. In this respect he was far behind his time.

While Pascal in 1654 was working on his *Conics*, his friend, the Chevalier **24** de Méré, raised with him questions such as the following: In eight throws of a die a player is to attempt to throw a one, but after three unsuccessful trials the game is interrupted. How should he be indemnified? Pascal wrote to Fermat on this, and their resulting correspondence became the effective starting point for the modern theory of probability, the thoughts of Cardan of a century before having been overlooked.[13] Although neither Pascal nor Fermat wrote up their results, Huygens in 1657 published a little tract, *De ratiociniis in ludo aleae* ("On Reasoning in Games of Dice") which was prompted by the correspondence of the Frenchmen. Pascal meanwhile had connected the study of probability with the arithmetic triangle, carrying the discussion so far beyond the work of Cardan that the triangular arrangement has ever since been known as Pascal's triangle. The triangle itself was more

[12] For further information on this and other aspects of Pascal's work see Henri Bosmans, "Sur l'oeuvre mathématique de Blaise Pascal," *Mathesis*, **38** (1924), Supplement, pp. 1–59. See also C. B. Boyer, "Pascal: The Man and the Mathematician," *Scripta Mathematica*, **26** (1963), 283–307, and René Taton, "L' 'essay pour les coniques' de Pascal," *Revue d'Histoire des Sciences*, **8** (1955), 1–18.

[13] See Oystein Ore, "Pascal and the Invention of Probability Theory," *American Mathematical Monthly*, **47** (1960), 409–419.

than 600 years old, but Pascal disclosed some new properties, such as the following:

> In every arithmetic triangle, if two cells are contiguous in the same base, the upper is to the lower as the number of cells from the upper to the top of the base is to the number of those from the lower to the bottom inclusive.

(Pascal called positions in the same vertical column, in Fig. 17.9, "cells of the same perpendicular rank," and those in the same horizontal row "cells of the same parallel rank"; cells in the same upward-sloping diagonal he called "cells of the same base.") The method of proof of this property is of more significance than the property itself, for here in 1654 Pascal gave an eminently clear-cut explanation of the method of mathematical induction. Indications of the method can be found earlier in work by Maurolycus; but Pascal had exceptional ability in clarifying concepts, hence he shares, with Fermat and others, in the development of reasoning by recurrence. The name "mathematical induction" seems to have originated much later in De Morgan's article on "Induction (Mathematics)" in the Penny Cyclopaedia of 1838.[14]

1	1	1	1	1	1	1
1	2	3	4	5	6	
1	3	6	10	15		
1	4	10	20			
1	5	15				
1	6					
1						

FIG. 17.9

Fermat hoped to interest Pascal in the theory of numbers, and in 1654 he sent him a statement of one of his most beautiful theorems (unproved until the nineteenth century):

> Every integer is composed of one, two, or three triangular numbers, of one, two, three, or four squares, of one, two, three, four, or five pentagons, of one, two, three, four, five, or six hexagons, and thus to infinity.

Pascal, however, was a mathematical dilettante, as well as a virtuoso, and did not pursue this problem. He did, nevertheless, consider a problem in

---

[14] For an extensive historical account of mathematical induction see W. H. Bussey, "Origin of Mathematical Induction," *American Mathematical Monthly*, **24** (1917), 199–207. Cf. Florian Cajori, "Origin of the Name 'Mathematical Induction,'" *American Mathematical Monthly*, **25** (1918), 197–201, and Hans Freudenthal, "Zur Geschichte der vollständigen Induktion," **6** (1953), 17–37.

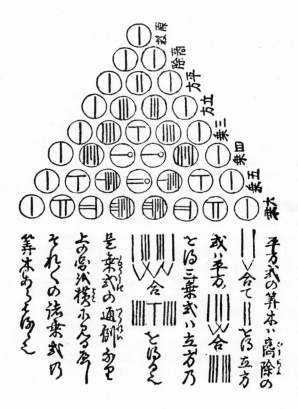

Pascal Triangle in Japan. From Murai Chūzen's *Sampō Dōshi-mon* (1781), showing also the sangi forms of the numerals.

number theory much discussed at the time—a formula for the sum of the *m*th powers of the first *n* consecutive integers—for this he related to the arithmetic triangle, to reasoning by recurrence, and to infinitesimal analysis. The formula, as usual with Pascal, is expressed verbally, but in modern symbolism it is equivalent to

$$_{m+1}C_1 \sum i^m + {}_{m+1}C_2 \sum i^{m-1} + \cdots + {}_{m+1}C_m \sum i = (n+1)^{m+1} - (n+1)$$

where the sums are taken from $i = 1$ to $i = n$. From this formula[15] Pascal

[15] For details see C. B. Boyer, "Pascal's Formula for the Sums of the Powers of the Integers," *Scripta Mathematica*, **9** (1943), 237–244.

easily derived the equivalent of the well-known calculus formula

$$\int_0^a x^n \, dx = \frac{a^{n+1}}{n+1}$$

25   On the night of November 23, 1654, from 10:30 to about 12:30, Pascal experienced a religious ecstasy which caused him to abandon science and mathematics for theology. The result was the writing of the *Lettres provinciales* and the *Pensées*; for only one brief period, in 1658–1659, did Pascal return to mathematics. One night in 1658 toothache or illness prevented him from falling asleep, and as a distraction from the pain he turned to the study of the cycloid. Miraculously, the pain eased, and Pascal took this as a sign from God that the study of mathematics was not displeasing to Him. Having found certain areas, volumes, and centers of gravity associated with the cycloid, Pascal proposed half a dozen such questions to the mathematicians of his day, offering first and second prizes for their solution—and naming Roberval as one of the judges. Publicity and timing were so poor that only two sets of solutions were submitted, and these contained at least some errors in computation. Pascal therefore awarded no prize; but he did publish his own solutions, along with other results, all preceded by a *Histoire de la roulette* (the name usually used for the curve in France), in a series of *Lettres de A. Dettonville* (1658–1659). (The name Amos Dettonville was an anagram of Louis de Montalte, the pseudonym used in the *Lettres provinciales*.) The contest questions and the *Lettres de A. Dettonville* brought interest in the cycloid to a focus, but they also stirred up a hornets' nest of controversy. The two finalists, Antoine de Lalouvère and John Wallis, both capable mathematicians, were disgruntled that prizes were withheld; and the Italian mathematicians were indignant that Pascal's "History of the Cycloid" gave virtually no credit to Torricelli, priority in discovery being conceded only to Roberval.

Much of the material in the *Lettres de A. Dettonville*, such as the equality of the arcs of spirals and parabolas, as well as the cycloid contest questions, had been known to Roberval and Torricelli; but some of this appeared in print for the first time. Among the new results was the equality of the arc length of an arch of the generalized cycloid $x = aK\phi - a \sin \phi, y = a - a \cos \phi$ and the semicircumference of the ellipse $x = 2a(1 + K) \cos \phi$, $y = 2a(1 - K) \sin \phi$. The theorem was expressed rhetorically rather than symbolically, and it was demonstrated in an essentially Archimedean manner, as were most of the demonstrations of Pascal in 1658–1659.

In connection with an integration of the sine function in his 1658 *Traité des sinus du quart de cercle* ("Treatise on the Sines of a Quadrant of a Circle") Pascal came remarkably close to a discovery of the calculus—so close that

Leibniz later wrote that it was upon reading this work by Pascal that a light suddenly burst upon him. Had Pascal not died, like Torricelli, shortly after his thirty-ninth birthday, or had he been more single-mindedly the mathematician, or had he been more attracted by algorithmic methods than by geometry and speculations on the philosophy of mathematics, there is little doubt that he would have anticipated Newton and Leibniz in their greatest discovery. Pascal was without doubt the greatest might-have-been in the history of mathematics; yet he is one of the important connecting links in mathematical development. In this respect, of course, he was not alone. In the next chapter we look at the work of the more immediate precursors of Newton and Leibniz.

BIBLIOGRAPHY

Bosmans, Henri, "Sur l'oeuvre mathématique de Blaise Pascal," *Mathesis*, **38** (1924), Supplement, pp. 1–59.

Boyer, C. B., *History of Analytic Geometry* (New York: Scripta Mathematica, 1956).

Boyer, C. B., *The History of the Calculus* (paperback ed., New York: Dover, 1959).

Boyer, C. B., "Pascal: The Man and the Mathematician," *Scripta Mathematica*, **26** (1963), 283–307.

Brassine, E., *Précis des oeuvres mathématiques de P. Fermat* (Paris, 1853).

Castelnuovo, G., *Le origini del calcolo infinitesimale nell'era moderna* (Bologna: Nicola Zanichelli, 1938).

Coolidge, J. L., *A History of Geometrical Methods* (New York: Oxford University Press, 1940; paperback ed., New York: Dover, 1963).

David, F. N., *Games, Gods and Gambling* (New York: Hafner, 1962).

Descartes, René, *Oeuvres*, ed. by Charles Adam and Paul Tannery (Paris: L. Cerf, 1897–1913, 12 vols. and supplement).

Duhamel, J. M. C., "Mémoire sur la méthode des maxima et minima de Fermat et sur les méthodes des tangentes de Fermat et Descartes", *Mémoires de l'Académie des Sciences de l'Institut Impérial de France*, **32** (1864), 269–330.

Fermat, Pierre de, *Oeuvres*, ed. by Paul Tannery and Charles Henry (Paris: Gauthier Villars, 1891–1922, 4 vols. and supplement).

Genty, Abbé Louis, *L'Influence de Fermat sur son siècle, relativement au progrès de la haute géométrie et du calcul* (Orleans, 1784).

Henry, Charles, "Recherches sur les manuscrits de Pierre de Fermat," *Bullettino di Bibliografia e di Storia delle Scienze Matematiche e Fisiche*, **12** (1879), 477–568, 619–740; **13** (1880), 437–470.

Loria, Gino, *Storia delle matematiche* (Torino: Sten, 1929–1933. 3 vols.).

Ore, Oystein, "Pascal and the Invention of Probability Theory," *American Mathematical Monthly*, **47** (1960), 409–419.

Pascal, Blaise, *Oeuvres*, ed. by Léon Brunschvicg and Pierre Boutroux (Paris: Hachette, 1904–1914, 14 vols.).

Scriba, C. J., "Zur Lösung des 2. Debeauneschen Problems durch Descartes," *Archive for History of Exact Sciences*, **1** (1961), 406–419.

Smith, D. E., and Marcia L. Latham, eds., *The Geometry of René Descartes* (paperback ed., New York: Dover, 1954).

Smith, D. E., *A Source Book in Mathematics* (New York: McGraw-Hill, 1929; paperback ed, New York: Dover, 1959, 2 vols.).

Todhunter, Isaac, *A History of Probability, from the Time of Pascal to that of Laplace* (reprint, New York: Chelsea, 1949).

Toeplitz, Otto, *The Calculus, a Genetic Approach* (Chicago: University of Chicago Press, 1963).

Torricelli, Evangelista, *Opere*, ed. by Gino Loria (Florence: G. Montanari, 1919–1944, 4 vols.).

Turnbull, H. W., *The Great Mathematicians* (New York: New York University Press, 1961).

Walker, Evelyn, *A Study of the Traité des Indivisibles of Gilles Persone de Roberval* (New York: Teachers College, 1932).

Wallner, C. R., "Die Wandlungen des Indivisibilienbegriffs von Cavalieri bis Wallis," *Bibliotheca Mathematica* (3), **4** (1903), 28–47.

Zeuthen, H. G., *Geschichte der Mathematik im 16. und 17. Jahrhundert*, trans. by R. Meyer (Leipzig, 1903).

EXERCISES

1. What possible causes can you suggest for the French supremacy in mathematics in the middle third of the seventeenth century?
2. Is systematic doubt, such as that of Descartes, a help or a hindrance in the development of mathematics? Explain.
3. Compare the influence of Descartes with that of Fermat in the development of mathematics?
4. How do you account for the fact that the projective geometry of Desargues and Pascal met with so little response on the part of their contemporaries?
5. Verify the Descartes-Euler polyhedral formula for each of the five regular solids.
6. Justify fully the construction Descartes used for solving $z^2 = az + b^2$.
7. Devise a construction like Descartes' for $z^2 = az - b^2$ and for $z^2 = b^2 - ay$.
8. Is our equality sign better or worse than that of Descartes? Explain.
9. Sketch the Cartesian parabola or trident for $a = 1$.
10. Verify Torricelli's rectification of $r = ae^{b\theta}$.
11. Find a necessary relation among the coefficients of Descartes' equation $y^2 = ay - bxy + cx - dx^2$ for the curve to be a parabola.
12. Using the method of Descartes, find the normal to $y^2 = 4x$ at the point $(1, 2)$.
13. Using Fermat's method, find the tangent to $x^2 = 4y$ at the point $(2, 1)$.
14. Find an equation of the Cartesian oval such that twice the distance of any point (x, y) from $(-1, 0)$ added to the distance of (x, y) from $(1, 0)$ will be 4.
15. Sketch the folium of Descartes for $a = 1$.
16. Show that by a translation of axes Fermat's equation $xy + a^2 = bx + cy$ can be reduced to the form $xy = K^2$.
17. Prove Fermat's theorem that if the sum of the squares of the distances of a variable point P from any number of fixed lines is a constant, the locus of P is a conic section.
18. Use Fermat's method to find the maximum and minimum values of $(x + 1)(2x^2 + 5x - 7)$.
19. Use the *method* of Fermat (but *not* his formula) to find $\int_0^4 x^5\,dx$.

20. Use Fermat's method (but not his formula) to find $\int_2^\infty 1/x^3 \, dx$.
21. Use Fermat's method (but not his formula) to find $\int_0^1 x^{\frac{3}{2}} \, dx$.
22. Which of the primes between 10 and 30 can be written as a sum of two squares? Find the two squares in cases in which such a sum is possible.
23. Find $2^{2^n} + 1$ for $n = 0, 1, 2, 3$, and verify that these are prime numbers.
24. Verify Fermat's lesser theorem for four cases, using the prime 3.
25. Using straightedge and compasses, construct by Roberval's method a tangent to the cycloid at a point on it.
*26. Verify the result of Roberval and Torricelli on the area under one arch of the cycloid.
*27. Solve the following problem set by Pascal: The area bounded by the above cycloid, the ordinate through the vertex, and a line parallel to the base, is revolved about the x-axis. Find the volume generated.
*28. Verify Roberval's theorem on the length of arc of the spiral and the parabola.
*29. Using Pascal's theorem, draw a tangent to an ellipse at a point on the ellipse.
**30. Verify Torricelli's discovery that the revolution of an infinite area may result in a finite volume.
*31. Using Desargues' complete quadrangle, show how to construct the tangents to a conic from an external point.
*32. Using the methods of differential equations, solve Debeaune's equation $y' = x - y$ for the curve through the origin.
*33. Using $1/(\sqrt{2} - 1) = \sqrt{2} + 1$ and Fermat's method of "infinite descent," prove that $\sqrt{2}$ is not rational.

CHAPTER XVIII

A Transitional Period

Mathematics—the unshaken Foundation of Sciences, and the plentiful Fountain of Advantage to human affairs.

Isaac Barrow

1 With the death of Desargues in 1661, of Pascal in 1662, and of Fermat in 1665, a great period in French mathematics came to a close. It is true that Roberval lived about another decade, but his contributions were no longer significant and his influence was limited by his refusal to publish. About the only mathematician of stature in France at the time was Philippe de Lahire (1640–1718), a disciple of Desargues and, like his master, an architect. Pure geometry obviously appealed to him, and his first work on conics in 1673 was synthetic, but he did not break with the analytic wave of the future. Lahire kept an eye out for a patron; hence in his *Nouveaux élémens des sections coniques* of 1679, dedicated to Colbert, the methods of Descartes came to the fore. The approach is metric and two-dimensional, proceeding, in the case of the ellipse and hyperbola, from the definitions in terms of the sum and difference of focal radii and, in the case of the parabola, from the equality of distances to focus and directrix. But Lahire carried over into analytic geometry some of Desargues' language. The axis of abscissas was the "trunk," points on it were "knots," and ordinates were "branches." Of his analytic language only the term "origin" has survived. Perhaps it was because of his terminology that contemporaries did not give proper weight to a significant point in his *Nouveaux élémens*—Lahire provided one of the first examples of a surface given analytically through an equation in three unknowns—which was the first real step toward solid analytic geometry. He, like Fermat and Descartes, had only a single reference point or origin O on a single line of reference or axis OB, to which he now added the reference or coordinate plane OBA (Fig. 18.1). Lahire found that then the equation of the locus of a point P such that its perpendicular distance PB from the axis shall exceed the distance OB (the abscissa of P) by a fixed quantity a, with respect to his coordinate system is $a^2 + 2ax + x^2 = y^2 + v^2$ (where v is the coordinate that is now generally designated by z). The locus is, of course, a cone.

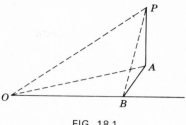

FIG. 18.1

In 1685 Lahire returned to synthetic methods in a book with the simple title *Sectiones conicae.*[1] This might be described as a version by Lahire of the Greek *Conics* of Apollonius translated into Latin from the French language of Desargues. The harmonic properties of the complete quadrangle, poles and polars, tangents and normals, and conjugate diameters are among the familiar topics treated from a projective point of view.

It is interesting to note that today Lahire's name is attached not to anything in his synthetic or analytic treatises on conics, but to a theorem from a paper of 1706 on "roulettes" in the *Mémoires* of the Académie des Sciences. Here he showed that if a smaller circle rolls without slipping along the inside of a larger circle with diameter twice as great, then (1) the locus of a point on the circumference of the smaller circle is a line segment (a diameter of the larger circle), and (2) the locus of a point which is not on the circumference but which is fixed with respect to the smaller circle is an ellipse. As we have seen, the first part of this theorem was known to Nasir Eddin and the second to Copernicus.[2] The name of Lahire deserves to be remembered, but it is a pity that it should be attached to a theorem he was not first to discover. In a sense history has been unkind to Lahire. He was the first modern specialist in geometry, both synthetic and analytic; but geometry was in a state of decline from which it did not revive for about a century.

Lahire was not the only geometer of the time to be unappreciated. In 1672 **2** the Danish mathematician Georg Mohr (1640–1697) published an unusual book entitled *Euclides danicus* in which he showed that any pointwise construction that can be performed with compasses and straightedge (that is, any "plane" problem) can be carried out with compasses alone. Despite all the insistence by Pappus, Descartes, and others on the principle of parsimony, many of the classical constructions were shown by Mohr to have violated

[1] A good but short description of this is given by J. L. Coolidge in *A History of the Conic Sections and Quadric Surfaces* (1945). For a fuller account see Ernst Lehmann, *La Hire und seine Sectiones conicae* (1888).
[2] See C. B. Boyer, "Note on Epicycles and the Ellipse from Copernicus to Lahire," *Isis* **38** (1947), 54–56.

this principle through the use of two instruments where one would suffice! Obviously one cannot draw a straight line with compasses; but if one regards the line as known whenever two distinct points on it are known, then the use of a straightedge in Euclidean geometry is superfluous. So little attention did mathematicians of the time pay to this amazing discovery that geometry using compasses only, without the straightedge, bears the name not of Mohr but of Mascheroni, who rediscovered the principle 125 years later. Mohr's book disappeared so thoroughly that not until 1928, when a copy was accidentally found by a mathematician browsing in a Copenhagen bookstore, did it become known that Mascheroni had been anticipated in proving the supererogation of the straightedge.

3   The year of Mohr's stillborn *Euclides danicus*, 1672, marked the publication in Italy of yet another work on circle-squaring, *Il problema della quadratura del circolo*, by Pietro Mengoli (1625–1686), a third unappreciated mathematician of the time. Mengoli, a clergyman, had grown up under the influence of Cavalieri (whose successor he was at Bologna), Torricelli, and Gregory of St. Vincent. Continuing their work on indivisibles and the area under hyperbolas, Mengoli learned how to handle such problems through a device the usefulness of which now began to be apparent almost for the first time— the use of infinite series. Mengoli saw, for example, that the sum of the alternating harmonic series $\frac{1}{1} - \frac{1}{2} + \frac{1}{3} - \frac{1}{4} + \cdots + (-1^n)/n + \cdots$ is ln 2. He had rediscovered Oresme's conclusion, arrived at by a grouping of terms, that the ordinary harmonic series does not converge, a theorem usually attributed to Jacques Bernoulli in 1689; he also showed the convergence of the reciprocals of the triangular numbers, a result for which Huygens usually is given credit. Mengoli tried unsuccessfully to find the sum of the reciprocals of the squares, and of other powers, a summation achieved a century later by Euler. The *Quadratura del circolo* included the infinite product for π that had been given by Wallis (described later in this chapter). Mengoli's preoccupation with infinite sums and products was an important step toward future developments in mathematics, but none of his countrymen were ready to follow his lead.

4   We have considered three unappreciated mathematicians working in the 1670s, and one reason they were not adequately recognized was that the center of mathematics was not in their countries. France and Italy, once the leaders, were mathematically in decline, and Denmark remained outside the main current. During the period that we are considering—the interval between Descartes and Fermat on the one hand and Newton and Leibniz on the other—there were two regions in particular in which mathematics was thriving: Great Britain and the Low Countries. Here we find not isolated figures, as in France, Italy, and Denmark, but a handful of prominent Britons and another handful of Dutch and Flemish mathematicians.

We have already noted that Descartes had spent a score of years in Holland, and his mathematical influence was decisive in that analytic geometry took root there more quickly than elsewhere in Europe. At Leyden in 1646 Frans van Schooten (1615–1660) had succeeded his father as professor of mathematics, and it was chiefly through the younger van Schooten and his pupils that the rapid development of Cartesian geometry took place. Descartes' *La géométrie* had not originally been published in Latin, the universal language of scholars, and the exposition had been far from clear; both of these handicaps were overcome when van Schooten printed a Latin version in 1649, together with supplementary material. Van Schooten's *Geometria a Renato Des Cartes* ("Geometry by René Descartes") appeared in a greatly expanded two-volume version in 1659–1661, and additional editions were published in 1683 and 1695. Thus it is probably not too much to say that although analytic geometry was introduced by Descartes, it was established by Schooten.

The need for explanatory introductions to Cartesian geometry had been recognized so promptly that an anonymous "Introduction" to it had been composed, but not published, by a "Dutch gentleman" within a year of its appearance. In another year Descartes received and approved a more extensive commentary on the *Geometry*, this one by Debeaune under the title *Notae breves*. The ideas of Descartes were here explained, with greater emphasis upon loci represented by simple second-degree equations, much in the manner of Fermat's *Isagoge*. Debeaune showed, for example, that $y^2 = xy + bx$, $y^2 = -2dy + bx$, and $y^2 = bx - x^2$ represent hyperbolas, parabolas, and ellipses respectively. This work by Debeaune received wide publicity through its inclusion in the 1649 Latin translation of the *Geometria*, together with further commentary by Schooten.

A more extensive contribution to analytic geometry was composed in 1658 by one of Schooten's associates, Jan De Witt (1629–1672), the well-known Grand Pensionary of Holland. De Witt had studied law at Leyden, but he had acquired a taste for mathematics while living in Schooten's house. He led a hectic life while directing the affairs of the United Provinces through periods of war in which he opposed the designs of Louis XIV. When in 1672 the French invaded the Netherlands, De Witt was dismissed from office by the Orange party and seized by an infuriated mob that tore him to pieces. Although he had been a man of action, he had found the time in his earlier years to compose a work entitled *Elementa curvarum*. This is divided into two parts, the first of which gives various kinematic and planimetric definitions of the conic sections. Among these are the focus-directrix ratio definitions;

our word "directrix" is due to him. Another construction of the ellipse that he gave is through the now familiar use of two concentric circles with the eccentric angle as parameter. Here the treatment is largely synthetic; but Book II by contrast makes such systematic use of coordinates that it has been described, with some justification, as the first textbook on analytic geometry. Descartes' *Géométrie* had not been a textbook in any real sense, and the exposition of Fermat had not been published until 1679, whereas De Witt's *Elementa curvarum* appeared as part of the 1659–1661 edition of Schooten's *Geometria a Renato Des Cartes*. The purpose of De Witt's work is to reduce all second-degree equations in x and y to canonical form through translation and rotation of axes. He knew how to recognize when such an equation represented an ellipse, when a parabola, and when a hyperbola, according as the so-called characteristic is negative, zero, or positive.[3]

Only a year before his tragic death De Witt combined the aims of the statesman with the views of a mathematician in his *A Treatise on Life Annuities* (1671), motivated perhaps by the little essay by Huygens on probabilities. In this *Treatise* De Witt expressed what now would be described as the notion of mathematical expectation; and in his correspondence with Hudde, he considered the problem of an annuity based on the last survivor of two or more persons.

6   In 1656–1657 Schooten had published a work of his own, *Exercitationes mathematicae*, in which he gave new results in the application of algebra to geometry. Included are discoveries made also by his most capable disciples, such as Johann Hudde (1629–1704), patrician who served for some thirty years as burgomaster of Amsterdam. Hudde corresponded with Huygens and De Witt on the maintenance of canals and on problems of probability and life expectancy; in 1672 he directed the work of inundating Holland to obstruct the advance of the French army.[4] In 1656 Hudde had written on the quadrature of the hyperbola by means of infinite series, as had Mengoli; but the manuscript has been lost. In Schooten's *Exercitationes* there is a section by Hudde on a study of coordinates of a fourth-degree surface, an anticipation of solid analytic geometry antedating even that of Lahire, although less explicitly described. Moreover, it appears that Hudde was the first mathematician to permit a literal coefficient in an equation to represent any real number, whether positive or negative. This final step in the process of

---

[3] For further details see C. B. Boyer, *History of Analytic Geometry* (1956), pp. 114–117, or J. L. Coolidge, *The Mathematics of Great Amateurs* (1949), pp. 119–131.

[4] For further details see the article, "Johannes Hudde, heer van Waveren en Sloterdijk," in *Nieuw Nederlandsch Biografisch Woordenboek* (Leiden, 1911), Vol. I, cols. 1172–1176. See also Joy B. Easton, "Johan De Witt's Kinematic Constructions of the Conics," *Mathematics Teacher*, **56** (1963), 632–635, and Karlheinz Haas, "Die mathematischen Arbeiten von Johann Hudde," *Centaurus*, **4** (1956), 235–284.

generalizing the notations of Viète in the theory of equations was made in a work by Hudde entitled *De reductione aequationum,* which also formed part of the 1659–1661 Schooten edition of Descartes' *Geometry.*

The two most popular subjects in Hudde's day were analytic geometry and mathematical analysis, and the burgomaster-to-be contributed to both. In 1657–1658 Hudde had discovered two rules pointing clearly toward algorithms of the calculus:

1. If r is a double root of the polynomial equation

$$a_0 x^n + a_1 x^{n-1} + \cdots + a_{n-1} x + a_n = 0$$

and if $b_0, b_1, \ldots, b_{n-1}, b_n$ are numbers in arithmetic progression, then r is a root also of

$$a_0 b_0 x^n + a_1 b_1 x^{n-1} + \cdots + a_{n-1} b_{n-1} x + a_n b_n = 0$$

2. If for $x = a$ the polynomial

$$a_0 x^n + a_1 x^{n-1} + \cdots + a_{n-1} x + a_n$$

takes on a relative maximum or minimum value, then a is a root of the equation

$$n a_0 x^n + (n-1) a_1 x^{n-1} + \cdots + 2 a_{n-2} x^2 + a_{n-1} x = 0$$

The first of these "Hudde's rules" is a camouflaged form of the modern theorem that if r is a double root of $f(x) = 0$, then r is also a root of $f'(x) = 0$. The second is a slight modification of Fermat's theorem that today appears in the form that if $f(a)$ is a relative maximum or minimum value of a polynomial $f(x)$, then $f'(a) = 0$. Note that not only did area and tangent problems antedate the calculus of Newton and Leibniz, but also the play on coefficients and exponents so familiar in elementary rules of the calculus.

The rules of Hudde were widely known, for they were published by **7** Schooten in 1659 in Volume I of *Geometria a Renato Des Cartes.* A few years earlier a similar rule for tangents had been used by another representative from the Low Countries, the canon René François de Sluse (1622–1685), a native of Liége who came of a distinguished Walloon family. He had studied in Lyons and Rome, where he may have become familiar with the work of Italian mathematicians. Possibly through Torricelli, perhaps independently, Sluse arrived in 1652 at a routine for finding the tangent to a curve whose equation is of the form $f(x, y) = 0$, where f is a polynomial. The rule, not published until 1673, when it appeared in the *Philosophical Transactions* of the Royal Society, may be stated as follows:[5] the subtangent will be the quotient obtained by placing in the numerator all the terms containing y, each multiplied by the exponent of the power of y appearing in it, and placing in the denominator all the terms containing x, each multiplied by the exponent of the power of x appearing in it and then divided by x. This is, of course, equivalent to forming the quotient now written as yf_y/f_x, a result known in about 1659 also to Hudde. Such instances show how discoveries in the calculus were crowding upon each other even before the work of Newton.

Sluse, sharing in the tradition of the Low Countries, was quite active also in promoting Cartesian geometry, even though he preferred the A and E of Viète and Fermat to the x and y of Descartes. In 1659 he published a popular book, *Mesolabum* ("Of Means"), in which he pursued the familar topic on the geometric constructions of the roots of equations. He showed that given any conic, one can construct the roots of any cubic or quartic equation through the intersection of the conic and a circle. The name of Sluse is attached also to a family of curves that he introduced in his correspondence with Huygens and Pascal in 1657–1658. These so-called "pearls" of Sluse, so named by Pascal, are curves given by equations of the form $y^m = kx^n(a - x)^b$. Sluse mistakenly thought that such cases as $y = x^2(a - x)$ were pearl-shaped, for, negative coordinates not then being understood, Sluse assumed symmetry with respect to the axis (of abscissas). However, Christiaan Huygens (1692–1695), who had the reputation of being Schooten's best pupil, found the maximum and minimum points and the point of inflection and was able to sketch the curve correctly for both positive and negative coordinates. Points of inflection had been found by many men before Huygens, including Fermat and Roberval.

8   Huygens was a scientist of international reputation who is recalled for the principle that bears his name in the wave theory of light, the observation of the rings of Saturn, and the effective invention of the pendulum clock. It was in connection with his search for improvements in horology that he made his most important mathematical discovery. He knew that the oscillations of a simple pendulum are not strictly isochronous, but depend upon the magnitude of the swing. To phrase it differently, if an object is placed on the side of a smooth hemispherical bowl and released, the time it takes to reach the lowest point will be almost, but not quite, independent of the height from which it is released. Now it happened that Huygens invented the pendulum clock at just about the time of the Pascal cycloid contest, in 1658, and it occurred to him to consider what would happen if one were to replace the hemispherical bowl by one whose cross section is an inverted cycloidal arch. He was

[5] See L. Rosenfeld, "René-François de Sluse et le problème des tangents," *Isis*, **10** (1928), 416–434.

Christiaan Huygens.

delighted to find that for such a bowl the object will reach the lowest point in exactly the same time, no matter from what height on the inner surface of the bowl the object is released. That is, the cycloid is the truly isochronous curve; on an inverted cycloidal arch an object will slide from any point to the bottom in exactly the same time, no matter what the starting point. But a big question remained. How does one get a pendulum to oscillate in a cycloidal, rather than a circular arc? Here Huygens made a further beautiful discovery. If one suspends from a point P at the cusp between two inverted cycloidal semi-arches PQ and PR (Fig. 18.2) a pendulum the length of which is equal to the length of one of the semiarches, the pendulum bob will swing in an arc that is an arch of a cycloid QSR of exactly the same size and shape as the cycloid of which arcs PQ and PR are parts. In other words, if the pendulum of the clock oscillates between cycloidal jaws, it will be truly isochronous.

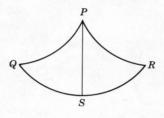

FIG. 18.2

Huygens made some pendulum clocks with cycloidal jaws, but he found that in operation they were no more accurate than those depending on the oscillations of an ordinary simple pendulum, which are nearly isochronous for very small swings. However, Huygens in this investigation had made a discovery of capital mathematical significance—the involute of a cycloid is a similar cycloid, or inversely, the evolute of a cycloid is a similar cycloid. This theorem, and further results on involutes and evolutes for other curves, were proved by Huygens in an essentially Archimedean and Fermatian manner by taking neighboring points and noting the result when the interval

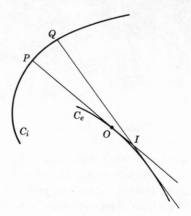

FIG. 18.3

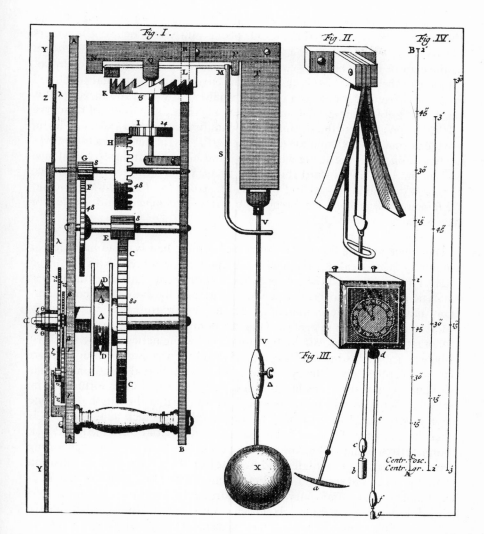

Diagrams from Huygens' *Horologium oscillatorium* (1673). The one labeled Fig. II shows the cycloidal jaws that caused the pendulum to swing in a cycloidal arc.

vanishes. Descartes and Fermat had used his device for normals and tangents to a curve, and now Huygens applied it to find what we call the radius of curvature of a plane curve. If at neighboring points P and Q on a curve (Fig. 18.3) one finds the normals and their point of intersection I, then as Q approaches P along the curve, the variable point I tends toward a fixed point

O, which is called the center of curvature of the curve for the point P, and the distance OP is known as the radius of curvature. The locus of the centers of curvature O for points P on a given curve C_i lie on a second curve C_e known as the evolute of C_i; and any curve C_i of which C_e is the evolute is called an involute of the curve C_e. It is clear that the envelope of the normals to C_i will by C_e, a curve tangent to each of the normals. In Fig. 18.2 the curve QPR is the evolute of the curve QSR and the curve QSR is an involute of the curve QPR. The positions of the string, as the pendulum bob swings back and forth, are the normals to QSR and the tangents to QPR. As the pendulum bob moves farther to one side, the string winds more and more about the cycloidal jaw, and as the bob falls toward the lowest point S, the string unwinds. Hence Huygens described the cycloid QSR as *ex evolutione descripta*, the cycloid QPR being the *evoluta*. (In French the terms *développante* and *développée* have since been adopted.)

9   The concepts of radius of curvature and evolute had been adumbrated in the purely theoretical work on *Conics* of Apollonius, but only with Huygens' interest in horology did the concepts find a permanent place in mathematics. Analytic geometry had been a product of essentially theoretical considerations, but Huygens' development of the idea of curvature was prompted by practical concerns. An interplay of the two points of view, the theoretical and the practical, often proves to be fruitful in mathematics, as the work of Huygens aptly illustrates. His cycloidal pendulum presented him with an obvious rectification of the cycloid, a result that Roberval had found earlier but had not published. The fact that the arc QS (in Fig. 18.2) is formed as the pendulum string winds about the curve QP shows that the length of the line PS is exactly equal to the length of the arc QP. Inasmuch as the line PS is twice the diameter of the circle that generates the cycloid QSR, the length of a complete arch of the cycloid must be four times the diameter of the generating circle. The theory of involutes and evolutes similarly led to the rectification of many other curves, and the Peripatetic-Cartesian dogma of the non-rectifiability of algebraic curves came more seriously into question. In 1658 one of Huygens' associates, Heinrich van Heuraet (1633–1660?), also a protégé of Schooten, discovered that the semicubical parabola $ay^2 = x^3$ can be rectified by Euclidean means, thus ending the uncertainty. The disclosure appeared in 1659 as one of the more important aspects of Schooten's *Geometria a Renato Des Cartes*. This result had been reached independently a little earlier by the Englishman William Neil (1637–1670), and was known independently a little later to Fermat in France, constituting another striking case of virtual simultaneity of discovery.

Of all Fermat's discoveries in mathematics it was only the rectification of the semicubical parabola, usually known as Neil's parabola, that was published by him. The solution appeared in 1660 as a supplement in the *Veterum geometria promota in septem de cycloide libris* ("Geometry of the Ancients Promoted in Seven Books on the Cycloid") by Antoine de Lalouvère (1600–1664), the circle-squarer who had striven for Pascal's prize. The rectification of Fermat was found by comparing a small arc of a curve with the circumscribed figure made up of tangents at the extremities of the arc. Van Heuraet's method was based on the rate of change in the arc, expressed in modern notation by the equation $ds/dx = \sqrt{1 + (y')^2}$.

The rectification by Neil depended on the recognition, already noted by Wallis in *Arithmetica infinitorum*, that a small arc is virtually the hypotenuse of a right triangle whose sides are the increments in the abscissa and the ordinate—that is, on the equivalent of the modern formula $ds = \sqrt{dx^2 + dy^2}$. Neil's rectification was published in 1659 by John Wallis in a treatise entitled *Tractatus duo, prior de cycloide, posterior de cissoide* ("Two Treatises, the First on the Cycloid, the Second on the Cissoid"). This work followed by a few months the work of Pascal on the cycloid, indicating the extent to which cycloid fever had seized mathematicians just before the invention of the calculus.

Huygens' work on involutes and evolutes was not published until 1673, when it appeared in his celebrated *Horologium oscillatorium*. This treatise on pendulum clocks is a classic that served as an introduction to Newton's *Principia* a little more than a decade later. It contained the law of centripetal force for circular motion, Huygens' law for pendular motion, the principle of the conservation of kinetic energy, and other important results in mechanics. The book was published at Paris, for Huygens had been in touch with the work of Pascal and Fermat and in 1666 had gone to Paris as a member of the newly established Académie des Sciences. He remained there until 1681, when the threatened revocation of the Edict of Nantes (effected in 1685) prompted him, a Protestant, to leave Catholic France; the death of Colbert in 1683 confirmed his decision not to return. He had earlier visited London, and throughout his life he maintained a broad interest in all things mathematical, but especially in higher plane curves. He rectified the cissoid and studied the tractrix; whereas Galileo had thought that the catenary was a parabola, Huygens showed that it is a nonalgebraic curve. In 1656 he had applied infinitesimal analysis to the conics, reducing the rectification of the parabola to the quadrature of the hyperbola (that is, to finding a logarithm). By the next year Huygens had become the first one to find the surface area of a segment of a paraboloid of revolution (the "conoid" of Archimedes), showing that the complanation can be achieved by elementary means.[6]

---

[6] There is no good summary in English of Huygens' mathematical work, but all that he did will be found in the sumptuous edition of his *Oeuvres complètes* (1888–1950).

10   Van Schooten died in 1660, the year in which the Royal Society was
founded in England (the charter, however, was granted in 1662), and the date
can be taken as marking a new shift in the mathematical center of the world.
The Leyden group, gathered about Schooten, was losing its momentum, and
it suffered a further blow when Huygens left for Paris in 1666. Meanwhile
a vigorous development in mathematics had been taking place in Great
Britain, and this was further encouraged by the formation of the Royal
Society, one of the oldest scientific organizations still in existence. (The
Accademia dei Lincei, founded in 1603 at Rome, seems to be the oldest.)
William Oughtred had died in 1660, but he left behind a brilliant student in
the person of John Wallis (1616–1703), the most influential English predecessor of Newton. Oughtred, an Episcopal minister, had given free lessons in
mathematics, and Wallis was one who profited most from the instruction.
Wallis, too, entered Holy Orders, but he spent most of his time as a mathematician. He had been educated at Cambridge, but in 1649 he was appointed
Savilian professor of geometry at Oxford, holding the chair that had first
been filled by Briggs when it was established in 1619. Wallis was known to
be a Royalist, although the regime of Cromwell was not averse to using his
services in the deciphering of secret codes; and when Charles II was restored
to the throne, Wallis became the king's chaplain. Wallis was a charter member of the Royal Society, which he had helped to organize. Earlier he had
published, in 1655, two very important books, one in analytic geometry, the
other in infinite analysis. These were the two leading branches of mathematics
at the time, and the genius of Wallis was well suited to advance them.

11   The *Tractatus de sectionibus conicis* of Wallis did for analytic geometry
in England what De Witt's *Elementa curvarum* had done for the subject on
the Continent. Wallis complained, in fact, that De Witt's work was an imita-
tion of his own *Tractatus*, but De Witt's treatise, although published four
years after that of Wallis, had actually been composed before 1655. The books
of both men may be described as the completion of the arithmetization of
conic sections that had been begun by Descartes. Wallis in particular replaced
geometrical concepts by numerical ones wherever possible. Even proportion,
the stronghold of ancient geometry, Wallis held to be an arithmetic concept.
In this his attitude represented the tendency of mathematics for at least the
following century, but it should be remarked that such a movement was
without a solid foundation, since real numbers had not been defined. The
work of Wallis is a good illustration of the fact, so often seen in the history
of mathematics, that an occasional disregard of the demands of logical rigor
can have a salutary effect on progress.

The *Conics* of Wallis opened by paying lip service to the generation of the
curves as sections of a cone, yet he deduced all of the familiar properties

through plane coordinate methods from the three standard forms $e^2 = ld - ld^2/t$, $p^2 = ld$, and $h^2 = ld + ld^2/t$, where e, p, and h are the ordinates of the ellipse, parabola, and hyperbola respectively, corresponding to abscissas d measured from a vertex at the origin, and where l and t are the latus rectum and "diameter" or axis. Later still he took these equations as the definitions of the conic sections, considered "absolutely," that is, without reference to the cone. Here he was closer even than Fermat to the modern definition of a conic as the locus of points on a plane coordinate system whose coordinates satisfy an equation of second degree in two variables, a fact of which Descartes had been aware but which he had not emphasized.

Had Wallis' *Conics* not appeared, the loss would not have been serious, **12** for De Witt's work appeared only four years later. However, there was no substitute for the *Arithmetica infinitorum* of Wallis, which also was published in 1655. Here Wallis arithmetized the *Geometria indivisibilibus* of Cavalieri, as he had arithmetized the *Conics* of Apollonius. Whereas Cavalieri had arrived at the result

$$\int_0^a x^m \, dx = \frac{a^{m+1}}{m+1}$$

through a laborious pairing of geometric indivisibles in a parallelogram with those in one of the two triangles into which a diagonal divides it, Wallis abandoned the geometrical background after having associated the infinitely many indivisibles in the figures with numerical values. If, for example, one wishes to compare the squares of the indivisibles in the triangle with the squares of the indivisibles in the parallelogram, one takes the length of the first indivisible in the triangle as zero, the second as one, the third as two, and so on up to the last, of length $n - 1$, if there are n indivisibles. The ratio of the squares of the indivisibles in the two figures would then be

$$\frac{0^2 + 1^2}{1^2 + 1^2} \quad \text{or} \quad \frac{1}{2} = \frac{1}{3} + \frac{1}{6}$$

if there were only two indivisibles in each; or

$$\frac{0^2 + 1^2 + 2^2}{2^2 + 2^2 + 2^2} = \frac{5}{12} = \frac{1}{3} + \frac{1}{12}$$

if there were three; or

$$\frac{0^2 + 1^2 + 2^2 + 3^2}{3^2 + 3^2 + 3^2 + 3^2} = \frac{14}{36} = \frac{1}{3} + \frac{1}{18}$$

if there were four. For $n + 1$ indivisibles the result is

$$\frac{0^2 + 1^2 + 2^2 + \cdots + (n-1)^2 + n^2}{n^2 + n^2 + n^2 + \cdots + n^2 + n^2} = \frac{1}{3} + \frac{1}{6n}$$

and if n is infinite, the ratio obviously is $\frac{1}{3}$. (For n infinite, the remainder term $1/6n$ becomes $1/\infty$ or zero. Wallis here was the first one to use the now familiar "love knot" symbol for infinity.) This is, of course, the equivalent of saying that $\int_0^1 x^2 \, dx = \frac{1}{3}$; Wallis extended the same procedure to higher integral powers of x. By incomplete induction he concluded that

$$\int_0^1 x^m \, dx = \frac{1}{m+1}$$

for all integral values of m.

Fermat rightly criticized Wallis' induction, for it lacks the rigor of the method of complete induction that Fermat and Pascal frequently used. Moreover, Wallis followed a still more questionable principle of interpolation under which he assumed that his result held for fractional values of m also, as well as for negative values (except $m = -1$). He even had the hardihood to assume that the formula held for irrational powers—the earliest statement in the calculus concerning what now would be called "a higher transcendental function." The use of exponential notation for fractional and negative powers was an important generalization of suggestions made earlier, as by Oresme and Stevin, but Wallis did not give a sound basis for his extension of the Cartesian exponentiation. He merely gave some particular instances of various cases—that a term or number with index -2 multiplied by the same term or number with index -3 is the term with index -5; or a term with index -3 multiplied by one with index 2 is one with index -1. Then he all too casually concluded; "And the same thing will happen in any other cases whatsoever of this sort, and hence the proposition is proved."[7] Wallis was long on discovery, but short on rigor, as the French eagerly pointed out.

Wallis was a chauvinistic Englishman, and when he later (in 1685) published his *Treatise of Algebra, Both Historical and Practical*, he belittled the work of Descartes, arguing, very unfairly, that most of it had been taken from Harriot's *Artis analyticae praxis*. The fact that his solutions of the Pascal contest questions had been rejected as not worthy of the prize evidently did not ameliorate his anti-Gallic bias. Wallis seems all too readily to have suspected others of ill will. In his *Treatise of Algebra* he wrote:

That it [algebra] was in use of old among the *Grecians*, we need not doubt; but studiously concealed (by them) as a great Secret. Examples we have of it in *Euclid*, at least in *Theo* upon him; who ascribes the invention of it (amongst them) to Plato.

One who has read our chapters on Greece will see that Wallis was far better

---

[7] See D. E. Smith, *A Source Book in Mathematics*, pp. 217–218.

as a mathematician than as a historian, for he equates algebra (or the analytics of Viète) with the ancient geometrical analysis.

At the time that Wallis sent in his reply to the Pascal challenge, Christopher **13** Wren (1632–1723) sent Pascal his rectification of the cycloid. Wren was educated at Oxford and later held there the Savilian professorship in astronomy. He, too, was elected to the Royal Society, of which he was president for a few years; had not the great fire of 1666 destroyed much of London, Wren might now be known as a mathematician, rather than as architect of St. Paul's Cathedral and some fifty other churches. The mathematical circle to which Wren and Wallis belonged in 1657–1658 evidently was applying the equivalent of the formula for arc length $ds^2 = dx^2 + dy^2$ to various curves, and was meeting with brilliant success. We mentioned earlier that William Neil when only twenty years old succeeded first in rectifying his curve in 1657; Wren found the length of the cycloid a year later. Both discoveries were incorporated, with due credit to the discoverers, by Wallis in his *Tractatus duo* of 1659, a book on infinitesimal problems related to the cycloid and cissoid. Neil seems not to have made other contributions to mathematics before his untimely death at the age of thirty-two. Wren's interests soon turned to physics and then to architecture; but in 1669 he published in the *Philosophical Transactions* the discovery that on the hyperboloid of revolution of one sheet there are two families of generating lines.

It is a pity that the geometry of surfaces and curves in three dimensions was then attracting so little attention that almost a century later solid analytic geometry still was virtually undeveloped. Wallis in his *Algebra* of 1685 included a study of a surface that belonged to the class now known as conoids (not, of course, in the Archimedean sense). Wallis' surface, which he called the "cono-cuneus" (or conical wedge), can be described as follows: Let C be a circle, let L be a line parallel to the plane of C, and let P be a plane perpendicular to L. Then the cono-cuneus is the totality of lines that are parallel to P and pass through points of L and C. Wallis suggested other conoidal surfaces obtained by replacing the circle C by a conic; and in his *Mechanica* of 1670 he had noted the parabolic sections on Wren's hyperboloid (or "hyperbolic cylindroid"). However, Wallis did not give equations for the surfaces, nor did he arithmetize geometry of three dimensions as he had plane geometry.

Wallis, undoubtedly the leading English mathematician before Newton, **14** made his most important contributions in infinitesimal analysis. Among these was one in which, while evaluating $\int_0^1 \sqrt{x - x^2}\, dx$, he anticipated some of the work of Euler on the gamma or factorial function. From the work of Cavalieri, Fermat, and others Wallis knew that this integral represents the area under the semicircle $y = \sqrt{x - x^2}$ and that this area therefore is $\pi/8$; but how can one obtain the answer through a direct evaluation of the integral by infinitesimal devices? Wallis could not answer this question, but his method of induction and interpolation produced an interesting result. After an evaluation of $\int_0^1 (x - x^2)^n \, dx$ for several positive integral values of n, Wallis arrived by incomplete induction at the conclusion that the value of this integral is $(n!)^2/(2n + 1)!$ Assuming that the formula holds for fractional values of n as well, Wallis concluded that

$$\int_0^1 \sqrt{x - x^2} \, dx = (\tfrac{1}{2}!)^2/2!$$

hence $\pi/8 = \tfrac{1}{2}(\tfrac{1}{2}!)^2$ or $\tfrac{1}{2}! = \sqrt{\pi}/2$. This is a special case of the Eulerian beta function, $B(m, n) = \int_0^1 x^{m-1}(1 - x)^{n-1} \, dx$, where $m = \tfrac{3}{2}$ and $n = \tfrac{3}{2}$.

Thomas Hobbes (1588–1679) was foremost among those who criticized Wallis' arithmetization of geometry, objecting strenuously to "the whole herd of them who apply their algebra to geometry" and referring to the *Arithmetica infinitorum* as "a scab of symbols." Hobbes, however, had more mathematical conceit than ability, insisting that he had squared the circle and had solved the other ancient geometrical problems. Wallis could well afford to disregard Hobbes and go on to further discoveries. Among his best known results is the infinite product

$$\frac{2}{\pi} = \frac{1 \cdot 3 \cdot 3 \cdot 5 \cdot 5 \cdot 7 \cdots}{2 \cdot 2 \cdot 4 \cdot 4 \cdot 6 \cdot 6 \cdots}$$

This expression can readily be obtained from the modern theorem

$$\lim_{n \to \infty} \frac{\displaystyle\int_0^{\pi/2} \sin^n x \, dx}{\displaystyle\int_0^{\pi/2} \sin^{n+1} x \, dx} = 1$$

and the formulas

$$\int_0^{\pi/2} \sin^m x \, dx = \frac{(m - 1)!!}{m!!}$$

for m an odd integer and

$$\int_0^{\pi/2} \sin^m x \, dx = \frac{(m - 1)!!}{m!!} \frac{\pi}{2}$$

for m even. (The symbol $m!!$ represents the product $m(m - 2)(m - 4) \ldots$ which terminates in 1 or 2 according as m is odd or even.) Hence the above expressions for $\int_0^{\pi/2} \sin^m x \, dx$ are known as Wallis' formulas. However, the method that Wallis actually used to achieve his product for $2/\pi$ was in reality again based on his principles of induction and interpolation, applied this time to $\int_0^1 \sqrt{1 - x^2} \, dx$, which he was unable to evaluate directly for lack of the binomial theorem.[8]

The binomial theorem for integral powers had been known in Europe at **15** least since 1527, but Wallis was unable, surprisingly, to apply his method of interpolation here. It looks as though this result may have been known to the young Scotsman James Gregory (1638–1675), a predecessor of Newton who died when only thirty-six. Gregory evidently had come in contact with the mathematics of several countries. His great-uncle Alexander Anderson (1582–1620?) had edited Viète's works, and James Gregory had studied mathematics not only at school in Aberdeen, but also with his older brother, David Gregory (1627–1720); a wealthy patron had introduced him to John Collins (1625–1683), librarian of the Royal Society. Collins was to British mathematicians what Mersenne had been to the French a generation earlier— the correspondent extraordinary. In 1663 Gregory went to Italy where the patron introduced him to the successors of Torricelli, especially Stefano degli Angeli (1623–1697). The many works of Angeli, protégé of Cardinal Michelangelo Ricci (1619–1682) who had been a close friend of Torricelli, were almost all on infinitesimal methods, with emphasis on the quadrature of generalized spirals, parabolas, and hyperbolas. Gregory studied with Angeli for several years (1664–1668) before returning to London, and it is likely that it was in Italy, through Mengoli and Angeli, that Gregory came to appreciate the power of infinite series expansions of functions and of infinite processes in general. In 1667, consequently, he published at Padua a work entitled *Vera circuli et hyperbolae quadratura*, containing very significant results in infinitesimal analysis.

For one thing, Gregory extended the Archimedean algorithm to the quadrature of ellipses and hyperbolas. He took an inscribed triangle of area a_0 and a circumscribed quadrilateral of area A_0; by successively doubling the number of sides of these figures he formed the sequence $a_0, A_0, a_1, A_1, a_2, A_2, a_3, A_3, \ldots$ and showed that a_n is the geometric mean of the two terms immediately preceding and A_n the harmonic mean of the two preceding terms. Thus he had two sequences—that of the inscribed areas and that of the circumscribed areas—both converging to the area of the conic; he used these to get very good approximations to elliptic and hyperbolic sectors. Incidentally, the word "converge" was here used by Gregory in this sense for the first time. Through this infinite process Gregory sought,

---

[8] For other aspects of his work see J. F. Scott, *The Mathematical Works of John Wallis, D.D., F.R.S. (1616–1703)* (1938), and C. J. Scriba, *Studien zur Mathematik des John Wallis (1616–1703)* (1966).

unsuccessfully, to prove the impossibility of squaring the circle by algebraic means. Huygens, regarded as the leading mathematician of the day, believed that π could be expressed algebraically, and a dispute arose over the validity of Gregory's methods. The question of π's transcendence was a difficult one, and it was to be another two centuries before it was resolved in Gregory's favor.

16   In 1668 Gregory published two more works, bringing together results from France, Italy, Holland, and England, as well as new discoveries of his own. One of these, *Geometriae pars universalis* ("The Universal Part of Geometry"), was published at Padua; the other, *Exercitationes geometricae* ("Geometrical Exercises"), at London.[9] As the title of the first book implies, Gregory broke from the Cartesian distinction between "geometrical" and "mechanical" curves. He preferred to divide mathematics into "general" and "special" groups of theorems, rather than into algebraic and transcendental functions. Gregory did not wish to distinguish even between algebraic and geometric methods, and consequently his work appeared in an essentially geometric garb that is not easy to follow. Had he expressed his work analytically, he might have anticipated Newton in the invention of the calculus, for virtually all the fundamental elements were known to him by the end of 1668. He was thoroughly familiar with quadratures and rectifications and probably saw that these are the inverses of tangent problems. He even knew the equivalent of $\int \sec x \, dx = \ln (\sec x + \tan x)$. He had found independently the binomial theorem for fractional powers, a result known earlier to Newton (but as yet unpublished), and he had, through a process equivalent to successive differentiation, discovered the Taylor series more than forty years before Taylor published it.[10] The Maclaurin series for $\tan x$ and $\sec x$ and for $\arctan x$ and $\operatorname{arcsec} x$ were all known to him, but only one of these, the series for $\arctan x$, bears his name. He could have learned in Italy that the area under the curve $y = 1/(1 + x^2)$, from $x = 0$ to $x = x$, is $\arctan x$; and a simple long division converts $1/(1 + x^2)$ to $1 - x^2 + x^4 - x^6 + \cdots$. Hence it is at once apparent from Cavalieri's formula that

$$\int_0^x \frac{dx}{1 + x^2} = \arctan x = x - \frac{x^3}{3} + \frac{x^5}{5} - \frac{x^7}{7} + \cdots$$

This result is still known as "Gregory's series."

[9] A thorough account of Gregory's work is given in H. W. Turnbull, *James Gregory Tercentenary Memorial Volume* (1939).

[10] The history of the Taylor series is very complicated indeed. Some anticipation of it is found in India before 1550. See C. T. Rajagopal and T. V. Vedamurthi, "On the Hindu Proof of Gregory's Series," *Scripta Mathematica*, **17** (1951), 65–74; see also **15** (1949), 201–209, and **18** (1952), 25–30.

A result somewhat analogous to Gregory's series was derived at about the same time by Nicolaus Mercator (1620–1687) and published in his *Logar-ithmotechnia* of 1668. Mercator (real name Kaufmann) was born at Holstein in Denmark, but he lived in London for a long time and became one of the first members of the Royal Society. In 1683 he went to France and designed the fountains at Versailles; he died at Paris four years later. The first part of Mercator's *Logarithmotechnia* is on the calculation of logarithms by methods stemming from those of Napier and Briggs; the second part contains various approximation formulas for logarithms, one of which is essentially that now known as "Mercator's series." From the work of Gregory of St. Vincent it had been known that the area under the hyperbola $y = 1/(1 + x)$, from $x = 0$ to $x = x$, is $\ln(1 + x)$. Hence, using Gregory's method of long division followed by integration, we have

$$\int_0^x \frac{dx}{1 + x} = \int_0^x (1 - x + x^2 - x^3 + \cdots)\, dx = \ln(1 + x)$$

$$= \frac{x}{1} - \frac{x^2}{2} + \frac{x^3}{3} - \frac{x^4}{4} + \cdots$$

Mercator took over from Mengoli the name "natural logarithms" for values that are derived by means of this series. Although the series bears Mercator's name, it appears that it was known earlier to both Hudde and Newton, though not published by them.

During the 1650s and 1660s a wide variety of infinite methods were developed, including the infinite continued fraction method for π that had been given by William Brouncker (1620?–1684), the first president of the Royal Society. The first steps in continued fractions had been taken long before in Italy, where Pietro Antonio Cataldi (1548–1626) of Bologna had expressed square roots in this form. Such expressions are easily obtained as follows: Let $\sqrt{2}$ be desired and let $x + 1 = \sqrt{2}$. Then $(x + 1)^2 = 2$ or $x^2 + 2x = 1$ or $x = 1/(2 + x)$. If on the right-hand side one continues to replace x as often as it appears by $1/(2 + x)$, one finds that

$$x = \cfrac{1}{2 + \cfrac{1}{2 + \cfrac{1}{2 + \cdots}}} = \sqrt{2} - 1$$

Through manipulation of Wallis' product for $2/\pi$, Brouncker was led somehow[11] to the expression

$$\frac{4}{\pi} = 1 + \cfrac{1}{2 + \cfrac{9}{2 + \cfrac{25}{2 + \cfrac{49}{2 + \cdots}}}}$$

Brouncker and Gregory found also certain infinite series for logarithms, but these were overshadowed by the greater simplicity of the Mercator series. Gregory also studied the curve $y = \ln x$, which he derived from the equiangular spiral $r = e^\theta$ by a geometrical transformation equivalent to letting the radius vector r of a point become the abscissa x and turning the arc θ into the ordinate. This may have been suggested by the comparison, so popular in Italy, of the parabola with the spiral of Archimedes. It is sad to report, however, that Gregory did not have an influence commensurate with his achievement. He returned to Scotland to become professor of mathematics, first at St. Andrews in 1668 and then at Edinburgh in 1674, where he became blind and died a year later. After his three treatises of 1667–1668 had appeared, he no longer published, and many of his results had to be rediscovered by others.

18   Newton could have learned much from Gregory, but the young Cambridge student evidently was not well acquainted with the work of the Scot. Instead it was two Englishmen, one at Oxford and the other at Cambridge, who made a deeper impression on him. They were John Wallis and Isaac Barrow (1630–1677). Barrow, like Wallis, entered Holy Orders but taught mathematics. In 1662 he was professor of geometry at Gresham College in London, and in 1664 be became the Lucasian professor of geometry at Cambridge, being the first to fill the chair established by Henry Lucas (1610?–1663) and later occupied by Newton, who succeeded Barrow. A mathematical conservative, Barrow disliked the formalisms of algebra, and in this respect his work is antithetical to that of Wallis. He thought that algebra should be part of logic rather than of mathematics, a view scarcely conducive to analytic discoveries. An admirer of the ancients, he edited the works of Euclid, Apollonius, and Archimedes, as well as publishing his own *Lectiones opticae* (1669) and *Lectiones geometriae* (1670), in the editing of both of which Newton assisted. The date 1668 is important for the fact that Barrow was giving his geometrical lectures at the same time that Gregory's *Geometria pars universalis* and Mercator's *Logarithmotechnia* appeared, as well as a

---

[11] It is not known how Brouncker reached this result, but a proof based on Euler's work is given in the chapter on Brouncker in J. L. Coolidge, *The Mathematics of Great Amateurs* (1949).

revised edition of Sluse's *Mesolabum*. Sluse's book included a new section dealing with infinitesimal problems and containing a method of maxima and minima. Wishing his *Lectiones geometriae* to take account of the state of the subject at the time, Barrow included an especially full account of the new discoveries. Tangent problems and quadratures were all the rage, and they figure prominently in Barrow's 1670 treatise. Here Barrow preferred the kinematic views of Torricelli to the static arithmetic of Wallis, and he liked to think of geometrical magnitudes as generated by a steady flow of points. Time, he said, has many analogies with a line; yet he viewed both as made up of indivisibles. Although his reasoning is much more like Cavalieri's than like Wallis' or Fermat's, there is one point at which algebraic analysis obtrudes prominently. At the end of Lecture X Barrow writes:

> Supplementary to this we add, in the form of appendices, a method for finding tangents by calculation frequently used by us, although I hardly know, after so many well-known and well-worn methods of the kind above, whether there is any advantage in doing so. Yet I do so on the advice of a friend [later shown to have been Newton]; and all the more willingly because it seems to be more profitable and general than those which I have discussed.

Then Barrow went on to explain a method of tangents which is virtually identical with that used in the differential calculus.[12] It is much like that of Fermat, but it makes use of two quantities—instead of Fermat's single letter E—quantities that are equivalent to the modern Δx and Δy. Barrow explained his tangent rule essentially as follows. If M is a point on a curve given (in modern notation) by a polynomial equation $f(x, y) = 0$ and if T is the point of intersection of the desired tangent MT with the x-axis, then Barrow marked off "an indefinitely small arc, MN, of the curve." He then drew the ordinates at M and N and through M a line MR parallel to the x-axis (Fig. 18.4). Then, designating by m the known ordinate at M by t the desired subtangent PT, and by a and e the vertical and horizontal sides of the triangle MRN, Barrow pointed out that the ratio of a to e is equal to the ratio of m to t. As we should now express it, the ratio of a to e for infinitely close points is the slope of the curve. To find this ratio Barrow proceeded much as Fermat had. He replaced x and y in $f(x, y) = 0$ by $x + e$ and $y + a$ respectively, then in the resulting equation he disregarded all terms not containing a or e (since these by themselves equal zero) and all terms of degree higher than the first degree in a and e, and finally he replaced a by m and e by t. From this the subtangent is found in terms of x and m, and if x and m are known, the quantity t is determined.

[12] For further details on Barrow's work see his *Geometrical Lectures*, ed. by J. M. Child (1916), and *The Mathematical Works of Isaac Barrow*, ed. by W. Whewell (1860). See also the forthcoming article on Barrow, by D. T. Whiteside, to appear in *Dictionary of Scientific Biography* (New York: Scribner's).

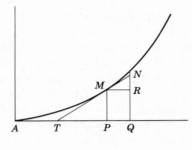

FIG. 18.4

Barrow apparently did not know directly of Fermat's work, for he nowhere mentioned his name; but the men to whom he referred as sources of his ideas include Cavalieri, Huygens, Gregory of St. Vincent, James Gregory, and Wallis, and it may be that Fermat's method became known to Barrow through them. Huygens and James Gregory in particular made frequent use of the procedure, and Newton, with whom Barrow was working, recognized that Barrow's algorithm was only an improvement of Fermat's.

Of all the mathematicians who anticipated portions of the differential and integral calculus, none approached more closely to the new analysis than Barrow. He seems to have recognized clearly the inverse relationship between tangent and quadrature problems. But his conservative adherence to geometric methods evidently kept him from making effective use of the relationship, and his contemporaries found his *Lectiones geometricae* difficult to understand. Fortunately, Barrow knew that at that very time Newton himself was working on the same problems, and the older man entreated his young associate to collect and publish his own results. Barrow in 1669 was called to London as chaplain to Charles II, and Newton, on Barrow's suggestion, succeeded him in the Lucasian chair at Cambridge. That the succession was most felicitous will become apparent in the next chapter.

BIBLIOGRAPHY

Agostini, Amedeo, "L'opera mathematica di Pietro Mengoli," *Archives Internationales d'Histoire des Sciences* (3), **29** (1950), 816–834.
Barrow, Isaac, *Geometrical Lectures*, ed. by J. M. Child (Chicago: Open Court, 1916).
Boyer, C. B., *History of Analytic Geometry* (New York: Scripta Mathematica, 1956).
Boyer, C. B., *History of the Calculus* (paperback ed., New York: Dover, 1959).
Cantor, Moritz, *Vorlesungen über Geschichte der Mathematik* (Leipzig: B. G. Teubner, 1892–1908, 4 vols.).

Castelnuovo, G., *Le origini del calcolo infinitesimale nell'era moderna* (Bologna: Nicola Zanichelli, 1938).

Chasles, Michel, *Aperçu historique sur l'origine et le développement des méthodes en géométrie* (new ed., Paris, 1875).

Coolidge, J. L., *A History of the Conic Sections and Quadric Surfaces* (Oxford: Clarendon, 1945).

Coolidge, J. L., *The Mathematics of Great Amateurs* (Oxford: Clarendon, 1949).

Haas, Karlheinz, "Die mathematischen Arbeiten von Johann Hudde," *Centaurus* **4** (1956), 235–284.

Huygens, Christiaan, *Oeuvres complètes* (The Hague: M. Nijhoff, 1888–1950, 22 vols.).

Lehmann, Ernst, *La Hire und seine Sectiones conicae* (Leipzig, 1888).

Loria, Gino, *Storia delle matematiche* (Torino: Sten, 1929–1933, 3 vols.).

Montucla, Etienne, *Histoire des mathématiques* (new ed., Paris, 1799–1802, 4 vols.).

Reiff, R., *Geschichte der Unendlichen Reihen* (Tübingen, 1889).

Rigaud, S. P., *Correspondence of Scientific Men of the Seventeenth Century* (Oxford, 1841, 2 vols.).

Rosenfeld, L., "René-François de Sluse et le problème des tangents," *Isis*, **10** (1928), 416–434.

Schooten, Frans van, *Geometria a Renato Des Cartes*, 2nd ed. (Amsterdam, 1659–1661).

Scott, J. F., *The Mathematical Works of John Wallis, D.D., F.R.S. (1616–1703)* (London: Taylor and Francis, 1938).

Scriba, C. J., *Studien zur Mathematik des John Wallis (1616–1703)* (Wiesbaden: Steiner, 1966).

Smith, D. E., ed., *A Source Book in Mathematics* (New York: McGraw-Hill, 1929; paperback ed., New York: Dover, 1959, 2 vols.).

Turnbull, H. W., *James Gregory Tercentenary Memorial Volume* (London: G. Bell, 1939).

Whewell, William, ed., *The Mathematical Works of Isaac Barrow* (Cambridge, 1860).

Zeuthen, H. G., *Geschichte der Mathematik im 16. und 17. Jahrhundert*, trans. by R. Meyer (Leipzig: B. G. Teubner, 1903).

EXERCISES

1. What reasons can you give for the fact that so little attention was given in the seventeenth century to solid analytic geometry?
2. Explain how the concept of curvature arose in the work of Huygens.
3. Why were curve rectifications discovered so much later than the quadratures of curves?
4. Explain clearly the differences between mathematical induction and Wallis' ordinary induction.
5. What were three of the chief sources of support of mathematicians in the seventeenth century? Give some illustrations of each.
6. Mention several mathematicians of the seventeenth century who were churchmen and compare the level of their work with that of several who were nonchurchmen.
7. Prove by mathematical induction that

$$\frac{1}{1 \cdot 2} + \frac{1}{2 \cdot 3} + \cdots + \frac{1}{n(n + 1)} = \frac{n}{n + 1}$$

and use this fact to find the sum of the reciprocals of the triangular numbers.

8. Use the method of Gregory and Mercator to find the first three terms of the series for arcsin x from

$$\int_0^x \frac{dx}{\sqrt{1 - x^2}}$$

9. Given two concentric circles of radii a and b (with $a > b$), show how to construct points on the ellipse $x = a \cos \theta$, $y = b \sin \theta$.

10. Justify Hudde's theorems on double roots of a polynomial equation and on relative maxima and minima.

11. Sketch the "pearl of Sluse" $y^2 = x(8 - x)$.

12. Find the length of $y^2 = x^3$ from $x = 0$ to $x = 1$.

13. Write $\sqrt{3}$ as a continued fraction.

14. Write the positive root of the equation $x^2 + 3x = 4$ as a continued fraction.

15. Use the first five terms in Mercator's series to find ln 1.1 approximately.

16. Use Barrow's method to find the subtangent to the curve $y = x^2 + 2x^3$ at $(2, 20)$.

17. Verify Wallis' formula

$$\frac{0^3 + 1^3 + 2^3 + \cdots + n^3}{n^3 + n^3 + n^3 + \cdots + n^3} = \frac{1}{4} + \frac{1}{4n}$$

for $n = 1, 2, 3,$ and 4.

18. Verify Wallis' formula

$$\int_0^1 (x - x^2)^n \, dx = \frac{(n!)^2}{(2n + 1)!}$$

for $n = 1, 2, 3,$ and 4.

*19. Prove that the evolute of a cycloid is another cycloid of the same size.

*20. Derive Wallis' infinite product for $2/\pi$ by the method indicated in Chapter 18.

*21. .Find the evolute of the parabola $y^2 = 2x$.

*22. Verify Van Heuraet's theorem for neighboring points that the ratio of the change in arc to the change in the abscissa is equal to the ratio of the normal to the ordinate.

*23. Find by Wallis' formula the length of one arch of the cycloid $x = a(\theta - \sin \theta)$, $y = a(1 - \cos \theta)$.

*24. Find direction numbers of two lines on the hyperboloid $x^2 + y^2 - z^2 = 1$ through the point $(1, 2, 2)$.

CHAPTER XIX

Newton and Leibniz

> Taking mathematics from the beginning of the world
> to the time of Newton, what he has done is much the
> better half.
>
> *Leibniz*

Isaac Newton, Barrow's successor, was born prematurely on Christmas Day **1** of 1642, the year of Galileo's death. His father had died before the sickly Isaac was born, and his mother married again when her son was three years old. The boy was brought up by his grandmother while he attended the neighborhood school, and a maternal uncle who was a Cambridge graduate recognized unusual ability in his nephew and persuaded Isaac's mother to enter the boy at Cambridge. Young Newton therefore enrolled at Trinity College in 1661, probably with no thought of being a mathematician, for he had made no particular study of the subject. Chemistry at first seemed to be his chief interest, and he retained a strong concern for it throughout his life. Early in his first year, however, he bought and studied a copy of Euclid, and shortly thereafter he read Oughtred's *Clavis*, the Schooten *Geometria a Renato Des Cartes*, Kepler's *Optics*, the works of Viète, and, perhaps most important of all, Wallis' *Arithmetica infinitorum*. Moreover, to this training we must add the lectures that Barrow gave as Lucasian professor, which Newton attended, after 1663. He also became acquainted with work of Galileo, Fermat, Huygens, and others. It is no wonder that Newton later wrote to Hooke, "If I have seen farther than Descartes, it is because I have stood on the shoulders of giants."

By the end of 1664 Newton seems to have reached the frontiers of mathematical knowledge and was ready to make contributions of his own. His first discoveries, dating from the early months of 1665, resulted from his ability to express functions in terms of infinite series—the very thing that Gregory was doing in Italy at about that time, although Newton could scarcely have known of this. Newton also began thinking, in 1665, of the rate of change, or fluxion, of continuously varying quantities, or fluents—such as lengths, areas, volumes, distances, temperatures. From that time on Newton linked together these two problems—of infinite series and of rates of change—as "my method."

Sir Isaac Newton.

During much of 1665–1666, immediately after Newton had earned his A.B. degree, Trinity College was closed because of the plague, and Newton went home to live and think. The result was the most productive period of mathematical discovery ever reported, for it was during these months, Newton later averred, that he had made four of his chief discoveries: (1) the binomial theorem, (2) the calculus, (3) the law of gravitation, and (4) the nature of colors. The first of these seems so obvious to us now that it is difficult to see why its discovery was so long delayed. For at least half a millennium the binomial coefficients for integral powers had been known. Cardan and Pascal, among others, were well aware of the rule of succession for coefficients; but they did not make use of the exponential notation of Descartes, hence could not make the relatively simple transition from an integral power to a fractional one. Stevin and Girard had suggested fractional powers, but did not really use them. Hence it was only with Wallis that fractional exponents came into common use, and we have seen that even Wallis, the great interpolater, was unable to write down an expansion for $(x - x^2)^{\frac{1}{2}}$ or $(1 - x^2)^{\frac{1}{2}}$. It remained for Newton to supply the expansions as part of his method of infinite series.

The binomial theorem, discovered in 1664 or 1665, was described in two **2** letters of 1676 from Newton to Henry Oldenburg (1615?–1677), secretary of the Royal Society, and published by Wallis (with credit to Newton) in Wallis' *Algebra* of 1685. The form of expression given by Newton (and Wallis) strikes the modern reader as awkward, but it indicates that the discovery was not just a simple replacement of an integral power by a fraction; it was the result of much trial and error on Newton's part in connection with divisions and radicals involving algebraic quantities. Finally Newton discovered that

The Extractions of Roots are much shortened by the Theorem

$$\overline{P + PQ}\Big|\frac{m}{n} = P\frac{m}{n} + \frac{m}{n}AQ + \frac{m-n}{2n}BQ + \frac{m-2n}{3n}CQ + \frac{m-3n}{4n}DQ + \text{etc.}$$

where $P + PQ$ stands for a Quantity whose Root or Power or whose Root of a Power is to be found, P being the first term of that quantity, Q being the remaining terms divided by the first term and m/n the numerical Index of the powers of $P + PQ \ldots$ Finally, in place of the terms that occur in the course of the work in the Quotient, I shall use A, B, C, D, etc. Thus A stands for the first term $P(m/n)$; B for the second term $(m/n)AQ$; and so on.[1]

This theorem was first announced by Newton in a letter of June 13, 1676, sent to Oldenburg but intended for Leibniz. In a second letter of October 24 of the same year Newton explained in detail just how he had been led to this binomial series. He wrote that toward the beginning of his study of mathematics he had happened on the work of Wallis on finding the area (from $x = 0$ to $x = x$) under curves whose ordinates are of the form $(1 - x^2)^n$.

[1] See D. E. Smith, *Source Book in Mathematics* (1929), pp. 224–228.

Upon examining the areas for exponents n equal to 0, 1, 2, 3, and so on, he found the first term always to be x, the second term to be $\frac{-0}{3}x^3$ or $-\frac{1}{3}x^3$ or $-\frac{2}{3}x^3$ or $\frac{-3}{3}x^3$, according as the power of n is 0 or 1 or 2 or 3, and so on. Hence, by Wallis' principle of "intercalculation" Newton assumed that the first two terms in the area for $n = \frac{1}{2}$ should be

$$x - \frac{\frac{1}{2}x^3}{3}$$

In the same fashion, proceeding by analogy, he found further terms, the first five being

$$x - \frac{\frac{1}{2}x^3}{3} - \frac{\frac{1}{8}x^5}{5} - \frac{\frac{1}{16}x^7}{7} - \frac{\frac{5}{128}x^9}{9}$$

He then realized that the same result could have been found by first deriving $(1 - x^2)^{\frac{1}{2}} = 1 - \frac{1}{2}x^2 - \frac{1}{8}x^4 = \frac{1}{16}x^6 - \frac{5}{128}x^8 - \cdots$, by interpolation in the same manner, and then finding the area through integration of the terms in this series. In other words, Newton did not proceed directly from the Pascal triangle to the binomial theorem, but indirectly from a quadrature problem to the binomial theorem.

3   It is likely that Newton's indirect approach was fortunate for the future of his work, for it made clear to him that one could operate with infinite series in much the same way as with finite polynomial expressions. The generality of this new infinite analysis was then confirmed for him when he derived the same infinite series through the extraction of the square root of $1 - x^2$ by the usual algebraic process, finally verifying the result by multiplying the infinite series by itself to recover the original radicand $1 - x^2$. In the same way Newton found that the result obtained for $(1 - x^2)^{-1}$ by interpolation (that is, the binomial theorem for $n = -1$) agreed with the result found by long division. Through these examples Newton had discovered something far more important than the binomial theorem; he had found that the analysis by infinite series had the same inner consistency, and was subject to the same general laws, as the algebra of finite quantities. Infinite series were no longer to be regarded as approximating devices only; they were alternative forms of the functions they represented. As Wallis expressed the view in his *Algebra* in describing Newton's binomial theorem, these infinite series or converging series "intimate the designation of some particular quantity by a regular Progression or rank of quantities, continually approaching to it; and which, if infinitely continued, must be equal to it."

Newton himself never published the binomial theorem, nor did he prove it; but he wrote out and ultimately published several accounts of his infinite analysis. The first of these, chronologically, was the *De analysi per aequationes numero terminorum infinitas*, composed in 1669 on the basis of ideas acquired in 1665–1666, but not published until 1711. In this he wrote:

And whatever the common Analysis [that is, algebra] performs by Means of Equations of a finite number of Terms (provided that can be done) this new method can always perform the same by Means of infinite Equations. So that I have not made any Question of giving this the Name of *Analysis* likewise. For the Reasonings in this are no less certain than in the other; nor the Equations less exact; albeit we Mortals whose reasoning Powers are confined within narrow Limits, can neither express, nor so conceive all the Terms of these Equations as to know exactly from thence the Quantities we want . . . . To conclude, we may justly reckon that to belong to the *Analytic Art*, by the help of which the Areas and Lengths, etc. of Curves may be exactly and geometrically determined.[2]

From then on, encouraged by Newton, men no longer tried to avoid infinite processes, as had the Greeks, for these now were regarded as legitimate in mathematics.

Newton's *De analysi* contained more, of course, than some work on infinite series; it is of great significance also as the first systematic account of Newton's chief mathematical discovery—the calculus. Barrow, the most important of Newton's mentors, was primarily a geometer, and Newton himself often has been described as an exponent of pure geometry; but the earliest manuscript drafts of his thoughts show that Newton made free use of algebra and a variety of algorithmic devices and notations. He had not, by 1666, developed his notation of fluxions, but he had formulated a systematic method of differentiation[3] that was not far removed from that published in 1670 by Barrow. It is only necessary to replace Barrow's *a* by Newton's *qo* and Barrow's *e* by Newton's *po* to arrive at Newton's first form for the calculus. Evidently Newton regarded *o* as a very small interval of time and *op* and *oq* as the small increments by which *x* and *y* change in this interval. The ratio *q/p*, therefore, will be the ratio of the instantaneous rates of change of *y* and *x*—that is, the slope of the curve $f(x, y) = 0$. The slope of the curve $y^n = x^m$, for example, is found from $(y + oq)^n = (x + op)^m$ by expanding both sides by the binomial theorem, dividing through by *o*, and disregarding terms still containing *o*, the result being

$$\frac{q}{p} = \frac{m}{n} \frac{x^{m-1}}{y^{n-1}} \quad \text{or} \quad \frac{q}{p} = \frac{m}{n} x^{m/n-1}$$

[2] The *De analysi* is included in the *Opera quae exstant omnia*, ed. by Samuel Horsley (1779–1785).

[3] For some of Newton's earliest work see *Unpublished Scientific Papers of Isaac Newton*, ed. by A. R. Hall and M. R. Hall (1962), and the *Correspondence of Newton*, ed. by H. W. Turnbull (1959–1961). By far the most important source for Newton's early work is *The Mathematical Papers of Isaac Newton* (vol. I, 1664–1666, ed. by D. T. Whiteside, Cambridge University Press, 1967). In this valuable work the editor shows that Newton was at first more strongly influenced by Descartes, Schooten, and Hudde than has been generally recognized.

Fractional powers no longer bothered Newton, for his method of infinite series had given him a universal algorithm.

When dealing later with an explicit function of x alone, Newton dropped his p and q and used o as a small change in the independent variable, a notation that was used also by Gregory. In the *De analysi*, for example, Newton proved as follows that the area under the curve $y = ax^{m/n}$ is given by

$$\frac{ax^{m/n + 1}}{m/n + 1}$$

Let the area be z and assume that

$$z = \frac{n}{m + n} ax^{(m + n)/n}$$

Let the moment or infinitesimal increase in the abscissa be o. Then the new abscissa will be $x + o$ and the augmented area will be

$$z + oy = \frac{n}{m + n} a(x + o)^{(m + n)/n}$$

If here one applies the binomial theorem, cancels the equal terms

$$z \text{ and } \frac{n}{m + n} ax^{(m + n)/n}$$

divides through by o, and discards the terms still containing o, the result will be $y = ax^{m/n}$. Conversely, if the curve is $y = ax^{m/n}$, then the area will be

$$z = \frac{n}{m + n} ax^{(m + n)/n}$$

This seems to be the first time in the history of mathematics that an area was found through the inverse of what we call differentiation, although the possibility of such a procedure evidently was known to Barrow and Gregory, and perhaps also to Torricelli and Fermat. Newton became the effective inventor of the calculus because he was able to exploit the inverse relationship between slope and area through his new infinite analysis. This is why in later years he frowned upon any effort to separate his calculus from his analysis by infinite series.

4   It is well known that in Newton's most popular presentation of his infinitesimal methods he looked upon x and y as flowing quantities, or fluents, of which the quantities p and q (above) were the fluxions or rates of change; when he wrote up this view of the calculus in about 1671, he replaced p and q by the "pricked letters" $\dot{x}$ and $\dot{y}$. The quantities or fluents of which x and y are the fluxions he designated by $\dot{x}$ and $\dot{y}$. By doubling the dots and dashes he was able to represent fluxions of fluxions or fluents of fluents. It should be noted that the title of the work, when published long afterward in 1742 (although an English translation appeared earlier in 1736), was not simply the method of fluxions, but *Methodus fluxionum et serierum infinitorum.*

In 1676 Newton wrote still a third account of his calculus, under the title *De quadratura curvarum*, and this time he sought to avoid both infinitely small quantities and flowing quantities, replacing these by a doctrine of "prime and ultimate ratios." He found the "prime ratio of nascent augments" or the "ultimate ratio of evanescent increments" as follows. Let the ratio of the changes in x and x^n be desired. Let o be the increment in x and $(x + o)^n - x^n$ the corresponding increment in x^n. Then the ratio of the increments will be

$$1 : \left[nx^{n-1} + \frac{n(n-1)}{2} ox^{n-2} + \cdots \right]$$

To find the prime and ultimate ratio one lets o vanish, obtaining the ratio $1 : (nx^{n-1})$. Here Newton is very close indeed to the limit concept, the chief objection being the use of the word "vanish." Is there really a ratio between increments that have vanished? Newton did not clarify this question, and it continued to distract mathematicians throughout the eighteenth century.[4]

Newton discovered his method of infinite series and the calculus in 1665– **5** 1666, and within the next decade he wrote at least three substantial accounts of the new analysis. The *De analysi* circulated among friends, including John Collins (1625–1683) and Isaac Barrow, and the infinite binomial expansion was sent to Oldenburg and Leibniz; but Newton made no move to publish his results, even though he knew that Gregory and Mercator in 1668 had disclosed their work on infinite series. The first account of the calculus that Newton put into print appeared in 1687 in *Philosophiae naturalis principia mathematica*, the most admired scientific treatise of all times. This book generally is described as presenting the foundations of physics and astronomy in the language of pure geometry. It is true that a large part of the work is in synthetic form, but there is also a large admixture of analytic passages. Section I of Book I is, in fact, entitled "The method of first and last ratios of

---

[4] For more details see C. B. Boyer, *History of the Calculus* (1939, Dover reprint 1959). Newton's chief treatises on the calculus have been collected and edited by D. T. Whiteside in *Mathematical Works* (1964), Vol. I. Newton's *Mathematical Papers* are being edited by Dr. Whiteside in a monumental work, estimated at eight volumes, the first volume of which was published in 1967 by the Cambridge University Press.

quantities, by the help of which we demonstrate the propositions that follow," including Lemma I:

> Quantities, and the ratios of quantities, which in any finite time converge continually to equality, and before the end of that time approach nearer to each other than by any given difference, become ultimately equal.[5]

This is, of course, an attempt at a definition of limit of a function. Lemma VII in Section I postulates that "the ultimate ratio of the arc, chord, and tangent, any one to any other, is the ratio of equality"; other lemmas assume the similarity of certain "evanescent triangles." Every now and then in Book I the author has recourse to an infinite series. However, calculus algorithms do not appear until in Book II, Lemma II, we come to the cryptic formulation:

> The moment of any genitum is equal to the moments of each of the generating sides multiplied by the indices of the powers of those sides, and by their coefficients continually.

Newton's explanation shows that by the word "genitum" he has in mind what we call a "term" and that by the "moment" of a genitum he means the infinitely small increment. Designating by a the moment of A and by b the moment of B, Newton proves that the moment of AB is $aB + bA$, that the moment of A^n is naA^{n-1}, and that the moment of $1/A$ is $-a/(A^2)$. Such sibylline expressions, which are the equivalents of the differential of a product, a power, and a reciprocal respectively, constitute Newton's first official pronouncement on the calculus, making it easy to understand why so few mathematicians of the time mastered the new analysis in terms of Newtonian language.

Newton was not the first one to differentiate or to integrate, nor to see the relationship between these operations in the fundamental theorem of the calculus. His discovery consisted in the consolidation of these elements into a general algorithm applicable to all functions, whether algebraic or transcendental. This was emphasized in a scholium that Newton published in the *Principia* immediately following Lemma II:

> In a letter of mine to Mr. J. Collins, dated December 10, 1672, having described a method of tangents, which I suspected to be the same with Sluse's method, which at that time was not made public, I added these words. This is one particular, or rather a Corollary, of a general method, which extends itself, without any troublesome calculation, not only to the drawing of tangents to any curved lines whether geometrical or mechanical... but also to the resolving other abstruser kinds of problems about the crookedness, areas, lengths, centres of gravity of curves, etc.; nor is it (as Hudden's

[5] *Mathematical Principles of Natural Philosophy*, trans. by Andrew Motte, translation revised by Florian Cajori (1934). Passages in translation of the *Principia* here are from this edition.

method de maximis et minimis) limited to equations which are free from surd quantities. This method I have interwoven with that other of working in equations by reducing them to infinite series.

In the first edition of *Principia* Newton admitted that Leibniz was in possession of a similar method, but in the third edition of 1726, following the bitter

Gottfried Wilhelm Leibniz

quarrel between adherents of the two men concerning the independence and priority of the discovery of the calculus, Newton deleted the reference to the calculus of Leibniz. It is now fairly clear that Newton's discovery antedated that of Leibniz by about ten years, but that the discovery by Leibniz was independent of that of Newton. Moreover, Leibniz is entitled to priority of publication, for he printed an account of his calculus in 1684 in the *Acta Eruditorum*, a sort of "scientific monthly" that had been established only two years before.

6   Gottfried Wilhelm Leibniz (1646–1716) was born at Leipzig, where at fifteen he entered the university and at seventeen earned his bachelor's degree. He studied theology, law, philosophy, and mathematics at the university, and he sometimes is regarded as the last scholar to achieve universal knowledge. By the time he was twenty, he was prepared for the degree of doctor of laws, but this was refused because of his youth. He thereupon left Leipzig and took his doctorate at the University of Altdorf in Nuremberg, where he was offered a professorship in law, which he declined. He then entered the diplomatic service, first for the elector of Mainz, then for the Brunswick family, and finally for the Hanoverians, whom he served for forty years. Among the electors of Hanover whom Leibniz served was one who, as great-grandson of James I of England, succeeded Queen Anne in 1714 as King George I. As an influential governmental representative Leibniz traveled widely. In 1672 he went to Paris, hoping to divert French acquisitorial designs against Germany through a "holy war" directed against Egypt (a suggestion later adopted by Napoleon). There he met Huygens, who suggested that if he wished to become a mathematician, he should read Pascal's treatises of 1658–1659. In 1673 a political mission took him to London, where he bought a copy of Barrow's *Lectiones geometricae*, met Oldenburg and Collins, and became a member of the Royal Society. It is largely around this visit that the later quarrel over priority centered, for Leibniz could have seen Newton's *De analysi* in manuscript; but it is doubtful that at this stage he would have derived much from it, for Leibniz was not yet well prepared in geometry or analysis.

In 1676 Leibniz again visited London, bringing with him his calculating machine; it was during these years between his two London visits that the differential calculus had taken shape. As was the case with Newton, infinite series played a large role in the early work of Leibniz. Huygens had set him the problem of finding the sum of the reciprocals of the triangular numbers— that is, $2/n(n + 1)$. Leibniz cleverly wrote each term as the sum of two fractions, using

$$\frac{2}{n(n + 1)} = 2\left(\frac{1}{n} - \frac{1}{n + 1}\right)$$

from which it is obvious, on writing out a few terms, that the sum of the first n terms is

$$2\left(\frac{1}{1} - \frac{1}{n+1}\right)$$

hence that the sum of the infinite series is 2. From this success he ingenuously concluded that he would be able to find the sum of almost any infinite series.

The summation of series again came up in the harmonic triangle, whose analogies with the arithmetic (Pascal) triangle fascinated Leibniz.

Arithmetic triangle	Harmonic triangle
1 1 1 1 1 1 1 $\cdots$	$\frac{1}{1}$ $\frac{1}{2}$ $\frac{1}{3}$ $\frac{1}{4}$ $\frac{1}{5}$ $\frac{1}{6} \cdots$
1 2 3 4 5 6 $\cdots$	$\frac{1}{2}$ $\frac{1}{6}$ $\frac{1}{12}$ $\frac{1}{20}$ $\frac{1}{30} \cdots$
1 3 6 10 15 $\cdots$	$\frac{1}{3}$ $\frac{1}{12}$ $\frac{1}{30}$ $\frac{1}{60} \cdots$
1 4 10 20 $\cdots$	$\frac{1}{4}$ $\frac{1}{20}$ $\frac{1}{60} \cdots$
1 5 15 $\cdots$	$\frac{1}{5}$ $\frac{1}{30} \cdots$
1 6 $\cdots$	$\frac{1}{6} \cdots$
1 $\cdots$	

In the arithmetic triangle each element (which is not in the first column) is the difference of the two terms directly below it and to the left; in the harmonic triangle each term (which is not in the first row) is the difference of the two terms directly above it and to the right. Moreover, in the arithmetic triangle each element (not in the first row or column) is the sum of all of the terms in the line above it and to the left, whereas in the harmonic triangle each element is the sum of all of the terms in the line below it and to the right. Because the number of terms in the latter case is infinite, Leibniz had much practice in summing infinite series. The series in the first line is the harmonic series, which diverges; for all other lines the series converge. The numbers in the second line are one-half the reciprocals of the triangular numbers, and Leibniz knew that the sum of this series is 1. The numbers in the third line are one-third the reciprocals of the pyramidal numbers

$$\frac{n(n+1)(n+2)}{1\cdot 2\cdot 3}$$

and the harmonic triangle indicates that the sum of this series is $\frac{1}{2}$; the numbers in the fourth line are one-fourth the reciprocals of the figurate numbers corresponding to the four-dimensional analogue of the tetrahedron, and the sum of these is $\frac{1}{3}$; and so on for the succeeding rows in the harmonic triangle. The numbers in the nth diagonal row in this triangle are the reciprocals of the numbers in the corresponding nth diagonal row of the arithmetic triangle divided by n.

7   From his studies on infinite series and the harmonic triangle Leibniz turned to reading Pascal's works on the cycloid and other aspects of infinitesimal analysis. In particular, it was on reading the letter of Amos Dettonville on *Traité des sinus du quart de cercle* that Leibniz reported that a light burst upon him. He then realized, in about 1673, that the determination of the tangent to a curve depended on the ratio of the *differences* in the ordinates and abscissas, as these became infinitely small, and that quadratures depended on the sum of the ordinates or infinitely thin rectangles making up the area. Just as in the arithmetic and harmonic triangles the processes of summing and differencing are oppositely related, so also in geometry the quadrature and tangent problems, depending on sums and differences respectively, are inverses of each other. The connecting link seemed to be through the infinitesimal or "characteristic" triangle, for where Pascal had used it to find the quadrature of sines, Barrow had applied it to the tangent problem. A comparison of the triangle in Barrow's diagram (Fig. 18.4) with that in Pascal's figure (Fig. 19.1) will disclose the marked similarity that

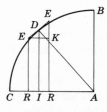

FIG. 19.1

evidently struck Leibniz so forcibly.[6] If EDE is tangent at D to the unit quarter circle BDC (Fig. 19.1), then, Pascal saw, AD is to DI as EE is to RR or EK. For a very small interval RR the line EE can be considered to be virtually the same as the arc of the circle intercepted between the ordinates ER. Hence, in the notation that Leibniz developed a few years later, we have $1/\sin \theta = d\theta/dx$, where θ is the angle DAC. Since $\sin \theta = \sqrt{1 - \cos^2 \theta}$ and $\cos \theta = x$, we have $d\theta = dx/\sqrt{1 - x^2}$. By the square-root algorithm and long division (or by the binomial theorem that Newton communicated to Leibniz, through Oldenburg, in 1676) it is a simple matter to find $d\theta = (1 + x^2/2 + \frac{3}{8}x^4 + \frac{5}{16}x^6 + \cdots)\,dx$. By use of the usual method of quadratures, as found in Gregory and Mercator, one obtains arcsin $x = x + x^3/6 +$

---

[6] See *The Early Mathematical Manuscripts of Leibniz*, trans. by C. I. Gerhardt, ed. by J. M. Child (1920), especially pp. 15–16.

$3x^5/40 + 5x^7/112 + \cdots$ (or, allowing for the negative slope and the constant of integration, arccos $= \pi/2 - x - x^3/6 - 3x^5/40 - 5x^7/112 - \cdots$). Newton, too, had arrived at this result earlier and by a similar method. From this it was possible to find the series for sin x by the process known as reversion, a scheme apparently first used by Newton but rediscovered by Leibniz. If we let $y = \arcsin x$ or $x = \sin y$ and for x assume a power series of the form $x = a_1 y + a_2 y^2 + a_3 y^3 + \cdots + a_n y^n + \cdots$, then, on replacing each x in the power series for arcsin x by this series in y, we have an identity in y. From this the quantities $a_1, a_2, a_3, \ldots, a_n, \ldots$ are determined by equating coefficients of terms of like degree. The resulting series, sin $y = y - y^3/3! + y^5/5! - \cdots$, was therefore known both to Newton and to Leibniz; and through $\sin^2 y + \cos^2 y = 1$ the series for cos y was obtained. The quotient of the sine and cosine series provides the tangent series, and their reciprocals give the other three trigonometric functions as infinite series. In the same way, through reversion of Mercator's series, Newton and Leibniz found the series for e^x.

Leibniz by 1676 had arrived at the same conclusion that Newton had reached several years earlier—that he was in possession of a method that was highly important because of its generality. Whether a function was rational or irrational, algebraic or transcendental (a word that Leibniz coined), his operations of finding sums and differences could always be applied. It therefore was incumbent upon him to develop an appropriate language and notation for the new subject. Leibniz always had a keen appreciation of the importance of good notations as a help to thought, and his choice in the case of the calculus was especially happy. After some trial and error he fixed on dx and dy for the smallest possible differences (differentials) in x and y, although initially he had used instead x/d and y/d to indicate the lowering of the degree. At first he wrote simply omn. y (or "all y's") for the sum of the ordinates under a curve, but later he used the symbol $\int y$, and still later $\int y\, dx$, the integral sign being an enlarged letter s for sum. Finding tangents called for the use of the *calculus differentialis*, and finding quadratures required the *calculus summatorius* or the *calculus integralis*; from these phrases arose our words "differential calculus" and "integral calculus."

The first account of the differential calculus was published by Leibniz in 1684 under the long but significant title of *Nova methodus pro maximis et minimis, itemque tangentibus, qua nec irrationales quantitates moratur* ("A New Method for Maxima and Minima, and also for Tangents, which is not Obstructed by Irrational Quantities"). Here Leibniz gave the formulas $dxy = x\, dy + y\, dx$, $d(x/y) = (y\, dx - x\, dy)/y^2$, and $dx^n = nx^{n-1}\, dx$ for products, quotients, and powers (or roots), together with geometrical applications. These formulas were derived by neglecting infinitesimals of higher order. If, for example, the smallest differences in x and y are dx and dy respectively, then dxy or the smallest difference in xy is $(x + dx)(y + dy) - xy$. Inasmuch as dx and dy are infinitely small, the term $dx\,dy$ is infinitely infinitely small and can be disregarded, giving the result $dxy = x\,dy + y\,dx$.

Two years later, again in the *Acta Eruditorum*, Leibniz published an explanation of the integral calculus in which quadratures are shown to be special cases of the inverse method of tangents. Here Leibniz emphasized the inverse relationship between differentiation and integration in the fundamental theorem of the calculus; he pointed out that in the integration of familiar functions "is included the greatest part of all transcendental geometry." Where Descartes' geometry had once excluded all nonalgebraic curves, the calculus of Newton and Leibniz showed how essential is the role of these in their new analysis. Were one to exclude transcendental functions from the new analysis, there would be no integrals of such algebraic functions as $1/x$ or $1/(1 + x^2)$. Moreover, Leibniz seems to have appreciated, as did Newton, that the operations in the new analysis can be applied to infinite series as well as to finite algebraic expressions. In this respect Leibniz was less cautious than Newton, for he argued that the infinite series $1 - 1 + 1 - 1 + 1 - \cdots$ is equal to $\frac{1}{2}$. In the light of recent work on divergent series, we cannot say that it is necessarily "wrong" to assign the "sum" $\frac{1}{2}$ in this case. It is nevertheless clear that Leibniz allowed himself to be carried away by the very success of his algorithms and was not deterred by uncertainty over concepts. Newton's reasoning was far closer to the modern foundations of the calculus than was that of Leibniz, but the plausibility of the Leibnizian view and the effectiveness of the differential notation made for a readier acceptance of differentials than of fluxions.

Newton and Leibniz both developed their new analysis rapidly to include differentials and fluxions of higher order, as in the case of the formula for curvature of a curve at a point. It probably was lack of clarity on Leibniz' part about higher orders of infinitesimals that led him to the mistaken conclusion than an osculating circle has four "consecutive" or coincident points of contact with a curve, rather than the three that determine the circle of curvature. The formula for the nth derivative (to use the modern language) of a product, $(uv)^{(n)} = u^{(n)}v^{(0)} + nu^{(n-1)}v^{(1)} + \cdots + nu^{(1)}v^{(n-1)} + u^{(0)}v^{(n)}$, a development paralleling the binomial expansion of $(u + v)^n$, bears the name of Leibniz. (In the Leibnizian theorem the exponents in parentheses indicate orders of differentiation rather than powers.) Also named for Leibniz is the rule, given in a memoir of 1692, for finding the envelope of a one-parameter family of plane curves $f(x, y, c) = 0$ through the elimination of c from the simultaneous equations $f = 0$ and $f_c = 0$, where f_c is the result of differentiating f partially with respect to c.

Newton retained his extraordinary mathematical ability to the very last;

when Leibniz in 1716 (the last year of his life) challenged Newton to find the orthogonal trajectories of a one-parameter family of plane curves, Newton within a few hours solved the problem and gave a method for finding trajectories in general. (Earlier, in 1696, Newton had been challenged to find the brachistochrone, or curve of quickest descent, and the day after receiving the problem he gave the solution, showing the curve to be a cycloid.) The name of Leibniz usually is attached also to the infinite series $\pi/4 = \frac{1}{1} - \frac{1}{3} + \frac{1}{5} - \frac{1}{7} + \cdots$, one of his first discoveries in mathematics. This series, which arose in his quadrature of the circle, is only a special case of the arctangent expansion that had been given earlier by Gregory. The fact that Leibniz was virtually self-taught in mathematics accounts in part for the frequent cases of rediscovery that appear in his work.

The great contribution of Leibniz to mathematics was the calculus, but **9** other aspects of his work deserve mention. The generalization of the binomial theorem to the multinomial theorem—the expansion of such expressions as $(x + y + z)^n$—is attributed to him, as is also the first reference in the Western world to the method of determinants. In letters of 1693 to L'Hospital, Leibniz wrote that he occasionally used numbers indicating rows and columns in a set of simultaneous equations:

$$10 + 11x + 12y = 0 \qquad\qquad 1_0 + 1_1 x + 1_2 y = 0$$
$$20 + 21x + 22y = 0 \quad \text{or} \quad 2_0 + 2_1 x + 2_2 y = 0$$
$$30 + 31x + 32y = 0 \qquad\qquad 3_0 + 3_1 x + 3_2 y = 0$$

We would write this as

$$a_1 + b_1 x + c_1 y = 0$$
$$a_2 + b_2 x + c_2 y = 0$$
$$a_3 + b_3 x + c_3 y = 0$$

If the equations are consistent, then

$$\begin{matrix} 1_0 \cdot 2_1 \cdot 3_2 & & 1_0 \cdot 2_2 \cdot 3_1 \\ 1_1 \cdot 2_2 \cdot 3_0 & = & 1_1 \cdot 2_0 \cdot 3_2 \\ 1_2 \cdot 2_0 \cdot 3_1 & & 1_2 \cdot 2_1 \cdot 3_0 \end{matrix}$$

which is equivalent to the modern statement that

$$\begin{vmatrix} a_1 & b_1 & c_1 \\ a_2 & b_2 & c_2 \\ a_3 & b_3 & c_3 \end{vmatrix} = 0$$

This anticipation by Leibniz of determinants went unpublished until 1850 and had to be rediscovered more than half a century later.

Leibniz' comments in the letter show that he was very conscious of the power in analysis of "characteristic" or notation that properly displays the elements of a given situation.[7] Evidently he thought highly of this contribution to notation because of its easy generalization, and he boasted that he showed that "Viète and Descartes hadn't yet discovered all of the mysteries" of analysis. Leibniz was, in fact, one of the greatest of all notation builders, being second only to Euler in this respect. He was the first mathematician of prominence to use systematically the dot for multiplication and to write proportions in the form $a:b = c:d$. The Leibnizian use of : for division is still widely employed. Moreover, it was in large part due to Leibniz and Newton that the = sign of Recorde triumphed over the symbol ∞ of Descartes. To Leibniz we owe also the symbols $\sim$ for "is similar to" and $\simeq$ for "is congruent to." Nevertheless, Leibniz' symbols for differentials and integrals remain his greatest triumphs in the field of notation. Leibniz was not responsible for the modern functional notation, but it is to him that the word "function," in much the same sense as it is used today, is due.

Among relatively minor contributions by Leibniz were his comments on complex numbers, at a time when they were almost forgotten, and his noting of the binary system of numeration. He factored $x^4 + a^4$ into

$$(x + a\sqrt{\sqrt{-1}})(x - a\sqrt{\sqrt{-1}})(x + a\sqrt{-\sqrt{-1}})(x - a\sqrt{-\sqrt{-1}})$$

and he showed that $\sqrt{6} = \sqrt{1+\sqrt{-3}} + \sqrt{1-\sqrt{-3}}$, an imaginary decomposition of a positive real number that surprised his contemporaries. However, Leibniz did not write the square roots of complex numbers in standard complex form, nor was he able to prove his conjecture that $f(x + \sqrt{-1}y) + f(x - \sqrt{-1}y)$ is real if $f(z)$ is a real polynomial. The ambivalent status of complex numbers is well illustrated by the remark of Leibniz, who was also a prominent theologian, that imaginary numbers are a sort of amphibian, halfway between existence and nonexistence, resembling in this respect the Holy Ghost in Christian theology. His theology obtruded itself again in his view of the binary system in arithmetic (in which only two symbols, unity and zero, are used) as a symbol of the creation in which God, represented by unity, drew all things from nothingness. He was so pleased with the idea that he wrote about it to the Jesuits, who had missionaries in China, hoping that they might use the analogy to convert the scientifically inclined Chinese emperor to Christianity.

---

[7] See Thomas Muir, *The Theory of Determinants in the Historical Order of Development* (1960), I, 6–10.

Leibniz was as much a philosopher as he was a mathematician; hence his **10** most significant mathematical contribution, other than the calculus, was in logic. In the calculus it was the element of universality that impressed him, and so it was with his other efforts. He hoped to reduce all things to order; to reduce logical discussions to systematic form, he wished to develop a universal characteristic that would serve as a sort of algebra of logic. His first mathematical paper had been a thesis on combinatorial analysis in 1666, and even at this early date he had visions of a formal symbolic logic. Universal symbols or ideograms were to be introduced for the small number of fundamental concepts needed in thought, and composite ideas were to be made up from this "alphabet" of human thoughts just as formulas are developed in mathematics. The syllogism itself was to be reduced to a sort of calculus expressed in a universal symbolism intelligible in all languages. Truth and error would then be simply a matter of correct or erroneous calculation within the system, and there would be an end to philosophical controversies. Moreover, new discoveries could be derived through correct but more or less routine operations on the symbols according to the rules of the logical calculus. Leibniz was justifiably proud of this idea, but his own enthusiasm for it was not matched by that of others. Perhaps his contemporaries looked upon it as too metaphysical, like the harmonies of his monads in the best of all possible worlds which were so ruthlessly and effectively satirized by Voltaire in *Candide*. Leibniz was noted in his day for an unbounded optimism that envisioned not only a universal language, but also a universal church through the union of Catholics and Protestants. In these respects the optimism of Leibniz today appears to have been unwarranted; but his suggestion of an algebra of logic was revived in the nineteenth century, and it has played a very effective role indeed in mathematics during the past century.

Leibniz was a scientist as well as a philosopher, and he and Huygens **11** developed the notion of kinetic energy, which ultimately, in the nineteenth century, became part of the broader concept of energy in general—one that Leibniz would most certainly have applauded for its universality. However, in science the contributions of Leibniz were overshadowed by those of Newton, which included the grandest mathematical formulation up to that time—the law of gravitation. In the opening sections of the *Principia* Newton had so generalized and clarified Galileo's ideas on motion that ever since we refer to this formulation as "Newton's laws of motion." Then Newton went on to combine these laws with Kepler's laws in astronomy and Huygens' law of centripetal force in circular motion to establish the great unifying principle that any two particles in the universe, whether two planets or two mustard seeds, or the sun and a mustard seed, attract each other with a force that varies inversely as the square of the distance between them. In the statement of this law Newton had been anticipated by others, including Robert Hooke (1638–1703), professor of geometry at Gresham College and Oldenburg's successor as secretary of the Royal Society. But Newton was the first to convince the world of the truth of the law because he was able to handle the mathematics required in the proof.

For circular motion the inverse square law is easily derived from Newton's $f = ma$, Huygens' $a = v^2/r$, and Kepler's $T^2 = Kr^3$ simply by noting that $T \propto r/v$ and then eliminating T and v from the equations, to arrive at $f \propto 1/r^2$. To prove the same thing for ellipses, however, required considerably more mathematical skill. Moreover, to prove that the distance is to be measured from the center of the bodies was so difficult a task that it evidently was this integration problem that induced Newton to lay the work on gravitation aside for almost twenty years following his discovery of the law in the plague year of 1665–1666. When in 1684 his friend Edmund Halley (1656–1742), a mathematician of no mean ability who also had guessed at the inverse square law, pressed Newton for a proof, the result was the exposition in the *Principia*. So impressed was Halley with the quality of this book that he had it published at his own expense.

The *Principia*, of course, contains far more than the calculus, the laws of motion, and the law of gravitation. It includes, in science, such things as the motions of bodies in resisting media[8] and the proof that, for isothermal vibrations, the velocity of sound should be the speed with which a body would strike the earth if falling without resistance through a height that is half that of a uniform atmosphere having the density of air at the surface of the earth and exerting the same pressure.[9] From his calculations Newton concluded that the speed of sound should be about 979 ft/sec, whereas he knew from experimental values that it actually is close to 1142 ft/sec. This puzzle in the *Principia* was not unraveled until almost a century later when Laplace explained the discrepancy as due to the fact that the vibrations of sound are to be considered as adiabatic. Another of the scientific conclusions in the *Principia* is a mathematical proof of the invalidity of the prevailing cosmic scheme—the Cartesian theory of vortices—for Newton showed, at the close of Book II, that, according to the laws of mechanics, planets in vortical motion would move more swiftly in aphelion than in perihelion, which contradicts the astronomy of Kepler. Nevertheless, it took about forty years before the Newtonian gravitational view of the universe, popularized by Maupertuis and Voltaire, displaced the Cartesian vortical cosmology in France.

---

[8] See, for example, Book II, Propositions IX–XVIII.
[9] See Book II, Propositions XLV–L.

One who reads only the headings of the three books in the *Principia* will **12** get the erroneous impression that it contains nothing but physics and astronomy, for the books are entitled, respectively, I. The Motion of Bodies, II. The Motion of Bodies (in Resisting Mediums), and III. The System of the World. However, the treatise contains also a great deal of pure mathematics, especially concerning the conic sections. In Lemma XIX of Book I, for example, the author solves the Pappus four-line locus problem, adding that his solution is "not an analytical calculus but a geometrical composition, such as the ancients required," an oblique and pejorative reference, apparently, to the treatment of the problem given by Descartes.

Newton throughout the *Principia* gave preference to a geometrical approach, probably because in his hands it provided elegant and convincing demonstrations of a universally familar language; but we have seen that where he found it expedient to do so, he did not hesitate to appeal to his method of infinite series and the calculus. Most of Section II of Book II, for example, is analytic. On the other hand, Newton's handling of the properties of conics is almost exclusively synthetic, for Newton here had no need to resort to analysis. Following the Pappus problem, he gave a couple of organic generations of conics through intersections of moving lines, and then he used these in half a dozen succeeding propositions to show how to construct a conic satisfying five conditions—passing through five points, for example, or tangent to five lines, or through two points and tangent to three lines. Propositions XIX through XXIX of Book I comprise virtually a little treatise on the organic description of conics and include some beautiful theorems. In Proposition XXVII, for instance, Newton related the properties of the complete quadrilateral to the conics. Using the fact that the centers of conics tangent to four lines lie on a straight line (now known as Newton's line) through the midpoints of the three diagonals, he found the conic touching five lines.

The *Principia* is the greatest monument to Newton, but it is by no means **13** the only one. Newton was so sensitive to criticism that after attacks by Hooke and others on his paper in the *Philosophical Transactions* for 1672 concerning the nature of color[10] he determined to publish nothing further. This article was of great import to physics for it was here that Newton announced what he regarded as one of the oddest of all of the operations of nature—that white light was merely a combination of rays of varying color, each color having its own characteristic index of refraction. That is, when a spectrum is formed on passing white light through a prism,

---

[10] For a facsimile reproduction of this paper, and of other Newtonian material, see *Isaac Newton's Papers and Letters on Natural Philosophy*, ed. by I. Bernard Cohen (Cambridge, Mass.: Harvard University Press, 1958).

the role of the prism is simply to sort out the rays according to the varying degrees of refrangibility. Such a revolutionary view was not easy for his contemporaries to accept, and the ensuing controversy upset Newton. For fifteen years he published nothing further until the urging of Halley induced him to write and publish the *Principia*. Meanwhile, the three versions of his calculus that he had written from 1669 to 1676, as well as a treatise on optics that he had composed, remained in manuscript form.

About fifteen years after the *Principia* appeared, Hooke died, and then, finally, Newton's aversion to publication seems to have abated somewhat. The *Opticks* appeared in 1704, and appended to it were two mathematical works: the *De quadratura curvarum*, in which an intelligible account of the Newtonian methods in the calculus finally appeared in print, and a little treatise entitled *Enumeratio linearum tertii ordinis* ("Enumeration of Curves of Third Degree").[11] The *Enumeratio* also had been composed by 1676, and it is the earliest instance of a work devoted solely to graphs of higher plane curves in algebra. Newton noted seventy-two species of cubics (half a dozen are omitted), and a curve of each species is carefully drawn. For the first time two axes are systematically used, and there is no hesitation about negative coordinates. Among the interesting properties of cubics indicated in this treatise are the fact that a third-degree curve can have not more than three asymptotes (as a conic can have no more than two) and that as all conics are projections of the circle, so all cubics are projections of a "divergent parabola" $y^2 = ax^3 + bx^2 + cx + d$.

14   The *Enumeratio* was not Newton's only contribution to analytic geometry. In the *Method of Fluxions*, written in Latin about 1671, he had suggested eight new types of coordinate system. One of these, Newton's "Third Manner" of determining a curve, was through what now are called bipolar coordinates. If x and y are the distances of a variable point from two fixed points or poles, then the equations $x + y = a$ and $x - y = a$ represent ellipses and hyperbolas respectively and $ax + by = c$ are ovals of Descartes. This type of coordinate system is infrequently used today, but that given by Newton as his "Seventh Manner; For Spirals" is now familiarly known under the name of polar coordinates. Using x where we now use θ or ϕ and y where we use r or ρ, Newton found the subtangent to the spiral of Archimedes $by = ax$, as well as to other spirals. Having given the formula for radius of curvature in rectangular coordinates, $R = (1 + \dot{y}^2)\sqrt{1 + \dot{y}^2/\dot{z}}$, where $z = \dot{y}$, he wrote the corresponding formula in polar coordinates as

$$R \sin \psi = \frac{y + yzz}{1 + zz - \dot{z}}$$

[11] See W. W. Rouse Ball, "On Newton's Classification of Cubic Curves," *Proceedings of the London Mathematical Society*, **22** (1890–1891), 104–143, for a good account of this work.

where $z = \dot{y}/y$ and ψ is the angle between the tangent and the radius vector.[12] Newton also gave equations for the transformation from rectangular to polar coordinates, expressing these as $xx + yy = tt$ and $tv = y$, where t is the radius vector and v a line representing the sine of the vectorial angle associated with the point (x, y) in Cartesian coordinates.

In the *Method of Fluxions*, as well as in *De analysi*, we find "Newton's **15** method" for the approximate solution of equations. If the equation to be solved is $f(x) = 0$, one first locates the desired root between two values $x = a_1$ and $x = b_1$ such that in the interval (a_1, b_1) neither the first nor the second derivative vanishes, or fails to exist. Then for one of the values, say $x = a_1, f(x)$ and $f''(x)$ will have the same sign. In this case the value $x = a_2$ will be a better approximation if

$$a_2 = a_1 - \frac{f(a_1)}{f'(a_1)}$$

and this procedure can be applied iteratively to obtain as precise an approximation a_n as may be desired. If $f(x)$ is of the form $x^2 - a^2$, the successive approximations in the Newton method are the same as those found in the Old Babylonian square-root algorithm; hence this ancient procedure sometimes is unwarrantedly called "Newton's algorithm." If $f(x)$ is a polynomial, Newton's method is in essence the same as the Chinese-Arabic method named for Horner; but the great advantage of the Newtonian method is that it applies equally to equations involving transcendental functions.

The *Method of Fluxions* contained also a diagram that later became known as "Newton's parallelogram," useful in developments in infinite series and in the sketching of curves. For a polynomial equation $f(x, y) = 0$ one forms a grid or lattice the intersection points of which are to correspond to terms of all possible degrees in the equation $f(x, y) = 0$; on this "parallelogram" one connects by straight-line segments those intersections that correspond to terms actually appearing in the equation and then forms a portion of a polygon convex toward the point of zero degree. In Figure 19.2 we have

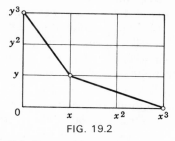

FIG. 19.2

[12] For further details see C. B. Boyer, "Newton as an Originator of Polar Coordinates," *American Mathematical Monthly*, **16** (1949), 73–78.

drawn the diagram for the folium of Descartes, $x^3 + y^3 - 3axy = 0$. Then the equations that are obtained by equating to zero in turn the totality of terms from the given equation whose lattice points line on each of the segments will be approximating equations for branches of the curve through the origin. In the case of the folium of Descartes, the approximating curves are $x^3 - 3axy = 0$ (or the parabola $x^2 = 3ay$) and $y^3 - 3axy = 0$ (or the parabola $y^2 = 3ax$); the graphing of portions of these parabolas near the origin will aid in the rapid sketching of the given equation $f(x, y) = 0$.

16 The three Newtonian books that are best known today are the *Principia*, the *Method of Fluxions*, and the *Opticks*; there is also a fourth work which in the eighteenth century appeared in a greater number of editions[13] than did the other three, and it, too, contained valuable contributions. This was the *Arithmetica universalis*, a work composed between 1673 and 1683, perhaps for Newton's lectures at Cambridge, and first published in 1707. This influential treatise contains the formulas, usually known as "Newton's identities," for the sums of the powers of the roots of a polynomial equation. Cardan had known that the sum of the roots of $x^n + a_1 x^{n-1} + \cdots + a_{n-1}x + a_n = 0$ is $-a$, and Viète had carried the relations between roots and coefficients somewhat further. Girard in 1629 has shown how to find the sum of the squares of the roots, or the sum of the cubes or of the fourth powers; but it was Newton who generalized this work to cover all powers. If $K \le n$, the relationships

$$S_K + a_1 S_{K-1} + \cdots + a_K K = 0$$

$$S_K + a_1 S_{K-1} + \cdots + a_K S_0 + a_{K+1} S_{-1} + \cdots + a_n S_{K-n} = 0$$

both hold; if $K > n$, the relationship

$$S_K + a_1 S_{K-1} + \cdots + a_{n-1} S_{K-n+1} + a_n S_{K-n} = 0$$

holds, where S_i is the sum of the ith powers of the roots. Using these relationships recursively, the sums of the powers of the roots can readily be found for any integral power. In the *Arithmetica universalis* there is also a theorem generalizing Descartes' rule of signs to determine the number of imaginary roots of a polynomial, as well as a rule for an upper bound for the positive roots. Another theorem asserts that if a cubic is cut by a variable line that moves parallel to itself, the locus of the barycenter of the three points of intersection is a straight line.

Despite his own contributions to the subject of algebra, Newton seems to have preferred the geometric analysis of the ancients. Consequently, the

---

[13] There were at least five Latin editions (1707, 1722, 1732, 1752, 1761) and three English editions (1720, 1728, 1769). A French translation appeared in 1802.

longest section in the *Arithmetica universalis* is that on the resolution of geometric questions. Here the solution of cubic equations is carried out with the help of a given conic section, for Newton regarded geometric constructions through curves other than the line and circle as part of algebra rather than of geometry:

> Equations are Expressions of Arithmetical Computation and properly have no place in Geometry .... Therefore the conic sections and all other Figures must be cast out of plane Geometry, except the right Line and the Circle. Therefore all these descriptions of the Conicks *in plano*, which the Moderns are so fond of, are foreign to Geometry.

Newton's conservatism here is in sharp contrast to his radical views in analysis—and to pedagogical views of the mid-twentieth century.

The *Principia* was the first of Newton's mathematical treatises to be **17** published, but it was the last in order of composition. Fame had come to him relatively promptly, for he had been elected to the Royal Society in 1672, four years after he had constructed his reflecting telescope (the idea for which had occurred also to Gregory even earlier). The *Principia* met with enthusiastic approval, and in 1689 Newton was elected to represent Cambridge in the British Parliament. Despite the generous recognition he received, Newton in 1692 became depressed and suffered a nervous breakdown. Perhaps feeling that continued scientific research was a strain, in 1696 he accepted appointment as Warden of the Mint, becoming Master of the Mint three years later. Evidently Newton felt comfortable and successful in his position there, partly, perhaps, because he had throughout much of his life been devoted to chemical research, with a special interest in alchemy. Theology and chronology also attracted his attention. In belief he seems to have been a crypto-Unitarian, while conforming outwardly to the Trinitarian views of the time. His *Chronology of Ancient Kingdoms Amended* and *Observations upon the Prophecies of Daniel and the Apocalypse of St. John* were published after his death.

Honors were heaped upon Newton in his later years. In 1699 he was elected a foreign associate of the Académie des Sciences, in 1703 he became president of the Royal Society, holding the post for the rest of his life, and in 1705 he was knighted by Queen Anne. Nevertheless, one event cast a cloud over Newton's life after 1695. In that year Wallis told him that in Holland the calculus was regarded as the discovery of Leibniz. In 1699 Nicolas Fatio de Duillier (1664–1753), an obscure Swiss mathematician who had moved to England, implied in a paper to the Royal Society that Leibniz may have taken his ideas on the calculus from Newton. At this affront Leibniz in the *Acta Eruditorum* for 1704 insisted that he was entitled to priority in publication and protested to the Royal Society against the imputation of plagiarism.

In 1705 Newton's *De quadratura curvarum* was unfavorably reviewed (by Leibniz?) in the *Acta Eruditorum*; and in 1708 John Keill (1671–1721), an Oxford professor, vigorously supported Newton's claims against those of Leibniz in a paper in the *Philosophical Transactions*. The repeated appeals of Leibniz to the Royal Society for justice finally led the Society to appoint a committee to study the matter and to report. The committee's report, under the title *Commercium epistolicum*, was published in 1712; but it left matters unimproved. It reached the banal conclusion that Newton was the first inventor, a point that had not been questioned seriously in the first place. Implications of plagiarism were supported by the committee in terms of documents that they assumed Leibniz had seen, but which we now know he had not received. The bitterness of national feeling reached such a point that in 1726, a decade after Leibniz had died, Newton deleted from the third edition of the *Principia* all reference to the fact that Leibniz had possessed a method in the calculus similar to the Newtonian.

As a consequence to the disgraceful priority dispute, British mathematicians were to some extent alienated from workers on the Continent throughout much of the eighteenth century. A penalty for the unfairness of followers of Newton toward Leibniz was thus visited on the next generations of mathematicians in England, with the result that British mathematics fell behind that of Continental Europe. Upon his death, Newton was buried in Westminster Abbey with such pomp that Voltaire, who attended the funeral, said later, "I have seen a professor of mathematics, only because he was great in his vocation, buried like a king who had done good to his subjects." Nevertheless, despite the recognition accorded mathematical achievement in England, development of mathematics there failed to match the rapid strides taken elsewhere in Europe during the eighteenth century.

BIBLIOGRAPHY

Ball, W. W. Rouse, "On Newton's Classification of Cubic Curves," *Proceedings of the London Mathematical Society*, **22** (1890–1891), 104–143.

Boyer, Carl B., *History of Analytic Geometry* (New York: Scripta Mathematica, 1956).

Boyer, Carl B., *History of the Calculus* (New York: Columbia University Press, 1939; paperback ed., New York: Dover, 1959).

Cajori, Florian, "The History of Notations of the Calculus," *Annals of Mathematics* (2), **25** (1924), 1–46.

Cajori, Florian, *A History of Mathematical Notations* (Chicago: Open Court, 1928–1929, 2 vols).

Child, J. M., ed., *The Early Mathematical Manuscripts of Leibniz*, trans. by C. I. Gerhardt (Chicago: Open Court, 1920).

De Morgan, Augustus, *Essays on the Life and Work of Newton* (Chicago and London: Open Court, 1914).

Gerhardt, C. I., *Die Entdeckung der differentialrechnung durch Leibniz* (Halle, 1848).

Gerhardt, C. I., *Die Geschichte der höheren Analysis.* Part I, *Die Entdeckung der höheren Analysis* (Halle, 1855).

Hall, A. R., and M. B. Hall, eds., *Unpublished Scientific Papers of Isaac Newton* (Cambridge: Cambridge University Press, 1962).

Hofmann, J. E., *Die Entwicklungsgeschichte der Leibnizschen Mathematik während des Aufenthaltes in Paris (1672–76)* (Munich: Leibniz Verlag, 1949).

Hofmann, J. E., "Zur Entdeckungsgeschichte der höheren Analysis im 17. Jahrhundert," *Mathematisch-Physikalische Semesterberichte*, **1** (1950).

Leibniz, G. W., *Mathematische Schriften*, ed. by C. I. Gerhardt, in *Gesammelte Werke*, ed. by G. H. Pertz, 3rd series, *Mathematik* (Halle, 1849–1863, 7 vols.).

Mahnke, Dietrich, "Neue Einblicke in die Entdeckungsgeschichte der höheren Analysis," *Abhandlungen der Preussische Akademie der Wissenschaften*, Physikalisch-Mathematische Klasse, **1** (1925), 1–64.

Mahnke, Dietrich, "Zur Keimesgeschichte der Leibnizschen Differentialrechnung," *Sitzungsberichte der Gesellschaft zur Beförderung der Gesamten Naturwissenschaften zu Marburg*, **47** (1932), 31–69.

More, L. T., *Isaac Newton, a Biography* (New York: Dover reprint, 1962).

Muir, Thomas, *The Theory of Determinants in the Historical Order of Development* (New York: Dover reprint, 1960, 4 vols.).

Newton, Sir Isaac, *Opera quae exstant omnia*, ed. by Samuel Harsley (London, 1779–1785, 5 vols.).

Newton, Sir Isaac, *Mathematical Principles of Natural Philosophy*, trans. by Andrew Motte, ed. by Florian Cajori (Berkeley, Calif.: University of California Press, 1934).

Newton, Sir Isaac, *Mathematical Works*, ed. by D. T. Whiteside (New York: Johnson Reprint, 1964–1967, 2 vols.).

Newton, Sir Isaac, *Mathematical Papers*, ed. by D. T. Whiteside (Cambridge: Cambridge University Press, 1967), Vol. I; Vols. II–VIII in press.

Rosenthal, A. "The History of Calculus," *American Mathematical Monthly*, **58** (1951), 75–86.

Sergescu, P., *Les recherches sur l'infini mathématique jusqu'à l'etablissement de l'analyse infinitésimale* (Paris, 1949).

Smith, D. E., ed., *Source Book in Mathematics* (New York: McGraw-Hill, 1929; paperback ed., New York: Dover, 1959, 2 vols.).

Toeplitz, Otto, *The Calculus, a Genetic Approach* (Chicago: University of Chicago Press, 1963).

Turnbull, H. W., *Mathematical Discoveries of Newton* (Glasgow: Blackie, 1945).

Turnbull, H. W., ed., *Correspondence of Newton* (Cambridge: Cambridge University Press, 1959–1961, 3 vols.).

Wieleitner, H., *Geschichte der Mathematik* (Leipzig: Walter de Gruyter, 1911, 2nd part, first half).

Zeuthen, H. G., *Geschichte der Mathematik im 16. und 17. Jahrhundert*, German ed. by Raphael Meyer (Leipzig: B. G. Teubner, 1903).

EXERCISES

1. Compare the lives of Newton and Leibniz with respect to means of support and opportunity for travel.
2. Compare the contributions of Newton and Leibniz to mathematical notations.
3. Was Newton more indebted to Wallis than to Barrow? Explain.
4. Was Leibniz more indebted to Pascal than to Barrow? Explain.
5. Are imaginary numbers halfway between existence and non-existence, as Leibniz asserted? Explain.
6. What relative advantages in mathematical research did Newton have over Leibniz?
7. Compare Newton's form of the binomial theorem with that used today, showing that the two forms are equivalent.
8. Discuss the extent to which Newton's prime and ultimate ratio is the same as the modern limit.
9. Use Newton's method of series to find $\int_0^1 \sqrt[3]{1 - x^3}\, dx$ approximately.
10. Find by Newton's first method, using o, p, and q, the slope of $y^2 = x^3 + 6x - 6$ at the point $(1, 1)$.
11. Show that if A changes from $A - a/2$ to $A + a/2$ and if B changes from $B - b/2$ to $B + b/2$, then AB will change by $aB + bA$.
12. Show how to derive the moment $3aA^2$ of A^3 by using the result of Exercise 11.
13. Show how to derive the moment of $1/A$ by using the result of Exercise 11.
14. Write down the numbers in the next diagonal for the arithmetic and harmonic triangles as given in the text.
15. Verify that the sum of the numbers in the third row of the harmonic triangle is $\frac{1}{2}$.
16. Using Leibniz' method, derive
$$d\left(\frac{x}{y}\right) = \frac{y\, dx - x\, dy}{y^2}$$
17. Using Leibniz' formula, find the sixth derivative of $x^2 e^{2x}$.
18. Using Leibniz' rule, find the envelope of the one-parameter family of lines $y = cx + c^2$.
19. Explain what is wrong in Leibniz' argument that the sum of the infinite alternating series $1 - 1 + 1 - 1 + \cdots$ is $\frac{1}{2}$.
20. Use an equivalent of Leibniz' method of determinants to see if the system of equations $2x - 3y = 6$, $3x + 4y = 7$, $x - 27y = 9$ is consistent.
21. By Newton's method find to the nearest hundredth the root of the equation $x^3 - 2x = 5$ which lies between 2 and 3.
22. By Newton's method find to the nearest tenth the root of the equation $2 \sin x = x$ which lies between $\pi/2$ and π.
23. Show graphically why Newton's method may fail if $f(a_1)$ and $f''(a_1)$ are not of the same sign.
24. Describe the curve given in bipolar coordinates by the equation $y = mx$, where m is a constant.
*25. Verify Leibniz' statement that $\sqrt{6} = \sqrt{1 + \sqrt{-3}} + \sqrt{1 - \sqrt{-3}}$.
*26. How high a uniform atmosphere did Newton assume in arriving at a speed of 979 ft/sec for sound?
*27. Use the Newton parallelogram to find approximating curves at the origin for $y^3 - 2xy + x^4 = 0$, sketching the approximating curves.
*28. Verify that Newton's formulas for radius of curvature are equivalent to those used today.
*29. Using Newton's method for reversion of series, derive the first four terms in the e^x series from
$$\ln(1 + x) = x - \frac{x^2}{2} + \frac{x^3}{3} - \frac{x^4}{4} + \cdots$$
*30. Use Newton's identities to find the sum of the fifth powers of the roots of $x^4 - x^3 - 7x^2 + x + 6 = 0$.

CHAPTER XX

The Bernoulli Era

He who can digest a second or third fluxion, a second
or third difference, need not, methinks, be squeamish
about any point in Divinity.

George Berkeley

The discoveries of a great mathematician, such as Newton, do not auto- **1**
matically become part of the mathematical tradition. They may be lost to
the world unless other scholars understand them and take enough interest
to look at them from various points of view, clarify and generalize them, and
point out their implications. Newton, unfortunately, was hypersensitive and
did not communicate freely, and consequently the method of fluxions was
not well known outside of England. Leibniz, on the other hand, found devoted
disciples who were eager to learn about the differential and integral calculus
and to transmit the knowledge to others. Foremost among the enthusiasts
were two Swiss brothers, Jacques Bernoulli (1654–1705) and Jean Bernoulli
(1667–1748), often known also by the Anglicized forms of their names, James
and John (or by the German equivalents, Jakob and Johann), each as quick
to offend as to feel offended.

No family in the history of mathematics has produced as many celebrated
mathematicians as did the Bernoulli family, which, unnerved by the Spanish
Fury in 1576, had fled to Basel from the Catholic Spanish Netherlands in
1583. Some dozen members of the family (see the genealogical chart) achieved
distinction in mathematics and physics, and four of them were elected foreign
associates of the Académie des Sciences. The first to attain prominence in
mathematics was Jacques Bernoulli. He was born and died at Basel, but he
traveled widely to meet scholars in other countries. His interest had been
directed toward infinitesimals by works of Wallis and Barrow, and the papers
of Leibniz in 1684–1686 enabled him to master the new methods. By 1690,
when he suggested the name "integral" to Leibniz, Jacques Bernoulli was
himself contributing papers on the subject to the *Acta Eruditorum*. Among
other things, he pointed out that at a maximum or minimum point the
derivative of the function need not vanish, but can take on an "infinite value"
or assume an indeterminate form. He was early interested in infinite series,

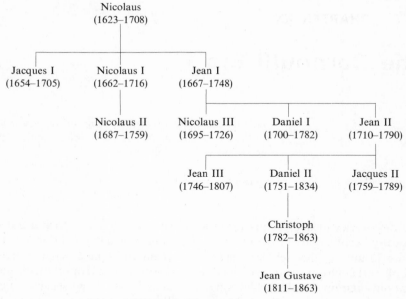

The mathematical Bernoullis: a genealogical chart.

and in his first paper on the subject in 1689 he gave the well-known "Bernoulli inequality" $(1 + x)^n > 1 + nx$ where x is real and $x > -1$ and $x \neq 0$ and n is an integer greater than one, but this can be found earlier in the seventh lecture of Barrow's *Lectiones geometriae* of 1670. To him frequently is attributed also the demonstration that the harmonic series is divergent, for most men were unaware of anticipations by Oresme and Mengoli. Jacques Bernoulli believed, in fact, that his brother had been the first to observe the divergence of the harmonic series.

Jacques Bernoulli was fascinated by the series of reciprocals of the figurate numbers, and although he knew that the series of reciprocals of the perfect squares is convergent, he was unable to find the sum of the series. Inasmuch as the terms of

$$\frac{1}{1^2} + \frac{1}{2^2} + \frac{1}{3^2} + \cdots + \frac{1}{n^2} + \cdots$$

are, term for term, less than or equal to those of

$$\frac{1}{1} + \frac{1}{1 \cdot 2} + \frac{1}{2 \cdot 3} + \frac{1}{3 \cdot 4} + \cdots + \frac{1}{n(n-1)} + \cdots$$

and the latter series was known to converge to 2, it was clear to Bernoulli that the former must converge.

A frequent correspondent with other mathematicians of the day, Jacques Bernoulli was *au courant* with the popular problems, many of which he solved independently. Among these were finding the equations of the catenary, the tractrix, and the isochrone, all of which had been treated by Huygens and Leibniz. The isochrone called for the equation of a plane curve along which an object would fall with uniform vertical velocity, and Bernoulli showed that the required curve is the semicubical parabola. It was in connection with such problems that the Bernoulli brothers discovered the power of the calculus, and they remained in touch with Leibniz on all aspects of the new subject. Jacques Bernoulli in his work on the isochrone in the *Acta Eruditorum* for 1690 used the word "integral," and a few years later Leibniz agreed that *calculus integralis* would be a better name than *calculus summatorius* for the inverse of the *calculus differentialis*. To differential equations Jacques Bernoulli contributed the study of the "Bernoulli equation" $y' + P(x)y = Q(x)y^n$ which he and Leibniz and Jean Bernoulli solved—Jean by reducing it to a linear equation through the substitution $z = y^{1-n}$. It was especially in connection with a problem from the calculus of variations that the Bernoulli brothers came into sharp conflict with each other. (Jacques was the fifth child in the family, Jean was the tenth; the younger man perhaps resented what he regarded as condescension on the part of his older brother.) Leibniz and the Bernoullis were seeking a solution to the brachistochrone problem—to find the curve along which a particle will slide in the shortest time from one given point to a second lower given point not directly beneath the first point. Jean had found an incorrect proof that the curve is a cycloid, and he challenged his brother to discover the required curve. After Jacques correctly proved that the curve sought is a cycloid, Jean tried to substitute his brother's proof for his own.

Jacques Bernoulli was fascinated by curves and the calculus, and one **2** curve bears his name—the "lemniscate of Bernoulli," given by the polar equation $r^2 = a \cos 2\theta$. The curve was described in the *Acta Eruditorum* of 1694 as resembling a figure eight or a knotted ribbon (lemniscus). However, the curve that most caught his fancy was the logarithmic spiral. This curve had been mentioned by Descartes and rectified by Torricelli, but Bernoulli showed that it had several striking properties not noted before: (1) the evolute of a logarithmic spiral is an equal logarithmic spiral; (2) the pedal curve of a logarithmic spiral with respect to its pole (that is, the locus of the projections of the pole upon the tangents to the given curve) is an equal logarithmic spiral; (3) the caustic by reflection for rays emanating from the pole (that is, the envelope of the rays reflected at points on the given curve) is an equal logarithmic spiral; and (4) the caustic by refraction for rays emanating from the pole (that is, the envelope of rays refracted at points on the curve) is an equal logarithmic spiral. It is easy to appreciate the feeling that led Bernoulli to request that the *spira mirabilis* be engraved on his tombstone together with the inscription *Eadem mutata resurgo* ("Though changed, I arise again the same").

Jacques Bernoulli had been led to spirals of a different type when he repeated Cavalieri's procedure in bending half of the parabola $x^2 = ay$ about the origin to produce a spiral of Archimedes; but whereas Cavalieri had studied the transformation by essentially synthetic methods, Bernoulli used rectangular and polar coordinates. Newton had used polar coordinates earlier—perhaps as early as 1671—but priority in publication seems to go to Bernoulli who in the *Acta Eruditorum* for 1691 proposed measuring abscissas along the arc of a fixed circle and ordinates radially along the normals. Three years later, in the same journal, he proposed a modification that agreed with the system of Newton. The coordinate y now was the length of the radius vector of the point, and x was the arc cut off by the sides of the vectorial angle on a circle of radius a described about the pole as center. These coordinates are essentially what we would now write as $(r, a\theta)$. Bernoulli, like Newton, was interested primarily in applications of the system to the calculus; hence he, too, derived formulas for arc length and radius of curvature in polar coordinates. For his "parabolic spiral" $r^2 = a\theta$ he noted that the question of arc length leads, through $ds = \sqrt{dr^2 + r^2 \, d\theta^2}$, to the integral of the square root of a quartic polynomial, the first specific instance of what now is known as an elliptic integral.

3   The mathematical contributions of the Bernoullis, like those of Leibniz, are found chiefly in articles in journals, especially the *Acta Eruditorum*; but Jacques Bernoulli wrote also a classical treatise entitled *Ars conjectandi* (or "Art of Conjecturing"), published in 1713, eight years after the author's death. This is the earliest substantial volume in the theory of probability, for Huygens' *De ludo aleae* had been only a brief introduction. The treatise of Huygens is, in fact, reproduced as the first of the four parts of the *Ars conjectandi*, together with commentary by Bernoulli. The second part of the *Ars conjectandi* includes a general theory of permutations and combinations, facilitated by the binomial and multinomial theorems. Here we find the first adequate proof of the binomial theorem for positive integral powers. The proof is by mathematical induction, a method of approach that Bernoulli had rediscovered while reading the *Arithmetic infinitorum* of Wallis and which he had published in the *Acta Eruditorum* in 1686. He gave Pascal credit for the binomial theorem with general exponent, but this attribution appears to be gratuitous. Newton seems to have first stated the theorem in general form for any rational exponent, although he gave no proof, this being supplied later by Euler. In connection with the expansion of $(1 + 1/n)^n$ Jacques

Bernoulli proposed the problem of the continuous compounding of interest—that is finding $\lim_{n \to \infty} (1 + 1/n)^n$. Since

$$\left(1 + \frac{1}{n}\right)^n < 1 + \frac{1}{1} + \frac{1}{1 \cdot 2} + \cdots + \frac{1}{1 \cdot 2 \ldots n} < 1 + 1 + \frac{1}{2} + \frac{1}{2^2} + \frac{1}{2^{n-1}} < 3$$

it was clear to him that the limit existed.

The second part of the *Ars conjectandi* contains also the "Bernoulli numbers." These arose as coefficients in a recursion formula for the sums of the powers of the integers, and they now find many applications in other connections. The formula was written by Bernoulli as follows:[1]

$$\int n^c = \frac{1}{c+1} n^{c+1} + \frac{1}{2} n^c + \frac{c}{2} A n^{c-1} + \frac{c(c-1)(c-2)}{2 \cdot 3 \cdot 4} B n^{c-3}$$

$$+ \frac{c(c-1)(c-2)(c-3)(c-4)}{2 \cdot 3 \cdot 4 \cdot 5 \cdot 6} C n^{c-5} \ldots$$

where $\int n^c$ means the sum of the cth powers of the first n positive integers and the letters $A, B, C, \ldots$ (the Bernoulli numbers) are the coefficients of the term in n (the last term) in the corresponding expressions for $\int n^2, \int n^4, \int n^6, \ldots$. (The numbers also can be defined as $n!$ times the coefficients of the even-powered terms in the Maclaurin expansion of the function $x/(e^x - 1)$.) The Bernoulli numbers are useful in writing the infinite series expansions of trigonometric and hyperbolic functions. The first three of the numbers are readily verified as $A = \frac{1}{6}$, $B = -\frac{1}{30}$, and $C = \frac{1}{42}$.

The third and fourth parts of the *Ars conjectandi* are devoted primarily to problems illustrating the theory of probability. The fourth and last part contains the celebrated theorem that now bears the author's name, and on which Bernoulli and Leibniz had corresponded—the so-called "Law of large numbers." This states that if p is the probability of an event, if m is the number of occurrences of the event in n trials, if ϵ is an arbitrarily small positive number, and if P is the probability that the inequality $|m/n - p| < \epsilon$ is satisfied, then $\lim_{n \to \infty} P = 1$.

Appended to the *Ars conjectandi* is a long memoir on infinite series. Besides the harmonic series and the sum of the reciprocals of the perfect squares, Bernoulli considered the series

$$\frac{1}{\sqrt{1}} + \frac{1}{\sqrt{2}} + \frac{1}{\sqrt{3}} + \frac{1}{\sqrt{4}} + \cdots$$

[1] See D. E. Smith, *Source Book in Mathematics* (1929), pp. 85–90, for a translation of the relevant pages of *Ars conjectandi* dealing with the Bernoulli numbers. For bibliographic references and further details on Jacques Bernoulli's work, see the article on Bernoulli by J. E. Hofmann to appear in Vol. I of the forthcoming *Dictionary of Scientific Biography* (New York: Scribner's).

He knew (by comparing the terms with those in the harmonic series) that this diverges, and he pointed to the paradox that the ratio of the "sum" of all the odd terms to the "sum" of all the even terms is as $\sqrt{2} - 1$ is to 1, from which the sum of all the odd terms appears to be less than the sum of all the even terms; but this is impossible because term for term the former are larger than the latter.

4 The father of the famous Bernoulli brothers, Nicolaus (1623–1708), had had very definite plans for the futures of his sons, and he had put obstacles in the way of their becoming mathematicians. Jacques, the older, had been destined for the ministry, and Jean was to have been a merchant or a physician. The younger brother did, in fact, write his doctoral dissertation in 1690 on effervescence and fermentation, but the following year he became so deeply interested in the calculus that during 1691–1692 he composed two little textbooks on the differential and integral calculus, although neither was published until long afterward. While he was in Paris in 1692, he instructed a young French marquis, G. F. A. de L'Hospital (1661–1704), in the new Leibnizian discipline; and Jean Bernoulli signed a pact under which in return for a regular salary, he agreed to send L'Hospital his discoveries in mathematics, to be used as the marquis might wish. The result was that one of Bernoulli's chief contributions, dating from 1694, has ever since been known as L'Hospital's rule on indeterminate forms. Jean Bernoulli had found that if $f(x)$ and $g(x)$ are functions differentiable at $x = a$ such that $f(a) = 0$ and $g(a) = 0$ and

$$\lim_{x \to a} \frac{f'(x)}{g'(x)}$$

exists, then

$$\lim_{x \to a} \frac{f(x)}{g(x)} = \lim_{x \to a} \frac{f'(x)}{g'(x)}$$

This well-known rule was incorporated by L'Hospital in the first textbook on the differential calculus to appear in print—*Analyse des infiniment petits*, published at Paris in 1696. This book, the influence of which dominated most of the eighteenth century, is based on two postulates: (1) that one can take as equal two quantities which differ only by an infinitely small quantity; and (2) that a curve can be considered as made up of infinitely small straight line segments that determine, by the angles that they make with each other, the curvature of the curve. These would today scarcely be regarded as acceptable, but L'Hospital considered them "so self-evident as not to leave the least scruple about their truth and certainty on the mind of an attentive reader." The basic differential formulas for algebraic functions are derived in the manner of Leibniz, and applications are made to tangents, maxima and minima, points of inflection, curvature, caustics, and indeterminate forms. The book closes with a statement indicating that the methods are general and can be extended also to transcendental curves.[2] L'Hospital was an exceptionally effective writer, for his *Traité analytique des sections coniques*, published posthumously in 1707, did for analytic geometry of the eighteenth century what the *Analyse* did for the calculus. The *Traité* was not especially original, but it had a pedagogical quality that made it a standard treatise on conics throughout most of the century.[3]

L'Hospital's *Analyse* met with great success and appeared in numerous 5 editions throughout the next century. In the preface the author had admitted that he owed much to Leibniz and the Bernoullis, especially to "the young professor at Groningen" (where Jean had been appointed in 1695). Jean Bernoulli wrote to L'Hospital to thank him for mentioning him in the volume, but after the death of the marquis in 1704 Bernoulli in letters to others virtually accused the author of plagiarism. Contemporaries regarded the claims of Bernoulli as unfounded, but the recent publication of the Bernoulli-L'Hospital correspondence[4] indicates that much of the work evidently was due to Bernoulli. Bernoulli did not publish his own textbook on the differential calculus (which finally was printed in 1924), and the text on the integral calculus appeared fifty years after it had been written—in his *Opera omnia* of 1742. In the interval Jean Bernoulli wrote prolifically on many advanced aspects of analysis—the isochrone, solids of least resistance, the catenary, the tractrix, trajectories, caustic curves, isoperimetric problems—achieving a reputation that led to his being called to Basel in 1705 to fill the chair left vacant by his brother's death. His correspondence with Leibniz was very active, and he espoused the Leibnizian cause against Newton with unwarranted aggressiveness. One might call him Leibniz' bulldog, for he did for the calculus what Huxley later accomplished for the Darwinian theory of evolution, His tactlessness led to bitter controversy with his brother, and his jealous nature led him to drive from home a son, Daniel, for having won a prize of the Académie des Sciences for which Jean, too,

---

[2] For further description of the *Analyse* see C. B. Boyer, "The First Calculus Textbooks," *The Mathematics Teacher*, **39** (1946), 159–167. Some of the material in the *Analyse* undoubtedly was the result of L'Hospital's independent work, for he was a capable mathematician. The rectification of the logarithmic curve, for example, seems to have appeared for the first time in 1692 in a letter from L'Hospital to Leibniz. An unusually extensive account of L'Hospital's work is found in J. L. Coolidge, *The Mathematics of Great Amateurs* (1949). Chapter 12. pp. 147–170.

[3] For a description of the contents of the *Traité*, see C. B. Boyer, *History of Analytic Geometry* (1956), pp. 150–154.

[4] See *Der Briefwechsel von Johann Bernoulli*, ed. by Otto Spiess (1955), Vol. I. Cf. Otto Spiess, "Une édition de l'oeuvre des mathématiciens Bernoulli," *Archives Internationales d'Histoire des Sciences*, **1** (1947), 356–362.

had competed; but he was a most inspiring teacher and an indefatigable research worker. He frequently is regarded as the inventor of the calculus of variations, because of his proposal in 1696–1697 of the problem of the brachistochrone; and he contributed to differential geometry through his work on geodesic lines on a surface. To him often is ascribed also the exponential calculus, for he studied not only the simple exponential curves $y = a^x$, but general exponentials such as $y = x^x$. For the area under the curve $y = x^x$ from $x = 0$ to $x = 1$ he found the striking infinite series representation

$$\frac{1}{1^1} - \frac{1}{2^2} + \frac{1}{3^3} - \frac{1}{4^4} + \cdots$$

This result he obtained by writing $x^x = e^{x \ln x}$, expanding this in the exponential series and integrating term by term, using integration by parts.

6   Jean and Jacques Bernoulli rediscovered the series for $\sin n\theta$ and $\cos n\theta$, in terms of $\sin \theta$ and $\cos \theta$, which Viète had known, and they extended them, uncritically, to include fractional values of n. Jean was aware also of relationships between inverse trigonometric functions and imaginary logarithms, discovering in 1702, through differential equations, the relationship

$$\arctan z = \frac{1}{i} \ln \sqrt{\frac{1 + iz}{1 - iz}}$$

He corresponded with other mathematicians on the logarithms of negative numbers, but here he mistakenly believed that $\log(-n) = \log n$. He tended to develop trigonometry and the theory of logarithms from an analytic point of view, and he experimented with several notations for a function of x, the one nearest to the modern being ϕx. His vague notion of a function was expressed as "a quantity composed in any manner of a variable and any constants." Among his numerous controversies was one with British mathematicians over whether or not the well-known series of Brook Taylor (1685–1731), published in the *Methodus incrementorum* of 1715, was a plagiarism of the Bernoulli series

$$\int y \, dx = yx - \frac{x^2}{2!} \frac{dy}{dx} + \frac{x^3}{3!} \frac{d^2 y}{dx^2} - \cdots$$

Neither Bernoulli nor Taylor was aware that both had been anticipated by Gregory in the discovery of "Taylor's series."

7   Jean Bernoulli maintained a zeal for mathematics as lively as was his persistence in controversy. Moreover, he was the father of three sons, Nicholas (1695–1726), Daniel (1700–1782), and Jean II (1710–1790), all of whom at some stage filled a position as professor of mathematics: Nicholas and Daniel at St. Petersburg and Daniel and Jean II at Basel. (Another Nicolaus (1687–1759), cousin of the one above, for a time held the chair in mathematics at Padua that Galileo once had filled.) There were still other Bernoullis who attained some eminence in mathematics,[5] but of these none achieved fame comparable to that of the original two brothers, Jacques and Jean. The most celebrated of the younger generation was Daniel, whose work in hydrodynamics is recalled in "Bernoulli's principle." In mathematics he is best known for his distinction, in the theory of probability, between mathematical expectation and "moral expectation," or between "physical fortune" and "moral fortune." He assumed that a small increase in a person's material means causes an increase in satisfaction that is inversely proportional to the means. In the form of an equation, $dm = K(dp/p)$, where m is the moral fortune, p is the physical fortune, and K is a constant of proportionality. This leads to the conclusion that as the physical fortune increases geometrically, the moral fortune increases arithmetically. This hypothesis by Bernoulli appears in the *Commentarii* of the Academy of Sciences at St. Petersburg for 1730–1731 (published 1738), for Daniel Bernoulli had spent the years 1725–1733 in Russia before returning to Basel. His work on probability took on varied aspects, including applications to business, medicine, and astronomy. In 1734 he and his father shared the prize offered by the Académie des Sciences for an essay on probabilities related to the inclinations of the orbital planes of the planets; in 1760 he read to the Paris Académie a paper on the application of probability theory to the question of the advantage of inoculation against smallpox.

When Daniel Bernoulli went to St. Petersburg in 1725, his older brother also was called there as professor of mathematics; in the discussions of the two men there arose a problem that has come to be known as the Petersburg paradox, probably because it first appeared in the *Commentarii* of the Academy there.[6] The problem is as follows: Suppose that Peter and Paul agree to play a game based on the toss of a coin. If a head is thrown on the first toss, Paul will give Peter one crown; if the first toss is tail, but a head appears on the second toss, Paul will give Peter two crowns; if a head appears for the first time on the third toss, Paul will give Peter four crowns; and so on, the amount to be paid if head appears for the first time on the nth toss

[5] An account of these is included in Smith, *History of Mathematics* (1958), I, 431–433. Among the lesser Bernoullis not already mentioned were three sons of Jean II: Jean III (1744–1807), Daniel II (1751–1834), and Jacques II (1759–1789). The first of these became a professor of mathematics at the Berlin Academy at the early age of nineteen. Two other significant members of the family were Christoph Bernoulli (1782–1863), a son of Daniel II, and Johann Gustav Bernoulli (1811–1863), a son of Christoph. Cf. J. O. Fleckenstein, *L'école mathématique baloise des Bernoulli à l'aube du XVIIIe siècle* (ca. 1958).

[6] The full title is *Commentarii Academiae Scientiarum Imperialis Petropolitanae*. This journal contains many articles by the younger Bernoullis and their Swiss colleague, Euler.

being 2^{n-1} crowns. What should Peter pay Paul for the privilege of playing the game? Peter's mathematical expectation, given by

$$\frac{1}{2} \cdot 1 + \frac{1}{2^2} \cdot 2 + \frac{1}{2^3} \cdot 2^2 + \cdots + \frac{1}{2^n} \cdot 2^{n-1} + \cdots$$

evidently is infinite, yet common sense suggests a very modest finite sum. When Georges Louis Leclerc, Comte de Buffon (1707–1788), made an empirical test of the matter, he found that in 2084 games Paul would have paid Peter 10,057 crowns. This indicates that for any one game Paul's expectation, instead of being infinite, is actually something less than 5 crowns! The paradox raised in the Petersburg problem was widely discussed during the eighteenth century, with differing explanations being given. Daniel Bernoulli sought to resolve it through his principle of moral expectation, in accordance with which he replaced the amounts 1, 2, 2^2, 2^3, ... 2^n, ... by $1^{\frac{1}{2}}$, $2^{\frac{1}{4}}$, $4^{\frac{1}{8}}$, $8^{\frac{1}{16}}$, ... Others preferred, as a solution of the paradox, to point out that the problem is inherently impossible in view of the fact that Paul's fortune is necessarily finite, hence he could not pay the unlimited sums that might be required in the case of a long delay in the appearance of a head.[7]

8   The theory of probability had many devotees during the early eighteenth century, and one of the most important of these was Abraham De Moivre (1667–1754). He had been born a French Huguenot, but shortly after the revocation of the Edict of Nantes he went to England, where he made the acquaintance of Newton and Halley and became a private teacher of mathematics. In 1697 he was elected to the Royal Society and subsequently to the Academies of Paris and Berlin.[8] He hoped to obtain a university position in mathematics, but this he never secured, partly because of his non-British origin; and Leibniz tried in vain to secure a professional position for him in Germany. Nevertheless, despite the long hours of tutoring necessary to support himself, De Moivre produced a considerable quantity of research. In 1711 he contributed to the *Philosophical Transactions* a long memoir on the laws of chance, and this he expanded into a celebrated volume, the *Doctrine of Chances*, that appeared in 1718 (and in later editions). The memoir and the volume contained numerous questions on dice, the problem of points (with unequal chances of winning), drawing balls of various colors from a bag, and other games. Some of the problems had appeared in Jacques Bernoulli's *Ars conjectandi*, the publication of which was earlier than the *Doctrine of Chances* but later than De Moivre's memoir. In the preface to the

---

[7] See I. Todhunter, *A History of the Mathematical Theory of Probability* (Cambridge, 1861) for extensive discussion of the problem.

[8] For an account of his life see Helen M. Walker, "Abraham de Moivre," *Scripta Mathematica*, **2** (1934), 316–333. See also the article on "Moivre, Abraham de (1667–1754)," by Agnes M. Clerke in *Dictionary of National Biography*, **38** (1894), 116–117.

Doctrine of Chances the author referred to the work on probability of Jacques, Jean, and Nicolaus Bernoulli. The various editions of the volume contain more than fifty problems on probability, as well as questions relating to life annuities. In general, De Moivre derived the theory of permutations and combinations from the principles of probability, whereas now it is customary to reverse the roles. For example, to find the number of permutations of two letters chosen from the six letters *a, b, c, d, e,* and *f,* he argues that the probability that a particular letter will be first is $\frac{1}{6}$ and the probability that another specific letter will be second is $\frac{1}{5}$. Hence the probability that these two letters will appear in that order is $\frac{1}{6} \cdot \frac{1}{5} = \frac{1}{30}$, from which one concludes that the number of all possible permutations, two at a time, is 30. De Moivre is often credited with the principle, published in the *Doctrine of Chances*, that the probability of a compound event is the product of the probabilities of its components, but this had been implied in earlier works.

De Moivre was especially interested in developing for probability general procedures and notations that he thought of as a new "algebra."[9] A generalization of a problem given earlier by Huygens usually is called, appropriately, De Moivre's problem: to find the probability of throwing a given number on a throw of *n* dice each having *m* faces. Some of his contributions to probability were published in a further volume, the *Miscellanea analytica* of 1730. In a supplement to this work De Moivre included some results that appeared also in the *Methodus differentialis* of James Stirling (1692–1770), published in the same year as the *Miscellanea analytica*. Among these is the approximation $n! \approx \sqrt{2\pi n}(n/e)^n$, which usually is known as Stirling's formula, although known earlier to De Moivre, and a series, also named for Stirling, relating ln *n*! and the Bernoulli numbers.

De Moivre apparently was the first one to work with the probability formula $\int_0^\infty e^{-x^2}\, dx = \sqrt{\pi}/2$, a result that appeared unobtrusively in a privately printed pamphlet of 1733 entitled *Approximatio ad summam terminorum binomii* $(a + b)^n$ *in seriem expansi*. This work, representing the first appearance of the law of errors or the distribution curve, was translated by De Moivre and included in the second edition (1738) of his *Doctrine of Chances*.[10] Many aspects of probability attracted De Moivre, including actuarial problems. In his work on *Annuities upon Lives*, which formed a part of the *Doctrine of Chances* and was separately reprinted in more than half a dozen editions, he adopted a rough-and-ready rule, known as "De Moivre's hypothesis of equal decrements," that annuities can be computed on the assumption that the number of a given group that die is the same during each year.

[9] The work of De Moivre on probability is fully described in Todhunter, *A History of the Mathematical Theory of Probability*, Chapter 9, pp. 135–193.
[10] It is reprinted in D. E. Smith, *Source Book in Mathematics*, pp. 566–575.

9   The *Miscellanea analytica* is important not only in probability, but also in the development of the analytic side of trigonometry. The well-known De Moivre's theorem, $(\cos\theta + i\sin\theta)^n = \cos n\theta + i\sin n\theta$, is not explicitly given, but it is clear from the work on cyclometry and other contexts that the author was quite familiar with this relationship, probably as early as 1707. In a paper in the *Philosophical Transactions* for 1707 De Moivre wrote that

$$\tfrac{1}{2}(\sin n\theta + \sqrt{-1}\cos n\theta)^{1/n} + \tfrac{1}{2}(\sin n\theta - \sqrt{-1}\cos n\theta)^{1/n} = \sin\theta$$

and in his *Miscellanea analytica* of 1730, he expressed the equivalent of

$$(\cos\theta \pm i\sin\theta)^{1/n} = \cos\frac{2K\pi \pm \theta}{n} \pm i\sin\frac{2K\pi \pm \theta}{n}$$

which he used to factor $x^{2n} + 2x\cos n\theta + 1$ into quadratic factors of the form $x^2 + 2x\cos\theta + 1$. Again in a *Philosophical Transactions* paper of 1739 he found[11] the nth roots of the "impossible binomial" $a + \sqrt{-b}$ through the procedure that we now use in taking the nth root of the modulus, dividing the amplitude by n, and adding multiples of $2\pi/n$.

De Moivre, dealing with imaginary numbers and the circular functions in *Miscellanea analytica*, came close to recognizing the hyperbolic functions in extending theorems on sectors of circles to analogous results on sectors of the rectangular hyperbola. In view of the breadth and depth of his results, it was natural that Newton in his later years should have told those who came to him with questions on mathematics, "Go to Mr. De Moivre; he knows these things better than I do." It is not surprising that De Moivre was one of the partisan commissioners appointed by the Royal Society in 1712 to report on the claims of Newton and Leibniz to the invention of the calculus.

In the *Philosophical Transactions* for 1697–1698 De Moivre had written on the "infinitonome"—that is, an infinite polynomial or infinite series—including the process of finding a root of such an expression, and it was largely in recognition of this paper that he had been elected a member of the Royal Society. The interest of De Moivre in infinite series and probability is reminiscent of the Bernoullis. De Moivre carried on an extensive and cordial correspondence with Jean Bernoulli during the decade 1704 to 1714, and it was the former who proposed the latter for election to the Royal Society in 1712.

10   One of the motives that had led De Moivre to be concerned with the factoring of $x^{2n} + ax^n + 1$ into quadratic factors was the desire to complete some of the work of Roger Cotes (1682–1716) on the integration of rational fractions through decompositions into partial fractions. The life of Cotes

[11] See D. E. Smith: *Source Book in Mathematics*, pp. 440–450, for a full account of this work.

is another tragic instance of a very promising career cut short by premature death. As Newton remarked, "If Cotes had lived, we might have known something." A student and later a professor at Cambridge, the young man had spent much of the time from 1709 to 1713 preparing the second edition of Newton's *Principia*, and three years later he had died, leaving behind him some significant but uncompleted work. Much of this was collected and published posthumously in 1722 under the title *Harmonia mensurarum*. The title is derived from the following theorem:

> If through a fixed point O a variable straight line is drawn cutting an algebraic curve in points $Q_1, Q_2, \ldots, Q_n$, and if a point P is taken on the line such that the reciprocal of OP is the arithmetic mean of the reciprocals of $OQ_1, OQ_2, \ldots, OQ_n$, then the locus of P is a straight line.

Most of the treatise, however, is devoted to the integration of rational fractions, including decomposition into quadratic factors of $x^n - 1$, work completed later by De Moivre. The *Harmonia mensurarum* is among the early works to recognize the periodicity of the trigonometric functions, cycles of the tangent and secant functions appearing here in print for perhaps the first time. It is one of the earliest books with a thorough treatment of the calculus as applied to the logarithmic and circular functions, including a table of integrals depending on these. In this connection the author gave what is known in trigonometry books as "Cotes' property of the circle," a result closely related to De Moivre's theorem, which allows one to write such expressions as

$$x^{2n} + 1 = \left(x^2 - 2x\cos\frac{\pi}{2n} + 1\right)\left(x^2 - 2x\cos\frac{3\pi}{2n} + 1\right)\cdots$$

$$\left(x^2 - 2x\cos\frac{(2n-1)\pi}{2n} + 1\right)$$

This result is readily confirmed if, having plotted on the unit circle the roots of -1 of order $2n$, one forms the products of conjugate imaginary pairs. Cotes apparently was among the earliest of mathematicians to anticipate the relationship $\ln(\cos\theta + i\sin\theta) = i\theta$, an equivalent of which had been given by him in a *Philosophical Transactions* article in 1714 and which was reprinted in the *Harmonia mensurarum*. This theorem is usually attributed to Euler, who first gave it in modern exponential form.

British mathematics boasted an impressive number of capable contributors **11** during the earlier part of the eighteenth century, of whom De Moivre, Cotes, and Stirling in particular were friends of Newton. Stirling published in 1717 a work entitled *Lineae tertii ordinis Neutonianae* in which he completed the classification of cubic curves, drawn up by Newton in 1704, by adding some cubics that Newton missed and by adding demonstrations that had been lacking in the original *Enumeratio*. Stirling showed, among other things, that if the y-axis is an asymptote of a curve of order n, the equation of the curve cannot contain a term in y^n and an asymptote cannot cut the curve in more than $n - 2$ points. For graphs of rational functions $y = f(x)/g(x)$ he found the vertical asymptotes by equating $g(x)$ to zero.[12] For conic sections Stirling gave a full treatment in which the axes, vertices, and asymptotes are found analytically from the general second-degree equation with respect to oblique coordinates.

12   The work of Newton and Stirling on plane curves was continued by Colin Maclaurin (1698–1746), perhaps the outstanding British mathematician of the generation after Newton. Born in Scotland and educated at Glasgow, which he entered at the age of eleven, he became professor of mathematics at Aberdeen when nineteen, and half a dozen years later taught at the University of Edinburgh. It is interesting to note that in Great Britain, Switzerland, and the Low Countries, the leading mathematicians in the seventeenth and eighteenth centuries were connected with universities, whereas in France, Germany and Russia they were more likely to be associated with the academies established by the absolute rulers.

Maclaurin had begun contributing papers to the *Philosophical Transactions* before he was twenty-one, and in 1720 he published two treatises on curves: *Geometrica organica* and *De linearum geometricarum proprietatibus*. The former in particular was a well-known work which extended the results of Newton and Stirling on conics, cubics, and higher algebraic curves. Among the propositions is one often known as the theorem of Bézout (in honor of the man who later gave an imperfect proof)—a curve of order m intersects a curve of order n in general in mn points. In connection with this theorem Maclaurin noticed a difficulty which is usually known as Cramer's paradox, in honor of a later rediscoverer. A curve of order n generally is determined, as Stirling had indicated, by $n(n + 3)/2$ points. Thus a conic is uniquely determined by five points and a cubic should be determined by nine points. By the Maclaurin-Bézout theorem, however, two curves of degree n intersect in n^2 points, so that two different cubics intersect in nine points. Hence it is obvious that $n(n + 3)/2$ points do not always uniquely determine a single curve of order n. The answer to the paradox did not appear until a century later when it was explained in the work of Plücker (see below).

The *Geometrica organica* contained interesting propositions on conics, including various organic constructions similar to some given by Newton in the *Principia*. We find also Pascal's theorem on a hexagon inscribed in a

[12] For further summary see C. Tweedie: *James Stirling, Sketch of His Life and Works* (1922).

conic, deduced from the properties of a quadrilateral inscribed in a conic. Extending this type of work to curves of third degree, Maclaurin showed that if a quadrilateral is inscribed in a cubic, and if the points of intersection of the opposite sides also lie on the curve, the tangents to the cubic at any two opposite vertices of the quadrilateral will meet on the curve.

In view of the striking results of Maclaurin in geometry, it is ironic that **13** today his name is recalled almost exclusively in connection with a portion of analysis in which he had been anticipated by some half dozen earlier workers. The so-called Maclaurin series, which appeared in his *Treatise of Fluxions* of 1742, is only a special case of the more general Taylor series, published by Brook Taylor (1685–1731) in 1715 in his *Methodus incrementorum directa et inversa*. Taylor was a Cambridge graduate, an enthusiastic admirer of Newton, and secretary of the Royal Society. He was much interested in perspective; on this subject he published two books in 1715 and 1719, in the second of which he gave the first general statement of the principle of vanishing points.[13] However, his name today is recalled almost exclusively in connection with the series

$$f(x + a) = f(a) + f'(a)x + f''(a)\frac{x^2}{2!} + f'''(a)\frac{x^3}{3!} + \cdots + f^{(n)}(a)\frac{x^n}{n!} + \cdots$$

which appeared in the *Methodus incrementorum*. This series becomes the familiar Maclaurin series upon substituting zero for *a*. The general Taylor series had been known long before to James Gregory, and in essence also to Jean Bernoulli; but Taylor was unaware of this. Moreover, the Maclaurin series had appeared in the *Methodus differentialis* of Stirling more than a dozen years before it was published by Maclaurin. Clio, the muse of history, often is fickle in the matter of attaching names to theorems!

The *Methodus incrementorum* contained also a number of other familiar **14** parts of calculus, such as formulas relating the derivative of a function to the derivative of the inverse function—for example, $d^2x/dy^2 = -d^2y/dx^2/(dy/dx)^3$ —singular solutions of differential equations, and an attempt to find an equation for a vibrating string. After 1719 Taylor gave up the pursuit of mathematics, but the young Maclaurin was then just beginning his fruitful career. His *Treatise of Fluxions* was not just another book on the techniques of the calculus, but an effort to establish the subject on a sound basis similar to that of the geometry of Archimedes. The motive here was to defend the subject from attacks that had been launched, especially by Bishop George Berkeley (1685–1753) in a tract of 1734 entitled *The Analyst*. Berkeley did

---

[13] The title of the 1715 work was *Linear Perspective*, that of the 1719 treatise was *Principles of Linear Perspective*, both published at London.

not deny the utility of the techniques of fluxions nor the validity of the results obtained by using these; but he had been nettled on having a sick friend refuse spiritual consolation because Halley had convinced the friend of the untenable nature of Christian doctrine. Hence the subtitle of the *Analyst* reads:

Or a Discourse Addressed to an Infidel Mathematician [presumably Halley]. Wherein It Is Examined Whether the Object, Principles, and Inferences of the Modern Analysis are More Distinctly Conceived, or More Evidently Deduced, than Religious Mysteries and Points of Faith. "First Cast the Beam Out of Thine Own Eye; and Then Shalt Thou See Clearly to Cast Out the Mote Out of Thy Brother's Eye."

Berkeley's account of the method of fluxions was quite fair, and his criticisms were well taken. He pointed out that, in finding either fluxions or the ratios of differentials, mathematicians first assume that increments are given to the variables and then take the increments away by assuming them to be zero. The calculus, as then explained, seemed to Berkeley to be only a compensation of errors. Thus, "by virtue of a twofold mistake you arrive, though not at science, yet at the truth." Even Newton's explanation of fluxions in terms of prime and ultimate ratios was condemned by Berkeley, who denied the possibility of a literally "instantaneous" velocity in which distance and time increments have vanished to leave the meaningless quotient 0/0. As he expressed it,

And what are these fluxions? The velocities of evanescent increments. And what are these same evanescent increments? they are neither finite quantities, nor quantities infinitely small, nor yet nothing. May we not call them ghosts of departed quantities?[14]

It was to answer such criticisms that Maclaurin wrote his *Treatise of Fluxions* in the rigorous manner of the ancients; but in doing so, he used a geometric approach that is less suggestive of the new developments that were to feature the analysis of Continental Europe. Perhaps this is not unrelated to the fact that Maclaurin was almost the last significant mathematician in Great Britain during the eighteenth century, a time when analysis, rather than geometry, was on the crest of the wave. Nevertheless, the *Treatise of Fluxions* contained a number of relatively new results, including the integral test for convergence of infinite series (given earlier by Euler in 1732 but generally overlooked).

15   If the name of Maclaurin today is recalled in connection with a series of which he was not the first discoverer, this is compensated for by the fact that a contribution he made bears the name of someone else who discovered and

[14] For further discussion of the *Analyst* controversy see Florian Cajori: *A History of the Conceptions of Limits and Fluxions in Great Britain, from Newton to Woodhouse* (1919). Cf. *The Works of George Berkeley*, ed. by A. C. Fraser (Oxford, 1901, 4 vols.), especially Vol. III.

printed it later. The well-known Cramer's rule, published in 1750 by Gabriel Cramer (1704–1752), probably was known to Maclaurin as early as 1729, the time when he was composing an algebra intended as a commentary on Newton's *Arithmetica universalis*. The Maclaurin *Treatise of Algebra* was published in 1748, two years after the author had died, and in it the rule for solving simultaneous equations by determinants appeared, two years earlier than in Cramer's *Introduction à l'analyse des lignes courbes algebriques*. The solution for *y* in the system

$$\begin{cases} ax + by = c \\ dx + ey = f \end{cases}$$

is given as

$$y = \frac{af - dc}{ae - db}$$

and the solution for *z* in the system

$$\begin{cases} ax + by + cz = m \\ dx + ey + fx = n \\ gx + hy + Kz = p \end{cases}$$

is expressed as

$$z = \frac{aep - ahn + dhm - dbp + gbn - gem}{aek - ahf + dhc - dbk + gbf - gec}$$

Maclaurin explained that the denominator consists, in the former case, of "the Difference of the Products of the opposite Coefficients taken from the Orders that involve the two unknown Quantities" and, in the latter case, "of all the Products that can be made of the three opposite Coefficients taken from the Orders that involve the three unknown Quantities." (He had earlier explained that he would call those quantities of the "same order that are prefixt to the same unknown Quantities in the different Equations . . . and those . . . that affect no unknown Quantity. But those are called opposite Coefficients that are taken each from a different Equation, and from a different Order of Coefficients.") The numerators in Maclaurin's patterns differ from the denominators merely in the substitution in the former of the constant terms for the coefficients of the terms in the unknown sought. Maclaurin told how to write out the solution similarly for four equations in four unknowns, "prefixing contrary signs to those that involve the Products of two opposite Coefficients." This statement shows that Maclaurin had in mind a rule for alternations in sign akin to that now ordinarily described in terms of the inversion principle.[15]

Maclaurin's *Treatise of Algebra* enjoyed an even wider popularity than his other works, with a sixth edition appearing at London in 1796. The world seems nevertheless to have learned of the solution of simultaneous equations by determinants more through Cramer than through Maclaurin, mainly, we suspect, because of the superiority of Cramer's notation, in which superscripts were attached to literal coefficients to facilitate the determination of signs. Then, too, mathematics in Great Britain was on the downgrade by the time Maclaurin's *Algebra* appeared, and Continental mathematicians paid relatively little attention to British authors. Conversely, English mathematicians displayed an indifference to the work of Continental analysts, thus accentuating the disparity in achievement after Maclaurin's time.

Maclaurin took an active part in opposing "Bonnie Prince Charlie" when the Young Pretender in 1745 marched against Edinburgh with an army of Highlanders. The forty-seven-year-old professor of mathematics escaped when the city finally was taken; but exposure in trench warfare and the flight to York were too much for him, and he died in 1746. De Moivre died eight years later in his eighty-eighth year, and British mathematics thereafter suffered an eclipse.

16   Continental Europe had not escaped controversy over the foundations of the calculus, but there the effect was less felt than in England. As early as in Leibniz' day objections to the new analysis had been raised by a Saxon nobleman, Count Ehrenfried Walter von Tschirnhaus (1651–1708). His name is still perpetuated in the "Tschirnhaus transformations" in algebra, by which he hoped to find a general method for solving equations of higher degree. A Tschirnhaus transformation of a polynomial equation $f(x) = 0$ is one of the form $y = g(x)/h(x)$, where g and h are polynomials and h does not vanish for a root of $f(x) = 0$. The transformations by which Cardan and Viète solved the cubic were special cases of such transformations. In the *Acta Eruditorum* of 1683 Tschirnhaus (or Tschirnhausen) showed that a polynomial of degree $n > 2$ can be reduced by his transformations to a form in which the coefficients of the terms of degrees $n - 1$ and $n - 2$ are both zero; for the cubic he found a transformation of the form $y = x^2 + ax + b$ which reduced the general cubic to the form $y^3 = K$. Another such transformation reduced the quartic to $y^4 + py^2 + q = 0$, thus adding new methods of solving the cubic and quartic.

---

[15] For references see C. B. Boyer, "Colin Maclaurin and Cramer's Rule," *Scripta Mathematica*, **27** (1966), 377–379. On other aspects of Maclaurin's work see C. Tweedie, "A Study of the Life and Writings of Colin Maclaurin," *Mathematical Gazette*, **8** (1915–1916), 132–151; **9** (1919), 303–305; and H. W. Turnbull, *Bi-centenary of the Death of Colin Maclaurin* (1951).

Tschirnhaus hoped to develop similar algorithms that would reduce the general equation of nth degree to a "pure" equation of nth degree containing only the terms of degree n and degree zero. His transformations constituted the most promising contribution to the solution of equations during the seventeenth century; but his elimination of the second and third coefficients by means of such transformations was far from adequate for the solution of the quintic. Even when the Swedish mathematician E. S. Bring (1736–1798) in 1786 showed that a Tschirnhaus transformation can be found that reduces the general quintic to the form $y^5 + py + q = 0$, the solution still remained elusive. In 1834 G. B. Jerrard (†1863), a Briton, showed that a Tschirnhaus transformation can be found that will eliminate the terms of degrees $n - 1$ and $n - 2$ and $n - 3$ from any polynomial equation of degree $n > 3$; but the power of the method is limited[16] by the fact that equations of fifth and higher degree are not solvable algebraically. Jerrard's belief that he could solve all algebraic equations was illusory.

Tschirnhaus was a man of wide acquaintance and interests. He had studied at Leyden and for a while served in the Dutch army; later he spent some time in England. He was for a time host to Georg Mohr, the "Danish Euclid"; he visited Paris several times, where in 1682 he was elected to the Académie des Sciences; he also set up a glassworks in Italy to further his experiments in light. He is noted as the discoverer of caustics by reflection (catacaustics) which bear his name. It was his report on these curves, the envelopes of a family of rays from a point source and reflected in a curve, that resulted in his election to the Paris Académie; and interest in caustics and similar families was continued by Leibniz, L'Hospital, Jacques and Jean Bernoulli, and others. His name is attached also to the "Tschirnhaus cubic" $a = r\cos^3\theta/3$, a form generalized later by Maclaurin to $r^n = a\cos n\theta$ for n rational. Sometimes Tschirnhaus is referred to as "the discoverer of porcelain," for he was one of the men who helped to establish the pottery works at Dresden for the Elector of Saxony in the early eighteenth century; but porcelain had been produced in China long before it was made in Europe.

Tschirnhaus had been in touch with Oldenberg and Leibniz during the formative years of the calculus, and he also had contributed many mathematical articles to the *Acta Eruditorum* after its establishment in 1682. Some of Tschirnhaus' work, however, was hastily composed and published prematurely, and the Bernoulli brothers and others pointed out errors. At one point Tschirnhaus rejected the basic concepts of the calculus and of infinite series, insisting that algebraic methods would suffice. In Holland objections to the calculus of Leibniz had been raised in 1694–1696 by the physician and geometer Bernard Nieuwentijt (1654–1718). In three separate treatises

[16] See L. E. Dickson, *Modern Algebraic Theories* (New York: B. H. Sanborn, 1930) pp. 186–197, for applications of the Tschirnhaus transformation.

published during these years at Amsterdam he admitted the correctness of the results, but he criticized the vagueness of Newton's evanescent quantities and the lack of clear definition in Leibniz' differentials of higher order.

17   Leibniz in 1695 had defended himself in the *Acta Eruditorum* from his "overprecise" critic,[17] and in 1701 a more detailed refutation of Nieuwentijt came from Switzerland from the pen of Jacob Hermann (1678–1733), a devoted pupil of Jacques Bernoulli. Illustrating the mobility of mathematicians during the early eighteenth century, Hermann taught mathematics at the Universities of Padua, Frankfort on the Oder, and St. Petersburg, before concluding his career at the University of Basel, his hometown. In the *Commentarii Academiae Petropolitanae* for the years 1729–1733 Hermann made contributions to solid analytic geometry and to polar coordinates in continuation of results made by the older Bernoulli brothers. Where Jacques Bernoulli had rather hesitantly applied polar coordinates to spirals, Hermann gave polar equations of algebraic curves as well, together with equations of transformation from rectangular to polar coordinates. Hermann's use of space coordinates also was bolder than that of Jean Bernoulli, who as early as 1692 had first referred to the use of coordinates as "Cartesian geometry." Bernoulli had rather timidly suggested an extension of Cartesian geometry to three dimensions, but Hermann applied space coordinates effectively to planes and several types of quadratic surfaces. He made a beginning in the use of direction angles by showing that the sine of the angle that the plane $az + by + cx = c^2$ makes with the xy-plane is given by $\sqrt{b^2 + c^2}/\sqrt{a^2 + b^2 + c^2}$.

18   In France, as well as in England, Germany and Holland, there was a group in the Académie des Sciences, especially shortly after 1700, who questioned the validity of the new infinitesimal methods as presented by L'Hospital. Among these was Michel Rolle (1652–1719), whose name is recalled in connection with Rolle's theorem, published in 1691 in an obscure book on geometry and algebra entitled *Méthode pour résoudre les égalitéz*: if a function is differentiable in the interval from a to b, and if $f(a) = 0 = f(b)$, then $f'(x)$ has at least one real root between a and b. The theorem, now so important in the calculus, was given only incidentally by Rolle in connection with an approximate solution of equations.[18]

Rolle's attack on the calculus, which he described as a collection of ingenious fallacies, was answered vigorously by Pierre Varignon (1654–1722), Jean Bernoulli's "best friend in France" and one who also had been

---

[17] For further details see C. B. Boyer, *History of the Calculus* (1959), pp. 213–215.

[18] A translation from the French of the section on Rolle's theorem appears in D. E. Smith, *Source Book in Mathematics*, pp. 251–260.

corresponding with Leibniz. Bernoulli simply told Rolle that he did not understand the subject, but Varignon sought to clarify the situation by showing indirectly that the infinitesimal methods could be reconciled with the geometry of Euclid. Most of the group opposing the calculus were admirers of the ancient synthetic geometry, and the controversy in the Académie des Sciences reminds one of the then contemporary literary controversy on "ancients vs. moderns."

Varignon, like the Bernoullis, had not at first expected to be a mathematician, being intended for the church; but when he accidentally came across a copy of Euclid's *Elements*, he changed his mind and held professorships in mathematics in Paris, becoming a member of the Académie. In the *Memoirs* of the Académie des Sciences for 1704 he continued and extended Jacques Bernoulli's use of polar coordinates, including an elaborate classification of spirals obtained from algebraic curves, such as the parabolas and hyperbolas of Fermat, by interpreting the ordinate as a radius vector and the abscissa as a vectorial arc. Varignon, one of the first French scholars to appreciate the calculus, had prepared a commentary on L'Hospital's *Analyse*, but this appeared only in 1725, after both men had died, under the title *Eclaircissemens sur l'analyse des infiniments petits*. Varignon was a more careful writer than L'Hospital, and he warned that infinite series were not to be used without investigation of the remainder term. Hence he had been rather worried about the attacks on the calculus, and in 1701 he had written to Leibniz about his differences with Rolle:

> The Abbé Galloys, who is really behind the whole thing, is spreading the report here [in Paris] that you have explained that you mean by the "differential" or the "infinitely small" a very small, but nevertheless constant and definite, quantity .... I, on the other hand, have called a thing infinitely small, or the differential of a quantity, if that quantity is inexhaustible in comparison with the thing.[19]

The view that Varignon expressed here is far from clear, but at least he recognized that a differential is a variable rather than a constant. Leibniz' reply from Hanover in 1702 seeks to avoid metaphysical quarrels, but his use of the phrase "incomparably small quantities" for differentials was scarcely more satisfactory than Varignon's explanation. Varignon's defense of the calculus nevertheless seems to have won Rolle's approval.

Rolle also had raised embarrassing questions about analytic geometry, especially concerning the Cartesian graphical solution of equations, so popular at the time. To solve $f(x) = 0$, for example, one arbitrarily chose a curve $g(x, y) = 0$ and, on combining it with $f(x) = 0$, obtained a new curve $h(x, y) = 0$ the intersections of which with $g(x, y) = 0$ furnish the solutions

---

[19] See Herbert Meschkowski, *Ways of Thought of Great Mathematicians* (San Francisco, 1964) pp. 57–58.

of $f(x) = 0$. Rolle saw that extraneous solutions may be introduced through this procedure. In his best-known work, the *Traité d'algèbre* of 1690, Rolle seems to have been first to state that there are n values for the nth root of a number, but he was able to prove this only for $n = 3$, for he died before the relevant works of Cotes and De Moivre appeared. Rolle was the most capable mathematician in the group from the Académie des Sciences that criticized the calculus. When he was convinced by Varignon of the essential soundness of the new analysis, opposition collapsed and the subject entered a century of unimpeded and rapid development on the Continent of Europe.

19   While the Bernoullis and their associates were defending and espousing developments in analytic geometry, the calculus, and probability, mathematics in Italy flowed along more or less unobtrusively with some preference for geometry. No outstanding figure appeared there, although several men left results important enough to be noted. Giovanni Ceva (1648–1734) is recalled today for the theorem that bears his name: A necessary and sufficient condition that lines from the vertices A, B, C of a triangle to points X, Y, Z on the opposite sides be concurrent is that

$$\frac{AZ \cdot BX \cdot CY}{ZB \cdot XC \cdot YA} = -1$$

This is closely related to the theorem of Menelaus which had been forgotten but was rediscovered and published also by Ceva in 1678.

More closely related to the interests of the Bernoullis were the contributions of Jacopo Riccati (1676–1754), who made Newton's work known in Italy. Riccati is remembered especially for his extensive study of the differential equation $dy/dx = A(x) + B(x)y + C(x)y^2$, now bearing his name, although Jacques Bernoulli had earlier studied the special case $dy/dx = x^2 + y^2$. Riccati may have known of this study, for Nicolas Bernoulli taught at Padua, where Riccati had been a student of Angeli and where Riccati came in contact with both Nicolas Bernoulli and Hermann. The work of the Bernoullis was well known in Italy. Count G. C. Fagnano (1682–1766) followed up work on the lemniscate of Bernoulli to show, around 1717–1718, that the rectification of this curve leads to an elliptic integral, as does the arc length of the ellipse, although certain arcs are rectifiable by elementary means. Fagnano's name still is attached to the ellipse $x^2 + 2y^2 = 1$ which presents certain analogies to the equilateral or rectangular hyperbola. The eccentricity of this ellipse, for example, is $1/\sqrt{2}$, whereas the eccentricity of the rectangular hyperbola is $\sqrt{2}$.

20   Italian mathematicians during the eighteenth century made few, if any, fundamental discoveries. The nearest approach to such a discovery undoubtedly was that of Girolamo Saccheri (1667–1733), a Jesuit who taught at colleges of his order in Italy. In the very year in which he died, he published a book entitled *Euclides ab omni naevo vindicatus* ("Euclid Cleared of Every Flaw") in which he made an elaborate effort to prove the parallel postulate. Saccheri had known of Nasir Eddin's efforts to prove the postulate almost half a millennium earlier, and he determined to apply the method of *reductio ad absurdum* to the problem. He began with a birectangular isosceles quadrilateral, now known as a "Saccheri quadrilateral"—one having sides AD and BC equal to each other and both perpendicular to the base AB. Without using the parallel postulate he easily showed that the "summit" angles C and D are equal and that there are, then, just three possibilities for these angles, described by Saccheri as (1) the hypothesis of the acute angle, (2) the hypothesis of the right angle, and (3) the hypothesis of the obtuse angle. By showing that hypotheses 1 and 3 lead to absurdities, he thought by indirect reasoning to establish hypothesis 2 as a necessary consequence of Euclid's postulates other than the parallel postulate. Saccheri had little trouble disposing of hypothesis 3, for he implicitly assumed a straight line to be infinitely long. From hypothesis 1 he derived theorem after theorem without encountering difficulty. We know now that he was here building up a perfectly consistent non-Euclidean geometry; but Saccheri was so thoroughly imbued with the conviction that Euclid's was the only valid geometry that he permitted this preconception to interfere with his logic. Where no contradiction existed, he twisted his reasoning until he thought that hypothesis 1 led to an absurdity. Hence he lost credit for what would undoubtedly have been the most significant discovery of the eighteenth century—non-Euclidean geometry. As it was, his name remained unsung for another century, for the importance of his work was overlooked by those who followed him.[20]

Saccheri had as his student another Italian mathematician who perhaps **21** deserves brief mention—Guido Grandi (1671–1742), whose name is remembered in the rose-petal curves so familiar in polar coordinates through the equations $r = a \cos n\theta$ and $r = a \sin n\theta$. These are known as "roses of Grandi" in recognition of his study of them. Grandi also is recalled as one who had corresponded with Leibniz on the question of whether or not the sum of the alternating infinite series $1 - 1 + 1 - 1 + 1 - 1 + \cdots$ can be taken to be $\frac{1}{2}$. This is suggested not only as the arithmetic mean of the two values of the partial sums of the first n terms, but also as the value when $x = +1$ of the generating function $1/(1 + x)$ from which the series $1 - x + x^2 - x^3 + x^4 - \cdots$ is obtained through division. In this correspondence

[20] Accounts of Saccheri's work are available in many places. An excellent description is given in Roberto Bonola, *Non-Euclidean Geometry, A Critical and Historical Study of Its Developments* (1955), pp. 22–44. See also D. E. Smith, *Source Book in Mathematics* (1929), pp. 351–359.

Grandi suggested that here one has a paradox comparable to the mysteries of Christianity, for on grouping terms in pairs one reaches the result

$$1 - 1 + 1 - 1 + 1 - \cdots = 0 + 0 + 0 + \cdots = \tfrac{1}{2}$$

paralleling the creation of the world out of nothing. Continuing such un-critical ideas to the integral of the generating function $1/(1 + x)$, Leibniz and Jean Bernoulli had corresponded on the nature of the logarithms of negative numbers. The series $\ln (1 + x) = x - x^2/2 + x^3/3 - x^4/4 + \cdots$, however, is of little help here since the series diverges for $x < -1$. Leibniz argued that negative numbers do not have real logarithms; but Bernoulli, believing the logarithmic curve to be symmetric with respect to the function axis, held that $\ln (-x) = \ln (x)$, a view that seems to be confirmed by the fact that $d/dx \ln (-x) = d/dx \ln (+x) = 1/x$. The question of the nature of logarithms of negative numbers was not definitively resolved by either of the correspondents, but rather by Bernoulli's most brilliant student. Jean Bernoulli had continued to exert an encouraging enthusiasm, through his correspondence, during the first half of the eighteenth century, for he outlived his older brother by forty-three years. Nevertheless, long before his death in 1748, as an octogenarian, his influence had become far less felt than that of his famous pupil, Euler, whose contributions to analysis, including the logarithms of negative numbers, were the essential core of mathematical developments during the middle years of the eighteenth century.

BIBLIOGRAPHY

Berkeley, George, *Works*, ed. by A. C. Fraser (Oxford: Clarendon, 1901, 4 vols.).

Bernoulli, Jacques, *Opera* (Geneva, 1744, 2 vols.).

Bernoulli, Jean, *Opera omnia* (Lausanne and Geneva, 1742, 4 vols.).

Bernoulli, Jean, *Die Differentialrechnung aus dem Jahre 1691/92* (Ostwald's Klassiker, No. 211, Leipzig: W. Engelmann, 1924).

Bonola, Roberto, *Non-Euclidean Geometry, A Critical and Historical Study of Its Developments* (paperback ed., New York: Dover, 1955).

Boyer, C. B., *History of Analytic Geometry* (New York: Scripta Mathematica, 1956).

Boyer, C. B., *History of the Calculus* (paperback ed., New York: Dover, 1959).

Cajori, Florian, *A History of the Conceptions of Limits and Fluxions in Great Britain, from Newton to Woodhouse* (Chicago: Open Court, 1919).

Coolidge, J. L., *The Mathematics of Great Amateurs* (New York: Oxford University Press, 1949; paperback ed., New York: Dover, 1963).

Edleston, J., *Correspondence of ... Newton and ... Cotes* (Cambridge, 1850).

Engel, Fr., and P. Stäckel, *Die Theorie der Parallellinien von Euklid bis auf Gauss* (Leipzig, 1895, 2 vols.).

Fedel, J., *Der Briefwechsel Johann (I) Bernoulli–Pierre Varignon aus den Jahren 1692 bis 1702* (Heidelberg, 1932).

Fleckenstein, J. O., *L'école mathématique baloise des Bernoulli à l'aube du XVIIIe siècle* (Paris: Palais de la Découverte, ca. 1958).

Hoffmann, J. E., *Classical Mathematics* (New York: Philosophical Library, 1959).

Jurin, James, *Geometry No Friend to Infidelity: or, a Defence of Sir Isaac Newton and the British Mathematicians* (London, 1734).

Leibniz, G. W., and Jean Bernoulli, *Commercium philosophicum et mathematicum* (Lausanne et Geneva, 1745, 2 vols.).

Maclaurin, Colin, *A Treatise of Fluxions* (Edinburgh, 1742, 2 vols.).

Meschkowski, Herbert, *Ways of Thought of Great Mathematicians* (San Francisco: Holden-Day, 1964).

Montucla, Etienne, *Histoire des mathématiques*, new ed. (Paris, 1799–1802, 4 vols.).

Reiff, R., *Geschichte der unendlichen Reihen* (Tübingen, 1889).

Smith, D. E., *Source Book in Mathematics* (New York: McGraw-Hill, 1929; paperback ed., New York; Dover, 1959, 2 vols.).

Smith, D. E., *History of Mathematics* (paperback ed., New York: Dover, 1958, 2 vols.).

Spiess, Otto, ed., *Der Briefwechsel von Johann Bernoulli* (Basel: Birkhäuser, 1955), Vol. I.

Struik, D. J., *A Concise History of Mathematics* (3rd ed., New York: Dover, 1967).

Todhunter, Isaac, *A History of the Mathematical Theory of Probability from the Time of Pascal to that of Laplace* (reprinted, New York: Chelsea, 1949).

Turnbull, H. W., *Bi-centenary of the Death of Colin Maclaurin* (Aberdeen: Aberdeen University Press, 1951).

Tweedie, Ch., *James Stirling. Sketch of His Life and Works* (Oxford: Clarendon, 1922).

Walker, Helen M., "Abraham de Moivre," *Scripta Mathematica*, **2** (1934). 316–333.

Weissenborn, Hermann, *Die Principien der höheren Analysis in ihrer Entwickelung von Leibniz bis auf Lagrange* (Halle, 1856).

Wieleitner, Heinrich, *Geschichte der Mathematik* (Leipzig: Walter de Gruyter, 1911–1921, 2nd part, 2 vols.).

EXERCISES

1. Mention some British and Swiss mathematicians who held positions in universities. Suggest some possible reasons why few prominent French mathematicians held university chairs.

2. Why did Russia have no outstanding native mathematician during the seventeenth and eighteenth centuries?

3. Mention at least four journals of the early eighteenth century that published articles on mathematics. What were the most popular branches of mathematics at the time?

4. What were the most important centers of mathematics during the time of the Bernoullis? In each case cite one or more mathematicians who were active there at the time.

5. How should British mathematicians have answered the objections of Berkeley to the method of fluxions?

6. Write $\cos 11\theta$ in terms of powers of $\sin \theta$ and $\cos \theta$.

7. Derive the Taylor series (through successive differentiation or otherwise).

8. Prove De Moivre's theorem (by mathematical induction or otherwise.)

9. Factor $x^6 + a^6$ and $x^6 - a^6$ into real quadratic factors.

10. Express d^3x/dy^3 in terms of derivatives of y with respect to x.

11. Express the answers for y in the systems of simultaneous linear equations used by Maclaurin.

12. Show that either of the Tschirnhaus transformations $y = x^2 + 4x + 11$ and $y = x^2 + 7$ will reduce the equation $x^3 + 3x^2 + 15x + 13 = 0$ to the pure two-term form. (Eliminate x from the pair of equations.)

13. Show that the substitution $y = x^2 + 3x$ will reduce $x^4 + 12x - 5 = 0$ to $y^4 + 98y^2 - 200 = 0$, hence will solve the quartic in x.

14. Give an example of a continuous curve for which $f(a) = f(b) = 0$ and for which $f'(x)$ is never zero between $x = a$ and $x = b$. Does this contradict Rolle's theorem? Explain.

15. Verify De Moivre's statement that the odds against throwing at least two aces in eight tosses of a single die are "about 3 to 2."

16. Prove that the "summit angles" of a Saccheri quadrilateral are equal.

17. Sketch the "rose of Grandi" given by the equation $r = \cos \frac{3}{2}\theta$.

*18. Show that the substitution $y = z^{1-n}$ reduces the Bernoulli equation $y' + P(x)y = Q(x)y^n$ to a linear differential equation.

*19. Show that if a particular solution $f(x)$ of the Riccati equation $y' = A(x) + B(x)y + C(x)y^2$ is known, this equation can be reduced to a Bernoulli equation through the substitution $y = z + f$.

*20. Show that the eccentricity of Fagnano's ellipse $x^2 + 2y^2 = 1$ is the reciprocal of the eccentricity of the equilateral hyperbola $x^2 - y^2 = 1$.

*21. Prove one of the properties of the logarithmic spiral $r = ae^{b\theta}$ mentioned in connection with the work of Jacques Bernoulli.

*22. Show that the length of the spiral $r^2 = a\theta$ from $\theta = 0$ to $\theta = a$ is given by an elliptic integral.

*23. Verify, by one of the methods described in the text, that the first two Bernoulli numbers are $\frac{1}{6}$ and $-\frac{1}{30}$.

*24. Justify the L'Hospital rule.

*25. Use L'Hospital's rule to find

$$\lim_{x \to 0} \frac{a^x - b^x}{c^x - d^x}$$

where $c \neq d$.

*26. Carry out the details to show that the area under $y = x^x$ from $x = 0$ to $x = 1$ is given by the infinite series

$$\sum \frac{(-1)^{n-1}}{n^n}$$

*27. Show that a curve of order n generally is determined by $n(n + 3)/2$ points.

CHAPTER XXI

The Age of Euler

> Algebra is generous; she often gives more than is
> asked of her.
>
> *D'Alembert*

The history of mathematics during the modern period is unlike that of **1** antiquity or the medieval world in at least one respect: no national group remained the leader for any prolonged period. In ancient times Greece stood head and shoulders over all other peoples in mathematical achievement; during much of the Middle Ages the level of mathematics in the Arabic world was higher than elsewhere. From the Renaissance to the eighteenth century the center of mathematical activity had shifted repeatedly—from Germany to Italy to France to Holland to England. Had religious persecution not driven the Bernoulli family from Antwerp, Belgium might have had its turn; but the family emigrated to Basel, and as a result Switzerland was the birthplace of many of the leading figures in the mathematics of the early eighteenth century. We have already mentioned the work of four of the mathematicians of the Bernoulli clan, as well as that of Hermann, one of their Swiss protégés. But the most significant mathematician to come from Switzerland during that time—or any time—was Leonhard Euler (1707–1783), who was born at Basel.

Euler's father was a clergyman who, like Jacques Bernoulli's father, hoped that his son would enter the ministry. However, the young man studied under Jean Bernoulli and associated with his sons, Nicolaus and Daniel, and through them discovered his vocation. The elder Euler also was adept in mathematics, having been a pupil under Jacques Bernoulli, and helped to instruct the son in the elements of the subject, despite his hope that Leonhard would pursue a theological career. At all events, the young man was broadly trained, for to the study of mathematics he added theology, medicine, astronomy, physics, and oriental languages. This breadth stood him in good stead when in 1727 he heard from Russia that there was an opening in medicine in the St. Petersburg Academy, where the young Bernoullis had gone as professors of mathematics. This important institution had been established only a few years earlier by Catherine I along lines laid down by her late husband,

Peter the Great, with the advice of Leibniz. On the recommendation of the Bernoullis, two of the brightest luminaries in the early days of the Academy, Euler was called to be a member of the section on medicine and physiology; but on the very day that he arrived in Russia, Catherine died. The fledgling Academy very nearly succumbed with her, because the new rulers showed less sympathy for learned foreigners than had Peter and Catherine.

The Academy somehow managed to survive, and Euler, in 1730, found himself in the chair of natural philosophy rather than in the medical section. His friend Nicolaus Bernoulli had died, by drowning, in St. Petersburg the year before Euler arrived, and in 1733 Daniel Bernoulli left Russia to occupy the chair in mathematics at Basel. Thereupon Euler at the age of twenty-six became the Academy's chief mathematician. He married and settled down to pursue in earnest mathematical research and rear a family that ultimately included thirteen children. The St. Petersburg Academy had established a research journal, the *Commentarii Academiae Scientiarum Imperialis Petropolitanae*, and almost from the start Euler contributed a spate of mathematical articles. The editors did not have to worry about a shortage of material as long as the pen of Euler was busy. It was said by the French academician François Arago that Euler could calculate without any apparent effort, "just as men breathe, as eagles sustain themselves in the air." As a result, Euler composed mathematical memoirs while playing with his children. In 1735 he had lost the sight of his right eye—through overwork, it is said—but this misfortune in no way diminished the rate of output of his research. He is supposed to have said that his pencil seemed to surpass him in intelligence, so easily did memoirs flow; and he published more than 500 books and papers during his lifetime. For almost half a century after his death, works by Euler continued to appear in the publications of the St. Petersburg Academy. A bibliographical list of Euler's works, including posthumous items, contains 886 entries; and it is estimated that his collected works, now being published under Swiss auspices, will run close to seventy-five substantial volumes.[1] His mathematical research during his lifetime averaged about 800 pages a year; no mathematician has ever exceeded the output of this man whom Arago characterized as "Analysis Incarnate."

Euler early acquired an international reputation; even before leaving Basel he had received an honorable mention from the Parisian Académie des Sciences for an essay on the masting of ships. In later years he frequently entered essays in the contests set by the Académie, and twelve times he won

[1] See Gustav Eneström, "Verzeichnis der Schriften Leonhard Eulers," *Jahresbericht der Deutschen Mathematiker-Vereinigung, Ergänzungsband IV*, Leipzig, 1910–1913. See also N. Fuss, *Lobrede auf Herrn Leonhard Euler* (1786), and Condorcet's *Eloge de M. Euler* for the Académie des Sciences, 1785. Cf. J. O. Fleckenstein, "L'état actuel de l'édition des oeuvres d'Euler," *XI Congrès International d'Histoire des Sciences*, Warsaw, 1965.

the coveted biennial prize. The topics ranged widely, and on one occasion, in 1724, Euler shared with Maclaurin and Daniel Bernoulli a prize for an essay on the tides.[2] (The Paris prize was won twice by Jean Bernoulli and ten times by Daniel Bernoulli.) Euler was never guilty of false pride, and he wrote works on all levels, including textbook material for use in the Russian schools. He generally wrote in Latin, and sometimes in French, although German was his native tongue. Euler had unusual language facility, as one should expect of a person with a Swiss background. This was fortunate, for one of the distinguishing marks of eighteenth-century mathematics was the readiness with which scholars moved from one country to another, and here Euler encountered no language problems. In 1741 Euler was invited by Frederick the Great to join the Berlin Academy, and the invitation was accepted. (Jean and Daniel Bernoulli also were invited from Switzerland, but they declined.) Euler spent twenty-five years at Frederick's court, but during this period he continued to receive a pension from Russia, and he submitted numerous papers to the St. Petersburg Academy, as well as to the Prussian Academy.

Euler's stay at Berlin was not entirely happy, for Frederick preferred a scholar who scintillated, as did Voltaire. The monarch, who valued philosophers above geometers, referred to the unsophisticated Euler as a "mathematical cyclops," and relationships at the court became intolerable for Euler. Catherine the Great was only too eager to have the prolific mathematician resume his place in the St. Petersburg Academy, and in 1766 Euler returned to Russia. During this year Euler learned that he was losing by cataract the sight of his remaining eye, and he prepared for ultimate blindness by practicing writing with chalk on a large slate and by dictating to his children. An operation was performed in 1771, and for a few days Euler saw once more; but success was short-lived and Euler spent almost all of the last seventeen years of his life in total darkness. Even this tragedy failed to stem the flood of his research and publication, which continued unabated until in 1783, at the age of seventy-six, he suddenly died while sipping tea and enjoying the company of one of his grandchildren.

From 1727 to 1783 the pen of Euler had been busy adding to knowledge in virtually every branch of pure and applied mathematics, from the most elementary to the most advanced. Moreover, in most respects Euler wrote in the language and notations we use today, for no other individual was so largely responsible for the form of college-level mathematics today as was Euler, the most successful notation-builder of all times. Upon his arrival in Russia in 1727, he had been engaged in experiments in the firing of canon;

---

[2] All three prize essays are included in Vol. III of the "Jesuits' edition" of Newton's *Principia* published at Geneva in 1760.

and in a manuscript account of his results, written probably in 1727 or 1728, Euler had used the letter e more than a dozen times to represent the base of the system of natural logarithms. The concept behind this number had been well known ever since the invention of logarithms more than a century before; yet no standard notation for it had become common. In a letter to Goldbach in 1731 Euler again used his letter e for "that number whose hyperbolic logarithm $= 1$"; it appeared in print for the first time in Euler's *Mechanica* of 1736, a book in which Newtonian dynamics is presented for the first time in analytic form. This notation, suggested perhaps by the first letter of the word "exponential," soon became standard.[3] The definitive use of the Greek letter π for the ratio of circumference to diameter in a circle also is largely due to Euler, although a prior occurrence is found in 1706, the year before Euler was born—in the *Synopsis Palmariorum Matheseos, or A New Introduction to the Mathematics*, by William Jones (1675-1749). It was Euler's adoption of the symbol π in 1737, and later in his many popular textbooks, that made it widely known and used. The symbol i for $\sqrt{-1}$ is another notation first used by Euler, although in this case the adoption came near the end of his life, in 1777. This use came so late probably because in his earlier works he had used i to represent an "infinite number," somewhat as Wallis had used ∞. Thus Euler wrote $e^x = (1 + x/i)^i$ where we should prefer $e^x = \lim_{h \to \infty} (1 + x/h)^h$. In fact, although Euler used i for $\sqrt{-1}$ in a manuscript dated 1777, this was published only in 1794. It was the adoption of the symbol by Gauss in his classic *Disquisitiones arithmeticae* of 1801 that resulted in its secure place in mathematical notations. The three symbols e, π, and i, for which Euler was in large measure responsible, can be combined with the two most important integers, 0 and 1, in the celebrated equality $e^{\pi i} + 1 = 0$ which contains the five most significant numbers (as well as the most important relation and the most important operation) in all of mathematics. The equivalent of this equality, in generalized form, had been included by Euler in 1748 in his best known textbook, *Introductio in Analysin infinitorum*; but the name of Euler today is not generally attached to any one of the symbols in this relationship. The so-called "Eulerian constant," often represented by the Greek letter γ, is a sixth important mathematical constant,[4] the number defined as $\lim_{n \to \infty} (1 + \frac{1}{2} + \frac{1}{3} + \cdots + 1/n - \ln n)$, a transcendental number that has been calculated to hundreds of decimal places, of which the first ten are 0.5772156649.

It is not only in connection with designations for important numbers that

---

[3] See D. E. Smith, *Source Book in Mathematics* (New York: McGraw-Hill, 1929), pp. 95–98 for the early uses of Euler's e notation.

[4] See J. W. L. Glaisher, "On the History of Euler's Constant," *Messenger of Mathematics*, **1** (1871), 25–30.

today we use notations introduced by Euler. In geometry, algebra, trigono-metry, and analysis we find ubiquitous use of Eulerian symbols, terminology, and ideas. The use of the small letters a, b, c for the sides of a triangle and of the corresponding capitals A, B, C for the opposite angles stems from Euler, as does the application of the letters r, R, and s for the radius of the inscribed and circumscribed circles and the semiperimeter of the triangle respectively. The beautiful formula $4rRs = abc$ relating the six lengths also is one of the many elementary results attributed to him, although equivalents of this result are implied by ancient geometry. The designation lx for logarithm of x, the use of the now-familiar Σ to indicate a summation, and, perhaps most important of all, the notation $f(x)$ for a function of x (used in the Petersburg *Commentaries* for 1734–1735) are other Eulerian notations related to ours. Our notations today are what they are more on account of Euler than of any other mathematician in history.

In evaluating developments in mathematics we must always bear in mind **3** that the ideas behind the notations are by far the better half; in this respect also the work of Euler was epoch-making. It may fairly be said that Euler did for the infinite analysis of Newton and Leibniz what Euclid had done for the geometry of Eudoxus and Theaetetus, or what Viète had done for the algebra of al-Khowarizmi and Cardan. Euler took the differential calculus and the method of fluxions and made them part of a more general branch of mathematics which ever since has been known as "analysis"—the study of infinite processes. If the ancient *Elements* was the cornerstone of geometry and the medieval *Al jabr wa'l muqābalah* was the foundation stone of algebra, then Euler's *Introductio in analysin infinitorum* can be thought of as the keystone of analysis. This important two-volume treatise of 1748 served as a fountainhead for the burgeoning developments in mathematics throughout the second half of the eighteenth century. From this time onward the idea of "function" became fundamental in analysis. It had been adumbrated in the medieval latitude of forms, and it had been implicit in the analytic geometry of Fermat and Descartes, as well as in the calculus of Newton and Leibniz. The fourth paragraph of the *Introductio* defines function of a variable quantity as "any analytic expression whatsoever made up from that variable quantity and from numbers or constant quantities." (Sometimes Euler thought of a function less formally and more generally as the relation-ship between the two coordinates of points on a curve drawn freehand in a plane.) Today such a definition is unacceptable, for it fails to explain what an "analytic expression" is. Euler presumably had in mind primarily the algebraic functions and the elementary transcendental functions; the strictly analytical treatment of the trigonometric functions was, in fact, in large measure established by the *Introductio*. The sine, for example, was no longer a line segment; it was simply a number or a ratio—the ordinate of a point on a unit circle, or the number defined by the series $z - z^3/3! + z^5/5! - \cdots$ for some value of z. From the infinite series for e^x, sin x, and cos x it was a short step to the "Euler identities"

$$\sin x = \frac{e^{\sqrt{-1}x} - e^{-\sqrt{-1}x}}{2\sqrt{-1}}$$

$$\cos x = \frac{e^{\sqrt{-1}x} + e^{-\sqrt{-1}x}}{2}$$

and

$$e^{\sqrt{-1}x} = \cos x + \sqrt{-1} \sin x$$

relationships that had in essence been known to Cotes and De Moivre but which in Euler's hands became familiar tools of analysis.

Euler had used imaginary exponents in 1740 in a letter to Jean Bernoulli in which he wrote $e^{x\sqrt{-1}} + e^{-x\sqrt{-1}} = 2 \cos x$; the familiar Euler identities appeared in the influential *Introductio* of 1748. The elementary transcendental functions—trigonometric, logarithmic, inverse trigonometric, and exponential—were written and thought of in much the form in which they are treated today. The abbreviations sin., cos., tang., cot., sec., and cosec. that were used by Euler in the Latin *Introductio* are closer to the present English forms than are the corresponding abbreviations in the Romance languages. Moreover, Euler was among the first to treat logarithms as exponents, in the manner now so familiar.

4   The first volume of the *Introductio* is concerned from start to finish with infinite processes—infinite products and infinite continued fractions, as well as innumerable infinite series. In this respect the work is the natural generalization of the views of Newton, Leibniz, and the Bernoullis, all of whom were fond of infinite series. However, Euler was very incautious in his use of such series. Although upon occasion he warned against the risk in working with divergent series, he himself used the binomial series $1/(1 - x) = 1 + x + x^2 + x^3 + \cdots$ for values of $x \geq 1$. In fact, by combining the two series $x/(1 - x) = x + x^2 + x^3 + \cdots$ and $x/(x - 1) = 1 + 1/x + 1/x^2 + \cdots$ Euler concluded that $\ldots 1/x^2 + 1/x + 1 + x + x^2 + x^3 + \cdots = 0$.

Despite his hardihood, through manipulations of infinite series Euler achieved results that had baffled his predecessors. Among these was the summation of the reciprocals of the perfect squares—$1/1^2 + 1/2^2 + 1/3^2 + 1/4^2 + \cdots$. Oldenburg, in a letter to Leibniz in 1673, had asked for the sum of this series, but Leibniz failed to answer; in 1689 Jacques Bernoulli had admitted his own inability to find the sum. Euler began with the familiar series $\sin z = z - z^3/3! + z^5/5! - z^7/7! + \cdots$. Then $\sin z = 0$ can be thought of as the infinite polynomial equation $0 = 1 - z^2/3! + z^4/5! - z^6/7! + \cdots$ (obtained by dividing through by z), or, if z^2 is replaced by w, as the equation $0 = 1 - w/3! + w^2/5! - w^3/7! + \cdots$. From the theory of algebraic equations it is known that the sum of the reciprocals of the roots is the negative of the coefficient of the linear term—in this case $1/3!$. Moreover, the roots of the equation in z are known to be π, 2π, 3π, and so on; hence the roots of the equation in w are π^2, $(2\pi)^2$, $(3\pi)^2$, and so on. Therefore

$$\frac{1}{6} = \frac{1}{\pi^2} + \frac{1}{(2\pi)^2} + \frac{1}{(3\pi)^2} + \cdots \qquad \text{or} \qquad \frac{\pi^2}{6} = \frac{1}{1^2} + \frac{1}{2^2} + \frac{1}{3^2} + \cdots$$

Through this carefree application to polynomials of infinite degree of algebraic rules valid for the finite case Euler had achieved a result that had baffled the older Bernoulli brothers; Euler in later years repeatedly made discoveries in similar fashion. When Jean Bernoulli learned of Euler's triumph, he wrote:

And so is satisfied the burning desire of my brother who, realizing that the investigation of the sum was more difficult than anyone would have thought, openly confessed that all his zeal had been mocked. If only my brother were alive now.

Euler's summation of the reciprocals of the squares of the integers seems to date from about 1736, and it is likely that it was to Daniel Bernoulli that he promptly communicated the result.[5] His interest in such series always was strong, and in later years he published the sums of the reciprocals of other powers of the integers. Using the cosine series instead of the sine series, Euler similarly found the result

$$\frac{\pi^2}{8} = \frac{1}{1^2} + \frac{1}{3^3} + \frac{1}{5^2} + \cdots$$

hence the corollary summation

$$\frac{\pi^2}{12} = \frac{1}{1^2} - \frac{1}{2^2} + \frac{1}{3^2} - \frac{1}{4^2} + \cdots$$

Many of these results appeared also in the *Introductio* of 1748, including the sums of reciprocals of even powers from $n = 2$ through $n = 26$. The series of reciprocals of odd powers are so intractable that it still is not known whether or not the sum of the reciprocals of the cubes of the positive integers

[5] See Paul Stäckel, "Eine vergessene Abhandlung Leonhard Eulers über die Summe der reziproken Quadrate der natürlichen Zahlen," *Bibliotheca Mathematica* (3), **8** (1907–1908), 37–60.

is a rational multiple of π^3, whereas Euler knew that for the 26th power the sum of the reciprocals is

$$\frac{2^{24} \cdot 76977927\pi^{26}}{1 \cdot 2 \cdot 3 \cdots 27}$$

5 Euler's imaginative treatment of series led him to some striking relationships between analysis and the theory of numbers. He showed, in a relatively easy proof, that the divergence of the harmonic series implies the Euclidean theorem on the infinitude of primes. If there were only K primes—p_1, $p_2, \ldots, p_K$—then every number n would be of the form $n = p_1{}^{\alpha_1}p_2{}^{\alpha_2} \ldots p_K{}^{\alpha_K}$. Let α be the greatest of the exponents α_i for the number n and form the product

$$P = \left(1 + \frac{1}{p_1} + \frac{1}{p_1{}^2} + \cdots + \frac{1}{p_1{}^\alpha}\right)\left(1 + \frac{1}{p_2} + \frac{1}{p_2{}^2} + \cdots \frac{1}{p_2{}^\alpha}\right) \cdots$$

$$\left(1 + \frac{1}{p_K} + \frac{1}{p_K{}^2} + \cdots + \frac{1}{p_K{}^\alpha}\right)$$

In this product the terms $\frac{1}{1}, \frac{1}{2}, \ldots, 1/n$ are bound to appear, as well as others, hence the product P cannot be smaller than $\frac{1}{1} + \frac{1}{2} + \cdots + 1/n$. From the formula for the sum of a geometric progression we see that the factors in the product are respectively smaller than

$$\frac{1}{1 - 1/p_1}, \qquad \frac{1}{1 - 1/p_2}, \qquad \frac{1}{1 - 1/p_3}$$

and so on. Hence

$$\frac{1}{1} + \frac{1}{2} + \frac{1}{3} + \cdots + \frac{1}{n} < \frac{p_1}{p_1 - 1} \cdot \frac{p_2}{p_2 - 1} \cdot \frac{p_3}{p_3 - 1} \cdots \frac{p_K}{p_K - 1}$$

for all values of n. Therefore if K, the number of primes, were finite, the harmonic series would necessarily be convergent.[6] In a considerably more involved analysis, Euler showed that the infinite series made up of the reciprocals of the primes is itself divergent, the sum S_n being asymptotic to $\ln \ln n$ for increasing values of the integer n.

Euler delighted in relationships between the theory of numbers and his rough and ready manipulations of infinite series. Heedless of the dangers lurking in alternating series, he found such results as $\pi = 1 + \frac{1}{2} + \frac{1}{3} + \frac{1}{4} - \frac{1}{5} + \frac{1}{6} + \frac{1}{7} + \frac{1}{8} + \frac{1}{9} - \frac{1}{10} + \cdots$. Here the sign of a term, after the first two, is determined as follows: if the denominator is a prime of form $4m - 1$, a

---

[6] This proof, and other work by Euler and the Bernoullis, appears in Gerhard Kowalewski, *Die klassischen Probleme der Analysis des Unendlichen* (1910).

minus sign is used, if the denominator is a prime of form $4m + 1$, a plus sign is used, and if the denominator is a composite number, the sign indicated by the product of the signs of its components is used. Operations on infinite series were handled with great abandon. From the result $\ln 1/(1 - x) = x + x^2/2 + x^3/3 + x^4/4 + \cdots$ Euler concluded that $\ln \infty = 1 + \frac{1}{2} + \frac{1}{3} + \frac{1}{4} + \cdots$, hence that $1/\ln \infty = 0 = 1 - \frac{1}{2} - \frac{1}{3} - \frac{1}{5} + \frac{1}{6} - \frac{1}{7} + \frac{1}{10} - \cdots$, where the last series is made up of all the reciprocals of primes (in which case the terms are taken as positive) and reciprocals of products of two distinct primes (in which case the terms are negative). The *Introductio* is replete with such series and with related infinite products, such as $0 = \frac{1}{2} \cdot \frac{2}{3} \cdot \frac{4}{5} \cdot \frac{6}{7} \cdot \frac{10}{11} \cdot \frac{12}{13} \cdot \frac{16}{17} \cdot \frac{18}{19} \cdots$ and $\infty = \frac{2}{1} \cdot \frac{3}{2} \cdot \frac{5}{4} \cdot \frac{7}{6} \cdot \frac{11}{10} \cdot \frac{13}{12} \cdot \frac{17}{16} \cdot \frac{19}{18} \cdots$. The symbol ∞ is freely regarded[7] as denoting the reciprocal of the number 0.

To the subject of logarithms Euler contributed not only the definition in terms of exponents that we use today, but also the correct view with respect to the logarithms of negative numbers. The notion that $\log(-x) = \log(+x)$ was upheld by the leading mathematician in France during the mid-eighteenth century, who died in the same year as Euler—Jean Le Rond d'Alembert (1717–1783). D'Alembert's cognomen was taken from the church of St. Jean Baptiste le Rond, near Notre-Dame de Paris, on the steps of which he had been abandoned as an infant. His mother was later discovered to be the aristocratic and vivacious Madame de Tencin, an eloquent writer and sister of a cardinal, and his father was the Chevalier Destouches, an artillery general. The foundling was brought up by the wife of a glazier; in later years when he became celebrated as a mathematician d'Alembert spurned the overtures of his mother, preferring to be recognized as the son of his impoverished foster parents. The surname d'Alembert was adopted, for reasons not known, when he was a young man.

Like Euler and the Bernoullis, d'Alembert, too, was broadly educated—in law, medicine, science and mathematics—a background that served him well when, from 1751 to 1772, he collaborated with Denis Diderot (1713–1784) in the twenty-eight volumes of the celebrated *Encyclopédie* or *Dictionnaire raisonné des sciences, des arts, et des métiers*. For the *Encyclopédie* d'Alembert wrote the much-admired "Discours préliminaire," as well as most of the mathematical and scientific articles. The *Encyclopédie*, despite d'Alembert's Jansenist education, showed strong tendencies toward the secularization of learning so characteristic of the Enlightenment, and it met with strong attack from Jesuits. Through his defense of the project d'Alembert became known as "the fox of the Encyclopedia" and incidentally played a significant role in the expulsion of the Jesuit order from France. As a result of his activities, and of his friendships with Voltaire and others among the

---

[7] See especially *Introductio* (new ed., Lyons, 1797), I, 229 ff.

"philosophes," he was one of those who paved the way for the French Revolution.[8] At the early age of twenty-four he had been elected to the Académie des Sciences, and in 1754 he became its *secrétaire perpetuel*, and as such perhaps the most influential scientist in France. Toward the close of Euler's residency in Berlin, Frederick the Great of Prussia invited d'Alembert to head the Prussian Academy; d'Alembert declined, arguing that it would be most inappropriate to place any contemporary in a position of academic superiority over the great Euler. D'Alembert was invited also by Catherine the Great of Russia to serve as tutor to her son, but this offer he likewise declined, despite the princely salary he was offered.

While Euler was busy with mathematical research in Berlin, d'Alembert was active in Paris; until 1757, when controversy over the problem of vibrating strings brought estrangement, correspondence between the two was frequent and cordial,[9] for their interests were much the same. Statements such as $\log(-1)^2 = \log(+1)^2$, equivalent to $2\log(-1) = 2\log(+1)$ or to $\log(-1) = \log(+1)$, had puzzled the best mathematicians of the earlier part of the eighteenth century, but by 1747 Euler was able to write to d'Alembert explaining correctly the status of logarithms of negative numbers. The result should really have been apparent to Jean Bernoulli and others who were more or less familiar with the formula $e^{i\theta} = \cos\theta + i\sin\theta$ even before Euler clearly enunciated it. This identity holds for all angles (in radian measure); in particular it leads, for $\theta = \pi$, to $e^{i\pi} = -1$—that is, to the statement that $\ln(-1) = \pi i$. Logarithms of negative numbers therefore are not real, as Jean Bernoulli and d'Alembert had thought, but pure imaginaries.

Euler called attention also to another property of logarithms that became apparent from his identity. Any number, positive or negative, has not one logarithm, but infinitely many. From the relationship $e^{i(\theta \pm 2K\pi)} = \cos\theta + i\sin\theta$, one sees that if $\ln a = c$, then $c \pm 2K\pi i$ are also natural logarithms of a. Moreover, from Euler's identity one sees that logarithms of complex numbers, real or imaginary, also are complex numbers. If, for example, one wishes a natural logarithm of $a + bi$, one writes $a + bi = e^{x+iy}$. One obtains $e^x \cdot e^{iy} = a + bi = e^x(\cos y + i\sin y)$. The solution of the simultaneous equations $e^x\cos y = a$ and $e^x\sin y = b$ (obtained by equating real and imaginary parts of the complex equation) yields the values $y = \arctan b/a$ and $x = \ln(b\csc\arctan b/a)$ [or $x = \ln(a\sec\arctan b/a)$].

7   D'Alembert had spent much of his time and effort attempting to prove the theorem conjectured by Girard and known today as the fundamental

---

[8] For an excellent nonmathematical account of his life and work, see Ronald Grimsley, *Jean d'Alembert (1717–1783)* (1963).

[9] See R. E. Langer, "The Life of Leonard Euler," *Scripta Mathematica*, **3** (1935), 61–66, 131–138.

theorem of algebra—that every polynomial equation $f(x) = 0$, having complex coefficients and of degree $n \geq 1$, has at least one complex root. So earnest were his efforts to prove the theorem (especially in a prize essay on "The General Cause of Winds" published in the *Memoirs* of the Berlin Academy for 1746) that in France today the theorem is widely known as the theorem of d'Alembert. If we think of the solution of such a polynomial equation as a generalization of the explicit algebraic operations, we can say that in essence d'Alembert wished to show that the result of any algebraic operation performed on a complex number is in turn a complex number. In a sense, then, Euler did for elementary transcendental operations what d'Alembert had tried to do for algebraic operations. Through the Euler identities it is not difficult to find, for example, such quantities as $\sin(1 + i)$ or $\arccos i$, expressed in standard complex-number form. In the former case one writes

$$\sin(1 + i) = \frac{e^{i(1+i)} - e^{-i(1+i)}}{2i}$$

from which one finds that $\sin(1 + i) = a + bi$, where $a = [(1 - e^2)\sin 1]/2e$ and $b = [(e^2 - 1)\cos 1]/2e$. In the latter case one writes $\arccos i = x + iy$ or $i = \cos(x + iy)$ or

$$i = \frac{e^{i(x+iy)} + e^{-i(x+iy)}}{2} = \frac{1 + e^{2y}}{2e^y}\cos x + i\frac{(i - e^{2y})}{2e^y}\sin x$$

Equating real and imaginary parts, one sees that $\cos x = 0$ and $x = +\pi/2$. Hence

$$\frac{1 - e^{2y}}{2e^y} = \pm 1 \qquad \text{or} \qquad e^y = \mp 1 \pm \sqrt{2}$$

Inasmuch as both x and y must be real, we see that $x = \pm\pi/2$ and $y = \ln(\mp 1 + \sqrt{2})$. In a similar manner one can carry out other elementary transcendental operations on complex numbers, the results being complex numbers. That is, the work of Euler showed that the system of complex numbers is closed under the elementary transcendental operations, whereas d'Alembert had suggested that the system of complex numbers is closed under algebraic operations.

Euler similarly showed that, surprisingly, an imaginary power of an imaginary number can be a real number. In a letter to Christian Goldbach (1690–1764) in 1746 he gave the remarkable result $i^i = e^{-\pi/2}$. From $e^{i\theta} = \cos\theta + i\sin\theta$ we have, for $\theta = \pi/2$, $e^{\pi i/2} = i$; hence

$$(e^{\pi i/2})^i = e^{\pi i^2/2} = e^{-\pi/2}$$

There are, in fact, infinitely many real values for i^i, as Euler later showed, given by $e^{-\pi/2 \pm 2K\pi}$, where K is an integer. In the *Memoirs* of the Berlin Academy for 1749 Euler showed that any complex power of a complex number, $(a + bi)^{c+di}$, can be written[10] as a complex number $p + qi$. This aspect of Euler's work was overlooked, and the real values of i^i had to be rediscovered in the nineteenth century.

D'Alembert likewise considered the expression $(a + bi)^{c+di}$, and at one point he took the base $a + bi$ in this combination to be a variable and differentiated the function—an anticipation of the theory of complex variables that was developed in the nineteenth century. D'Alembert assumed that a calculus of complex variables would follow a pattern similar to that for algebraic combinations of real variables, so that the expression $f(x + iy) d(x + iy)$ would always be expressible in the form $dp + i\,dq$ in which real and imaginary parts are separated, a result that he was unable to prove. In a paper of 1752 on the resistance of fluids he arrived at the so-called Cauchy-Riemann equations that loom so large in complex analysis. If the analytic function $f(x + iy) = u + iv$, then $f(x - iy) = u - iv$, $\partial u/\partial x = \partial v/\partial y$, and $\partial u/\partial y = -\partial v/\partial x$.

8   D'Alembert was an unusual combination of caution and boldness in his view of mathematical developments. He regarded Euler's use of divergent series as open to suspicion (1768), despite the successes achieved. Moreover, d'Alembert objected to the Eulerian assumption that differentials are symbols for quantities that are zero and yet qualitatively different. Inasmuch as Euler restricted himself to well-behaved functions, he had not become involved in the subtle difficulties that later were to make his naïve position untenable. Meanwhile, d'Alembert believed that the "true metaphysics" of the calculus was to be found in the idea of a limit. In the article on the "differential" that he wrote for the *Encyclopédie*, d'Alembert stated that "the differentiation of equations consists simply in finding the limits of the ratio of finite differences of two variables included in the equation." Opposing the views of Leibniz and Euler, d'Alembert insisted that "a quantity is something or nothing: if it is something, it has not yet vanished; if it is nothing, it has literally vanished. The supposition that there is an intermediate state between these two is a chimera."[11] This view would rule out the vague notion of differentials as infinitely small magnitudes, and d'Alembert held that the differential notation is merely a convenient manner of speaking that depends for its justification on the language of limits. His *Encyclopédie* article on the differential referred to Newton's *De quadratura curvarum*, but d'Alembert interpreted Newton's phrase "prime and ultimate

---

[10] See articles on i^i by H. S. Uhler and R. C. Archibald in *American Mathematical Monthly*, **28** (1921), 114–121.

[11] See his *Mélanges de littérature, d'histoire et de philosophie* (1767), pp. 249–250.

ratio" as a limit, rather than as a first or last ratio of two quantities just springing into being. In the article on "Limit" which he composed for the *Encyclopédie*, he called one quantity the limit of a second [variable] quantity if the second can approach the first nearer than by any given quantity (without actually coinciding with it.)[12] The imprecision in this definition was removed in the works of nineteenth-century mathematicians.

Euler had thought of an infinitely large quantity as the reciprocal of an infinitely small magnitude; but d'Alembert, having outlawed the infinitesimal, defined the indefinitely large in terms of limits. A line, for example, is said to be infinite with respect to another if their ratio is greater than any given number. He went on to define indefinitely large quantities of higher order in a manner similar to that used by mathematicians today in speaking of orders of infinity with respect to functions. D'Alembert denied the existence of the actually infinite, for he was thinking of geometrical magnitudes rather than of the theory of aggregates proposed a century later. D'Alembert's formulation of the limit concept lacked the clear-cut phraseology necessary to make it acceptable to his contemporaries. Continental textbook writers of the later eighteenth century therefore generally continued to use the language and views of Leibniz and Euler, rather than those of d'Alembert.

D'Alembert, a man of wide interests, today is perhaps best known for **9** what is referred to as d'Alembert's principle—the internal actions and reactions of a system of rigid bodies in motion are in equilibrium. This appeared in 1743 in his celebrated treatise, *Traité de dynamique*. Other treatises by d'Alembert concerned music, the three-body problem, the precession of the equinoxes, motion in resisting media, and lunar perturbations. In studying the problem of vibrating strings he was led to the partial differential equation $\partial^2 u/\partial t^2 = \partial^2 u/\partial x^2$, for which in 1747 he gave (in the *Memoirs* of the Berlin Academy) the solution $u = f(x + t) + g(x - t)$, where f and g are arbitrary functions. The theory of ordinary differential equations had been well developed before this time, but the more difficult subject of the solution of partial differential equations was then a field for pioneers. Euler made further progress in this branch of analysis by giving for the more general equation $\partial^2 u/\partial t^2 = a^2(\partial^2 u/\partial x^2)$ the solution $u = f(x + at) + g(x - at)$.

The solution of ordinary differential equations had in a sense begun as soon as the inverse relationship between differentiation and integration had been recognized. But most differential equations cannot easily be reduced to simple quadratures, requiring instead ingenious substitutions or algorithms for their solution. One of the achievements of the eighteenth century

[12] See A. Robinson, *Non-standard Analysis* (Amsterdam: North Holland, 1966), pp. 267–268.

was the discovery of groups of differential equations that are solvable by means of fairly simple devices. The Bernoulli equations, noted in the preceding chapter, form one such group. Another type was identified by the precocious mathematician Alexis Claude Clairaut (1713–1765) and is named for him— the family of equations of the form $y = xy' + f(y')$. In this case the substitution $p = y'$, followed by differentiation of the terms of the equation with respect to x, will lead to an equation in x, p, and dp/dx which is of first order and solvable, the general solution being $y = cx + f(c)$. The Clairaut differential equation has also a singular solution, among the first of this type to be found, Taylor having earlier given one such solution. D'Alembert found the singular solution of the somewhat more general type of differential equation $y = xf(y') + g(y')$, hence this is known as d'Alembert's equation.

10   Alexis Claude Clairaut was one of the most precocious of mathematicians, outdoing even Blaise Pascal in this respect. At the age of ten he was reading the textbooks of L'Hospital on conics and the calculus, when he was thirteen he read a paper on geometry to the Académie des Sciences, and when only eighteen he was admitted, through special dispensation with respect to age requirements, to membership in the Académie. (D'Alembert was elected to the Académie at the age of twenty-four.) In the year of his election Clairaut published a celebrated treatise, *Recherches sur les courbes à double courbure*, the substance of which had been presented to the Académie two years earlier. Like the *Géométrie* of Descartes, the *Recherches* of Clairaut appeared without the name of the author on the title page, although in this case, too, the authorship was generally known. The treatise of Clairaut carried out for space curves the program that Descartes had suggested almost a century before—their study through projections on two coordinate planes. It was, in fact, this method that suggested the name given by Clairaut to gauche or twisted curves inasmuch as their curvature is determined by the curvatures of the two projections. In the *Recherches* numerous space curves are determined through intersections of various surfaces, distance formulas for two and three dimensions are explicitly given, an intercept form of the plane is included, and tangent lines to space curves are found. This book by the teen-age Clairaut constitutes the first treatise on solid analytic geometry.[13] Clairaut, one of a family of twenty children only one of whom survived the father, was important for other contributions to analysis. He observed that the mixed second-order partial derivatives f_{xy} and f_{yx} of a function $f(x, y)$ are in general equal (we know now that this holds if these derivatives are

[13] An excellent source of information on Clairaut is found in Pierre Brunet, "La vie et l'oeuvre de Clairaut," *Revue d'Histoire des Sciences et de leurs Applications,* **4** (1951), 13–40, 109–153; **5** (1952), 334–349; **6** (1953), 1–17. See also C. B. Boyer, "Clairaut and the Origin of the Distance Formula," *American Mathematical Monthly,* **55** (1948), 556–557.

continuous at the point in question), and he used this fact in the test $M_y \equiv N_x$, familiar in differential equations, for exactness of the differential expression $M(x, y) \, dx + N(x, y) \, dy$. In celebrated works on applied mathematics, such as *Théorie de la figure de la terre* (1743) and *Théorie de la lune* (1752), he made use of potential theory. His textbooks, *Eléments de géométrie* (1741) and *Eléments d'algèbre* (1746), the former composed for the Marquise du Châtelet, were part of a plan, reminiscent of those of our own day, to improve the teaching of mathematics.

Incidentally, Clairaut had a younger brother who rivaled him in precocity, for at the age of fifteen the brother, known to history only as "le cadet Clairaut," published in 1731 (the same year as that in which the older brother's *Recherches* had appeared) a book on calculus entitled *Traité de quadratures circulaires et hyperboliques*. This virtually unknown genius died tragically of smallpox during the next year.[14] The father of the two Clairaut brothers was himself a capable mathematician, but today he is recalled primarily through the work of his sons, two of the most precocious mathematicians of all times.

One of the interesting differential equations of the eighteenth century is that called by d'Alembert the Riccati equation—$y' = p(x)y^2 + q(x)y + r(x)$. This equation had been studied by a number of mathematicians, including several of the Bernoullis, as well as by Giacomo Riccati (1676–1754) and his son Vincenzo (1707–1775). But it was Euler who first called attention to the fact that if a particular solution $v = f(x)$ is known, then the substitution $y = v + 1/z$ converts the Riccati equation into a linear differential equation in z, so that a general solution can be found. In the Petersburg *Commentarii* for 1760–1763 Euler also pointed out that if two particular solutions are known, then a general solution is expressible in terms of a simple quadrature.

Euler was, without any doubt, the individual most responsible for methods used today in introductory college courses in the solution of differential equations, and even many of the specific problems appearing in current textbooks can be traced back to the great treatises Euler wrote on the calculus—*Institutiones calculi differentialis* (Petersburg, 1755) and *Institutiones calculi integralis* (Petersburg, 1768–1770, 3 vols.). The use of integrating factors, the systematic methods of solving linear equations of higher order with constant coefficients, and the distinction between linear homogeneous and nonhomogeneous equations, and between particular and general solutions, are among his contributions to the subject, although on some points credit must be shared with others. Daniel Bernoulli, for example, had solved the equation $y'' + Ky = f(x)$ independently of Euler and at about the same time in 1739–1740, and d'Alembert as well as Euler had general methods, about 1747, for solving complete linear equations. To some extent our ubiquitous indebtedness to Euler in the field of differential equations is betokened in the fact that a type of linear equation with variable coefficients bears his name. The Euler equation $x^n y^{(n)} + a_1 x^{n-1} y^{(n-1)} + \cdots + a_n y^{(0)} = f(x)$ (where exponents included within parentheses indicate orders of differentiation) is easily reduced, through the substitution $x = e^t$, to a linear equation having constant coefficients.

Euler's four volumes of *Institutiones* contain by far the most exhaustive treatment of the calculus up to that time. Besides the elements of the subject and the solution of differential equations, we find such things as "Euler's theorem on homogeneous functions"—if $f(x, y)$ is homogeneous of order n, then $x f_x + y f_y = nf$—a development of the calculus of finite differences, standard forms for elliptic integrals (in which field d'Alembert also was active), and the theory of the beta and gamma (or factorial) functions based on the "Eulerian integrals" $\Gamma(p) = \int_0^\infty x^{p-1} e^{-x} \, dx$ and $B(m, n) = \int_0^1 x^{m-1}(1 - x)^{n-1} \, dx$ and related through such formulas as $B(m, n) = \Gamma(m)\Gamma(n)/\Gamma(m + n)$. Wallis had anticipated some of the properties of these integrals, but through the organization of Euler these higher transcendental functions became an essential part of advanced calculus and of applied mathematics. About a century later the integral in the beta function was generalized by Pafnuti L. Tchebycheff (1821–1894), who demonstrated that the "Tchebycheff integral" $\int x^p (1 - x)^q \, dx$ is a higher transcendental function unless p or q or $p + q$ is an integer.

12   One of the characteristics of the Age of Enlightenment was the tendency to apply to all aspects of society the quantitative methods that had been so successful in the physical sciences. In this respect it is not surprising to find both Euler and d'Alembert writing on problems of life expectancy, the value of an annuity, lotteries, and other aspects of social science. Probability, after all, had been among the chief interests of Euler's friends, Daniel and Nicolaus Bernoulli. According to Euler's calculations, published in the *Memoirs* of the Berlin Academy for 1751, a payment of 350 crowns should purchase for a newborn infant a deferred annuity of 100 crowns to commence at age twenty and to continue for life. Among the lottery problems that he published in the Berlin Academy *Memoirs* for 1765, the following is one of the simplest. Let n tickets be numbered consecutively from 1 to n and let three tickets be drawn at random. Then the probability that a sequence of three consecutive numbers will be drawn is

$$\frac{2 \cdot 3}{n(n - 1)}$$

the probability that two consecutive numbers (but not three) will be drawn is

$$\frac{2 \cdot 3(n - 3)}{n(n - 1)}$$

and the probability that no consecutive numbers will be drawn is

$$\frac{(n - 3)(n - 4)}{n(n - 1)}$$

No new concepts are required for the solution, but, as we might anticipate, Euler contributed to notations here as he had elsewhere. He wrote that he found it useful to represent the expression

$$\frac{p(p - 1) \cdots (p - q + 1)}{1 \cdot 2 \cdots q}$$

by

$$\left[\frac{p}{q} \right]$$

a form essentially equivalent to the modern notation

$$\binom{p}{q}$$

D'Alembert, unlike Euler, is noted, in the history of probability, chiefly for his opposition to opinions generally received. For example, in the article on "Croix ou Pile" published in 1754 in the *Encyclopédie*, d'Alembert suggests that the probability of throwing head in the course of two throws with a coin should be $\frac{2}{3}$, rather than the $\frac{3}{4}$ commonly accepted, inasmuch as the game is finished if head appears on the first throw. A Geneva mathematician pointed out to d'Alembert that his three cases (H, TH, TT) are not equally likely, but d'Alembert remained skeptical of the common argument. In the article above he had referred to the status of the Petersburg paradox as a scandal; evidently this encouraged him to look upon the first principles of probability as unsound. In view of this situation he suggested that where possible probabilities should be determined by experiment. In this he had the approval of the Comte de Buffon (1707–1788), author of a celebrated multi-volume *Histoire naturelle*.

To scientists in general Buffon is known as an iconoclastic who proposed an estimate of about 75,000 years for the age of the earth, instead of the commonly accepted figure of approximately 6000 years. To mathematicians, Buffon is known for two contributions—a translation into French of Newton's *Method of Fluxions* and the "Buffon needle problem" in the theory of probability. Buffon, too, had been impressed by the "Petersburg paradox," and in an "Essai d'arithmétique morale," published in 1777 in the fourth volume of a supplement to his *Histoire naturelle*, he gave several reasons for regarding the game as inherently impossible. Buffon suggested also, in the same "Essai," what was essentially a new branch of probability— problems involving geometrical considerations. He proposed that a large plane area be ruled with equidistant parallel straight lines and that a thin needle be thrown at random upon the plane area.[15] The probability that the needle will fall across one of the lines he correctly gave as $2l/\pi d$, where d is the distance between the lines and l is the length of the needle, and $l < d$. The "Essai" contained also a collection of tables, covering the years 1709 to 1766 at Paris, on births, marriages, and deaths, as well as results on life expectancies to which d'Alembert took exception.

It was during the eighteenth century that the practice of variolation— inoculation with a weakened form of smallpox in order to develop immunity against the disease—was introduced into Europe from the Levant; this provoked controversy among those who sought to apply the theory of probability to the affairs of life. In 1760 Daniel Bernoulli read before the Académie des Sciences at Paris an essay on the advantages of inoculation, but before the Essai was published in the *Memoirs* of the Académie d'Alembert already had raised objections. D'Alembert did not deny the advantages, but he argued that Bernoulli had overstated them. Part of the argument centered on the distinctions d'Alembert insisted must be made between the "mean life" and the "probable life" of an individual. The "probable life" of an infant was about eight years (that is, half of the infants of the time died before they were eight years old), whereas his "mean life" or average life span was about twenty-six years. (A comparison of these figures with corresponding data of today makes vivid the appallingly low state of medical research of past centuries.) Controversies over the probability that variolation would be advantageous were effectively terminated at the end of the century when vaccination against smallpox was discovered by Dr. Edward Jenner.

13   D'Alembert shared interests with Euler in many aspects of mathematics, especially in analysis and applied mathematics, but there was one direction in which Euler made great strides without rivalry on the part of d'Alembert. This was in the theory of numbers, a subject that has held strong attraction for many of the greatest mathematicians, such as Fermat and Euler, but no appeal for others, including Newton and d'Alembert. Euler did not publish a

---

[15] An account of this is found in Isaac Todhunter, *History of the Mathematical Theory of Probability from the Time of Pascal to that of Laplace* (1865). See also N. T. Gridgeman, "Geometric Probability and the Number π," *Scripta Mathematica*, **25** (1960), 183–195.

treatise on the subject, but he wrote letters and articles on various aspects of the theory of numbers. It will be recalled that Fermat had asserted, among other things, (1) that numbers of the form $2^{2^n} + 1$ apparently are always prime and (2) that if p is prime and a is an integer not divisible by p, then $a^p - a$ is divisible by p. The first of these conjectures Euler exploded in 1732 through his uncanny ability for computation, showing that $2^{25} + 1 = 4,294,967,297$ is factorable into $6,700,417 \times 641$. Today the Fermat conjecture has been so thoroughly deflated that mathematicians incline to the contrary opinion—that there are no prime Fermat numbers beyond the number 65,537 corresponding to $n = 4$.

In the same way that Euler, by means of a counterexample, had upset one of Fermat's conjectures, the twentieth century has disproved a suggestion made by Euler. If n is greater than two, Euler believed, at least n nth powers are required to provide a sum that is itself an nth power:[16] but in 1966 it was shown that the sum of only four fifth powers can be a fifth power,[17] for $27^5 + 84^5 + 110^5 + 133^5 = 144^5$. It should be noted, however, that in the latter case it required two centuries and the services of a high-speed computing device to detect the inadvertence.

For the second of the conjectures, known as Fermat's lesser theorem, Euler was the first one to publish a proof (although Leibniz had left an earlier demonstration in manuscript). The proof of Euler, which appeared in the Petersburg *Commentarii* for 1736, is so surprisingly elementary that we describe it here. The proof depends on an induction on a. If $a = 1$ the theorem obviously holds. We now show that if the theorem holds for any positive integral value of a, such as $a = k$, then it necessarily holds for $a = k + 1$. To show this, we use the binomial theorem to write $(k + 1)^p$ as $k^p + mp + 1$, where m is an integer. On subtracting $k + 1$ from both sides, we see that $(k + 1)^p - (k + 1) = mp + (k^p - k)$. Inasmuch as the last term on the right is divisible by p, by hypothesis, the right-hand side of the equation, hence also the left-hand side, obviously is divisible by p. The theorem therefore holds, through mathematical induction, for all values of a provided that a is prime to p.

Having proved Fermat's lesser theorem, Euler demonstrated a somewhat more general statement in which he used what has been called "Euler's ϕ-function."[18] If m is a positive integer greater than one, the function $\phi(m)$ is defined as the number of integers less than m which are prime to m (but including the integer one in each case). It is customary to define $\phi(1)$ as 1; for $n = 2$, 3, and 4, for example, the values of $\phi(n)$ are 1, 2, and 2 respectively.

[16] L. E. Dickson, *History of the Theory of Numbers* (1952), II, 648.

[17] L. J. Lander and T. R. Parkin, "Counterexample to Euler's Conjecture on Sums of Like Powers," *Bulletin of the American Mathematical Society*, **72** (1966), 1079.

[18] A good account of this is included in Oystein Ore, *Number Theory and its History* (1948).

If p is a prime, then clearly $\phi(p) = p - 1$. It can be proved that

$$\phi(m) = m\left(1 - \frac{1}{p_1}\right)\left(1 - \frac{1}{p_2}\right)\cdots\left(1 - \frac{1}{p_r}\right)$$

where $p_1, p_2, \cdots p_r$ are the distinct prime factors of m. Using this result Euler showed that $a^{\phi(m)} - 1$ is divisible by m if a is relatively prime to m.

Euler settled two of Fermat's conjectures, but did not dispose of "Fermat's last theorem," although he did prove the impossibility of integer solutions of $x^n + y^n = z^n$ for the case $n = 3$. In 1747 Euler added to the three pairs of amicable numbers known to Fermat, bringing the list up to thirty pairs; later he increased this to more than sixty. Euler also gave a proof that all even perfect numbers are of the form given by Euclid—$2^{n-1}(2^n - 1)$, where $2^n - 1$ is prime. Whether or not there can be an odd perfect number remains an open question. Also unresolved to this day is a question raised in correspondence between Euler and Christian Goldbach (1690–1764). In writing to Euler in 1742 Goldbach suggested that every even integer is the sum of two primes. This so-called Goldbach's theorem appeared in print (without proof) in 1770 in England in the *Meditationes algebraicae* of Edward Waring (1734–1793). Waring was senior wrangler at Cambridge in 1757 and Lucasian professor of mathematics there from 1760. His works contained many important results. They were, however, poorly written and not widely read, so that the familiar ratio test for the convergence of infinite series more frequently is known as Cauchy's test, despite the fact that it had been given by Waring as early as 1776. The *Meditationes algebraicae* contains not only the Goldbach conjecture, but also a complementary conjecture that every odd integer is a prime or the sum of three primes. Among other unproved assertions is one known as Waring's theorem or Waring's problem. Euler had proved that every positive integer is the sum of not more than four squares; Waring surmised that every positive integer is the sum of not more than nine cubes, or the sum of not more than nineteen fourth powers. The first half of this bold guess was proved in the early twentieth century; the second part is still unproved, beyond the fact that Hilbert in 1909 showed that every positive integer is expressible as the sum of not more than N positive nth powers, where N is some function of n. Waring published also in the *Meditationes algebraicae* a theorem named for his friend and pupil, John Wilson (1741–1793)—if p is a prime, then $(p - 1)! + 1$ is a multiple of p. Wilson, too, was a senior wrangler at Cambridge, but he left mathematics for law, where he rose to a judgeship and to knighthood.

14   The leading Continental mathematicians of the mid-eighteenth century were primarily analysts, but we have seen that their contributions were not

limited to analysis. D'Alembert had given an imperfect proof of the fundamental theorem of algebra, and Clairaut in 1740 had published a textbook,
Élémens d'algèbre, which was so popular it went through a sixth edition in
1801. Euler not only contributed to the theory of numbers, but also composed
a popular algebra textbook that appeared in German and Russian editions
at St. Petersburg in 1770–1772, in French (under the auspices of d'Alembert)
in 1774, and in numerous other versions, including American editions in
English. The exceptionally didactic quality of Euler's *Algebra* is attributed to
the fact that it was dictated by the blind author through a relatively untutored
domestic. The textbooks of Clairaut and Euler were not widely used in
England, in part because of British mathematical isolationism during the
later eighteenth century and in part because Maclaurin and others had
composed good textbooks on an elementary level. Maclaurin's *Treatise of
Algebra* went through half a dozen editions from 1748 to 1796. A rival
Treatise of Algebra by Thomas Simpson (1710–1761) boasted at least
eight editions at London from 1745 to 1809; another, *Elements of Algebra*,
by Nicholas Saunderson (1682–1739), enjoyed five editions between 1740
and 1792. Simpson was a self-taught genius who won election to the Royal
Society in 1745 but whose turbulent life ended in failure half a dozen years
later. His name nevertheless is preserved in the so-called Simpson's rule,
published in his *Mathematical Dissertations on Physical and Analytical
Subjects* (1743), for approximate quadratures using parabolic arcs; but this
result had appeared in somewhat different form in 1668 in the *Exercitationes
geometricae* of James Gregory. Saunderson's life, by contrast, was an example
of personal triumph over an enormous handicap—total blindness from the
age of one, resulting from an attack of smallpox.

Algebra textbooks of the eighteenth century illustrate a tendency toward
increasingly algorithmic emphasis, while at the same time there remained
considerable uncertainty about the logical bases for the subject. Most
authors felt it necessary to dwell at length on the rules governing multiplications of negative numbers, and some rejected categorically the possibility of
multiplication of two negative numbers. The century was, par excellence, a
textbook age in mathematics, and never before had so many books appeared
in so many editions. Simpson's *Algebra* had a companion volume, *Elements
of Plane Geometry*, which went through five editions from 1747 to 1800, but
among the host of textbooks of the time few achieved quite the record of the
edition by Robert Simson (1687–1768) of the *Elements of Euclid*. This work,
by a man trained in medicine who became professor in mathematics at the
University of Glasgow, first appeared in 1756, and by 1834 it boasted a
twenty-fourth English edition, not to mention translations into other
languages nor geometries more or less derived from it, for most modern
English versions of Euclid are heavily indebted to it.

15   Simson sought to revive ancient Greek geometry, and in this connection he published "restorations" of lost works, such as Euclid's *Porisms* and the *Determinate Sections* of Apollonius. Partly as a result of Simson's enthusiasm for antiquity, England throughout the eighteenth century remained a stronghold of synthetic geometry, and analytic methods made little headway in geometry. This may be one of the reasons that progress in analysis in Britain lagged far behind that on the Continent. It is customary to place much of the blame for this backwardness on the supposedly clumsy method of fluxions as compared with that of the differential calculus, but such a view is not easily justified. Fluxional notations even today are conveniently used by physicists, and they are readily adapted to analytic geometry; but no calculus, whether differential or fluxional, is appropriately wedded to synthetic geometry. Hence the British predilection for pure geometry seems to have been a far more effective deterrent to research in analysis than was the notation of fluxions. Nor is it fair to place the blame for British geometrical conservatism largely on the shoulders of Newton. After all, Newton's *Method of Fluxions* was replete with analytic geometry, and even the *Principia* contained more analysis than generally is recognized. Perhaps it was an excessive insistence on logical precision that had led the British into a narrow geometrical view. We noted previously the arguments of Berkeley against the mathematicians, and Maclaurin had felt that the most effective way to meet these on a rational basis was to return to the rigor of classical geometry. Almost 2000 years earlier, in Greece, an insistence on rigor seems to have hampered the development of a numerical algebra; in England in the eighteenth century the situation was somewhat similar. On the Continent, on the other hand, the feeling was akin to the advice that d'Alembert is said to have given to a hesitating mathematical friend: "Just go on ahead, and faith will soon return." It is easy to criticize the logic of Euler and d'Alembert, but it is unthinkable that anyone should question their immensely significant roles in the development of mathematics.

Synthetic geometry was not entirely forgotten on the Continent, for in 1741 Clairaut published an *Élémens de géométrie* which also boasted some half dozen editions, but this was an insipid textbook with little solidity and less rigor. Euler and d'Alembert contributed little to the field, despite the fact that today the line containing the circumcenter, the orthocenter, and the barycenter of a triangle is known as the Euler line of the triangle. That these centers of a triangle are collinear seems to have been known earlier to Simson, whose name has been attached to another line related to a triangle.[19] Such minor additions to pure geometry pale into insignificance, however,

[19] See, for example, R. A. Johnson, *Modern Geometry* (Boston: Houghton Mifflin, 1929); reprinted as *Advanced Euclidean Geometry* (New York: Dover paperback, 1960), pp. 137 ff, 206 ff.

when compared to Continental contributions to analytic geometry during the mid-eighteenth century.

We have described the analytic geometry of Clairaut, especially in connection with developments in three dimensions, but the material in the second volume of Euler's *Introductio* was more extensive, more systematic, and more effective. As early as 1728 Euler contributed to the Petersburg *Commentarii* papers on the use of coordinate geometry in three-space, giving general equations for three broad classes of surfaces—cylinders, cones, and surfaces of revolution. He recognized that the equation of a cone with vertex at the origin is necessarily homogeneous. He showed also that the shortest curve (geodesic) between two points on a conical surface would become the straight line between these points if the surface were flattened out into the form of a plane—one of the earliest theorems concerning developable surfaces. Euler's awareness of the significance of making work as general as possible is seen especially in the second volume of his *Introductio*. This book did more than any other to make the use of coordinates, in both two and three dimensions, the basis of a systematic study of curves and surfaces. Instead of concentrating on the conic sections, Euler gave a theory of curves in general, based on the function concept that had been central in the first volume. Transcendental curves are not given short shrift, as had been customary, so that here, practically for the first time, graphical study of trigonometric functions formed a part of analytic geometry. The other common transcendental curves also are included, as well as some not so common, such as $y = x^x$, $y^x = x^y$, and $y = (-1)^x$.

The *Introductio* includes also two accounts of polar coordinates which **16** are so thorough and systematic that the system frequently, but erroneously, is attributed to Euler. Whole classes of curves, both algebraic and transcendental, are considered; for the first time the equations for transformations from rectangular to polar coordinates appear in strictly modern trigonometric form. Moreover, Euler made use of the general vectorial angle and of negative values for the radius vector, so that the spiral of Archimedes, for example, appeared in its dual form, symmetric with respect to the 90° axis. D'Alembert evidently was influenced by this work when he wrote the article on "Géométrie" for the *Encyclopédie*. Euler's *Introductio* also was chiefly responsible for the systematic use of what is called the parametric representation of curves—that is, an expression of each of the Cartesian coordinates as a function of an auxiliary independent variable. For the cycloid, for example, Euler used the form

$$\begin{cases} x = b - b \cos \dfrac{z}{a} \\ y = z + b \sin \dfrac{z}{a} \end{cases}$$

A long and systematic appendix to the *Introductio* is perhaps Euler's most significant contribution to geometry, for it represents virtually the first textbook exposition of solid analytic geometry. Surfaces, both algebraic and transcendental, are considered in general and then are subdivided into categories. Here we find, evidently for the first time, the notion that surfaces of second degree constitute a family of quadrics in space analogous to the conic sections in plane geometry. Beginning with the general ten-term quadratic equation $f(x, y, z) = 0$, Euler noted that the aggregate of terms of second degree, when equated to zero, gives the equation of the asymptotic cone, real or imaginary. More importantly, he used the equations for translation and rotation of axes (in the form that, incidentally, still bears Euler's name) to reduce the equation of a nonsingular quadric surface to one of the canonical forms corresponding to the five fundamental types—the real ellipsoid, the hyperboloids of one and two sheets, and the elliptic and hyperbolic paraboloids. One aspect of modern courses in analytic geometry that is not found in the *Introductio* (or in other books of the time) is a systematic study of the loci of elementary geometry, the line and the circle, the plane and the sphere. Nevertheless, the work of Euler comes closer to modern textbooks than did any other book before the French Revolution.

17   Many mathematicians of all ages have fancied themselves also as philosophers. Euler and d'Alembert were among these, but both of them missed an opportunity that another philosophically inclined mathematician tried to exploit. This was Johann Heinrich Lambert (1728–1777), a Swiss-German writer on a wide variety of mathematical and nonmathematical themes, who for a couple of years was an associate of Euler in the Berlin Academy. It is said that when Frederick the Great asked him in which science he was most proficient, Lambert curtly replied, "All." He might be better known today if he had not tried, immodestly, to master all fields of science, for he was indeed a man of exceptional ability.

We have seen that Saccheri had believed that he had demolished the possibilities that the sum of the angles of a plane triangle might be more or less than two right angles. Lambert called attention to the well-known fact that on the surface of a sphere the angle sum of a triangle is indeed more than two right angles, and he suggested that a surface might be found on which the triangle angle-sum falls short of two right angles. In trying to complete what Saccheri had attempted—a proof that denial of Euclid's parallel postulate leads to a contradiction—Lambert, in 1766, wrote *Die Theorie der Parallellinien*, although this appeared, posthumously, only in 1786. Instead of beginning with a Saccheri quadrilateral, he adopted as his starting point a quadrilateral having three right angles (now known as a Lambert quadrilateral) and then considered for the fourth angle the three possibilities, namely, that it might be acute, right, or obtuse. Corresponding to these three cases he showed, in the manner of Saccheri, that the angle sum of a triangle would be respectively less than, equal to, or greater than two right angles. Going beyond Saccheri, he demonstrated that the extent to which the sum falls short of, or exceeds, two right angles is proportional to the area of the triangle. In the obtuse-angled case this situation is similar to a classical theorem in spherical geometry—that the area of a triangle is proportional to its spherical excess—and Lambert speculated that the hypothesis of the acute angle might correspond to a geometry on a novel surface, such as a sphere of imaginary radius. In 1868, it was shown by Eugenio Beltrami (1835–1900) that Lambert had indeed been correct in his conjecture of the existence of some such surface. It turned out to be, however, not a sphere with an imaginary radius, but a real surface known as a pseudosphere—a surface of constant negative curvature generated by revolving the tractrix above its axis.[20]

Although Lambert, like Saccheri, tried to prove the parallel postulate, he seems to have been aware of his lack of success. He wrote:

Proofs of the Euclidean postulate can be developed to such an extent that apparently a mere trifle remains. But a careful analysis shows that in this seeming trifle lies the crux of the matter; usually it contains either the proposition that is being proved or a postulate equivalent to it.

No one else came so close to the truth without actually discovering non-Euclidean geometry.

Lambert is known today also for other contributions. One of these is the first proof, presented to the Berlin Academy in 1761, that π is an irrational number. (Euler in 1737 had shown that e is irrational.) Lambert showed that if x is a nonzero rational number, then tan x cannot be rational. Inasmuch as tan $\pi/4 = 1$, a rational number, it follows that $\pi/4$ cannot be a rational number, hence neither can π. This did not, of course, dispose of the circle-squaring question, for quadratic irrationalities are constructible; at about this time circle-squarers had become so numerous that the Academy at Paris in 1775 passed a resolution that no purported solutions of the quadrature problem would be officially examined. As another contribution of Lambert to mathematics we should recall that he did for the hyperbolic functions what Euler had done for the circular functions, providing the modern view and notation. Comparisons of the ordinates of the circle $x^2 + y^2 = 1$ and of the hyperbola $x^2 - y^2 = 1$ had fascinated mathematicians for a century, and by 1757 Vincenzo Riccati, an Italian, had suggested a development of hyperbolic functions. It remained for Lambert to introduce

---

[20] The work of Lambert on non-Euclidean geometry is fully described in F. Engel and P. Stäckel: *Die Theorie der Parallellinien von Euklid bis auf Gauss* (1895). A shorter account is found, for example, in Roberto Bonola, *Non-Euclidean Geometry* (1912).

the notations sinh x, cosh x, and tanh x for the hyperbolic equivalents of the circular functions of ordinary trigonometry and to popularize the new hyperbolic trigonometry that modern science finds so useful. Corresponding to Euler's three identities for sin x, cos x, and e^{ix}, there are three similar relationships for the hyperbolic functions expressed by the equations

$$\sinh x = \frac{e^x - e^{-x}}{2}, \qquad \cosh x = \frac{e^x + e^{-x}}{2}$$

and

$$e^x = \cosh x + \sinh x$$

Lambert also wrote on cosmography, descriptive geometry, map making, logic, and the philosophy of mathematics, but his influence did not match that of Euler or d'Alembert.

18   Euler and d'Alembert died in the same year, 1783; this was also the year of death of Etienne Bézout (1730–1783), a mathematician who represents a characteristic aspect of the subject at that time. We have mentioned that the eighteenth century produced many enormously successful textbooks; we might add that it was the second half of the century that produced also the genre often known as a *Cours d'analyse*—a multivolume work covering the subject matter of mathematics from the lowest to the highest level. One of the most successful of all of these was Bézout's *Cours de mathématique*, a six-volume work that first appeared in 1764–1769, which was almost immediately issued in a new edition of 1770–1772 and which boasted many versions in French and other languages. (The first American textbook in analytic geometry, incidentally, derived in 1826 from Bézout's *Cours*.) It was through such compilations, rather than through the original works of the authors themselves, that the mathematical advances of Euler and d'Alembert became widely known. Bézout himself was no mere hack, and his name is familiar today in connection with the use of determinants in algebraic elimination. In a memoir of the Paris Academy for 1764, and more extensively in a treatise of 1779 entitled *Théorie générale des équations algébriques*, Bézout gave artificial rules, similar to Cramer's, for solving n simultaneous linear equations in n unknowns. He is best known for an extension of these to a system of equations in one or more unknowns in which it is required to find the condition on the coefficients necessary for the equations to have a common solution. To take a very simple case, one might ask for the condition that the equations $a_1 x + b_1 y + c_1 = 0$, $a_2 x + b_2 y + c_2 = 0$, $a_3 x + b_3 y + c_3 = 0$ have a common solution. The necessary condition is that the eliminant

$$\begin{vmatrix} a_1 & b_1 & c_1 \\ a_2 & b_2 & c_2 \\ a_3 & b_3 & c_3 \end{vmatrix}$$

here a special case of the "Bézoutiant," should be 0. Somewhat more complicated eliminants arise when conditions are sought for two polynomial equations of unequal degree to have a common solution. Bézout also was the first one to give a satisfactory proof of the theorem, known to Maclaurin and Cramer, that two algebraic curves of degrees m and n respectively intersect in general in $m \cdot n$ points; hence this is often called Bézout's theorem. Euler also had contributed to the theory of elimination, but less extensively than did Bézout.

During the eighteenth century the French universities were not outstanding in mathematics. It was the academies and military schools that produced a substantial number of mathematicians, and a *Cours de mathématique* like that of Bézout was likely to be used at institutions such as these. Bézout himself taught at a military school and was an examiner for the navy, hence he was in touch with the curricula of the time. However, within a few years of the deaths of the mathematicians featured in this chapter (Buffon died only a year before the fall of the Bastille in 1789) the system of higher education in France was to undergo a drastic revision as a result of the upheaval produced by the French Revolution. During this short but significant period France became once more the mathematical center of the world, as she had been during the middle of the seventeenth century. The next chapter is devoted to a group of mathematicians who lived and worked in the city of Paris during some of her most trying days.

BIBLIOGRAPHY

Bell, E. T., *Men of Mathematics* (New York: Simon & Schuster, 1937).

Bell, E. T., *Development of Mathematics*, 2nd ed. (New York: McGraw-Hill, 1945).

Bonola, Roberto, *Non-Euclidean Geometry* (New York, 1912; reprinted, New York: Dover, 1955).

Boyer, C. B., *History of Analytic Geometry* (New York: Scripta Mathematica, 1956).

Boyer, C. B., *History of the Calculus* (paperback ed., New York: Dover, 1959).

Brunet, Pierre, "La vie et l'oeuvre de Clairaut," *Revue d'Histoire des Sciences et de leurs Applications*, **4** (1951), 13–40, 109–153; **5** (1952), 334–349; **6** (1953). 1–17.

Cajori, Florian, *History of Mathematical Notations* (Chicago: Open Court, 1928–1929, 2 vols.).

Coolidge, J. L., *History of the Conic Sections and Quadric Surfaces* (Oxford: Clarendon, 1945).

D'Alembert, J., *Encyclopédie* (Paris, 1751–1765).

D'Alembert, J., *Mélanges de littérature, d'historie, et de philosophie*, 4th ed. (Amsterdam, 1767, 5 vols.).

Dickson, L. E., *History of the Theory of Numbers* (New York: Chelsea reprint, 1952, 3 vols.).

Dugas, René, *A History of Mechanics* (New York: Central Book Co., ca. 1955).

Engel, F., and P. Stäckel, *Die Theorie der Parallellinien von Euklid bis auf Gauss* (Leipzig, 1895, 2 vols.).

Eneström, Gustav, "Verzeichnis der Schriften Leonhard Eulers," *Jahresbericht der Deutschen Mathematiker-Vereinigung, Ergänzungsband IV*, Leipzig, 1910–1913.

Euler, Leonhard, *Opera omnia*, ed. by F. Rudio and others (Leipzig and Lausanne: B. G. Teubner, 1911–   ).

Fuss, N., *Lobrede auf Herrn Leonhard Euler* (Basel, 1786).

Grimsley, Ronald, *Jean d'Alembert (1717–83)* (Oxford: Clarendon, 1963).

Hofmann, J. E., *Classical Mathematics. A Concise History of the Classical Era in Mathematics*, trans. by Henrietta O. Midonick (New York: Philosophical Library, 1959).

Kowalewski, Gerhard, *Die klassischen Probleme der Analysis des Unendlichen* (Leipzig: W. Engelmann, 1910).

Ore, Oystein, *Number Theory and its History* (New York: McGraw-Hill, 1948).

Reiff, R., *Geschichte der unendlichen Reihen* (Tübingen, 1889).

Stäckel, Paul, "Eine vergessene Abhandlung Leonhard Eulers über die Summe der reziproken Quadrate der natürlichen Zahlen," *Bibliotheca Mathematica* (3), **8** (1907–1908), 37–60.

Struik, D. J., *Concise History of Mathematics*, 3rd ed. (New York: Dover, 1967).

Todhunter, Isaac, *History of the Mathematical Theory of Probability from the Time of Pascal to that of Laplace* (Cambridge, 1865).

Toeplitz, Otto, *The Calculus, a Genetic Approach* (Chicago: University of Chicago Press, 1963).

Truesdell, C., *The Rational Mechanics of Flexible or Elastic Bodies*, Introduction to *Leonhardi Euleri Opera Omnia* (Zurich: Orell Füssli, 1960, 2nd series, Vols. X–XI).

Wieleitner, Heinrich, *Geschichte der Mathematik* (Berlin and Leipzig: Walter de Gruyter, 1921, Vol. II, 2nd half).

EXERCISES

1. Describe the chief sources of support of mathematicians during the eighteenth century, giving specific instances.
2. Which branches of mathematics were most actively developed during the middle of the eighteenth century? Give examples to support your answer.
3. Name four periodicals that published mathematical articles during the eighteenth century, mentioning at least one contributor in each case.
4. Which mathematical treatise of the mid-eighteenth century do you regard as having been most influential? Give reasons for your answer.
5. How do you account for the fact that Russia for the first time was an important mathematical center in the eighteenth century?
6. In the eighteenth century a number of well-known mathematicians moved from one country to another. Mention several of them, indicating the circumstances surrounding the change.
7. Describe the most important contributions made by Euler to mathematical notations.

8. Prove the three Euler identities:

$$\sin x = \frac{e^{ix} - e^{-ix}}{2i}$$

$$\cos x = \frac{e^{ix} + e^{-ix}}{2}$$

and

$$e^{ix} = \cos x + i \sin x$$

9. Use the Euler identities to find, in $a + bi$ form, a natural logarithm of $1 + i$.
10. Write $\sin (1 + i)$ as a complex number in $a + bi$ form.
11. Write e^{p+qi} as a complex number in $a + bi$ form.
12. Write one value of i^{i} as a complex number in $a + bi$ form, both exactly and in terms of decimal approximations.
13. Verify that if one adds the three probabilities of Euler given above in connection with a lottery problem, the sum is one, and indicate why this should be so.
14. Sketch Euler's curve $y = (-1)^x$.
15. Find the minimum point of the curve $y = x^x$ and sketch the curve.
16. Sketch the spiral of Archimedes $r = a\theta$ as Euler did—that is, for positive and negative values of θ.
17. Identify, with respect to Euler's five fundamental types of quadric surfaces, the following: $a^2x^2 + b^2y^2 = 2z$; $a^2x^2 - b^2y^2 = 2z$; $a^2x^2 + b^2y^2 + c^2z^2 = d^2$; $a^2x^2 - b^2y^2 + c^2z^2 = d^2$; $a^2x^2 - b^2y^2 - c^2z^2 = d^2$.
18. Derive, in the manner of Euler, the sum of the terms in the series

$$\frac{1}{1^2} + \frac{1}{3^2} + \frac{1}{5^2} + \cdots + \frac{1}{(2n-1)^2} + \cdots$$

19. Verify the hyperbolic function identity $\cosh^2 x - \sinh^2 x \equiv 1$.
20. Verify that $\sinh ix = i \sin x$ and $\cosh ix = \cos x$.
21. If the hyperbolic tangent and hyperbolic cotangent are defined by the identities

$$\tanh x \equiv \frac{\sinh x}{\cosh x} \quad \text{and} \quad \coth x \equiv \frac{\cosh x}{\sinh x}$$

sketch the curves $y = \tanh x$ and $y = \coth x$.
22. Show that $\sinh x$ is a monotonically increasing function.
23. Use Clairaut's test to see if the following are exact differentials: $(\sin y - y \sin x) \, dx + (x \cos y + \cos x) \, dy$ and $2xy^3 \, dx - 3x^2y^2 \, dy$.
*24. Find the singular solution of the Clairaut differential equation $y = xy' + (y')^2$ and sketch this solution on the same set of axes with a few of the particular solutions.
*25. Show that if $v = f(x)$ is a particular solution of the Ricatti equation $y' = p(x)y^2 + q(x)y + r(x)$, then the substitution $y = v + z$ converts the equation to a Bernoulli equation and the substitution $y = v + 1/z$ converts the equation to a linear equation.
*26. Solve the Euler equation $x^2y'' + 4xy' + 2y = 0$.
*27. Show that $\int x^p(1 - x)^q \, dx$ is an elementary function if p or q or $p + q$ is an integer.
*28. In a letter to Goldbach of 1741 Euler pointed out that $2^{-1+i} + 2^{-1-i}$ is nearly equal to $\frac{10}{13}$. Verify this approximation.
*29. Show that if m is a power α of a prime p (that is, $m = p^{\alpha}$), then $\phi(m) = p^{\alpha}(1 - 1/p)$, where ϕ is the "Euler ϕ-function."

CHAPTER XXII

Mathematicians
of the French Revolution

> The advancement and perfection of mathematics are
> intimately connected with the prosperity of the State.
>
> *Napoleon I*

1 The eighteenth century had the misfortune to come after the seventeenth and before the nineteenth. How could any period that followed the "Century of Genius" and which preceded the "Golden Age" of mathematics be looked upon as anything but a prosy interlude? Analytic geometry and the calculus were invented in the seventeenth century; the rise of mathematical rigor and the flowering of geometry are associated with the nineteenth. There is a scholarly history of mathematics of the sixteenth and seventeenth centuries and one (incomplete) for the nineteenth century:[1] but there is no comparable history of mathematics in the eighteenth century, nor do we readily look to the eighteenth century for significant trends in mathematics. This is in marked contrast to what is true in other fields. For Americans the date 1776 was decisive; in France the year 1789 was crucial. Nor was the Age of Revolutions confined to the sphere of politics. The Industrial Revolution changed the whole fabric of Western society, and the thermotic revolution during the same years laid the foundations of modern chemistry. Can it be that mathematics during these stirring events was enjoying a nap? This chapter will show that mathematicians of France at the time of the Revolution not only contributed handsomely to the fund of knowledge, but that they were in large measure responsible for the chief lines of development in the explosive proliferation of mathematics during the succeeding century. We are even tempted to add to the already impressive list of revolutions of the time two more: a "geometrical revolution" and an "analytical revolution."[2]

---

[1] See H. G. Zeuthen, *Geschichte der Mathematik im XVI und XVII. Jahrhundert* (German ed., Leipzig, 1903), and Felix Klein, *Vorlesungen über die Entwicklung der Mathematik im 19. Jahrhundert* (Berlin, 1926–1927), 2 vols. The last volume and a half of Moritz Cantor, *Vorlesungen über Geschichte der Mathematik* (4 vols., Leipzig, 1880–1908) by themselves would constitute an ample history of eighteenth-century mathematics.

[2] In developing this argument considerable use is made of Niels Nielsen's *Géomètres français sous la révolution* (Copenhagen, 1929); but whereas Nielsen has catalogued alphabetically the contributions of some four score French mathematicians, it is the object of this chapter to furnish a synthesis of the mathematical milieu of the time with emphasis on only a handful of individuals. The chapter is an amplification of an article on "Mathematicians of the French Revolution" that appeared in *Scripta Mathematica*, **25** (1960), 11–31.

Every age is inclined to think of itself as one of revolution—a period of tremendous change. But almost every age of rapid change has been preceded by a long period in which preparations for the revolution are made, sometimes consciously, more often unconsciously. Among the heralds of the French Revolution were Voltaire, Rousseau, d'Alembert, and Diderot—not one of whom lived to see the fall of the Bastille (Voltaire and Rousseau died in 1778, d'Alembert in 1783, and Diderot a year later)—and their associate Condorcet, who fell a victim in the holocaust that he helped to father. In mathematics six men who were to show the way—Monge, Lagrange, Laplace, Legendre, Carnot, and Condorcet—were to be in the midst of the turmoil, and it is with these men that this chapter is chiefly concerned.

Our half dozen mathematicians were almost of an age: Lagrange, the oldest, was born in 1736; Condorcet was born in 1743; Monge in 1746; Laplace in 1749; Legendre in 1752; Carnot, the youngest, was born in 1753. With the exception of Condorcet, who died a suicide in prison, these mathematicians all lived to be septuagenarians, and one, Legendre, an octogenarian.

In France of the eighteenth century, universities were not the mathematical **2** foci that they are today, and one is hard put to it to name even one eighteenth-century mathematician at, say, the University of Paris. During the fourteenth century Paris had been one of the scientific centers of the world (the other being at Oxford), but it had long since lost this position. It was behind the times: when Europe turned to Cartesianism, Paris clung to Peripatetic Scholasticism; and when most of the scientific world had turned to Newtonionism, Paris fought a rearguard action for Cartesianism. Most of the French mathematicians of the eighteenth century were associated not with the universities, but with either the church or the military; others found royal patronage or became private teachers. Lagrange (1736–1813), the only one of our group who was not strictly a Frenchman, was born at Turin of once prosperous parents with French and Italian backgrounds. Joseph-Louis, the youngest of eleven children and the only one to survive beyond infancy, was educated there and as a young man became professor of mathematics in the military academy of Turin; but later he found successive royal patrons in Frederick the Great of Prussia and Louis XVI of France. The family of Condorcet (1743–1794) included influential members in the cavalry and the church, hence his education presented no problem. At Jesuit schools and later at the Collège de Navarre he made an enviable reputation in mathematics; but instead of becoming a captain of cavalry, as his family had hoped, he lived the life of a scholar in much the same sense as Voltaire, Diderot, and d'Alembert. The third of our sextet, Gaspard Monge (1746–1818), was the son of a poor tradesman. However, through the influence of a lieutenant colonel who had been struck by the boy's ability, Monge was permitted to attend some courses at the École Militaire de Mézières; he so impressed those in authority that he soon became a member of the teaching staff—the only one of our group of six who was primarily a teacher, perhaps the most influential mathematics teacher since the days of Euclid. Laplace (1749–1827) also was born without wealth; like Monge, he found influential friends who saw that he obtained an education—again in a military academy. Legendre (1752–1833) experienced no difficulty in securing an education; but even he was not a university teacher in the strict sense, although for five years he taught in the École Militaire at Paris. The youngest of our group, Lazare Carnot (1753–1823), was sufficiently above bourgeois standing to be permitted to attend the École Militaire at Mézières, where Monge was one of his teachers. Upon graduation Carnot entered the army, although, lacking a title, he could not, under the *ancien régime*, aspire to a rank above that of captain.[3] This must have rankled in his mind—as it did in the case of so many others that the proverb arose at examination time that "the competent were not noble and the noble were not competent." The economic wastefulness of the government may have been the immediate cause of the French Revolution, but it was far from the only one. The enormous waste of human resources was also an important factor, and symptomatic of this was the failure at first of the men of our group to win positions commensurate with their ability; not one of the six expressed regret later when the old order passed away.

Of the mathematical encyclopedias of the late eighteenth century the most successful, judging from repeated editions, was that by Bézout, instructor in the school at Mézières that both Monge and Carnot attended. The *Cours de mathématique* of Bézout was, during the first third of the nineteenth century, still a very influential work, especially in America where parts of it appeared in English translation at West Point and other academies. The fourth part of Bézout's *Cours*—the principles of mechanics—is the *raison d'être* of the program. The emphasis given to mechanics and to the closing section on navigation is in keeping with the use of the *Cours de mathématiques* as a text in a military academy. The mathematical preeminence of France (and, indeed, of Continental Europe as a whole) in the eighteenth century was based in large measure on the application of analysis to mechanics as taught in technical schools, and it was under this influence that the mathematicians of the French Revolution had been brought up, in marked contrast to the situation in England. One should naturally expect the contrast in mathematical spirit to become sharper during the Revolution, for France had greater need for technical training, and England became more thoroughly isolated from the Continent.

---

[3] For further details on his life see C. B. Boyer, "The Great Carnot," *The Mathematics Teacher*, **49** (1956), 7–14.

Every one of the six men we have named as the mathematical leaders **3** during the Revolution had produced abundantly before 1789. Lagrange had published his *Mécanique analytique* (1788), as well as frequent papers on algebra, analysis, and geometry. Condorcet, perhaps the most interesting of the six because of the breadth of his interests, had published *De calcul intégral* as early as 1765 and *Essai sur l'application de l'analyse à la probabilité des décisions rendues à la pluralité des voix* in 1785. A firm believer in the perfectibility of man, a basic tenet of the Philosophes, Condorcet was the only one of our six who can be said to have played an anticipatory role in the events leading to 1789. (It is ironic to note that of our mathematical sextet the one who did most to bring about the Revolution was the only one to lose his life through it, although two others, Carnot and Monge, were not always safe from the guillotine.) Monge had contributed numerous mathematical articles to the *Mémoires of the Académie des Sciences*. Inasmuch as he succeeded Bézout as examiner for the School of the Marine, Monge was urged by those in authority to do what Bézout had done—write a *Cours de mathématiques* for the use of candidates. Monge, however, was interested in teaching and research rather than in writing textbooks, and he completed only one volume of the project—*Traité élémentaire de statique* (Paris, 1788). He was attracted not only to both pure and applied mathematics, but also to physics and chemistry. In particular, he participated with Lavoisier in experiments, including those on the composition of water, which led to the chemical revolution of 1789. Through his numerous activities Monge had become, at the time of the revolution, one of the best known of French scientists. In fact, his reputation as a physicist and chemist was perhaps greater than that as a mathematician, for his geometry had not been properly appreciated. His chief work, the *Géométrie descriptive*, had not been published because his superiors felt that it was in the interests of national defense to keep it confidential. (Classified material is not a monopoly of the mid-twentieth century!) Laplace and Legendre were regular contributors to learned periodicals, and Carnot by 1786 had published a second edition of his *Essai sur les machines en général*, as well as some verses and a work on fortifications.

In looking at the achievements of these six men, one is struck by a lack of **4** utilitarian motive in their work. Carnot's *Essai* would appear, from the title, to be technically oriented, but a glance at the book shows that it deals with broad principles, not with technology. The *Mécanique* of Lagrange likewise is concerned with a postulational treatment of the subject, far removed from criteria of practicability. The beauty of Lagrange's work is apparent not to the engineer but to the pure mathematician; even in the more elementary portions of his work there is an aesthetic quality. It is primarily to him that we owe such compact forms, though somewhat differently expressed, as

$$\frac{1}{2!} \begin{vmatrix} x_1 & y_1 & 1 \\ x_2 & y_2 & 1 \\ x_3 & y_3 & 1 \end{vmatrix} \quad \text{and} \quad \frac{1}{3!} \begin{vmatrix} x_1 & y_1 & z_1 & 1 \\ x_2 & y_2 & z_2 & 1 \\ x_3 & y_3 & z_3 & 1 \\ x_4 & y_4 & z_4 & 1 \end{vmatrix}$$

for the area of a triangle and for the volume of a tetrahedron, respectively, results that appeared in a paper, "Solutions analytiques de quelques problèmes sur les pyramides triangulaires," delivered in 1773 and published in 1775.[4] Such work looks pretty, but inconsequential; yet it contained an idea that was to become, through the educational reforms of the Revolution, very important. As Lagrange expressed it, "I flatter myself that the solutions which I am going to give will be of interest to geometers as much for the methods as for the results. These solutions are purely analytic and can even be understood without figures." True to his promise, there is not a single diagram throughout the work. Monge, too, although he used diagrams and models in descriptive and differential geometry, seems somehow to have come to the conclusion that one should avoid the use of diagrams in elementary analytic geometry. Perhaps Carnot felt somewhat the same way, for his *Essai*, antedating the *Mécanique* of Lagrange, contains not a single diagram.

Laplace, of all the members of our sextet, came closest to being an applied mathematician, but even in his case we must interpret the phrase in a very broad sense. After all, how "practical" in those days was the theory of probability or celestial mechanics? We can safely conclude that, in spite of their education in predominantly technical schools, the great figures in mathematics just before the Revolution had shown remarkable "purity" of interest.

5   The fall of the Bastille in 1789 found our six men divided into two categories: the three L's (Lagrange, Laplace, and Legendre) took no significant part in shaping the political events that were to follow; the other three (Carnot, Condorcet, and Monge) welcomed the changed outlook and played definite roles in revolutionary activities. Men from both groups, however, participated in at least one mathematical project during the Revolution.

The reform of the system of weights and measures is an especially appropriate example of the way in which mathematicians patiently persisted in their efforts in spite of confusion and politican difficulties. As early in the Revolution as 1790 Talleyrand proposed the reform of weights and measures. The problem was referred to the Académie des Sciences, in which a committee,

[4] See his *Oeuvres*, III, 658–692.

of which Lagrange and Condorcet were two of the members, was established to draw up a proposal. Legendre should have been a member, for he had achieved quite a reputation for his triangulation of France; revolutionary politics seem to have been responsible for his being overlooked. The Committee agreed on a decimal system, although there appear to have been some earnest supporters of a duodecimal scheme. Lagrange firmly supported the decimalists against the duodecimalists, for he was not greatly impressed by the argument about divisibility. (He is reported to have almost regretted not adopting as a base for the system some *prime* number, such as eleven, but it has been suggested that he may have done this simply to obstruct the duodecimalists.)

As is well known, the Committee considered two alternatives for the basic length in the new system. One was the length of the pendulum which should beat seconds. The equation for the pendulum being $T = 2\pi\sqrt{l/g}$, this would make the standard length g/π^2. But the Committee was so impressed by the accuracy with which Legendre and others had measured the length of a terrestrial meridian that in the end the meter was defined to be the ten-millionth part of the distance between the equator and the pole. The resulting metric system was ready in most respects in 1791, but there was confusion and delay in establishing it. The National Convention in 1793 suppressed the Académie des Sciences, while the Jardin des Plantes was greatly expanded. This inconsistency seems to have been the result of political forces. The Académie was led by older and more conservative men, the Jardin by younger scientists who were eager in their support of the new government. There was, moreover, quite a cult of Robespierre which represented a back-to-nature attitude derived in part from Rousseau. Evidently there was in France an attitude toward physical science something like Goethe's belligerency toward Newtonian physics. The Jardin des Plantes represented "safe" science, that of the Académie was suspect.

The closing of the Académie was a blow to mathematics; but the Convention continued the Committee on Weights and Measures, although it purged the Committee of some members, such as Lavoisier, and enlarged it by adding others, including Monge. At one point Lagrange was very nearly lost to the Committee, for the provincially minded Convention had banned foreigners from France; but Lagrange was specifically exempted from the decree and remained to serve as head of the Committee. Still later the Committee was made responsible to the Institut National that had replaced the Académie des Sciences; Lagrange, Laplace, Legendre, and Monge all served on the Committee at this stage. By 1799 the work of the Committee had been completed, and the metric system as we have it today became a reality. It will be noted that five of our group of six revolutionary mathematicians took active part in this project, only Carnot being unconnected with it; but we shall find that Carnot was engaged in many other essential activities, both political and mathematical. The metric system is, of course, one of the more tangible mathematical results of the Revolution, but in terms of the development of our subject it cannot be compared in significance with other contributions.

6   Condorcet, a physiocrat, a philosophe, and an encyclopedist, belonged to the circle of Voltaire and d'Alembert. He was a capable mathematician who had published books on probability and the integral calculus, but he was also a restless visionary and idealist who was interested in anything related to the welfare of mankind. He, like Voltaire, had a passionate hatred of injustice; although he held the title of marquis, he saw so many inequalities in the *ancien régime* that he wrote and worked toward reform. With implicit faith in the perfectibility of mankind and believing that education would eliminate vice, he argued for free public education, an admirably forward-looking view, especially for those days.[5] Condorcet is perhaps best remembered mathematically as a pioneer in social mathematics, especially through the application of probability and statistics to social problems. When, for example, conservative elements (including the Faculty of Medicine and the faculty of Theology) had attacked those who advocated inoculation against smallpox, Condorcet (together with Voltaire and Daniel Bernoulli) had come to the defense of variolation.

With the opening of the Revolution, Condorcet's thoughts turned from mathematics to administrative and political problems. The educational system had collapsed under the effervescence of the Revolution, and Condorcet saw that this was the time to try to introduce the reforms he had in mind. He presented his plan to the Legislative Assembly, of which he became President, but agitation over other matters precluded serious consideration of it. Condorcet published his scheme in 1792, but the provision for free education became a target of attack. Not until years after his death did France achieve Condorcet's ideal of free public instruction.

Condorcet had had high hopes for the Revolution—until extremists seized control. He then boldly denounced the Septembrists, and was ordered arrested for his pains. He sought hiding, and during the long months of concealment he composed the celebrated *Sketch for a Historical Picture of the Progress of the Human Mind*,[6] indicating nine steps in the rise of mankind from a tribal stage to the founding of the French Republic, with a prediction of the bright tenth stage that he believed the Revolution was about to usher in.

[5] The extent of his influence can well ge gauged by the fact that an impressive tome of 891 pages was published by Franck Alengry in 1904 with the title *Condorcet: Guide de la révolution française*. This book, however, is on his social and legal philosophy, rather than on his mathematics. See also G. G. Granger, *La mathématique sociale du Marquis de Condorcet* (1956).

[6] A convenient English translation by June Barraclough appeared in 1955 (New York: Noonday Press).

Shortly after completing this work (in 1794), and believing that his presence endangered the lives of his hosts, he left his hiding place. Promptly recognized as an aristocrat, he was arrested; the following morning he was found dead on the floor of his prison, presumably a suicide.

Condorcet had been sympathetic to the moderate Gironde wing of the **7** Revolution. Monge was plebian and an important member of the more radical Jacobin Club; but he, too, was to have some trouble, even though he was an enthusiastic partisan and joined patriotic organizations. He was assigned a role in the reform of weights and measures, ordered by the Constituent Assembly in 1790, but his post as examiner for the navy had kept him from Paris for a couple of years. On his return to the city in 1792 he was named Ministre de la Marine, apparently on the suggestion of Condorcet, and it was in his capacity as Minister of the Navy that to Monge fell the task of signing the official record of the trial and execution of the King. The French fleet, however, was so poorly organized and so ineffectual that Monge was unable to achieve anything significant, and within a year he demanded that he be replaced. He nevertheless remained active in politics and governmental operations, and he devoted an enormous amount of energy to meeting the needs for gunpowder of the revolutionary arsenal. At the instance of the Committee of Public Safety he published also a *Description de l'art de fabriquer les canons*. Throughout the Revolution Monge found himself in a precarious position, for he was too liberal for the conservatives and too conservative for the extremists.

More important for the future of mathematics were the efforts of Monge, after the crisis of foreign invasion had subsided, to establish a school for the preparation of engineers. As Condorcet had been the guiding spirit in the Committee on Instruction, so Monge was the leading advocate of institutions of higher learning. The result was the formation in 1794 of a Commission of Public Works, of which Monge was an active member, charged with the establishment of an appropriate institution. The school was the famous École Polytechnique, which took form so rapidly that students were admitted in the following year. At all stages of its creation the role of Monge was essential, both as administrator and as teacher. It is gratifying to note that the two functions are not incompatible, for Monge was eminently successful in both. He was even able to overcome his reluctance to write textbooks, for in the reform of the mathematics curriculum the need for suitable books was acute.

Monge found himself lecturing on two subjects both essentially new to a university curriculum. The first of these was known as stereotomy, now more commonly called descriptive geometry. Monge gave a concentrated course in the subject to 400 students, and a manuscript outline of the syllabus survives.

This shows that the course was of wider scope, both on the pure and the applied side, than is now usual. Besides the study of shadow, perspective, and topography, attention was paid to the properties of surfaces, including normal lines and tangent planes, and to the theory of machines. Among the problems set by Monge, for example, was that of determining the curve of intersection of two surfaces each of which is generated by a line that moves so as to intersect three skew lines in space. Another was the determination of a point in space equidistant from four lines. Such problems point up a change in mathematical education which was sponsored primarily by the French Revolution. As long ago as the Golden Age of Greece Plato had pointed out that the state of solid geometry was deplorable, and the medieval decline in mathematics had hit solid geometry harder than it had plane geometry. One who could not cross the *pons asinorum* could scarcely be expected to reach the study of three dimensions. The inventors of analytic geometry, Descartes and Fermat, had been well aware of the fundamental principle of solid analytic geometry—that every equation in three unknowns represents a surface, and conversely—but they had not taken steps to develop it. One can say that whereas the seventeenth century was the century of curves—the cycloid, the limaçon, the catenary, the lemniscate, the equiangular spiral, the hyperbolas, parabolas, and spirals of Fermat, the pearls of Sluse, and many others—the eighteenth was the century that really began the study of surfaces.[7] It was Euler (see above) who called attention to the quadric surfaces as a family analogous to the conics, and his *Introductio* in a sense established the subject of solid analytic geometry (although we must perforce mention Clairaut as a precursor); but Euler was not a proselytizer, hence his subject found no place in the school curriculum. One reason may have been that, like Descartes, he did not begin with the simplest rectilinear cases. Lagrange, influenced perhaps by his calculus of variations, manifested interest in problems in three dimensions and emphasized their analytic solution. He was first, for example, to give the formula

$$D = \frac{ap + bq + cr - d}{\sqrt{a^2 + b^2 + c^2}}$$

for the distance D from a point (p, q, r) to the plane $ax + by + cz = d$. But Lagrange did not have a geometer's heart, nor did he have enthusiastic disciples. Monge, by contrast, was a specialist in geometry—almost the first since Apollonius—as well as a superior teacher and a curriculum builder. (Parenthetically it may be mentioned that Monge had two brothers who also were professors of mathematics, thus putting the name of Monge in

[7] J. L. Coolidge, "The Beginnings of Analytic Geometry in Three Dimensions," *American Mathematical Monthly*, **55** (1948), 76–86, called attention to the slow progress made in solid analytic geometry pefore the days of Euler.

a class with that of the Bernoullis, the Cassinis, the Clairauts, and the Pascals as designating a family of mathematicians.) The rise of solid geometry consequently was due in part to the mathematical and revolutionary activities of Gaspard Monge. Had he not been politically active, the École Polytechnique might never have come into being; had he not been an inspiring teacher, the revival of geometry in three dimensions might not have taken place.

The École Polytechnique was not the only school created at the time. The École Normale had been hastily opened to some 1400 or 1500 students, less carefully selected than those at the École Polytechnique, and it boasted a mathematical faculty of high calibre, Monge, Lagrange, Legendre, and Laplace being among the instructors. It was the lectures of Monge at the École Normale in 1794–1795 that finally were published as his *Géométrie descriptive*;[8] but administrative difficulties made the school short-lived. The idea behind the new descriptive geometry, or method of double orthographic projection, is essentially very easy to understand. One simply takes two planes at right angles to each other, one vertical, the other horizontal, and then projects the figure to be represented orthogonally on these planes, the projections of all edges and vertices being clearly indicated. The projection on the vertical plane is known as the "elevation," the other projection is called the "plan." Finally, the vertical plane is folded or rotated about the line of intersection of the two planes until it also is horizontal. The elevation and plan thus provide one with a diagram in two dimensions of the three-dimensional object. This simple procedure, now so common in mechanical drawing, produced in the days of Monge almost a revolution in military engineering design.

Descriptive geometry was not the only contribution of Monge to three- **8** dimensional mathematics, for at the École Polytechnique he taught also a course in "application of analysis to geometry." Just as the abbreviated title "analytic geometry" had not yet come into general use, so also there was no "differential geometry," but the course given by Monge was essentially an introduction to this field. Here, too, no textbook was available, and so Monge found himself compelled to compose and print his *Feuilles d'analyse* (1795) for the use of students. Here the analytic geometry of three dimensions really came into its own; it was this course, required of all students at the École Polytechnique, that formed the prototype of the present program in solid analytic geometry. Students, however, evidently found the course difficult, for the lectures skimmed very rapidly over the elementary forms of the line and plane, the bulk of the material being on the applications of the

---

[8] An account of the methods in this work is given in W. H. Roever, *The Mongean Method of Descriptive Geometry* (1933); a history of the methods is given in Gino Loria, *Storia della geometria descrittiva* (1921).

calculus to the study of curves and surfaces in three dimensions. Monge was ever reluctant to write textbooks on the elementary level, or to organize material that was not primarily his own. However, he found collaborators ready to edit material that he included in his course; and so in 1802 there appeared in the *Journal de l'École Polytechnique* an extensive memoir by Monge and Jean-Nicolas-Pierre Hachette (1769–1834) on *Application d'algèbre à la géométrie*. Its first theorem is typical of a more elementary approach to the subject. It is the well-known eighteenth-century generaliza- tion of the Pythagorean theorem: The sum of the squares of the projections of a plane figure upon three mutually perpendicular planes is equal to the square of the area of the figure. Monge and Hachette proved the theorem just as in modern courses; in fact, the whole volume could serve without difficulty as a text in the twentieth century. Equations for transformations of axes, the usual treatment of lines and planes, the determination of the principal planes of a quadric are treated fully. It is in the analytic geometry of Monge, rather than that of Clairaut and Euler, that we first find a systematic study of the straight line in three dimensions. Monge showed that if, for instance, the line is given by the intersection of the planes $ax + by + cz + d = 0$ and $a'x + b'y + c'z + d = 0$, a plane through a point (x', y', z') that is orthogonal to the line has the form $A(x - x') + B(y - y') + c(z - z') = 0$, where A, B, and C are respectively the expressions (now called direction numbers of the line) $bc' - b'c, ca' - c'a$, and $ab' - a'b$. Other formulas give the distance from a point to a line and the shortest distance between two skew lines. For the latter, Monge wrote the given lines in the projection form

$$\begin{cases} y = Ax + B \\ z = cx + D \end{cases} \quad \text{and} \quad \begin{cases} y = A'x + B' \\ z = c'x + D' \end{cases}$$

and the equations of the desired common normal as

$$\begin{cases} y = \alpha x + \beta \\ z = \gamma x + \delta \end{cases}$$

Inasmuch as the common normal intersects both given lines, we know that

$$(\gamma - C)(\beta - B) = (\alpha - A)(\delta - D) \text{ and } (\gamma - C')(\beta - B') = (\alpha - A')(\delta - D')$$

From the fact that the common normal is perpendicular to each of the given lines we have $1 + A\alpha + C\gamma = 0$ and $1 + A'\alpha + C'\gamma = 0$. Solving the four simultaneous equations for α, β, γ, and δ, the equations of the desired normal are known.

Most of the results of Monge on the analytic geometry of the line and plane were given in memoirs dating from 1771. In his systematic arrangement of the material in the *Feuilles d'analyse* of 1795, and especially in the 1802 memoir with Hachette, we find most of the solid analytic geometry and the elementary differential geometry that are included in undergraduate college textbooks. One thing that might be missed is the explicit use of determinants, for this was the work of the nineteenth century. Nevertheless, we might, as in the case of Lagrange, look upon Monge's use of symmetric notations as an anticipation of determinants, but without the now customary arrangement (due to Cayley).

Among the new results given by Monge are two theorems that bear his name: (1) The planes drawn through the midpoints of the edges of a tetrahedron perpendicular to the opposite edges meet at a point M (which has since been called the "Monge point" of the tetrahedron). M turns out to be the midpoint of the segment joining the centroid and the circumcenter. (2) The locus of the vertices of the trirectangular angle whose faces are tangent to a given quadric surface is a sphere, known as the "Monge sphere" or director sphere of the quadric. The equivalent of this locus in two dimensions leads to what is called the "Monge circle" of a conic, even though the locus had been given a century earlier in synthetic form by Lahire. In 1809 Monge proved in various ways that the centroid of a tetrahedron is the point of concurrency of the lines joining the midpoints of opposite edges; he gave also the analogue of the Euler line in three-space, showing that for the orthocentric tetrahedron the centroid is twice as far from the orthocenter as from the circumcenter. Lagrange was so impressed by the work of Monge that he is said to have exclaimed; "With his application of analysis to geometry this devil of a man will make himself immortal."[9]

As already indicated, Monge possessed a quite unusual combination of **9** talents, for he was at once a capable administrator, an imaginative research mathematician, and an inspiring teacher. The one trait of a pedagogue he might have had, but lacked, was that of a textbook compilator. But if Monge here showed a deficiency, it was more than made up for by his young and eager students. We can say with no fear of contradiction that the pupils of Monge let loose a spate of elementary textbooks on analytic geometry such as has never been equaled—not even in our own day, deluged as we are with new books. If we judge from the sudden appearance of so many analytic geometries beginning with 1798, a revolution had taken place in mathematical instruction. Analytic geometry, which for a century and more had been overshadowed by the calculus, suddenly achieved a recognized place in the schools; this "analytical revolution" can be credited primarily to Monge. Between the years 1798 and 1802 four elementary analytic geometries appeared from the pens of Sylvestre François Lacroix (1765–1843), Jean-Baptiste Biot (1774–1862), Louis Puissant (1769–1843), and F. L. Lefrançais,

[9] For an excellent comprehensive account of the work of Monge, see René Taton, *L'oeuvre scientifique de Monge* (1951).

all directly inspired by the lectures at the Ecole Polytechnique; Polytechnicians were responsible for as many books again in the next decade. Most of these were eminently successful texts, appearing in numerous editions. The volume by Biot achieved a fifth edition in less than a dozen years; that by Lacroix, student and colleague of Monge, appeared in twenty-five editions within ninety-nine years! Perhaps we should speak instead of the "textbook revolution," for Lacroix's other textbooks were almost as spectacularly successful, his *Arithmetic* and his *Geometry* appearing in 1848 in the twentieth and sixteenth editions, respectively. The twentieth edition of his *Algebra* was published in 1859, and the ninth edition of his *Calculus* in 1881. These figures do not include translations into other languages.

10   Monge is known to most readers as a founder of modern *pure* geometry. Through Poncelet and other *anciens élèves* of the École Polytechnique, pure or *synthetic* geometry did indeed undergo a glorious renaissance, largely through the inspiration of Monge; but there is an aspect of Monge's work that is less well known. Virtually without exception, the textbook writers in *analytic* geometry ascribe the inspiration for their work to Monge, although Lagrange occasionally is mentioned as well. Lacroix most clearly expressed the point of view as follows:

> In carefully avoiding all geometric constructions, I would have the reader realize that there exists a way of looking at geometry which one might call *analytic geometry*, and which consists in deducing the properties of extension from the smallest possible number of principles by purely analytic methods, as Lagrange has done in his mechanics with regard to the properties of equilibrium and movement.[10]

Lacroix held that algebra and geometry "should be treated separately, as far apart as they can be; and that the results in each should serve for mutual clarification, corresponding, so to speak, to the text of a book and its translation." Lacroix pointed to Lagrange's work on the tetrahedron as an instance of this point of view, but he believed that Monge "was the first one to think of presenting in this form the application of algebra to geometry." (The historian of astronomy J. B. J. Delambre likewise ascribed to Monge the "resurrection of the alliance of algebra and geometry.") His own section on solid analytic geometry Lacroix admitted to be almost entirely the work of Monge. Perhaps teachers today can take satisfaction in the thought that analytic geometry as presented by Fermat and Descartes, a lawyer and a philosopher, respectively, remained ineffectual, and that only when it was given a new form by genuine pedagogues—Monge and those of his students who in turn became teachers at the École Polytechnique—did it show vitality.

[10] See the preface to his *Traité de calcul* (Paris, 1797).

It is interesting to note that Lacroix declined to use the name "analytic geometry" as a title for this textbook, and edition after edition carried the ponderous title *Traité élémentaire de trigonométrie rectiligne et sphérique et application de l'algèbre à la géométrie*. Although the phrase "analytic geometry" had appeared every now and then during the eighteenth century, it seems first to have been used as the title of a textbook by Lefrançais in an edition of his *Essais de géométrie* of 1804 and by Biot in an 1805 edition of his *Essais de géométrie analytique*, the latter of which, translated into English as well as other languages, was used for many years at West Point. We need not look in detail at the contents of the texts of Lacroix, Lefrançais, Biot, and others, for they resemble very closely the books of the early twentieth century in this country.

Monge was an outstanding figure of the Revolution; yet the mathematician whose name was on the tongue of every Frenchman during the Revolution was not Monge but Carnot. It was Lazare Carnot, who, when the success of the Revolution was threatened by confusion within and invasion from without, organized the armies and led them to victory. As ardent a republican as Monge, he nevertheless shunned all political cliques; having a high sense of intellectual honesty, he tried to be impartial in reaching decisions. After investigation he absolved the royalists of the infamous charge that they had mixed powdered glass in flour intended for the Revolutionary armies, but he felt bound by conscience to vote for the death of the king. (The American Tom Paine, sometimes regarded in his country as dangerously radical, voted *against* the execution of the king.) Reasoned impartiality, however, is difficult to maintain in times of crisis, and Robespierre, whom Carnot had antagonized, threatened that Carnot would lose his head at the first military disaster. Had Carnot been merely a mathematician and a politician, like Monge and Condorcet, he might well have gone to the guillotine. But Carnot had won the admiration of his countrymen for his remarkable military successes; and when a voice in the Convention proposed his arrest, the deputies spontaneously rose to his defense, acclaiming him the "Organizer of Victory." Hence it was instead the head of Robespierre that fell, and Carnot survived to take an active part in the formation of the École Polytechnique. Carnot was greatly interested in education at all levels, even though he seems never to have taught a class. His son Hippolyte served as minister of public instruction in 1848. (Another son, Sadi, became a celebrated physicist; and a grandson, also named Sadi, became the fourth president of the Third French Republic. See the genealogical chart.)

Carnot led a charmed political life until 1797. He had gone from the National Assembly to the Legislative Assembly, to the National Convention, to the powerful Committee of Public Safety, to the Council of Five Hundred

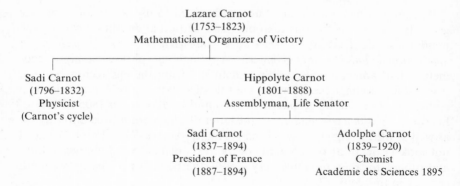

Celebrated Carnots: a genealogical chart.

and the Directorate. In 1797, however, he refused to join a partisan coup d'état and was promptly ordered deported. His name was stricken from the roles of the Institut and his chair of geometry was voted unanimously to General Bonaparte. Even Monge, fellow republican and mathematician, approved the intellectual outrage. About the only thing that can be said in extenuation of his action is that Monge seems to have been mesmerized by Napoleon. Monge followed his idol through thick and thin, his devotion being such that he became literally sick every time Napolean lost a battle. This is in contrast to Carnot, who initially was responsible for Bonaparte's rise to power through his appointment to the Italian campaign, but who did not hesitate to oppose the Frankenstein he had created, although it nearly cost him his life.

Mathematically, Carnot's proscription turned out to be a good thing, for it gave him an opportunity, while in exile, to complete a work that had been on his mind for some time. One should expect that a man engaged in affairs of enormous practical exigency, as was Carnot, would tend to think in terms of immediate practicality. Trajectories would appear to be a more likely subject for study than abstract metaphysical reflections. But the work that Carnot had been planning during his politically busy days was, *mirable dictu*, the *Réflexions sur la métaphysique du calcul infinitésimal*,[11] which appeared in 1797. This was not a work on applied mathematics; it came closer to philosophy than physics, and in this respect it adumbrated the period of rigor and concern for foundations so typical of the next century. Carnot's *Réflexions* became very popular and ran through a number of editions in several

[11] Two English translations were published. One appeared under the title "Reflections on the Theory of the Infinitesimal Calculus," trans. by W. Dickson, *Philosophical Magazine*, **8** (1800), 222–240, 335–352; **9** (1801), 39–56. The other was published as *Reflexions on the Metaphysical Principles of the Infinitesimal Analysis*, trans. by W. R. Browell (Oxford, 1832).

languages, proving that even in times that try men's souls pure mathematics finds many devotees.

Throughout the second half of the eighteenth century there was enthu- **12** siasm for the results of the calculus, but confusion about its basic principles. No one of the usual approaches, whether by the fluxions of Newton, the differentials of Leibniz, or the limits of d'Alembert, seemed to be satisfying. Hence Carnot, considering the conflicting interpretations, sought to show "in what the veritable spirit" of the new analysis consisted. In his selection of the unifying principle, however, he made a most deplorable choice. He concluded that "the true metaphysical principles" are "the principles of the compensation of errors." Infinitesimals, he argued, are "quantités inappréciables" which, like imaginary numbers, are introduced only to facilitate the computation, and are eliminated in reaching the final result. "Imperfect equations" are made "perfectly exact," in the calculus, by eliminating the quantities, such as infinitesimals of higher order, the presence of which occasioned the errors. To the objection that vanishing quantities either are or are not zero, Carnot responded that "what are called infinitely small quantities are not simply any null quantities at all, but rather null quantities assigned by a law of continuity which determines the relationship" —an argument that is strongly reminiscent of Leibniz. The divers approaches to the calculus, he claimed, were nothing but simplifications of the ancient method of exhaustion, reducing this in various ways to a convenient algorithm.

Carnot's *Réflections* enjoyed a wide popularity, appearing in many languages and editions. Unsuccessful though its synthesis of views was, it undoubtedly helped to make mathematicians dissatisfied with the "abominable little zeroes" of the eighteenth century and to lead toward the age of rigor in the nineteenth. Carnot's reputation today, however, depends primarily on other works. In 1801 he published *De la correlation des figures de géométrie*, again a work characterized by its high degree of generality. In it Carnot sought to establish for pure geometry a universality comparable to that enjoyed by analytic geometry. He showed that several of Euclid's theorems can be regarded as specific instances of a more inclusive theorem for which a single demonstration suffices. We find in the *Elements*, for instance, the theorem that if two chords AD and BC in a circle intersect in a point K, the product of AK by KD is equal to the product of BK by KC (Fig. 22.1). Later we run across the theorem that if KDA and KCB are secants to a circle, the product of AK by KD is equal to the product of BK by KC. These two theorems Carnot would regard merely as special cases, correlated through the use of negative quantities, of a general property of lines and circles. If we note that for the chords $CK = CB - BK$, whereas for the secants

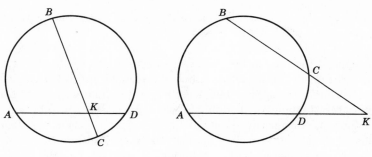

FIG. 22.1

$CK = BK - CB$, the relationship $AK \cdot KD = CK \cdot KB$ can be carried over from the one case to the other simply by a change of sign. And tangency is only another case in which B and C, say, coincide, so that $BC = 0$. Although the graphical representation of complex numbers had not yet come into general use, Carnot did not hesitate to suggest also a correlation of figures through imaginary numbers. He cited as an example the fact that the circle $y^2 = a^2 - x^2$ is related to the hyperbola $y^2 = x^2 - a^2$ through the identity $x^2 - a^2 = (\sqrt{-1})^2(a^2 - x^2)$.

13   Carnot greatly expanded his correlation of figures in his *Géométrie de position* in 1803, a book that placed him beside Monge as a founder of modern pure geometry. The development of mathematics has been characterized by a striving for ever higher and higher degrees of generality, and it is this quality that gives significance to the work of Carnot. His penchant for generalization led him to beautiful analogues of well-known theorems of plane geometry. The equivalent of the familiar law of cosines in trigonometry, $a^2 = b^2 + c^2 - 2bc \cos A$, had been known at least as far back as the days of Euclid; Carnot extended this ancient theorem to an equivalent form, $a^2 = b^2 + c^2 + d^2 - 2cd \cos B - 2bd \cos C - 2bc \cos D$, for a tetrahedron, where a, b, c, d are the areas of the four faces and B, C, D are the angles between the faces of areas c and d, b and d, and b and c respectively. The passion for generality that is found in his work has been the driving force of modern mathematics, especially in the twentieth century. Topology in particular, concerned as it is with the properties of figures that remain invariant under a continuous deformation, would delight Carnot, if he could return today, for he would recognize it as going far beyond his correlation of figures.

The *Géométrie de position* is a classic in pure geometry, but it contains also significant contributions to analysis. Although analytic geometry had completely overshadowed synthetic geometry for more than a century, its supremacy had been won in terms of two coordinate systems, rectangular and polar. In the rectangular system the coordinates of a point P in a plane are the distances of P from two mutually perpendicular lines or axes; in the polar system one of the coordinates of P is the distance of P from a fixed point O (the pole), and the other is the angle that line OP makes with a fixed line (polar axis) through O. Carnot saw that coordinate systems could be modified in many ways. For example, the coordinates of P may be the distances of P from two fixed points O and Q; or one coordinate may be the distance OP and the other the area of the triangle OPQ. In such generalizations Carnot simply rediscovered and extended a suggestion that Newton had made, but which had been generally overlooked; but Carnot's thought characteristically carried him further. In all of the cases so far considered, the equation of a curve depends on the particular coordinate frame of reference that is used; yet the properties of a curve are not bound to any one choice of pole or axes. It should be possible, Carnot reasoned, to find coordinates that do not "depend on any particular hypothesis or on any basis of comparison taken in absolute space." Thus he initiated the search for what now are known as intrinsic coordinates. One of these he found in the familiar radius of curvature of a curve at a point. For the other he introduced a quantity to which he gave no name but which since has come to be called aberrancy or angle of deviation. This is an extension of the ideas of tangency and curvature. The tangent to a curve at a point P is the limiting position of a secant line PQ as Q approaches P along the curve; the circle of curvature is the limiting position of the circle though the points P, Q, and R as Q and R approach P along the curve. If, now, one passes a parabola through points P, Q, R, and S and finds the limiting position of this parabola as the points Q, R, and S approach P along the curve, the aberrancy at P is the angle between the axis of this parabola and the normal to the curve. Aberrancy is related to the third derivative of a function in much the same sense that slope and curvature are related to the first and second derivatives, respectively; and it turns out that just as the slope of a line and the curvature of a circle are constant, so is the aberrancy of a conic section the same for all points.[12]

Carnot's name is known among mathematicians for a theorem that bears **14** his name, which appeared in 1806 in an *Essai sur le théorie des transversales*. This again is an extension of an ancient result. Menelaus of Alexandria had shown that if a straight line intersects the sides AB, BC, and CA of a triangle (or these sides extended) in points P, Q, and R respectively, and if $a' = AP$, $b' = BQ$, $c' = CR$ and $a'' = AR$, $b'' = BP$, $c'' = CQ$, then $a'b'c' = a''b''c''$ (Fig. 22.2). Carnot showed that if the straight line in the theorem of Menelaus

---

[12] For further details see C. B. Boyer, "Carnot and the Concept of Deviation," *American Mathematical Monthly*, **61** (1954), 459–463.

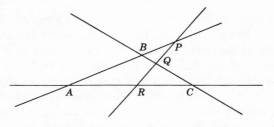

FIG. 22.2

is replaced by a curve of order n which intersects AB in the (real or imaginary) points $P_1, P_2, P_3, \ldots, P_n$, BC in the points $Q_1, Q_2, Q_3, \ldots, Q_n$, and CA in the points $R_1, R_2, R_3, \ldots, R_n$, then the theorem of Menelaus holds if one takes a' as the product of the n distances $AP_1, AP_2, AP_3, \ldots, AP_n$, with similar definitions for b' and c' and analogous definitions for a'', b'', and c'' (Fig. 22.3). The theory of transversals is only a small part of a work that contains other interesting generalizations. From the familiar formula of Heron of Alexandria for the area of a triangle in terms of its three sides, Carnot went on to a corresponding result for the volume of the tetrahedron in terms of its six edges; finally he derived a formula, comprising 130 terms, for finding the tenth of the ten segments joining five points at random in space if the other nine are known.

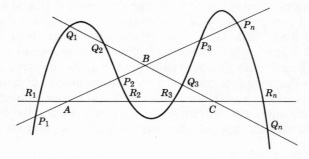

FIG. 22.3

Carnot was a soldier, a politician, a poet, and a geometer; but he was also a speculator. The failure of colonial ventures, in which he had invested far too heavily, resulted in financial ruin in 1809, at which point the emperor magnanimously granted him a position.[13]

---

[13] There are several books on the life of Carnot. See, for example, Marcel Reinhard, *Le grand Carnot* (1950–1952), and S. J. Watson, *Carnot* (1954).

Carnot was not the only one of our Revolutionary group who felt the need **15** for greater rigor in mathematics. We have mentioned the lamentable state of geometry as portrayed by Bézout's *Cours de mathématiques*. This prompted Legendre, who was, after all, primarily an analyst, to revive some of the intellectual quality of Euclid. The result was the *Éléments de géométrie* which appeared in 1794, the year of the Terror. Here, too, we see the very antithesis of what generally is regarded as practical. As Legendre says in the preface, his object is to present a geometry that shall satisfy *l'espirit*. The result of Legendre's efforts was a remarkably successful textbook—one of the mathematical products of the Revolution that had pervasive influence, for twenty editions appeared within the author's lifetime. Legendre wrote that his object was "to compose a very rigorous elements" of geometry, but he did not wax pedantic to the point of making a fetish of rigor at the expense of clarity.

Often we are inclined to think of American mathematics as influenced primarily by German scholarship, for a generation ago one went to Göttingen to be in touch with the foremost scholars in the field. We are prone to forget that during much of the nineteenth century it was French mathematics that dominated American teaching, and this was primarily through the work of the men whom we have been considering. Textbooks by Lacroix, Biot, and Lagrange were published in America for use in the schools, but perhaps the most influential of all was the geometry of Legendre. *Davies' Legendre* became almost a synonym for geometry in America. As late as 1885 Dean Van Amringe of Columbia wrote in the preface of still another edition:

It is believed that in clearness and precision of definition, in general simplicity and rigor of demonstration, in orderly and logical development of the subject, and in compactness of form, *Davies' Legendre* is superior to any work of its grade for the general training of the logical powers of pupils, and for their instruction in the great body of elementary geometric truth.

The success of Legendre's *Éléments* should not lead one to think of the **16** author as a geometer. The fields in which Legendre made significant advances were numerous, but chiefly nongeometrical—differential equations, calculus, theory of functions, theory of numbers, and applied mathematics. He composed a three-volume treatise, *Exercises du calcul intégral* (1811–1819), which rivaled that of Euler for comprehensiveness and authoritativeness; later he expanded aspects of this in another three volumes comprising the *Traité des fonctions elliptiques et des intégrales eulériennes* (1825–1832). In these important treatises, as well as in earlier memoirs, Legendre introduced the name "Eulerian integrals" for the beta and gamma functions. More importantly, he provided some basic tools of analysis, so helpful to mathematical physicists, which bear his name. Among these are the Legendre

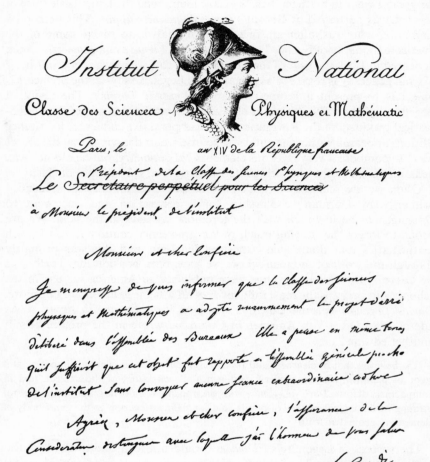

Autograph letter of Legendre. In some of his letters the form "Le Gendre" appears, as in this case. In general the name is spelled Legendre.

functions, which are solutions of the Legendre differential equation $(1 - x^2)y'' - 2xy' + n(n + 1)y = 0$. Polynomial solutions for positive integral values of n are known as Legendre polynomials.

Legendre spent much effort in reducing elliptic integrals (quadratures of the form $\int R(x, s)\, dx$, where R is a rational function and s is the square root of a polynomial of third or fourth degree) to three standard forms that have since borne his name. The elliptic integrals of first and second kind in Legendre's form are

$$F(K, \phi) = \int_0^\phi \frac{dt}{\sqrt{1 - K^2 \sin^2 t}}$$

and

$$E(K, \phi) = \int_0^\phi \sqrt{1 - K^2 \sin^2 t}\, dt$$

respectively, where $K^2 < 1$; those of the third form are somewhat more complicated. Tables of these integrals, tabulated for given K and varying values of ϕ, can be found in most comprehensive handbooks, for the integrals arise in many problems. Legendre's elliptic integral of the first kind arises naturally in solving the differential equation for the motion of a simple pendulum; that of the second kind appears in seeking the length of arc of an ellipse. Elliptic integrals arose also in Legendre's earlier memoirs, especially in one of 1785 on the gravitational attraction of an ellipsoid, a problem in connection with which there appeared what are known as zonal harmonics or "Legendre's coefficients"—functions used effectively by Laplace in potential theory.

Legendre was an important figure in geodesy, and in this connection he developed the statistical method of least squares. A simple case of the method of least squares may be described as follows. If observations have led to *three* or more approximate equations in two variables—say, $a_1x + b_1y + c_1 = 0$, $a_2x + b_2y + c_2 = 0$, and $a_3x + b_3y + c_3 = 0$—one adopts as the "best" values of x and y the solution of the *two* simultaneous equations

$$(a_1{}^2 + a_2{}^2 + a_3{}^2)x + (a_1b_1 + a_2b_2 + a_3b_3)y + (a_1c_1 + a_2c_2 + a_3c_3) = 0$$
$$(a_1b_1 + a_2b_2 + a_3b_3)x + (b_1{}^2 + b_2{}^2 + b_3{}^2)y + (b_1c_1 + b_2c_2 + b_3c_3) = 0$$

The *Memoirs* of the Institut contain also one of Legendre's attempts to **17** prove the parallel postulate, but of all his contributions to mathematics Legendre was most pleased with the works on elliptic integrals and the theory of numbers. He published a two-volume *Essai sur la théorie des nombres* (1797–1798), the first treatise to be devoted exclusively to the subject. The famous "last theorem of Fermat" attracted him, and in about 1825 he gave a proof of its unsolvability for $n = 5$. Almost equally famous is a theorem on congruences which Legendre published in the treatise of 1797–1798. If, given integers p and q, there exists an integer x such that $x^2 - q$ is divisible by p, then q is known as a quadratic residue of p; we now write (following a notation introduced by Gauss) $x^2 \equiv q \pmod{p}$, reading this as "x^2 is congruent to q modulo p." Legendre rediscovered a beautiful theorem, given earlier in less modern form by Euler, known as the law of quadratic reciprocity: if p and q are primes, then the congruences $x^2 \equiv q \pmod{p}$ and $x^2 \equiv p \pmod{q}$ are either *both* solvable or *both* unsolvable, unless both p and q are of the form $4n + 3$, in which case one is solvable and the other is not. For example, $x^2 \equiv 13 \pmod{17}$ has the solution $x = 8$, and $x \equiv 17$ (mod 13) has the solution $x = 11$; and it can be shown that $x^2 \equiv 5 \pmod{13}$ and $x^2 \equiv 13 \pmod 5$ have no solution. On the other hand, $x^2 \equiv 19 \pmod{11}$ is not solvable, whereas $x^2 = 11 \pmod{19}$ has the solution $x = 7$. The theorem is here stated in the customary modern form. In the exposition of Legendre it becomes

$$(p/q)(q/p) = (-1)^{(p-1)(q-1)/4}$$

where the Legendre symbol (p/q) denotes 1 or -1 according as $x^2 \equiv p \pmod q$ is, or is not, solvable for x.

Ever since the days of Euclid it had been known that the number of primes is infinite; yet it is obvious that the density of prime numbers decreases as we move on to ever larger integers. Hence it became one of the most famous problems to describe the distribution of primes among the natural numbers. Mathematicians were looking for a rule, known as the prime number theorem,[14] which should express the number of primes less than a given integer n as a function of n, usually written $\pi(n)$. In his well-known treatise of 1797–1798 Legendre conjectured, on the basis of a count of a large number of primes, that $\pi(n)$ approaches $n/(\ln n - 1.08366)$ as n increases indefinitely. This conjecture comes close to the truth, but a precise statement of the theorem that $\pi(n) \to n/\ln n$, suggested several times during the following century, was not proved until 1896. Legendre showed that there is no rational algebraic function that always gives primes, but he noted that $n^2 + n + 17$ is prime for all values of n from 1 to 16 and $2n^2 + 29$ is prime for values of n from 1 to 28. (Euler earlier had shown that $n^2 - n + 41$ is prime for values of n from 1 to 40.)

18   If Carnot and Legendre were disciples of clear and rigorous thought, Lagrange was the high priest of the cult. At the height of the Terror, Lagrange

---

[14] For an historical sketch of these attempts see two books by Edmund Landau, *Handbuch der Lehre von der Verteilung der Primzahlen* (Leipzig, 1909), and *Vorlesungen über Zahlentheorie* (Leipzig, 1927).

had thought seriously of leaving France; but just at this critical juncture the École Normale and the École Polytechnique were established, and Lagrange was invited to lecture on analysis. Lagrange seems to have welcomed the opportunity to teach, although it had been many years since he had done lecturing at Turin. In the interval he had been under the patronage of sovereigns, but during the Revolution he did not take sides for or against the king or the second estate. Perhaps this was the result of political apathy, possibly it was due to Lagrange's mental depression at the time.[15] At all events, his appointment to the newly established schools woke him from his lethargy. The new curriculum called for new lecture notes, and these Lagrange supplied for various levels. For students at the École Normale in 1795 he prepared and delivered lectures that today would be appropriate for a high school class in advanced algebra or for a course in college algebra; the material in these notes enjoyed a popularity that extended to America, where they were published as *Lectures on Elementary Mathematics.*[16] For scholars on the higher level of the École Polytechnique, Lagrange lectured on analysis and prepared what has ever since been regarded as a classic in mathematics. The results, in his *Théorie des fonctions analytiques*, appeared in the same year as Carnot's *Réflexions*, and together they make 1797 a banner year for the rise of rigor.

Lagrange's function theory, which developed some ideas that he had presented in a paper about twenty-five years earlier, certainly was not useful in the narrower sense, for the notation of the differential was far more expeditious and suggestive than the Lagrangian "derived function," from which our name "derivative" comes. The whole motive of the work was not to try to make the calculus more utilitarian, but to make it more logically satisfying. The key idea is easy to describe. The function $f(x) = 1/(1 - x)$, when expanded by long division, yields the infinite series $1 + 1x + 1x^2 + 1x^3 + \cdots + 1x^n \cdots$. If the coefficient of x^n is multiplied by $n!$, Lagrange called the result the value of the nth derived function of $f(x)$ for the point $x = 0$—with suitable modification for expansions of functions about other points. To this work by Lagrange we owe the commonly used notation for derivatives of various orders, $f'(x), f''(x), \ldots, f^n(x) \ldots$. Lagrange thought that through this device he had eliminated the need for limits or infinitesimals, although he continued to use the latter side by side with his derived functions; but, alas, there are flaws in his fine new scheme. Not every function can be so expanded, for there were lapses in Lagrange's putative proof of the expandability; moreover, the question of the convergence of the infinite series

[15] See George Sarton, "Lagrange's Personality", *Proceedings of the American Philosophical Society*, **88** (1944), 457–496; also E. T. Bell, *Men of Mathematics* (1937), Chapter 10.
[16] Translated by T. J. McCormack (1901). The French text is found in Vol. VII of the *Oeuvres* of Lagrange.

brings back the need for the limit concept. But the work of Lagrange during the Revolution can be said to have had a broader influence through the initiation of a new subject which has ever since been the center of attention in mathematics—the theory of functions of a real variable.

19   Lagrange generally is regarded as the keenest mathematician of the eighteenth century, only Euler being a close rival, and there are aspects of his work that are not easily described in an elementary historical survey. Among these is Lagrange's first, and perhaps his greatest, contribution—the calculus of variations. This was a new branch of mathematics, the name of which originated from notations used by Lagrange from about 1760. In its simplest form the subject seeks to determine a functional relationship $y = f(x)$ such that an integral $\int_a^b g(x, y)\,dx$ shall be a maximum or minimum. Problems of isoperimetry or of quickest descent are special cases in the calculus of variations. In 1755 Lagrange had written to Euler about the general methods that he had developed for handling problems of this type, and Euler generously held up publication of somewhat related work of his own in order that the younger man should receive full credit for the newer methods that Euler regarded as superior.[17]

From the time of his first publications in the *Miscellanea* of the Turin Academy in 1759–1761, the reputation of Lagrange was established. When in 1766 Euler and d'Alembert advised Frederick the Great on Euler's successor at the Berlin Academy, they both urged the appointment of Lagrange. Frederick then presumptuously wrote Lagrange that it was necessary that the greatest geometer of Europe should live near the greatest of kings. Lagrange assented; he remained in Berlin for twenty years, leaving only after Frederick's death, three years before the start of the French Revolution.

It was during his days at Berlin that Lagrange published important memoirs on mechanics, the three-body problem, his early ideas on derived functions, and important work on the theory of equations. In 1767 he published a memoir on the approximation of roots of polynomial equations by means of continued fractions; in another paper in 1770 he considered the solvability of equations in terms of permutations on their roots. It was the latter work that was to lead to the enormously successful theory of groups and to the proofs by Galois and Abel of the unsolvability, in the usual terms, of equations of degree greater than four. The name of Lagrange is today attached to what is perhaps the most important theorem of group theory—if o is the order of a subgroup g of a group G of order O, then o is a factor of O. Finding that the resolvent of a quintic equation, far from being of degree less

[17] See C. Carathéodory, "The Beginning of Research in the Calculus of Variations," *Osiris*, **3** (1938), 224–240; also Robert Woodhouse, *A History of the Calculus of Variations in the Eighteenth Century* (reprinted, New York: Chelsea, n.d.).

than five, as one should have expected, was a sextic, Lagrange conjectured that polynomial equations above fourth degree are not solvable in the usual sense.

Ever on the lookout for generality and elegance in the treatment of **20** problems, Lagrange was responsible for the method of variation of parameters in the solution of nonhomogeneous linear differential equations. That is, if $c_1u_1 + c_2u_2$ is a general solution of $y'' + a_1y' + a_2y = 0$ (where u_1 and u_2 are functions of x), he replaced the parameters c_1 and c_2 by undetermined variables v_1 and v_2 (functions of x) and determined the latter so that $v_1u_1 + v_2u_2$ should be a solution of $y'' + a_1y' + a_2y = f(x)$. In the determination of maxima and minima of a function such as $f(x, y, z, w)$ subject to constraints $g(x, y, z, w) = 0$ and $h(x, y, z, w) = 0$, he suggested the use of Lagrange multipliers to provide an elegant and symmetric algorithm. Under this method one introduces two undetermined constants λ and μ, forms the function $F \equiv f + \lambda g + \mu h$, from the six equations $F_x = 0$, $F_y = 0$, $F_z = 0$, $F_w = 0$, $g = 0$, and $h = 0$ eliminates the multipliers λ and μ, and solves for the desired values of x, y, z, and w.

Like so many of the leading modern mathematicians, Lagrange had a deep interest in the theory of numbers. Although he did not use the language of congruences, Lagrange showed, in 1768, the equivalent of the statement that for a prime modulus p the congruence $f(x) \equiv 0$ can have not more than n distinct solutions, where n is the degree (except for the trivial case in which all coefficients of $f(x)$ are divisible by p). Two years later he published a demonstration of the theorem, for which Fermat claimed to have had a proof, that every positive integer is the sum of at most four perfect squares; hence this theorem often is known as Lagrange's four-square theorem. At the same time he gave also the first proof of a result known as Wilson's theorem, which had appeared in Waring's *Meditationes algebraicae* of the same year[18]—for any prime p, the integer $(p - 1)! + 1$ is divisible by p. Lagrange contributed also to the theory of probability, but in this branch he took second place to Laplace, who was younger.

We have said little so far about Laplace, who in his day was regarded, as **21** a mathematician, as highly as Lagrange. There are two reasons for the relative neglect. First, Laplace took virtually no part in revolutionary activities. He seems to have had a strong sense of intellectual honesty in science, but in politics he was without convictions. This does not mean that he was timid, for he seems to have associated freely with those of his scientific colleagues who were suspect during the period of crisis. It has been said that he too would

[18] For the history of this and other theorems in number theory, see L. E. Dickson, *History of the Theory of Numbers* (1919–1923).

have been in danger of the guillotine except for his contributions to science, but this statement seems to be questionable, since he often appeared as a brazen opportunist. He played a role in the Committee on Weights and Measures, but this was not of great significance. He naturally was a professor at the École Normale and the École Polytechnique but, unlike Monge and Lagrange, he did not publish lecture notes. His publications were primarily on celestial mechanics, in which he stands preeminent in the period since Newton. Laplace did have one fling at political administration some years later, when Napoleon, a great admirer of men of science, appointed him Minister of the Interior—a post that Carnot also had held for a while under Napoleon. But it is well known that Laplace, unlike Carnot, showed no aptitude for the office, and Napoleon quipped that he "carried the spirit of the infinitely small into the management of affairs." A second reason for our failure to emphasize the work of Laplace is that it did not have the immediate and persistent influence that can be traced to others in our group. His compilations represent in a sense the end of an era rather than the beginning of a new period—although we must make an exception in the case of his work in probability and potential theory.

The theory of probability owes more to Laplace than to any other mathematician. From 1774 on he wrote many memoirs on the subject, the results of which he embodied in the classic *Théorie analytique des probabilités* of 1812. He considered the theory from all aspects and at all levels,[19] and his *Essai philosophique des probabilités* of 1814 is an introductory account for the general reader. Laplace wrote that "at the bottom the theory of probabilities is only common sense expressed in numbers"; but his *Théorie analytique* shows the hand of a master analyst who knows his advanced calculus. It is replete with integrals involving beta and gamma functions; and Laplace was among the earliest to show that $\int_{-\infty}^{\infty} e^{-x^2}\, dx$, the area under the probability curve, is $\sqrt{\pi}$. The method by which he achieved this result was somewhat artificial, but it is not far removed from the modern device of transforming

$$\int_0^\infty e^{-x^2}\, dx \cdot \int_0^\infty e^{-y^2}\, dy = \int_0^\infty \int_0^\infty e^{-(x^2 + y^2)}\, dx\, dy$$

to polar coordinates as

$$\int_0^\pi \int_0^\infty re^{-r^2}\, dr\, d\theta$$

[19] By far the longest chapter of Isaac Todhunter, *A History of the Mathematical Theory of Probability* (1949), is devoted exclusively to the work of Laplace.

which is easily evaluated and leads to

$$\int_0^\infty e^{-x^2}\,dx = \frac{\sqrt{\pi}}{2}$$

Among the many things to which Laplace called attention in his *Théorie analytique* was the calculation of π through Buffon's needle problem which had been all but forgotten for thirty-five years. This sometimes is known as the Buffon–Laplace needle problem,[20] inasmuch as Laplace extended the original problem to a crisscross of two mutually perpendicular sets of equidistant parallel lines. If the distances are a and b, the probability that a needle of length l (less than a and b) will fall on one of the lines is

$$p = \frac{2l(a+b) - l^2}{\pi ab}$$

Laplace also rescued from oblivion the work of the Rev. Thomas Bayes (†1761) on inverse probability. We find in the book also the theory of least squares, invented by Legendre, together with a formal proof that Legendre had failed to give. The *Théorie analytique* also contains the Laplace transform which is so useful in differential equations. If $f(x) = \int_0^\infty e^{-xt} g(t)\,dt$, the function $f(x)$ is said to be the Laplace transform of the function $g(x)$.

The works of Laplace involve a considerable application of higher **22** mathematical analysis. Typical was his study of the conditions for the equilibrium of a rotating fluid mass, a subject that he had considered in connection with the nebular hypothesis of the origin of the solar system. The hypothesis had been presented in a popular form in 1796 in *Exposition du système du monde*, a book that bears the same relation to the *Mécanique céleste* (1799–1825, 5 vols.) as does the *Essai philosophique des probabilités* to the *Théorie analytique*. According to the theory of Laplace the solar system evolved from an incandescent gas rotating about an axis. As it cooled, the gas contracted, causing ever more rapid rotation, according to the conservation of angular momentum, until successive rings broke off from the outer edge to condense and form planets. The rotating sun constitutes the remaining central core of the nebula. The idea behind this hypothesis was not entirely original with Laplace, for it had been proposed in qualitative skeletal form by Thomas Wright and Immanuel Kant, but the quantitative fleshing out of

[20] For a good account see N. T. Gridgman, "Geometric Probability and the Number π," *Scripta Mathematica*, **25** (1960), 183–195.

the theory forms part of the multivolume *Mécanique céleste*. It is in this classic also that we find, in connection with the attraction of a spheroid on a particle, the Laplacian use of the idea of potential and the Laplace equation. In a highly technical paper of 1782 on "Théorie des attractions des sphéroïdes et de la figure des planètes," included also in the *Mécanique céleste*, Laplace developed the very useful concept of potential—a function whose directional derivative at every point is equal to the component of the field intensity in the given direction. Also of fundamental importance in astronomy and mathematical physics is the so-called Laplacian of a function $u = f(x, y, z)$. This is simply the sum of the second-order partial derivatives of u—$u_{xx} + u_{yy} + u_{zz}$—often abbreviated $\nabla^2 u$ (read "del-squared of u) where ∇^2 is called Laplace's operator. The function $\nabla^2 u$ is independent of the particular coordinate system that is used; under certain conditions gravitational, electrical, and other potentials satisfy the Laplace equation $u_{xx} + u_{yy} + u_{zz} = 0$. Euler had run across this equation somewhat incidentally in 1752 in studies on hydrodynamics, but Laplace made it a standard part of mathematical physics.

The publication of the *Mécanique céleste* of Laplace commonly is regarded as marking the culmination of the Newtonian view of gravitation. Accounting for all the perturbations in the solar system, Laplace showed the motions to be secular, so that the system could be regarded as stable. There no longer appeared to be any need for occasional divine intervention. Napoleon is said to have commented to Laplace on the latter's failure to mention God in his monumental work, whereupon Laplace is reported to have replied, "I have no need for that hypothesis." Lagrange, being told about this, is quoted as saying, "Ah, but it is a beautiful hypothesis."

Laplace completed not only the gravitational portion of Newton's *Principia*, but also some points in the physics. Newton had computed a velocity of sound on purely theoretical grounds, only to find that the calculation resulted in too small a value for the speed. Laplace in 1816 was the first one to point out that the lack of agreement between calculated and observed speeds was due to the fact that the computations in the *Principia* were based on the assumption of isothermal compressions and expansions, whereas in reality the oscillations for sound are so rapid that compressions are adiabatic, thereby increasing the elasticity of the air and the speed of sound.

The minds of Laplace and Lagrange, the two leading mathematicians of the Revolution, were in many ways direct opposites. For Laplace nature was the essence, and mathematics was only a kit of tools that he handled with extraordinary skill; for Lagrange mathematics was a sublime art that was its own excuse for being. The mathematics of the *Mécanique céleste* has often been described as difficult, but no one calls it beautiful; the *Mécanique analytique*, on the other hand, has been admired as "a scientific poem" in the perfection and grandeur of its structure.

This chapter should, in a sense, close with the date 1799, for at that time **23** Bonaparte seized power and one can regard the period of the Revolution as ended. However, under Napolean the favorable conditions for the growth of mathematics persisted. Moreover, this date was far from the end of the activity of the five survivors in our group, for every one of them continued, as we have seen, to make contributions to mathematics, and some to politics as well. Honors came to all of them, Monge, Carnot, and Lagrange being named counts of the empire and Laplace achieving the title of marquis. Of our group of six, only Legendre seems never to have borne a title. Mathematically the chapter has a happy ending, for our scholars were able to continue their work until the end. Politically, however, two were to suffer defeat. Carnot and Monge had strong political convictions, and both of them had voted for the death of Louis XVI. Carnot, the more consistent of the two, ever was opposed to dictators, and in 1804 he was the only Tribune with sufficient courage and conviction to vote against naming Napoleon emperor. Yet later, when he felt that the welfare of France demanded, Carnot willingly served under Napoleon, both in the army and in governmental administration. Monge, on the other hand, supported his idol from the revolutionary corporal to the despotic emperor. He and Fourier accompanied Bonaparte on the Italian and Egyptian campaigns, and it was Monge who executed the delicate task of determining what works of art were to be brought back to Paris as war booty.

Following the restoration of the French monarchy, Carnot was forced to seek exile in Magdeburg, and Monge was banished and stripped of all his honors, including his place in the École Polytechnique and Institut National. The turn in events was accepted courageously by Carnot, who continued his scholarly activities, but it broke the spirit of Monge, who died shortly afterward. Lagrange had died a few years before the Napoleonic crisis. Legendre seems to have remained politically neutral throughout the changes, for he was shy and retiring; but he produced a steady stream of publications on elliptic integrals and the theory of numbers, as well as contributions to other parts of mathematics. Toward the end of his life he, too, suffered politically. Because he resisted the move of the government to dictate to the Académie des Sciences, he was deprived of his pension. Laplace, on the other hand, made peace with each regime as it came along, including in editions of his works glowing tributes to whichever side happened to be in power. Posterity, as a result, has admired Laplace for his mathematics while disdaining his political maneuvering.

It is now more than a century and a half since the days of which we have been speaking, and we can look back on the period dispassionately. One lesson that can be drawn from the survey is that the things that really count in mathematics, and have lasting influence, are not those that immediate practicality dictates. Even in times of crisis it is things of the "spirit" (in the French sense) that count most, and this spirit is perhaps best imparted by great teachers. But perhaps more important than this is the moral that, like Carnot, one should never lose heart, no matter how disillusioning the political or intellectual outlook may be.

BIBLIOGRAPHY

Alengry, Franck, *Condorcet: Guide de la révolution française* (Paris: Giard & E. Brière, 1904).

Arago, François, "Biographies of Distinguished Scientific Men" (on Laplace), *Smithsonian Institution, Annual Report*, 1874, pp. 129–168.

Bell, E. T., *Men of Mathematics* (New York: Simon and Schuster, 1937).

Boyer, C. B., "Carnot and the Concept of Deviation," *American Mathematical Monthly*, **61** (1954), 459–463.

Boyer, C. B., "The Great Carnot," *The Mathematics Teacher*, **49** (1956), 7–14.

Burlingame, Anne E., *Condorcet, the Torch Bearer of the French Revolution* (Boston: Stratford, 1930).

Cajori, Florian, *A History of Mathematics*, rev. ed. (New York: Macmillan, 1919).

Cantor, Moritz, *Vorlesungen über Geschichte der Mathematik* (Leipzig: B. G. Teubner, 1880–1908, 4 vols.).

Carnot, Hippolyte, *Mémoires sur Carnot par son fils* (Paris, 1861–1863, 2 vols.).

Carnot, L. N. M., "Reflections on the Theory of the Infinitesimal Calculus," trans. by W. Dickson, *Philosophical Magazine*, **8** (1800), 222–240, 335–352; **9** (1801), 39–56.

Carnot, L. N. M., *Reflexions on the Metaphysical Principles of the Infinitesimal Analysis*, trans. by W. R. Browell (Oxford, 1832).

Chasles, Michel, *Aperçu historique sur l'origine et de développement des méthodes en géométrie* (Bruxelles, 1837; 2nd ed., Paris, 1875).

Condorcet, M. J. A. N. C. de, *Oeuvres* (Paris, 1847–1849, 12 vols.).

Condorcet, M. J. A. N. C. de, *Sketch for a Historical Picture of the Progress of the Human Mind*, trans. by June Barraclough (New York: Noonday Press, 1955).

Dickson, L. E., *History of the Theory of Numbers* (Washington, D.C.: Carnegie Institution, 1919–1923, 3 vols.).

École Polytechnique, *Livre du centenaire, 1794–1894* (Paris, 1894–1897, 3 vols.).

École Normale, *Le centenaire de l'École Normale, 1795–1895* (Paris, 1895).

Granger, G. G., *La mathématique sociale du Marquis de Condorcet* (Paris: Presses Universitaires de France, 1956).

Lagrange, J. L., *Oeuvres* (Paris, 1867–1892, 14 vols.).

Lagrange, J. L., *Lectures on Elementary Mathematics*, trans. by T. J. McCormack, 2nd ed. (Chicago; Open Court, 1901).

Laplace, P. S., *The System of the World*, trans. by H. H. Harte (London: Longmans-Green, 1830, 2 vols.).

Laplace, P. S., *Oeuvres complètes* (Paris: Gauthier-Villars, 1878–1912, 14 vols. in 15).

Laplace, P. S., *A Philosophical Essay on Probabilities* (New York: Wiley, 1902).

Legendre, A. M., *Essai sur la théorie des nombres*, 2nd ed. (Paris: Courcier, 1808).

Legendre, A. M., *Elements of Geometry and Trigonometry*, trans. by David Brewster, revised by Charles Davies (New York: A. S. Barnes, 1851).

Loria, Gino, *Storia della geometria descrittiva* (Milan: Hoepli, 1921).

Loria, Gino, *Storia delle matematiche* (Turin: Sten, 1929–1933, 3 vols.).

Muir, T. P., *The Theory of Determinants in the Historical Order of Development* (London, 1906–1923, 4 vols.; reprinted, New York: Dover, 1960).

Nielsen, Niels, *Géomètres français sous la révolution* (Copenhagen: Levin & Munksgaard, 1929).

Reinhard, Marcel, *Le grand Carnot* (Paris: Hachette, 1950–1952, 2 vols.).

Roever, W. H., *The Mongean Method of Descriptive Geometry* (New York: Macmillan, 1933).

Sarton, George, "Lagrange's Personality," *Proceedings of the American Philosophical Society*, **88** (1944), 457–496.

Smith, D. E., *A Source Book in Mathematics* (New York: McGraw-Hill, 1929).

Smith, D. E., *The Poetry of Mathematics and Other Essays* (New York: Scripta Mathematic, 1934).

Taton, René, *L'oeuvre scientifique de Monge* (Paris: Presses Universitaires de France, 1951).

Todnunter, Isaac, *A History of the Mathematical Theory of Probability from the Time of Pascal to that of Laplace* (New York: Chelsea reprint, 1949).

Watson, S. J., *Carnot* (London: Bodley Head, 1954).

EXERCISES

1. Did the exigencies of the French Revolution lead to increased emphasis on applied, rather than pure, mathematics? Explain, citing specific instances.

2. What were the chief means of support of French mathematicians before the Revolution? After the Revolution? Explain, citing specific cases.

3. Mention three outstanding mathematicians in France who supported the Revolution and describe their activities in this connection.

4. What alternatives did the Revolutionary Committee on Weights and Measures consider for determining the length of a meter? What well-known mathematicians served on the Committee?

5. Mention five families of the seventeenth and eighteenth centuries, each of which produced several mathematicians. In the case of each of three of these families describe the chief contributions of the most distinguished member of the family.

6. Describe the book that, of those published during the French Revolution, exerted the greatest influence in America.

7. Prove the two theorems of elementary geometry associated with the diagrams in Fig. 22.1.

8. Prove that the area of a triangle with vertices (x_1, y_1), (x_2, y_2), and (x_3, y_3) is equal to

$$\frac{1}{2} \begin{vmatrix} x_1 & y_1 & 1 \\ x_2 & y_2 & 1 \\ x_3 & y_3 & 1 \end{vmatrix}$$

9. Using the method of Monge's descriptive geometry, draw the "elevation" and the "plan" of the line segment from the point (1, 2, 3) to the point (4, 1, 7). Find the lengths of the segment and of each of the projections.

10. Show, using the parametric form $x = a\cos\theta$, $y = b\sin\theta$, that the circumference of the ellipse is given by an elliptic integral.

11. Using long division, expand the function $1/(1 + 2x)$ in a Taylor's series to five terms; hence, without differentiating, find the fourth derivative of this function for $x = 0$.

12. Verify Legendre's assertion that $n^2 + n + 17$ is prime for positive integral values of n less than 17.

13. Use the method of least squares to find the "best" values of x and y satisfying the system

$$x - y + 1 = 0$$

$$2x - y = 0$$

$$3x - 2y - 2 = 0$$

14. Find $\pi(n)$ for $n = 100$ and compare the result with $n/(\ln n - 1)$.

15. Verify for $p = 11$ that $(p - 1)! + 1$ is divisible by p if p is prime.

16. Show that the integral

$$\int_0^a \frac{dx}{\sqrt{(1 - x^2)(1 - K^2 x^2)}}$$

is reduced to one of Legendre's forms through the substitution $x = \sin t$.

17. Show that the probability formula for Buffon's needle problem (Chapter 21) can be derived from that of Laplace (see text) by letting one of the distances, say b, increase indefinitely.

*18. Prove that the centroid of a tetrahedron is the point of concurrency of the lines joining the midpoints of opposite edges.

*19. Prove the following theorem of Monge: planes drawn through the midpoints of the edges of a tetrahedron perpendicular to the opposite edges meet at a point which is the midpoint of the segment joining the centroid of the tetrahedron to the circumcenter of the tetrahedron.

*20. Find the equation of the plane through the point (1, 2, 3) which is perpendicular to the line $x + 2y - 2z - 6 = 0 = 2x - y + 2z$.

*21. Find the perpendicular distance from the point to the line in the preceding question.

*22. Prove the formula of Lagrange

$$D = \frac{ap + bq + cr - d}{\sqrt{a^2 + b^2 + c^2}}$$

for the perpendicular distance from the point (p, q, r) to the plane $ax + by + cz = d$.

*23. Prove that the sum of the squares of the projections of a triangle upon three mutually perpendicular planes is equal to the square of the area of the triangle.

*24. Prove that the volume of a tetrahedron with vertices (x_1, y_1, z_1), (x_2, y_2, z_2), (x_3, y_3, z_3), and (x_4, y_4, z_4) is equal to

$$\frac{1}{6}\begin{vmatrix} x_1 & y_1 & z_1 & 1 \\ x_2 & y_2 & z_2 & 1 \\ x_3 & y_3 & z_3 & 1 \\ x_4 & y_4 & z_4 & 1 \end{vmatrix}$$

*25. Complete the details in the proof that

$$\int_{-\infty}^{\infty} e^{-x^2}\, dx = \sqrt{\pi}$$

*26. Use the method of Lagrange multipliers to prove that the largest rectangular parallelepiped with a given surface area is a cube.

*27. Use Lagrange's method of variation of parameters to solve the equation $y'' + a^2 y = \sec x$, knowing that $\sin ax$ and $\cos ax$ are solutions of the reduced equation.

CHAPTER XXIII

The Time
of Gauss and Cauchy

> Mathematics is the queen of the sciences and number
> theory the queen of mathematics.
>
> ⨍   *Gauss*

1 The leading mathematicians during the French Revolution had been, almost without exception, French, but with the beginning of the nineteenth century France again had to share honors with other lands. The greatest mathematician of the time—perhaps of all times—was so German that he never left Germany, even on a visit. Carl Friedrich Gauss (1777–1855), unlike any of the men discussed in the preceding chapter, was an infant prodigy. His father was a hardworking Brunswick laborer, stubborn in his views, who tried to keep the son from receiving an appropriate education; but his mother, uneducated herself, encouraged her son in his studies and took great pride in his achievement until her death at the age of ninety-seven. As a youth Carl attended the local school, where the teacher was known as a taskmaster. One day, in order to keep the class occupied, the teacher had the students add up all the numbers from one to a hundred, with instructions that each should place his slate on a table as soon as he had completed the task. Almost immediately Carl placed his slate on the table, saying, "There it is"; the teacher looked at him scornfully while the others worked diligently. When the instructor finally looked at the results, the slate of Gauss was the only one to have the correct answer, 5050, with no further calculation. The ten-year-old boy evidently had computed mentally the sum of the arithmetic progression $1 + 2 + 3 + \cdots + 99 + 100$, presumably through the formula $m(m + 1)/2$. At fifteen Gauss attended college in Brunswick, through the help of the Duke of Brunswick, and in 1795, with the Duke's continuing assistance, he entered Göttingen. He was undecided at the time whether to become a philologist or a mathematician, although he already had thought up and justified the method of least squares—a decade before Legendre's publication of the device. On March 30, 1796, he made up his mind in favor of mathematics, for on that day, when he was still a month short of being

544

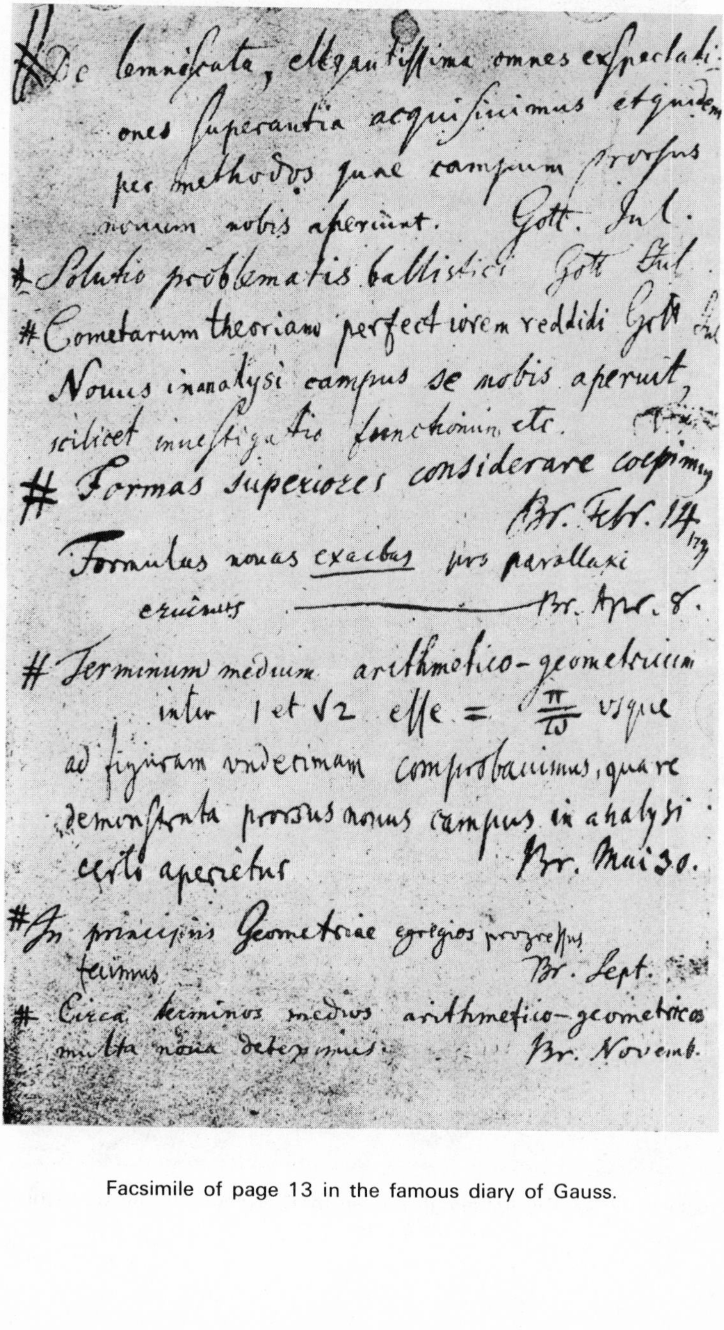

Facsimile of page 13 in the famous diary of Gauss.

Carl Friedrich Gauss. The full-length portrait by R. Wimmer in the Deutsches Museum, Munich (1925).

nineteen years old, he made a brilliant discovery. For more than 2000 years men had known how to construct, with compasses and straightedge, the equilateral triangle and the regular pentagon (as well as certain other regular polygons, the numbers of whose sides are multiples of two, three, and five), but no other polygon with a prime number of sides. On the critical day in 1796, Gauss constructed according to Euclidean rules the regular polygon of seventeen sides. On the same day he began to keep a diary in which he recorded some of his greatest discoveries, the first entry being that on the seventeen-sided regular polygon.

On July 10, 1796, Gauss confided to his diary the discovery that every integer is the sum of at most three triangular numbers. This diary of only nineteen pages is perhaps the most precious document in the whole history of mathematics, for in it are recorded 146 brief statements of results, the last being dated July 9, 1814. Through the diary it is possible to verify Gauss's earlier discovery in cases of disputed priority, as well as to trace the development of his genius, for some of his most original thoughts were never published during his lifetime. (His reluctance to publish was matched only by that of his modern rival in mathematical fame—Isaac Newton.) The diary, forming a little booklet of nineteen pages, had remained hidden among family papers until 1898, when it was found in the possession of a grandson in Hamlin. In 1901 its contents were published by the mathematician Felix Klein in a volume celebrating the sesquicentennial of the Göttingen Scientific Society,[1] and the grandson agreed to have the diary preserved among the Gauss archives, which are chiefly at Brunswick and Göttingen.

Gauss's seal bore the motto, *pauca sed matura*—"few, but ripe"—and his mind so teemed with original ideas that he did not have time to see that all of them ripened to the point of perfection on which he insisted before publication. His decisive discovery of March 30, 1796, however, he announced publicly in a literary journal; so proud of it did he remain that, in the manner of Archimedes, an ancient rival in greatness, he expressed the wish that a seventeen-sided regular polygon be carved on his tombstone. This wish was never carried out because the stonemason insisted that the resulting figure would be indistinguishable from a circle; but on a monument to Gauss in Brunswick the egregious polygon is indeed carved.[2]

2

[1] Felix Klein, "Gauss' wissenschaftliches Tagebuch 1796–1814," *Festschrift zur Feier des Hundertfünfzigjährigen Bestehens der Königlichen Gesellschaft der Wissenschaften zu Göttingen* (Berlin, 1901), pp. 1–44. The content appeared again in 1903 in *Mathematische Annalen*, **57** (1903). 1–34, and, with extensive notes, in Gauss, *Werke* (1917), Vol. X, pp. 483–574. There is a French translation by P. Eymard and J. P. Lafon, "Le journal mathématique de Gauss," *Revue d'Histoire des Sciences et de leurs Applications*, **9** (1956), 21–51.

[2] Details on his life, more than on his work, are available in G. Waldo Dunnington, *Carl Friedrich Gauss, Titan of Science; A Study of His Life and Work* (1955). A less substantial but eminently readable account is available in William L. Schaaf, *Carl Friedrich Gauss* (1964).

During short periods Gauss left Göttingen for the University of Helmstädt, and it is from the latter institution that he received his doctorate in 1798. The thesis, published at Helmstädt in 1799, bears in Latin the ponderous title, "New Demonstration of the Theorem That Every Rational Integral Algebraic Function in One Variable Can Be Resolved into Real Factors of First or Second Degree." This statement, which Gauss later referred to as "the fundamental theorem of algebra," is essentially the proposition known in France as d'Alembert's theorem; but Gauss showed that all previously attempted demonstrations, including some by Euler and Lagrange, were inadequate. The graphical representation of complex numbers already had been discovered in 1797 by Caspar Wessel (1745–1818) and published in the transactions of the Danish academy for 1798; but the work of Wessel went virtually unnoticed, hence the plane of complex numbers today usually is referred to as the Gaussian plane, even though Gauss did not publish his views until some thirty years later. Ever since the days of Girard it had been generally known that the real numbers—positive, negative, and zero—can be pictured as corresponding to points on a straight line. Wallis had even suggested that pure imaginary numbers should be represented by a line perpendicular to the axis of real numbers. Oddly enough, however, no one before Wessel and Gauss took the obvious step of thinking of the real and imaginary parts of a complex number $a + bi$ as rectangular coordinates of points in a plane. Taking this simple step made mathematicians feel much more comfortable about imaginary numbers, for these now could be visualized in the sense that every point in the plane corresponds to a complex number, and vice versa. Seeing is believing, and the old ideas about the nonexistence of imaginary numbers were generally abandoned.[3]

3   The doctoral thesis of Gauss proved that every polynomial equation $f(x) = 0$ has at least one root, whether the coefficients are real or imaginary. We cannot here go into the details of the proof, but an illustration will at least indicate the lines of his thought. We shall solve the equation $z^2 - 4i = 0$ graphically, showing that there is a complex value of $z = a + bi$ which will satisfy the equation. Replacing z by $a + bi$ and separating real and imaginary parts in the equation, we have $a^2 - b^2 = 0$ and $ab - 2 = 0$. Interpreting a and b as variable quantities and sketching these equations on the same set of axes, one for the real part a, the other for the imaginary part b, we have two curves; one consists of the lines $a + b = 0$ and $a - b = 0$, the other of the rectangular hyperbola $ab = +2$ (Fig. 23.1). It is clear that the curves have a point of intersection P in the first quadrant (and, incidentally, another

---

[3] See Ernest Nagel, "Impossible Numbers," *Studies in the History of Ideas*, (ed. by Department of Philosophy, Columbia University), **3** (1935), 427–474; and Hermann Hankel, *Theorie der complexen Zahlensysteme* (Leipzig, 1867).

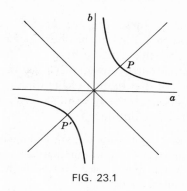

FIG. 23.1

P' in the third). We should note in particular that one branch of the first curve moves away from the origin along the directions $\theta = 1\pi/4$ and $\theta = 3\pi/4$ and that a branch of the second curve moves asymptotically toward the directions $\theta = 0\pi/4$ and $\theta = 2\pi/4$; the point of intersection lies between the last two directions, $\theta = 0$ and $\theta = \pi/2$. The a and b coordinates of this point of intersection are the real and imaginary parts of the complex number which is a solution of the equation $z^2 - 4i = 0$. Had our original polynomial equation been of third degree, instead of second degree, there would have been a branch of one curve approaching the directions $\theta = 1\pi/6$ and $\theta = 3\pi/6$ and the other curve would have been approaching the directions $\theta = 0\pi/6$ and $\theta = 2\pi/6$. The branches are in each case continuous; hence they are bound to intersect somewhere in the interval from θ to 0 to $\theta = \pi/3$. For an equation of degree n there will be a branch of one curve having asymptotic directions $\theta = 1\pi/2n$ and $\theta = 3\pi/2n$, while a branch of the other curve will have asymptotic directions $\theta = 0\pi/2n$ and $\theta = 2\pi/2n$; these branches necessarily intersect in the interval from $\theta = 0$ to $\theta = \pi/n$, and the a and b coordinates of the point of intersection will be the real and imaginary parts of the complex number satisfying the equation. Hence we see that, no matter what the degree of a polynomial equation, it is bound to have at least one complex root. From this result we can easily demonstrate the thesis of Gauss that the polynomial can be factored into real linear and quadratic factors.

The proof of the fundamental theorem of algebra given by Gauss in his thesis is based in part on geometrical considerations. Years after, in 1816, Gauss published two new demonstrations, as well as another in 1850, striving to find a proof that is entirely algebraic.[4]

[4] The proofs are collected in Vol. III of his *Werke* (1870–1930). An English translation of the second proof is included in the *Source Book in Mathematics*, ed. by David Eugene Smith (1958), pp. 292–306. It is now generally believed that the fundamental theorem of algebra depends essentially on topological considerations. See Hans Zassenhaus, "On the Fundamental Theorem of Algebra," *American Mathematical Monthly*, **74** (1967), 485–497.

4   Only two years after his thesis had appeared, Gauss published his best known book, a treatise in Latin on the theory of numbers with the title *Disquisitiones arithmeticae*, dedicated to his patron, the Duke of Brunswick. This celebrated work[5] is chiefly responsible for the development of the language and notations of that branch of the theory of numbers known as the algebra of congruences, an early instance of equivalence classes. The exposition opens with the definition:

> If a number a divides the difference between two numbers b and c, then b and c are said to be congruent, otherwise incongruent; and a itself is called the modulus. Either of the numbers is called a residue of the other, in the former case, a nonresidue in the latter case.

The notation Gauss adopted was that in use today—$b \equiv c \pmod{a}$—and he proceeded to build an algebra for the relationship denoted by $\equiv$ similar to the familiar common algebra expressed in the language of equality. Some but not all of the rules of ordinary algebra can be carried over into the new algebra. For example, in ordinary algebra if $ax = ay$, where $a \neq 0$, then $x = y$. That this cancellation law does not hold for congruences is clear from an example: if $a = 3$, $x = 4$, and $y = 7$, it is indeed true that $3 \cdot 4 \equiv 3 \cdot 7 \pmod 9$, but it is not true that $4 \equiv 7 \pmod 9$. The divisor a, in the cancellation law, must be relatively prime to the modulus in order to have the law apply to congruences. Then, too, if $x \cdot y = 0$, we can conclude in ordinary algebra that either x or y (or both) must be zero. That this is not the case for congruences becomes apparent from the fact that $6 \cdot 5 \equiv 0 \pmod{15}$, but neither 6 nor 5 is congruent to zero (mod 15). For the rule to hold in the case of congruences, the modulus and the integers x and y must have no common factor. Again, in ordinary algebra the linear equation $ax = b$ (where a, b, and x are integers and $a \neq 0$) can have not more than one solution. The linear congruence, on the other hand, can have several distinct solutions, as we see from the fact that $x = 1$, $x = 4$, and $x = 7$ are solutions of the congruence $6x \equiv 15 \pmod 9$. Only if a and m are relatively prime can we be sure that $ax \equiv b \pmod m$ will have one and only one solution smaller than m. On the other hand, the relationship $\equiv$ has the same three simple properties of reflexivity, symmetry, and transitivity as does the relationship $=$: (1) $a \equiv a \pmod m$; (2) if $a \equiv b \pmod m$, then $b \equiv a \pmod m$; and (3) if $a \equiv b \pmod m$ and $b \equiv c \pmod m$, then $a \equiv c \pmod m$. That is, both $=$ and $\equiv$ are equivalence relations.

5   At several points the work of Gauss overlapped with that of Legendre, and the latter came to have a jealous dislike of the younger and more brilliant

[5] An English translation by A. A. Clarke appeared in 1966 (Yale University Press). A French translation had appeared more promptly at Paris in 1807.

man. In the *Disquisitiones*, for example, appears the law on quadratic reciprocity which Legendre had published a couple of years earlier. Gauss called this the *theorema aureum*, or the gem of arithmetic. In later work Gauss sought to find comparable theorems for congruences $x^n \equiv p \pmod q$ for $n = 3$ and 4; but for these cases he found it necessary to extend the meaning of the word integer to include the so-called Gaussian integers—that is, numbers of the form $a + bi$, where a and b are integers. The Gaussian integers form an integral domain like that of the real integers, but more general. Problems in divisibility become more complicated, for 5 no longer is a prime, being factorable into the product of the two "primes" $1 + 2i$ and $1 - 2i$. In fact, no real prime of the form $4n + 1$ is a "Gaussian prime," whereas real primes of the form $4n - 1$ remain primes in the generalized sense. In the *Disquisitiones* Gauss included the Fundamental Theorem of Arithmetic, one of the basic principles that continues to hold in the integral domain of Gaussian integers. In fact, any integral domain for which factoring is unique is known today as a Gaussian integral domain. One of the contributions of the *Disquisitiones* was a rigorous proof of the theorem, known since the days of Euclid, that any positive integer can be represented in one and only one way (except for the order of the factors) as a product of primes.

Not all that Gauss discovered about prime numbers is contained in the *Disquisitiones*. On the back page of a copy of a table of logarithms that he had obtained as a boy of fourteen is written cryptically in German:

$$\text{Primzahlen unter } a\ (= \infty)\ \frac{a}{1a}$$

This is a statement of the celebrated prime number theorem—the number of primes less than a given integer a approaches asymptotically the quotient $a/\ln a$ as a increases indefinitely.[6]

Legendre had come close to anticipating this theorem, as we have seen; but the odd thing is that if Gauss wrote this, as we presume he did, he kept this beautiful result to himself. We do not know whether or not he had a proof of the theorem, or even when the statement was written. The distribution of primes has had a fascination for mathematicians. In 1845, when Gauss was an old man, a Parisian professor, Joseph L. F. Bertrand (1822-1900), guessed that if $n > 3$, there always is at least one prime between n and $2n$ (or, more precisely, $2n-2$) inclusive. This conjecture, known as Bertrand's postulate, was proved in 1850 by Pafnuti Tchebycheff (or Chebychev or Chebichev or Tschebytschew) of the University of St. Petersburg. Tchebycheff was a rival of Lobachevsky as the leading Russian mathematician of his day and became a foreign associate of the Institut de France and of the Royal

[6] See Hans Reichart, ed., *C. F. Gauss. Leben und Werk* (1960).

Society of London. Tchebycheff, evidently unaware of Gauss' work on primes, was able to show that if $\pi(n)(\ln n)/n$ approaches a limit as n increases indefinitely, this limit must be one; but he could not demonstrate the existence of a limit. Not until two years after Tchebycheff's death was a proof generally known; then, in 1896, two mathematicians, working independently, came up with demonstrations in the same year. One was the Belgian mathematician C. J. de la Vallée-Poussin (1866–1962), who lived to be almost ninety-six; the other was a Frenchman, Jacques Hadamard (1865–1963), who was almost ninety-eight when he died.

6   Problems on the number and distribution of primes have fascinated many mathematicians from Euclid's day to our own. What may be regarded as a deep and difficult corollary to Euclid's theorem on the infinity of primes was proved by a friend of Gauss who in 1855 was to succeed him at Göttingen. This was Peter Gustav Lejeune Dirichlet (1805–1859), the man who did more than anyone else to amplify the *Disquisitiones*. The Dirichlet theorem states not only that the number of prime numbers is infinite, but that if one considers only those integers in an arithmetic progression $a, a + b, a + 2b,$ $\ldots, a + nb, \ldots$ in which a and b are relatively prime, then even in this relatively more sparse subset of the integers there still will be infinitely many primes. The proof Dirichlet gave required complicated tools from analysis, where Dirichlet's name is again preserved in the Dirichlet test for uniform convergence of a series. Among other contributions of Dirichlet was the first proof of the theorem known as Bertrand's postulate. We cannot go into these ever more specialized intricacies of nineteenth-century number theory, but it should be noted that Dirichlet's theorem showed that the discrete domain of the theory of numbers cannot be studied in isolation from the branch of mathematics dealing with continuous variables—that is, that number theory required the aid of analysis. Gauss himself, in the *Disquisitiones*, had given a striking example of the fact that the properties of prime numbers intrude in the most unexpected ways even into the realm of geometry.

Toward the end of the *Disquisitiones* Gauss included the first important discovery he had made in mathematics—the construction of the regular polygon of seventeen sides; he carried the topic to its logical conclusion by showing which of the infinitely many possible regular polygons can be constructed and which cannot. General theorems, such as that which Gauss now proved, are of ever so much more value than a single case, no matter how spectacular this may be. It will be recalled that Fermat had believed that numbers of the form $2^{2^n} + 1$ are primes, a conjecture that Euler had shown to be incorrect. The number $2^{2^2} + 1 = 17$ is indeed prime, as are also $2^{2^3} + 1 = 257$ and $2^{2^4} + 1 = 65,537$. Gauss already had shown the polygon of seventeen sides to be constructible, and the question naturally arises whether a regular polygon of 257 or 65,537 sides can be constructed with Euclidean tools. In the *Disquisitiones* Gauss answered the question in the affirmative, showing that a regular polygon of N sides can be constructed with Euclidean tools if and only if the number N is of the form $N = 2^m p_1 p_2 p_3 \ldots p_r$, where m is any positive integer and the p's are distinct Fermat primes. There remains one aspect of the problem that Gauss did not answer, and which has not yet been answered. Is the number of Fermat primes finite or infinite? For $n = 5, 6, 7, 8,$ and 9 it is known that the Fermat numbers are *not* prime, and it appears possible that there are five and only five constructible regular polygons of a prime number of sides, two that were known in antiquity and the three that were discovered by Gauss. A friend whom Gauss much admired, Ferdinand Gotthold Eisenstein (1823–1852), professor of mathematics at Berlin, added a new conjecture about prime numbers when he hazarded the thought, unverified up to the present time, that numbers of the form $2^2 + 1$, $2^{2^2} + 1$, $2^{2^{2^2}} + 1$, and so on, are primes. To Gauss is attributed the remark that "There have been only three epoch-making mathematicians, Archimedes, Newton, and Eisenstein." Whether, given a normal span of years, Eisenstein might have fulfilled such a glowing prediction is a matter of conjecture, for the young man died when not yet thirty years old, having remained a Privatdozent.

Gauss had planned the *Disquisitiones* as a two-volume work, but he never **7** got around to writing the second volume. He once described mathematics as the queen of the sciences and arithmetic (that is, theory of numbers) as the queen of mathematics; but Gauss in the early nineteenth century virtually abandoned his Queen of Mathematics, his attention distracted by too many other subjects. Later in life he expressed regret that he had abandoned his first love. Among the other fields that caught his fancy was astronomy, to which he was drawn almost by accident. On the very first day of the nineteenth century a new planet or asteroid, Ceres, had been discovered, but a few weeks afterward the tiny body was lost to sight. Gauss realized that he had most unusual computational ability, as well as the added advantage of the method of least squares, and he took up the challenge to compute, from the few recorded observations of the planet, the orbit in which it moved. For the task of computing orbits from a limited number of observations he devised a scheme, known as Gauss's method, which still is used to track satellites. The result was a resounding success, the planet being rediscovered at the end of the year in very nearly the position indicated by his calculations. From now on astronomy and statistics remained among his serious interests, and he spent much of the next score of years in astronomical calculations. This activity seems to be at variance with the view Gauss expressed that

"All the measurements in the world are not the equivalent of a single theorem that produces a significant advance in our greatest of sciences." Nevertheless, his reputation now was clearly established in astronomy, as well as in mathematics; in 1807 he was appointed director of the Göttingen Observatory, a post he held for some forty years. He could have had a professorship in mathematics, and it is likely that he chose the position at the observatory less from preference for astronomy than from dislike of teaching. He enjoyed helping a few brilliant devotees, but classroom instruction seems to have repelled him.[7]

8   The position of Gauss at the observatory nevertheless did not deter him from making further important contributions to mathematics. In 1811 he informed an astronomer friend, F. W. Bessel (1784–1846), of a discovery he had made in what was soon to become a new subject in the hands of the leading French mathematician, Augustin-Louis Cauchy (1789–1867) and which today bears the latter's name. The theory of functions of a real variable had been developed by Lagrange, but the theory of functions of a complex variable awaited the efforts of Cauchy; yet Gauss perceived a theorem of fundamental significance in the as yet unworked field. If in the complex or Gaussian plane one draws a simple closed curve, and if a function $f(z)$ of the complex variable $z = x + iy$ is analytic (that is, has a derivative) at every point on the curve and within the curve, then the line integral of $f(z)$ taken along the curve is zero.

The unpublished memoranda of Gauss hung like a sword of Damocles over mathematics of the first half of the nineteenth century. When an important new development was announced by others, it frequently turned out that Gauss had had the idea earlier, but had permitted it to go unpublished. Among the striking instances of this situation was the disclosure of elliptic functions, a discovery in which four outstanding figures are involved. One of these was, of course, Legendre, who had spent some forty years studying elliptic integrals almost single-handedly. He had developed a great many formulas, some of them resembling relationships among inverse trigonometric functions (a number of which had been known long before to Euler). This was not surprising inasmuch as the elliptic integral

includes
$$\int \frac{dx}{\sqrt{(1 - K^2 x^2)(1 - x^2)}}$$

$$\int \frac{dx}{\sqrt{1 - x^2}}$$

---

[7] For an excellent overall summary of the views of Gauss, see the chapter on "The Prince of Mathematicians" in E. T. Bell, *Men of Mathematics* (1937). This chapter is reprinted in James Newman, *The World of Mathematics* (New York: Simon and Schuster, 1956, 4 vols.).

as the special case for which $K = 0$. However, it remained for Gauss and two younger contemporaries to take full advantage of a point of view that greatly facilitates the study of elliptic integrals. If

$$u = \int_0^v \frac{dx}{\sqrt{1 - x^2}}$$

then $u = \arcsin v$. Here u is expressed as a function of the independent variable v (x being only the dummy variable for integrating), but it turns out to be more felicitous to reverse the roles of u and v by choosing u as the independent variable. In this case we have $v = f(u)$, or, in the language of trigonometry, $v = \sin u$. The function $v = \sin u$ is more expeditiously manipulated, and it has a striking property that $u = \arcsin v$ does not have— it is periodic. The private papers of Gauss show that perhaps as early as 1800 he had discovered the double periodicity of elliptic (or lemniscatic) functions. It was not until 1827–1828, however, that this remarkable property was disclosed by one of the most brilliant mathematicians of all times, Niels Henrik Abel (1802–1829), in a memoir in Crelle's *Journal*. The young Abel died when he was only twenty-six years old, leaving behind him profound results in algebra and the theory of functions, some of which we shall describe before returning to Gauss.

Abel's life illustrates vividly how closely poverty can be related to tragedy. **9** He was born into a large family, the son of the pastor of the little village of Findö in Norway. When he was sixteen, his teacher urged him to read the great books in mathematics, including the *Disquisitiones* of Gauss. In his reading Abel noted that Euler had proved the binomial theorem for rational powers only, and so he filled the gap by giving a proof valid for the general case. When Abel was eighteen, his father died, and much of the care of the family fell on his young and weak shoulders; yet within the next year he made a remarkable mathematical discovery. Ever since the cubic and quartic equations had been solved in the sixteenth century, men had studied the quintic. Abel at first thought he had hit upon a solution; but in 1824 he published a memoir, "On the Algebraic Resolution of Equations," in which he reached the opposite conclusion: he gave the first proof that no solution is possible, thus putting an end to the long search. There can be no general formula, expressed in explicit algebraic operations on the coefficients of a polynomial equation, for the roots of the equation if the degree of the equation is greater than four. An earlier proof, less satisfactory and generally overlooked, of the unsolvability of the quintic had been published in 1799 by

Paolo Ruffini (1765–1822), and hence the result now is referred to as the "Abel–Ruffini theorem."

The impossibility proof on the quintic, one of the most celebrated theorems in mathematics, was produced by Abel when he was only nineteen; yet in 1826, when he visited Legendre, Cauchy, and other mathematicians in Paris, he still had not been offered an appropriate academic post. On this visit to Paris, Abel wrote to a friend: "Every beginner has a great deal of difficulty in getting noticed here. I have just finished an extensive treatise on a certain class of transcendental functions, ... but Mr. Cauchy scarcely deigned to glance at it." By this time, unlike the eighteenth century, most openings in mathematics were professorships at universities, and Abel was hoping for one of these. Consequently he left the memoir with Cauchy for his considera-tion. It contained novel results of extraordinary generality, but Cauchy proceeded to mislay it, with the result that it did not appear in print until it was much too late. Abel in 1829 died of tuberculosis, generally assumed to have been brought on by his poverty and his responsibilities for his dependent family. Two days after his death a letter was written to inform him that he was to be appointed as professor of mathematics at the University of Berlin, a position that had been secured for him by Crelle. In the same year the German mathematician Carl Gustav Jacob Jacobi (1804–1851) wrote to Legendre to inquire about the memoir Abel had left with Cauchy, for Jacobi had intimations that it touched on his outstanding discovery. Upon looking into the matter, Cauchy in 1830 dug up the manuscript, which Legendre later described as "a monument more lasting than bronze," and it was published in 1841 by the French Institut among the memoirs presented by foreigners. It contained an important generalization of Abel's work on elliptic functions. If

$$u = \int_0^v \frac{dx}{\sqrt{(1 - K^2 x^2)(1 - x^2)}}$$

u is a function of v, $u = f(v)$, the properties of which had been very extensively described by Legendre in his treatise on elliptic integrals. What Legendre had missed, and what Gauss, Abel, and Jacobi saw, was that by inverting the functional relationship between u and v, one obtains a more useful and more beautiful function, $v = f(u)$. This function, usually written $v = \operatorname{sn} u$ and read "v is the sine amplitude of u," together with others defined in a somewhat similar manner, are known as elliptic functions.[8]

The most striking property of these new higher transcendental functions was, as their three independent discoverers saw, that in the theory of complex

---

[8] Some historical confusion has arisen because Legendre used the phrase "fonctions ellip-tiques" in the title of his treatise on elliptic integrals. As used by Legendre, however, the phrase referred to elliptic integrals, and not to what are now known as elliptic functions.

variables they have a *double* periodicity, that is, there are two complex numbers m and n such that $v = f(u) = f(u + m) = f(u + n)$. Whereas the trigonometric functions have a real period only (a period of 2π) and the function e^x has an imaginary period only ($2\pi i$), the elliptic functions have two distinct periods. So impressed was Jacobi with the simplicity achieved through a simple inversion of the functional relationship in elliptic integrals that he regarded the advice, "You must always invert," as the secret of success in mathematics.

Priority, with respect to the discovery of double periodicity, is not easily established. Jacobi's classic treatise, *Fundamenta nova theoriae functionum ellipticarum*, appeared in 1829, the year of Abel's death, and Abel's work was published in 1827–1828. But Gauss seems to have been by far the earliest to make the discovery, which had lain dormant among his papers for a quarter of a century before Abel and Jacobi again hit upon it. Jacobi deserves credit also for several critical theorems related to elliptic functions. In 1834 he proved that if a single-valued function of one variable is doubly periodic, the ratio of the periods can not be real, and that it is impossible for a single-valued function of a single independent variable to have more than two distinct periods. To him we owe also a study of the "Jacobi theta functions," a class of quasi doubly-periodic entire functions of which the elliptic functions are quotients.

The fateful misplaced memoir by Abel contained the hint of something **10** even more general than the elliptic functions. If one replaces the elliptical integral by

$$u = \int_0^v \frac{dx}{\sqrt{P(x)}}$$

where $P(x)$ is a polynomial the degree of which may exceed four, and if one again inverts the relationship between u and v to obtain $v = f(u)$, this function is a special case of what is known as an "Abelian function." It was Jacobi, however, who in 1832 first demonstrated that the inversion can be carried out not only for a single variable, but for functions of several variables.

The name of Abel is perpetuated also in the Abelian group for any group that is commutative. The name of Jacobi, too, is known today, but chiefly in an entirely different connection. The history of determinants is an oddly chequered one, with sporadic bits of the subject appearing in ancient China, some work by Leibniz, a rule of Cramer, and some symmetric results of Lagrange. Not until the nineteenth century did a sustained development take place, initiated to a considerable extent, at least on the Continent, by Cauchy and Jacobi. This is one of the few branches in which the role of Gauss was slight, although it was from the terminology of Gauss in a somewhat different context that Cauchy derived the name "determinant" for what he otherwise described as a class of alternating symmetric functions—such as $a_1 b_2 - b_1 a_2$. A good case could be made out for having the definitive history of determinants begin in 1812, when Cauchy read to the Institut a long memoir on the subject,[9] although in doing so one would fail to do justice to some pioneer work as early as 1772 by Laplace and Vandermonde. Cauchy was at the time a military engineer with Ponts et Chaussées. He had been born in Paris a few weeks after the fall of the Bastille, and he had been educated at the École Polytechnique, where both Lagrange and Laplace had taken an interest in his progress. Incidentally, it was Cauchy who took the seat of Monge in the Academy when in the restoration of 1816 both Carnot and Monge were expelled. At the same time Cauchy was made professor at the École Polytechnique; but a reversal came with the revolution in 1830. Cauchy throughout his life remained a devout Catholic and a convinced reactionary. Consequently, he vigorously defended the Jesuit Order which d'Alembert had attacked; and when "his king," Charles X, went into exile, Cauchy also left Paris, not to return until 1838. However, no matter where he was, whether at Turin or Prague, or under what government, Cauchy continued to pour out such a stream of books and memoirs that he was second only to Euler in extent of output. But whereas Euler had been willing to write with questionable logic on almost any aspect of mathematics, pure or applied, Cauchy followed in the tradition of Lagrange in his preference for pure mathematics in elegant form with due attention to rigorous proofs. His 1812 paper on determinants, to be followed by many others from him on the same topic, was in this tradition in giving emphasis to the symmetries of notation with which it abounds.

In the pedagogical approach to determinants today it is customary to begin with the square array and then to attach a meaning or value to this through an expansion in terms of transpositions or permutations. In the memoir of Cauchy the author did the opposite. He began with the n elements or numbers, $a_1, a_2, a_3, \ldots a_n$, and formed the product of these by all the differences of distinct elements—$a_1 a_2 a_3 \ldots a_n (a_2 - a_1)(a_3 - a_1) \ldots (a_n - a_1)$ $(a_3 - a_2) \ldots (a_n - a_2) \ldots (a_n - a_{n-1})$. He then defined the determinant as the expression obtained upon changing every indicated power into a second subscript, so that $a_r{}^s$ becomes $a_{r \cdot s}$; he wrote this as $S(\pm a_{1 \cdot 1} a_{2 \cdot 2} a_{3 \cdot 3} \ldots a_{n \cdot n})$. Then he arranged the n^2 different quantities in this determinant in a square

---

[9] In the same year a shorter and less definitive memoir was read before the Institut by Binet. For full accounts of these and other works on determinants, see Thomas Muir, *The Theory of Determinants in the Historical Order of Development*. The first four volumes, covering the work up to 1900, have been reprinted by Dover Publications. New York, 1960.

array not unlike that used today:

$$a_{1\cdot 1}, a_{1\cdot 2}, a_{1\cdot 3}, \ldots a_{1\cdot n}$$

$$a_{2\cdot 1}, a_{2\cdot 2}, a_{2\cdot 3}, \ldots a_{2\cdot n}$$

$$\ldots \ldots \ldots \ldots \ldots \ldots$$

$$a_{n\cdot 1}, a_{n\cdot 2}, a_{n\cdot 3}, \ldots a_{n\cdot n}$$

As thus arranged, the n^2 quantities in his determinant were said to form "a symmetric system of order n." He defined conjugate terms as elements the orders of whose subscripts are reversed, and he called terms that are self-conjugate principal terms; the product of the terms in what we call the main diagonal or the principal diagonal he called the principal product. Later in the memoir Cauchy gave other rules for determining the sign of a term in the expansion, using circular substitutions.

Cauchy's eighty-four page memoir of 1812 was not his only work on the subject of determinants; from then on he found many opportunities to use them in a variety of situations. In a memoir of 1815, on wave propagation, he applied the language of determinants to a problem in geometry and also to one in physics. Cauchy asserted that if A, B, C are the lengths of three edges of a parallelepiped, and if the projections of these on the x, y, and z axes of a rectangular coordinate system are

$$A_1, B_1, C_1$$

$$A_2, B_2, C_2$$

$$A_3, B_3, C_3$$

then the volume of the parallelepiped will be $A_1B_2C_3 - A_1B_3C_2 + A_2B_3C_1 - A_2B_1C_3 + A_1B_2C_3 - A_3B_2C_1 = S(\pm A_1B_2C_3)$. In the same memoir, in connection with the propagation of waves, he applied his determinant notation to partial derivatives, replacing a condition that required two lines for its expression by the simple abbreviation

$$S\left(\pm \frac{dx}{da} \frac{dy}{db} \frac{dz}{dc} \right) = 1$$

The left-hand side of this is obviously what now is called the "Jacobian" of x, y, z with respect to a, b, c. The name of Jacobi is attached to functional determinants of this form not because he was the first to use them, but because he was an algorithm builder who was especially enthusiastic about the possibilities inherent in determinant notations. It was not until 1829 that Jacobi first used the functional determinants that bear his name.

11   The early years of Jacobi's life were in a way the antithesis of those of Abel, for Jacobi's father was a prosperous banker, whose children never felt the pinch of want. C. G. J. Jacobi, the second son, secured a good education at the University of Berlin, with concentrations in philology and mathematics. Like Gauss, he finally settled for the latter, but, unlike Gauss, Jacobi was a born teacher who delighted in instructing others. The most celebrated results of his research were those in elliptic functions, published in 1829, which brought him the praise of Legendre. By means of this new analysis Jacobi later proved again the four-square theorem of Fermat and Lagrange. In 1829 Jacobi also published a paper in which he made extensive and general use of Jacobians, expressing these in a more modern form than had Cauchy:

$$\frac{\partial u}{\partial x}, \quad \frac{\partial u}{\partial x_1}, \quad \frac{\partial u}{\partial x_2}, \quad \dots \quad \frac{\partial u}{\partial x_{n-1}}$$

$$\frac{\partial u_1}{\partial x}, \quad \frac{\partial u_1}{\partial x_1}, \quad \frac{\partial u_1}{\partial x_2}, \quad \dots \quad \frac{\partial u_1}{\partial x_{n-1}}$$

$$\dots \dots \dots \dots \dots \dots \dots \dots \dots \dots \dots$$

$$\frac{\partial u_{n-1}}{\partial x}, \quad \frac{\partial u_{n-1}}{\partial x_1}, \quad \frac{\partial u_{n-1}}{\partial x_2}, \quad \dots \quad \frac{\partial u_{n-1}}{\partial x_{n-1}}$$

Jacobi became so enamoured of functional determinants that he insisted on thinking of ordinary numerical determinants as Jacobians of n linear functions in n unknowns.

Jacobi's use of functional determinants in a paper on algebra in 1829 was only incidental, as had been that of Cauchy. Had this been all from the pen of Jacobi, his name would not have been attached to the particular determinant that we are considering. In 1841, however, he published a long memoir "De determinantibus functionalibus," specifically devoted to the Jacobian. He pointed out, among other things, that this functional determinant is in many ways an analogue, for functions of several variables, to the differential quotient of a function of a single variable; and of course he called attention to its role in determining whether or not a set of equations or functions are independent. He showed that if a set of n functions in n variables are functionally related, the Jacobian must vanish identically; if the functions are mutually independent, the Jacobian cannot be identically zero.

12   The papers of Jacobi that we have mentioned, as well as many others by Jacobi, Abel, and Dirichlet, appeared in what is usually known as *Crelle's Journal.* This was one of the specifically mathematical periodicals that were a feature of the nineteenth century. Before 1794 there had been scientific journals, but none primarily devoted to serious mathematics. The initiative for the establishment of mathematical periodicals came from the École Polytechnique when it began publishing its *Journal*. Shortly thereafter, in 1810, the first privately established mathematical periodical was begun by an artillery officer who was an *ancien élève* of the École Polytechnique. This was the *Annales de Mathématiques Pures et Appliquées*, edited by Joseph-Diaz Gergonne (1771–1859). The editor was a thoroughly capable mathematician who contributed articles to his journal, but these will be described in the next chapter. In Germany a periodical similar to Gergonne's *Annales*, and even more successful, was begun in 1826 by August Leopold Crelle (1780–1855) under the title *Journal für die reine und angewandte Mathematik*. Again the editor was primarily an engineer, but so heavily weighted were the articles in the direction of pure (*reine*) mathematics (notably those by Abel, five of which appeared in the very first volume) that wags suggested the title might be more appropriate if the two German words *und angewandte* ("and applied") were replaced by the single word *unangewandte* ("unapplied"). The *Annales* of Gergonne did not long survive, but in 1836 the lacuna was filled by a French periodical paralleling that of Crelle in motive as well as title—*Journal de Mathématiques Pures et Appliquées*, founded and edited by Joseph Liouville (1809–1882). Liouville was an active research mathematician whose work will be described in the chapter on the arithmetization of analysis. Editors of both the French and the German *Journal* have changed frequently since that time, but these periodicals, continuing publication to this day, generally still are known in common parlance as Crelle's *Journal* and Liouville's *Journal*. These two venerable periodicals have since, of course, been joined by a host of others, especially within the last hundred years, some of them organs of mathematical organizations. Before 1865 quite a number of scientific societies had been established, but there was no large-scale organization devoted primarily to mathematics. In that year the London Mathematical Society was organized, and it initiated publication of its *Proceedings*. Since then almost every large nation has come to boast of one or more mathematical organizations within its borders, some of them supporting several periodicals. So extensive has the publication of mathematical research become that today a single volume of an abstracting journal, such as *Mathematical Reviews*, sponsored by the American Mathematical Society (organized in 1888), is several times the size of the journals of Crelle and Liouville combined.[10]

Gauss is sometimes described as the last mathematician to know everything in his subject. Such a generalization is bound to be inexact, but it does **13**

[10] For the dates of founding of some of the mathematical organizations, see George Sarton, *The Study of the History of Mathematics* (1936), p. 62.

emphasize the breadth of interests Gauss displayed. In 1806 he published an account of his astronomical methods in a book with the title *Theoria motus*, where he referred to his discovery of the principle of least squares. Legendre believed this principle to have been his own property and virtually accused Gauss of plagiarism, sharing his indignation with Jacobi. We know now that Gauss was correct; but one cannot help but wish that he had been more prompt in making disclosure of his discoveries, or more gentle in breaking the news to later rediscoverers. Gauss, like Newton, tried to be scrupulously fair to others, but he sometimes appeared to be less than cordial to the work of others, especially Cauchy, his nearest rival in mathematical achievement and one to whose work Gauss did not once refer.

Cauchy in one respect was quite unlike Gauss: he leaped into print as soon as he had achieved something. Perhaps this is one of the reasons that the chief characteristic of nineteenth-century mathematics—the introduction of rigor—is attributed to Cauchy, rather than to Gauss, despite the high standard of logical precision that Gauss set for himself. Possibly also it was the pedagogical tradition of the École Polytechnique that played a role here, for there was far more of the pedagogue in Cauchy, who enjoyed teaching, than in Gauss, who hated it. Gauss had latent theorems on complex variables written down here and there in a diary or in memoranda, but it was Cauchy who kept filling the *Journal* of the École Polytechnique and the *Comptes Rendus* of the Académie with ever longer memoirs. These were on a variety of topics, but especially on the theory of functions of a complex variable, a field in which, from 1814 on, Cauchy became the effective founder. In 1806 Jean Robert Argand (1768–1822) of Geneva had published an account of the graphical representation of complex numbers. Although at first this went almost as unnoticed as the work of Wessel, by the end of the second decade of the nineteenth century most of Europe was familiar, through Cauchy, not only with the Wessel-Argand-Gaussian diagram for a complex number, but with the fundamental properties of complex functions as well. In the eighteenth century problems in complex variables occasionally had arisen in connection with the physics of Euler and d'Alembert, but now they became a part of pure mathematics. Inasmuch as two dimensions are required for a pictorial representation of the independent variable alone, it would take four dimensions to portray graphically a functional relationship between two complex variables, $w = f(z)$. Of necessity, therefore, complex variable theory entails a higher degree of abstraction and complexity than does the study of functions of a real variable. Definitions and rules of differentiation, for example, cannot readily be carried over from the real case to the complex, and the derivative in the latter case is no longer pictured as the slope of the tangent to a curve. Without the crutch of visualization, one is likely to require more precise and careful definitions of concepts. To supply this need was one of

Cauchy's contributions to the calculus—both for real variables and for complex variables.

The first teachers in the École Polytechnique had set a precedent according **14** to which even the greatest of mathematicians are not above writing textbooks on all levels, and Cauchy followed in this tradition. In three books—*Cours d'analyse de l'École Polytechnique* (1821), *Résumé des leçons sur le calcul infinitésimal* (1823), and *Leçons sur le calcul différentiel* (1829)—he gave to elementary calculus the character that it bears today. Rejecting the Taylor's theorem approach of Lagrange, he made the limit concept of d'Alembert fundamental, but he gave it an arithmetic character of greater precision. Dispensing with geometry and with infinitesimals or velocities, he gave a relatively clear-cut definition of limit:

> When the successive values attributed to a variable approach indefinitely a fixed value so as to end by differing from it by as little as one wishes, this last is called the limit of all the others.[11]

Where many earlier mathematicians had thought of an infinitesimal as a very small fixed *number*, Cauchy defined it clearly as a dependent *variable*:

> One says that a variable quantity becomes infinitely small when its numerical value decreases indefinitely in such a way as to converge toward the limit zero.

In the calculus of Cauchy the concepts of function and limit of a function were fundamental. In defining the derivative of $y = f(x)$ with respect to x, he gave to the variable x an increment $\Delta x = i$ and formed the ratio

$$\frac{\Delta y}{\Delta x} = \frac{f(x + i) - f(x)}{i}$$

The limit of this difference quotient as i approaches zero he defined as the derivative $f'(x)$ of y with respect to x. The differential he relegated to a subsidiary role, although he was aware of its operational facility. If dx is a finite quantity, the differential dy of $y = f(x)$ is defined simply as $f'(x)\,dx$. Cauchy also gave a satisfactory definition of a continuous function. The function $f(x)$ is continuous within given limits if between these limits an infinitely small increment i in the variable x produces always an infinitely small increment, $f(x + i) - f(x)$, in the function itself. When we bear in

[11] This and other definitions can be found in Vol. III of Cauchy's *Oeuvres complètes* (Paris 1882–1932, 25 vols.). See also C. B. Boyer, *Concepts of the Calculus* (New York: Dover paperback, 1959), pp. 271 ff; and A. Robinson, *Non-Standard Analysis* (Amsterdam: North-Holland, 1966), pp. 269 ff.

mind Cauchy's definition of infinitely small quantities in terms of limits, his definition of continuity parallels that used today.

During the eighteenth century integration had been treated as the inverse of differentiation. Cauchy's definition of derivative makes it clear that the derivative will not exist at a point for which the function is discontinuous; yet the integral may afford no difficulty. Even discontinuous curves may determine a well-defined area. Hence Cauchy defined the definite integral in terms of the limit of the integral sums in a manner not very different from that used in elementary textbooks today, except that he took the value of the function always at the left-hand endpoint of the interval. If $S_n = (x_1 - x_0)f(x_0) + (x_2 - x_1)f(x_1) \cdots + (X - x_{n-1})f(x_{n-1})$, then the limit S of this sum S_n, as the magnitudes of the intervals $x_i - x_{i-1}$ decrease indefinitely, is the definite integral of the function $f(x)$ for the interval from $x = x_0$ to $x = X$. It is from Cauchy's concept of the integral as a limit of a sum, rather than from the antiderivative, that the many fruitful modern generalizations of the integral have arisen.

Having defined the integral independently of differentiation, it was necessary for Cauchy to prove the usual relation between the integral and the antiderivative, and this he accomplished through use of the theorem of mean value. If $f(x)$ is continuous over the closed interval $[a, b]$ and differentiable over the open interval (a, b), then there will be some value x_0 such that $a < x_0 < b$ and $f(b) - f(a) = (b - a)f'(x_0)$. This is a fairly obvious generalization of Rolle's theorem, which was known a century earlier. The mean-value theorem, however, did not attract serious attention until the days of Cauchy, but it has since continued to play a basic role in analysis. It is with justice, therefore, that a still more general form,

$$\frac{f(b) - f(a)}{g(b) - g(a)} = \frac{f'(x_0)}{g'(x_0)}$$

with suitable restrictions on $f(x)$ and $g(x)$, is known as Cauchy's mean-value theorem.

15 The history of mathematics teems with cases of simultaneity and near simultaneity of discovery, some of which have already been noted. The work by Cauchy that we have just described is another case in point, for similar views were developed at about the same time by Bernhard Bolzano (1781–1848), a Czechoslovakian priest whose theological views were frowned upon by his church and whose mathematical work was most undeservedly overlooked by his lay and clerical contemporaries. Cauchy, during his exile, for a time lived at Prague, where Bolzano was born and died; yet there is no indication that the men met. The similarity in their arithmetization of the calculus, of their definitions of limit, derivative, continuity, and convergence was only a coincidence. Bolzano in 1817 had published a book, *Rein analytischer Beweis*, devoted to a purely arithmetic proof of the location theorem in algebra, and this had required a nongeometrical approach to the continuity of a curve or function. Going considerably further in his unorthodox ideas, he disclosed some important properties of infinite sets in a posthumous work of 1850, *Paradoxien des Unendlichen.*[12]

From Galileo's paradox on the one-to-one correspondence between integers and perfect squares, Bolzano went on to show that similar correspondences between the elements of an infinite set and a proper subset are commonplace. For example, a simple linear equation, such as $y = 2x$, establishes a one-to-one correspondence between the real numbers y in the interval from 0 to 2, for example, and the real numbers x in half this interval. That is, there are just as many real numbers between 0 and 1 as between 0 and 2, or just as many points in a line segment 1 inch long as in a line segment 2 inches long. Bolzano seems even to have recognized, by about 1840, that the infinity of real numbers is of a type different from the infinity of integers, being nondenumerable. In such speculations on infinite sets the Bohemian philosopher came closer to parts of modern mathematics than had his better-known contemporaries. Both Gauss and Cauchy seem to have had a kind of *horror infiniti*, insisting that there could be no such thing as a completed infinite in mathematics. Their work on "orders of infinity" in reality was far removed from the concepts of Bolzano, for to say, as Cauchy in essence did, that a function y is infinite of order n with respect to x if $\lim_{x \to \gamma} y/x^n = K \neq 0$ is quite different from making a statement about correspondences between sets.

Bolzano was a "voice crying in the wilderness," and many of his results **16** had to be rediscovered later. Among these was the recognition that there are pathological functions that do not behave as mathematicians had always expected them to behave. Newton, for instance, had assumed that curves are generated by smooth and continuous motions. There might be occasional abrupt changes in direction or even some discontinuities at isolated points; but throughout the first half of the nineteenth century it was generally assumed that a continuous real function must have a derivative at most points. In 1834, however, Bolzano had thought up a function continuous for an interval but, despite physical intuition to the contrary, having no derivative at any point in the interval.[13] The example given by Bolzano unfortunately

---

[12] A good English translation appeared a century later. See Bernhard Bolzano, *Paradoxes of the Infinite*, ed. by D. A. Steele (1950).

[13] For a full description of the function see Gerhard Kowalewski, "Über Bolzanos nicht-differenzierbare stetige Funktion," *Acta Mathematica*, **44** (1923), 315–319. A brief account is contained in C. B. Boyer, *Concepts of the Calculus* (1959), pp. 269–270.

did not become known; hence credit for building the first continuous but nowhere differentiable function generally goes to Weierstrass about a third of a century later. Similarly it is the name of Cauchy, rather than that of Bolzano, that is attached to an important test of convergence for an infinite series or sequence. Occasionally before their time there had been warnings about the need to test an infinite series for convergence. Gauss as early as 1812, for example, used the ratio test to show that his hypergeometric series

$$1 + \frac{\alpha\beta}{\gamma}x + \frac{\alpha\beta(\alpha + 1)(\beta + 1)}{1 \cdot 2\gamma(\gamma + 1)}x^2 + \cdots$$

$$+ \frac{\alpha\beta(\alpha + 1)(\beta + 1) \cdots (\alpha + n - 1)(\beta + n - 1)}{1 \cdot 2 \cdots (n - 1)\gamma(\gamma + 1) \cdots (\gamma + n - 1)}x^n + \cdots$$

converges for $|x| < 1$ and diverges for $|x| > 1$. This test seems to have been first used long before, in England, by Edward Waring although it generally bears the name of d'Alembert or, more occasionally, that of Cauchy.

The name of Cauchy appears today in connection with a number of theorems on infinite series, for, despite some efforts on the part of Gauss and Abel, it was largely through Cauchy that the mathematician's conscience was pricked concerning the need for vigilance with regard to convergence. Having defined a series to be convergent if, for increasing values of n, the sum S_n of the first n terms approaches a limit S, called the sum of the series, Cauchy proved that a necessary and sufficient condition that an infinite series converge is that, for a given value of p, the magnitude of the difference between S_n and S_{n+p} tends toward zero as n increases indefinitely. This condition for "convergence within itself" has come to be known as Cauchy's criterion, but it was known earlier to Bolzano (and possibly still earlier to Euler).

Cauchy also announced in 1831 the theorem that an analytic function of a complex variable $w = f(z)$ can be expanded about a point $z = z_0$ in a power series that is convergent for all values of z within a circle having z_0 as center and passing through the singular point of $f(z)$ nearest to z_0. From this time on the use of infinite series became an essential part of the theory of functions of both real and complex variables. Several tests for convergence bear Cauchy's name, as does a particular form of the remainder in the Taylor series expansion of a function, the more usual form being attributed to Lagrange. The period of rigor in mathematics was taking hold rapidly. It is said that when Cauchy read to the Académie his first paper on the convergence of series, Laplace hurried home to verify that he had not made use of any divergent series in his *Mécanique céleste*. Toward the end of his life Cauchy became aware of the important notion of "uniform convergence," but here, too, he was not alone, having been anticipated by the physicist G. G. Stokes (1819–1903).

The prolific Cauchy contributed to almost as many fields as did his contemporary, Gauss. Although in the theory of numbers his work is less well known than that of Legendre and Gauss, it is to Cauchy that we owe the first general proof of one of the most beautiful and difficult theorems of Fermat—that every positive integer is the sum of at most three triangular numbers or four square numbers or five pentagonal numbers or six hexagonal numbers, and so on indefinitely. This proof is a fitting climax to the study of figurate numbers initiated by the Pythagoreans some 2300 years earlier.

Cauchy evidently was little attracted to geometry in its various forms. In **17** 1811, however, in one of his very earliest memoirs, he presented a generalization of the Descartes-Euler polyhedral formula $E + 2 = F + V$, where E, F, and V are the number of edges, faces, and vertices of the polyhedron; we have noted a case of his application of determinants in finding the volume of a tetrahedron. Gauss, too, was not especially fond of geometry, yet he thought about the subject sufficiently to do two things: (1) to arrive, by 1824, at an important unpublished conclusion on the parallel postulate and (2) to publish in 1827 a classic treatise which generally is regarded as the cornerstone of a new branch of geometry. Gauss, while still a student at Göttingen, had tried to prove the parallel postulate, as had also his intimate friend Wolfgang (or Farkas) Bolyai (1775–1856). Both men continued to look for a proof, the latter giving up in despair, the former eventually coming to the conviction that not only was no proof possible, but that a geometry quite different from that of Euclid might be developed. Had Gauss developed and published his thoughts on the parallel postulate, he would have been hailed as the inventor of non-Euclidean geometry, but his silence on the subject resulted in credit going to others, as we shall see in the next chapter.

The new branch of geometry that Gauss initiated in 1827 is known as differential geometry, and it belongs perhaps more to analysis than to the traditional field of geometry. Ever since the days of Newton and Leibniz men had applied the calculus to the study of curves in two dimensions, and in a sense this work constituted a prototype of differential geometry. Euler and Monge had extended this to include an analytic study of surfaces; hence they sometimes are regarded as the fathers of differential geometry. Nevertheless, not until the appearance of the classical treatise of Gauss, *Disquisitiones circa superficies curvas*,[14] was there a comprehensive volume devoted

---

[14] An English translation is available in *General Investigations of Curved Surfaces*, trans. by Adam Hiltebeitel and James Morehead (1965).

entirely to the subject. Roughly speaking, ordinary geometry is interested in the totality of a given diagram or figure, whereas differential geometry concentrates on the properties of a curve or a surface in the immediate neighborhood of a point on the curve or the surface. In this connection Gauss extended the work of Huygens and Clairaut on the curvature of a plane or gauche curve at a point by defining the curvature of a surface at a point—the "Gaussian curvature" or "the total curvature." If at a point P on a well-behaved surface S one erects a line N normal to S, the pencil of planes through N will cut the surface S in a family of plane curves each of which will have a radius of curvature. The directions of the curves with the largest and smallest radii of curvature, R and r, are called the principal directions on S at P, and they happen always to be perpendicular to each other. The quantities R and r are known as the principal radii of curvature of S at P, and the Gaussian curvature of S at P is defined as $K = 1/rR$. (The quantity $K_m = 1/r + 1/R$, known as the mean curvature of S at P, also turns out to be useful.) Gauss gave formulas for K in terms of the partial derivatives of the surface with respect to various coordinate systems, curvilinear as well as Cartesian; he also discovered what even he regarded as "remarkable theorems" about properties of families of curves, such as geodesics, drawn on the surface. It was in continuing studies like these in differential geometry that the mathematicians of the nineteenth century paved the way for the scientific theories of the twentieth.

18   The concentration in this chapter on the pure mathematics of Gauss and Cauchy may leave the reader with a false impression of their work as a whole. Both men contributed far more heavily than we have indicated to the physical sciences. Gauss, for example, published works on astronomy and geodesy, on capillarity and crystallography. So significant were his achievements in terrestrial magnetism and in magnetic devices that the standard unit of magnetic intensity today is known as the gauss. (For this reason a ship that has been protectively prepared against magnetic mines in naval parlance is said to have been degaussed.) It was Gauss the mathematician who in 1833–1834 collaborated with Wilhelm Weber to produce the first successful electromagnetic telegraph. Whereas the bulk of the nonmathematical work by Gauss came later in life, in the case of Cauchy it came chiefly during his earlier years. Gauss's treatise on refraction (*Dioptrische Untersuchungen*) was published in 1840; the memoir on optical wave propagation (*Mémoire sur la théorie des ondes*) by Cauchy, the younger man, appeared in 1815, during the very period in which the wave theory of light was being established by Young and Fresnel. Cauchy was a founder also of the mathematical theory of elasticity and contributed to celestial mechanics, fields in which an overshadowed contemporary, Siméon-Denis Poisson (1781–1840), also was doing work at the time.

Poisson was the son of a small-town administrator who took charge of local affairs when the Revolution broke out, and the child was reared under republican principles; but he later became a staunch Legitimist and in 1825 was rewarded with the title of baron. In 1837, under Louis Philippe, he became a peer of France. Relatives at first had hoped that the young man would become a physician, but strong mathematical interests led him in 1798 to enter the École Polytechnique, where on graduation he became successively lecturer, professor, and examiner. He is said to have once remarked that life is good for only two things: to do mathematics and to teach it. Consequently he published almost 400 works, and he enjoyed a reputation as an excellent instructor. The direction of his research is indicated in part by a sentence from a letter written in 1826 by Abel concerning the mathematicians in Paris: "Cauchy is the only one occupied with pure mathematics; Poisson, Fourier, Ampère, etc., busy themselves exclusively with magnetism and other physical subjects."[15] This should not be taken too literally, but Poisson, in memoirs of 1812, did help to make electricity and magnetism a branch of mathematical physics, as did Gauss, Cauchy, and Green. Poisson was also a worthy successor to Laplace in studies on celestial mechanics and the attraction of spheroids. The Poisson integral in potential theory, the Poisson brackets in differential equations, the Poisson ratio in elasticity, and Poisson's constant in electricity indicate the importance of his contributions to various fields of applied mathematics. Two of his best known treatises were the *Traité de mécanique* (2 vols., 1811, 1833) and *Recherches sur la probabilité des jugements* (1837). In the latter appears the familiar Poisson distribution or Poisson's law of large numbers. In the binomial distribution $f(x) = (p + q)^n$, (where $p + q = 1$ and n is the number of trials), as n increases indefinitely the binomial distribution ordinarily tends toward a normal distribution; but if, as n increases indefinitely, p approaches zero, the product np remaining constant, the limiting case of the binomial distribution is the Poisson distribution.

Gauss and Cauchy died within two years of each other, the former in 1855, the latter in 1857. Honors of various forms had been bestowed freely on both of them, as had been the case earlier with Lagrange, Carnot, and Laplace. Lagrange and Carnot had been given the title of count, and on Laplace the emperor had bestowed the title of marquis. Cauchy, as a reward for his faithfulness, had been made a baron by Charles X. Gauss never bore a rank of nobility in the legal sense, but posterity by common consent has hailed him as the Prince of Mathematicians.

[15] See E. T. Bell, *Men of Mathematics* (1937), p. 318. See also Maximilien Marie, *Histoire des sciences mathématiques et physiques* (Paris, 1883–1888, 12 vols.), XI, 174–191.

BIBLIOGRAPHY

Bell, E. T., *Men of Mathematics* (New York: Simon and Schuster, 1937).

Bolzano, Bernhard, *Paradoxes of the Infinite*, trans. by D. A. Steele (London: Routledge and Kegan Paul, 1950).

Bonconpagni, B., "La Vie et les travaux du Baron Cauchy... par C.-A. Valson," *Bullettino di Bibliografia e di Storia delle Scienze Matematiche e Fisiche*, 2 (1896, 1–95).

Boyer, C. B., *Concepts of the Calculus* (New York: Dover paperback, 1959).

Brill, A., and M. Noether, "Die Entwickelung der Theorie der algebraischen Funktionen in älterer and neurer Zeit," *Jahresbericht, Deutsche Mathematiker-Vereinigung*, 3 (1892–1893), 107–566.

Dunnington, G. Waldo, *Carl Friedrich Gauss, Titan of Science; A Study of His Life and Work* (New York: Exposition Press, 1955).

Gauss, C. F., *Werke* (Leipzig, 1866–1933, 12 vols.).

Gauss, C. F., *Inaugural Lecture on Astronomy and Papers on the Foundations of Mathematics*, trans. by G. W. Dunnington (Baton Rouge, La.: Louisiana State University, 1937).

Gauss, C. F., *Theory of the Motion of Heavenly Bodies* (New York: Dover reprint, 1963).

Gauss, C. F., *General Investigations of Curved Surfaces*, trans. by Adam Hiltebeitel and James Morehead (New York: Raven Press, 1965).

Gauss, C. F., *Disquisitiones arithmeticae*, English translation by A. A. Clarke (New Haven, Conn.: Yale University Press, 1966).

Jourdain, P. E. B., "The Theory of Functions with Cauchy and Gauss," *Bibliotheca Mathematica* (3), 6 (1905), 190–207.

Jourdain, P. E. B., "The Origin of Cauchy's Conceptions of a Definite Integral and of the Continuity of a Function," *Isis*, 1 (1913), 661–703.

Klein, Felix, *Vorlesungen über die Entwicklung der Mathematik im 19. Jahrhundert* (Berlin: Springer, 1926–1927, 2 vols.).

Merz, J. T., *A History of European Thought in the Nineteenth Century* (Edinburgh and London, 1896–1914, 4 vols.; New York: Dover paperback, 1965).

Muir, Thomas, *The Theory of Determinants in the Historical Order of Development* (London, 1906–1930, 4 vols. and Supplement; reprinted, New York: Dover, 1960).

Pierpont, James, "Mathematical Rigor, Past and Present," *Bulletin, American Mathematical Society*, 34 (1928), 23–53.

Pringsheim, A., and J. Molk, "Principes fondamentaux de la théorie des fonctions," *Encyclopédie des Sciences Mathématiques*, Vol. II, Part I, Fascicule 1.

Reichardt, Hans, ed., *C. F. Gauss. Leben und Werk* (Gauss Gedenkband, Berlin: Haude & Spener, 1960).

Sarton, George, *The Study of the History of Mathematics* (Cambridge, Mass., 1936; New York: Dover paperback, 1957).

Schaaf, William L., *Carl Friedrich Gauss* (New York: Watts, 1964).

Smith, D. E., *Source Book in Mathematics* (New York: Dover reprint, 1958, 2 vols.).

Stolz, Otto, "B. Bolzanos Bedeutung in der Geschichte der Infinitesimalrechnung," *Mathematische Annalen*, 18 (1881), 255–279.

Valson, C.-A., *La Vie et les travaux du baron Cauchy*, with a preface by M. Hermite (Paris, 1868), Vol. I.

EXERCISES

1. Which were established first, mathematical organizations or mathematical journals? Give instances to support your answer.

2. Describe several respects in which the mathematics of Gauss and Cauchy departed from the eighteenth-century tradition.

3. Was Cauchy or Gauss the more influential in contributing to mathematics of the present-day undergraduate level in college? Explain.

4. Explain clearly why Gauss has been described as "the Prince of Mathematicians".

5. What are elliptic functions? Why are they so called and how do they differ from circular (trigonometric) and hyperbolic functions?

6. Describe some differences in the backgrounds, temperaments, and interests of Gauss and Cauchy.

7. Describe two respects in which the algebra of congruences resembles that of equality and two respects in which it differs from that of equality, giving illustrations in each case.

8. Graph the numbers $1 - i$, $3i$, and $i + 7$ on a Gaussian diagram.

9. For the fifty regular polygons having not more than fifty-two sides, list those that are constructible with straightedge and compasses only.

10. Is a regular polygon of 26,214 sides, or of 17,990 sides, constructible with straightedge and compasses only? Explain.

11. Show that the equation $x^2 + 2ix + i = 0$ has a root of the form $a + bi$.

12. Find a value x_0 that satisfies the mean-value theorem $f(b) - f(a) = (b - a)f'(x_0)$ for $f(x) = x^3$, $a = -1$, and $b = 2$.

13. If $f(x) = x^{\frac{1}{3}}$, $a = -1$, and $b = 2$, is there a value x_0 that satisfies the mean-value theorem of Exercise 12? Explain.

14. If $f(x) = 1/x$, $a = -1$, and $b = 2$, is there a value of x_0 that satisfies the mean-value theorem of Exercise 12? Explain.

15. If $f(x) = x^3$, $g(x) = x^2$, $a = 1$, and $b = 3$, find a value x_0 that satisfies the Cauchy generalized mean-value theorem

$$\frac{f(b) - f(a)}{g(b) - g(a)} = \frac{f'(x_0)}{g'(x_0)}.$$

16. Use the Cauchy generalized mean-value theorem to prove L'Hospital's rule for indeterminate forms.

17. Show that the Jacobian of the functions $\sin xy$ and $\cos xy$ is identically zero and find a functional relationship between the two functions.

18. Show that the functions e^{xy} and $e^x e^y$ are not functionally dependent.

19. Compare the value of $\pi(n)$ with that of $n/\ln n$ for $n = 100$.

20. Verify Bertrand's postulate for $n = 42$.

*21. Solve the equation $ix^3 + 8 = 0$ graphically by making the substitution $x = u + iv$, then equating the real parts and the imaginary parts on both sides of the equation, and sketching on the same set of axes the resulting equations in u and v. Verify the solution by solving $ix^3 + 8 = 0$ by De Moivre's theorem.

*22. Show that $4 \cdot 5 \equiv 4 \cdot 8 \pmod{12}$ and that $5 \not\equiv 8 \pmod{12}$.

*23. Find three solutions, each less than 9, of the congruence $3x \equiv 51 \pmod 9$.

*24. Prove that if $a \equiv b \pmod m$ and $b \equiv c \pmod m$, then $a \equiv c \pmod m$.

*25. Factor 13 into the product of two Gaussian primes.

*26. Use Cauchy's principle and the singularities of arctan x and its derivative to show that the expansion of arctan x in terms of powers of x converges for $|x| < 1$.

CHAPTER XXIV

The Heroic Age
in Geometry

> There is no branch of mathematics, however abstract, which may not some day be applied to phenomena of the real world.
>
> *Lobachevsky*

1 Geometry, of all the branches of mathematics, has been most subject to changing tastes from age to age. In classical Greece it had climbed to the zenith, only to fall to its nadir at about the time that Rome fell. It had recovered some lost ground in Arabia and in Renaissance Europe; in the seventeenth century it stood on the threshold of a new era, only to be all but forgotten, at least by research mathematicians, for nearly two more centuries, languishing in the shade of the ever-proliferating branches of analysis. Britain, especially throughout the later eighteenth century, had fought a losing battle to restore Euclid's *Elements* to its once glorious position, but she had done little to advance research in the subject. Through the efforts of Monge and Carnot, there were some stirrings of revival in pure geometry during the period of the French Revolution, but the almost explosive rediscovery of geometry as a living branch of mathematics came chiefly with the dawn of the nineteenth century. As one might have anticipated, the École Polytechnique had a hand in this movement, for there the well-known Brianchon theorem was discovered by a student and was published in 1806 in the *Journal of l'École Polytechnique*. Charles Julien Brianchon (1785–1864) had entered the school only the year before, studying under Monge and reading Carnot's *Géométrie de position*. The twenty-one-year-old student, later an artillery officer and teacher, first reestablished the long-forgotten theorem of Pascal, which Brianchon expressed in the modern form: In any hexagon inscribed in a conic section, the three points of intersection of the opposite sides always lie on a straight line. Continuing through some other demonstrations, he came to the one that bears his name: "In any hexagon circumscribed about a conic section, the three diagonals cross each other in the same point." As Pascal had been impressed by the number of corollaries that he had been able to derive from his theorem, so Brianchon remarked that his own theorem "is pregnant with curious consequences."[1] The theorems of Pascal and Brianchon are, in fact, fundamental in the projective study of conics. They form, in addition, the first clear-cut instance of a pair of significant "dual" theorems in geometry, that is, theorems that remain valid (in plane geometry) if the words point and line are interchanged. If we let the phrase "a line is tangent to a conic" be read as "a line is on a conic," the two theorems can be expressed in the following combined form:

The six $\begin{cases} \text{vertices} \\ \text{sides} \end{cases}$ of a hexagon lie on a conic if

and only if the three $\begin{cases} \text{points} \\ \text{lines} \end{cases}$ common to the three pairs of

opposite $\begin{cases} \text{sides} \\ \text{vertices} \end{cases}$ have a $\begin{cases} \text{line} \\ \text{point} \end{cases}$ in common.

Such relationships between points and lines on conics were later exploited effectively by another alumnus of the École Polytechnique, the man who became the effective founder of projective geometry. This was Jean-Victor Poncelet (1788–1867), who also studied under Monge, who entered the army corps of engineers just in time to take part in Napoleon's ill-fated 1812 campaign in Russia, and who spent several years in a Moscow prison. Upon his return to France he became perhaps the most important figure in the revival in pure geometry. Among his earlier discoveries was one that he shared with Brianchon, and the two men published a joint paper on it in Gergonne's *Annales* for 1820–1821. The title of the paper, "Recherches sur la détermination d'une hyperbole équilatère," gives no clue to a beautiful theorem it contained—that concerning the now well-known nine-point circle. Here Brianchon and Poncelet gave an elementary proof that

> The circle which passes through the feet of the perpendiculars, dropped from the vertices of any triangle on the sides opposite them, passes also through the midpoints of these sides as well as through the midpoints of the segments which join the vertices to the point of intersection of the perpendiculars.[2]

This theorem generally is named for neither Brianchon nor Poncelet, but for an independent mathematician, Karl Wilhelm Feuerbach (1800–1834), who in 1822 published this and some related theorems in a little German book with a long-winded and diffuse title. That the nine-point circle should

---

[1] See *Source Book in Mathematics*, ed. by D. E. Smith (1959), pp. 331–336.
[2] The proof by Brianchon and Poncelet is included in the *Source Book in Mathematics*, ed. by D. E. Smith, pp. 337–338. It appears that some properties of the nine-point circle were known earlier to Euler. See N. A. Court, *College Geometry*, 2nd ed. (New York: Barnes and Noble, 1952), p. 299.

be known as Feuerbach's circle is justified not by priority of publication, but through other properties disclosed by Feuerbach. The center of the nine-point circle, he showed, lies on the Euler line and is midway between the orthocenter and the circumcenter. More remarkable still is the property contained in what is now known as Feuerbach's theorem: the nine-point circle of any triangle is tangent internally to the inscribed circle and tangent externally to the three escribed circles, possibly "the most beautiful theorem in elementary geometry that has been discovered since the time of Euclid."[3] In connection with the proof Feuerbach made use of the fact that the radii of the inscribed and escribed triangles are given by

$$\frac{2K}{\pm a \pm b \pm c}$$

where not more than one minus sign is used.

2   The work of Feuerbach, who died when only thirty-four-years old, is typical of the numerous new results in the modern geometry of triangles and circles that were disclosed throughout the nineteenth century. No country had a monopoly in this development, but Germany had perhaps the largest role, for she nurtured Jakob Steiner (1796–1863), one of the many independent rediscoverers of the nine-point circle and, more importantly, the man who generally is regarded as the greatest geometer of modern times, as Apollonius was of ancient times. Steiner was born in Switzerland, but he was educated at Heidelberg and Berlin, and through Jacobi he obtained a professorship at Berlin that he held until his death. In his hands synthetic geometry made strides comparable to those made earlier in analysis, and he vied with Abel in the number of articles that he contributed to Crelle's *Journal.* His name is recalled in many connections, including the properties of the Steiner points: if one joins in all possible ways the six points on a conic in Pascal's mystic hexagram, one obtains sixty Pascal lines that intersect three by three in twenty Steiner points. Steiner proved also that all Euclidean constructions can be performed with straightedge alone, provided that one is given also a single fixed circle. He intensely disliked analytic methods and demonstrated by synthetic methods alone, in a paper in Crelle's *Journal,* a striking theorem that appears naturally to belong to analysis— that a surface of third order contains only twenty-seven lines.

[3] See J. L. Coolidge, "The Heroic Age of Geometry," *Bulletin of the American Mathematical Society,* **35** (1929), 19–37, especially p. 21. For the history of this theorem see John S. Mackay, "History of the Nine-point Circle," *Proceedings of the Edinburgh Mathematical Society,* **11** (1892), 19; see also **5** (1886–1887), 62. The theorem itself is contained, in translation, in *Source Book in Mathematics,* ed. by D. E. Smith, pp. 339–345.

The history of geometry in the nineteenth century is replete with cases of independent rediscovery, and Steiner was involved in a number of these. In 1822 Poncelet, inspired by the work of Mascheroni, had suggested that all Euclidean plane constructions can be carried out with a straightedge if in addition one has given in the plane a single circle and its center, a theorem proved in 1833 by Steiner. That is, the Poncelet–Steiner theorem shows that one cannot, in Euclidean geometry, dispense entirely with the compasses, but that having used them to draw one circle, one can thereafter discard them in favor of the straightedge alone, somewhat as Mascheroni had used compasses alone.[4]

Steiner reminds one of Gauss in that ideas and discoveries thronged through his mind so rapidly that he could scarcely reduce them to order on paper. As early as 1824 he had discovered the fruitful geometrical transformation known as inversive geometry.[5] If two points P and P' lie on a ray from the center O of a circle C of radius $r \neq 0$, and if the product of the distances OP and OP' is r^2, then P and P' are said to be inverse to each other with respect to C. To every point P outside the circle there is a corresponding point inside the circle. Inasmuch as there is no outside point P' corresponding to P when P coincides with the center O, one has in a sense a paradox similar to that of Bolzano—the inside of every circle, no matter how small, contains, as it were, one more point than the portion of the plane outside the circle. In an exactly analogous manner one readily defines the inverse of a point in three-dimensional space with respect to a sphere.

A host of theorems in plane or solid inversive geometry are readily proved by either analytic or synthetic methods. In particular, it is easy to show that a circle not passing through the center of inversion is transformed, under a plane inversion, into a circle, whereas a circle through the center of inversion goes into a straight line not passing through the center of inversion (with analogous results holding for spheres and planes in three-dimensional inversive geometry). Somewhat more difficult to establish is the more significant result that inversion is a conformal transformation—that is, angles between curves are preserved in this geometry. That angle-preserving transformations are far from usual is clear from a theorem proved by Joseph Liouville that in space the only ones that are conformal are inversions and similarity and congruency transformations.[6] Steiner did not publish his ideas on inversion, and the transformation was rediscovered several times by other mathematicians of the century, including Lord Kelvin (or William

---

[4] Further details are given in Howard Eves, *An Introduction to the History of Mathematics*, rev. ed. (New York: Holt, 1964), pp. 99–100.

[5] See N. A. Court, "Notes on Inversion," *The Mathematics Teacher*, **55** (1962), 655–657.

[6] D. J. Struik, "Outline of a History of Differential Geometry," *Isis*, **19** (1933), 92–120; **20** (1933), 161–191.

Thompson, 1824–1907), who in 1845 arrived at it through physics and who applied it to problems in electrostatics.

If the center O of the circle of inversion of radius a is at the origin of a plane Cartesian coordinate system, the coordinates x' and y' of the inverse P' of a point $P(x, y)$ are given by the equations

$$x' = \frac{a^2 x}{x^2 + y^2}, \qquad y' = \frac{a^2 y}{x^2 + y^2}$$

These equations later suggested to Luigi Cremona (1830–1903), a professor of geometry successively at Bologna, Milan, and Rome, the study of the much more general transformation $x' = R_1(x, y)$, $y' = R_2(x, y)$, where R_1 and R_2 are rational algebraic functions. Such transformations, of which those for inversion are only a special case, are now known as Cremona transformations, in honor of the man who in 1863 published an account of them and who later generalized them for three dimensions.

3   A chief characteristic of geometry during the second half of the nineteenth century was the enthusiasm with which men studied a wide variety of transformations. One of the most popular of these was a group of transformations forming what is known as projective geometry. Hints of such a geometry had been contained in the work of Pascal and Desargues, but not until the early years of the nineteenth century was there a systematic development of this, especially by Poncelet. During the years 1813–1814 of his imprisonment in Russia, Poncelet had composed a treatise on analytic geometry, *Applications d'analyse et de géométrie*, which was based on the principles he had learned at the École Polytechnique. This work, however, was not published until about half a century later (2 vols., 1862–1864), despite the fact that it originally was intended to serve as an introduction to the author's far more celebrated *Traité des propriétés projectives des figures* of 1822. The latter work differed sharply from the former in that it was synthetic, rather than analytic, in style. Poncelet's tastes had changed on his return to Paris, and from that time on he was a staunch advocate of synthetic methods. He realized that the advantage that analytic geometry had appeared to have lay in its generality, and he therefore sought to make statements in synthetic geometry as general as possible. To further this design he formulated what he called the "principle of continuity" or "the principle of permanence of mathematical relations." This he described as follows:

> The metric properties discovered for a primitive figure remain applicable, without other modifications than those of change of sign, to all correlative figures which can be considered to spring from the first.

As an example of the principle Poncelet cited the theorem of the equality of the products of the segments of intersecting chords in a circle, which becomes, when the point of intersection lies outside the circle, an equality of the products of the segments of secants. If one of the lines is tangent to the circle, the theorem nevertheless remains valid upon replacing the product of the segments of the secant by the square of the tangent. Cauchy was inclined to scoff at Poncelet's principle of continuity, for it appeared to him to be nothing more than a bold induction. In a sense this principle is not unlike the view of Carnot, but Poncelet carried it further to include the points at infinity that Kepler and Desargues had suggested. Thus one could say of two straight lines that they always intersected—either in an ordinary point or (in the case of parallel lines) in a point at infinity, called an ideal point. In order to achieve the generality of analysis, Poncelet found it necessary to introduce into synthetic geometry not only ideal points, but also imaginary points, for only thus could he say that a circle and a straight line always intersect. Among his striking discoveries was that all circles whatsoever, drawn in a plane, have two points in common. These are two ideal imaginary points, known as the circular points at infinity and usually designated as I and J (or, more informally, as Isaac and Jacob).

Poncelet argued that his principle of continuity, which presumably had been suggested by analytic geometry, was properly a development of synthetic geometry, and he quickly became a champion of the latter against the analysts. During the second half of the eighteenth century there had been some controversy, especially in Germany, about the relative merits of analysis and synthesis. In 1759 the mathematician and historian A. G. Kaestner (1719–1800), professor at Leipzig and Göttingen, had argued that analysis was superior as a heuristic approach to problems, affording power and economy of thought; one of his students, G. S. Klügel (1739–1812), in 1767 wrote that he suspected the English of seeking, through the difficulty of their synthetic proofs, to enhance their reputations. During the early nineteenth century interest in the rival methodologies in France was such that a prize was offered in 1813 by the Bordeaux Scientific Society for the best essay characterizing synthesis and analysis and the influence that each had exerted. The winning essay, by a teacher at Versailles,[7] closed with the hope that there might be a reconciliation between the two camps; but half a dozen years later the controversy broke out again and became increasingly bitter. The two chief rivals, interestingly, were Poncelet and Gergonne, both

---

[7] For more details see C. B. Boyer, "Analysis: Notes on the Evolution of a Subject and a Name," *The Mathematics Teacher*, **47** (1954), 450–462, especially p. 459. An extensive discussion is available in G. Fano and S. Carrus, "Exposé parallèle du développement de la géométrie synthétique et de la géométrie analytique pendant le 19ième siècle," *Encyclopédie des sciences mathématiques*, III, 3, 185–259.

students of Monge, a man who had been equally at home in analytic and synthetic geometry. At first the rivalry was friendly, and both men in 1818, the year Monge died, published papers in Gergonne's *Annales*, Poncelet arguing for the superiority of synthetic geometry and Gergonne defending analytic methods. But by 1826 there arose a priority controversy over the newly discovered principle of duality. We saw earlier how the theorems of Pascal and Brianchon were related through a simple interchange of the words point and line, and Gergonne had become convinced that analytic methods would show that such an interchange is universally valid. That is, for any theorem of plane geometry involving points and lines, Gergonne confidently assumed that the dual of this theorem, obtained by interchanging the words point and line, also will be valid, and he began publishing pairs of dual theorems in parallel columns in his *Annales*. Poncelet argued that he had discovered duality first, and that the principle was a consequence of the relationships in pure geometry between a pole and its polar line with respect to a conic.

4   Priority in recognition of the principle of duality is not easily determined, but in 1829 the logical justification of the principle was firmly established by Julius Plücker (1801–1868) through an important new point of view in analytic geometry. As Monge had been perhaps the first modern specialist in geometry in general, so Plücker became the first specialist in analytic geometry in particular. His earliest publications in Gergonne's *Annales* in 1826 had been largely synthetic, but inadvertently he became so embroiled in controversy with Poncelet that he forsook the camp of the synthesists and became the most prolific of all analytic geometers. Algebraic methods, he came to believe firmly, were much to be preferred to the purely geometric approach of Poncelet and Steiner. That his name survives in coordinate geometry in what is called Plücker's abridged notation is a tribute to his influence, although in this case the phrase does him more than justice. During the early nineteenth century a number of men, including Gergonne, had recognized that analytic geometry was burdened by awkwardness in algebraic computation; hence they began to abbreviate notations drastically. The family of all circles through the intersection of the two circles $x^2 + y^2 + ax + by + c = 0$ and $x^2 + y^2 + a'x + b'y + c' = 0$, for instance, was written by Gabriel Lamé (1795–1870) in 1818 simply as $mC + m'C' = 0$, using two parameters or multipliers m and m'. Gergonne and Plücker preferred a single Greek multiplier, the former writing $C + \lambda C' = 0$, from which we have the word "lambdalizing," and the latter using $C + \mu C' = 0$, resulting in the phrase "Plücker's μ." Lamé seems to have been the initiator in the study in analytic geometry of one-parameter families through abridged notation, but it was Plücker who, especially during the years 1827–1829, carried this study furthest. Incidentally, that the linear pencil of circles $C + \mu C' = 0$ form an interesting "radical family," whether or not $C_1 = 0$ and $C_2 = 0$ intersect, had been recognized some fifteen years earlier by L. Gaultier in connection with pure geometry; hence the radical axis (the straight-line member of the family $C + \mu C' = 0$) is sometimes known as the "line of Gaultier." This line has the property that from any point on it the tangents drawn to members of the radical family of circles are equal—or, as Steiner expressed it, the "power" of a point on the radical axis with respect to members of the family is the same for all circles in the family.

Among the many uses Plücker made of abridged notation was one of 1828, in Gergonne's *Annales*, in which he explained the Cramer–Euler paradox. If, for example, one has fourteen random points in a plane, the quartic curve through these points can be written as $Q + \mu Q' = 0$, where $Q = 0$ and $Q' = 0$ are distinct quartics through the same thirteen of the fourteen given points. Let μ be so determined that the coordinates of the fourteenth point satisfy $Q + \mu Q' = 0$. Then $Q = 0$, $Q' = 0$, and $Q + \mu Q' = 0$ all have in common not only the original *thirteen* points, but also all *sixteen* points of intersection of $Q = 0$ and $Q' = 0$. Hence with any set of thirteen points there are three additional points dependent upon, or associated with, the original thirteen, and no set of fourteen or more points selected from the combined set of sixteen dependent points will determine a unique quartic curve, despite the fact that a random set of fourteen points will in general determine a quartic curve uniquely. More generally, any given set of

$$\frac{n(n + 3)}{2} - 1$$

random points will determine a concomitant set of

$$n^2 - \left[\frac{n(n + 3)}{2} - 1\right] = \frac{(n - 1)(n - 2)}{2}$$

additional "dependent" points such that any curve of degree n through the given set of points will pass also through the dependent points.[8] Plücker gave also a dual of his theorem on the paradox, as well as generalizations to surfaces in three dimensions.

It was Plücker who, in the first volume of his *Analytisch-geometrische* **5** *Entwicklungen* (1828), elevated the abridged notation of Lamé and Gergonne to the status of a principle; in the second volume of this influential work (1831) Plücker effectively rediscovered a new system of coordinates that had been independently invented three times before. This was what we now call

[8] See Charlotte A. Scott, "On the Intersections of Plane Curves," *Bulletin of the American Mathematical Society*, **4** (1897–1898), 260–273.

homogeneous coordinates, of which Feuerbach was one inventor.[9] Another discoverer was A. F. Möbius (1790–1860) who published his scheme in 1827 in a work with the title *Barycentrische Calcul*. The author of this classic is best known, however, for the one-sided surface that bears his name—the Möbius strip or band obtained by joining the ends of a segment of ribbon after one end has been turned upside down. Still another inventor of homogeneous coordinates was Étienne Bobillier (1797–1832), a graduate of the École Polytechnique who published his new coordinate system in Gergonne's *Annales* for 1827–1828. The notations and patterns of reasoning of the four inventors of homogeneous coordinates differed somewhat,[10] but they all had one thing in common—they made use of *three* coordinates instead of *two* to locate a point in a plane. The systems were equivalent to what are also known as trilinear coordinates. Plücker, in fact, at first specifically took his three coordinates x, y, and t of a point P in a plane to be the three distances of P from the sides of a triangle of reference. Later in Volume II of his *Analytisch-geometrische Entwicklungen* he gave the more usual definition of homogeneous coordinates as any set of ordered number triples (x, y, t) related to the Cartesian coordinates (X, Y) of P such that $x = Xt$ and $y = Yt$. It will be apparent immediately that the homogeneous coordinates of a point P are not unique, for the triple (x, y, t) and (kx, ky, kt) correspond to the same Cartesian pair $(x/t, y/t)$. This lack of uniqueness, however, causes no more difficulty than does the lack of uniqueness in polar coordinates or the lack of uniqueness of form in the case of equal fractions. The name homogeneous stems, of course, from the fact that when one uses the equations of transformation to convert the equation of a curve $f(X, Y) = 0$ in rectangular Cartesian coordinates to the form $f(x/t, y/t) = 0$, the new equation will contain terms all of the same degree in the variables x, y, and t. More importantly, it will be noted that there is in the system of Cartesian coordinates no number pair corresponding to a homogeneous plane number triple of the form $(x, y, 0)$. Such a triple (provided that x and y are not both zero) designates an ideal point, or a "point at infinity." At long last the infinite elements of Kepler, Desargues, and Poncelet had been tied down to a coordinate system of ordinary numbers. Moreover, just as any ordered triple of real numbers (not all zero) in homogeneous coordinates corresponds to a point in a plane, so also does every linear equation $ax + by + ct = 0$ (provided that a, b, and c are not all zero) correspond to a straight line in the plane. In particular, all the "points at infinity" in the plane obviously lie on the line given by the equation $t = 0$, known as the line at infinity or the ideal line in the plane.

[9] See Albert Keifer, *Die Einführung der homogenen Koordinaten durch* K. W. Feuerbach: M. D. Schauberg (Strassburg, 1910).

[10] For further details and references see C. B. Boyer, *History of Analytic Geometry* (1956), pp. 241–244, 249–252.

It is obvious that this new system of coordinates is ideally suited to the study of projective geometry, which up to this time had been approached almost exclusively from the point of view of pure geometry.

Homogeneous coordinates were a big step in the direction of the arithmetization of geometry, but in 1829 Plücker contributed to Crelle's *Journal* a paper with a revolutionary point of view that broke completely with the old Cartesian view of coordinates as line segments. The equation of a straight line in homogeneous coordinates has the form $ax + by + ct = 0$. The three coefficients or parameters (a, b, c) determine a unique straight line in the plane, just as the three homogeneous coordinates (x, y, t) correspond to a unique point in the plane. Inasmuch as coordinates are numbers, hence not unlike coefficients, Plücker saw that one could modify the usual language and call (a, b, c) the homogeneous *coordinates* of a line. If, finally, one reverses the Cartesian convention so that letters at the beginning of the alphabet designate variables and those near the end of the alphabet constants, the equation $ax + by + ct = 0$ represents a pencil of lines through the fixed point (x, y, t), rather than a pencil of points on the fixed line (a, b, c). If, now, one considers the noncommittal equation $pu + qv + rw = 0$, it is clear that one can consider this indifferently as the totality of points (u, v, w) lying on the fixed line (p, q, r) or as the totality of lines (p, q, r) through the fixed point (u, v, w).

Plücker had discovered the immediate analytic counterpart of the geometric principle of duality, about which Gergonne and Poncelet had quarreled; it now became clear that the justification that pure geometry had sought in vain was here supplied by the algebraic point of view. The interchange of the words "point" and "line" corresponds merely to an interchange of the words "constant" and "variable" with respect to the quantities p, q, r and u, v, w. From the symmetry of the algebraic situation it is clear that every theorem concerning $pu + qv + rw = 0$ appears immediately in two forms, one the dual of the other. Moreover, Plücker showed that every curve (other than a straight line) can be regarded as having a dual origin: it is a locus generated by a moving point and enveloped by a moving line, the point moving continuously along the line while the line continues to rotate about the point. Oddly enough, the degree of a curve in point coordinates (the "order" of the curve) need not be the same as the degree of the curve in line coordinates (the "class" of the curve), and one of Plücker's great achievements, published in Crelle's *Journal* for 1834, was the discovery of four equations, bearing his name, that relate the class and order of a curve with the singularities of the curve:

$$m = n(n - 1) - 2\delta - 3\kappa \quad \text{and} \quad n = m(m - 1) - 2\tau - 3\iota$$

$$\iota = 3n(n - 2) - 6\delta - 8\kappa \quad \text{and} \quad \kappa = 3m(m - 2) - 6\tau - 8\iota$$

where m is the class, n the order, δ the number of nodes, κ the number of cusps, ι the number of stationary tangents (points of inflection), and τ the number of bitangents. From these equations it is clear at a glance that a conic (of order two) can have no singularities and thus must also be of class two.

In a paper in Crelle's *Journal* for 1831 Plücker had extended the principle of duality to three dimensions, where the relationships between homogeneous coordinates (a, b, c, d) of a plane and the homogeneous coordinates (x, y, z, t) of a point showed that the dual of a theorem in three-space is obtained through an interchange of the words "point" and "plane," the word "line" remaining unchanged. The French geometer Michel Chasles (1793–1880) claimed to have had the idea of line and plane coordinates at about the same time as Plücker, adding still another instance of simultaneity of discovery in nineteenth-century geometry.[11] In later papers and volumes Plücker extended his work to include imaginary Cartesian and homogeneous coordinates. It was now a trivial matter to justify Poncelet's theorem that all circles have in common two imaginary points at infinity, for the points $(1, i, 0)$ and $(i, 1, 0)$ both satisfy the equation $x^2 + y^2 + axt + byt + ct^2 = 0$, no matter what values a, b, c may take on. Plücker showed also that the foci of conics have the property that the imaginary tangents from these points to the curve pass through the above two circular points; he therefore defined a focus of a higher plane curve as a point having this property.

7   During the days of Descartes and Fermat, and again during the time of Monge and Lagrange, France had been the center for the development of analytic geometry, but with the work of Plücker leadership in the field crossed the Rhine to Germany. Nevertheless, Plücker was to a considerable extent the proverbial prophet without honor in his own country. There Steiner was inordinately admired; and Steiner had taken an intense dislike to analytic methods. The term analysis implies a certain amount of technique or machinery. Analysis often is referred to as a tool, a term never applied to synthesis; and Steiner objected to all kinds of tools or "props" in geometry. Calculation, he argued, replaces thinking, whereas geometry should stimulate thinking. Möbius remained neutral in the analysis-synthesis controversy, but Jacobi, despite the fact that he himself was an algorithm-builder, joined Steiner in polemically opposing Plücker.[12] Discouraged, Plücker in 1847 turned from geometry to physics, where he published a series of papers on magnetism and spectroscopy.

---

[11] See C. B. Boyer, *History of Analytic Geometry*, p. 251, note 65.

[12] An excellent account of Plücker's work is found in Wilhelm Ernst, *Julius Plücker* (1933). See also Alfred Clebsch, "Notice sur les travaux de Jules Plücker," trans. by Paul Mansion, *Bullettino di Bibliografia e di Storia delle Scienze Matematiche e Fisiche*, **5** (1872), 183–212.

It is one of the ironies of history that Plücker's work found support where one would least have expected it. England throughout the eighteenth century had been a stronghold of synthetic geometry; while Monge and Lagrange were engineering the analytical revolution in France, coordinate geometry in England had scarcely gone beyond the work of Newton. Even Wallis' *Conics* had fallen into disuse at Cambridge, where interest in mathematics was at a low ebb at the opening of the nineteenth century.[13] The rapid rise in French and German mathematical research had made little impression in Great Britain. Faced by this situation, a group of young Cambridge mathematicians in 1812 formed what they called the Analytical Society. In the words of Charles Babbage (1792–1871), one of the leaders, the aim of the Society was to promote "the principles of pure *d*-ism as opposed to the *dot*-age of the university." (A second aim of the Society was "to leave the world wiser than they found it.") This was, of course, a reference to the continued refusal of the English to abandon the dotted fluxions of Newton for the differentials of Leibniz; more generally it also implied a desire to take advantage of the great strides in mathematics that had been made on the Continent.[14] In 1816, as a result of the Society's inspiration, an English translation of Lacroix's one-volume *Calculus* was published, and within a few years British mathematicians were in a position to vie with their contemporaries on the Continent. For example, George Green (1793–1841), a self-educated miller's son, in 1828 published for private circulation an essay on electricity and magnetism that contained the important theorem bearing his name: if $P(x, y)$ and $Q(x, y)$ have continuous partial derivatives over a region R of the xy-plane bounded by a curve C, then $\int_C P \, dx + Q \, dy = \iint_R (Q_x - P_y) \, dx \, dy$. This theorem, or its analogue in three dimensions, also is known as Gauss's theorem, for Green's results were largely overlooked until rediscovered by Lord Kelvin in 1846. The theorem meanwhile had been discovered also by Michel Ostrogradski (1801–1861), and in Russia it bears his name to this day.

The awakening of mathematics in England is typified by the fact that in **8** 1839 the *Cambridge Mathematical Journal* was founded. Shortly after this England produced one of the most prolific mathematicians of all times, a man who was rivaled in productivity only by Euler and Cauchy. This was Arthur Cayley (1821–1895), a brilliant student at Cambridge who won most of the prizes in mathematics and who from an early age contributed papers to the *Cambridge Mathematical Journal* and its successors. Cayley was primarily an algebraist, rather than a specialist in geometry; but it was on the

[13] See W. W. R. Ball, *A History of the Study of Mathematics at Cambridge* (Cambridge, 1889).
[14] See J. M. Dubbey, "The Introduction of the Differential Notation to Great Britain," *Annals of Science*, **19** (1963), 37–48.

algebraic side that Plücker had been weakest. One notes with surprise that Plücker had not taken advantage of developments in determinants, possibly because of his feud with Jacobi; this may have been why he did not systematically develop an analytic geometry of more than three dimensions. Plücker had come close to this notion through his observation in 1846 that the four parameters determining a line in three-dimensional space can be thought of as four coordinates; but only long afterward, in 1865, did he return to analytic geometry and develop the idea of a "new geometry of space"—a four-dimensional space in which straight lines, rather than points, were the basic elements. Meanwhile, Cayley in 1843 had initiated the ordinary analytic geometry of n-dimensional space, using determinants as an essential tool. In this notation, using homogeneous coordinates, the equations of the line and plane, respectively, can be written as

$$\begin{vmatrix} x & y & t \\ x_1 & y_1 & t_1 \\ x_2 & y_2 & t_2 \end{vmatrix} = 0 \quad \text{and} \quad \begin{vmatrix} x & y & z & t \\ x_1 & y_1 & z_1 & t_1 \\ x_2 & y_2 & z_2 & t_2 \\ x_3 & y_3 & z_3 & t_3 \end{vmatrix} = 0$$

Cayley pointed out that the corresponding fundamental $(n - 1)$ dimensional element in n-dimensional space can be expressed in homogeneous coordinates by a determinant, similar to those above, of order $n + 1$. Many of the simple formulas for two and three dimensions, when properly expressed, can easily be generalized to n dimensions. In 1846 Cayley published a paper in Crelle's *Journal* in which he again extended some theorems from three dimensions to a space of four dimensions; in 1847 Cauchy in the *Comptes Rendus* published an article in which he considered "analytical points" and "analytical lines" in space of more than three dimensions.[15]

9   Only a year after the publication of Cayley's first paper on the geometry of higher dimensions, somewhat similar views appeared in Germany in the *Ausdehnungslehre* of Hermann Grassmann (1809–1877). In forbiddingly novel notation and obscure exposition, Grassmann tried to build a calculus of "extensive magnitudes" involving an indefinite number of elements or dimensions, but his efforts to build a kind of vector analysis for n dimensions met with little recognition from his contemporaries.[16] Perhaps one reason for this was that the towering figure of Steiner continued to champion pure

---

[15] For excerpts from these and other works see *Source Book in Mathematics*, ed. by D. E. Smith, pp. 524–545.

[16] A summary account of Grassmann's ideas is found in J. L. Coolidge, *A History of Geometrical Methods* (1963), pp. 252–257, and a translation of a portion of Grassmann's *Ausdehnungslehre* appears in the Smith *Source Book in Mathematics*, pp. 684–696.

geometry, Steiner, in his *Systematische Entwicklungen* of 1832, had produced a treatment of projective geometry based on metrical considerations. Some years later pure geometry found another German devotee in K. G. C. von Staudt (1798–1867), whose *Geometrie der Lage* of 1847 built up projective geometry without reference to magnitude or number. In France the work of Poncelet had been continued by Chasles, also a graduate of the École Polytechnique, where he became professor of geometry. To Chasles was due the emphasis on the six cross-ratios or anharmonic ratios

$$\frac{c-a}{c-b} \Big/ \frac{d-a}{d-b}$$

of four collinear points or four concurrent lines, and the invariance of these under projective transformations. His *Traité de géométrie supérieure* (1852) was influential also in establishing the use of directed line segments in pure geometry. Chasles, who is noted also for his *Aperçu historique sur l'origine et la développement des méthodes en géométrie* (1837), was one of the last great projective geometers in France, and it was chiefly in Germany that his work was continued by such men as Steiner and von Staudt. To the latter in particular is due much of the form that synthetic projective geometry has taken.[17]

It has been difficult to present a picture of geometrical developments **10** in the first half of the nineteenth century because of the crosscurrents and interrelations of the multifarious aspects of the subject, but there was one aspect, the rise of non-Euclidean geometry, that developed independently of the movements described. Nevertheless, in this case also, we find a startling case of simultaneity of discovery, for similar notions occurred, during the first third of the nineteenth century, to three men, one German, one Hungarian, and one Russian. We already have noted that Gauss during the second decade of the century had come to the conclusion that the efforts to prove the parallel postulate made by Saccheri, Lambert, Legendre, and his Hungarian friend Farkas Bolyai were in vain and that geometries other than Euclid's were possible. However, he had not shared this view with others; he had simply elaborated the idea, as he said, "for himself." Hence efforts to prove the parallel postulate continued, and among those attempting such a proof was young Nicolai Ivanovitch Lobachevsky (1793–1856), the son of a minor government official who died when Nicolai was only seven. Lobachevsky attended the University of Kazan, despite the straitened circumstances of the family, and there he came in touch with distinguished professors whom the university had attracted from Germany, including J. M. Bartels (1769–

[17] For further details see J. L. Coolidge, *A History of Geometrical Methods*, pp. 92–101.

1836), a man under whom Gauss had earlier studied. At the early age of twenty-one Lobachevsky became a member of the teaching staff, and by 1827 he had been appointed Rector of the university. There he remained, as teacher and administrator, to the end of his days, despite the fact that blindness and lack of appreciation of his work saddened his last years.

Lobachevsky and Ostrogradsky were both eminent Russian mathematicians, but they differed sharply on things mathematical and political. Ostrogradsky had studied extensively at Paris, where he came under the influence of the French analysis of Cauchy—a conventional subject in which rapid progress was being made. Lobachevsky, on the other hand, had been brought up with a more German and geometrical background, where the frontiers and direction of progress were more controversial. Moreover, Ostrogradsky came from a prosperous, aristocratic, and conservative social background, whereas Lobachevsky, constantly faced by poverty and privation, never enjoyed social position and often espoused unpopular liberal causes. Thus it was that in their days Ostrogradsky enjoyed an esteem that Lobachevsky did not; but today the name of Ostrogradsky is known, if at all, in connection with a single theorem, whereas Lobachevsky is regarded as the "Copernicus of geometry," the man who revolutionized the subject through the creation of a whole new branch, Lobachevskian geometry, showing thereby that Euclidean geometry was not the exact science or absolute truth it previously had been taken to be. In a sense the discovery of non-Euclidean geometry dealt a devastating blow to Kantian philosophy comparable to the effect on Pythagorean thought resulting from the disclosure of incommensurable magnitudes. Through the work of Lobachevsky it became necessary to revise fundamental views of the nature of mathematics; but Lobachevsky's colleagues were too close to the situation to see it in proper perspective, and the trailblazer had to pursue his thoughts in lonely isolation.

11   Lobachevsky's revolutionary view seems not to have come to him as a sudden inspiration. In an outline of geometry that he drew up in 1823, presumably for classroom use, Lobachevsky said of the parallel postulate simply that "no rigorous proof of the truth of this has ever been discovered."[18] Apparently he did not then exclude the possibility that such a proof might yet be discovered. Three years later at Kazan University he read in French a paper (now lost) on the principles of geometry, including "une démonstration rigoureuse du théorème des paralleles." The year 1826 in which this paper was delivered may be taken as the unofficial birthdate of Lobachevskian geometry, for it was then that the author presented many of the characteristic

[18] See V. Kagan, *N. Lobachevsky and his Contribution to Science* (1957), p. 33; also Alexander Vucinich, "Nikolai Ivanovich Lobachevskii: The Man behind the First Non-Euclidean Geometry," *Isis*, **53** (1962), 465–481.

theorems of the new subject. Another three years later, in the *Kazan Messenger* for 1829, Lobachevsky published an article, "On the Principles of Geometry," which marks the official birth of non-Euclidean geometry. Between 1826 and 1829 he had become thoroughly convinced that Euclid's fifth postulate cannot be proved on the basis of the other four, and in the paper of 1829 be became the first mathematician to take the revolutionary step of publishing a geometry specifically built on an assumption in direct conflict with the parallel postulate: Through a point *C* lying outside a line *AB* there can be drawn more than one line in the plane and not meeting *AB*. With this new postulate Lobachevsky deduced a harmonious geometrical structure having no inherent logical contradictions. This was in every sense a valid geometry, but so contrary to common sense did it appear, even to Lobachevsky, that he called it "imaginary geometry."

Lobachevsky was well aware of the significance of his discovery of "imaginary geometry," as is clear from the fact that during the score of years from 1835 to 1855 he wrote out three full accounts of the new geometry. In 1835–1838 his *New Foundations of Geometry* appeared in Russian; in 1840 he published *Geometrical Investigations on the Theory of Parallels* in German; and in 1855 his last book, *Pangeometry*, was published simultaneously in French and Russian. (All have since been translated into other languages, including English.) From the second of the three works Gauss learned of Lobachevsky's contributions to non-Euclidean geometry, and it was on his recommendation that Lobachevsky in 1842 was elected to the Göttingen Scientific Society. In letters to friends Gauss praised Lobachevsky's work, but he never gave it support in print, for he feared the jibes of "the Bœotians." Partly for this reason the new geometry became known only very slowly.

The Hungarian friend of Gauss, Farkas Bolyai, had spent much of his **12** life trying to prove the parallel postulate, and when he found that his own son Janos Bolyai (1802–1860) was absorbed in the problem of parallels, the father, a provincial mathematics teacher, wrote to the son, a dashing army officer:

> For God's sake, I beseech you, give it up. Fear it no less than sensual passions because it, too, may take all your time, and deprive you of your health, peace of mind, and happiness in life.

The son, not dissuaded, continued his efforts until in about 1829 he came to the conclusion reached only a few years before by Lobachevsky. Instead of attempting to prove the impossible, he developed what he called the "Absolute Science of Space," starting from the assumption that through a point not on a line infinitely many lines can be drawn in the plane, each parallel to the given line. Janos sent his reflections to his father, who published them in the form of an appendix to a treatise that he had completed, bearing a long Latin title beginning with *Tentamen*. The elder Bolyai's *Tentamen* bears an imprimatur dated 1829, the year of Lobachevsky's *Kazan Messenger* article, but it did not actually appear until 1832.

The reaction of Gauss to the "Absolute Science of Space" was similar to that in the case of Lobachevsky—sincere approval, but lack of support in print. When Farkas Bolyai wrote to ask for an opinion on the unorthodox work of his son, Gauss replied that he could not praise Janos' work, for this would mean self-praise, inasmuch as he had held these views for many years. The temperamental Janos was understandably disturbed, fearing that he would be deprived of priority. Continued lack of recognition, as well as the publication of Lobachevsky's work in German in 1840, so upset him that he published nothing more. The lion's share of the credit for the development of non-Euclidean geometry consequently belongs to Lobachevsky.[19]

13   Non-Euclidean geometry continued for several decades to be a fringe aspect of mathematics until it was thoroughly integrated through the remarkably general views of G. F. B. Riemann (1826–1866). The son of a village pastor, Riemann was brought up in very modest circumstances, always remaining frail in body and shy in manner. He nevertheless secured a good education, first at Berlin and later at Göttingen, where he took his doctorate with a thesis in theory of functions of a complex variable. It is here that we find the so-called Cauchy-Riemann equations, $u_x = v_y, u_y = -v_x$, which an analytic function $w = f(z) = u + iv$ of a complex variable $z = x + iy$ must satisfy, although this requirement had been known even in the days of Euler and d'Alembert.[20] The thesis also led to the concept of a Riemann surface, anticipating the part that topology ultimately was to play in analysis.

In 1854 Riemann became *Privatdozent* at the University of Göttingen, and according to custom he was called upon to deliver a *Habilitationschrift* before the faculty. The result in Riemann's case was the most celebrated probationary lecture in the history of mathematics, for it presented a deep and broad view of the whole field of geometry. The thesis bore the title "Über die Hypothesen welche der Geometrie zu Grunde liegen" (On the Hypotheses Which Lie at the Foundation of Geometry), but it did not present a specific example. It urged instead a global view of geometry as a study of manifolds of any number of dimensions in any kind of space.

---

[19] Portions of the works of both Lobachevsky and Bolyai appear in English in many places, including the Smith *Source Book*. A very full account is found in Friedrich Engel and Paul Stäckel, *Urkunden zur Geschichte der nichteuklidischen Geometrie* (1898–1913). Among the more extensive accounts in English is that in Roberto Bonola, *Non-Euclidean Geometry* (1955), which contains translations of Lobachevsky's *Theory of Parallels* and Bolyai's *Science Absolute of Space*.

[20] See E. T. Bell, *Development of Mathematics* (New York: McGraw-Hill, 1940), p. 465.

His geometries are non-Euclidean in a far more general sense than is Lobachevskian, where the question is simply how many parallels are possible through a point. Riemann saw that geometry should not even necessarily deal with points or lines or space in the ordinary sense, but with sets of ordered n-tuples that are combined according to certain rules.

Among the most important rules in any geometry, Riemann saw, is that for finding the distance between two points that are infinitesimally close together. In ordinary Euclidean geometry this "metric" is given by $ds^2 = dx^2 + dy^2 + dz^2$; but infinitely many other formulas can be used as a distance formula, and of course the metric used will determine the properties of the space or the geometry. A space whose metric is of the form

$$ds^2 = g_{11}\, dx^2 + g_{12}\, dx\, dy + g_{13}\, dx\, dz$$
$$+ g_{21}\, dy\, dx + g_{22}\, dy^2 + g_{23}\, dy\, dz$$
$$+ g_{13}\, dz\, dx + g_{23}\, dz\, dy + g_{33}\, dz^2$$

where the g's are constants or, more generally, functions of x, y, and z, is known as a Riemannian space. Thus (locally) Euclidean space is only the very special case of a Riemannian space in which $g_{11} = g_{22} = g_{33} = 1$ and all the other g's are zero. Riemann even developed from the metric a formula for the Gaussian curvature of a "surface" in his "space." It is no wonder that after Riemann's lecture, and for almost the only time in his long career, Gauss expressed enthusiasm for the work of someone else.[21]

There is a more restricted sense in which we today use the phrase Riemannian geometry: the plane geometry that is deduced from Saccheri's hypothesis of the obtuse angle if the infinitude of the straight line is also abandoned. A model for this geometry is found in the interpretation of the "plane" as the surface of a sphere and of a "straight line" as a great circle on the sphere. In this case the angle-sum of a triangle is greater than two right angles, whereas in the geometry of Lobachevsky and Bolyai (corresponding to the hypothesis of the acute angle) the angle-sum is less than two right angles. This use of Riemann's name, however, fails to do justice to the fundamental change in geometrical thought that his 1854 *Habilitationschrift* (not published until 1867) brought about. It was Riemann's suggestion of the general study of curved metric spaces, rather than of the special case equivalent to geometry on the sphere, that ultimately made the theory of general relativity possible. Riemann himself contributed heavily to theoretical physics in a number of directions, and it was therefore fitting that in 1859 he should have been appointed as successor to Dirichlet in the chair at Göttingen that Gauss had filled.

[21] See E. T. Bell, *Men of Mathematics* (New York: Simon and Schuster, 1937), pp. 484–509.

In showing that non-Euclidean geometry with angle-sum greater than two right angles is realized on the surface of a sphere, Riemann essentially verified the consistency of the axioms from which the geometry is derived. In much the same sense Eugenio Beltrami (1835–1900), a colleague of Cremona at Bologna and later professor at Pisa, Pavia, and Rome, showed that there was at hand a corresponding model for Lobachevskian geometry. This is the surface generated through the revolution of a tractrix about its asymptote, a surface known as a pseudosphere inasmuch as it has constant negative curvature, as the sphere has constant positive curvature. If we define the "straight line" through two points on the pseudosphere as the geodesic through the points, the resulting geometry will have the properties resulting from the Lobachevskian postulates. Inasmuch as the plane is a surface with constant zero curvature, Euclidean geometry can be regarded as an intermediary between the two types of non-Euclidean geometry.

14   The unification of geometry that Riemann had achieved was especially relevant in the microscopic aspect of differential geometry, or geometry "in the small." Analytic geometry, or geometry "in the large," had not been much changed. In fact, Riemann's lecture was given at about the midpoint of Plücker's self-imposed geometrical retirement, during which there had been something of a lull in analytic geometric activity in Germany. In 1865 Plücker again resumed mathematical publication, this time in British publications instead of in Crelle's *Journal*, probably because Cayley had shown interest in Plücker's work. In this year he published a paper in the *Philosophical Transactions* (often known simply as *Phil. Trans.*), expanded three years later into a book, on a "New Geometry of Space." Here he explicitly formulated a principle at which he had hinted about twenty years before. A space, he argued, need not be thought of as a totality of points; it can equally well be visualized as composed of lines. In fact, any figure that formerly had been thought of as a locus or totality of points can itself be taken as a space *element*, and the dimensionality of the space will correspond to the number of parameters determining this element. If our ordinary three-space is considered a "cosmic haystack of infinitely thin, infinitely long straight straws," rather than an "agglomeration of infinitely fine birdshot,"[22] it is four-dimensional rather than three-dimensional. In 1868, the year of Plücker's book based on this theme, Cayley developed analytically in the *Phil. Trans.* the notion of the ordinary two-dimensional Cartesian plane as a space of five dimensions, the elements of which are conics. There are in Plücker's *Neue Geometrie des Raumes* also other new ideas. The geometrical representation of a single equation $f(x, y, z) = 0$ in

---

[22] See E. T. Bell, *Men of Mathematics*, p. 400.

point coordinates is called a surface, two simultaneous equations correspond to a curve, and three equations determine one or more points. In the "new geometry" of his four-dimensional line-space Plücker called the "figure" represented by a single equation $f(r, s, t, u) = 0$ in the four coordinates of his line-space a "complex," two equations designated a "congruence," and three a "range." He found that the quadratic line complex has properties similar to those of the quadric surface, but he did not live to complete the extensive study he planned. He died in 1868, the year in which his *New Geometry* appeared, edited by one of his students, Felix Klein (1849–1925).

Klein had been Plücker's assistant at the University of Bonn during the **15** latter's return to geometry, and in a sense he was Plücker's successor in his enthusiasm for analytic geometry. However, the young man's work in the field took a different direction—one that served to bring some element of unity into the diversity of new results of research. The new view may have been in part the result of visits to Paris, where Lagrange's hints of group theory had been developed, especially through substitution groups, into a full-blown branch of algebra. Klein was deeply impressed by the unifying possibilities in the group concept, and he spent much of the rest of his life in developing, applying, and popularizing the notion. In some of this work he collaborated with the Norwegian mathematician Sophus Lie (1842–1899), fellow student with Klein at Göttingen who discovered contact transformations and wrote a ponderous three-volume treatise on the theory of transformation groups (1888–1893). Lie's contact transformations, systematized by Klein, set up a one-to-one correspondence between the lines and spheres in Euclidean space in such a way that intersecting lines correspond to tangent spheres.[23] (In conformity with Plücker's view, the lines and spheres in Euclidean three-space each constitute a four-dimensional space.) In general, contact transformations are analytic transformations that carry tangent surfaces into tangent surfaces.

A set of elements are said to form a group with respect to a given operation if (1) the set of elements is closed under the operation, (2) the set contains an identity element with respect to the operation, (3) for every element in the set there is an inverse element with respect to the operation, and (4) the operation is associative. The elements can be numbers (as in arithmetic), points (in geometry), transformations (in algebra or geometry), or anything at all. The operation can be arithmetic (such as addition or multiplication) or geometric (as a rotation about a point or an axis), or any other rule for combining two elements of a set (such as two transformations) to form a third element in the set. The generality of the group concept is readily apparent;

---

[23] See Coolidge. *History of Geometrical Methods.* pp. 298 ff.

Klein, in a celebrated inaugural address in 1872, when he became professor at Erlangen, showed how it could be applied as a convenient means of characterizing the various geometries that had appeared during the century.

The address that Klein gave, which became known as the *Erlanger Programm*, described geometry as the study of those properties of figures that remain invariant under a particular group of transformations. Hence any classification of groups of transformations becomes a codification of geometries. Plane Euclidean geometry, for example, is the study of such properties of figures, including areas and lengths, as remain invariant under the group of transformations made up of translations and rotations in the plane—the so-called rigid transformations, equivalent to Euclid's unstated axiom that figures remain unchanged when moved about in a plane. Analytically the rigid plane transformations can be written in the form

$$\begin{cases} x' = ax + by + c \\ y' = dx + ey + f \end{cases}$$

where $ae - bd = 1$; these form the elements of a group. The "operation" that "combines" two such elements is simply that of performing the transformations in order. It is easy to see that if the transformation above is followed by a second,

$$\begin{cases} x'' = Ax' + By' + C \\ y'' = Dx' + Ey' + F \end{cases}$$

the result of the two operations performed successively is equivalent to some single operation of this type that will carry the point (x, y) into the point (x'', y'').

If in this transformation group one replaces the restriction that $ae - bd = 1$ by the more general requirement that $ae - bd \neq 0$, the new transformations also form a group. However, lengths and areas do not necessarily remain the same, but a conic of given type (ellipse, parabola, or hyperbola) will, under these transformations, remain a conic of the same type. Such transformations, studied earlier by Möbius, are known as affine transformations; they characterize a geometry known as "affine geometry," so called because a finite point goes into a finite point under any such transformation. It is clear, then, that Euclidean geometry, in Klein's view, is only a special case of affine geometry. Affine geometry in its turn becomes only a special case of a still more general geometry—projective geometry. A projective transformation can be written in the form

$$x' = \frac{ax + by + c}{dx + ey + f}, \qquad y' = \frac{Ax + By + c}{dx + ey + f}$$

It is clear that if $d = 0 = e$ and $f = 1$, the transformation is affine. Interesting properties of projective transformations include the fact that (1) a conic is transformed into a conic and (2) cross ratio remains invariant. Pappus had been aware of these properties a millennium and a half earlier, but he had no inkling of the group concept that made possible such neat classifications of geometries. In fact, for Pappus, there had been only *one* geometry, for the ideal points of projective geometry would have been unthinkable in antiquity. The *Erlanger Programm* of Klein was so clearly a product of the nineteenth century that it could not meaningfully be transferred to *any* earlier age. At first it had only a limited circulation, but before the end of the century it came to enjoy a wide influence throughout the international mathematical world.[24] The continuing influence of the *Erlanger Programm* today can be seen in almost any modern college survey of geometry.[25]

The work of Klein is in a sense a fitting climax to "The Heroic Age in Geometry," for he taught and lectured for half a century. So contagious was his enthusiasm that some late-nineteenth-century figures were willing to prophesy that not only geometry, but all of mathematics, ultimately would be comprised within the theory of groups. Nevertheless, not all of Klein's work was concerned with groups. His classic history of mathematics in the nineteenth century (published posthumously)[26] shows how familiar he was with all aspects of the subject; his name is also recalled today in topology in the one-sided surface known as the Klein bottle. He was much concerned with non-Euclidean geometry, to which he contributed the names "elliptic geometry" and "hyperbolic geometry" for the hypotheses of the obtuse and acute angle respectively; for the latter he proposed a simple model as an alternative to that of Beltrami. Let the hyperbolic plane be pictured as the points interior to a circle C in the Euclidean plane, let the hyperbolic "straight line" through two points P_1 and P_2 be that portion of the Euclidean line P_1P_2 that lies within C, and let the "distance" between the two points P_1 and P_2 within the circle be defined as

$$\ln \frac{P_2Q_1 \cdot P_1Q_2}{P_1Q_1 \cdot P_2Q_2}$$

where Q_1 and Q_2 are the points of intersection of the line P_1P_2 with the circle

16

---

[24] An English translation, under the title "A Comparative Review of Recent Researches in Geometry," appeared in 1893 in the second volume of the *Bulletin of the New York Mathematical Society* (now the *Bulletin of the American Mathematical Society*).

[25] See, for instance, Annita Tuller, *A Modern Introduction to Geometries* (Princeton, N. Y.: D. Van Nostrand Co., 1967).

[26] *Vorlesungen über die Entwicklung der Mathematik im 19. Jahrhundert* (1926–1927).

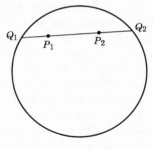

FIG. 24.1

C (Fig. 24.1). With an appropriate definition of "angle" between two "lines," the "points," "lines," and "angles" in Klein's hyperbolic model have properties similar to those in Euclidean geometry, except for the parallel postulate.

Not since Monge had there been a more influential teacher, for in addition to giving inspiring lectures, Klein was concerned with the teaching of mathematics at many levels and exerted a strong influence in pedagogical circles. In 1886 he became professor of mathematics at Göttingen, and under his leadership the university became the Mecca to which students from many lands, including America, flocked. In his later years Klein played very effectively the role of an "elder statesman" in the realm of mathematics.[27] Thus the golden age of modern geometry that had begun so auspiciously in France at the École Polytechnique, with the work of Lagrange, Monge, and Poncelet, reached its zenith in Germany, at the University of Göttingen, through the research and inspiration of Gauss, Riemann, and Klein.

BIBLIOGRAPHY

Bolyai, Johann, *Geometrische Untersuchungen*, ed. by Paul Stäckel (Leipzig: B. G. Teubner, 1913).

Bonola, Roberto, *Non-Euclidean Geometry*, trans. by H. S. Carslaw (New York: Dover reprint, 1955).

Boyer, C. B., *History of Analytic Geometry* (New York: Scripta Mathematica, 1956).

Cajori, Florian, *History of Mathematics*, 2nd ed. (New York: Macmillan, 1931).

Chasles, Michel, *Aperçu historique sur l'origine et la développement des méthodes en géométrie* (Bruxelles, 1837; 2nd ed., Paris, 1875).

Clebsch, Alfred, "Notice sur les travaux de Jules Plücker," trans. by Paul Mansion, *Bullettino di Bibliografia e di Storia delle Scienze Matematiche e Fisiche*, **5** (1872) 183–212.

[27] See, for example, G. A. Miller, "Felix Klein and the History of Mathematics," *Proceedings of the National Academy of Sciences*, **13** (1927), 611–613.

Coolidge, J. L., "The Heroic Age of Geometry," *Bulletin of the American Mathematical Society*, **35** (1929), 19–37.

Coolidge, J. L., *A History of Geometrical Methods* (New York: Dover reprint, 1963).

DeVries, Hk., "How Analytic Geometry Became a Science," *Scripta Mathematica*, **14** (1948), 5–15.

Engel, Friedrich, and Paul Stäckel, *Urkunden zur Geschichte der nichteuklidischen Geometrie* (Leipzig: B. G. Teubner, 1898–1913, 2 vols.).

Ernst, Wilhelm, *Julius Plücker* (Bonn, 1933).

Fano, G., "Gegensatz von synthetischer und analytischer Geometrie in seiner historischen Entwicklung im XIX. Jahrhundert," *Encyklopädie der Mathematischen Wissenschaften*, Vol. III, Part 1, first half, pp. 223–288; French translation in *Encyclopédie des Sciences Mathématiques*, Tome III, Vol. 1, pp. 185–259.

Kagan, V., N. *Lobachevsky and his Contribution to Science* (Moscow: Foreign Languages Publishing House, 1957).

Klein, Felix, *Vorlesungen über die Entwicklung der Mathematik im 19. Jahrhundert* (Berlin: Springer, 1926–1927, 2 vols.).

Klein, Felix, *Elementary Mathematics from an Advanced Standpoint*, Vol. II, *Geometry* (New York: Dover paperback, 1939).

Kötter, Ernst, "Die Entwickelung der synthetischen Geometrie, von Monge bis auf Staudt, 1847," *Jahresbericht der Deutscher Mathematiker-Vereinigung*, **5**, part 2 (1898–1901).

Loria, Gino, *Storia delle matematiche* (Turin: Sten, 1929–1935, 3 vols.).

Loria, Gino, *Il passato e il presente delle principali teorie geometriche*, 4th ed. Padua: Cedam, 1931).

Nagel, Ernest, "The Formation of Modern Conceptions of Formal Logic in the Development of Geometry," *Osiris*, **7** (1939), 142–224.

Patterson, Boyd C., "The Origins of the Geometric Principle of Inversion," *Isis*, **19** (1933), 154–180.

Pierpont, James, "The History of Mathematics in the Nineteenth Century," *Bulletin of the American Mathematical Society*, **11** (1904), 136–159.

Schmidt, Franz, and Paul Stäckel, eds., *Briefwechsel zwischen Carl Friedrich Gauss und Wolfgang Bolyai* (Leipzig, 1899).

Simon, Max, "Über die Entwicklung der Elementargeometrie im XIX. Jahrhundert," *Jahresbericht der Deutschen Mathematiker-Vereinigung, Ergänzungsbände*, **1** (1906).

Smith, D. E., *Source Book in Mathematics* (New York: Dover reprint, 1959).

Sommerville, D. M. Y., *Bibliography of Non-Euclidean Geometry* (London: Harrison, 1911).

Vucinich, Alexander, "Nikolai Ivanovich Lobachevskii, "The Man behind the First Non-Euclidean Geometry," *Isis*, **53** (1962), 465–481.

EXERCISES

1. Was the role of the École Polytechnique, an engineering school, a help or a hindrance in the nineteenth-century revival of pure geometry? Explain.

2. Cite half a dozen cases of independent and nearly simultaneous discovery in geometry during the first half of the nineteenth century, mentioning the parts played by the discoverers.

3. Name three leading geometers in France and three in Germany during the nineteenth century, citing some of their chief contributions and indicating whether they were primarily analysts or synthesists.

4. Was it chance or a natural consequence that non-Euclidean geometry first arose in Germany, Russia, and Hungary, rather than in France or England? Give reasons for your answer.

5. Were the discovery and the justification of the principle of duality the result of developments in synthetic geometry or in analytic geometry? Explain.

6. How do you account for the fact that England, a stronghold of geometry during the eighteenth century, failed to take the lead in the revival of pure geometry in the nineteenth century?

7. Describe several respects in which coordinate geometry in the nineteenth century differed from that of Fermat and Descartes.

8. Find the radical axis, or "line of Gaultier," of the pair of circles $x^2 + y^2 + 2x - 4y = 0$ and $(x - 2)^2 + (y - 4)^2 - 1 = 0$. Sketch several members of the radical family together with the radical axis.

9. Show that with Klein's definition of distance, in his model of hyperbolic geometry, three collinear points in the order P, Q, R have the property that $\overline{PQ} + \overline{QR} = \overline{PR}$.

10. Show that in Klein's model the line is infinitely long in the sense that the distance between two points on it increases indefinitely as one of the points is held fixed and the other tends toward a point of intersection of the line with the bounding circle.

11. For the three-four-five right triangle find the radii of the inscribed and escribed circles and the radius of the nine-point circle.

12. Construct, with straightedge and compasses, the inscribed circle, the three escribed circles, and the nine-point circle of a triangle.

13. Show that the nine-point circle of an isosceles right triangle is tangent to both the circumscribed and the inscribed circles.

14. Show that in plane homogeneous coordinates the inverse $P'(x', y', t')$ of $P(x, y, t)$ with with respect to the circle $x^2 + y^2 = a^2t^2$ is given by the equations $x' = a^2xt$, $y' = a^2yt$, $t' = x^2 + y^2$.

15. Show that the inverse of a parabola is not a parabola.

*16. Prove that the center of the nine-point circle of any triangle is midway between the circumcenter and the orthocenter.

*17. Prove that if the base angles of a quadrilateral are right angles, and if the two sides perpendicular to the base are equal, the summit angles are equal. Are these summit angles acute, right, or obtuse? Explain.

*18. Prove that the plane inverse of a circle not passing through the center of inversion is a circle not passing through the center of inversion. What is the inverse of a circle that passes through the center of inversion?

*19. Does an inversion of three-dimensional space with respect to a sphere transform a plane into a plane? Explain fully.

*20. Prove that a circle orthogonal to the circle of inversion is transformed into itself, but that the inverse of the center is not the center of the inverted circle.

*21. If $C_1 = 0$ and $C_2 = 0$ are two circles in a plane, describe the members of the family $K_1C_1 + K_2C_2 = 0$ for the following three cases: (a) $C_1 = 0$ and $C_2 = 0$ intersect; (b) $C_1 = 0$ and $C_2 = 0$ are tangent; (c) $C_1 = 0$ and $C_2 = 0$ have no point in common.

*22. Verify Green's theorem, using both a line integral and a double integral, for the case in which $P = xy$ and $Q = x^2 + y^2$ and the region R is that portion of the plane which is bounded by the unit square with vertices $(0, 0)$, $(1, 0)$, $(1, 1)$, and $(0, 1)$.

*23. Show that a conic under inversion is transformed into a cubic or a quartic according as the center of the inversion lies on the conic or not.

*24. What would be the dimensionality of a plane geometry in which the fundamental elements are not points, but (a) straight lines, or (b) circles, or (c) parabolas, or (d) conic sections, or (e) algebraic curves of third degree?

CHAPTER XXV

The Arithmetization
of Analysis

> In most sciences one generation tears down what
> another has built, and what one has established another
> undoes. In mathematics alone each generation builds a
> new story to the old structure.
>
> *Hermann Hankel*

1 Analysis, the study of infinite processes, had been understood by Newton
and Leibniz to be concerned with continuous magnitudes, such as lengths,
areas, velocities, and accelerations, whereas the theory of numbers clearly
has as its domain the discrete set of natural numbers. We have nevertheless
seen that Bolzano tried to give purely arithmetic proofs of propositions, such
as the location theorem in elementary algebra, that seemed to depend on
properties of continuous functions; and Plücker had thoroughly arithmetized
analytic geometry. The theory of groups had originally been concerned with
discrete sets of elements, but Klein envisioned a unification of both discrete
and continuous aspects of mathematics under the group concept. The
nineteenth century was indeed a period of correlation in mathematics, and
the arithmetization of analysis, a phrase coined by Klein in 1895, was one
aspect of this tendency.

The key word in analysis is, of course, "function," and it was especially
in the clarification of this term that the arithmetizing trend arose. Differences
of opinion on the representation of functions had developed before the
middle of the eighteenth century when d'Alembert and Euler had given
solutions for the problem of a vibrating string in so-called "closed form,"
using two arbitrary functions, whereas Daniel Bernoulli had found a solution
in terms of an infinite series of trigonometric functions. Since the latter
solution seemed to imply periodicity, whereas the arbitrary functions of
d'Alembert and Euler were not necessarily periodic, it looked as though
Bernoulli's solution was less general. That this was not the case was shown
in 1824 by J. B. J. Fourier (1768–1830).

Joseph Fourier was the son of a tailor in Auxerre; he received his education
through the Benedictine Order, in which he at one point intended to become

a priest. He became instead a teacher of mathematics, first at the local military school and later at the École Normale and the École Polytechnique. In 1798 he joined Monge in Napoleon's Egyptian adventure, becoming secretary of the Institut d'Egypte and compiling the *Description de l'Egypte.* On his return to France he held a number of administrative posts, but he had opportunity nevertheless to continue scholarly pursuits. He is best known today for his celebrated *Théorie analytique de la chaleur* of 1822. This book, described by Kelvin as "a great mathematical poem," was a development of ideas that ten years earlier had won him the Académie prize for an essay on the mathematical theory of heat. Lagrange, Laplace, and Legendre, the referees, had criticized the essay for a certain looseness of reasoning; the later clarification of Fourier's ideas was to some extent the reason that the nineteenth century come to be called the age of rigor.[1]

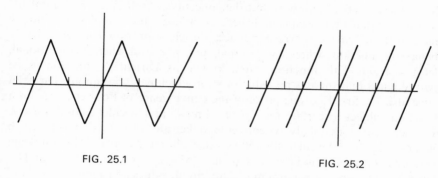

FIG. 25.1   FIG. 25.2

The chief contribution of Fourier and his classic in mathematics was the idea, adumbrated by Daniel Bernoulli, that any function $y = f(x)$ can be represented by a series of the form

$$y = \tfrac{1}{2}a_0 + a_1 \cos x + a_2 \cos 2x + \cdots + a_n \cos nx + \cdots$$
$$+ b_1 \sin x + b_2 \sin 2x + \cdots + b_n \sin nx + \cdots$$

now known as a Fourier series. Such a series representation affords considerably greater generality in the type of functions that can be studied than does the Taylor series. Even if there are many points at which the derivative does not exist (as in Fig. 25.1) or at which the function is not continuous (as in Fig. 25.2), the function may still have a Fourier expansion; this expansion is easily found on noting that

$$a_0 = \frac{1}{\pi} \int_{-\pi}^{\pi} f(x)\,dx, \qquad a_n = \frac{1}{\pi} \int_{-\pi}^{\pi} f(x) \cos nx\,dx$$

[1] See P. E. B. Jourdain, "Note on Fourier's Influence on the Conceptions of Mathematics," *International Congress of Mathematicians* (Cambridge, 1912), Vol. II, pp. 526–527.

and

$$b_n = \frac{1}{\pi} \int_{-\pi}^{\pi} f(x) \sin nx \, dx$$

Fourier, like Monge, had fallen from grace when the Bourbon restoration followed the exile of Napoleon in 1815, but his work has ever since been fundamental both in physics and in mathematics. Functions no longer needed to be of the well-behaved form with which mathematicians had been familiar. Lejeune Dirichlet, for instance, in 1837 suggested[2] a very broad definition of function: if a variable y is so related to a variable x that whenever a numerical value is assigned to x, there is a rule according to which a unique value of y is determined, then y is said to be a function of the independent variable x. This comes close to the modern view of a correspondence between two sets of numbers, but the concepts of "set" and "real number" had not at that time been established. To indicate the completely arbitrary nature of the rule of correspondence, Dirichlet proposed a very "badly behaved" function: when x is rational, let $y = c$, and when x is irrational, let $y = d \neq c$. This function, often known as Dirichlet's function, is so pathological that there is no value of x for which it is continuous. Dirichlet gave also the first rigorous proof of the convergence of Fourier series for a function subject to certain restrictions, known as Dirichlet's conditions. A Fourier series does not always converge to the value of a function from which it is derived, but Dirichlet in Crelle's *Journal* for 1828 proved the following theorem. If $f(x)$ is periodic of period 2π, if for $-\pi < x < \pi$ the function $f(x)$ has a finite number of maximum and minimum values and a finite number of discontinuities, and if $\int_{-\pi}^{\pi} f(x) \, dx$ is finite, then the Fourier series converges to $f(x)$ at all points where $f(x)$ is continuous, and at jump-points it converges to the arithmetic mean of the right-hand and left-hand limits of the function. Useful also is another theorem known as Dirichlet's test: if the terms in the series $a_1 b_1 + a_2 b_2 + \cdots + a_n b_n + \cdots$ are such that the b's are positive and monotonically tending toward zero, and if there is a number M such that $|a_1 + a_2 + \cdots + a_m| < M$ for all values of m, then the series converges.

The name of Dirichlet arises in many other connections in pure and applied mathematics. Especially important in thermodynamics and electrodynamics is the Dirichlet problem: Given a region R bounded by a closed curve C and a function $f(x, y)$ continuous on C, find a function $F(x, y)$ continuous in R and on C that satisfies the Laplace equation in R and is equal to f on C. In pure mathematics Dirichlet is well known for his application of analysis to the theory of numbers, in connection with which he introduced the Dirichlet series—$\Sigma a_n e^{-\lambda_n S}$—where the Dirichlet coefficients

[2] See Dirichlet, *Werke* (1889–1897) I, 135.

a_n are complex numbers, the Dirichlet exponents λ_n are real monotonically increasing numbers, and S is a complex variable.

Dirichlet's successor at Göttingen, Bernhard Riemann, also arrived at **2** deep theorems relating number theory and classical analysis. Euler had noted connections between prime-number theory and the series

$$\frac{1}{1^s} + \frac{1}{2^s} + \frac{1}{3^s} + \cdots + \frac{1}{n^s} + \cdots$$

where s is an integer—a special case of the Dirichlet series. Riemann studied the same series for s a complex variable, the sum of the series defining a function $\zeta(s)$, which has since been known as Riemann's zeta function. One of the tantalizing suggestions that mathematicians have not yet been able to prove or disprove is the famous Riemann conjecture that all of the imaginary zeroes $s = \sigma + i\tau$ of the zeta function[3] have real part $\sigma = \frac{1}{2}$. Probably no branch of mathematics has bequeathed so many unsolved problems as has the theory of numbers. Riemann was a many-sided mathematician with a fertile mind, contributing not only to geometry and the theory of numbers, but also to analysis. In analysis he is recalled for his part in the refinement of the definition of the integral, for emphasis on the Cauchy-Riemann equations, and for the Riemann surfaces. These surfaces are an ingenious scheme for uniformizing a function, that is, representing a one-to-one mapping of a complex function that in the ordinary Gaussian plane would be multivalued. Here we see the most striking aspect of Riemann's work—a strongly intuitive and geometrical backbround in analysis that contrasts sharply with the arithmetizing tendencies of the Weierstrassian school. His approach has been called "a method of discovery," whereas that of Weierstrass was, as we shall see, "a method of demonstration,"[4] and his results were so significant that Bertrand Russell described him as "logically the immediate predecessor of Einstein." It was Riemann's intuitive genius in physics and mathematics that produced such concepts as that of the curvature of a Riemannian space or manifold, without which the theory of general relativity could not have been formulated.[5]

---

[3] For properties of the zeta function see E. Landau, *Handbuch der Lehre von der Verteilung der Primzahlen* (Leipzig, 1909). An account of the Riemann hypothesis is found in E. T. Bell, *The Development of Mathematics* (1940), p. 293.

[4] Henri Poincaré, "L'oeuvre mathématique de Weierstrass," *Acta Mathematica*, **22** (1898–1899), 1–18.

[5] In E. T. Bell, *Men of Mathematics* (New York: Simon and Schuster, 1937), there is a warmly sympathetic chapter, entitled "Anima Candida," devoted entirely to Riemann and his work (pp. 484–509).

3   The theory of numbers deals primarily with the integers or, more generally, with ratios of integers—the so-called rational numbers. Such numbers always are roots of a linear equation $ax + b = 0$ with integral coefficients. Real analysis deals with a more general type of number that may be either rational or irrational. It had been known in essence to Euclid that the roots of $ax^2 + bx + c = 0$, where a, b, c are integral multiples of a given length, can be constructed geometrically with straightedge and compasses. If the coefficients of $ax^n + bx^{n-1} + cx^{n-2} + \cdots + px + q = 0$, where n and a, b, $c, \cdots, q$ are integers and $n > 2$, the roots of the equation generally are not constructible with Euclidean tools. The roots of such an equation, for $n > 0$, are known as algebraic numbers, to indicate the manner in which they are defined. Inasmuch as every rational number is a root of such an equation for $n = 1$, the question naturally arises whether or not every irrational number is a root of such an equation for some $n \geq 2$. The negative of this question was finally established in 1844 by Liouville, who in that year constructed an extensive class of nonalgebraic real numbers. The particular class that he developed are known as Liouville numbers, the more comprehensive set of nonalgebraic real numbers being called transcendental numbers. Liouville's construction of transcendental numbers is quite involved, but if one does not insist on a proof of transcendentalism, some simple examples of transcendental numbers can be given—such as $0.1001000100001\ldots$, or numbers of the form

$$\sum_{n=1}^{\infty} \frac{1}{10^{n!}}$$

To prove that any particular real number, such as e or π, is not algebriac is usually quite difficult. Liouville, for example, was able to show, in his *Journal* for 1844, that neither e nor e^2 could be the root of a quadratic equation with integral coefficients; hence, given a unit line segment, lines of length e and e^2 are not constructible by Euclidean tools. But it was almost thirty years before another French mathematician, Charles Hermite (1822–1901), pursuing the views of Liouville, was able to show in 1873 in an article in the *Comptes Rendus* of the Académie that e could not be the root of *any* polynomial equation with integral coefficients—that is, that e is transcendental.[6]

The name "Hermite's theorem" often is given to the statement that e is a transcendental number. Hermite, like many of his distinguished predecessors, was educated at the École Polytechnique, where later he taught and where he was generally regarded as the foremost French writer in the theory of functions since the days of Cauchy. Among his notable achievements was a

---

[6] For a full account of the historical background see U. G. Mitchell and Mary Strain, "The Number e," *Osiris*, **1** (1936), 476–496. See also D. E. Smith, *Source Book in Mathematics* (New York: Dover reprint, 1959), pp. 99–106.

solution of the general quintic equation by means of elliptic functions.[7] Liouville also is noted for a variety of other contributions. In analysis his work is recalled in Liouville's theorem—if $f(z)$, an entire analytic function of the complex variable z, is bounded over the complex plane, then $f(z)$ is a constant. From this theorem the fundamental theorem of algebra can be deduced as a simple corollary as follows: if $f(z)$ is a polynomial of degree greater than zero, and if $f(z)$ were nowhere zero in the complex plane, then its reciprocal $F(z) = 1/f(z)$ would satisfy the conditions of the Liouville theorem. Consequently $F(z)$ would have to be a constant, which obviously it is not. Therefore the equation $f(z) = 0$ is satisfied by at least one complex value $z = z_0$. In plane analytic geometry there is another "Liouville theorem"—the lengths of the tangents from a point P to a conic C are proportional to the cube roots of the radii of curvature of C at the corresponding points of contact.[8]

The status of the number π baffled mathematicians for nine years longer than did the number e. Lambert in 1770 and Legendre in 1794 had shown that both π and π^2 are irrational, but this proof had not put an end to the age-old question of the squaring of the circle. The matter was finally put to rest in 1882 in a paper in the *Mathematische Annalen* by C. L. F. Lindemann (1852–1939) of Munich. The article, entitled "Über die Zahl π," showed conclusively, in extending the work of Liouville and Hermite, that π also is a transcendental number. Lindemann in his proof first demonstrated that the equation $e^{ix} + 1 = 0$ cannot be satisfied if x is algebraic. Inasmuch as Euler had shown that the value $x = \pi$ does satisfy the equation, it must follow that π is not algebraic. Here, finally, was the answer to the classical problem of the quadrature of the circle. In order for the quadrature of the circle to be possible with Euclidean tools, the number π would have to be the root of a second degree equation and therefore, a fortiori, algebraic. Since π is not algebraic, the circle cannot be squared according to the classical rules.[9] Emboldened by his success, Ferdinand Lindemann later published several purported proofs of Fermat's last theorem, but they were shown by others to be invalid.

---

[7] See E. Picard, "L'oeuvre scientifique de Charles Hermite," in École Normal Supérieur, *Annales Scientifiques* (3), **18** (1901), 9–34, or the preface to Charles Hermite, *Oeuvres*, ed. by Emile Picard (1905–1917).

[8] For this and other aspects of his work see Gino Loria, "J. Liouville and His Work," *Scripta Mathematica*, **4** (1936), 147–154, 257–262; 301–305; or, in French, *Archeion*, **18** (1936), 117–139.

[9] For an exceptionally extensive account of the later history of the three classical problems see Felix Klein, *Famous Problems of Elementary Geometry*, trans. by Beman and Smith (reprinted in New York, 1955). See also E. W. Hobson, *Squaring the Circle*, and D. E. Smith, "The History and Transcendence of π," in J. W. A. Young, *Monographs on Topics of Modern Mathematics* (New York, 1915), pp. 387–416. Cf. Hermann von Baravalle, "The Number π," *The Mathematics Teacher*, **45** (1952), 340–348, or **60** (1967), 479–487.

4   The year 1872 was a red-letter year not only in geometry, but more particularly in analysis. In that year crucial contributions toward the arithmetization of analysis were made by no fewer than five mathematicians, one French, the others German. The Frenchman was H. C. R. (Charles) Méray (1835–1911) of Burgundy; the four Germans were Karl Weierstrass (1815–1897) of the University of Berlin, his student H. E. Heine (1821–1881) of Halle, Georg Cantor (1845–1918), also of Halle, and J. W. R. Dedekind (1831–1916) of Braunschweig. These men in a sense represented the climax in half a century of investigation into the nature of function and number that had begun in 1822 with Fourier's theory of heat and with an attempt made in that year by Martin Ohm (1792–1872) to reduce all of analysis to arithmetic in *Versuch eines vollständig konsequenten Systems der Mathematik*. There were two chief causes of uneasiness in this fifty-year interval. One was the lack of confidence in operations performed on infinite series. It was not even clear whether or not an infinite series of functions— of powers, or of sines and cosines, for example—always converges to the function from which it was derived. A second cause for concern was occasioned by the lack of any definition of the phrase "real number" that lay at the very heart of the arithmetization program. Bolzano by 1817 had been so fully aware of the need for rigor in analysis that Klein referred to him as the "father of arithmetization"; but Bolzano had been less influential than Cauchy, whose analysis was still encumbered with geometrical intuition. Even Bolzano's continuous nondifferentiable function of about 1830 was overlooked by successors, and the example of such a function given by Weierstrass (in classroom lectures in 1861 and in a paper to the Berlin Academy in 1872) was generally thought to be the first illustration of it.

Riemann meanwhile had exhibited a function $f(x)$ that is discontinuous at infinitely many points in an interval and yet the integral of which exists and defines a continuous function $F(x)$ that for the infinity of points in question fails to have a derivative. Riemann's function is in a sense less pathological than are those of Bolzano and Weierstrass, but it made clear that the integral required a more careful definition than that of Cauchy, which had been guided largely by geometrical feeling for the area under a curve. The present-day definition of the definite integral over an interval in terms of upper and lower sums generally is known as the Riemann integral, in honor of the man who gave necessary and sufficient conditions that a bounded function be integrable. The Dirichlet function, for instance, does not have a Riemann integral for any interval. Still more general definitions of the integral, with weaker conditions on the function, were proposed in the next century, but the definition of the integral used in most undergraduate courses in the calculus still is that of Riemann.

There was a gap of some fifty years between the work of Bolzano and that **5**
of Weierstrass, but the unity of effort in this half century and the need for
rediscovering Bolzano's work were such that there is a celebrated theorem
that bears the name of both men, the Bolzano-Weierstrass theorem: a
bounded set S containing infinitely many elements (such as points or numbers)
contains at least one limit-point. This theorem was proved by Bolzano and
apparently was known also to Cauchy, but it was the work of Weierstrass
that made it familar to mathematicians.

Skepticism about Fourier's series had been expressed by Lagrange, but
Cauchy in 1823 thought he had proved the convergence of the general
Fourier series. Dirichlet had shown that Cauchy's proof was inadequate and
had provided sufficient conditions for the convergence. It was in seeking to
liberalize Dirichlet's conditions for the convergence of a Fourier Series that
Riemann developed his definition of the Riemann integral; in this connection
he showed that a function $f(x)$ may be integrable in an interval without being
representable by a Fourier series.[10] It was the study of infinite trigonometric
series that led also to the theory of sets of Cantor, to be described later.

Dirichlet died in the critical year 1872. Only a year later there also died **6**
at the early age of thirty-four a young man who had given promise of significant contributions both to mathematics and to its history. This was Hermann
Hankel (1839–1873), a student of Riemann and professor of mathematics at
Leipzig. In 1867 he had published a book, *Theorie der komplexen Zahlensysteme*, in which he pointed out that "the condition for erecting a universal
arithmetic is therefore a purely intellectual mathematics, one detached from
all perceptions." We have seen that the revolution in geometry took place
when Gauss, Lobachevsky, and Bolyai freed themselves from preconceptions
of space. In somewhat the same sense the thoroughgoing arithmetization of
analysis became possible only when, as Hankel foresaw, mathematicians
understood that the real numbers are to be viewed as "intellectual structures"
rather than as intuitively given magnitudes inherited from Euclid's geometry.
The view of Hankel was not really new; for a generation, as we shall see in
the next chapter, algebraists, especially in Great Britain, had been developing
a universal arithmetic and multiple algebras. The implications for analysis,
however, had not been widely recognized. Bolzano during the early 1830s
had made an attempt to develop a theory of real numbers as limits of rational
number sequences,[11] but this had gone unnoticed and unpublished until

[10] For further details see Jerome H. Manheim, *The Genesis of Point Set Topology* (1964),
Chapters 3 and 4. Work related to the analysis of Riemann and Weierstrass is extensively treated
in Felix Klein, *Vorlesungen über die Entwicklung der Mathematik im 19. Jahrhundert* (1926–
1927), Vol. I.

[11] See B. von Rootselaar, "Bolzano's Theory of Real Numbers," *Archive for History of Exact
Sciences*, **2** (1964–1965), 168–180.

1962. Sir William Rowan Hamilton (1805–1865) perhaps had felt some such need, but his appeal to time rather than space was a change in language, although not in logical form, from the usual geometrical background. The crux of the matter was first effectively seized upon and published by the quintet of 1872 mentioned earlier.

Méray was prompt to present his thoughts, for as early as 1869 he had published an article calling attention to a serious lapse in reasoning of which mathematicians from the time of Cauchy had been guilty. Essentially the *petitio principii* consisted in defining the limit of a sequence as a real number and then in turn defining a real number as a limit of a sequence (of rational numbers). It will be recalled that Bolzano and Cauchy had attempted to prove that a sequence that "converges within itself"—that is, one for which S_{n+p} differs from S_n (for a given integer p and for n sufficiently large) by less than any assigned magnitude ϵ—also converges in the sense of external relations to a real number S, the limit of the sequence. Méray, in his *Nouveau préçis d'analyse infinitésimale* of 1872, cut the Gordian knot by not invoking the external condition of convergence or the real number S. Using only the Bolzano-Cauchy criterion, where n, p, and ϵ are rational numbers, convergence can be described without reference to irrational numbers. In a broad sense he regarded a converging sequence as *determining* either a rational number as a limit or a "fictitious number" as a "fictitious limit." These "fictitious numbers" can, he showed, be ordered, and in essence they are what we know as the irrational numbers. Méray was somewhat vague as to whether or not his converging sequence *is* the number. If it is, as seems to be implied, then his theory is equivalent to one developed at the same time by Weierstrass.

7   Weierstrass sought to separate the calculus from geometry and to base it upon the concept of number alone; like Méray, he also saw that to do this it was necessary to give a definition of irrational number that is independent of the limit concept, inasmuch as the latter had up to this point presupposed the former. To correct Cauchy's logical error Weierstrass settled the question of the *existence* of a limit of a convergent sequence by making the sequence itself the number or limit. The scheme of Weierstrass is too subtle to present in detail here, but in considerably oversimplified form we may say that the number $\frac{1}{3}$ is not the *limit* of the series $\frac{3}{10} + \frac{3}{100} + \frac{3}{1000} + \cdots + \frac{3}{10^n} + \cdots$; it *is* the sequence associated with this series. (Actually, in Weierstrass' theory, the irrational numbers are more broadly defined as *aggregates* of the rationals, rather than more narrowly as *ordered sequences* of rationals as we have implied.)

Weierstrass did not publish his views on the arithmetization of analysis, but they were made known by students, such as Ferdinand Lindemann and

Eduard Heine, who had attended his lectures. In 1871 Cantor had initiated a third program of arithmetization, similar to those of Méray and Weierstrass. Heine suggested simplifications that have led to the so-called Cantor-Heine development, published by Heine in Crelle's *Journal* for 1872 in the article "Die Elemente der Funktionenlehre." We cannot go into this in detail, but in essence the scheme resembled that of Méray in that convergent sequences that fail to converge to rational numbers are taken by fiat to define irrational numbers. A thoroughly distinct approach to the same problem, and the one that today is best known, was given in the same year by Dedekind in a celebrated book, *Stetigkeit und die Irrationalzahlen* ("Continuity and Irrational Numbers").[12]

Dedekind's attention had been directed to the problem of irrational numbers as early as 1858, when he found himself lecturing on the calculus. The limit concept, he concluded, should be developed through arithmetic alone, without the usual guidance from geometry, if it were to be rigorous. Instead of simply seeking a way out of Cauchy's vicious circle, Dedekind asked himself, as the title of his book implies, what there is in continuous geometrical magnitude that distinguishes it from the rational numbers. Galileo and Leibniz had thought that the "continuousness" of points on a line was the result of their density—that between any two points there is always a third. However, the rational numbers have this property, yet they do not form a continuum. Upon pondering this matter, Dedekind came to the conclusion that the essence of the continuity of a line segment is not due to a vague hang-togetherness, but to an exactly opposite property—the nature of the division of the segment into two parts by a point on the segment. In any division of the points of the segment into two classes such that each point belongs to one and only one class, and such that every point of the one class is to the left of every point in the other, there is one and only one point that brings about the division. As Dedekind wrote, "By this commonplace remark the secret of continuity is to be revealed." Commonplace the remark may have been, but its author seems to have had some qualms about it, for he hesitated for some years before committing himself in print.

Dedekind saw that the domain of rational numbers can be extended to form a continuum of real numbers if one assumes what now is known as the Cantor-Dedekind axion—that the points on a line can be put into one-to-one correspondence with the real numbers. Arithmetically expressed, this means that for every division of the rational numbers into two classes A and B such that every number of the first class, A, is less than every number of the

---

[12] An English translation of this is found under the title *Essays on the Theory of Numbers*, trans. by W. W. Beman (Chicago, 1901.). This contains also the translation of Dedekind's *Was sind und was sollen die Zahlen* of 1888.

second class, B, there is one and only one real number producing this *Schnitt*, or Dedekind cut. If A has a largest number, or if B contains a smallest number, the cut defines a rational number; but if A has no largest number and B no smallest, then the cut defines an irrational number. If, for example, we put in A all negative rational numbers and also all positive rational numbers whose squares are less than two, and in B all positive rational numbers whose squares are more than two, we have subdivided the entire field of rational numbers in a manner defining an irrational number—in this case the number that we usually write as $\sqrt{2}$. Now, Dedekind pointed out, the fundamental theorems on limits can be proved rigorously without recourse to geometry. It was geometry that had pointed the way to a suitable definition of continuity, but in the end it was excluded from the formal arithmetic definition of the concept. The Dedekind cut in the rational number system, or an equivalent construction of real number, now has replaced geometrical magnitude as the backbone of analysis.

The definitions of real number are, as Hankel indicated they should be, intellectual constructions on the basis of the rational numbers, rather than something imposed on mathematics from without. Of the definitions above, one of the most popular has been that of Dedekind. Early in the twentieth century a modification of the Dedekind cut was proposed by Bertrand Russell (1872–). He noted that since either of Dedekind's two classes A and B is uniquely determined by the other, one alone suffices for the determination of a real number. Thus $\sqrt{2}$ can be defined simply as that segment or subclass of the set of rational numbers made up of all positive rational numbers whose squares are less than two and also of all negative rational numbers; similarly every real number is nothing more than a segment of the rational number system.

9   Weierstrass, as part of an arithmetization program, not only contributed to a satisfactory definition of real number, but also to an improved definition of the limit concept. The definition of Cauchy had made use of such phrases as "successive values" or "approach indefinitely" or "as little as one wishes." Although these are suggestive, and probably pedagogically comforting, they nevertheless lack the precision that is generally expected of mathematicians. In his lectures, therefore, Weierstrass emphasized what has sometimes been called the "static theory of the variable." Heine, in his *Elemente* of 1872, influenced by Weierstrass' lectures, defined the limit of the function $f(x)$ at x_0 as follows:

> If, given any ϵ, there is an η_0 such that for $0 < \eta < \eta_0$ the difference $f(x_0 \pm \eta) - L$ is less in absolute value than ϵ, then L is the limit of $f(x)$ for $x = x_0$.

In this cold and precise definition there is no suggestion of flowing entities generating magnitudes of higher dimension, no recourse to moving points or lines, no dropping of infinitely small quantities. There is nothing left but real numbers, the operation of addition (and its inverse, subtraction), and the relationship "less than." The unequivocal language and symbolism of Weierstrass and Heine banished from the calculus the notion of variability and rendered unnecessary the persistent resort to fixed infinitesimals. The "Age of Rigor" had truly arrived, replacing the older heuristic devices and intuitive views by critical logical precision. Today the η of Weierstrass frequently is replaced by another Greek letter, δ, but the definitions of limit of a function found in current textbooks are in essence those introduced by Weierstrass and Heine almost a century ago. So-called delta-and-epsilon proofs, or epsilontics, now are part of the mathematician's stock-in-trade.

Weierstrass had been brought up in a devout but liberal Catholic family, his **10** father having been converted from Protestantism. Karl, the eldest son, had a brother and two sisters, but not one of the four children married, possibly because of the father's domineering attitude; and Karl had at least one other eccentricity—a dislike of music. He did so well in school that his father insisted that he prepare for public service by studying law at the University of Bonn, where he became an expert in drinking and fencing, rather than in law or mathematics, and left without a degree. He then prepared himself at Münster for secondary school teaching, where an instructor, Christoph Gudermann (1798–1851), took Weierstrass under his wing. Gudermann was especially interested in elliptic and hyperbolic functions, where his name is still recalled in the Gudermannian: If u is a function of x satisfying the equation $\tan u = \sinh x$, then u is known as the Gudermannian of x, written as $u = \operatorname{gd} x$. More important to mathematics than this minor contribution were the time and inspiration that the teacher gave to his student, who was destined in turn to become the greatest mathematics teacher of the mid-nineteenth century—at least as measured in terms of the number of successful research workers he produced. Gudermann had impressed upon the young Weierstrass what a useful tool the power series representation of a function was, and it was in this connection that Weierstrass produced his greatest work, following in the footsteps of Abel.

Weierstrass earned his teacher's certificate at the late age of twenty-six, and for more than a dozen years he taught at various secondary schools. In 1854, however, a paper on Abelian functions, appearing in Crelle's *Journal*, brought him such recognition that shortly thereafter he was offered, and accepted, a professorship at the University of Berlin. Weierstrass was then almost forty, making him a striking exception to the common notion that a great mathematician must make his mark early in life. Despite his delayed start, he was freely acknowledged, during the last third of the century, to be the leading analyst in the world.[13]

It had been generally assumed, before the middle of the nineteenth century, that if an infinite series converges for some interval to a continuous and differentiable function $f(x)$, then a second series obtained by differentiating the original series term by term necessarily will converge, for the same interval, to $f'(x)$. Several mathematicians showed that this is not necessarily the case and that term-by-term differentiation can be trusted only if the series is *uniformly* convergent for the interval—that is, if a single N can be found such that for every value of x in the interval the partial sums $S_n(x)$ will differ from the sum $S(x)$ of the series by less than a given ϵ for all $n > N$. Weierstrass showed that for a uniformly convergent series term-by-term integration also was permissible. In the matter of uniform convergence Weierstrass was far from alone, for the concept was hit upon independently at about the same time by at least three other men—Cauchy in France (perhaps by 1853), Sir G. G. Stokes at Cambridge (in 1847), and P. L. V. Seidel (1821–1896) in Germany (in 1848).[14] However, perhaps no one is more deserving to be known as the father of the critical movement in analysis than is Weierstrass.[15] From 1857 until his retirement in 1890 he urged a generation of students to use infinite series representations with care, and one of them, Heine, in 1870 proved that the Fourier series development of a continuous function is unique if one imposes the condition that it be uniformly convergent. In this respect he was smoothing out difficulties in the work of Dirichlet and Riemann on Fourier series.

One of the important contributions of Weierstrass to analysis is known as analytic continuation. Weierstrass had shown that the infinite power series representation of a function $f(x)$, about a point P_1 in the complex plane, converges at all points within a circle C_1 whose center is P_1 and which passes through the nearest singularity. If, now, one expands the same function about a second point P_2 other than P_1, but within C_1, this series will be convergent within a circle C_2 having P_2 as center and passing through the singularity nearest to P_2. This circle may include points outside C_1, hence one has extended the area of the plane within which $f(x)$ is defined analytically by a power series; the process can be continued with still other circles. Weierstrass therefore defined an analytic function as one power series together with all those that are obtainable from it by analytic continuation. The importance of work such as that of Weierstrass is felt particularly in mathematical physics,

---

[13] On his life see E. T. Bell, *Men of Mathematics* (New York: Simon and Schuster, 1937), Chap. 22, or Ganesh Prasad, *Some Great Mathematicians of the Nineteenth Century* (1933–1937), Vol. I, Chap. 5.

[14] See E. T. Bell, *The Development of Mathematics*, p. 270.

[15] See James Pierpont, "Mathematical Rigor, Past and Present," *Bulletin of the American Mathematical Society*, **34** (1928), 23–53

in which solutions of differential equations are rarely found in any form other than as an infinite series.

In some respects the life of Dedekind was similar to that of Weierstrass: he, too, was one of four children, and he, too, never married; and both men lived into their eighties. On the other hand, the members of Dedekind's family were Lutheran; he also made an earlier start in mathematics than had Weierstrass, entering Göttingen at the age of nineteen and earning his doctorate three years later with a thesis on the calculus which elicited praise from Gauss. Dedekind stayed at Göttingen for a few years, teaching and listening to lectures by Dirichlet, and then he took up secondary school teaching, chiefly at Brunswick, for the rest of his life. Dedekind lived so long after his celebrated introduction of "cuts" that the famous publishing house of Teubner had listed his death in its *Calendar for Mathematicians* as September 4, 1899. This amused Dedekind, who lived more than a dozen years longer, and he wrote to the editor that he had passed the day in question in stimulating conversation with his friend Georg Cantor.

The life of Cantor was tragically different from that of his friend Dedekind.[16]   **11**
Cantor was born in St. Petersburg of parents who had migrated from Denmark, but most of his life was spent in Germany, the family having moved to Frankfurt when he was eleven. His parents were Christians of Jewish background—his father had been converted to Protestantism, his mother had been born a Catholic. The son Georg took a strong interest in the finespun arguments of medieval theologians concerning continuity and the infinite, and this militated against his pursuing a mundane career in engineering as suggested by his father. In his studies at Zurich, Göttingen, and Berlin the young man consequently concentrated on philosophy, physics, and mathematics—a program that seems to have fostered his unprecedented mathematical imagination. He took his doctorate at Berlin in 1867 with a thesis on the theory of numbers, but his early publications show an attraction to Weierstrassian analysis. This field prompted the revolutionary ideas that sprang to his mind in his late twenties. We have already noted the work of Cantor in connection with the prosaic phrase, "real number"; but his most original contributions centered about the provocative word "infinity."

Even since the days of Zeno men had been talking about infinity, in theology as well as in mathematics, but no one before 1872 had been able to tell precisely what he was talking about. All too frequently in discussions of the infinite the examples cited were such things as unlimited power or indefinitely large magnitudes. Occasionally attention had been focused instead, as in the work of Galileo and Bolzano, on the infinitely many elements in a

---

[16] See Bell, *Men of Mathematics*, Chap. 29, or Prasad, *Great Mathematicians*, Vol. II, Chap. 7.

collection—for example, the natural numbers or the points in a line segment. Cauchy and Weierstrass saw only paradox in attempts to identify an actual or "completed" infinity in mathematics, believing that the infinitely large and small indicated nothing more than the potentiality of Aristotle—an incompleteness of the process in question. While under the influence of Weierstrass' analysis, two of his students nevertheless came to a contrary conclusion. The first of these was Dedekind, who saw in Bolzano's paradoxes not an anomaly, but a universal property of infinite sets which he took as a precise definition:

> A system S is said to be *infinite* when it is similar to a proper part of itself; in the contrary case S is said to be a finite system.

In somewhat more modern terminology, a set S of elements is said to be infinite if the elements of a proper subset S' can be put into one-to-one correspondence with the elements of S. That the set S of natural numbers is infinite, for instance, is clear from the fact that the subset S' made up of all triangular numbers is such that to each element n of S there corresponds an element of S' given by $n(n + 1)/2$. This positive definition of a "completed infinite" set is not to be confused with the negative statement sometimes written with Wallis' symbol as $1/0 = \infty$. This last "equation" simply indicates that there is no real number that multiplied by zero will produce the number one.

12   Dedekind's definition of an infinite set appeared in 1872 in his *Stetigkeit und irrationale Zahlen*. (In 1888 Dedekind amplified his ideas in another important treatise, *Was sind und was sollen die Zahlen*.) Two years later Cantor married, and on the honeymoon he took his bride to Interlaken, where they met Dedekind. In the same year, 1874, Cantor published in Crelle's *Journal* one of his most revolutionary papers.[17] He, like Dedekind, had recognized the fundamental property of infinite sets, but, unlike Dedekind, Cantor saw that not all infinite sets are the same. In the finite case, sets of elements are said to have the same (cardinal) number if they can be put into one-to-one correspondence. In a somewhat similar way, Cantor set out to build a hierarchy of infinite sets according to the *Mächtigkeit* or "power" of the set. The set of perfect squares or the set of triangular numbers has the same power as the set of all the positive integers, for the groups can be put into one-to-one correspondence. These sets seem to be much smaller than the set of all rational fractions, yet Cantor showed that the latter set also is

---

[17] A very thorough account of Cantor's work is found in the Introduction to an English translation of two of Cantor's papers of 1895 and 1897 published under the title *Contributions to the Founding of the Theory of Transfinite Numbers*, ed. by P. E. B. Jourdain (1915). See also Herbert Meschkowski, *Ways of Thought of Great Mathematicians* (San Francisco: Holden-Day, 1964), pp. 91–104.

countable or denumerable—that is, it too can be put into one-to-one correspondence with the positive integers, hence has the same power. To show this, we merely follow the arrows in Figure 25.3, "counting" the fractions along the way.

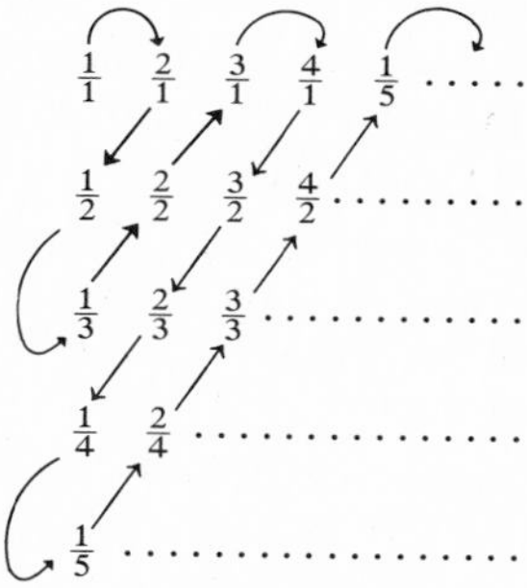

FIG. 25.3

The rational fractions are so dense that between any two of them, no matter how close, there always will be another; yet Cantor's arrangement[18] showed that the set of fractions has the same power as does the set of integers. One begins to wonder if all sets of numbers have the same power, but Cantor proved conclusively that this is not the case. The set of all real numbers, for example, has a higher power than does the set of rational fractions. To show this, Cantor used a *reductio ad absurdum*. Assume that the real numbers between 0 and 1 are countable, are expressed as nonterminating decimals (so that $\frac{1}{3}$, for example, appears as 0.333..., $\frac{1}{2}$ as 0.499..., and so on), and are arranged in denumerable order:

$$a_1 = 0.a_{11}a_{12}a_{13}\cdots$$

$$a_2 = 0.a_{21}a_{22}a_{23}\cdots$$

$$a_3 = 0.a_{31}a_{32}a_{33}\cdots$$

$$\cdots\cdots\cdots\cdots\cdots$$

where a_{ij} is a digit between 0 and 9 inclusive. To show that not all of the real numbers between 0 and 1 are included above, Cantor exhibited an infinite decimal different from all of those listed. To do this, simply form the decimal $b = 0.b_1b_2b_3\ldots$ where $b_K = 9$ if $a_{KK} = 1$ and $b_K = 1$ if $a_{KK} \neq 1$. This real number will be between 0 and 1 and yet it will be unequal to any one of those in the arrangement that was presumed to contain *all* of the real numbers between 0 and 1.

---

[18] Cantor proved the denumerability of rational numbers in his paper of 1874, but he there used a different type of proof. Later he gave the demonstration as given above.

13   The real numbers can be subdivided into two types in two different ways: (1) as rational and irrational or (2) as algebraic and transcendental. Cantor showed that even the class of algebraic numbers, which is far more general than that of rational numbers, nevertheless has the same power as that of the integers. Hence it is the transcendental numbers that give to the real number system the "density" that results in a higher power. That it is fundamentally a matter of density that determines the power of a set is suggested in the fact that the power of the set of points on an infinitely extended line is just the same as the power of the set of points in any segment of the line, however small. To show this, let RS be the infinitely extended line and let PQ be any finite segment (Fig. 25.4). Place the segment so that it intersects RS at a point O but is not perpendicular to RS and does not lie on RS. If the points M and N are so chosen that PM and QN are parallel to RS, and MON is perpendicular to RS, then by drawing lines through M intersecting both OP and OR, and lines through N intersecting OQ and OS, a one-to-one correspondence is easily established.

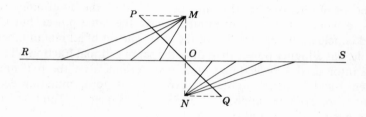

FIG. 25.4

More surprising still is the fact that dimensionality is not the arbiter of the power of a set. The power of the set of points in a unit line setment is just the same as that of the points in a unit area or in a unit volume—or, for that matter, all of three-dimensional space. (Dimensionality, however, retains some measure of authority in that any one-to-one mapping of points in a space of unlike dimensionality is necessarily a discontinuous mapping.) So paradoxical were some results in point-set theory that Cantor himself on one occasion in 1877 wrote to Dedekind, "I see it, but I don't believe it"; and he asked his friend to check the proof.[19] Publishers, too, were very hesitant about accepting his papers, and several times the appearance of articles by Cantor in Crelle's *Journal* was delayed by editorial indecision

[19] An especially readable account of Cantor's work is found in Herbert Meschkowski, *Evolution of Mathematical Thought* (1965), Chap. 5.

and concern lest error lurk in the unconventional approach to mathematical concepts.

Cantor's amazing results led him to the establishment of the theory of **14** sets as a full-fledged mathematical discipline, known as *Mengenlehre* (theory of assemblages) or *Mannigfaltigkeitslehre* (theory of manifolds), a branch that in the mid-twentieth century was to have profound effects on the teaching of mathematics. At the time of its founding Cantor spent much effort in convincing his contemporaries of the validity of his results, for there was considerable *horror infiniti*, and mathematicians were reluctant to accept the *eigentlich Unendlich* or completed infinity. In piling evidence upon evidence, Cantor in the end built a whole transfinite arithmetic. The "power" of a set became the "cardinal number" of the set. Thus the "number" of the set of integers was the "smallest" transfinite number, E, and the "number" of the set of real numbers or of points on a line is a "larger" number, C, the number of the continuum. Still unanswered is the question whether or not there are transfinite numbers between E and C. Cantor himself showed that there are indefinitely many transfinite numbers beyond C, for he proved that the set of subsets of a set always is of a higher power than the set itself. Hence the "number" of the set of subsets of C is a third transfinite number, the set of subsets of this set of subsets determines a fourth number, and so on indefinitely. As there are infinitely many natural numbers, so also are there infinitely many transfinite numbers.

The transfinite numbers described above are *cardinal* numbers, but Cantor developed also an arithmetic of transfinite ordinal numbers. Ordering relations are a ticklish matter in mathematics, and so it turns out that transfinite ordinal arithmetic differs strikingly from finite ordinal arithmetic. For finite cases the rules for ordinal numbers are essentially the same as for cardinal numbers. Thus $3 + 4 = 4 + 3$, whether these digits represent cardinal or ordinal numbers. However, if one designates by ω the ordinal number of the "counting numbers," then $\omega + 1$ is not the same as $1 + \omega$, for $1 + \omega$ obviously is the same as ω. Moreover, one can show that $\omega + \omega = \omega$ and $\omega \cdot \omega = \omega$, properties unlike those of finite ordinals but resembling those of transfinite cardinals.

Dedekind and Cantor were among the most capable mathematicians, and **15** certainly the most original, of their day; yet neither man secured a top-ranking professional position. Dedekind spent almost a lifetime teaching on the secondary school level, and Cantor spent most of his career at the University of Halle, a small school without particular reputation. Cantor had hoped to achieve the distinction of a professorship at the University of Berlin, and he blamed Leopold Kronecker (1823–1891) for his lack of success. Kronecker, like Cantor, was born of Jewish parents, but, again like Cantor, he preferred Protestant Christianity.[20] At the University of Berlin he came in contact with Weierstrass, Dirichlet, Jacobi, and Steiner, taking his doctorate in 1845 with a thesis in algebraic number theory. Like Weierstrass, he approved of the universal arithmetization of analysis, but he demanded that the arithmetic be finite, and here he came into sharp conflict with Cantor. Reverting to ancient Pythagorean views, Kronecker insisted that arithmetic and analysis be based on the whole numbers. "God made the integers," he used to say, "and all the rest is the work of man." He categorically rejected the real number constructions of his day on the ground that they cannot be achieved through finite processes only, and he called for an arithmetical revolution that would ban the irrational numbers as nonexistent. In analysis Kronecker did little but openly criticize his contemporaries in lectures and conversation. He is said to have asked Lindemann of what use was the proof that π is not algebraic inasmuch as irrational numbers are nonexistent. In algebra Kronecker made significant contributions, but analysts of the time regarded his views as excessively metaphysical. Sometimes it is reported that his movement died of inanition;[21] we shall see later that it can be said to have reappeared in a new form in the work of Poincaré and Brouwer.

Kronecker was for most of his life a very prosperous businessman, but he was strongly associated with scholars at the University of Berlin, where he finally accepted a professorship in 1883. His finitism obviously embarrassed Weierstrass, but it was Cantor whom he wounded most seriously. Not only did Kronecker stand in the way of a position for Cantor at Berlin, but he sought to undermine the branch of mathematics that Cantor was creating. Cantor, in turn, wrote a vigorous defense in 1883 in his *Grundlagen einer allgemeinen Mannigfaltigkeitslehre* ("Foundations of a General Theory of Manifolds"), holding that "definite numerations can be undertaken with infinite sets just as well as with finite." He had no fear of falling into what he described as an "abyss of transcendentals," yet occasionally he did lapse into arguments of theological type. Kronecker continued his attacks on the hypersensitive and temperamental Cantor, and in 1884 Cantor suffered the first of the nervous breakdowns that were to recur throughout the remaining thirty-three years of his life. Fits of depression sometimes led him to doubt his own work, although he was comforted to some extent by the support of men such as Hermite. Toward the end he did earn recognition for his achievements, but his death in 1918 in a mental institution in Halle is a reminder that

---

[20] For his life and work see Bell, *Men of Mathematics*, Chap. 25, or Prasad, *Great Mathematicians*, Vol. II, Chap. 3.

[21] Pierpont, "Mathematical Rigor, Past and Present," *Bulletin of the American Mathematical Society*, **34** (1928), 23–53, pp. 38–40.

genius and madness sometimes are closely related. The tragedy of his personal life is mitigated by the paen of praise of one of the leading mathematicians of the early twentieth century, David Hilbert, who described the new transfinite arithmetic as "the most astonishing product of mathematical thought, one of the most beautiful realizations of human activity in the domain of the purely intelligible." Where timid souls had hesitated, Hilbert exclaimed, "No one shall expel us from the paradise which Cantor has created for us."[22]

BIBLIOGRAPHY

Bell, E. T., *The Development of Mathematics* (New York: McGraw-Hill, 1940).

Bourbaki, Nicolas, *Éléments d'histoire des mathématiques* (Paris: Hermann, 1960).

Boyer, Carl B., *The Concepts of the Calculus* (New York: Columbia University Press, 1939; reprinted by Dover, 1959).

Brill, A., and M. Noether, "Die Entwicklung der Theorie der algebraischen Funktionen in älterer und neurer Zeit," *Jahresbericht der Deutsche Mathematiker Vereinigung*, **3** (1892–1893), 107–566.

Cajori, Florian, *History of Mathematics*, 2nd ed. (New York: Macmillan, 1931).

Cantor, Georg, *Contributions to the Founding of the Theory of Transfinite Numbers*, trans. by P. E. B. Jourdain (Chicago and London: Open Court, 1915; New York: Dover paperback, n.d.).

Dantscher, Victor, *Vorlesungen über die Weierstrassche Theorie der irrationalen Zahlen* (Leipzig and Berlin: B. G. Teubner, 1908).

Dedekind, Richard, *Essays on the Theory of Numbers*, trans. by W. W. Beman (Chicago: Open Court, 1901).

Dirichlet, G. L., *Werke*, ed. by L. Kronecker and L. Fuchs (Berlin, 1889–1897, 2 vols.).

Eves, Howard, and Carroll V. Newsom, *An Introduction to the Foundations and Fundamental Concepts of Mathematics* (New York: Rinehart, 1958).

Fourier, Joseph, *Oeuvres*, ed. by G. Darboux (Paris, 1888–1890, 2 vols.).

Heine, E., "Die Elemente der Funktionenlehre," *Journal für die Reine und Angewandte Mathematik*, **74** (1872), 172–188.

Hermite, Charles, *Oeuvres*, ed. by Émile Picard (Paris: Gauthier-Villars, 1905–1917, 4 vols.).

Hobson, E. W., "On the Infinite and the Infinitesimal in Mathematic Analysis," *Proceedings of the London Mathematical Society*, **35** (1903), 117–140.

Jourdain, P. E. B., "The Development of the Theory of Transfinite Numbers," *Archiv der Mathematik und Physik* (3), **10** (1906), 254–281; **17** (1908–1909), 287–311; **16** (1910), 21–43; **22** (1913–1914), 1–21.

Jourdain, P. E. B., "Note on Fourier's Influence on the Conceptions of Mathematics," *International Congress of Mathematicians* (Cambridge, 1912), Vol. II, pp. 526–527.

Jourdain, P. E. B., "On Isoid Relations and Theories of Irrational Number," *Proceedings of the International Congress of Mathematicians* (Cambridge, 1912), Vol. II, pp. 492–496.

[22] "Sur l'infini," *Acta Mathematica*, **48** (1926), 91–122, especially pp. 97–100; or "Über das Unendlich," *Mathematische Annalen*, **95** (1926), 161–190, especially pp. 167–170.

Klein, Felix, *On Riemann's Theory of Algebraic Functions and Their Integrals*, trans. by Frances Hardcastle (Cambridge: Cambridge University Press, 1893).

Klein, Felix, *Vorlesungen über die Entwicklung der Mathematik im 19. Jahrhundert* (Berlin: Springer, 1926–1927, 2 vols.).

Kronecker, Leopold, *Werke*, ed. by Kurt Hensel (Leipzig: B. G. Teubner, 1895–1931, 5 vols.).

Langer, Rudolph E., *Fourier Series, the Genesis and Evolution of a Theory* (Oberlin, Ohio: The Mathematical Association of America, 1947).

Loria, Gino, "Le mathématicien J. Liouville et ses oeuvres," *Archeion*, **18** (1936), 117–139; English version in *Scripta Mathematica*, vol. 4 (1936).

Manheim, Jerome H., *The Genesis of Point Set Topology* (New York: Pergamon Press, 1964).

Méray, Charles, *Nouveau précis d'analyse infinitésimale* (Paris, 1872).

Merz, J. T., *A History of European Thought in the Nineteenth Century* (Edinburgh and London, 1896–1914, 4 vols; reprinted, New York: Dover, 1965).

Meschkowski, Herbert, *Evolution of Mathematical Thought* (San Francisco: Holden-Day, 1965).

Pierpont, James, "The History of Mathematics in the Nineteenth Century," *Bulletin of the American Mathematical Society*, **11** (1904), 136–159.

Pierpont, James, "Mathematical Rigor, Past and Present," *Bulletin of the American Mathematical Society*, **34** (1928), 23–53.

Pincherle, Salvatore, "Saggio di una introduzione alla teoria delle funzioni analitiche secondo i principii del Prof. C. Weierstrass," *Giornale di Matematiche*, **18** (1880), 178–254, 317–357.

Poincaré, Henri, "L'oeuvre mathématique de Weierstrass," *Acta Mathematica*, **22** (1898–1899), 1–18.

Prasad, Ganesh, *Some Great Mathematicians of the Nineteenth Century* (Benares: Benares Mathematical Society, 1933–1934, 2 vols.).

Waismann, Friedrich, *Introduction to Mathematical Thinking* (New York: Harper reprint, 1959).

Weierstrass, Karl, *Mathematische Werke* (Berlin: Mayer & Müller, 1894–1927, 7 vols.).

EXERCISES

1. Explain why the nineteenth century has been called "a century of correlation" in mathematics, citing specific contributions to support this view.

2. To what extent were developments in analysis in the nineteenth century motivated by internal factors in mathematics, rather than by the needs and preferences of society? Give specific illustrations to support your answer.

3. Compare the level of rigor in analysis in the nineteenth century with that in the eighteenth century and with that of Archimedes' works, supporting your answer with specific examples.

4. Describe the importance of the year 1872 in the arithmetization of analysis.

5. Were leading mathematicians in the nineteenth century more or less likely to be teachers than was the case in the eighteenth century? Cite instances to support your answer.

6. Is Kronecker's assertion, that God made the integers and that all other numbers are the work of man, defensible or indefensible? Explain.

7. Show that the set of all real numbers between 3 and 7 inclusive can be put into one-to-one correspondence with the set of all real numbers between 1 and 11.

8. Define precisely the phrases "real number" and "irrational number." When and how was the need for irrational numbers first recognized and when and how did the need for precise definition arise? Explain clearly.

9. Compare the definition of "limit of a function" given by Weierstrass with that formulated earlier by Cauchy, indicating relative advantages or disadvantages.

10. Which has the higher transfinite number, the set of all integers or the class of all the nth roots of all the integers? Explain.

11. Given two line segments of unequal length, show that a one-to-one correspondence can be set up between the points on the two segments.

12. Write in the form of infinite decimals a dozen different transcendental numbers, in each case giving the rule of succession for the digits. Are any of these numbers rational? Explain.

13. Show that if $u = \text{gd } x$, then $\cos u = \text{sech } x$ and $\sin u = \tanh x$.

*14. In the Fourier expansion

$$f(x) = \tfrac{1}{2}a_0 + a_1 \cos x + a_2 \cos 2x + \cdots + b_1 \sin x + b_2 \sin 2x + \cdots$$

multiply both sides of the equation by $\cos 2x$ and then integrate both sides from $-\pi$ to $+\pi$ to show that

$$a_2 = \frac{1}{\pi}\int_{-\pi}^{\pi} f(x)\cos 2x \, dx$$

*15. As in Exercise 14, multiply both sides of the equation by $\sin 2x$ and integrate between $-\pi$ and $+\pi$ to obtain the analogous formula for b_2.

*16. Sketch the function $f(x) = 1$ for $n\pi \le x < (n + 1)\pi$ when n is even and $f(x) = 0$ for $n\pi \le x < (n + 1)\pi$ when n is odd, and find the Fourier expansion of the function.

CHAPTER XXVI

The Rise
of Abstract Algebra

It is no paradox to say that in our most theoretical
moods we may be nearest to our most practical
applications.

A. N. Whitehead

1 The nineteenth century, more than any preceding period, deserved to be
known as the Golden Age in mathematics. The additions to the subject
during these one hundred years far outweighed, both in quantity and quality,
the total combined productivity of all preceding ages. The century was also,
with the possible exception of the Heroic Age in ancient Greece, the most
revolutionary in the history of mathematics. In 1829 a new world in geometry
was discovered by Lobachevsky, a Russian who had had a German teacher,
and in 1874 the field of analysis had been startled by the mathematics of the
infinite which had been introduced by Cantor, a German who had been
born in Russia. No longer was France the clearly acknowledged center of the
mathematical world, although she provided the meteoric career of Évariste
Galois (1811–1832). The international character of the subject is seen in the
fact that in algebra the two most revolutionary contributions, in 1843 and
1847, were made by mathematicians who taught in Ireland. The first of these
was the work of Sir William Rowan Hamilton (1805–1865), the second came
from George Boole (1815–1864). The most prolific contributors to nineteenth-century algebra, however, were Englishmen who spent part of their
time in America—Arthur Cayley (1821–1895) and J. J. Sylvester (1814–1897)
—and it was chiefly from their alma mater, Cambridge, that the rise in
modern algebra stemmed.

Cambridge University in the earliest years of the nineteenth century was
scarcely the place to which one would have looked for new developments in
mathematics. It is true that, a hundred years earlier, it had been the alma
mater of Sir Isaac Newton; but chauvinism and the priority controversy
in the calculus had led to intellectual isolationism for which the British paid
dearly. Scottish universities of the eighteenth century had maintained closer
contact with the Continent than had those in England, but the former were

relatively weaker in mathematics than in biology and chemistry. When Jacobi visited Cambridge in 1842, he was asked who was the greatest living mathematician in England, and he replied, "There is none."[1] This was at the time an ungraciously harsh judgment. It is true, nevertheless, that a generation earlier in England in general, and at Cambridge in particular, there had been almost no awareness of the tremendous strides in analysis and geometry that had been made on the Continent. Perhaps it was for this reason that Britain, when she emerged from the web of provincialism, took the lead in the direction of algebra, for in this field Continental developments in the eighteenth century had been less striking.

The turning point in British mathematics came in 1815 with the formation **2** at Trinity College, Cambridge, of the Analytical Society, mentioned earlier, which was made up of three young Cantabrigians: the algebraist George Peacock (1791–1858), the astronomer John Herschel (1792–1871), and Charles Babbage (1792–1871) of "Calculating Engines" fame. The immediate purpose of the Society was to reform the teaching and notation of the calculus; and in 1817, when Peacock was appointed an examiner for the mathematical tripos, differential notation replaced fluxional symbols on the Cambridge examination. Peacock was himself a Cambridge graduate and teacher, the first of many Trinity College men who were to lead in the development of algebra. He graduated as second wrangler—that is, he took second place in the celebrated tripos examination (initiated in 1725) for undergraduates who had specialized in mathematics—the first wrangler being John Herschel, another of the founders of the Analytical Society. Peacock was a zealous administrator and reformer, taking an active part in the modification of the university statutes and in the establishment of the Astronomical Society of London, the Philosophical Society of Cambridge, and the British Association for the Advancement of Science, the last of which set the pattern for the American Association for the Advancement of Science. The last twenty years of his life were spent as dean of Ely cathedral.

Peacock did not produce any outstanding new results in mathematics, but he was of great importance in the reformation of the subject in Britain, especially with respect to algebra. There had been at Cambridge a tendency in algebra as conservative as that in geometry and analysis; whereas on the Continent mathematicians were developing the graphical representation of complex numbers, in England there were protests that not even negative numbers had validity. In an effort to justify the broader views in algebra, Peacock in 1830 published his *Treatise on Algebra*, in which he sought to give the subject a logical structure comparable to that of Euclid's *Elements*.

[1] Alexander Macfarlane, *Lectures on Ten British Mathematicians of the Nineteenth Century* (1916), p. 10. Cf. W. W. R. Ball, *A History of the Study of Mathematics at Cambridge* (1889).

Without using their modern names, he attempted, none too successfully by modern standards, to formulate the fundamental laws of arithmetic—the commutative and associative laws for addition and multiplication and the distributive law for multiplication over addition. This approach, amplified later into a two-volume work of 1842–1845, marks the beginnings of postulational thinking in arithmetic and algebra. In the first volume the author applied the rules to numbers, or, as he called the subject, "arithmetical algebra"; in the second volume, devoted to "symbolical algebra," he carried the rules over to the study of magnitudes in general. In Peacock's "arithmetical algebra" the symbols $+$ and $-$ are understood to have only their ordinary significance, so that the expression $a - b$ has meaning only if a is greater than b (for Peacock had the natural numbers in mind). In "symbolical algebra" such restrictions are removed, but the rules of the numerical algebra are nevertheless assumed to hold universally in the more abstract system:[2]

All the results of arithmetical algebra which are deduced by the application of its rules, and which are general in form though particular in value are results likewise of symbolical algebra, where they are general in value as well as in form.

The justification for such a bold extrapolation is not made clear; Peacock merely accepts this as a "principle of the permanence of equivalent forms" somewhat akin to the correlation principle that Carnot and Poncelet had used so fruitfully in geometry. However, the algebraic form of this fuzzy postulate in one respect served as a deterrent to progress, for it suggested that the laws of algebra are the same no matter what the numbers or objects within the algebra may be. Peacock, it appears, was thinking primarily of the number system of integers and the real magnitudes of geometry, and his distinction between the two types of algebra was not so different from that which Viète had made between "logistica numerosa" and "logistica speciosa." Hence it was that the subtitle for Peacock's second volume is *On Symbolical Algebra and its Applications to the Geometry of Position*, the last three words of which might imply that the author had been reading Carnot.

3   Peacock, the "Euclid of algebra," found support for his views in the work of Augustus De Morgan (1806–1871), a man who helped also to found the British Association for the Advancement of Science (1831) and who in a sense joined Peacock in constituting what might be called a "British School" in mathematics. De Morgan had been born in India, his father having been

---

[2] *Treatise on Algebra* (reprinted from the 1842–1845 edition, New York: Scripta Mathematica, 1940), Vol. I, pp. vi–viii. A life of Peacock is found in Macfarlane, *Ten British Mathematicians*, Chap. 1.

associated with the East India Company, but he attended Trinity College, graduating as fourth wrangler. He could not hold a fellowship at Cambridge or Oxford because he refused to submit to the necessary religious test, despite the fact that he had been brought up in the Church of England, in which his mother hoped he would become a minister. De Morgan consequently was appointed, at the early age of twenty-two, a professor of mathematics at the newly established London University, later University College of the University of London, where he continued to teach except for short periods following resignations prompted by cases of abridgement of academic freedom. He always remained a champion of religious and intellectual toleration, and he was equally a writer and teacher of exceptional ability. He was born blind in one eye, a handicap that might account for some of his innocuous eccentricities, such as his dislike of rural life, his refusal ever to vote in an election, and his failure to apply for membership in the Royal Society. He was a lover of conundrums and witticisms, many of which are collected in his well-known *Budget of Paradoxes*, a delightful satire on circle-squarers edited after his death by his widow.[3]

Peacock was something of a prophet in the development of abstract algebra, and De Morgan was to him somewhat as Elisha was to Elijah. In Peacock's *Algebra* the symbols were generally understood to be numbers or magnitudes, but De Morgan would keep them abstract. He left without meaning not only the letters that he used, but also the symbols of operation; letters such as A, B, C might stand for virtues and vices and $+$ and $-$ might mean reward and punishment. De Morgan insisted that, "with one exception, no word or sign of arithmetic or algebra has one atom of meaning throughout this chapter, the object of which is symbols and their laws of combination, giving a symbolic algebra which may hereafter become the grammar of a hundred distinct significant algebras." (The exception mentioned by De Morgan is the symbol of equality, for he thought that in $A = B$ the symbols A and B must "have the same resulting meaning, by whatever steps attained.") This idea, expressed as early as 1830 in his *Trigonometry and Double Algebra*, comes close to the modern recognition that mathematics deals with propositional functions, rather than with functions; but De Morgan seems not to have realized the entirely arbitrary nature of the rules and definitions of algebra. He was sufficiently close to Kantian philosophy to believe that the usual fundamental laws of algebra should apply to any algebraic system whatsoever. He saw that in going from the "single algebra" of the real number system to the "double algebra" of the complex numbers, the rules of operation remain the same. And De Morgan believed that these two forms

---

[3] An extensive account of De Morgan's life and work is included in Macfarlane, *Ten British Mathematicians*, Chap. 2. See also *Memoir of A. D. M. by his Wife Sophia Elizabeth De Morgan, with Selections from His Letters* (1882).

exhaust the types of algebra that are possible and that a triple or quadruple algebra could not be developed. In this important respect he was shown to be wrong by Hamilton, another man of Trinity, but this time not Trinity College, Cambridge, but Trinity College, Dublin. (Yet another mathematician of Trinity (Dublin) was George Salmon [1819–1904], who taught both mathematics and divinity there and was the author of excellent textbooks on conics, algebra, and analytic geometry.)

4 Hamilton's father, a practicing attorney, and his mother, said to have been intellectually gifted, both died while he was a boy; but even before he was orphaned, the young Hamilton's education had been determined by an uncle who was a linguist. An extremely precocious youngster, William read Greek, Hebrew, and Latin by the time he was five; at the age of ten he was acquainted with half a dozen oriental languages. A meeting with a lightning calculator a few years later perhaps spurred Hamilton's already strong interest in mathematics, as friendships with Wordsworth and Coleridge probably encouraged him to continue to produce the bad poetry he had been writing since boyhood.[4] Hamilton entered Trinity College, Dublin, and while still an undergraduate there, at the age of twenty-two, he was appointed Royal Astronomer of Ireland, Director of the Dunsink Observatory, and Professor of Astronomy. In the same year he presented to the Irish Academy a paper on systems of rays in which he expressed one of his favorite themes— that space and time are "indissolubly connected with each other." In a sense this view could be taken to presage the theory of relativity, but Hamilton drew from it a less fruitful conclusion: inasmuch as geometry is the science of space alone, algebra must be the science of pure time. Perhaps Hamilton here was following in algebra the lead of Newton who, when he had difficulty defining abstract concepts in the method of fluxions, felt more comfortable in appealing to the notion of time in the physical universe. Possibly he was simply concluding that, since geometry is the science of space, and space and time are the two aspects of sensuous intuition, algebra should be the science of time.[5]

Shortly after presenting his first paper, Hamilton's prediction of conical refraction in certain crystals was experimentally confirmed by physicists. This verification of a mathematical theory assured his reputation, and at the age of thirty he was knighted. Two years earlier, in 1833, he had presented a long and significant paper before the Irish Academy in which he introduced

---

[4] Accounts of his life and work are found in E. T. Bell, *Men of Mathematics* (1937), Chap. 19, and Alexander Macfarlane, *Ten British Mathematicians*, Chap. 3. In *Scripta Mathematica*, **10** (1944), there are a number of articles devoted to Hamilton's life and work. The most extensive study is R. P. Graves, *Life of Sir William Rowan Hamilton* (1882). See also C. Lanczos, "William Rowan Hamilton," *American Scientist*, **55** (1967), 129–143.

[5] E. T. Bell, *Men of Mathematics*, p. 359.

a formal algebra of real number couples the rules of combination of which are precisely those given today for the system of complex numbers. The important rule for multiplication of the couples is, of course,

$$(a, b)(\alpha, \beta) = (a\alpha - b\beta, a\beta + b\alpha)$$

and he interpreted this product as an operation involving rotation.[6] Here one sees the definitive view of a complex number as an ordered pair of real numbers, an idea that had been implied in the graphical representations of Wessel, Argand, and Gauss, but which now for the first time was made explicit.

Hamilton realized that his ordered pairs could be thought of as directed entities in the plane, and he naturally tried to extend the idea to three dimensions by going from the binary complex number $a + bi$ to ordered number triples $a + bi + cj$. The operation of addition created no difficulty, but for ten years he was baffled by multiplication of n-tuples for n greater than two. One day in 1843, as he was walking with his wife along the Royal Canal, he had a flash of inspiration: his difficulty would vanish if he used quadruples instead of triples and if he abandoned the commutative law for multiplication. It had been more or less clear that for number quadruples $a + bi + cj + dk$ one should take $i^2 = j^2 = k^2 = -1$; now Hamilton saw in addition that he should let $ij = k$, but $ji = -k$, and similarly $jk = i = -kj$ and $ki = j = -ik$. In other respects the laws of operation are as in ordinary algebra.

Just as Lobachevsky had created a new geometry consistent within itself, by abandoning the parallel postulate, so Hamilton created a new algebra, also consistent within itself, by discarding the commutative postulate for multiplication. He stopped in his walk, and with a knife he cut the fundamental formula $i^2 = j^2 = k^2 = ijk$ on a stone of Brougham Bridge; the same day, October 16, he asked the Royal Irish Academy for leave to read a paper on quaternions at the next session. The key discovery was sudden, but the discoverer had been working toward it for some fifteen years. Hamilton, quite naturally, always regarded the discovery of quaternions as his greatest achievement; in retrospect it is clear that it was not so much this particular type of algebra that was significant, but rather the discovery of the tremendous freedom that mathematics enjoys to build algebras that need not satisfy the restrictions imposed by the so-called "fundamental laws," which up to that time, supported by the vague principle of permanence of form, had been invoked without exception. For the last twenty years of his life Hamilton spent his energies on his favorite algebra, which he was inclined to imbue with cosmic significance, and which some British mathematicians regarded as a kind of Leibnizian *arithmetica universalis*. His *Lectures on Quaternions*

---

[6] An account of the paper in which Hamilton presented this view is given by C. C. MacDuffee in "Algebra's Debt to Hamilton," *Scripta Mathematica*, **10** (1944), 25–36.

appeared in 1853, and thereafter he devoted himself to the preparation of the enlarged *Elements of Quaternions*. This was not quite completed when he died in 1865, but it was edited and published by his son in the following year. The tragedy of a semi-invalid wife had dogged his later years, and occasional alcoholic intemperance led wags to say that while he might indeed be a master of pure time, he was not a master of sublunary time. Nevertheless, it is gratifying for Americans to recall that in those unhappy years of civil strife the newly established National Academy of Sciences named Sir William Rowan Hamilton its first foreign associate.

5   In 1844, the year following Hamilton's discovery of quaternion multiplication, somewhat similar views were published in Germany by Grassmann in his treatise entitled *Die lineale Ausdehnungslehre, ein neuer Zweig der Mathematik* ("The Theory of Linear Extension, a New Branch of Mathematics"). This is a very general calculus of vectors in any number of dimensions, and here, too, we find the development of the idea of noncommutative multiplication. In fact, multiplication in Grassmann's system is not necessarily associative. It is interesting to note that Grassmann, like Hamilton, was also a linguist, being a specialist in Sanskrit literature. Unlike Hamilton, he never occupied a position of prominence, but taught on the secondary school level. Moreover, the significance of his *Ausdehnungslehre* was slow to be recognized, for the book was not only unconventional, but difficult to read. One reason was that Grassmann, like Desargues before him, used a very unconventional terminology, but more fundamental was the novelty and extreme generality of the author's approach to the question of extension. Even Gauss, who expressed approval of Grassmann's work, seems to have found the philosophical abstraction excessive.[7]

Grassmann rewrote his *Ausdehnungslehre* in a second edition of 1862, and his influence then began to be felt more strongly. In particular, it resulted in the development in America, primarily through the efforts of a Yale University physicist, Josiah Williard Gibbs (1839–1903), of the more limited algebra of vectors in three-dimensional space. The algebra of vectors is again a multiple algebra in which the commutative law for multiplication fails to hold. In fact, it was proved in 1867 by Hankel that the algebra of complex numbers is, as De Morgan suspected, the most general algebra that is possible under the fundamental laws of arithmetic.[8] The *Vector Analysis* of Gibbs appeared in 1881 and again in 1884, and he published

---

[7] E. T. Bell, *The Development of Mathematics* (1940), p. 183. There is no adequate account in English of Grassmann and his work, but an extract in translation from the *Ausdehnungslehre* is included in *Source Book in Mathematics*, ed. by D. E. Smith (1929), pp. 684–685.

[8] See Kenneth O. May, "The Impossibility of a Division Algebra of Vectors in Three Dimensional Space," *American Mathematical Monthly*, 73 (1966), 289–291.

further articles throughout the decade. These works led to a spirited and not too genteel controversy with the proponents of quaternions over the relative merits of the two algebras. In 1895 a colleague of Gibbs at Yale organized an International Association for Promoting the Study of Quaternions and Allied Systems of Mathematics, of which the first president was a rabid supporter of quaternions. It was not long before allied systems, such as vectors and their generalization, tensors, for a time eclipsed quaternions,[9] but today they have a recognized place in algebra, as well as in quantum theory. Moreover, although Hamilton's name is infrequently linked with vectors, since Gibbs' notations came mostly from Grassmann, nevertheless the chief properties of vectors had been worked out in Hamilton's protracted investigations in multiple algebras.

By the middle of the nineteenth century German mathematicians had **6** stood head and shoulders above those of other nationalities in analysis and geometry, with the universities of Berlin and Göttingen in the lead and with publication centering in Crelle's *Journal*. Algebra, on the other hand, was for a while almost a British monopoly, with Trinity College, Cambridge, in the forefront and the *Cambridge Mathematical Journal* as the chief medium of publication. Peacock and De Morgan both were from Trinity, as was also Cayley, a heavy contributor to both algebra and geometry, who had graduated as senior wrangler. We have noted Cayley's work in analytic geometry, especially in connection with the use of determinants; but Cayley also was one of the first men to study matrices, another instance of the British concern for form and structure in algebra. This work grew out of a memoir of 1858 on the theory of transformations. If, for example, we follow the transformation

$$T_1 \begin{cases} x' = ax + by \\ y' = cx + dy \end{cases}$$

by another transformation

$$T_2 \begin{cases} x'' = Ax' + By' \\ y'' = Cx' + Dy' \end{cases}$$

the result (which had appeared earlier, for example, in the *Disquisitiones arithmeticae* of Gauss in 1801) is equivalent to the single composite transformation

---

[9] A book on the history of vector analysis, by Michael J. Crowe, is scheduled to appear shortly. Crowe also is preparing a translation of Grassmann's *Ausdehnungslehre*. There is an historical introduction in P. H. Moon and D. E. Spencer, *Vectors* (Princeton, N.J., 1965). See also G. J. Pawlikowski, "The Men Responsible for the Development of Vectors," *The Mathematics Teacher*, **60** (1967), 393–396.

$$T_2 T_1 \begin{cases} x'' = (Aa + Bc)x + (Ab + Bd)y \\ y'' = (Ca + Dc)x + (Cb + Dd)y \end{cases}$$

If, on the other hand, we reverse the order of T_1 and T_2, so that T_2 is the transformation

$$\begin{cases} x' = Ax + By \\ y' = Cx + Dy \end{cases}$$

and T_1 is the transformation

$$\begin{cases} x'' = ax' + by' \\ y'' = cx' + dy' \end{cases}$$

then these two, applied successively, are equivalent to the single transformation

$$T_1 T_2 \begin{cases} x'' = (aA + bC)x + (aB + bD)y \\ y'' = (cA + dC)x + (cB + dD)y \end{cases}$$

Reversing the order of the transformations in general gives a different result. Expressed in the language of matrices,

$$\begin{pmatrix} a & b \\ c & d \end{pmatrix} \cdot \begin{pmatrix} A & B \\ C & D \end{pmatrix} = \begin{pmatrix} aA + bC & aB + bD \\ cA + dC & cB + dD \end{pmatrix}$$

but

$$\begin{pmatrix} A & B \\ C & D \end{pmatrix} \cdot \begin{pmatrix} a & b \\ c & d \end{pmatrix} = \begin{pmatrix} Aa + Bc & Ab + Bd \\ Ca + Dc & Cb + Dd \end{pmatrix}$$

Inasmuch as two matrices are equal if and only if all corresponding elements are equal, it is clear that once again we have an instance of noncommutative multiplication.

The definition of multiplication of matrices is as indicated above, and the sum of two matrices (of the same dimensions) is defined as the matrix obtained by adding the corresponding elements of the matrices. Thus

$$\begin{pmatrix} a & b \\ c & d \end{pmatrix} + \begin{pmatrix} A & B \\ C & D \end{pmatrix} = \begin{pmatrix} a + A & b + B \\ c + C & d + D \end{pmatrix}$$

Multiplication of a matrix by a scalar K is defined by the equation

$$K \cdot \begin{pmatrix} a & b \\ c & d \end{pmatrix} = \begin{pmatrix} Ka & Kb \\ Kc & Kd \end{pmatrix}$$

The matrix

$$\begin{pmatrix} 1 & 0 \\ 0 & 1 \end{pmatrix}$$

which is usually denoted by I, leaves every square matrix of second order invariant under multiplication; hence it is called the identity matrix under multiplication. The only matrix leaving another such matrix invariant under addition is of course the zero matrix

$$\begin{pmatrix} 0 & 0 \\ 0 & 0 \end{pmatrix}$$

which consequently is the identity matrix under addition. With these definitions we can think of the operations on matrices as constituting an "algebra," a step that was taken by Cayley and the American mathematicians Benjamin Peirce (1809–1880) and his son Charles S. Peirce (1839–1914). The Peirces played somewhat the role in America that Hamilton, Grassmann, and Cayley had filled in Europe. The study of matrix algebra and of other noncommutative algebras has everywhere been one of the chief factors in the development of an increasingly abstract view of algebra,[10] especially in the twentieth century.

Shortly after receiving his degree at Trinity, Cayley took to the law for **7** fourteen years; this interfered little with his mathematical research, and he published several hundred papers during these years. Many of the papers were in the theory of algebraic invariants, a field in which he and his friend James Joseph Sylvester were preeminent. Cayley and Sylvester were a study in contrasts, the former being mild and even-tempered, the latter mercurial and impatient. Both were Cambridge men—Cayley at Trinity, Sylvester at St. John's—but Sylvester was ineligible for a degree because he was a Jew. For three years following 1838 Sylvester had taught at University College, London, where he was a colleague of his former teacher, De Morgan; after this he accepted a professorship at the University of Virginia. Discipline problems so upset the temperamental mathematician that he left precipitately after only three months. Upon returning to England he spent almost ten years in business and then turned to the study of law, in connection with which, in 1850, he first met Cayley. The two men were ever afterward friends and mathematicians, and ultimately both left the law. In 1854 Sylvester took a position at the Royal Military Academy at Woolwich, and in 1863 Cayley accepted the Sadlerian professorship at Cambridge. In 1876 Sylvester had

[10] See, for example, Nicolas Bourbaki, *Éléments d'histoire des mathématiques* (1960), pp. 120–128.

one more fling at teaching in America, this time at the newly established Johns Hopkins University, where he remained until he was almost seventy, when he accepted a professorship offered him by Oxford University. In 1881, while Sylvester was still at Johns Hopkins, Cayley accepted an invitation to deliver there a series of lectures on Abelian and theta functions. Although Cayley's papers, which rival those of Euler and Cauchy in number, are predominantly in algebra and geometry, he did contribute also to analysis, and his only book, published in 1876, is a *Treatise on Elliptic Functions*.[11]

Cayley's interests were divided, but Sylvester's loyalty to algebra was firm, and it is fitting that his name is attached to what is known as Sylvester's dialytic method in eliminating an unknown from two polynomial equations. The device is a simple one and consists in multiplying one or both of the two equations by the unknown quantity to be eliminated, repeating the process if necessary until the total number of equations is one greater than the number of powers of the unknown. From this set of $n + 1$ equations one can then eliminate all the n powers, thinking of each power as a different unknown. Thus, to eliminate x from the pair of equations $x^2 + ax + b = 0$ and $x^3 + cx^2 + dx + e = 0$, one multiplies the first by x and then multiplies the resulting equation, and also the second equation above, by x. Then, thinking of each of the four powers of x as a separate unknown, the determinant

$$\begin{vmatrix} 0 & 0 & 1 & a & b \\ 0 & 1 & a & b & 0 \\ 1 & a & b & 0 & 0 \\ 0 & 1 & c & d & e \\ 1 & c & d & e & 0 \end{vmatrix}$$

known as the resultant in Sylvester's method, when equated to zero, gives the result of the elimination.

8   More important than his work in elimination was Sylvester's collaboration with Cayley in the development of the theory of "forms" (or "quantics," as Cayley preferred to call them), through which the men came to be known as "invariant twins." Between 1854 and 1878 Sylvester published almost a dozen papers on forms—homogeneous polynomials in two or more variables—and their invariants.[12] The most important cases in analytic geometry

[11] His *Collected Mathematical Papers* (1889–1898) fill fourteen large volumes.

[12] Short summaries of some of Cayley's papers on quantics and other topics are included in Ganesh Prasad, *Some Great Mathematicians of the Nineteenth Century* (1933–1934), II, 1–33. See also R. W. Feldmann, "History of Elementary Matrix Theory," *The Mathematics Teacher*, **55** (1962), 482–484, 589–590, 657–659.

James Joseph Sylvester.

and physics are the quadratic forms in two and three variables, for, when equated to a constant, these represent conics and quadrics. In particular, the quantic or form $Ax^2 + 2Bxy + Cy^2$, when equated to a nonzero constant, represents an ellipse (real or imaginary), a parabola, or a hyperbola according as $B^2 - AC$ is less than, equal to, or greater than zero. Moreover, if the form is transformed under a rotation of axes about the origin into the new form $A'x^2 + 2B'xy + C'y^2$, then $(B')^2 - A'C' = B^2 - AC$—that is, the expression $B^2 - AC$, known as the characteristic of the form, is an invariant under such a transformation. The expression $A + C$ is another invariant. Still other important invariants associated with the form are the roots k_1 and k_2 of the characteristic equation

$$\begin{vmatrix} A - k & B \\ B & C - k \end{vmatrix} = 0 \quad \text{or} \quad \begin{vmatrix} A' - k & B' \\ B' & C' - k \end{vmatrix} = 0$$

These roots are, in fact, the coefficients of x^2 and y^2 in the canonical form $k_1 x^2 + k_2 y^2$ to which the form, if not of parabolic type, can be reduced through a rotation of axes. The effervescent Sylvester boasted that he had discovered and developed the reduction of binary forms to canonical form at one sitting, "with a decanter of port wine to sustain nature's flagging energies."[13]

If we designate by M the matrix of coefficients of the form and by I the identity matrix of order two, the characteristic equation can be written as $|M - kI| = 0$, where the vertical lines represent the determinant of the matrix. One of the important properties of the algebra of matrices is that a matrix M satisfies its characteristic equation, a result given in 1858 and known as the Hamilton–Cayley theorem. It sometimes is held that Cayley's algebra of matrices was an outcome of Hamilton's algebra of quaternions, but Cayley in 1894 specifically denied such a link. He admired the theory of quaternions, but he asserted that his development of matrices stemmed from that of determinants as a convenient mode of expressing a transformation.[14]

9   While the Trinity mathematicians Hamilton and Cayley (one from Dublin, the other from Cambridge) were developing two new types of algebra, a third and radically different form of algebra was being invented by an essentially self-taught Britisher, George Boole. Born into an impecunious lower-class tradesman's family at Lincoln, England, Boole had only a common school education; but he learned both Greek and Latin independently, believing that this knowledge would help him to rise above his station. During his early years as an elementary school teacher, Boole found that he had to learn more mathematics, and he began mastering the works of Laplace and Lagrange, as well as studying additional foreign languages. Having become friendly with De Morgan, he also took a keen interest in a controversy over logic that the Scottish philosopher Sir William Hamilton (1788–1856), not to be confused with the Irish mathematician Sir William

---

[13] Bell, *Men of Mathematics*, p. 398. In Chapter 21 there is a good general account of the lives of both Sylvester and Cayley. In Macfarlane, *Ten British Mathematicians*, there is a full chapter on each.

[14] E. T. Bell, *Development of Mathematics*, pp. 188–189.

Rowan Hamilton (1805–1865), had raised with De Morgan. (The Scottish Sir William was a baronet who had inherited his title, the Irish Sir William was a knight who had earned the title.) The result was that Boole in 1847 published a short work entitled *The Mathematical Analysis of Logic*, a little book that De Morgan recognized as epoch-making.

The history of logic may be divided, with some slight degree of over-simplification, into three stages: (1) Greek logic, (2) Scholastic logic, and (3) mathematical logic.[15] In the first stage, logical formulas consisted of words of ordinary language, subject to the usual syntactical rules. In the second stage, logic was abstracted from ordinary language but characterized by differentiated syntactical rules and specialized semantic functions. In the third stage, logic became marked by the use of an artificial language in which words and signs have narrowly limited semantic functions. Whereas in the first two stages logical theorems were *derived* from ordinary language, the logic of the third stage proceeds in a contrary manner—it first *constructs* a purely formal system, and only later does it look for an interpretation in everyday speech. Although Leibniz sometimes is regarded as a precursor of the latter point of view, its floruit date is really the year in which Boole's first book appeared, as well as De Morgan's *Formal Logic*. The work of Boole, in particular, emphasized that logic should be associated with mathematics, rather than with metaphysics, as the Scottish Sir William Hamilton had argued.

More important even than his mathematical logic was Boole's view of mathematics itself. In the Introduction to his *Mathematical Analysis of Logic* the author objected to the then-current view of mathematics as the science of magnitude or number (a definition still adopted in some of the weaker dictionaries). Espousing a far more general view, Boole wrote:

> We might justly assign it as the definitive character of a true Calculus, that it is a method resting upon the employment of Symbols, whose laws of combination are known and general, and whose results admit of a consistent interpretation.... It is upon the foundation of this general principle, that I propose to establish the Calculus of Logic, and that I claim for it a place among the acknowledged forms of Mathematical Analysis.[16]

Peacock's *Algebra* of 1830 had suggested that the symbols of objects in algebra need not stand for numbers, and De Morgan argued that interpretations of the symbols for operations also were arbitrary; Boole carried the formalism to its conclusion. No longer was mathematics to be limited to

---

[15] See I. M. Bochenski, *Formale Logik* (Amsterdam: North Holland, 1956; English translation 1961).

[16] George Boole, *The Mathematical Analysis of Logic* (1847) pp. 3–4. There is a reprint of this work—New York: Philosophical Library, 1948.

questions of number and continuous magnitude. Here for the first time the view is clearly expressed that the essential characteristic of mathematics is not so much its content as its form. If any topic is presented in such a way that it consists of symbols and precise rules of operation upon these symbols, subject only to the requirement of inner consistency, this topic is part of mathematics. Although the *Mathematical Analysis of Logic* did not achieve wide recognition, it probably was upon the weight of this work that Boole two years later was appointed professor of mathematics at the newly established Queens College in Cork.

A great mathematician and philosopher of the twentieth century, Bertrand Russell, has asserted that the greatest discovery of the nineteenth century was the nature of pure mathematics. He adds to this claim the words, "Pure Mathematics was discovered by Boole in a work which he called *The Laws of Thought*." In this assertion Russell is referring to Boole's best-known work, published in 1854. To be more accurate it might have been better to have cited the earlier book of 1847, in which much the same views had been presented.

10   Boole's *Investigation of the Laws of Thought* of 1854 is a classic in the history of mathematics, for it amplified and clarified the ideas presented in 1847, establishing both formal logic and a new algebra, known as Boolean algebra, the algebra of sets, or the algebra of logic. Boole used the letters $x, y, z, \ldots$ to represent objects of a subset of things—numbers, points, ideas, or other entities—selected from a universal set or universe of discourse, the totality of which he designated by the symbol or "number" 1. For example, if the symbol 1 represents all Europeans, x might stand for all Europeans who are French citizens, y might be all European men over twenty-one, and z might be all Europeans who are between five and six feet tall. The symbol or number 0 Boole took to indicate the empty set, containing no element of the universal set—what now is known as the null set. The sign $+$ between two letters or symbols, as $x + y$, he took to be the union of the subsets x and y—that is, the set made up of all the elements in x or y (or both). The multiplication sign $\times$ represented the intersection of sets, so that $x \times y$ means the elements or objects that are in the subset x and also in the subset y. In the example above, $x + y$ consists of all Europeans who are French citizens, or are men over twenty-one, or both; $x \times y$ (written also as $x \cdot y$ or simply as xy) is the set of French citizens who are men over twenty-one. (Boole, unlike De Morgan, used exclusive union, not permitting common elements in x and y, but modern Boolean algebra more conveniently takes $+$ to be the inclusive union of sets that may have elements in common.) The sign $=$ represents the relationship of identity. It is clear that the five fundamental laws of algebra now hold for this Boolean algebra, for

$x + y = y + x$, $xy = yx$, $x + (y + z) = (x + y) + z$, $x(yz) = (xy)z$, and $x(y + z) = xy + xz$. Nevertheless, not all of the rules of ordinary algebra continue to be valid. For example, $1 + 1 = 1$ and $x \cdot x = x$. (The second of these appears in the work of Boole, but not the first, since he used exclusive union.) The equation $x^2 = x$ has only the two roots, in ordinary algebra, $x = 0$ and $x = 1$; in this respect the algebra of logic and ordinary algebra are in agreement. The equation $x^2 = x$, when written in the form $x(1 - x) = 0$, also suggests that $1 - x$ should designate the complement of the subset x—that is, all the elements in the universal set that are not in the subset x. Although it is true in Boolean algebra that $x^3 = x$ or $x(1 - x^2) = 0$ or $x(1 - x)(1 + x) = 0$, the solution in ordinary algebra differs from that in Boolean algebra, in which there are no negative numbers. Boolean algebra differs from ordinary algebra also in that if $zx = zy$ (where z is not the null set), it does not follow that $x = y$; nor is it necessarily true that if $xy = 0$, then x or y must be 0.

Boole showed that his algebra provided an easy algorithm for syllogistic reasoning. The equation $xy = x$, for example, says very neatly that all x's are y's. If it is also given that all y's are z's, then $yz = y$. Upon substituting in the first equation the value of y given by the second equation, the result is $x(yz) = x$. Using the associative law for multiplication, the last equation can be written as $(xy)z = x$, and, upon replacing xy by x, we have $xz = x$, which is simply the symbolic way of saying that all x's are z's.

The *Mathematical Analysis of Logic* (1847) and, a fortiori, *The Laws of Thought* (1854) contain much more of the algebra of sets than we have indicated. In particular, the latter work includes applications to probability. Today Boolean algebra is used widely not only by pure mathematicians but also by others who apply it to problems in insurance and information theory. Notations have changed somewhat since Boole's day, so that union and intersection are generally indicated by $\cup$ and $\cap$, rather than $+$ and $\times$, and the symbol for the null set is ϕ rather than 0; but the fundamental principles are those that were laid down by Boole more than a century ago.

There is an aspect of Boole's work that is not closely related to his treatises in logic or the theory of sets, but which is familiar to every student of differential equations. This is the algorithm of differential operators, which he introduced in order to facilitate the treatment of linear differential equations. If, for example, we wish to solve the differential equation $ay'' + by' + cy = 0$, the equation is written in the notation $(aD^2 + bD + c)y = 0$. Then, regarding D as an unknown quantity rather than an operator, we solve the *algebraic* quadratic equation $aD^2 + bD + c = 0$. If the roots of the algebraic equation are p and q, then e^{px} and e^{qx} are solutions of the differential equation and $Ae^{px} + Be^{qx}$ is a general solution of the differential equation. There are many other situations in which Boole, in his *Treatise on Differential Equations* of

1859, pointed out parallels between the properties of the differential operator (and its inverse) and the rules of algebra. British mathematicians in the second half of the nineteenth century were thus again becoming leaders in algorithmic analysis, a field in which, fifty years before, they had been badly deficient.

Boole died in 1864, only ten years after publishing his *Laws of Thought*, but recognition, including an honorary degree from the University of Dublin, had come to him before his death.[17] It is curious to note that Cantor, who like Boole was one of the chief trailblazers of the century, was one of the few who declined to accept the work of Boole.

11   Among those who continued Boole's work after his death were De Morgan and Benjamin Peirce. These men hit independently on what is known as De Morgan's law of duality—for every proposition involving logical addition and multiplication, there is a corresponding proposition in which the words addition and multiplication are interchanged. In particular we have what is called De Morgan's formula: If x and y are subsets of a set S, then the complement of the union of x and y is the intersection of the complements of x and y, and the complement of the intersection of x and y is the union of the complements of x and y.

Benjamin Peirce was connected with Harvard College for more than fifty years, first as a student and later as a professor. His chief work was his paper on *Linear Associative Algebra*, read to the American Association for the Advancement of Science in 1864, but printed only in 1881 in the *American Journal of Mathematics*. Linear associative algebras include ordinary algebra, vector analysis, and quaternions as special cases, but are not restricted to the units $1, i, j, k$. Peirce worked out multiplication tables for 162 algebras, a far cry from the idea prevalent early in the century that there was only a single algebra! C. S. Peirce continued his father's work in this direction by showing that of all these algebras there are only three in which division is uniquely defined: ordinary real algebra, the algebra of complex numbers, and the algebra of quaternions. It was in connection with his work on linear associative algebra that Benjamin Peirce in 1870 gave the well-known definition, "Mathematics is the science which draws necessary conclusions." His son was in wholehearted agreement with this view, as a result of Boole's influence, but he stressed that mathematics and logic are not the same.

---

[17] There is no full-length biography of Boole. For further details see Macfarlane, *Ten British Mathematicians*, pp. 50–63, Bell, *Men of Mathematics*, Chap. 23, or a forthcoming article on Boole by T. A. A. Broadbent in *Dictionary of Scientific Biography* (to be published by Charles Scribner's Sons, New York). A very readable account of Boolean algebra is included also in Herbert Meschkowski, *Ways of Thought of Great Mathematicians* (1964), pp. 74–83. See also the article on "Logic, History of," in *The Encyclopedia of Philosophy*, ed. by Paul Edwards (8 vols., New York: Macmillan, 1967), IV, 513–571.

Charles Sanders Peirce.

"Mathematics is purely hypothetical: it produces nothing but conditional propositions. Logic, on the contrary, is categorical in its assertions."[18]

[18] For short excerpts from the work of both Benjamin Peirce and C. S. Peirce, see *The Treasury of Mathematics*, ed. by Henrietta O. Midonick (1965), pp. 610–642. Cf. *Studies in the Philosophy of Charles Sanders Peirce* (2nd series, Amherst, Mass.: University of Massachusetts Press, 1964).

This distinction was to be argued further throughout the mathematical world in the first half of the twentieth century.

In England somewhat similar ideas were pursued by William Kingdon Clifford (1845–1879), still another Trinity graduate, whose brilliant work, like that of an earlier Trinity graduate, Roger Cotes, was cut short by premature death in his thirty-fourth year. Clifford was extraordinary in several respects. For one thing, he was a gymnast who could pull himself up on the bar with either hand—a most unusual feat for anyone, and especially for a mathematician. Moreover, he won prizes for declamation, something almost unheard of for one who graduated as second wrangler. Also, like the Oxford mathematician C. L. Dodgson (1832–1898), better known as Lewis Carroll, author of *Alice in Wonderland*, he composed *The Little People*, a collection of tales for children. In 1870 Clifford wrote a paper "On the Space-Theory of Matter" in which he showed himself to be a staunch British supporter of the non-Euclidean geometry of Lobachevsky and Riemann.[19] In algebra Clifford also espoused the newer views, and his name is perpetuated today in the so-called Clifford algebras, of which octonians or biquaternions are special cases. These noncommutative algebras were used by Clifford to study motions in non-Euclidean spaces, certain manifolds of which are known as spaces of Clifford and Klein.[20] How different was the progressive British mathematics of the later part of the nineteenth century from the stultifyingly conservative views at the opening of the century!

12   The multiplicity of algebras invented in the nineteenth century might have given to mathematics a centrifugal tendency had it not been for the development of certain structural concepts. One of the most important of these was the notion of a group, the unifying role of which in geometry already has been indicated. In algebra the group concept was without doubt the most important force making for cohesiveness, and it was an essential factor in the rise of abstract views. No one person was responsible for the rise of the group idea, but the figure that loomed largest in this connection was that of the man who gave the concept its name, the young Évariste Galois, who died tragically before the age of twenty-one.

Galois was born just outside Paris in the village of Bourg-la-Reine, where his father served as mayor. His well-educated parents had not shown any particular aptitude for mathematics, but the young Galois did acquire from them an implacable hatred of tyranny. When he first entered school at the age of twelve, he showed little interest in Latin, Greek, or algebra,

[19] See Bell, *Men of Mathematics*, pp. 503–504, and D. J. Struik, *Concise History of Mathematics* (1967), pp. 171–173.

[20] An account of the life of Clifford will be found in Macfarlane, *Ten British Mathematicians*, pp. 78–91, but the judgments there expressed on Clifford's mathematics are not well founded.

but he was fascinated by Legendre's *Geometry*. Later he read with under-standing the algebra and analysis in the works of masters like Lagrange and Abel, but his routine classwork in mathematics remained mediocre, and his teachers regarded him as eccentric. By the age of sixteen Galois knew what his teachers had failed to recognize—that he was a mathematical genius. He hoped, therefore, to enter the school that had nurtured so many celebrated mathematicians, the École Polytechnique, but his lack of systema-tic preparation resulted in his rejection. This was but the first embittering failure. Nevertheless, Galois at the age of seventeen worked up his fundamen-tal discoveries in a paper, which he asked Cauchy to present to the Académie. Cauchy not only misplaced the paper, as he had misplaced one of Abel's important articles; he lost the paper! Now Galois hated not only examiners but also academicians. A failure in his second attempt at admission to the École Polytechnique heightened his bitterness; but the heaviest shock of all was yet to fall. Under attack because of clerical intrigues, his father felt himself persecuted and committed suicide.

Despite the blows that he had experienced, Galois entered the École Normale to prepare for teaching; he also continued his research, in 1830 submitting a memoir in competition for the Académie's prize in mathematics. Fourier, the secretary of the Académie took the paper home, died shortly afterward, and the paper was lost. Faced on all sides by tyranny and frustra-tion, Galois made the cause of the 1830 revolution his own. A blistering letter criticizing the indecision of the director of the École Normale resulted in Galois' expulsion; but once more he tried to submit a paper to the Académie, this time through Poisson. The paper contained important results now a part of what is known as Galois theory; but Poisson, the referee, returned it with the remark that it was "incomprehensible." Thoroughly disillusioned, Galois joined the National Guard. In 1831, at a gathering of republicans, he proposed a toast that was interpreted as a threat to the life of Louis Philippe, and he was arrested. Although released, he was again arrested some months later and sentenced to six months in jail. Shortly afterward he became involved with a coquette and, under a code of "honor," was unable to avoid a duel. In a letter to friends he wrote, "I have been challenged by two pat-riots—it was impossible for me to refuse." The night before the duel, with forebodings of death, Galois spent the hours jotting down, in a letter to a friend, notes for posterity concerning his discoveries.[21] He asked that the letter be published (as it was within the year) in the *Revue Encyclopédique* and expressed the hope that Jacobi and Gauss might publicly give their opinion as to the importance of the theorems. On the morning of May 30, 1832, Galois met his adversary in a duel with pistols. He was shot through

[21] The letter appears in translation in *Source Book in Mathematics*, ed. by D. E. Smith, pp. 278–285.

the intestines and lay where he fell until a passing peasant took him to a hospital, where he died of peritonitis the following morning. His funeral was attended by several thousand republicans.[22] He was only twenty years old at the time, the youngest mathematician ever to make such significant discoveries.

13   Galois had entrusted to the recipient of his last letter some manuscripts intended for the Académie and in 1846 Liouville edited some of these and published them in his *Journal*. Liouville had found the task difficult, for Galois had violated Descartes' advice, "When transcendental questions are under discussion, be transcendentally clear"; but Liouville felt rewarded when, after filling in gaps in the proof, he saw the method by which Galois had demonstrated this beautiful theorem:

> In order that an irreducible equation of prime degree be solvable by radicals, it is necessary and sufficient that all its roots be rational functions of any two of them.

The principal object of Galois' researches had been to determine when polynomial equations are solvable by radicals. Gauss, in his criteria for the constructibility of regular polygons, had in essence solved the question of the solvability of the equation $a_0 X^n + a_n = 0$ in terms of rational operations and square roots on the coefficients. Galois generalized the result to provide criteria for the solvability of $a_0 X^n + a_1 X^{n-1} + \cdots + a_{n-1} X + a_n = 0$ in terms of rational operations and nth roots on the coefficients. His approach to the problem, now known as Galois theory, was another of the highly original contributions to algebra in the nineteenth century. However, it has been said that Galois theory is like garlic in that there is no such thing as a little of it. One must make a substantial study of it to appreciate the reasoning —as Galois' experience with his contemporaries showed. Nevertheless, we can indicate in a general way what is behind Galois theory and why it has been important.

Galois began his investigations with some work of Lagrange on permutations on the roots of a polynomial equation. Any change in the ordered arrangement of n objects is called a permutation on these objects. If, for example, the order of the letters a, b, c is changed to c, a, b, this permutation is written succinctly as (acb), a notation in which each letter is taken into the letter immediately following, the first letter being understood to be the successor of the last letter. Thus the letter a was carried into c, c in turn was carried into b, and b went into a. The notation (ac) or (ac, b), however, means that a goes into c, c goes into a, and b goes into itself. If two permutations

---

[22] See George Sarton, "Evariste Galois," *Osiris*, **3** (1937), 241–259.

are performed successively, the resulting permutation is known as the product of the two permutation transformations. Thus the product of (*acb*) and (*ac*, *b*), written as (*acb*)(*ac*, *b*), is the permutation (*a*, *bc*). The identical permutation *I* takes each letter into itself—that is, it leaves the order *a*, *b*, *c* unchanged. The set of all permutations on the letters *a*, *b*, *c* clearly satisfies the definition of a group, given in Chapter XXIV on geometry; this group, containing six permutations, is known as the symmetric group on *a*, *b*, *c*. In the case of *n* distinct elements, $x_1, x_2, \ldots, x_n$, the symmetric group on these contains *n*! transformations. If these elements are the roots of an irreducible equation, the properties of the symmetric group provide necessary and sufficient conditions that the equation be solvable by radicals.

Inspired by Abel's proof of the unsolvability by radicals of the quintic equation, Galois discovered that an irreducible algebraic equation is solvable by radicals if and only if its group—that is, the symmetric group on its roots—is solvable. The description of a solvable group is quite complicated, involving as it does relationships between the group and its subgroups. The three permutations (*abc*), (*abc*)², and (*abc*)³ = *I* form a subgroup of the symmetric group on *a*, *b*, and *c*. Lagrange had already shown that the order of a subgroup must be a factor of the order of the group; but Galois went deeper and found relations between the factorability of the group of an equation and the solvability of the equation. Moreover, to him we owe the use in 1830 of the word "group" in its technical sense in mathematics.[23]

Galois theory can provide an algorithm for actually finding the roots of an equation, when these are expressible in radicals; but the emphasis in the Galois approach in the theory of equations generally is directed more toward algebraic structure than toward the handling of specific cases. Although his work was done before that of most of the British algebraists of the great period 1830–1850, his ideas were without influence until they were published in 1846. Algebra then was becoming so general and abstract that heuristic methods of finding roots were subordinated to logical questions concerning existence theorems. Galois was so close to modern attitudes in algebra and yet so inarticulate that he was unable to make himself understood by the teachers of his day. Today we hear much about the "New Math" in schools, but it is new only in the sense that the views of Galois are finally coming into their own, more than a century after fate had dealt so cruelly with him.

[23] See his *Oeuvres*, ed. by E. Picard (1897), p. 28. For details on Galois theory and the solvability of equations see Emile Artin, *Galois Theory* (Notre Dame Mathematical Lectures, No. 2, Notre Dame, Ind., 1959); or see a volume on modern algebra, such as Garrett Birkhoff and Saunders MacLane, *Survey of Modern Algebra* (New York: Macmillan, 1941), or L. E. Dickson, *Modern Algebraic Theories* (Chicago: B. H. Sanborn, 1926). A very concise explanation is found in E. T. Bell, *The Development of Mathematics*, pp. 218–220. Cf. Garrett Birkhoff, "Galois and Group Theory," *Osiris*, **3** (1937), 260–268.

14   The work of Galois was important not only in making the abstract notion of group fundamental in the theory of equations, but also led, through the contributions of Dedekind, Kronecker, and Kummer, to what may be called an arithmetical approach to algebra, somewhat akin to the arithmetization of analysis. This does not mean a return to the medieval and Renaissance view of algebra as an algorithm for finding an unknown number. It means rather the development of a careful postulational treatment of algebraic structure in terms of various number fields. The concept of field was implicit in work by Abel and Galois, but Dedekind in 1879 seems to have been the first one to give an explicit definition of a number field—a set of numbers that form an Abelian group with respect to addition and with respect to multiplication (except for the inverse of zero) and for which multiplication distributes over addition. Simple examples are the system of rational numbers, the real number system, and the complex number field. Kronecker in 1881 gave other instances through his domains of rationality. The set of numbers of the form $a + b\sqrt{2}$, where a and b are rational, form a field, as is easily verified. In this case the number of elements in the field is infinite. A field with a finite number of elements is known as a Galois field, and a simple instance of this is the field of integers modulo 5 (or any prime).

The concern for structure and the rise of new algebras, especially during the second half of the nineteenth century, led to broad generalizations in number and arithmetic. We have noted already that Gauss extended the idea of integer through the study of Gaussian integers of the form $a + bi$, where a and b are integers. Dedekind generalized further in the theory of "algebraic integers"—numbers satisfying a polynomial equation with integral coefficients having leading coefficient unity. Such systems of "integers" do not, of course, form a field, for inverses under multiplication are lacking. They do have something in common in that they satisfy the other requirements for a number field; they are thus said to form an "integral domain." Such generalizations of the word integer are, however, bought at a price—the loss of unique factorization. Therefore Dedekind and a contemporary mathematician, Ernst Eduard Kummer (1810–1893), introduced into arithmetic the concept of an "ideal," based on the notion of a "ring."

A set of elements is said to form a ring if (1) it is an Abelian group with respect to addition, (2) the set is closed under multiplication, and (3) multiplication is associative and is distributive over addition. (Hence a ring that is commutative under multiplication, has a unit element, and has no divisors of zero is an integral domain.) An ideal, then, is a subset I of elements of a ring R which (1) form an additive group and (2) are such that whenever x belongs to R and y belongs to I, then xy and yx belong to I. The set of even integers, for example, is an ideal in the ring of integers. It turns out that in the ring (or integral domain) R of algebraic integers, any ideal I of R can be represented uniquely (except for the order of the factors) as a product of prime ideals. That is, uniqueness of factorization can be saved through the theory of ideals.[24]

Kummer had been left fatherless at the age of three, but his mother saw to it that her son secured an education at the University of Halle, earning his doctorate at the age of twenty-one. After about a dozen years of teaching in gymnasia, he succeeded Dirichlet at Berlin when in 1855 the latter became the successor of Gauss at Göttingen; Kummer remained there until his retirement in 1883. Shortly after earning his degree, Kummer had become interested in Fermat's last theorem, for which Cauchy at one time mistakenly thought he had a proof. Kummer was able to prove the theorem for a large class of exponents, but a general proof eluded him. The stumbling block seems to have been the fact that in the factoring of $x^n + y^n$, through the solution of $x^n + y^n = 0$ for x in terms of y, the algebraic integers, or roots of the equation, do not necessarily satisfy the fundamental theorem of arithmetic—that is, they are not uniquely factorable. The result was that, although he failed to solve Fermat's theorem, in the attempt to do so he created in a sense a new arithmetic—the theory of ideals that he discovered in 1846, many years before the similar development by Dedekind. One of the lessons that the history of mathematics clearly teaches is that the search for solutions to unsolved problems, whether solvable or unsolvable, invariably leads to important discoveries along the way.

Mathematics often has been likened to a tree, for it grows through an ever **15** more widely spreading and branching structure above ground, while at the same time it sinks its roots ever deeper and wider in the search for a firm foundation. This double growth was especially characteristic of the development of analysis in the nineteenth century, for the rapid expansion of the theory of functions had been accompanied by the rigorous arithmetization of the subject from Bolzano to Weierstrass. In algebra the nineteenth century had been more notable for new developments than for attention to foundations, and Peacock's efforts to provide a sound basis were feeble in comparison with the precision of Bolzano in analysis. During the closing years of the century, however, there were several efforts to provide stronger roots for algebra. The complex number system is defined in terms of the real numbers, which are explained as classes of rational numbers, which in turn are ordered pairs of integers; but what, after all, are the integers? Everyone thinks that he knows, for example, what the number three is—until he tries to define or explain it—and the idea of equality of integers is assumed to be obvious. Not satisfied to leave the basic concepts of arithmetic, hence of

[24] For a good introduction to ideal theory, see N. H. McCoy, *Rings and Ideals* (Carus Mathematical Monographs, No. 8, The Mathematical Association of America, 1948).

algebra, in so vague a state, the German logician and mathematician F. L. G. Frege (1848–1925) was led to his well-known definition of cardinal number. The basis for his views came from the theory of sets of Boole and Cantor. It will be recalled that Cantor had regarded two infinite sets as having the same "power" if the elements of the sets can be put into one-to-one correspondence. Frege saw that this idea of the correspondence of elements is basic also in the notion of equality of integers. Two finite sets are said to have the same cardinal number—that is, to be equal—if the elements in either class can be put into one-to-one correspondence with the elements in the other. If, then, one were to begin with an initial set, such as the set of fingers on the normal human hand, and were to form the much more comprehensive set of all sets the elements of which can be put into one-to-one correspondence with the elements of the initial set, then this set of all such sets would constitute a cardinal number, in this case the number five. More generally, Frege's definition of the cardinal number of a given class, whether finite or infinite, is the class of all classes that are similar to the given class (where by "similar" one means that the elements of the two classes in question can be placed in one-to-one correspondence).

Frege's definition of cardinal number (amended later to avoid paradoxes) appeared in 1884 in a well-known book, *Die Grundlagen der Arithmetik* ("The Foundations of Arithmetic"), and from the definition he derived the properties of the whole numbers that are familiar in grade school arithmetic. During the succeeding years Frege amplified his views in the two-volume *Grundgesetze der Arithmetik* ("Basic Laws of Arithmetic"), the first volume of which appeared in 1893 and the second ten years later. Here the author undertook to derive the concepts of arithmetic from those of formal logic, for he disagreed with the assertion of C. S. Peirce that mathematics and logic are clearly distinct. Frege had been educated at the Universities of Jena and Göttingen, and he taught at Jena during a long career. Nevertheless, his program did not meet with much response until undertaken independently early in the twentieth century by Bertrand Russell, when it became one of the chief goals of mathematicians.[25] Frege was keenly disappointed by the poor reception of his work, but the fault lay in part in the excessively novel and philosophical form in which the results were cast. History shows that novelty in ideas is more readily accepted if couched in relatively conventional form.

16   Italy had taken somewhat less active part in the development of abstract algebra than had France, Germany, and England, but during the closing years

[25] There is an English translation of *The Foundations of Arithmetic* by J. L. Austin, 2nd ed. (New York: Philosophical Library, 1953). A portion of this translation is included in *The Treasury of Mathematics*, ed. by Henrietta O. Midonick. There is also an English version of the introductory parts of Vol. I and the Epilogue to Vol. II of *The Basic Laws of Arithmetic*, trans. by Montgomery Furth (Berkeley: University of California Press, 1964).

of the nineteenth century there were Italian mathematicians who took a deep interest in mathematical logic. Best known of these was Giuseppe Peano (1858–1932), whose name is recalled today in connection with the Peano axioms upon which so many rigorous constructions of algebra and analysis depend. His aim was similar to that of Frege, but it was at the same time more ambitious and yet more down-to-earth.[26] He hoped in his "Formulaire de mathématiques" (1894 et seq.) to develop a formalized language that should contain not only mathematical logic, but all the most important branches of mathematics. That his program attracted a large circle of collaborators and disciples resulted in part from his avoidance of metaphysical language and from his felicitous choice of symbols—such as $\in$ (belongs to the class of), $\cup$ (logical sum or union), $\cap$ (logical product or intersection), and $\supset$ (is contained in)—many of which are used even today. For his foundations of arithmetic he chose three primitive concepts (zero, number [that is, non-negative whole number], and the relationship "is the successor of") satisfying five postulates:

1. Zero is a number.
2. If a is a number, the successor of a is a number.
3. Zero is not the successor of a number.
4. Two numbers of which the successors are equal are themselves equal.
5. If a set S of numbers contains zero and also the successor of every number in S, then every number is in S.

The last requirement is, of course, the axiom of induction. The Peano axioms, first formulated in 1889 in *Arithmetices principia nova methodo exposita*, represent the most striking attempt of the century to reduce common arithmetic, hence ultimately most of mathematics, to the stark essentials of formal symbolism. (He expressed the postulates in symbols, rather than in the words that we have used.) Here the postulational method attained a new height of precision, with no ambiguity of meaning and no concealed assumptions. Peano also spent much effort in the development of symbolic logic, a favorite pursuit of the twentieth century.

A further contribution by Peano to mathematics should perhaps be mentioned, since it represented one of the disquieting discoveries of the time. The nineteenth century had opened with a recognition that curves and functions need not be of the well-behaved type that had theretofore preempted the field, and Peano in 1890 showed how thoroughly mathematics could outrage common sense when he constructed continuous space-filling curves—that is, curves given by parametric equations $x = f(t)$, $y = g(t)$, where f and g are continuous real functions in the interval $0 \leq t \leq 1$, the

[26] Nicolas Bourbaki, *Eléments d'histoire des mathématiques* (1960), p. 20.

points of which completely fill the unit square $0 \le x \le 1, 0 \le y \le 1$. This paradox, of course, is all of a piece with Cantor's discovery that there are no more points in a unit square than in a unit line segment,[27] and it was among the factors that caused the following century to devote much more attention to the basic structure of mathematics. Peano himself, however, in 1903 was distracted by his invention of the international language which he called "Interlingua" or "Latino sine flexione," with vocabulary drawn from Latin, French, English, and German. This movement turned out to be far more ephemeral than his axiomatic structure in arithmetic.

In retrospect we can admire the nineteenth century as a period of unparalleled achievement, whether in geometry, analysis, or algebra. In extent, imagination, rigor, abstraction, and generality no previous century could compare with it. Nevertheless, despite the rapid advance and the definitive formulations, there was little feeling that mathematical developments were destined to slow down. The *fin de siècle* pessimism that Lagrange had expressed at the close of the eighteenth century was conspicuously absent at the end of the nineteenth. The Victorian Era exuded nothing but optimism, as far as mathematics was concerned. In the concluding chapter we shall point to a few of the respects in which this sanguine expectation was amply fulfilled—but not before serious mistrust had assailed the serenity of mathematicians during the early years of the new century. Paradox was to follow paradox until the twentieth century resembled more a period of great doubts than one of great expectations. Fortunately, the principle of challenge and response seems to have operated; in mathematical achievements already recorded, this century easily leads all the rest.

BIBLIOGRAPHY

Ball, W. W. R., *A History of the Study of Mathematics at Cambridge* (Cambridge: Cambridge University Press, 1889).

Bell, E. T., *Men of Mathematics* (New York: Simon and Schuster, 1937).

Bell, E. T., *The Development of Mathematics* (New York: McGraw-Hill, 1940).

Bochenski, I. M., *Formale Logik* (Amsterdam: North Holland, 1956; English trans. 1961).

Boole, George, *The Laws of Thought*, reprinted as Vol. II in his *Collected Logical Works* (Chicago: Open Court, 1916).

Boole, George, *The Mathematical Analysis of Logic* (Cambridge: Macmillan, 1847; reprinted, New York: Philosophical Library, 1948).

Boole, George, *Studies in Logic and Probability* (London: Waters, 1952).

---

[27] A description of Peano's curve, as well as of his axioms, is contained in Ettore Carruccio, *Mathematics and Logic in History and in Contemporary Thought*, trans. by Isabel Quigly (1964), pp. 306–308. This book includes reference also to other aspects of Peano's work.

Boole, George, *Treatise on Differential Equations* (London: Macmillan, 1859).

Bourbaki, Nicolas, *Éléments d'histoire des mathématiques* (Paris: Hermann, 1960).

Carruccio, Ettore, *Mathematics and Logic in History and in Contemporary Thought*, trans. by Isabel Quigly (Chicago: Aldine, 1964).

Cayley, Arthur, *Collected Mathematical Papers* (Cambridge, 1889–1898, 14 vols.).

Crowe, Michael J., *A History of Vector Analysis* (in press).

De Morgan, Sophia Elizabeth, *Memoir of A. D. M. by his Wife Sophia Elizabeth De Morgan, with Selections from His Letters* (London, 1882).

Feldmann, R. W., "History of Elementary Matrix Theory," *The Mathematics Teacher*, **55** (1962), 482–484, 589–590, 657–659.

Galois, Évariste, *Oeuvres*, ed. by E. Picard (Paris, 1897).

Graves, R. P., *Life of Sir William Rowan Hamilton* (Dublin: Hodges, Figgis, 1882, 3 vols.).

Klein, Felix, *Vorlesungen über die Entwicklung der Mathematik im 19. Jahrhundert* (Berlin: Springer, 1926–1927, 2 vols.).

MacDuffee, C. C., "Algebra's Debt to Hamilton," *Scripta Mathematica*, **10** (1944), 25–36.

Macfarlane, Alexander, *Lectures on Ten British Mathematicians of the Nineteenth Century* (New York: Wiley, 1916).

Meschkowski, Herbert, *Ways of Thought of Great Mathematicians* (San Francisco: Holden-Day, 1964).

Midonick, Henrietta O., ed., *The Treasury of Mathematics* (New York: Philosophical Library, 1965).

Peacock, George, *Treatise on Algebra* (1840–1845, 2 vols.; reprinted, New York: Scripta Mathematica, 1940).

Prasad, Ganesh, *Some Great Mathematicians of the Nineteenth Century* (Benares: Benares Mathematical Society, 1933–1934, 2 vols.).

Prior, A. N., ed., "Logic, History of," *The Encyclopedia of Philosophy* (New York: Macmillan, 1967, 8 vols.), IV, 513–571.

Sarton, George, "Évariste Galois," *Osiris*, **3** (1937), 241–259.

Smith, D. E., ed., *Source Book in Mathematics* (New York: McGraw-Hill, 1929; reprinted, New York: Dover, 1959).

Struik, D. J., *Concise History of Mathematics*, 3rd ed. (New York: Dover paperback, 1967).

EXERCISES

1. Britain during the nineteenth century was strong in algebra, whereas in the eighteenth century she had given preference to synthetic geometry. Explain how certain factors may account for this, citing specific instances.

2. Did algebra in the nineteenth century contribute more strikingly novel views than analysis and geometry? Cite specific instances to support your answer.

3. As algebra became more abstract during the nineteenth century, did the means of support for mathematicians change materially? Describe the livelihoods of some of the leading figures.

4. Bertrand Russell claimed that it was the nineteenth century that discovered the nature of pure mathematics. Explain what he had in mind in making this statement.

5. To what extent were the careers of Hamilton and Boole similar, and in what ways were their lives and contributions at variance?

6. Cayley and Sylvester are sometimes known as "invariant twins." Explain why and indicate in what ways they were far from being twins.

7. Does algebra of the nineteenth century bear out the common view that most great mathematical discoveries are made by young men? Cite instances to justify your answer.

8. During the nineteenth century many leading mathematicians studied and taught at Trinity College, Cambridge, where Newton had studied and taught. Mention some of these, indicating their roles at Cambridge and their contributions to mathematics.

9. Explain in what sense the algebra of complex numbers is a "double algebra." Why did De Morgan combine this with elements of trigonometry?

10. De Morgan, born in the nineteenth century, proposed the following conundrum concerning his age; I was x years old in the year x^2. Solve the conundrum.

11. If A is the matrix

$$\begin{pmatrix} 1 & 2 \\ 3 & 1 \end{pmatrix}$$

and B is the matrix

$$\begin{pmatrix} 1 & 3 \\ 2 & 1 \end{pmatrix}$$

is $AB = BA$?

12. Show that for a quadratic polynomial in x and y the sum of the coefficients of the terms in x^2 and y^2 remains invariant under a rotation of axes about the origin.

13. Show by means of diagrams that De Morgan's formula for sets is justified.

14. If A is the permutation (abc) and B is the permutation (cba), is the product AB the same as the product BA? Explain.

15. Show that the Gaussian integer $1 + i$ has no inverse with respect to multiplication, hence that the Gaussian integers $a + bi$ do not form a field.

16. Using Hamilton's concept of the complex number $x + yi$ as nothing but the real number couple (x, y), write the quotient of $(a, b) \div (\alpha, \beta)$ as a real number couple.

17. From Hamilton's quaternion formula $i^2 = j^2 = k^2 = ijk = -1$, using the associative law for multiplication, show that $ij = -ji$, $ik = -ki$, and $jk = -kj$.

18. Using Boole's method of operators solve the differential equation $y'' + 3y' + 2y = 0$.

*19. Show that the set of all permutations on the letters a, b, c form a group. Is it an Abelian group? Did Abel know about this group? Explain.

*20. Show that the set of integers modulo 5 form a field.

*21. Show that for the field of integers modulo 5 every value of x satisfies the equation $x^5 = x$.

*22. Show that numbers of the form $a + b\sqrt{2}$, where a and b are rational, form a field.

*23. Prove that the relationship $B^2 - AC$ among the coefficients of the conic $Ax^2 + 2Bxy + Cy^2 + 2Dx + 2Ey + F = 0$ is invariant under a rotation of axes. Use this result to prove that the conic is an ellipse, a parabola, or a hyperbola according as $B^2 - AC$ is less than, equal to, or greater than zero.

CHAPTER XXVII

Aspects
of the Twentieth Century

> The golden age of mathematics—that was not the age
> of Euclid, it is ours.
>
> *C. J. Keyser*

One of the definitive contributions of the nineteenth century was the recogni- **1**
tion that mathematics is not a natural science, but an intellectual creation of
man. Bertrand Russell wrote in the *International Monthly* for 1901:

> The nineteenth century, which prides itself upon the invention of steam and evolution,
> might have derived a more legitimate title to fame from the discovery of pure mathe-
> matics.

T. H. Huxley (1825–1895), "Darwin's bulldog" in the defense of evolution,
had noted that "Mathematics is that subject which knows nothing of
observation, nothing of experiment, nothing of induction, nothing of
causation." On another occasion, when he had criticized Kelvin for what
he regarded as an underestimate of the age of the earth, Huxley made the
well-known comment:

> Mathematics may be compared to a mill of exquisite workmanship which grinds you
> stuff of any degree of fineness; but, nevertheless, what you get out depends on what you
> put in; and as the grandest mill in the world will not extract wheat-flour from peas-cods,
> so pages of formulae will not get a definite result out of loose data.

That is, toward the turn of the century it was generally recognized even by
nonmathematicians that mathematics is postulational thinking, in which
from arbitrary premises one draws valid conclusions. Whether or not the
postulates are true, in a scientific sense, is immaterial; in fact, the very words
in which the postulates are expressed are undefined terms. This led Bertrand
Russell to his facetious description of mathematics, in 1901, as the subject in
which no one knows what he is talking about, nor whether what he says is
true. Two years later, in the opening of his *Principles of Mathematics*, Russell
formulated a precise definition of mathematics:

Pure Mathematics is the class of all propositions of the form "p implies q," where p and q are propositions containing one or more variables, the same in the two propositions, and neither p nor q contains any constants except logical constants.[1]

This definition emphasizes that it is the logical structure that is the essential feature of mathematics, rather than any categorical statements it may contain referring to the world of the senses. Russell, in short, would equate mathematics with logic, but in this respect there has not been universal agreement among mathematicians. Sylvester had disagreed sharply with Huxley, arguing that mathematics springs

... direct from the inherent powers and activities of the human mind, and from continually renewed introspection of that inner world of thought of which the phenomena are as varied and require as close attention to discern as those of the outer physical world.

Sylvester, in other words, leaned toward what is now called the intuitionist's view of mathematics, for he regarded the objective of pure mathematics as "unfolding the laws of human intelligence," much as physics discloses the laws of the world, of the senses.[2] In this respect he represented a rejection of the formalizing tendencies of Boole, Dedekind, and Peano. Kronecker could perhaps be counted in Sylvester's camp, despite the arithmetization that he represented, for he regarded the integers of mathematics as having God-given meaning. More clearly intuitionistic was a mathematician who can be regarded as one of the two leading transition figures between the nineteenth and twentieth centuries—Henri Poincaré (1854–1912), a man whom Sylvester in old age admired as a prolific youth.

2   When Gauss died in 1855, it was generally thought that there never again would be a universalist in mathematics—one who is at home in all branches, pure and applied. If anyone has since proved this view wrong, it is Poincaré, for he took all mathematics as his province. In several respects, however, Poincaré differed fundamentally from Gauss. Gauss had been a calculating prodigy who throughout his life did not flinch from involved computations, whereas Poincaré was not especially early in showing mathematical promise and readily admitted that he had difficulty with simple arithmetic calculations. Poincaré's case shows that to be a great mathematician, one need not excel in number facility; there are other more advantageous aspects of innate mathematical ability. Also, whereas Gauss wrote relatively little, polishing his works, Poincaré wrote hastily and extensively, publishing more

[1] See Bertrand Russell, *Principles of Mathematics* (reprinted, New York: Norton, 1938), p. 3. See also pp. vii ff.

[2] *Collected Mathematical Papers*, ed. by H. F. Baker (Cambridge: Cambridge University Press, 1904–1912, 4 vols.), III, 424.

memoirs per year than any other mathematician. Moreover, Poincaré, especially in later life, wrote popular books with a philosophical flair, something that had not tempted Gauss. On the other hand, similarities between Poincaré and Gauss are numerous and fundamental. Both so teemed with ideas that it was difficult for them to jot the thoughts down on paper, both had a strong preference for general theorems over specific cases, and both contributed to a wide variety of branches of science.

Poincaré was born at Nancy,[3] a city that was to harbor quite a number of leading mathematicians in the twentieth century. The family achieved eminence in various ways; his cousin Raymond served as president of France during World War I. Henri was clumsily ambidextrous, and his ineptitude in physical exercise was legendary. He had poor eyesight and was very absent-minded, but, like Euler and Gauss, he had a remarkable capacity for mental exercises in all aspects of mathematical thought. Upon graduating from the École Polytechnique in 1875, he took a degree in mining engineering in 1879 and became attached to the Department of Mines for the rest of his life. In 1879 he earned also a doctorate in science at the University of Paris, where, until his death in 1912, he held several professorships in mathematics and science.

Poincaré's doctoral thesis had been on differential equations (not on methods of solution, but on existence theorems), which led to one of his most celebrated contributions to mathematics—the properties of automorphic functions; in fact, he was the virtual founder of the theory of these functions. An automorphic function $f(z)$ of the complex variable z is one which is analytic, except for poles, in a domain D and which is invariant under a denumerably infinite group of linear fractional transformations

$$z' = \frac{az + b}{cz + d}$$

Such functions are generalizations of trigonometric functions—as we see if $a = 1 = d$, $c = 0$, and b is of the form $2k\pi$—and of elliptic functions. Hermite had studied such transformations for the restricted case in which the coefficients a, b, c, and d are integers for which $ad - bc = 1$ and had discovered a class of elliptic modular functions invariant under these. But Poincaré's generalizations disclosed a broader category of functions, known as zeta-Fuchsian functions, which, Poincaré showed, could be used to solve the second-order linear differential equation with algebraic coefficients.

[3] For a bibliography of sources on Poincaré's life and work, see George Sarton, *The Study of the History of Mathematics* (Cambridge, Mass.: Harvard University Press, 1936), pp. 93–94. Chapter 28, appropriately titled "The Last Universalist," in E. T. Bell, *Men of Mathematics* (New York: Simon and Schuster, 1937), provides a vivid account of Poincaré's life and work.

3   Poincaré did not stay in any field long enough to round out his work; a contemporary said of him, "He was a conqueror, not a colonist." In his teaching at the Sorbonne he would lecture on a different topic each school year—capillarity, elasticity, thermodynamics, optics, electricity, telegraphy, cosmogony, and others; the presentation was such that in many cases the lectures appeared in print shortly after they had been delivered. In astronomy alone he published half a dozen volumes—*Les méthodes nouvelles de la mécanique céleste* (3 vols., 1892–1899) and *Leçons de mécanique céleste* (3 vols., 1905–1910)—being in this respect a worthy successor of Laplace. Especially important were the methods he used to attack the three-body problem and its generalizations. Significant also for cosmogony was a memoir of 1885 in which he showed that a pear-shape can be a figure of relative equilibrium assumed by a homogeneous fluid subject to Newtonian gravitation and rotating uniformly about an axis, and the question of a pear-shaped earth has continued to interest geodesists to our day. Sir George H. Darwin (1845–1912), son of Charles Darwin (1809–1882), wrote in 1909 that Poincaré's celestial mechanics would be a vast mine for researchers for half a century.

It is interesting that Poincaré, like Laplace, also wrote extensively on probability. In some respects his work is only a natural continuation of that of Laplace and the analysts of the nineteenth century; but Poincaré was Janus-faced and to some extent anticipated the great interest in topology that was to be so characteristic of the twentieth century. Topology was not the invention of any one man. Some topological problems are found in the work of Euler, Möbius, and Cantor, and even the word "topology" had been used in 1847 by J. B. Listing (1808–1882) in the title of a book, *Vorstudien zur Topologie* ("Introductory Studies in Topology"); but as a date for the beginning of the subject none is more appropriate than 1895, the year in which Poincaré published his *Analysis situs*. This book for the first time provided a systematic development.

Topology is now a broad and fundamental branch of mathematics, with many aspects; but it can be subdivided into two fairly distinct subbranches—combinatorial topology and point-set topology. Poincaré had little enthusiasm for the latter, and when in 1908 he addressed the International Mathematical Congress at Rome, he referred to Cantor's *Mengenlehre* as a disease from which later generations would regard themselves as having recovered.[4] Combinatorial topology, or analysis situs, as it was then generally called, is the study of intrinsic qualitative aspects of spatial configurations that remain invariant under continuous one-to-one transformations. It is often

---

[4] The published proceedings of such international congresses are available in many libraries, and they can very profitably be consulted for the development of mathematics in the twentieth century.

referred to popularly as "rubber-sheet geometry," for deformations of, say, a balloon, without puncturing or tearing it, are instances of topological transformations. A circle, for example, is topologically equivalent to an ellipse; the dimensionality of a space is a topological invariant, as is also the Descartes-Euler number $N_0 - N_1 + N_2$ for simple polyhedra. Among Poincaré's original contributions to topology was a generalization of the Descartes-Euler polyhedral formula for spaces of higher dimensionality, making use of what he called "Betti numbers," in honor of Enrico Betti (1823–1892), who had taught at the University of Pisa and had noted some of the properties of these topological invariants. Most of topology, nevertheless, deals with qualitative rather than quantitative aspects of mathematics, and in this respect it typifies a sharp break from the styles prevailing in nineteenth-century analysis. Poincaré's attention seems to have been directed toward analysis situs by attempts at qualitative integrations of differential equations. Poincaré, like Riemann, was especially adept at handling problems of a topological nature, such as finding out the properties of a function without worrying about its formal representation in the classical sense, for these men were intuitionists with sound judgment. Had Poincaré's interests in topology continued, he might have anticipated more of this branch of mathematics, one of the most favored and most fruitful lines of research in the twentieth century. His restless mind, however, was occupied with everything that was happening in physics and mathematics at the turn of the century, from hertzian waves and X-rays to quantum theory and the theory of relativity.

As an instance of Poincaré's many-sidedness, it is to him that we owe a suggestive model of Lobachevskian geometry within a Euclidean framework.[5] Suppose that a world is bounded by a large sphere of radius R and the absolute temperature at a point within the sphere is $R^2 - r^2$, where r is the distance from the center of the sphere; suppose also that the index of refraction of the pellucid medium is inversely proportional to $R^2 - r^2$. Moreover, assume that the dimensions of objects change from point to point, being proportional to the temperature at any given place. To inhabitants of such a world, the universe would appear to be infinite; and rays of light, or "straight lines," would not be rectilinear, but would be circles orthogonal to the limiting sphere and would appear to be infinite. "Planes" would be spheres orthogonal to the limiting sphere, and two such non-Euclidean "planes" would intersect in a non-Euclidean "line." The axioms of Euclid would hold, with the exception of the parallel postulate.

---

[5] For this and other contributions by Poincaré see the "Éloge historique d'Henri Poincaré" by Gaston Darboux in *Oeuvres Henri Poincaré* (1916–1956), Vol. II (1952), vii–lxxi. See also Vito Volterra et al, *Henri Poincaré: l'oeuvre scientifique, l'oeuvre philosophique* (1914).

4   Poincaré died at the height of his powers at the age of fifty-eight, having written more than any other mathematician of our century.[6] Klein compared him with Cauchy in versatility, and many regarded him as the leading mathematician of his day. His chief rival, David Hilbert (1862–1943), came from Germany and was of strikingly different temperament and views. Here was a second transition figure between the nineteenth and twentieth centuries; but whereas Poincaré seems perhaps to belong more to the earlier century, Hilbert clearly is more at home in the later, in view of his emphasis on structure. Hilbert, like Immanuel Kant (1724–1804), had been born at Königsberg in East Prussia, but unlike Kant he traveled widely, especially to attend the international congresses of mathematicians that have become so characteristic of this century. The first formal mathematical congress was held at Zurich in 1893, the second met at Paris in 1900, and since then congresses have been scheduled more or less regularly every four years, the Fifteenth International Congress being held in 1966 at Moscow. At the Paris Congress of 1900, Hilbert, a renowned professor at Göttingen, presented an address in which he attempted, on the basis of trends in mathematical research at the close of the glorious nineteenth century, to predict the direction of future advances. This he did by proposing twenty-three problems which he believed would be, or should be, among those occupying the attention of mathematicians in the twentieth century. "If we would obtain an idea of the probable development of mathematical knowledge in the immediate future," he said, "we must let the unsettled questions pass before our minds and look over the problems which the science of today sets and whose solution we expect from the future."[7]

Although he objected to the view that the concepts of arithmetic alone are susceptible of a fully rigorous treatment, he admitted that the development of the arithmetic continuum by Cauchy, Bolzano, and Cantor was one of the two most notable achievements of the nineteenth century—the other being the non-Euclidean geometry of Gauss, Bolyai, and Lobachevsky—and thus the first of the twenty-three problems concerned the structure of the real number continuum. The question is made up of two related parts: (1) is there a transfinite number between that of a denumerable set and the number of the continuum; and (2) can the numerical continuum be considered a well-ordered set? The second part asks whether the totality of all real numbers can be arranged in another manner so that every partial assemblage will have a first element. This is closely related with the axiom of choice named for

---

[6] See Ernest Lebon, *Henri Poincaré, bibliographie analytique des écrits* (1909).

[7] See the translation by Mary Winston Newson of Hilbert's "Mathematical Problems" in *Bulletin of the American Mathematical Society* (2), **8** (1902), 437–479. The original German had appeared in the *Göttinger Nachrichten* for 1900, pp. 253–297, and in *Archiv der Mathematik und Physik* (3), **1** (1901), 44–63, 213–237.

the German mathematician Ernst Zermelo (1871–1956) who formulated it in 1904. Zermelo's axiom asserts that, given any set of mutually exclusive nonempty sets, there exists at least one set that contains one and only one element in common with each of the nonempty sets.[8] As an illustration of a problem involving Zermelo's axiom, consider the set of all real numbers n such that $0 \leq n \leq 1$; let us call two of these real numbers equivalent if their difference is rational. There obviously are infinitely many classes of equivalent real numbers. If we form a set S made up of one number from each of these classes, is S denumerable or nondenumerable? The axiom of choice, indispensable in analysis, was in 1940 proved by Kurt Gödel (1906–   ) to be consistent with other axioms of set theory; but in 1963 it was demonstrated by Paul Cohen (1934–   ) that the axiom of choice is independent of the other axioms in a certain system of set theory, thus showing that the axiom cannot be proved within this system.[9] This seems to preclude a clear-cut solution to Hilbert's first problem.

Hilbert's second problem, also suggested by the nineteenth-century age of **5** rigor, asked whether it can be proved that the axioms of arithmetic are consistent—that a finite number of logical steps based upon them can never lead to contradictory results. A decade later there appeared the first volume of *Principia mathematica* (3 vols., 1910–1913), by Bertrand Russell and Alfred North Whitehead (1861–1947), the most elaborate attempt up to that time to develop the fundamental notions of arithmetic from a precise set of axioms. This work, in the tradition of Leibniz, Boole, and Frege and based on Peano's axioms, carried out in minute detail a program intended to prove that all of pure mathematics can be derived from a small number of fundamental logical principles. This would justify the view of Russell, expressed earlier, that mathematics is indistinguishable from logic. But the system of Russell and Whitehead, not entirely formalized, seems to have met with more approval among logicians than among mathematicians. Moreover, the *Principia* left unanswered the second query of Hilbert. Efforts to solve this problem led in 1931 to a surprising conclusion on the part of a young Austrian mathematician, Kurt Gödel, who had emigrated to the United States and became a member of the Institute for Advanced Study at Princeton. Gödel showed that within a rigidly logical system such as Russell and Whitehead had developed for arithmetic, propositions can be formulated that are undecidable or undemonstrable within the axioms of the system. That is, within the system there exist certain clear-cut statements that can be

[8] See Herman Rubin and Jean E. Rubin, *Equivalents of the Axiom of Choice* (Amsterdam: North-Holland, 1963).

[9] See P. J. Cohen, "The Independence of the Continuum Hypothesis," *Proceedings of the National Academy of Science*, **50** (1963), 1143–1148; **51** (1964), 105–110.

neither proved nor disproved.[10] Hence one cannot, using the usual methods, be certain that the axioms of arithmetic will not lead to contradictions. In a sense Gödel's theorem, sometimes regarded as the most decisive result in mathematical logic, seems to dispose negatively of Hilbert's second query. In its implications the discovery by Gödel of undecidable propositions is as disturbing as was the disclosure by Hippasus of incommensurable magnitudes, for it appears to foredoom hope of mathematical certitude through use of the obvious methods. Perhaps doomed also, as a result, is the ideal of science—to devise a set of axioms from which all phenomena of the natural world can be deduced. Nevertheless, mathematicians and scientists alike have taken the blow in stride and have continued to pile theorem upon theorem at a rate greater than ever before. Most assuredly no scholar of today would echo the assertion of Babbage in 1813 that "The golden age of mathematical literature is undoubtedly past."

The problems raised by Gödel's theorem have been approached from outside arithmetic itself through a new aspect of mathematical logic that arose toward the middle of the twentieth century and is known as meta-mathematics. This is not concerned with the symbolism and operations of arithmetic, but with the interpretation of these signs and rules. If arithmetic cannot lift itself from the quagmire of possible inconsistency, perhaps meta-mathematics, standing outside the arithmetical bog, can save the day by other means—such as transfinite induction. Some mathematicians would at least hope for a means of determining, for every mathematical proposition, whether it is true, false, or undecidable. In any case, even the discouragingly negative answer to Hilbert's second query has thus spurred, rather than daunted, mathematical creativity.

6   The next four questions in Hilbert's list are somewhat more technical and abstruse than the first two, but queries seven and eight are concerned with familiar notions. In problem seven it is asked whether the number α^β, where α is algebraic (and not zero or one) and β is irrational and algebraic, is transcendental. In alternative geometric form, Hilbert expressed this by asking whether in an isosceles triangle the ratio of the base to a side is transcendental if the ratio of the vertex angle to the base angles is algebraic and irrational.

This question was disposed of in 1934 when Aleksander Osipovich

---

[10] Kurt Gödel, "Über formal unentscheidbare Sätze der *Principia Mathematica* und verwandter Systeme," *Monatshefte der Mathematik und Physik*, **38** (1931), 173–198; or see his *Consistency of the Axiom of Choice and the Generalized Continuum Hypothesis with the Axioms of Set Theory* (Princeton, N.J.: Princeton University Press, 1940, rev. ed. 1951). Innumerable efforts have been made to explain the Gödel proof in language appropriate for the nonspecialist. See in particular Ernest Nagel and J. R. Newman, "Gödel's Proof," *The World of Mathematics* (New York: Simon and Schuster, 1956, 4 vols.), III, 1668–1695.

Gelfond (1906–   ) proved that Hilbert's conjecture, now known as Gelfond's theorem, was indeed correct—α^β is transcendental if α is algebraic and neither zero nor one, and β is algebraic and not rational. However, in mathematics an answer to one question merely raises others, and mathematicians are as yet unable to answer a question such as whether or not α^β is transcendental if α and β are transcendental. It is not known, for example, if e^e or π^π or π^e or the Eulerian constant γ is transcendental. It is, however, known that e^π is transcendental, for $e^\pi = 1/e^{-\pi} = 1/i^{2i}$, and i^{2i} is transcendental by Gelfond's theorem.[11]

Hilbert's eighth query simply renewed the call, familiar in the nineteenth century, for a proof of Riemann's conjecture that the zeros of the zeta function, except for the negative-integral zeros, all have real part equal to one half. A proof of this, he felt, might lead to a proof of Goldbach's conjecture on the infinity of prime pairs; but no demonstration has yet been given, although it is more than a century since Riemann hazarded the guess.

We cannot go into the other problems that Hilbert posed—problems in topology, differential equations, calculus of variations, and other fields— except to say that roughly half of them remain unsolved[12] and that, naturally, mathematics developed also in many directions not at all anticipated in 1900. We should note in addition that the first half of the twentieth century was not unlike the earlier periods in the history of mathematics in at least one respect—each old problem that was solved bequeathed to posterity several new problems. As Hilbert said in proposing his problems, "As long as a branch of science offers an abundance of problems, so long is it alive"; He closed with words of encouragement in the face of the virtually exponential growth in mathematics, pointing out that as the subject expands, tools are sharper and methods are simpler, hence a scholar can master the field despite its breadth.

Hilbert bequeathed to mathematics more than one set of problems. In **7** 1899, the year before his Paris address, he had published a small but celebrated volume entitled *Grundlagen der Geometrie* ("Foundations of Geometry"). This work, translated into the major languages,[13] exerted a strong influence on the mathematics of the twentieth century. Through the arithmetization of analysis and the axioms of Peano, most of mathematics, except for

[11] See Einar Hille, "Gelfond's Solution of Hilbert's Seventh Problem," *American Mathematical Monthly*, **49** (1942), 654–661. Cf. A. O. Gelfond, *Transcendental and Algebraic Numbers*, trans. by Leo F. Boron (New York: Dover, 1960).

[12] For the status of the problems after thirty years, see L. Bieberbach, "Über den Einfluss von Hilberts Pariser Vortrag über 'Mathematische Probleme' auf die Entwicklung der Mathematik in den letzten dreissig Jahren," *Naturwissenschaften*, **18** (1936), 1101–1111.

[13] An English translation by E. J. Townsend, *The Foundations of Geometry*, was published by Open Court, LaSalle, Ill., in 1902. An eighth German edition appeared in 1956.

geometry, had achieved a strict axiomatic foundation. Geometry in the nine-teenth century had flourished as never before, but it was chiefly in Hilbert's *Grundlagen* than an effort was first made to give it the purely formal character found in algebra and analysis. Euclid's *Elements* did have a deductive structure, to be sure, but it was replete with concealed assumptions, meaning-less definitions, and logical inadequacies. Hilbert understood that not all terms in mathematics can be defined and therefore began his treatment of geometry with three undefined objects—point, line, and plane—and six undefined relations—being on, being in, being between, being congruent, being parallel, and being continuous. In place of Euclid's five axioms (or common notions) and five postulates, Hilbert formulated for his geometry a set of twenty-one assumptions, since known as Hilbert's axioms. Eight of these concern incidence and include Euclid's first postulate, four are on order properties, five are on congruency, three are on continuity (assumptions not explicitly mentioned by Euclid), and one is a parallel postulate essentially equivalent to Euclid's fifth postulate.[14] Following the pioneer work by Hilbert, alternative sets of axioms have been proposed by others; and the purely formal and deductive character of geometry, as well as other branches of mathematics, has been thoroughly established since the beginning of the twentieth century.

8 Hilbert, through his *Grundlagen*, became the leading exponent of an "axiomatic school" of thought which has been influential in fashioning contemporary attitudes in mathematics and mathematical education.[15] The *Grundlagen* opened with a motto taken from Kant: "All human know-ledge begins with intuitions, proceeds to concepts, and terminates in ideas," but Hilbert's development of geometry established a decidedly anti-Kantian view of the subject. It emphasized that the undefined terms in geometry should not be assumed to have any properties beyond those indicated in the axioms. The intuitive-empirical level of the older geometrical views must be disregarded, and points, lines, and planes are to be understood merely as elements of certain given sets. Set theory, having taken over algebra and analysis, now was invading geometry. Similarly, the undefined relations are to be treated as abstractions indicating nothing more than a correspondence or mapping. Through analytic geometry the formal treatment of geometry had been associated with the axiomatization of algebra, and the ultimate outcome of the association was a degree of abstraction exceeding anything

[14] A list of the postulates can be found in Ralph G. Stanton and Kenneth D. Fryer, *Topics in Modern Mathematics* (Englewood Cliffs, N.J.: Prentice-Hall, 1964), pp. 167–170, as well as in the various editions of Hilbert's *Foundations*.

[15] See, for example, Rolf Nevanlinna, "Reform in Teaching Mathematics," *The American Mathematical Monthly*, **73** (1966), 451–464.

in the nineteenth century. Peano in 1888 had in essence defined a vector space over the field of real numbers; but in what has since been known as Hilbert space[16] the ideas of Hamilton, Grassmann, and Peano are further generalized. The elements are not Euclid's points, but infinite sequences of complex numbers $x_1, x_2, \ldots$ for which the sequence $x_1^2 + x_2^2 + \cdots$ converges. This is a special case of a vector space, more general forms of which were introduced later in the century by Stefan Banach (1892–1945) and others. A vector space is a set of elements called vectors, subject to the usual rules of combination of vectors and scalars. Vector spaces of varying types are determined according to the restrictions imposed on the vector elements. Hilbert space, for example, is a vector space whose elements have infinitely many components, subject to the condition that $(x_1^2 + x_2^2 + \cdots)$ be finite. Hilbert space, despite the extraordinary degree of abstraction that it represents, has found application in modern quantum theory. Banach space is a still more abstract vector space in which the elements need not be defined with respect to the complex number field. In technical language, a Banach space is a normed linear space complete in the metric determined by the norm; a Hilbert space is a Banach space whose norm $\|x\|$ has the parallelogram property $\|x + y\|^2 + \|x - y\|^2 = 2\|x\|^2 + 2\|y\|^2$. Banach was the brightest luminary in a great "Polish School" of mathematics that flourished between the two World Wars at the Universities of Lwow and Warsaw. Banach taught at Lwow, and the head of the group at Warsaw was Waclaw Sierpinski (1882–   ), a contributor to the theory of numbers, topology, and the theory of sets, and a founder in 1920 of *Fundamenta Mathematicae*, one of the world's most distinguished mathematical journals. Sierpinski was a remarkably successful teacher, and many of his students made names for themselves in American mathematics when the Polish circle was dispersed and Sierpinski was deported by the Germans.[17] With the end of the war Sierpinski returned to ruined Warsaw, and publication of *Fundamenta Mathematicae* was resumed; but the younger Banach had died very shortly after hostilities had ceased.[18]

Hilbert was interested in all aspects of pure mathematics, and his name is connected with a simple space-filling curve which is more easily described than is the somewhat similar curve given by Peano. The Hilbert curve is

---

[16] See P. R. Halmos, *Introduction to Hilbert Space* (New York: Chelsea, 1951), or S. K. Berberian, *Introduction to Hilbert Space* (New York: Oxford, 1961).

[17] See Matthew M. Fryde, "Waclaw Sierpinski—Mathematician," *Scripta Mathematica*, **27** (1964), 105–111.

[18] See Hugo Steinhaus, "Stefan Banach, 1892–1945," *Scripta Mathematica*, **26** (1963), 93–100. Steinhaus was also one of the distinguished members of the Polish group of mathematicians. For some additional, but inadequate, notes on Polish mathematicians, see Sister Mary Grace, "Poland's Contribution to Mathematics," *The Mathematics Teacher*, **60** (1967), 383–386. For those who read Polish there is a far more extensive account in a volume by Jadwiga Dianni and Adam Wachulka, *One Thousand Years of Polish Mathematics* (Warsaw: PZVS, 1963).

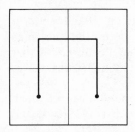 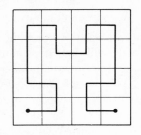 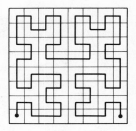

FIG. 27.1

generated by continuing indefinitely the process suggested in Fig. 27.1. Beginning with a unit square, we subdivide this into four square regions as shown, and then connect in order the four midpoints of the regions. Each of these four square regions in turn is subdivided into four parts, the centers of which are connected as shown, always beginning in the lower left box and ending in the lower right box. It is clear that the limiting curve in this process will pass through every point in the square; incidentally, this is another instance of a continuous curve that is nowhere differentiable. The nineteenth century had noted a few of the pathological cases that can arise in algebra, analysis, and geometry, but it was in the twentieth century that anomalies and paradoxes ran rampant. Among the other oddities was a continuous closed curve proposed in 1904 by Helge von Koch (1870–1924) of Stockholm and defined essentially as follows.[19] Beginning with an equilateral triangle having unit sides, we trisect each of the three unit segments, erect an equilateral triangle on the middle third or base, and then delete the base of each of these three new equilateral triangles (see Fig. 27.2). The result is a broken-line closed figure of twelve sides and having a total length of four units. Upon

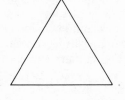 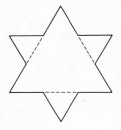 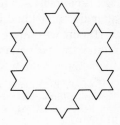

FIG. 27.2

[19] "Une méthode géométrique élémentaire pour l'étude de certaines questions de la théorie des courbes planes," *Acta Mathematica*, **30** (1906), 145–174. We have somewhat modified the von Koch curve for purposes of exposition.

trisecting each of the twelve sides, erecting twelve equilateral triangles on the middle thirds, and deleting the bases, we have a closed figure with forty-eight sides and a length of $\frac{16}{3}$. Upon continuing this procedure indefinitely, a limiting, crinkly curve, known as the von Koch curve or the snowflake curve, is determined. It not only has no tangent at any point, but it has the amazing property that, given any two points on the curve, the arc length between the two points is infinite.

Hilbert, like Poincaré, was a many-sided mathematician who contributed **9** to the theory of numbers, mathematical logic, differential equations, the three-body problem, and other aspects of mathematical physics. It was in connection with his work on the foundation of mathematics[20] that he became involved in the sharpest controversy of the century, which in a sense was a continuation of the earlier conflict between Cantor and Kronecker. Hilbert admired the Cantorian *Mengenlehre*, whereas Poincaré was sharply critical of it. The theories of Cantor, like the abstract spaces of Hilbert, seem to be far removed from an intuitive-empirical basis such as Poincaré and some of his contemporaries preferred. At the Paris Congress in 1900, at which Hilbert presented his problems, Poincaré delivered a paper in which he compared the roles of logic and intuition in mathematics. Mathematicians then and later came to be grouped into two or three schools of thought, depending upon their views regarding the foundations of their subject. Those adopting views akin to Poincaré's formed an indistinctly defined group with intuitive predilections. Hilbert came to be regarded as the leader of a "formalist" school of thought, which some of his successors carried to the conclusion that mathematics is nothing but a meaningless game played with meaningless marks according to certain formal rules agreed upon beforehand.

Related to the formalist group, but not identified with it, were a number of mathematicians who hesitated to subscribe to the entirely arbitrary nature of the rules of the game. Headed by Bertrand Russell, these men, often described as the "logicist" or "logicalist" school, would equate mathematics and logic, in opposition to C. S. Peirce, but in agreement with Frege. It was L. E. J. Brouwer (1882–1966) of the University of Amsterdam who was especially effective in rallying the opponents of the formalism of Hilbert and the logicism of Russell. He insisted that the elements and axioms of mathematics are considerably less arbitrary than would appear. In his doctoral dissertation of 1907 and in later articles Brouwer attacked the logical foundations of arithmetic and analysis, becoming generally known as the founder of a now clearly recognizable "intuitionist school." According to Brouwer, language and logic are not presuppositions for mathematics, a subject that has its source in intuition that makes its concepts and inferences immediately

[20] *Grundlagen der Mathematik*, with P. Bernays (Berlin: Springer, 1934–1939, 2 vols.).

clear to us;[21] a statement that an object exists having a given property means that there is a known method that enables the object to be found or constructed in a finite number of steps. In particular, he argued that the method of indirect proof, to which transfinite arithmetic had frequent recourse, is invalid. Ever since the time of Aristotle the three basic laws of logic had been held sacrosanct: (1) the law of identity, A is A; (2) the law of contradiction, A cannot simultaneously be B and not B; and (3) the law of the excluded middle (or *tertium non datur*), A is either B or not B, for there is no other alternative. Brouwer denied the last of these laws of logic and refused to accept results based on it. For example, he asked the formalists whether it is true or false that "the sequence of digits 1 2 3 4 5 6 7 8 9 occurs somewhere in the decimal representation of π." Since no known method exists for making a decision, one cannot apply here the law of the excluded middle and assert that the proposition is either true or false.

10   In 1918 the intuitionist cause was joined by Hermann Weyl (1885–1955), despite the fact that he had studied under Hilbert, whom he later succeeded at Göttingen in 1930. Weyl asserted that, in basing analysis on the arithmetic continuum, the formalists had constructed a house that "in an essential part is built on sand."[22] Hilbert compared the attacks of Brouwer and Weyl with the negativism of Kronecker in the preceding age, but he was not able to demolish their arguments. Weyl was one of the many outstanding mathematicians produced by the University of Göttingen, and he contributed to several branches of mathematics and to the two leading advances in science during the early years of the century. He had been a colleague of Albert Einstein (1879–1955) at Zurich in 1913, and in 1918 Weyl supported the theory of relativity in a widely read book, *Raum-Zeit-Materie* ("Space-Time-Matter"). During the next ten years he wrote a series of papers on the applications of group theory to quantum mechanics, to which Einstein also had made important contributions. At the height of his career, in 1933, Weyl resigned his post at Göttingen in protest against the Nazi dismissal of his colleagues, and the glorious period of mathematics at the university came to an abrupt close. Weyl went to America and became a member of the Institute for Advanced Study at Princeton, of which Einstein in 1933 had been made a life member. The close relationship between abstract mathematics and scientific theory that the work of Poincaré, Hilbert, Weyl, and Einstein represents has been especially characteristic of the twentieth century, and it was in no way disturbed by controversies within mathematics concerning the foundations of the subject.

[21] See the article, "Mathematics, Foundations of," by S. C. Kleene in *Encyclopaedia Britannica*, XV (1963), 82B–83. Cf. Edith H. Luchins and A. S. Luchins, "Logicism," *Scripta Mathematica*, **27** (1965), 223–243.

[22] See Hermann Weyl: *Philosophy of Mathematics and Natural Science*, based on a translation by Olaf Helmer (1940).

The formalists and logicists were particularly embarrassed by a number of paradoxes in set theory that had long been familiar, such as that of the town barber who shaves all those, and only those, who do not shave themselves. Is this barber included or not in the set of all those who shave themselves? Here, too, the law of the excluded middle appears to be inapplicable. Another instance is that usually known as "Russell's antinomy." Is the set of all sets that are not members of themselves a member of itself? Whether the answer is affirmative or negative, a contradiction results. Such paradoxes have raised serious doubts whether a program such as that of Russell and Whitehead, based as it is on the notion of set, can ever be successful. The Russell paradox was propounded in 1902, and the chagrin it caused among specialists in mathematical logic was well expressed by Frege in 1903 in an appendix to the second volume of his *Grundgesetze*:

Hardly anything more unwelcome can befall a scientific writer than that one of the foundations of his edifice be shaken after the work is finished. I have been placed in this position by a letter [containing the paradox] of Mr. Bertrand Russell just as the printing of this [second] volume was nearing completion. . . . *Solatium miseris, socios habuisse malorum.* I too have this solace, if solace it is; for everyone who in his proofs has made use of extensions of concepts, classes, sets [including Dedekind's systems] is in the same position. It is not just a matter of my particular method of laying the foundations, but of whether a logical foundation for arithmetic is possible at all.[23]

This recognition of three major views concerning the nature of mathematics should not lead to the conclusion that every mathematician is found in one or another of the three camps. Nothing could be further from the truth, and even within each school of thought there is great diversity of opinion. One might almost suggest that no two mathematicians of our day agree on the nature of their subject. Certainly the word "mathematics" has meant different things to the peoples of the world at different periods in history, and it would be unrealistic to expect wide agreement within a field that has become so vast. During the first half of the twentieth century conflict between factions was at times very sharp; but since then there has been more of a feeling, reminiscent of the view of d'Alembert some two hundred years earlier, that we should get on with the development of the subject, both in terms of foundations and in superstructure, without excessive concern about a particular creed.[24]

11   Not all leading mathematicians of the early twentieth century took active part in the formalist-intuitionist controversy. In particular, Henri Lebesgue (1875–1941), one of the most original and productive workers, revolutionized

[23] See Gottlob Frege, *The Basic Laws of Arithmetic: Exposition of the System*, trans. by Montgomery Furth (Berkeley, Calif.: University of California Press, 1964), p. 127.
[24] An intensive exposition of the conflicting views is presented in Max Black: *The Nature of Mathematics* (1933). See also Abraham S. Luchins and Edith H. Luchins, *Logical Foundations of Mathematics for Behavioral Scientists* (1965) for a more popular account.

an important aspect of analysis without subscribing to one of the chief orthodoxies. He took a position somewhat intermediate between intuitionists and formalists—one that might be described as French logical empiricism. But if on philosophical foundations Lebesgue occupied a middle of-the-road position, in his research he scandalized conventional analysts, such as Hermite, with his predilection for pathological types of functions. He had had the usual type of mathematical training, although he had shown exceptional irreverence in questioning statements made by his professors; but his dissertation, accepted at Nancy in 1902, was most unusual in virtually remaking the field of integration. His work was so great a departure from accepted views that Lebesgue, like Cantor, at first was assailed both by external criticism and by internal self-doubt; but the value of his views was increasingly recognized, and in 1910 he was appointed to the Sorbonne. However, he did not create a "school of thought," nor did he concentrate on the field that he had opened. Although his concept of the integral was in itself a striking case of generalization, Lebesgue feared that, "Reduced to general theories, mathematics would be a beautiful form without content. It would quickly die."[25] Later developments seem to indicate that his fears concerning the baneful influence of generality in mathematics were without foundation.

The Riemann integral had dominated studies in integration before Lebesgue became the "Archimedes of the extension period." But toward the close of the nineteenth century studies in trigonometric series and the *Mengenlehre* of Cantor had made mathematicians more keenly aware that the essential idea in functionality should be a pointwise correspondence or "mapping" in the newer sense, and not smoothness of variation. Cantor had even struggled with notions of measurable sets, but under his definition the measure of the union of two sets could be less than the sum of the measures of the sets. Defects in Cantor's definition were removed by Emile Borel (1871–1956), the immediate predecessor of Lebesgue in studies on measure theory. Borel, like Carnot, to some extent lived a double life, for from a professorship at Paris he turned to active participation in governmental affairs. From 1924 through 1936 he served in the Chamber of Deputies, and, before his arrest in 1940 under the Vichy regime, he had been minister of the navy. His record in mathematical publication before 1924 had been impressive, including more than half a dozen books. One of the earlier volumes had been on an unusual theme: *Leçons sur les séries divergentes* (1901). Here the author showed how for some divergent series a "sum" can be defined that will make sense in relationships and operations involving such series. For example, if the series is $\sum u_n$, then a "sum" can be defined as

[25] See Henri Lebesgue, *Measure and the Integral* (1966), p. 5. This volume contains a well-written "Biographical Sketch" by K. O. May.

$\int_0^\infty e^{-x} \sum_0^\infty a_n x^n/n! \, dx$, if this integral exists. During the first decades of this century there was lively interest in such definitions; but Borel's more lasting influence was in the application of the theory of sets to the theory of functions, where his name is recalled in the familiar Heine-Borel theorem:

> If a closed set of points on a line can be covered by a set of intervals so that every point of the set is an interior point of at least one of the intervals, then there exists a finite number of intervals with this covering property.

In somewhat different terminology this theorem had been expressed by Heine in 1872, but it had been overlooked until reenunciated in 1895 by Borel.[26] Borel's name is attached also to any set that can be obtained from closed and open sets on the real line by repeated applications of the operations of union and intersection to denumerable numbers of sets. Any Borel set is a measurable set in his sense.

Lebesgue, pondering Borel's work on sets, saw that Riemann's definition of the integral has the drawback of applying only in exceptional cases, for it assumes not more than a few points of discontinuity in the function. If a function $y = f(x)$ has many points of discontinuity, then as the interval $x_{i+1} - x_i$ becomes smaller, values of $f(x_{i+1})$ and $f(x_i)$ do not necessarily come closer together. Instead of subdividing the domain of the independent variable, Lebesgue therefore subdivided the range $\bar{f} - f$ of the function into subintervals Δy_i and within each subinterval selected a value η_i. Then he found the "measure" $m(E_i)$ of the set E_i of points on the x-axis for which values of $f(x)$ are approximately equal to η_i. As Lebesgue liked to express the difference informally, the earlier integrators had added indivisibles, large or small, in order from left to right, whereas he preferred to group together indivisibles of comparable size before adding.[27] That is, for the earlier Riemann sums $S_n = \sum f(x_i) \Delta x_i$ he substituted the Lebesgue-type sum $S_n = \eta_i m(E_i)$ and then let the intervals tend toward zero.

The Lebesgue integral that we have here described very roughly is in actuality defined far more precisely in terms of upper and lower bounds and the Lebesgue measure of a set, an abstruse concept that cannot be explained here, but an illustrative example may suggest how the Lebesgue procedures operate. Let it be granted that the Lebesgue measure of all rational numbers in the interval $[0, 1]$ is zero and that the Lebesgue measure of all irrational numbers in this interval is one; let the integral of $f(x)$ be required over this interval, where $f(x)$ is zero for all rational values of x and $f(x)$ is one for all irrational values of x. Inasmuch as $m(E_i) = 0$ for all values of i except $i = n$, where $\eta_n = 1$, we have $S_n = 0 + 0 + \cdots + \eta_n m(E_n) = 1 \cdot 1 = 1$; hence the

---

[26] See E. T. Bell, *The Development of Mathematics* (1940), p. 452 for references.

[27] For an excellent introduction to the Lebesgue integral see the words of Lebesgue himself, in English translation, in his *Measure and the Integral*, cited in footnote 25.

Lebesgue integral is 1. The Riemann integral of the same function over the same interval does not, of course, exist.

We have not defined the phrases "measure of a set" or "measurable function" since this is not easily done in a few elementary words. Moreover, the word "measure" can take on various meanings. When Lebesgue presented his new concept of the integral, he used the word in the specific sense now known as the Lebesgue measure. This was an extension of classical notions of length and area to sets more general than those associated with the usual curves and surfaces. Today the word "measure" is used more broadly still, a measure on a field R being simply a nonnegative function μ with the property $\mu(\sum A_i) = \sum \mu(A_i)$ for every countable disjoint class A_i contained in R. Not only does the new concept of integral cover a wider class of functions than does that of Riemann, but the inverse relationship between differentiation and integration (in Lebesgue's generalized sense) is subject to fewer exceptions. For example, if $g(x)$ is differentiable in $[a, b]$ and if $g'(x) = f(x)$ is bounded, then $f(x)$ is Lebesgue integrable and $g(x) - g(a) = {}_L\int_a^x f(t)\,dt$, whereas with the same restrictions on $g(x)$ and $g'(x)$ the Riemann integral ${}_R\int_a^x f(t)\,dt$ might not even exist.

Lebesgue's ideas date from the closing years of the nineteenth century, but they became widely known through his two classic treatises: *Leçons sur les séries trigonométriques* (1903) and *Leçons sur l'intégration et la recherche des fonctions primitives* (1904).[28] The revolutionary views they contained paved the way for further generalizations. Among these are the Denjoy integral and the Haar integral, proposed by a Frenchman, Arnaud Denjoy (1884–   ), and a Hungarian, Alfred Haar (1885–1933), respectively. Another well-known integral of the twentieth century is the Lebesgue-Stieltjes integral, a combination of the ideas of Lebesgue and the Dutch analyst T. J. Stieltjes (1856–1894). The work of these men and others has so altered the concept of the integral, through generalization, that it has been said that although integration is as old as the time of Archimedes, "the *theory* of integration was a creation of the twentieth century."[29]

12   The new theories of integration were closely allied with another pronounced characteristic of the twentieth century, the rapid growth of point set topology. Maurice Fréchet (1878–   ) at the University of Paris, in his

---

[28] There are many expositions in English. See J. Burkill, *The Lebesgue Integral* (Cambridge: Cambridge University Press, 1951); J. H. Williamson, *Lebesgue Integration* (New York: Holt, Rinehart, 1962); Stanislaw Hartman and J. Mikusinski, *The Theory of Lebesgue Measure and Integration*, trans. by Leo F. Boron (Oxford: Pergamon, 1961); L. Cesari, *Surface Area* (Princeton: Princeton University Press, 1956).

[29] E. T. Bell, *Development of Mathematics*, p. 448, Cf. Arnaud Denjoy, *Un demi-siècle (1907–1956) de notes communiquées aux académies* (Paris: Gauthier-Villars, 1957, 2 vols.); and *Introduction à la théorie des fonctions de variables réelles* (Paris: Hermann, 1937, 2 vols.). See also Leopoldo Nachbin, *The Haar Integral*, trans. by Lulu Bechtolsheim (Princeton: D. Van Nostrand, 1965).

doctoral dissertation of 1906, showed clearly that function theory no longer could do without a very general view of set theory. What Fréchet had in mind were not necessarily sets of numbers, but sets of elements of arbitrary nature, such as curves or points; upon such arbitrary sets he built a "functional calculus" in which a functional operation is defined on a set E when to each element A of E there corresponds a numerically determined value $U(A)$. His interest was not in a particular instance of a set E, but in those set-theoretical results that are independent of the nature of the set elements. In this very broad calculus the notion of limit is much broader than limits as previously defined, the latter being included in the former as special cases, just as the Lebesgue integral includes the integrals of Riemann and Cauchy. Probably no aspect of twentieth-century mathematics stands out more clearly than does the ever greater degree of generalization and abstraction. From the time of Hilbert and Fréchet the notions of abstract set and abstract space have been fundamental in research.

It is interesting to note that Hilbert and Fréchet came to their generalizations of the concept of space from somewhat differing directions. Hilbert had become interested, as had Poincaré, in the study of integral equations, especially through the work of Ivar Fredholm (1866–1927). In a sense an integral equation can be considered an extension of a system of n equations in n unknowns to a system of infinitely many equations in infinitely many unknowns, a topic that had been touched upon, in the form of infinite determinants, by von Koch. As he worked in integral equations from 1904 to 1910, Hilbert did not explicitly refer to infinite dimensional spaces, but he did develop the concept of continuity of a function of infinitely many variables. To what extent Hilbert formally constructed the "space" that later was named for him may be a moot point, but the basic ideas were there, and their impact on the mathematical world was great. During the years that Hilbert was concerned with integral equations, Hadamard was doing research in the calculus of variations, and his protégé Fréchet consciously sought in 1906 to generalize the methods in this field through what he called *functional calculus*. Whereas the ordinary calculus deals with functions, the functional calculus concerns functionals. Whereas a function is a correspondence between a set S_1 of numbers and another set S_2 of numbers, a functional is a correspondence between a class C_1 of functions and another class C_2 of functions. Fréchet formulated generalized definitions, corresponding roughly to terms such as limit, derivative, and continuity in the ordinary calculus, applicable to the function spaces he thus created, to a considerable extent introducing a new vocabulary for the new situation.[30]

[30] A detailed and perceptive account of the rise of function spaces through the work of Hilbert and Fréchet is given in Michael Bernkopf, "The Development of Function Spaces with Particular Reference to Their Origins in Integral Equation Theory," *Archive for History of Exact Sciences*, **3** (1966), 1–96.

Topology is said by some to have begun with the analysis situs of Poincaré; others claim that it dates from the set theory of Cantor, or perhaps from the development of abstract spaces. Still others regard Brouwer as the founder of topology, especially for his topological invariance theorems of 1911 and for his fusion of the methods of Cantor with those of analysis situs. At all events, with Brouwer there began the period of intensive evolution of topology that has continued to the present day. During this "golden age" of topology American mathematicians have been conspicuous contributors. It has been said that "topology began as much geometry and little algebra, but that now it is much algebra and little geometry."[31] Whereas once topology could be described as geometry without measurement, today algebraic topology threatens to dominate the field, a change that has resulted largely from leadership in the United States.

In 1913 Weyl lectured on Riemann surfaces[32] at Göttingen, where Hilbert had been given a chair on the recommendation of Klein, and he, too, emphasized the abstract nature of a surface, or a "two-dimensional manifold," as he preferred to call it. The concept of a manifold, he asserted, should not be tied to a point space (in the usual geometric sense), but given broader meaning. We merely begin with a collection of things called "points" (which can be any objects whatsoever) and introduce a concept of continuity through appropriate definition. The classical formulation of this view was given a year later by Felix Hausdorff (1868–1942), the "high priest" of point set topology.

The first portion of Hausdorff's *Grundzüge der Mengenlehre* ("Basic Features of Set Theory") of 1914 is a systematic exposition of the characteristic features of set theory, where the nature of elements is of no consequence; only the relations among the elements are important. In the latter portion of the book we find a clear-cut development of "Hausdorff topological spaces" from a set of axioms. By a topological space the author understands a set E of elements x and certain subsets S_x known as neighborhoods of x. The neighborhoods are assumed to satisfy the following four "Hausdorff axioms":

1. To each point x there corresponds at least one neighborhood $U(x)$, and each neighborhood $U(x)$ contains the point x.
2. If $U(x)$ and $V(x)$ are two neighborhoods of the same point x, there must exist a neighborhood $W(x)$ that is a subset of both.
3. If the point y lies in $U(x)$, there must exist a neighborhood $U(y)$ that is a subset of $U(x)$.

---

[31] See *Recent Soviet Contributions to Mathematics*, ed. by J. P. LaSalle and S. Lefschetz (1962), p. 13.

[32] Hermann Weyl, *The Concept of a Riemann Surface*, trans. by G. R. Maclane, 3rd ed. (Reading, Mass.: Addison-Wesley, 1955).

4. For two different points x and y there are two neighborhoods $U(x)$ and $U(y)$ with no points in common.[33]

Neighborhoods as so defined permitted Hausdorff to introduce the concept of continuity. Through additional axioms he developed the properties of various more restricted spaces, such as the Euclidean plane.

If any one book marks the emergence of point set topology as a separate discipline, it is Hausdorff's *Grundzüge*. It is interesting to note that although it was the arithmetization of analysis that began the train of thought that led from Cantor to Hausdorff, in the end the concept of number is thoroughly submerged under a far more general point of view. Moreover, although the word "point" is used in the title, the new subject has as little to do with the points of ordinary geometry as with the numbers of common arithmetic. Topology has emerged in the twentieth century as a subject that unifies almost the whole of mathematics, somewhat as philosophy seeks to coordinate all knowledge. Because of its primitiveness, topology lies at the basis of a very large part of mathematics, providing it with an unexpected cohesiveness.

The high degree of formal abstraction that had found its way into analysis, **13** geometry, and topology in the early twentieth century could not help but invade algebra. The result was a new type of algebra, sometimes inadequately described as "modern algebra," a product largely of the second third of the century. It is indeed true that a gradual process of generalization in algebra had developed during the nineteenth century, but in the twentieth century the degree of abstraction took a sharp turn upward.[34] No longer did x and y necessarily represent unknown numbers (real or imaginary) or lines, as in Descartes' work; they now could refer to elements of any type—substitutions, geometrical figures, matrices, and so on. When in the sixteenth century *cosa* was used for the unknown, the "thing" was understood, of course, to be a magnitude; now the literal meaning of the Italian and Spanish word (cosa = thing) is literally applicable, for there is no restriction on the nature of the elements of abstract algebra beyond those specifically postulated in the axioms. Some indication of the direction in which algebra had been moving can be gained by comparing the algebraic papers presented at the Fifth International Congress at Cambridge, England, in 1912, just before the interruption occasioned by World War I, and those read at the Eleventh International Congress at Cambridge, Massachussetts, in 1950, following the second break in the series caused by World War II. The transition from classical to abstract algebra during the interbellum era will be apparent from a perusal of the table of contents, as will be also the emergence of topology

[33] See Paul Alexandroff, *Elementary Concepts of Topology*, trans. by Alexis N. Obolensky (1965), p. 17, or Jerome H. Manheim, *The Genesis of Point Set Topology* (1964), pp. 126–127.
[34] See Oystein Ore, *L'algèbre abstraite* (Paris: Hermann, 1936).

as a rival to its mother, geometry. The enormous development of topology today arises in part from the fact that it is difficult to imagine an aspect of analysis or geometry that should not be based on a previous topological study; and despite the apparent vagueness that topology sometimes exhibits, it is closely linked with the most precise mathematical questions. Should the second half of the century continue in the direction in which it has opened, abstract algebra and topology will seize the lion's share in mathematical research.

14   If mathematics changed form between the wars, it is equally true that much of the mathematics following World War II represented a radically new departure heralding a new era.[35] Set theory and measure theory throughout the twentieth century have invaded an ever widening portion of mathematics, and few branches have been as thoroughly influenced by the trend as has the theory of probability, to which Borel had contributed his *Eléments de la théorie des probabilités* (1909). The opening year of the new century was auspicious for probability both in physics and in genetics, for in 1901 Gibbs published his *Elementary Principles in Statistical Mechanics*, and in the same year the *Biometrika* was founded by Karl Pearson (1857–1936). Francis Galton (1822–1911), precocious cousin of Charles Darwin and a born statistician, had studied regression phenomena; in 1900 Pearson, Galton Professor of Eugenics at the University of London, had popularized the chi-square test. One of Poincaré's titles had been "Professor of the Calculus of Probabilities," indicating the rising interest in the subject. In Russia the study of linked chains of events was initiated, especially in 1906–1907, by Andrei Andreyevich Markov (or Markoff, 1856–1922), student of Tchebycheff and coeditor of his teacher's *Oeuvres* (2 vols., 1899–1904). In the kinetic theory of gases and in many social and biological phenomena the probability of an event depends often on preceding outcomes, and especially since the middle of the twentieth century Markov chains of linked probabilities have been widely studied.[36] As mathematical foundations for the expanding theory of probability were sought, statisticians found the appropriate tool at hand, and today no rigorous presentation of probability theory is possible without using the notions of measurable functions and modern theories of integration. In Russia, for example, Andrei Nicolaevich Kolmogoroff (1903–   ) made important advances in Markov processes (1931) and satisfied in part Hilbert's sixth project, calling for axiomatic foundations of probability, through the

[35] Jean Dieudonné, "Recent Developments in Mathematics," *American Mathematical Monthly*, **71** (1964), 239–248.

[36] There is no adequate account of the history of recent probability theory, but an elementary presentation of some aspects is found in Amy C. King and C. B. Read, *Pathways to Probability* (1963). See also the historical-bibliographical notes at the end of E. B. Dynkin, *Markov Processes*, trans. from the Russian (New York: Springer, 1965, 2 vols.), II, 240–266.

use of Lebesgue measure theory. Classical analysis had been concerned with continuous functions, whereas probability problems generally involve discrete cases. Measure theory and the extensions of the integration concept were ideally suited to bring about a closer association of analysis and probability, especially after the middle of the century when Laurent Schwartz (1915–   ) of the University of Paris generalized the concept of differentiation through the theory of distributions (1950–1951).

The Dirac delta function of atomic physics had shown that the pathological functions that long had occupied mathematicians were useful also in science. In the more difficult cases, however, differentiability breaks down, with resulting problems in the solution of differential equations—one of the chief connecting links between mathematics and physics—especially where singular solutions are involved. To surmount this difficulty Schwartz introduced a broader view of differentiability, one made possible by the development, in the first half of the century, of general vector spaces by Banach, Fréchet, and others. A linear vector space is a set of elements $a, b, c \ldots$ satisfying certain conditions, including especially the requirement that if a and b are elements of L, and if α and β are complex numbers, then $\alpha a + \beta b$ is an element of L. If the elements of L are functions, the linear vector space is called a linear function space, and a mapping of this case is called a linear functional. By a "distribution" Schwartz meant a linear and continuous functional on the space of functions that are differentiable and satisfy certain other conditions. The Dirac measure, for example, is a special case of a distribution. Schwartz then developed an appropriate definition of the derivative of a distribution such that the derivative of a distribution always is itself a distribution. This provides a powerful generalization of the calculus, with immediate applications to probability theory and physics. Functional analysis, essentially a generalization of the calculus of variations, and the theory of distributions have also been important topics of mathematical research since the middle of the century.[37]

Probability and statistics in the twentieth century have been inextricably **15** linked not only with pure mathematics, but with a strikingly different characteristic of our times—an increasing dependency on high-speed computers. The subject of machine calculation was not really new, for Pascal and Leibniz had had modest success long before. In fact, the prophet of the elaborate computing machine had been Charles Babbage, an eccentric who had carried on a lifelong polemic against organ-grinders while he tried desperately to secure funds to complete his ambitious project of building a

---

[37] For a synoptic account see J.-P. Marchand, *Distributions: An Outline* (Amsterdam: North-Holland, 1962). Cf. Laurent Schwartz, *Théorie des distributions* (Paris: Hermann, 1950–1951, 2 vols.; 2nd ed., 1957).

"difference engine." This device, conceived in 1833, was for a while supported by the British government; when the Chancellor of the Exchequer in 1842 cut off financial support, Babbage bitterly compared him to the destroyer of the beautiful Temple of Ephesus. The machine envisioned by Babbage would have had much of the flexibility of modern machines, but without the latter's speed. It would have performed all arithmetic operations and would have stored information for later recall, using an elaborate pattern of gears, wheels, and levers. His "engine," a digital computer, was never completed. The modern era of mechanical computation may be said to have begun in about 1925 at the Massachusetts Institute of Technology where Vannevar Bush (1890–   ) and his associates built a large-scale analogue calculator, powered by electric motors, but otherwise mechanical. In 1939 the International Business Machines Corporation began construction of the MARK I, a fully automatic electromechanical device along the lines of Babbage's vision; but before it was completed in 1944 it had been outmoded by plans for ENIAC (Electronic Numerical Integrator and Calculator). The latter was the first fully electronic calculator, based on the flow of electrons through vacuum tubes. It had been initiated under pressure from military needs, and among those who served as consultants on the project was John von Neumann (1903–1957), who had been born in Budapest and had taught at Berlin and Hamburg before coming to America in 1930, where, along with Einstein, he became one of the first permanent members of the Institute for Advanced Study in 1933. Between 1944 and 1946 he had helped to draw up for the Army a report on computer capabilities, and in 1949 the first stored-program computer went into operation. Two years later UNIVAC I (Universal Automatic Calculator) was completed by Sperry Rand Corporation, but so rapidly does the field of machine computation change that this computer is now a museum piece in the Smithsonian Institution.[38]

Electricity has so altered our way of life that it often is said that we live in an electrical age; now electronic devices may be about to alter a large part of our mathematical development. Computers today have become so vast and intricate that they surpass the dreams of Babbage, who lived a century before his time. Problems that were hopelessly beyond the capabilities of mathematicians of earlier ages have recently been solved with the aid of high-speed computers. If, as Kepler said, the invention of logarithms doubled the life of an astronomer, how much more has the electronic computer expanded the careers of scientists and mathematicians! With its increased power has also come a proliferation of new fields and applications of

---

[38] A very informative account of the development of computers is found in Jeremy Bernstein, *The Analytical Engine: Computers —Past, Present and Future* (1963). See also *Babbage's Calculating Machine or Difference Engine*, ed. by Philip Morrison and Emily Morrison (New York, 1961).

mathematics—linear programming, game theory, operations research, and many more. Von Neumann, one of the most creative and versatile mathematicians of our century, was a pioneer in a new approach to mathematical economics. Econometrics had long made use of mathematical analysis, but it was especially through the *Theory of Games and Economic Behavior* of von Neumann and Oskar Morgenstern in 1944 that so-called finite mathematics came to play an increasing role in the social sciences. Interrelationships among the various branches of thought had become so complicated that Norbert Wiener (1894–1964), a mathematical prodigy and for many years professor of mathematics at the Massachusetts Institute of Technology, in 1948 published his *Cybernetics*, a book establishing a new subject devoted to the study of control and communication in animals and machines. Von Neumann and Wiener both were deeply involved also in quantum theory, and the former in 1955 was appointed to the Atomic Energy Commission; but it would be wrong to conclude that men like these were just applied mathematicians. They contributed at least as extensively to pure mathematics—to set theory, group theory, operational calculus, probability, and mathematical logic and foundations. It had been von Neumann, in fact, who in about 1929 had given Hilbert space its name, its first axiomatization, and its present highly abstract form.[39] Wiener had been important in the early twenties in the origins of the modern theory of linear spaces, and in particular in the development of Banach space. The remarkable expansion in applied mathematics in the twentieth century has in no way slowed the development of pure mathematics, nor has the rise of newer branches diminished the vigor of the older.

The fundamental concepts of modern (or abstract) algebra, topology, **16** and vector spaces were laid down between 1920 and 1940, but the next score of years saw a veritable upheaval in methods of algebraic topology that carried over into algebra and analysis. The result was a new subject known as homological algebra, the first book on which, by Henri Cartan (1904–   ) and Samuel Eilenberg (1913–   ), appeared in 1955, to be followed in the next dozen years by several other monographs. Homological algebra is a development of abstract algebra concerned with results valid for many different kinds of spaces—an invasion of algebraic topology in the domain of pure algebra. The rapidity with which this general and powerful cross between abstract algebra and algebraic topology has grown is apparent in the swift increase in the number of articles on homological algebra listed in *Mathematical Reviews*. Moreover, so widely applicable are results in the field that the older labels of algebra, analysis, and geometry scarcely fit the

---

[39] For an account of the work of von Neumann see a series of articles comprising a memorial number of the *Bulletin of the American Mathematical Society*, **64** (May 1958).

results of recent research. Never before has mathematics been so thoroughly unified as in our day. Most of the enormous development during the twenty years following World War II has had little to do with the natural sciences, being spurred on by problems within pure mathematics itself; yet within the same period the applications of mathematics to science have multiplied exceedingly. The explanation for this anomaly seems to be clear: abstraction and the discernment of patterns have been playing more important roles in the study of nature, just as they have in mathematics. Hence even in our day of hyperabstract thinking, mathematics continues to be the language of science, just as it was in antiquity.[40] That there is an intimate connection between experimental phenomena and mathematical structures seems to be fully confirmed in the most unexpected manner by the recent discoveries of contemporary physics, although the underlying reasons for the agreement remain obscure. "From the axiomatic point of view, mathematics appears thus as a storehouse of abstract forms—the mathematical structures; and it so happens—without our knowing why—that certain aspects of empirical reality fit themselves into these forms, as if through a kind of preadaptation."[41]

17   It has been repeatedly stressed here that mathematics of the twentieth century has seen an emphasis on abstraction and an increasing concern with the analysis of broad patterns. Perhaps nowhere is this more clearly apparent than in the mid-twentieth-century works that have emanated from the polycephalic mathematician known as Nicolas Bourbaki. This is a non-existent Frenchman with a Greek name that has appeared on the title pages of several dozen volumes in a continuing major work, *Éléments de mathématique*, which is intended to survey all of worthwile mathematics. The home of Bourbaki is given as Nancy, a city that has provided a number of leading mathematicians of this century; it may not be a coincidence that in Nancy there is a statue to the colorful and once very real General Charles Denis Sauter Bourbaki (1816–1897), who in 1862 was offered, but declined, the throne of Greece and whose role in the Franco-Prussian War was very tangible. Nicolas Bourbaki, nevertheless, is not a relative in any sense of the word; the name has simply been appropriated to designate a group of mathematicians, almost exclusively French, who form a sort of cryptic *société anonyme*.[42] As an institutional connection N. Bourbaki sometimes uses the University of Nancago, a playful reference to the fact that two of the moving spirits within the group were for a while connected with

[40] M. H. Stone, "The Revolution in Mathematics," *Liberal Education*, **47** (1961), 304–327. See especially p. 326.

[41] N. Bourbaki, "The Architecture of Mathematics," *American Mathematical Monthly*, **57** (1950), 221–232. This is a translation of an article that appeared in *Les grands courants de la pensée mathématique*, ed. by F. Le Lionnais (1962). See especially p. 231.

[42] Paul R. Halmos, "Nicolas Bourbaki," *Scientific American*, **196** (May 1957), 88–99.

universities in the Chicago area—André Weil (1906–   ) at the University of Chicago (more recently, however, at the Institute for Advanced Study at Princeton) and Jean Dieudonné (1906–   ) at Northwestern University (formerly at the University of Nancy, later at the University of Paris). The first volume of Bourbaki's *Éléments* appeared in 1939, the thirty-first in 1965; so far the work has not yet exhausted what is known as part I, *Les structures fondamentales de l'analyse*. This part contains half a dozen sub-headings: (1) Set Theory, (2) Algebra, (3) General Topology, (4) Functions of a Real Variable, (5) Topological Vector Spaces, and (6) Integration. These headings indicate that only a small portion of the mathematics contained in these volumes was in existence a century earlier. The presentation of the subject by Bourbaki is characterized by uncompromising adherence to the axiomatic approach and to a starkly abstract and general form that portrays clearly the logical structure. The Bourbakique approach to mathematics thus is somewhat analogous, on the highest research level, to the changes in mathematics that have taken place in grade school and high school curricula. The hope in both cases is that the emphasis on structure will effect a considerable economy of thought. For example, in the early nineteenth century the discovery that the structure of the complex number system was the same as that of points in the Euclidean plane showed that the properties of the latter, studied for over two millennia, could be applied to the former. The result was an exuberant proliferation in complex analysis. There is no reason why the current concern for similarities in structure should not, in the years to come, yield similar dividends.

The so-called new mathematics in the schools also shares with Bourbaki the desire to substitute ideas for calculations. Romantics in mathematics earlier in the century had feared a takeover of their subject by an arid formalism encouraged by logicism. By the middle of the century the feud between formalists and intuitionists had quieted, and Bourbaki sees no need to take sides in the controversy. "What the axiomatic method sets as its essential aim," he writes, "is exactly that which logical formalism by itself can not supply, namely the profound intelligibility of mathematics."[43] In the same vein one of the leaders of the group, generally regarded as an outstanding mathematician of the mid-twentieth century, wrote that "If logic is the hygiene of the mathematician, it is not his source of food."[44]

Poincaré had once remarked that in mathematics the "prophets of misfortune, ... the pessimists, have always been compelled to retreat"; this

---

[43] Bourbaki, "The Architecture of Mathematics," *American Mathematical Monthly*, **57** (1950), 223.

[44] André Weil, "The Future of Mathematics," *American Mathematical Monthly*, **57** (1950), 295–306; see p. 297. This is a translation of an article from *Les grands courants de la pensée mathematique*, ed. by F. Le Lionnais (1948; new ed. 1962).

optimism is present in the mathematics of today. Weil, echoing the view of Hilbert, has pointed to the multitude of problems at hand as a sure sign of the vitality of mathematics; of the future he has this to say: "The great mathematician of the future, as of the past, will flee the well-trodden path. It is by unexpected *rapprochements*, which our imagination would not have known how to arrive at, that he will solve, in giving them another twist, the great problems which we shall bequeath to him."[45] Looking ahead, Weil is confident also of one further thing: "In the future, as in the past, the great ideas must be simplifying ideas."[46]

From knowledge of the past one can anticipate in a very general sense what the future may hold. But if there is some element of truth in the aphorism "history repeats itself," the history of mathematics nevertheless has shown that the "repetitions" are so varied and unanticipated as to preclude any meaningful forecast of things to come. It has been asserted[47] that a graph representing the growth of science, including mathematics, closely approximates an exponential curve, and it is not unreasonable to hope that future developments in mathematics may conform to such a pattern. Nevertheless, folly and wisdom are so conjoined in human society that there now exists a very real possibility that man's mathematics may some day become the instrument of his own undoing.

BIBLIOGRAPHY

Alexandroff, Paul, *Elementary Concepts of Topology*, trans. by Alexis N. Obolensky (New York: Frederick Ungar, 1965).

Bell, E. T., *The Development of Mathematics* (New York: McGraw-Hill, 1940).

Bernkopf, Michael, "The Development of Function Spaces with Particular Reference to Their Origins in Integral Equation Theory," *Archive for History of Exact Sciences*, **3** (1966), 1–96.

Bernstein, Jeremy, *The Analytical Engine: Computers—Past, Present and Future* (New York: Random House, 1963).

Beth, E. W., *The Foundations of Mathematics* (Amsterdam: North Holland, 1959).

Black, Max, *The Nature of Mathematics* (New York: Harcourt, Brace, 1933).

Bochenski, I. M., *A History of Formal Logic*, trans. by Ivo Thomas (Notre Dame, Ind.: University of Notre Dame Press, 1961).

Bourbaki, N., "The Architecture of Mathematics," *American Mathematical Monthly*, **57** (1950), 221–232.

Bourbaki, N., *Éléments d'histoire des mathématiques* (Paris: Hermann, 1960).

---

[45] "L'avenir des mathématiques," in *Les grands courants de la pensée mathématique*, ed. by F. Le Lionnais (1962), pp. 307 320; see p. 317.

[46] "The Future of Mathematics," *American Mathematical Monthly*, **57** (1950), 304.

[47] D. J. Price, *Science since Babylon* (New Haven, Conn.: Yale University Press, 1961), Chap. 5. Cf. K. O. May, "Quantitative Growth of the Mathematical Literature," *Science*, **154** (1966), 1672–1673.

Delachet, André, *Contemporary Geometry*, trans. by H. G. Bergmann (New York: Dover Publications, 1962).

Dieudonné, Jean, "Recent Developments in Mathematics," *The American Mathematical Monthly*, **71** (1964), 239–248.

Enriques, F., *The Historic Development of Logic*, trans. by J. Rosenthal (New York: Holt, Rinehart, 1929).

Hilbert, David, "Mathematical Problems," trans. by Mary Winston Newson, in *Bulletin of the American Mathematical Society* (2), **8** (1902), 437–479.

Hilbert, David, *Foundations of Geometry*, trans. by E. J. Townsend, 2nd ed. (Chicago: Open Court, 1910).

Hilbert, David, *Gesammelte Abhandlungen* (Berlin: Springer, 1932–1935, 3 vols.).

King, Amy, C. and C. B. Read, *Pathways to Probability* (New York: Holt, Rinehart, 1963).

LaSalle, J. P., and S. Lefschetz, eds., *Recent Soviet Contributions to Mathematics* (New York: Macmillan, 1962).

Lebesgue, Henri, *Leçons sur l'intégration* (Paris: Gauthier-Villars, 1904).

Lebesgue, Henri, *Measure and the Integral*, ed. by K. O. May (San Francisco: Holden-Day, 1966).

Lebon, Ernest, *Henri Poincaré, bibliographie analytique des ecrits* (Paris: Gauthier-Villars, 1909).

Le Lionnais, F., ed. *Les grands courants de la pensée mathématique*, new ed. (Paris: Albert Blanchard, 1962).

Luchins, Abraham S., and Edith H. Luchins, *Logical Foundations of Mathematics for Behavioral Scientists* (New York: Holt, Rinehart, 1965).

Manheim, Jerome H., *The Genesis of Point Set Topology* (New York: Macmillan, 1964).

Nevanlinna, Rolf, "Reform in Teaching Mathematics," *American Mathematical Monthly*, **73** (1966), 451–464.

Nörlund, N. E., "Correspondence de Henri Poincaré et de Felix Klein," *Acta Mathematica*, **39** (1923), 94–132.

Picard, Émile, *Les sciences mathématiques en France depuis un demi-siècle* (Paris: Gauthier-Villars, 1917).

Pierpont, James, "Mathematical Rigor, Past and Present," *Bulletin of the American Mathematical Society*, **37** (1928), 23–53.

Poincaré, Henri, *Oeuvres* (Paris: Gauthier-Villars, 1916–1956, 11 vols.).

Prasad, Ganesh, *Mathematical Research in the Last Twenty Years* (Berlin: Walter de Gruyter, 1923).

Russell, Bertrand, *Principles of Mathematics*, 2nd ed. (New York: Norton, 1938).

Stone, M. H., "The Revolution in Mathematics," *Liberal Education*, **47** (1961), 304–327.

Volterra, Vito, et al., *Henri Poincaré: l'oeuvre scientifique, l'oeuvre philosophique* (Paris: Alcan, 1914).

Weil, André, "The Future of Mathematics," *American Mathematical Monthly*, **57** (1950), 295–306.

Weyl, Hermann, *Philosophy of Mathematics and Natural Science*, based on a translation by Olaf Helmer (Princeton, N.J.: Princeton University Press, 1940).

Wilder, R. L., "The Origin and Growth of Mathematical Concepts," *Bulletin of the American Mathematical Society*, **59** (1953), 423–448.

Wilder, R. L., "The Role of the Axiomatic Method," *American Mathematical Monthly*, **74** (1967), 115–127.

EXERCISES

1. Give three definitions or descriptions of mathematics from the nineteenth and twentieth centuries, explaining which you prefer and why.
2. Describe the views of the three chief nineteenth-century schools of thought concerning the foundations of mathematics, mentioning one or two outstanding figures in each school.
3. Describe several paradoxes and anomalies in mathematics of the twentieth century, indicating their significance.
4. Explain why the traditional distinctions between algebra and geometry became less pronounced during the twentieth century. Which field developed more rapidly in the century and why?
5. Were mathematical advances inspired more by science and technology in the twentieth century than in the nineteenth? Explain.
6. Compare the influence on mathematical attitudes of the Gödel theorem with that of the discovery of incommensurable magnitudes.
7. Has the rate of discovery in mathematics been increasing or decreasing during the twentieth century? How do you account for this?
8. Would ancient Greek mathematicians today be classed as formalists, intuitionists, or logicists? Explain.
9. Name three prominent foreign-born mathematicians who became members of the Institute for Advanced Study at Princeton (familiarly known as "The Princetitute") and briefly describe their chief contributions to mathematics.
10. Mention three respects in which the French city of Nancy was associated importantly with mathematicians of the twentieth century.
11. Describe some of the contributions of Poland to mathematics in the interval between the two World Wars.
12. What is the Lebesgue integral over $[0, 1]$ of the function $f(x)$, where $f(x) = 1$ for x rational and $f(x) = 0$ for x irrational? Give motivation for your answer.
13. Show that the product of two linear fractional transformations (in one variable) is a linear fractional transformation.
14. If a linear fractional transformation

$$z' = \frac{az + b}{cz + d}$$

satisfies the condition $ad - bc = 1$, show that the inverse transformation also satisfies this conditions.
15. If each of two linear fractional transformations satisfies the condition in Exercise 14, show that the product of the two transformations also satisfies this condition.
16. Which of the following are known to be transcendental: π^e, e^π, e^e, π^π, $(\sqrt{2})^e$, $\pi^{\sqrt{3}}$, $(\sqrt{2})^{\sqrt{3}}$, ln 1, tan $\pi/3$, i^i. Explain.
*17. Prove that $\log_{10} 3$ is irrational and use this result and Gelfond's theorem to show that $\log_{10} 3$ is transcendental. How many of the common logarithms of integers between 1 and 10 inclusive are algebraic?
*18. The first three steps in the definition of the von Koch or snowflake curve are shown in Figure 27.2. Find the perimeter and area of the configuration corresponding to the fourth step.
*19. The perimeter of the von Koch or snowflake curve is infinite, but the area bounded by it is finite. Find this area.

General Bibliography

Archibald, R. C., *Outline of the History of Mathematics* (Buffalo: Slaught Memorial Papers of the Mathematical Association of America, 1949). Especially valuable for a very extensive bibliography.

Ball, W. W. R., *A Short Account of the History of Mathematics* (London: Macmillan, 1888). One of the most popular histories of mathematics; it appeared in a 6th ed. in 1915 and was reprinted as a Dover paperback in 1960. Obsolescent, but still of interest.

Ball, W. W. R., *Mathematical Recreations and Problems of Past and Present Times* (London: Macmillan, 1892). Very popular; contains considerable history.

Bell, E. T., *Men of Mathematics* (New York: Simon and Schuster, 1937). Very readable biographical accounts assume relatively little mathematical background.

Bell, E. T., *Developement of Mathematics* (New York: McGraw-Hill, 1940). Excellent account, especially on modern mathematics, for a reader with good mathematical background.

Bochner, Salomon, *The Role of Mathematics in the Rise of Science* (Princeton, N.J.: Princeton University Press, 1966). Not a continuous narrative, but a collection of essays.

Bourbaki, Nicolas, *Éléments d'histoire des mathématiques* (Paris: Hermann, 1960). Not a connected history, but accounts of certain aspects, especially of modern times.

Braunmühl, Anton von, *Vorlesungen über Geschichte der Trigonometrie* (Leipzig: B. G. Teubner, 1900–1903, 2 vols.). Still the standard in the field.

Cajori, Florian, *A History of Mathematics*, 2nd ed. (New York: Macmillan, 1919). The most ambitious single-volume source in English.

Cajori, Florian, *A History of Mathematical Notations* (Chicago: Open Court, 1928–1929, 2 vols.). The definitive work on the subject.

Cantor, Moritz, *Vorlesungen über Geschichte der Mathematik* (Leipzig: Teubner, 1892–1908, 4 vols.). The most extensive history of mathematics so far published. Some volumes are in a 2nd ed., and the whole work is available in a reprint.

Carruccio, Ettore, *Mathematics and Logic in History and in Contemporary Thought*, trans. by Isabel Quigly (Chicago: Aldine, 1964). An eclectic survey. Italian authors predominate in the Bibliography.

Chasles, Michel, *Aperçu historique sur l'origine et le développement des méthodes en géométrie*, 2nd ed. (Paris, 1875). A classic work especially strong on synthetic geometry in the early nineteenth century.

Coolidge, J. L., *A History of Geometrical Methods* (Oxford: Clarendon, 1940). An excellent work presupposing considerable mathematical background. Available in a Dover paperback of 1963.

Dickson, L. E., *History of the Theory of Numbers* (Washington, D.C.: Carnegie Institution, 1919–1923). Definitive in its area. Available in reprint, New York: Stechert, 1934.

Encyclopédie des sciences mathématiques pures et appliquées (Paris, 1904–1914). This is essentially a partial translation of *Encyklopädie der mathematischen Wissenschaften* (Leipzig: Teubner, 1898–1935), but the French version contains many additional historical references.

Eves, Howard, *An Introduction to the History of Mathematics*, rev. ed. (New York: Holt, Rinehart and Winston, 1964). A notably successful textbook.

Gillispie, C. C., ed., *Dictionary of Scientific Biography* (to be published in about six volumes by Charles Scribner's Sons, New York, beginning in 1968).

Heath, T. L., *A History of Greek Mathematics* (Oxford: Clarendon, 1921, 2 vols.). The best treatment of the subject in English. Available in paperback in a somewhat abbreviated form as *A Manual of Greek Mathematics* (New York: Dover, 1963).

Hofmann, J. E., *Geschichte der Mathematik* (Berlin: Walter de Gruyter, 1953–1957, 3 vols.; Vol. I, new ed., 1963). The handy pocket-size volumes contain extraordinarily useful biobibliographical indexes. These indexes tragically were omitted from the English translation which appeared in two volumes (New York: Philosophical Library, 1957–1959) under the titles *The History of Mathematics* and *Classical Mathematics*, making the translation of very limited usefulness.

Itard, Jean, and Pierre Dedron, *Mathématiques et mathématiciens* (Paris: Magnard, 1959). Elementary but useful; contains excerpts from sources.

James, Glenn, and R. C. James, *Mathematics Dictionary*, 2nd ed. (Princeton, N.J.: D. Van Nostrand, 1959). Useful, but not so thoroughgoing as Naas and Schmid.

Kaestner, A. G., *Geschichte der Mathematik* (Göttingen, 1796–1800, 4 vols.). Especially useful for practical mathematics and science in the Renaissance.

Klein, Felix, *Vorlesungen über die Entwicklung der Mathematik im 19. Jahrhundert* (Berlin, 1926–1927, 2 vols.). Survey on a high level, left incomplete on the death of the author.

Kline, Morris, *Mathematics in Western Culture* (New York: Oxford, 1953). Attractively written on a popular level.

Klügel, G. S., *Mathematisches Wörterbuch* (Leipzig, 1803–1836, 7 vols.). Portrays the status of the subject a century and a half ago.

Loria, Gino, *Storia delle matematiche* (Turin: Sten, 1929–1933, 3 vols.). A sound and substantial work. A second edition, Milan: U. Hoepli, appeared in 1950.

Marie, Maximilien, *Histoire des sciences mathématiques* (Paris, 1883–1888, 12 vols.). Not a systematic history, but a series of biographies chronologically arranged and listing the chief works of the individuals.

Midonick, Henrietta O., ed., *The Treasury of Mathematics* (New York: Philosophical Library 1965). Useful, but selections emphasize non-European contributions.

Montucla, J. E., *Histoire des mathématiques*, new ed. (Paris, 1799–1802, 4 vols.). Still quite useful, especially for applications of mathematics to science. Available in reprint, Paris: A. Blanchard, 1960.

Moritz, R. E., *Memorabilia Mathematica or the Philomath's Quotation-Book* (New York, 1914). Contains more than two thousand quotations, arranged by subject and with an index. Available in Dover paperback under the title *On Mathematics and Mathematicians*.

Muir, Thomas, *The Theory of Determinants in the Historical Order of Development* (London, 1906–1930, 4 vols. and Supplement). By far the most comprehensive in the area. Available in Dover reprint.

Naas, Joseph, and H. L. Schmid, *Mathematisches Wörterbuch* (Berlin: Akademie, 1961, 2 vols.). An altogether exemplary dictionary containing extraordinarily many definitions and short biographies.

Newman, James, ed., *The World of Mathematics* (New York: Simon and Schuster, 1956, 4 vols.). Includes much material on the history of mathematics.

Read, C. B., "Articles on the History of Mathematics: A Bibliography of Articles Appearing in Six Periodicals," *School Science and Mathematics* (1959), 689–717. Especially useful for introductory material.

Sarton, George, *Introduction to the History of Science* (Baltimore: Carnegie Institution, 1927–1948, 3 vols. in 5). A monumental work and a standard tool in research in the history of science and mathematics up to 1400.

Sarton, George, *The Study of the History of Mathematics* (Cambridge, Mass.: Harvard University Press, 1936; New York: Dover paperback reprint, 1957). A slim but useful guide. See also Sarton's *Horus: A Guide to the History of Science* (New York: Ronald Press, 1952).

Schaaf, W. L., *A Bibliography of Mathematical Education* (Forest Hills, N.Y.: Stevinus Press, 1941). An index of periodical literature since 1920 containing more than four thousand items.

Schaaf, W. L., *Recreational Mathematics: A Guide to the Literature*, 3rd ed. (Washington, D.C.: National Council of Teachers of Mathematics, 1963). Contains between two and three thousand references to books and articles.

Scott, J. F., *A History of Mathematics* (London: Taylor and Francis, 1958). Good on British mathematicians, but not up-to-date on the pre-Hellenic period.

Smith, D. E., *History of Mathematics* (Boston: Ginn, 1923–1925, 2 vols.; Dover paperback, New York, 1958). Still very useful for biographical details, especially dates of birth and death, and for elementary aspects of mathematics.

Smith, D. E., *A Source Book in Mathematics* (New York: McGraw-Hill, 1929; Dover paperback, New York, 1959, 2 vols.). Useful, but selection is far from ideal.

Struik, D. J., *A Concise History of Mathematics*, 3rd ed., (New York: Dover, 1967). Quite brief, but reliable and on a very scholarly level, with many references.

Struik, D. J., *Source Book in Mathematics* (Cambridge, Mass: Harvard University Press, in press). Very good coverage in algebra, analysis, and geometry from 1200 to 1800.

Tannery, Paul, *Mémoires scientifiques* (Paris: Gauthier-Villars, 1912–1934, 13 vols.). These volumes contain many articles on history of mathematics, especially on Greek antiquity and on the seventeenth century, by one of the great authorities in the field.

Taton, René, ed., *Histoire générale des sciences* (Paris: Presses Universitaires de France, 1957–1961, 3 vols.; English trans., New York: Basic Books, 1964–1965). An excellent and authoritative account.

Taylor, Eva, G. R., *The Mathematical Practitioners of ... England* (Cambridge:

Cambridge University Press, 1954–1966, 2 vols.). Biographical details on individuals from 1485 to 1840.

Todhunter, Isaac, *History of the Calculus of Variations During the Nineteenth Century* (Cambridge, 1861). Old, but a standard work in the field.

Todhunter, Isaac, *History of the Mathematical Theory of Probability from the Time of Pascal to that of Laplace* (Cambridge, 1865). A thorough and standard work in the field.

Tropfke, Johannes, *Geschichte der Elementar-Mathematik*, 2nd ed. (1921–1924, 7 vols.). An important history for the elementary branches. Some volumes appeared in an incomplete 3rd ed.

Van der Waerden, B. L., *Science Awakening*, trans. by Arnold Dresden (New York: Oxford University Press, 1961). An account of pre-Hellenic and Greek mathematics, very attractively illustrated. A somewhat less attractive paperback edition (New York: Wiley, 1963) is available.

Wieleitner, Heinrich, *Geschichte der Mathematik* (Part II, from Descartes to about 1800, Leipzig, 1911–1921, 2 vols.). Very useful work on an intermediate mathematical level; not to be confused with the author's briefer two-volume Sammlung Göschen *Geschichte der Mathematik* (Berlin, 1939).

Youschkevitch, A. P., *Geschichte der Mathematik im Mittelalter* (Leipzig: Teubner, 1964). A substantial and authoritative account.

Zeller, Sister Mary Claudia, *The Development of Trigonometry from Regiomontanus to Pitiscus* (Ann Arbor, Mich.: University of Michigan, Ph.D. Dissertation, 1944). The nearest thing in English to a history of trigonometry.

Zeuthen, H. G., *Geschichte der Mathematik im XVI und XVII Jahrhundert* (German ed., Leipzig: Teubner, 1903). A sound and useful account.

Appendix

CHRONOLOGICAL TABLE

(Dates before −776 are approximations only)

	−5,000,000,000,000	Origin of the sun
	−5,000,000,000	Origin of the earth
	−600,000,000	Beginning of Paleozoic Age
	−225,000,000	Beginning of Mesozoic Age
	−60,000,000	Beginning of Cenozoic Age
	−2,000,000	Origin of man
	−50,000	Neanderthal Man
−50,000 Evidence of counting	−25,000	Paleolithic art; Cro-Magnon Man
−25,000 Primitive geometrical designs	−10,000	Mesolithic agriculture
	−5000	Neolithic civilizations
−4241 Hypothetical origin of Egyptian calendar	−4000	Use of metals
	−3500	Use of potter's wheel; writing
−3000 Hieroglyphic numerals in Egypt	−3000	Use of wheeled vehicles
	−2800	Great Pyramid
−2773 Probable introduction of Egyptian calendar		
−2400 Positional notation in Mesopotamia	−2400	Sumerian-Akkadian Empire
−1850 Moscow (Golenishev) papyrus; cipherization		
	−1800	Code of Hammurabi
	−1700	Hyksos domination of Egypt; Stonehenge in England
	−1600	Kassite rule in Mesopotamia; New Kingdom in Egypt
	−1400	Catastrophe in Crete

Chronological Table—*(contd.)*

−1100?	Chou-pei
−1350	Phonecian alphabet; use of iron; sundial; water clocks
−1200	Trojan War; Exodus from Egypt
−776	First Olympiad
−753	Traditional founding of Rome
−743	Era of Nabonassar
−740	Works of Homer and Hesoid (approx.)
−586	Babylonian Captivity
−585	Thales of Miletus; deductive geometry (?)
−540	Pythagorean arithmetic and geometry (approx.)
	Rod numerals in China (approx.)
	Indian *Sulvasūtras* (approx.)
−538	Persians took Babylon
−480	Battle of Thermopylae
−477	Formation of Delian League
−461	Beginning of Age of Pericles
−450	Spherical earth of Parmenides (approx.)
−430	Death of Zeno; works of Democritus
	Astronomy of Philolaus (approx.)
	Elements of Hippocrates of Chios (approx.)
−430	Hippocrates of Cos (approx.)
	Atomic doctrine (approx.)
−429	Death of Pericles; plague at Athens
−428	Birth of Archytas; death of Anaxagoras
−427	Birth of Plato
−420	Trisectrix of Hippias (approx.)
	Incommensurables (approx.)
−404	End of Peloponnesian War
−399	Death of Socrates; *Anabasis* of Xenophon
−369	Death of Theaetetus

−347	Death of Plato
−332	Alexandria founded
−323	Death of Alexander
−322	Deaths of Aristotle and Demosthenes
−311	Beginning of Seleucid Era in Mesopotamia
−306	Ptolemy I (Soter) of Egypt
−283	Pharos at Alexandria
−264	First Punic War opened
−232	Death of Asoka, the "Buddhist Constantine"
−210	Great Chinese Wall begun
−166	Revolt of Judas Maccabaeus
−146	Destruction of Carthage and Corinth
−121	Gaius Gracchus killed

−360	Eudoxus on proportion and exhaustion (approx.)
−350	Menaechmus on conic sections (approx.)
	Dinostratus on quadratrix (approx.)
−335	Eudemus: *History of Geometry* (approx.)
−330	Autolycus: *On the Moving Sphere* (approx.)
−320	Aristaeus: *Conics* (approx.)
−300	Euclid's *Elements* (approx.)
−260	Aristarchus' heliocentric astronomy (approx.)
−230	Seive of Eratosthenes (approx.)
−225	*Conics* of Apollonius (approx.)
−212	Death of Archimedes
−180	Cissoid of Diocles (approx.)
	Conchoid of Nicomedes (approx.)
	Hypsicles and 360° circle (approx.)
−150	Spires of Perseus (approx.)
−140	Trigonometry of Hipparchus (approx.)

Chronological Table—(contd.)

−60	Geminus on parallel postulate (approx.)	
+75	Works of Heron of Alexandria (approx.)	
100	Nicomachus: *Arithmetica* (approx.)	
	Menelaus: *Spherics* (approx.)	
125	Theon of Smyrna and Platonic mathematics	
150	Ptolemy: *The Almagest* (approx.)	
250	Diophantus: *Arithmetica* (approx.?)	
320	Pappus: *Mathematical Collections* (approx.)	
390	Theon of Alexandria (fl.)	
415	Death of Hypatia	
470	Tsu Ch'ung-chi's value of π (approx.)	
476	Birth of Aryabhata	
485	Death of Proclus	
520	Anthemius of Tralles and Isidore of Miletus	
524	Death of Boethius	
529	Closing of the schools at Athens	

−75	Cicero restored tomb of Archimedes
−60	Lucretius: *De rerum natura*
−44	Death of Julius Caesar
+79	Death of Pliny the Elder at Vesuvius
116	Trajan extends Roman Empire
122	Hadrian's Wall in Britain begun
180	Death of Marcus Aurelius
286	Division of Empire by Diocletian
324	Founding of Constantinople
378	Battle of Adrianople
455	Vandals sack Rome
476	Traditional "fall" of Rome
496	Clovis adopted Christianity
526	Death of Theodoric
529	Founding of the monastery at Monte Cassino
532	Building of Hagia Sophia by Justinian

560	Eutocius' commentaries on Archimedes (approx.)	590	Gregory the Great elected pope
		622	Hejira of Mohammed
628	Brahma-sphuta-siddhânta		
662	Bishop Sebokht mentioned Hindu numerals	641	Library at Alexandria burned
		732	Battle of Tours
735	Death of the Venerable Bede		
775	Hindu works translated into Arabic	814	Death of Charlemagne
830	Al-Khowarizmi: *Algebra* (approx.)	910	Benedictine abbey at Cluny
901	Death of Thabit ibn-Qurra	987	Accession of Hugh Capet
998	Death of abu'l-Wefa	999	Gerbert became Pope Sylvester II
		1028	School of Chartres
1037	Death of Avicenna	1066	Battle of Hastings
1039	Death of Alhazen	1096	First Crusade
1048	Death of al-Biruni	1100	Henry I of England crowned
1114	Birth of Bhaskara		
1123	Death of Omar Khayyam	1170	Murder of Thomas à Becket
1142	Adelard of Bath translated Euclid		
1202	Fibonacci: *Liber abaci*		

Chronological Table—*(contd.)*

1260	Campanus' trisection (approx.)	1204	Crusaders sack Constantinople
	Jordanus Nemorarius: *Arithmetica* (approx.)		Death of Maimonides
1270	Wm. of Moerbeke translated Archimedes (approx.)	1215	Magna Carta
1274	Death of Nasir Eddin	1265	"First" parliament in England
1303	Chu Shi-kië and the Pascal triangle	1271	Travels of Marco Polo; mechanical clocks (approx.)
1328	Bradwardine: *Liber de proportionibus*	1286	Invention of eyeglasses (approx.)
1336	Death of Richard of Wallingford		
1360	Oresme's latitude of forms (approx.)	1348	The Black Death
		1364	Death of Petrarch
		1431	Joan of Arc burned
1436	Death of al-Kashi	1440	Invention of printing
1464	Death of Nicholas of Cusa	1453	Fall of Constantinople
1472	Peurbach: *New Theory of the Planets*		
1476	Death of Regiomontanus	1473	Sistine Chapel
1482	First printed Euclid		
1484	Chuquet: *Triparty*	1483	Murder of the princes in the Tower
		1485	Henry VII, first Tudor

1492	Discovery of America by Columbus
1498	Execution of Savonarola
1517	Protestant Reformation
1520	Field of the Cloth of Gold
1534	Act of Supremacy
1543	Vesalius: *De fabrica*
	Ramus: *Reproof of Aristotle*
1553	Servetus burned at Geneva
1558	Accession of Elizabeth I
1564	Birth of Shakespeare; deaths of Vesalius and Michelangelo
1572	Massacre of St. Bartholomew
1584	Assassination of William of Orange
1588	Drake defeated the Spanish Armada
1598	Edict of Nantes
1603	Deaths of Wm. Gilbert and Elizabeth I
1609	Galileo's telescope

1489	Use of + and − by Widmann
1492	Use of decimal point by Pellos
1494	Pacioli: *Summa*
1525	Rudolff: *Coss*
1526	Death of Scipione dal Ferro
1527	Apian published the Pascal triangle
1543	Tartaglia published Moerbeke's *Archimedes*
	Copernicus: *De revolutionibus*
1544	Stifel: *Arithmetica integra*
1545	Cardan: *Ars magna*
1557	Recorde: *Whetstone of Witte*
1564	Birth of Galileo
1572	Bombelli: *Algebra*
1579	Viète: *Canon mathematicus*
1585	Stevin: *La disme*
	Harriot's report on "Virginia"
1595	Pitiscus: *Trigonometria*
1603	Death of Viète
1609	Kepler: *Astronomia nova*
1614	Napier's logarithms

Chronological Table—*(contd.)*

1620	Bürgi's logarithms
1629	Fermat's method of maxima and minima
1631	Harriot: *Artis analyticae praxis*
	Oughtred: *Clavis mathematicae*
1635	Cavalieri: *Geometria indivisibilibus*
1637	Descartes: *Discours de la méthode*
1639	Desargues: *Brouillon projet*
1640	*Essay pour les coniques* of Pascal
1642	Birth of Newton; death of Galileo
1647	Deaths of Cavalieri and Torricelli
1655	Wallis: *Arithmetica infinitorum*
1657	Neil rectified his parabola
1658	Huygens' cycloidal pendulum clock
1667	Gregory: *Geometriae pars universalis*
1668	Mercator: *Logarithmotechnia*
1670	Barrow: *Lectiones geometriae*
1672	Assassination of De Witt

1616	Deaths of Shakespeare and Cervantes
1620	Landing of the Pilgrims
1626	Deaths of Francis Bacon and Willebrord Snell
1628	Harvey: *De motu cordis et sanguinis*
1636	Harvard College founded
1643	Accession of Louis XIV
1644	Torricelli's barometer
1649	Charles I beheaded
1651	Hobbes: *Leviathan*
	Von Guericke's air pump
1660	The Restoration
1662	Royal Society founded
1666	Académie des Sciences founded

1678 Ceva's theorem	1679 Writ of Habeas Corpus
	1682 *Acta eruditorum* founded
	1683 Seige of Vienna
1684 Leibniz' first paper on the calculus	1685 Revocation of the Edict of Nantes
1687 Newton: *Principia*	1689 The Glorious Revolution
1690 Rolle: *Traité d'algèbre*	
1696 Brachistochrone (the Bernoullis)	1699 Death of Racine
L'Hospital's rule	1702 Opening of Queen Anne's War
1706 Use of π by William Jones	1711 Birth of Hume
1715 Taylor: *Methodus incrementorum*	1718 Fahrenheit's thermometer
1718 De Moivre: *Doctrine of Chances*	
1722 Cotes: *Harmonia mensurarum*	
1730 Stirling's formula	1730 Réaumur's thermometer
1731 Clairaut on skew curves	
1733 Saccheri: *Euclid Vindicated*	
1734 Berkeley: *The Analyst*	1737 Linnaeus: *Systema naturae*
	1738 Daniel Bernoulli: *Hydrodynamica*
	1740 Accession of Frederick the Great
1742 Maclaurin: *Treatise of Fluxions*	1742 Centigrade thermometer
1743 D'Alembert: *Traité de dynamique*	
1748 Euler: *Introductio*; Agnesi: *Istituzioni*	1749 Volume I of Buffon's *Histoire naturelle*

Chronological Table—*(contd.)*

Year		Year	
1750	Cramer's rule; Fagnano's ellipse	1751	Volume I of Diderot's *Encyclopédie*
		1752	Franklin's kite
1759	*Die freye Perspektive* of Lambert		
		1767	Watt's improved steam engine
1770	Hyperbolic trigonometry		
		1774	Discovery of Oxygen (Priestley, Scheele, Lavoisier)
1777	Buffon's needle problem	1776	American Declaration of Independence
1779	Bézout on elimination		
		1781	Discovery of Uranus by Herschel
		1783	Composition of water (Cavendish, Lavoisier)
1788	Lagrange: *Mécanique analytique*		
		1789	French Revolution
1794	Legendre: *Eléments de géométrie*	1794	Lavoisier guillotined
1795	Monge: *Feuilles d'analyse*	1795	École Polytechnique; École Normale
1796	Laplace: *Système du monde*	1796	Vaccination (Jenner)
1797	Lagrange: *Fonctions analytiques*		
	Mascheroni: *Geometria del compasso*		
	Wessel: *Essay on . . . direction*		
	Carnot: *Métaphysique du calcul*	1799	Metric system
		1800	Volta's battery
1801	Gauss: *Disquisitiones arithmeticae*	1801	Ceres discovered
1803	Carnot: *Géométrie de position*	1803	Dalton's atomic theory
		1804	Napoleon crowned emperor
1810	Volume I of Gergonne's *Annales*		
		1814	Fraunhofer lines

1815	"The Analytical Society" at Cambridge
1817	Bolzano: *Rein analytischer Beweis*
1822	Poncelet: *Traité*; Fourier series; Feuerbach's theorem
1826	Crelle's *Journal* founded
	Principle of Duality (Poncelet, Plücker, Gergonne)
	Elliptic functions (Abel, Gauss, Jacobi)
1827	Homogeneous coordinates (Möbius, Plücker, Feuerbach)
	Cauchy: *Calculus of Residues*
1828	Green: *Electricity and Magnetism*
1829	Lobachevskian geometry
	Death of Abel at age 26
1830	Peacock: *Algebra*
1832	Bolyai: *Absolute Science of Space*
	Death of Galois at age 20
1834	Steiner became professor at Berlin
1836	Liouville's *Journal* founded
1837	*Cambridge and Dublin Mathematical Journal*
1843	Hamilton's quaternions
1844	Grassmann: *Ausdehnungslehre*
1847	Von Staudt: *Geometrie der Lage*
1852	Chasles: *Traité de géométrie supérieure*

1815	Battle of Waterloo
1817	Optical transverse vibrations (Young and Fresnel)
1820	Oersted discovered electromagnetism
1826	Ampère's work in electrodynamics
1827	Ohm's law
1828	Synthesis of urea by Wöhler
1829	Death of Thomas Young
1830	Lyell: *Principles of Geology*
	Comte: *Cours de philosophie positive*
1831	Faraday's electromagnetic induction
1832	Babbage's analytical engine
1836	First telegraph
1842	Conservation of energy (Mayer and Joule)
1846	Discovery of Neptune (Adams and Leverrier)
	Use of anesthesia
1848	Marx: *Communist Manifesto*
1850	Dickens: *David Copperfield*

Chronological Table—*(contd.)*

1854	Riemann's Habilitationschrift	1858	Atlantic cable
	Boole: *Laws of Thought*	1859	Darwin: *Origin of Species*
1855	Dirichlet succeeded Gauss at Göttingen		Chemical spectroscopy (Bunsen and Kirchhoff)
		1868	Cro-Magnon caves discovered
1863	Cayley appointed at Cambridge	1869	Opening of Suez Canal
1864	Weierstrass appointed at Berlin		Mendeleef's periodic table
1872	Dedekind: *Stetigkeit und irrationale Zahlen*	1873	Maxwell: *Electricity and Magnetism*
	Heine: *Elemente*		
	Méray: *Nouveau précis*		
	Klein's Erlanger Programm		
1873	Hermite proved *e* transcendental	1876	Bell's telephone
1874	Cantor's *Mengenlehre*		
1877	Sylvester appointed at Johns Hopkins	1887	Discovery of herzian waves
1881	Gibbs: *Vector Analysis*	1888	Pasteur Institute founded
1882	Lindemann proved π transcendental		
1884	Frege: *Grundlagen der Arithmetik*	1895	Discovery of X-rays (Roentgen)
1888	Beginnings of American Mathematical Society	1896	Discovery of radioactivity (Becquerel)
1889	Peano's axioms		
1895	Poincaré: *Analysis situs*		
1896	Prime number theorem proved (Hadamard and De la Vallée-Poussin)		

1899	Hilbert: *Grundlagen der Geometrie*
1900	Hilbert's problems
	Volume I of Russell and Whitehead: *Principia*
1903	Lebesgue integration
1906	Functional calculus (Fréchet)
1907	Brouwer and intuitionism
1914	Hausdorff: *Grundzüge der Mengenlehre*
1916	Einstein's general theory of relativity
1917	Hardy and Ramanujan on theory of numbers
1923	Banach spaces
1930	Weyl succeeded Hilbert at Göttingen
1931	Gödel's theorem
1933	Weyl resigned at Göttingen
1934	Gelfond's theorem
1939	Volume I of Bourbaki: *Eléments*
1955	Homological algebra (Cartan and Eilenberg)
1963	Paul J. Cohen on continuum hypothesis
1966	15th International Congress of Mathematicians (Moscow)

1897	Discovery of electron (J. J. Thomson)
1898	Discovery of radium (Marie Curie)
1900	Freud: *Die Traumdeutung*
1901	Planck's quantum theory
1903	First powered air flight
1905	Special relativity (Einstein)
1914	Assassination of Austrian Archduke
1915	Panama Canal opened
1917	Russian Revolution
1927	Lindbergh flew the Atlantic
1928	Fleming discovered penicillin
1933	Hitler became chancellor
1941	Pearl Harbor
1945	Bombing of Hiroshima
1946	First meeting of U.N.
1963	Assassination of President Kennedy
1965	Death of Sir Winston Churchill

Index

697

7396